Sound and Music Computing

Special Issue Editors

Tapio Lokki
Stefania Serafin
Meinard Müller
Vesa Välimäki

MDPI • Basel • Beijing • Wuhan • Barcelona • Belgrade

MDPI

Special Issue Editors
Tapio Lokki
Aalto University
Finland

Stefania Serafin
Aalborg University
Denmark

Meinard Müller
Friedrich-Alexander Universität Erlangen-Nürnberg
Germany

Vesa Välimäki
Aalto University
Finland

Editorial Office
MDPI
St. Alban-Anlage 66
Basel, Switzerland

This edition is a reprint of the Special Issue published online in the open access journal *Applied Sciences* (ISSN 2076-3417) from 2017–2018 (available at: http://www.mdpi.com/journal/applsci/special_issues/Music_Computing).

For citation purposes, cite each article independently as indicated on the article page online and as indicated below:

Lastname, F.M.; Lastname, F.M. Article title. *Journal Name* **Year**, *Article number*, page range.

First Editon 2018

Cover photo courtesy of Mattia Bernardi.

ISBN 978-3-03842-907-4 (Pbk)
ISBN 978-3-03842-908-1 (PDF)

Table of Contents

About the Special Issue Editors

Tapio Lokki, associate professor, received his MSc and DSc(Tech.) degrees from the Helsinki University of Technology. Since 2012, he has been an associate professor (tenured) at the Aalto University, Espoo, Finland, where he leads the virtual acoustics team. The passion of Prof. Lokki is to understand how rooms modify the sounds that we hear. To pursue the understanding of room acoustics, his team is investigating auralization, spatial sound reproduction, binaural technology, and novel objective and subjective evaluation methods, as well as physically based room acoustics modeling methods. Particular interest has been devoted to concert halls, in which the team has developed new measurement techniques, analysis methods for spatial impulse responses, and sensory evaluation methods to understand perceptual differences between concert halls. Prof. Lokki has published over 200 scientific articles and has received research funding from the European Research Council and the Academy of Finland.

Stefania Serafin is Professor with special responsibilities in sound in multimodal environments at Aalborg University in Copenhagen. She previously was Associate Professor (2006–2012) and Assistant Professor (2003–2006) at Aalborg University Copenhagen. She received a Ph.D. in Computer-Based Music Theory and Acoustics from Stanford University in 2004 and a Master in Acoustics, Computer Science and Signal Processing Applied to Music from IRCAM (Paris) in 1997. She is currently principal investigator for the Nordic Sound and Music Computing Nordic University Hub, supported by Nordforsk. Her main research interests are sound models for interactive systems and multimodal interfaces and sonic interaction design.

Meinard Müller, Professor, received a Diploma degree in mathematics and a Ph.D. degree in computer science from the University of Bonn, Germany. In 2007, he finished his habilitation in the field of multimedia retrieval and became a member of the Saarland University and the Max-Planck Institut für Informatik. Since 2012, he holds a professorship for Semantic Audio Processing at the International Audio Laboratories Erlangen. His recent research interests include music processing, music information retrieval, audio signal processing, multimedia retrieval, and motion processing. He served in the IEEE Audio and Acoustic Signal Processing Technical Committee from 2010 to 2015 and has been a member of the Board of Directors of the International Society for Music Information Retrieval since 2009. Meinard Müller has coauthored more than 150 peer-reviewed scientific papers and wrote the monograph "Information Retrieval for Music and Motion" (Springer, 2007) and the textbook "Fundamentals of Music Processing" (Springer, 2015, www.music-processing.de).

Vesa Välimäki is Full Professor of audio signal processing and Vice Dean for research at the Aalto University School of Electrical Engineering, Espoo, Finland. He received his Master of Science in Technology and Doctor of Science in Technology degrees, both in electrical engineering, from the Helsinki University of Technology, Espoo, Finland, in 1992 and 1995, respectively. In 1996, he was a Postdoctoral Research Fellow at the University of Westminster, London, UK. In 2008–2009, he was a Visiting Scholar at the Center for Computer Research in Music and Acoustics (CCRMA), Stanford University, Stanford, CA, USA. He is a Fellow of the IEEE and a Fellow of the Audio Engineering Society. Prof. Välimäki was the General Chair of the International Conference on Digital Audio Effects, DAFx-08, in 2008, and was the General Chair of the Sound and Music Computing Conference, SMC-17, in 2017.

applied
sciences

MDPI

Editorial

Special Issue on "Sound and Music Computing"

Tapio Lokki [1,*], Meinard Müller [2], Stefania Serafin[3] and Vesa Välimäki [4]

1   Department of Computer Science, Aalto University, 02150 Espoo, Finland
2   International Audio Laboratories Erlangen, Friedrich-Alexander Universität Erlangen-Nürnberg,
    91058 Erlangen, Germany; meinard.mueller@audiolabs-erlangen.de
3   Department of Architecture, Design and Media Technology, Aalborg University,
    2450 Copenhagen SV, Denmark; sts@create.aau.dk
4   Department of Signal Processing and Acoustics, Aalto University, 02150 Espoo, Finland;
    vesa.valimaki@aalto.fi
*   Correspondence: tapio.lokki@aalto.fi

Received: 22 March 2018; Accepted: 27 March 2018; Published: 28 March 2018

1. Introduction

Sound and music computing is a young and highly multidisciplinary research field. It combines scientific, technological, and artistic methods to produce, model, and understand audio and sonic arts with the help of computers. Sound and music computing borrows methods, for example, from computer science, electrical engineering, mathematics, musicology, and psychology.

For this special issue, 44 manuscripts were submitted and were carefully reviewed. Finally, 29 high-quality articles were published, and we are very pleased with the outcome. Some of the articles are revised and extended versions of papers published earlier in related international conferences, such as in the 14th Sound and Music Computing Conference SMC-17 (Espoo, Finland), the 18th International Society for Music Information Retrieval Conference ISMIR-17 (Suzhou, China), or the 2017 New Interfaces for Musical Expression Conference NIME-17 (Copenhagen, Denmark).

This editorial briefly summarizes the published articles and guides you to read them in detail. The articles could be categorized in many ways, as such multidisciplinary field has a wide variety of topics. Here, we have organized the articles based on their application areas or special techniques applied in research. We hope that these articles will inspire researchers in sound and music computing to conduct more excellent research and spread the word about this vibrant, multidisciplinary field.

2. Sound and Music Computing Techniques

2.1. Audio Signal Processing

Cecchi et al. [1] have written the only review article for this special issue. Their long paper gives a complete overview of audio signal processing methods for the equalization of the loudspeaker-room response, which is a fundamental problem in sound reproduction. The increasing popularity of small mobile speakers having non-ideal properties makes this topic ever more important.

Necciari et al. [2] propose an improved auditory filter bank called *Audlet*, which allows perfect reconstruction. The new filter bank is compared with the gammatone filter bank, and its used in a single-channel audio source separation task is demonstrated.

Brandtsegg et al. [3] discuss approaches to real-time convolution with time-varying filters, which extends the convolution reverberation concept. For example, the sounds produced by two players can be convolved with each other to obtain exciting audio effects.

Damskägg and Välimäki [4] address a problem known as time-scale modification in which the objective is to temporally stretch or compress a given audio signal while preserving properties like pitch and timbre. To handle different signal characteristics, the main idea of the paper is to modify the phase of the signal's time-frequency bins in an adaptive fashion using an implicit bin-wise fuzzy classification based on three classes (sinusoid, noise, transient).

Esqueda et al. [5] present virtual analogue models of the Lockhart and Serge wavefolders. The input–output relationship of both circuits was digitally modeled using the Lambert-W function. Aliasing distortion is ameliorated using a first-order antiderivative method. An earlier version of this paper received a best paper award at the SMC-17 conference.

2.2. Machine and Deep Learning

Deep learning is a hot topic also in sound and music computing. In this special issue, there are several articles applying deep learning techniques to various problems. The article by Wang et al. [6] describes an automatic music transcription algorithm combining deep learning and spectrogram factorization techniques. It is applied to a specific piano, and the results outperform the earlier methods in note-level polyphonic piano music transcription. Blaauw and Bonada [7] describe a singing synthesizer based on deep neural networks called the Neural Parametric Singing Synthesizer (NPSS), which can generate high-quality singing when a musical score and lyrics are given as the input. The NPSS can learn the timbre and expressive features of a singer from a small set of recordings. Lee et al. [8] discuss a learning approach based on convolutional neural networks (CNNs) to derive meaningful feature representations directly from the waveform of an audio signals (rather than using frame-based input representations such as the short-time Fourier transform). Such approaches are interesting in view of end-to-end music classifications tasks including genre classification and auto tagging. As one main contribution, the authors discuss the properties of the learned sample-level filters and show how their CNN-based learning approach behaves under certain downsampling and normalization effects.

Machine learning is also traditionally applied by many researchers. Green and Murphy [9] report on spatial analysis of binaural room impulse responses. The results of this article indicate that neural networks are able to detect the direction of the direct sound, but are less accurate at predicting the direction of arrival of the reflections, even in quite simple cases. More work on this topic is needed, to be able to study room acoustics with machine learning. Lovedee-Turner and Murphy [10] have collected a database of spatial sound recordings for the purpose of classification of acoustic scenes as well as the material for machine learning algorithms. To validate the database they also introduce a classifier that performs better than a traditional Mel-frequency-cepstral-coefficient classifier. The article by Pesek et al. [11] introduces algorithmic concepts for modeling and detecting recurrent patterns in symbolically encoded music. Given a monophonic symbolic representation of a piece of music, the algorithm outputs a hierarchical representation of melodic patterns using an unsupervised learning procedure without the need of hard-coded rules from music theory. Also the comprehensive article by Bountouridis [12] is concerned with pattern analysis of symbolic music representations. Inspired by multiple sequence alignment techniques that are well known in bioinformatics, the authors show how such methods can be adapted to symbolic music analysis. In particular, sequence alignment and retrieval techniques are used for measuring melodic similarities and for detecting musically interesting relations within song families. Carabez et al. [13] study a brain-computer interface, which consists of headphones and an electroencephalography- or EEG-based measurement system, which registers the user's brain activity. Using machine learning techniques, they demonstrate promising results on reading from the user's mind the direction of arrival of sound stimuli.

2.3. Automatic Transcription and Programming

Mcleod et al. [14] address a central problem in music information retrieval known as music transcription: given an audio recording of a piece of music, the goal is to extract symbolic note parameters such as note onsets and pitches. In this article, the authors focus on a-cappella music recordings with four singers (bass, tenor, alto, soprano). Combining an acoustic model based on probabilistic latent component analysis (PLCA) and a music language model based on Hidden Markov

Models (HMM), the authors present an approach for jointly tackling the problems of multi-pitch transcription as well as voice assignment.

Lazzarini [15] presents a new framework for unit generator development for the computer music language Csound, introducing the concept of unit generators and opcodes, and its centrality with regards to music programming languages in general, and Csound in specific.

3. Sound and Music Computing Applications

3.1. Sound Synthesis and User Control

Sound synthesis and control of novel and computer-based instruments are one of the main areas in sound and music computing. In Selfridge et al. [16] several physical models of objects swinging through air are presented. Listening tests showed that the models were rated as plausible as recordings. Such models are particularly interesting when used in real-time audio-visual simulations. This is a revised and extended version of a paper winning a best paper award at SMC-17 Conference. Michon et al. [17] present two original concepts: mobile device augmentation and hybrid instruments. Several tools, techniques, as well as thoughtful considerations and useful advices on how to design such instruments are presented. This paper is an extension of a paper that won the best paper award at the NIME-17 conference. The paper by MacRitchie and Milne [18] investigates four different pitch layouts on the computer screen, and finds how easy or difficult it is to play melodies on each of them. Their results lead to novel design rules for such musical instruments. Kelkar and Jensenius [19] asked people to listen to short melodies and move their hands as if their movement was creating the sound. The authors found that people tend to use one of six different mapping strategies. They also observed an interesting gender difference, as one of the strategies was more often used by women than by men.

3.2. Audio Mixing and Audio Coding

Wilson and Fazenda [20] present a method to generate automatic audio mixes. The study concerns three audio processing activities: level-balancing, stereo-panning, and equalization. The presented work will pave the way to automatize the work of audio engineers, especially in object-based audio broadcasting.

Jia et al. [21] propose an efficient, psychoacoustic coding method for multiple sound objects in a spatial audio scene. This technology can be applied to 3-D movies, spatial audio communication systems, and virtual classrooms.

3.3. Games and Virtual Reality

Hansen and Hiraga [22] introduce and evaluate *Music Puzzle*, which is an audio-based game. Interestingly, they tested the game with different user groups. People with hearing loss had problems in a game that used speech, but less with a game based on music. In contrast, people with low engagement in music performed worse in a music game. Based on this study the authors could explain the impact of hearing acuity and musical experience on focused listening of different sounds.

Schaerlaeken et al. [23] investigate the impact of playing for a virtual audience, both from the perspective of the player and the audience. The study highlights the use of immersive virtual environments as a research tool and a training assistant for musicians who are eager to learn how to cope with their anxiety in front of an audience.

Yiyu et al. [24] discuss an audio processor architecture, which is suitable for rendering a virtual acoustic environment using a finite-difference approach. Such a system can be useful for providing realistic acoustic experiences for gaming or virtual reality.

Puomio et al. [25] present a perceptual study on the effect of virtual sound source positions in spatial audio rendering using headphones with head-tracking. A listening test was conducted comparing optimized and non-optimized virtual loudspeaker setups in the simulations of a small room and a concert hall. Their results suggest that the simulation of a small room benefits more from the optimization of virtual source positions than a large room.

3.4. Sonic Interaction, Musicology, and New Hardware

Verde et al. [26] investigate computational musicology for the study of tape music works, and existing computer vision techniques are applied to the analysis of such tracks.

Hayes and Stein [27] present an approach to incorporate environmental factors within the field of site-responsive sonic art using embedded audio and data processing techniques. The main focus is on the role of such systems within an ecosystemic framework, both in terms of incorporating systems of living organisms, as well as sonic interaction design.

In Yağanoğlu and Köse [28] a wearable vibration communication system for the deaf is presented. The wearable device proved to have a high success rate in localization tasks, which are problematic for deaf people.

Quintana-Suárez et al. [29] authored an article on a sensor device that enables to remotely monitor the activity and health of elderly people. Such technology is generally called Ambient Assisted Living, and this article in particular presents a low-cost acoustic sensor.

4. Conclusions

Overall the special issue shows the variety of topics researched by the sound and music community, ranging from spatial sound, sound processing, sonic interaction design, and music information retrieval to new interfaces for musical expression with applications in art, culture, gaming, and virtual and augmented reality.

Conflicts of Interest: The authors declare no conflict of interest.

References

1. Cecchi, S.; Carini, A.; Spors, S. Room Response Equalization—A Review. *Appl. Sci.* **2018**, *8*, 16.
2. Necciari, T.; Holighaus, N.; Balazs, P.; Průša, Z.; Majdak, P.; Derrien, O. Audlet Filter Banks: A Versatile Analysis/Synthesis Framework Using Auditory Frequency Scales. *Appl. Sci.* **2018**, *8*, 96.
3. Brandtsegg, Ø.; Saue, S.; Lazzarini, V. Live Convolution with Time-Varying Filters. *Appl. Sci.* **2018**, *8*, 103.
4. Damskägg, E.P.; Välimäki, V. Audio Time Stretching Using Fuzzy Classification of Spectral Bins. *Appl. Sci.* **2017**, *7*, 1293.
5. Esqueda, F.; Pöntynen, H.; Parker, J.D.; Bilbao, S. Virtual Analog Models of the Lockhart and Serge Wavefolders. *Appl. Sci.* **2017**, *7*, 1328.
6. Wang, Q.; Zhou, R.; Yan, Y. A Two-Stage Approach to Note-Level Transcription of a Specific Piano. *Appl. Sci.* **2017**, *7*, 901.
7. Blaauw, M.; Bonada, J. A Neural Parametric Singing Synthesizer Modeling Timbre and Expression from Natural Songs. *Appl. Sci.* **2017**, *7*, 1313.
8. Lee, J.; Park, J.; Kim, K.L.; Nam, J. SampleCNN: End-to-End Deep Convolutional Neural Networks Using Very Small Filters for Music Classification. *Appl. Sci.* **2018**, *8*, 150.
9. Green, M.C.; Murphy, D. EigenScape: A Database of Spatial Acoustic Scene Recordings. *Appl. Sci.* **2017**, *7*, 1204.
10. Lovedee-Turner, M.; Murphy, D. Application of Machine Learning for the Spatial Analysis of Binaural Room Impulse Responses. *Appl. Sci.* **2018**, *8*, 105.
11. Pesek, M.; Leonardis, A.; Marolt, M. SymCHM—An Unsupervised Approach for Pattern Discovery in Symbolic Music with a Compositional Hierarchical Model. *Appl. Sci.* **2017**, *7*, 1135.
12. Bountouridis, D.; Brown, D.G.; Wiering, F.; Veltkamp, R.C. Melodic Similarity and Applications Using Biologically-Inspired Techniques. *Appl. Sci.* **2017**, *7*, 1242.
13. Carabez, E.; Sugi, M.; Nambu, I.; Wada, Y. Identifying Single Trial Event-Related Potentials in an Earphone-Based Auditory Brain-Computer Interface. *Appl. Sci.* **2017**, *7*, 1197.

14. McLeod, A.; Schramm, R.; Steedman, M.; Benetos, E. Automatic Transcription of Polyphonic Vocal Music. *App. Sci.* **2017**, *7*, 1285.
15. Lazzarini, V. Supporting an Object-Oriented Approach to Unit Generator Development: The Csound Plugin Opcode Framework. *Appl. Sci.* **2017**, *7*, 970.
16. Selfridge, R.; Moffat, D.; Reiss, J.D. Sound Synthesis of Objects Swinging through Air Using Physical Models. *Appl. Sci.* **2017**, *7*, 1177.
17. Michon, R.; Smith, J.O.; Wright, M.; Chafe, C.; Granzow, J.; Wang, G. Mobile Music, Sensors, Physical Modeling, and Digital Fabrication: Articulating the Augmented Mobile Instrument. *Appl. Sci.* **2017**, *7*, 1311.
18. MacRitchie, J.; Milne, A.J. Exploring the Effects of Pitch Layout on Learning a New Musical Instrument. *Appl. Sci.* **2017**, *7*, 1218.
19. Kelkar, T.; Jensenius, A.R. Analyzing Free-Hand Sound-Tracings of Melodic Phrases. *Appl. Sci.* **2018**, *8*, 135.
20. Wilson, A.; Fazenda, B.M. Populating the Mix Space: Parametric Methods for Generating Multitrack Audio Mixtures. *Appl. Sci.* **2017**, *7*, 1329.
21. Jia, M.; Zhang, J.; Bao, C.; Zheng, X. A Psychoacoustic-Based Multiple Audio Object Coding Approach via Intra-Object Sparsity. *Appl. Sci.* **2017**, *7*, 1301.
22. Hansen, K.F.; Hiraga, R. The Effects of Musical Experience and Hearing Loss on Solving an Audio-Based Gaming Task. *Appl. Sci.* **2017**, *7*, 1278.
23. Schaerlaeken, S.; Grandjean, D.; Glowinski, D. Playing for a Virtual Audience: The Impact of a Social Factor on Gestures, Sounds and Expressive Intents. *Appl. Sci.* **2017**, *7*, 1321.
24. Yiyu, T.; Inoguchi, Y.; Otani, M.; Iwaya, Y.; Tsuchiya, T. A Real-Time Sound Field Rendering Processor. *Appl. Sci.* **2018**, *8*, 35.
25. Puomio, O.; Pätynen, J.; Lokki, T. Optimization of Virtual Loudspeakers for Spatial Room Acoustics Reproduction with Headphones. *Appl. Sci.* **2017**, *7*, 1282.
26. Verde, S.; Pretto, N.; Milani, S.; Canazza, S. Stay True to the Sound of History: Philology, Phylogenetics and Information Engineering in Musicology. *Appl. Sci.* **2018**, *8*, 226.
27. Hayes, L.; Stein, J. Desert and Sonic Ecosystems: Incorporating Environmental Factors within Site-Responsive Sonic Art. *Appl. Sci.* **2018**, *8*, 111.
28. Yağanoğlu, M.; Köse, C. Wearable Vibration Based Computer Interaction and Communication System for Deaf. *Appl. Sci.* **2017**, *7*, 1296.
29. Quintana-Suárez, M.A.; Sánchez-Rodríguez, D.; Alonso-González, I.; Alonso-Hernández, J.B. A Low Cost Wireless Acoustic Sensor for Ambient Assisted Living Systems. *Appl. Sci.* **2017**, *7*, 877.

applied
sciences

MDPI

Article

Stay True to the Sound of History: Philology, Phylogenetics and Information Engineering in Musicology

Sebastiano Verde, Niccolò Pretto *, Simone Milani and Sergio Canazza

Department of Information Engineering, University of Padova, via Gradenigo 6/B, 35131 Padova, Italy; sebastiano.verde@dei.unipd.it (S.V.); simone.milani@dei.unipd.it (S.M.); sergio.canazza@dei.unipd.it (S.C.)
* Correspondence: niccolo.pretto@dei.unipd.it; Tel.: +39-049-827-6465

Academic Editor: Stefania Serafin
Received: 3 November 2017; Accepted: 29 January 2018; Published: 1 February 2018

Abstract: This work investigates computational musicology for the study of tape music works tackling the problems concerning stemmatics. These philological problems have been analyzed with an innovative approach considering the peculiarities of audio tape recordings. The paper presents a phylogenetic reconstruction strategy that relies on digitizing the analyzed tapes and then converting each audio track into a two-dimensional spectrogram. This conversion allows adopting a set of computer vision tools to align and equalize different tracks in order to infer the most likely transformation that converts one track into another. In the presented approach, the main editing techniques, intentional and unintentional alterations and different configurations of a tape recorded are estimated in phylogeny analysis. The proposed solution presents a satisfying robustness to the adoption of the wrong reading setup together with a good reconstruction accuracy of the phylogenetic tree. The reconstructed dependencies proved to be correct or plausible in 90% of the experimental cases.

Keywords: tape music analysis; audio philology; digitized audio recordings; digital phylogeny; computational musicology; spectrogram alignment; audio forensics

---

1. Introduction

The interesting field of computational musicology is given in relation to the study of tape music, which represents a particular case of recorded sound art with important implications with respect to the preservation side, as well as the musicological analysis side. Tape music consists of the (processed) fragments, samples and speed manipulation of pre-recorded sounds used in modern composition. Since the 1950s, its peculiar working method was made popular by composers of the Columbia-Princeton Electronic Music Center and, in Europe, of the Studio di Fonologia Musicale of RAI Milan [1]. This music cannot be set in conventional notation: the musical text is non-existent, incomplete, insufficiently precise and transmitted in a non-traditional format. The performance of these music works is no longer the traditional one, in which one or more musicians are used to perform a score: the composer becomes also the luthier and the performer of the completed product, recorded on magnetic tape configured as a *unicum*. The uniqueness of the tape music works tackles a well-known problem in the visual arts field, such as the attribution and the generation of different versions (called witnesses, in the philology field).

In order to achieve scholarly analysis of the musical works, the audio signal, stored in analogue tapes, must be digitized, along with all ancillary information (e.g., text on the box, accompanying material, etc.). The output of this process is the preservation master, the bibliographic equivalent of which is the facsimile or the diplomaticcopy [2–7].

The production of a preservation master requires competences from engineering, archivistics and philology. The engineering side of the work consists of, but is obviously not limited to, the development of ad hoc tools that provide solutions both to problems of managerial and philological-documental characters. Van Huis [8] criticizes the general inertia of archival institutions in the face of new technologies, which ignores the potential to reach users familiar with the use of a Google-like search engine and/or a peer-to-peer network (simple, but not authoritative).

This work applied computational musicology to the study of tape music works tackling the problems concerning stemmatics (or the Latin, stemmata):

1. constructing the stemma codicum (recension, or the Latin recensio) starting with a set of the sources (all the different witnesses of that musical work);
2. selection (or selectio), where the original source is determined by examining variants, selecting the best ones [9].

These studies often use a general-purpose audio editor in order to compare sonograms or wave forms. This paper presents an innovative approach to this problem integrating methodologies typically used in the field of forensic science.

Recent years have witnessed a significant leap forward in sound trace analysis thanks to new processing tools derived from this research field. More precisely, multimedia forensics researchers have been investigating new accurate phylogenetic reconstruction strategies to be applied on unordered sets of similar digital audio/image/video contents [10–12]. Such availability, which has been fostered by the recent disposal of versatile acquisition, editing and sharing tools, poses the problem of discriminating the original file, identifying the owner or reconstructing the processing history of each copy. To these purposes, forensic researchers have been borrowing some of the analysis strategies from phylogenetic biology. The underlying idea is that multimedia contents can "mutate" as organisms evolve: a digital image or an image can change over time to slightly different versions of itself, which can generate other versions, as well [13]. These different versions are referred to as near-duplicates (ND). The generation process of sets of ND images or video sequences can be well described by means of a structure called an image or video phylogeny tree (IPT or VPT), and several algorithms have been recently proposed in the literature to reconstruct it [14–16]. Most of the proposed solutions analyze the relations between similar contents and infer the subset of links that correctly represents the chains of dependencies and transformations. Although, in the last few years, several works have been targeting the analysis of images and video contents, little effort has been put toward audio phylogenetic approaches. To the best of our knowledge, the only algorithms proposed in the literature are represented by [12,17]. Moreover, the phylogenetic analysis problems have been extensively investigated for digital multimedia contents since online material offers the largest datasets of near-duplicate contents. The investigation of phylogenetic approaches involving audio-visual contents stored on magnetic carriers is still at its earliest stages. In the end, multimedia phylogenetic strategies have been employed for copyright infringement detection and fake material identification; their use in cultural heritage and multimedia restoration is one of the novelty aspects presented by this paper.

The authors propose in the following sections an automated approach to stemmatics, applying a phylogenetic evolutionary framework to music digital philology. Different witnesses (audio files) are analyzed by software developed by the authors and then grouped according to their shared characteristics, listed in a tree in order to derive relationships between them (Figure 1). The current paper presents a novel methodology that automatizes the creation of such a dependency tree and proves to be sufficiently robust to acquisition errors (wrong reading speed or setup). The proposed solution relies on collecting the tapes to be analyzed into a set of digital audio tracks and representing them by means of two-dimensional spectrograms. After aligning them using computer vision strategies, it is possible to infer the most likely transformation interlying between them and to characterize it via a dissimilarity metric. Such dissimilarity metrics are then used to characterize the edge weights of a complete graph where nodes correspond to the acquired audio tracks. By running a minimum spanning tree (MST) algorithm, it is possible to estimate the phylogenetic tree that links the different contents.

(**a**) Original tape.

(**b**) Modified version of the original tape.

(**c**) Spectrogram of the original tape.

(**d**) Spectrogram of the modified version of the tape.

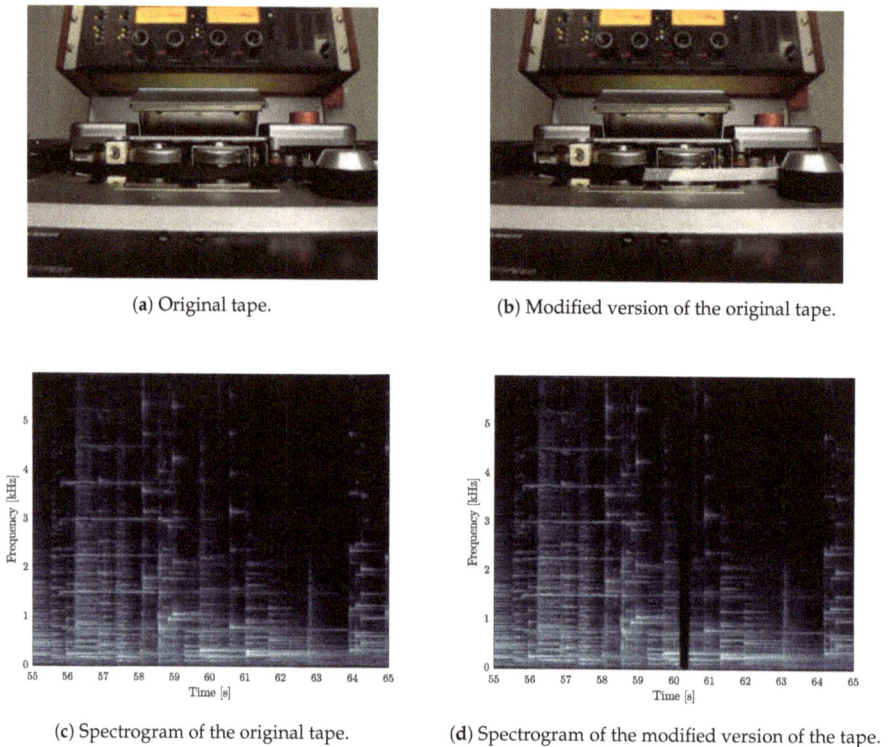

Figure 1. Example of near-duplicates (witnesses). In the middle of the tape (**a**) has been added a piece of leader-tape obtaining the modified version (**b**); The difference between the two versions can be clearly observed comparing the corresponding spectrograms (**c**,**d**).

Experimental data show that the proposed solution permits reconstructing the underlying story of each tape with good accuracy; moreover, the reconstruction process is not affected by digitization errors such as a different reading speed, wrong equalization and filtering. The results achieved show that this methodology gives a precise answer to the questions about the reliability of audio recordings as document witnesses, clarifying the concept of fidelity to the original.

In the following, the paper is organized as follows. Section 2 overviews some of the related works in the literature, while Section 3 describes how the phylogenetic reconstruction problem can be applied to magnetic tapes. Then, the full methodology is presented in Section 4, together with the reconstruction algorithm that permits estimating the audio phylogenetic tree (APT). This strategy is then evaluated on an experimental dataset (described in Section 5), and the obtained results are reported and discussed in Section 6. Section 7 draws the final conclusions.

2. Related Works

Many scholars ([18]; for an overview, see [19]) in the musicology field tackle the problems concerning stemmatics in their study of tape music works. As briefly outlined in the Introduction, these studies often use a general-purpose audio editor and are based on the comparison of sonograms or wave forms. However, other works based on computational musicology exist: Nicola Orio and co-workers [20–22] presented a tool to analyze the similarities and the differences of two witnesses of a music work. A graphical representation of the alignment curve, which matches pairs of points in the two signals in a bi-dimensional representation, gives a direct view of the main differences between two witnesses: by matching individual musical events, it is possible to compare the lengths of the events, the amplitude envelopes and the two spectral representations.

On the other hand, the audio phylogenetic field has been limitedly explored. Although several phylogenetic approaches for images and videos have been recently proposed, thus far, digital audio phylogeny research is taking the firsts steps.

A first approach was proposed by Nucci et al. in [12], where the authors designed a strategy to reconstruct the processing history of a set of near-duplicate (ND) audio tracks. According to the formulation given in [23], a near-duplicate is a transformed version of an object that remains recognizable; the audio tracks in [12] were generated via trim, fade and perceptual audio coding with a closed set of parameter values. A different set of values is then used in the analysis phase to compute the dissimilarity metric between couples of tracks. The proposed solution permits obtaining a good accuracy, but requires a significant computational effort; moreover, its efficiency is limited by the assumption of knowing the set of applied transformations.

A more flexible and computationally-efficient approach was proposed in [17]; in this latter approach, audio tracks are time- and frequency-aligned by representing each audio file with a spectrogram and using image registration techniques investigated in the field of computer vision. Moreover, the set of possible transformations includes time-frequency operations, as well. By employing highly-optimized computer vision libraries, the proposed algorithm requires a significantly lower computational load while not being constrained by a closed set of transformations. Such versatility suggested adopting that approach in this work, as well. The following sections will provide further details.

3. Problem Description

Tape music is a genre of electroacoustic music in which the artwork coincides with the tape on which the audio signal is recorded [6]. In some cases, the composer did not provide a score; hence, the tape could be considered the final product of the creative process. The carrier has therefore a prominent role and must be considered in the philological analysis [18,24]. The peculiarity of this type of analogue carrier is the possibility of editing the tape with several techniques. The main ones are introduced in [25], where the described techniques vary from straight recording to superimposition. Some of the main alterations considered in this paper are presented below.

The editing consists on the physical alterations of the tape that is cut in pieces and then recomposed to obtain the desired sound or effect. Every piece is joined with the rest of the tape in a splice by using a strip of plastic coated with a thermal or pressure-sensitive adhesive called splicing tape [26]. As recommended in [27], the tape has to be cut at an angle of $45°$ to $60°$, measured with respect to the tape edge in order to avoid electrical disturbance. Without this disturbance, it is very difficult to find the slice by analyzing the audio track, and it could only be hypothesized if the audio content suddenly changes.

In some cases, the splice does not join two pieces of magnetic tape, but one side consists of a leader-tape. It is a flexible plastic or paper strip that usually is spliced to either end of a roll of recording material [26]. In this case, the leader-tape extends the tape length in order to fasten the extremities of the tape to the hub of the flange and avoid wasting the magnetic tape that can be read entirely. In the creative process, the leader-tape is also important because it could be used inside the tape with several purposes, such as adding pauses or signaling new events or units.

Silence parts could be also obtained erasing previous recordings. The erasing head could be used to clean the tape before recording a new audio signal. Furthermore, as described in [25], a new signal could be recorded over the old one using superimposition. The result of these techniques is the sum of the two signals. In this paper, these techniques are generally referred to as overdubbing.

All these alterations are irreversible and create new versions of the opera, which in philology are called witnesses. As outlined in the Introduction, this paper presents an innovative approach to the musicological analysis using phylogenetic techniques, typically adopted in forensics, in order to reconstruct the stemma codicum, which can be considered as an audio phylogenetic tree (APT) [17].

Nevertheless, the analysis is performed in the digital domain and not in the analogue one. This introduces a variable to the phylogenetic problem described in [17]: the digitization. An analogue tape can be read and digitized several times, creating similar digital versions of a unique document. These versions can differ from each other in the configuration of the tape recorder, in the quality of the Analog-to-Digital (A/D) converters and in the digital format on which the signal is saved. This study considers recorder configuration only, as the other options are not connected to the analogue carrier and thus not necessary in order to create a suitable model that takes into account the audio tape peculiarities.

The most important parameter to be configured in the tape recorder during the listening and the digitization process is the replay speed. Six standard speeds are used: 30 ips (76.2 cm/s), 15 ips (38.1 cm/s), 7.5 ips (19.05 cm/s), 3.75 ips (9.525 cm/s), 15/8 ips (4.76 cm/s) and 15/16 ips (2.38 cm/s) [28]. The wrong choice of this parameter implies a time stretch and a pitch change that heavily distort the signal. To be thorough, this effect was used in electroacoustic music as a technique for altering the signal on purpose [25], but this aspect goes beyond the analysis of this work.

Another important parameter to be set in the tape recorder is the equalization: a post-equalization curve is applied during the reading, in order to compensate the pre-equalization curve applied during the recording, essentially acting as the integrator to make the overall transfer function nearly flat [29]. Several standards exist, and they are commonly called using the name of the association that proposed the standard. For 30 ips, the most diffused is the standard AES derived by Audio Engineering Society, whereas for 15 ips and 7.5 ips the most used standards are CCIR and NAB, from International Radio Consultative Committee (in French) and National Association of Broadcasters, respectively [30–32].

Applying the wrong post-equalization curve implicates the wrong frequency response and, thus, a non-flat overall transfer function.

Furthermore, the recording may be encoded with a noise reduction system. The most common are Dolby A and Dolby SR (professional), Dolby B and Dolby C (domestic) and dbxTypes I (professional) and II (domestic) [28]. When reading the tape, the same noise reduction system must be used in order to compensate the one adopted in the recording phase. Again, the lack of compensation or the wrong system choice deeply changes the signal. In this paper, only the former problem is tackled, together with the opposite case: the use of a noise reduction system in the decoding phase when the original system was not encoded.

Considering the combination of all these configurations, it is evident how many different digitized versions of the same tape could be obtained. The approach proposed in this work handles this aspect considering all the versions of the same tape as a single node in the phylogenetic tree.

The problem could be further extended considering multiple copies recorded and digitized with different tape recorders and the possibility of finding some audio documents obtained by pieces of tape recorded by different machines (professional or not). In this case, the analogue filters and characteristics of tape recorders could lead to differences in digitized copies. The same could happen with machines that were not correctly calibrated. A further variable that could be considered in the analysis is the possibility of having different copies of the same tape and the presence of damage and/or syndrome (such as SSS [7]) that impact the digitization results, creating dissimilarities and artifacts despite the original tape being the same. This extension has been provided to better explain the overall complexity of the problem, but it goes far beyond the scope of this article, which seeks to prove the effectiveness of this new approach with respect to the problem with a simplified model that nonetheless includes the main alterations.

4. Algorithms

In this work, we propose an innovative approach to tape music phylogeny, based on the application of computer vision techniques to the time-frequency representation of audio tracks. The core idea consists of mapping the digital audio signals obtained from the tapes into bi-dimensional images. Consequently, the employment of a robust feature extraction algorithm permits gathering a set of local spectral fingerprints, which can be exploited in order to align pairs of spectrogram images. This alignment makes it possible to compare different tracks and estimate which tape editing operation (if any) interlies between them.

A similar strategy has been adopted in [33,34] to identify and retrieve different digital copies of the same audio tracks. In these cases, spectrogram-based features are used to determine whether the track matches or not. The phylogenetic approach departs from such a problem since its aim is to parameterize the similarity between two audio tracks, i.e., how much they differ. As a matter of fact, it is necessary to design a correct and effective registration algorithm for the two analyzed signals, as well as accurate equalization techniques that permit compensating the dissimilarity associated with reading/writing operations. The wrong equalization leads to noisy dissimilarity values, implying the wrong reconstruction of the dependencies.

Given a set of N audio tracks, the core idea of the proposed algorithm is to characterize the dissimilarity between each couple of digital audio tracks. This procedure yields the creation of a $N \times N$ dissimilarity matrix $D = [d_{i,j}]$, where $d_{i,j}$ denotes the dissimilarity between the i-th and the j-th tracks. As a consequence, dissimilarity computation is repeated for each one of the $N \cdot (N-1)$ possible ordered pairs (i,j). Then, the algorithm builds a complete directed graph where nodes correspond to the analyzed set of tracks and edge weights are the computed dissimilarity values.

The description of the proposed strategy can be divided into the following steps or units (Figure 2): (i) pre-processing; (ii) leader-tape detection; (iii) spectrogram registration; (iv) overdub detection; (v) estimation of the phylogenetic tree.

The following paragraphs present a detailed description of each step.

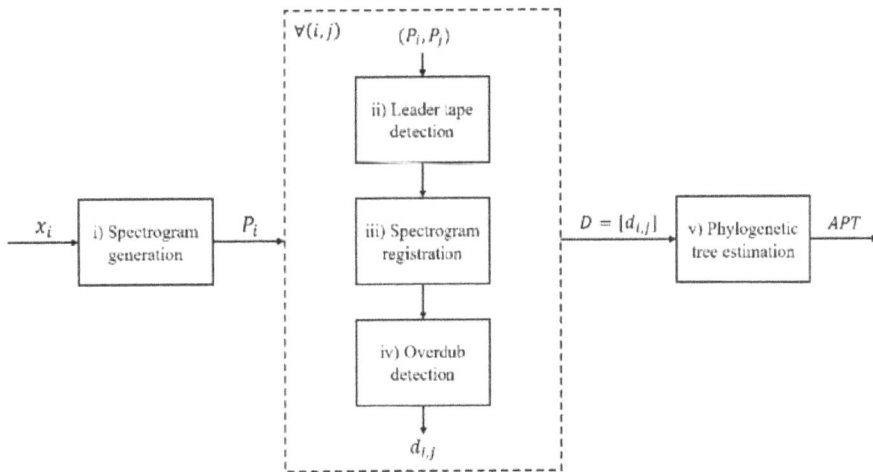

Figure 2. Block diagram of the proposed algorithm. The input consists of the digitalized audio tracks x_i, $i = 1, \ldots, N$, and the output is the estimated audio phylogeny tree (APT).

4.1. Audio Pre-Processing

At first, each audio track $x_i(n)$ is converted into the related spectrogram by computing the short-time Fourier transform:

$$X_i(f, m) = \sum_{n=-\infty}^{+\infty} x_i(n)\, w(n - mL)\, e^{-j2\pi f n} \tag{1}$$

where $w(\cdot)$ is a windowing function and L is the stride parameter. Coefficients $X_i(f, m)$ are computed for a finite set of N_f frequencies f and a finite set of M windows ($m = 0, \ldots, M-1$). In our experiments, we adopted Hamming-windowed frames of 4096 samples, with an overlap rate of 0.75, and a set of $N_f = 512$ linearly-spaced frequencies, ranging from 0 to 6 kHz. This latter choice allows one to reduce the computational burden (by reducing the size of images to process by a factor of four) without affecting the system performance, given that most of the spectral information is usually found at a low frequency.

By associating each spectrogram coefficient to the pixel of a grayscale image $P_i(u,v)$, we obtain a $N_f \times M$ gray level image, where the pixel intensity is obtained by converting the value $|X_i(f,m)|^2$ into an 8-bit integer. In order to remove part of the background noise, if $|X_i(f,m)|^2 < \delta$, the pixel $P_i(u,v)$ is set to zero.

From the obtained spectrogram image $P_i(u,v)$, a set of keypoints $\mathcal{K}_i = \{(u_k,v_k)\}$ with the related descriptors is computed by using the speeded-up robust features (SURF) algorithm [35].

After the pre-processing step, each image pair (P_i, P_j), $i = 1, \ldots, N, j = 1, \ldots, N, i \neq j$, is passed to the next modules.

4.2. Leader Tape Detection

Considering two spectrograms, (P_i, P_j), this step aims at detecting the presence of a leader-tape inserted into one of the two tapes. This detector relies on the fact that, whether or not a leader is present, there is not a single affine transformation that maps the keypoints found on P_i onto those of P_j. The reason is that keypoints lying in the portion of the spectrogram after a leader-tape insertion will carry an offset in their time coordinates with respect to those found on a spectrogram that does not contain such insertion.

The algorithm proceeds as follows.

1. From two sets of keypoints $(\mathcal{K}_i, \mathcal{K}_j)$, find a subset of matched pairs by comparing the related descriptors. Given the matched pairs $((u_k, v_k), (u'_h, v'_h))$, estimate the optimum geometric transform mapping P_i onto P_j with the RANSAC algorithm [36]. If a leader-tape is present, the set of inlier points returned by the algorithm will converge to a subset of keypoints belonging to only one of the two portions of the spectrogram separated by the leader (Figure 3).

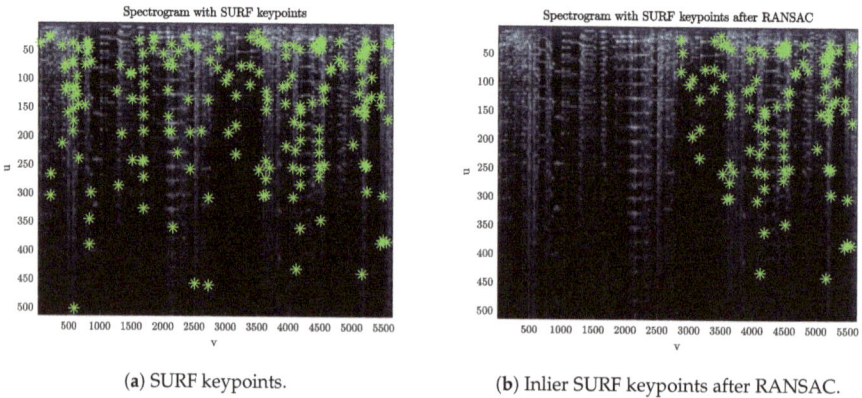

(a) SURF keypoints. (b) Inlier SURF keypoints after RANSAC.

Figure 3. Spectrogram image $P_i(u,v)$ of an audio track $x_i(n)$, with green asterisks representing the detected SURF keypoints. Subfigures show the SURF keypoints (a) and inlier keypoints after RANSAC (b). Note that the remaining inlier points are located to the right of the leader-tape.

2. Define a function $g_i(v)$ counting the number of keypoints detected in $P_i(u,v)$ for each image column v (in order to avoid strong oscillations, $g(v)$ is processed with a moving-average low-pass filter). Then, define $g'_i(v)$ as the number of inlier points left on $P_i(u,v)$ after the RANSAC

algorithm. In the presence of a leader insertion, distance $|g_i(v) - g_i'(v)|$ shows an evident step that can be detected by looking for gradient peaks.

3. Let v_l be the coordinate associated with the detected step. Define the following sets:

$$\mathcal{K}_i^{(L)} = \{(u_k, v_k) \in \mathcal{K}_i | v_k < v_l\},$$
$$\mathcal{K}_i^{(R)} = \{(u_k, v_k) \in \mathcal{K}_i | v_k > v_l\}, \tag{2}$$

i.e., the subsets of keypoints found on the left side (L) and on the right side (R) of the spectrogram with respect to the candidate leader location. Similarly, define $\mathcal{K}_j^{(L)}$ and $\mathcal{K}_j^{(R)}$.

4. Perform a new geometric transform estimation, on the left and right portion of the images separately, according to the subdivision defined in (2). The estimated models come in the form of 3×3 homography matrices, $H^{(L)}$ and $H^{(R)}$, from which it is possible to extract the translation components along the v direction, $t^{(L)}$ and $t^{(R)}$. The length of the candidate leader is then given by:

$$w_l = |t^{(L)} - t^{(R)}|. \tag{3}$$

If $w_l \neq 0$, the algorithm concludes that a leader-tape is present within the current spectrogram pair.

Finally, the algorithm tries to infer the correct phylogenetic relation that links P_i and P_j, namely whether P_j was derived from P_i by inserting a leader-tape or vice versa. This can be achieved by measuring the average spectral energy around the detected location, knowing that leader insertions are characterized by a very low-energy region in the related spectrogram. If we find a significant difference between the average energies measured in the two images (with respect to a suitable tolerance threshold), it is possible to conclude that the phylogenetic ancestor is the the one related to the highest energy content. Specifically, the algorithm distinguishes the two following cases.

If $\sum_u P_j(u, v_l) \gg \sum_u P_i(u, v_l)$, then P_j is assumed to be the phylogenetic ancestor. The dissimilarity matrix is updated with $d_{i,j} = +\infty$, indicating that a phylogenetic relation from i to j is not possible. The algorithm stops the analysis of the current image pair and switches to the next one.

Otherwise, if $\sum_u P_i(u, v_l) \ll \sum_u P_j(u, v_l)$ or $\sum_u P_i(u, v_l) \simeq \sum_u P_j(u, v_l)$, the algorithm proceeds with the next steps.

4.3. Spectrogram Registration

Spectrogram registration consists of warping P_i towards P_j according to the geometric transform estimated through their matched keypoints.

1. If a leader-tape has been detected in P_j, compensate it on P_i by adding a band of black pixels centered in v_l and with length w_l.
2. Estimate the global geometric transform H by running RANSAC on all keypoints.
3. Warp P_i towards P_j according to H, obtaining P_i'.
4. Compute the dissimilarity value $d_{i,j}$ as the MSE of P_i' and P_j:

$$d_{i,j} = \frac{1}{U \cdot V} \sum_{u,v} |P_j(u, v) - P_i'(u, v)|^2, \tag{4}$$

where U and V are the spectrograms' height and width in pixels.

4.4. Overdub Detection

This second detection step deals with the identification of an overdub in the analyzed tapes. It is positioned after the registration module, as it requires the spectrogram pair to be already aligned.

1. Compute the residual spectrogram as the pixel-wise absolute difference of P_i' and P_j (Figure 4a).

$$P_r(u, v) = |P_i'(u, v) - P_j(u, v)| \tag{5}$$

2. Define the function $e(v)$ representing the energy content of the residual spectrogram over time.

$$e(v) = \sum_u P_r(u, v), \qquad v = 1, \ldots, V \tag{6}$$

3. Look for strong variations in the residual energy by computing the first derivative $e'(v)$ and applying an outlier detector (three scaled MAD from the median, where MAD denotes the median absolute deviation), obtaining a set of points $\mathcal{O} = \{v_k\}$ (Figure 4b).
4. Process the points $v_k \in \mathcal{O}$ in order to obtain the interval $[v_1, v_2]$ corresponding to the candidate overdub. The employed criterion is that of selecting the couple of points which maximizes the average energy ratio between the regions inside and outside those points.

$$(v_1, v_2) = \arg \max_{(v_a, v_b) \in \mathcal{O}^2} \frac{\mathbb{E}\left[e(v)\right]_{v_a < v < v_b}}{\mathbb{E}\left[e(v)\right]_{v < v_a \vee v > v_b}} \tag{7}$$

where $\mathbb{E}\left[e(v)\right]_{\mathcal{I}}$ denotes the expectation of $e(v)$ for $v \in \mathcal{I}$.

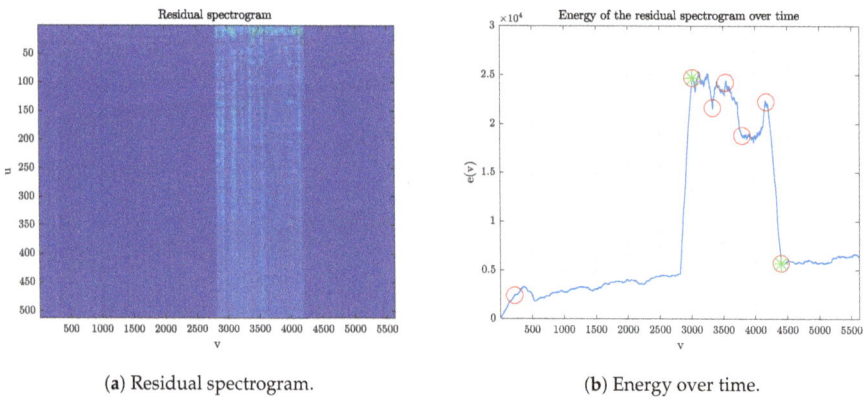

(a) Residual spectrogram.

(b) Energy over time.

Figure 4. Residual spectrogram and related energy-over-time associated with a track pair (i, j) containing an overdub, which appears in (**a**) as a bright region with clean edges. The red circles in (**b**) represent the detected outliers $v_k \in \mathcal{O}$, and the two points marked with green asterisks are the selected edges (v_1, v_2).

Given a detected overdub spanning from v_1 to v_2, the algorithm tries to infer the phylogenetic relation. Again, we compare energy statistics inside and outside the overdub region, but in this case, we consider P'_i and P_j, instead of P_r.

5. Scan through the spectrogram rows $u = 1, \ldots, U$. For each u, compute:

$$\begin{aligned} c_i(u) &= \left| \mathbb{E}\left[P_i\left(u, v\right)\right]_{v_1 < v < v_2} - \mathbb{E}\left[P_i\left(u, v\right)\right]_{v < v_1 \vee v > v_2} \right| \\ c_j(u) &= \left| \mathbb{E}\left[P_j(u, v)\right]_{v_1 < v < v_2} - \mathbb{E}\left[P_j(u, v)\right]_{v < v_1 \vee v > v_2} \right| \end{aligned} \tag{8}$$

which represent the discrepancies between average spectral energy inside and outside the overdubbed region, in P_i and P_j, for each frequency sub-band (row), u. The spectrogram presenting a higher $c(u)$ for the majority of rows u is assumed to be the overdubbed one, i.e., the phylogenetic descendant.

In a similar way to what is done in the leader detection step, if j is chosen as the ancestor of i, the algorithm sets $d_{i,j} = +\infty$. Otherwise, the dissimilarity value computed in (4) is kept.

4.5. Tree Estimation

Once the dissimilarity value $d_{i,j}$ has been computed for every (i, j), the algorithm analyzes the resulting dissimilarity matrix D in order to estimate the phylogenetic tree.

1. Starting from the matrix D, build an undirected graph $\mathcal{G} = \{\mathcal{V}, \mathcal{E}\}$ with N nodes, where the i-th node is associated with the audio track $x_i(n)$ and each edge (i, j) exists if and only if $d_{i,j} < +\infty$ and $d_{j,i} < +\infty$.
2. Run a maximal clique algorithm on $\mathcal{G}$, obtaining $\mathcal{C}_1, \ldots, \mathcal{C}_K \subseteq \mathcal{V}$.
3. Compute the $K \times K$ clique-dissimilarity matrix $D_\mathcal{C}$ as:

$$D_\mathcal{C}(p, q) = \frac{1}{|\mathcal{C}_p||\mathcal{C}_q|} \sum_{i \in \mathcal{C}_p, j \in \mathcal{C}_q} d_{i,j} \tag{9}$$

where $|\cdot|$ denotes the cardinality of a clique.

4. Starting from the matrix $D_\mathcal{C}$, build a complete directed graph $\mathcal{G}_\mathcal{C} = \{\mathcal{V}_\mathcal{C}, \mathcal{E}_\mathcal{C}\}$, with K nodes, where every node is a clique of the undirected graph $\mathcal{G}$ and each edge (p, q) has a weight equal to $D_\mathcal{C}$, corresponding to the average dissimilarity between the audio tracks belonging to the p-th and the q-th cliques.
5. Compute the phylogenetic tree as the minimum spanning arborescence $\hat{\mathcal{G}}_\mathcal{C} = \{\mathcal{V}_\mathcal{C}, \hat{\mathcal{E}}_\mathcal{C}\}$, i.e., the directed rooted spanning tree with minimum weight.

$$\hat{\mathcal{E}}_\mathcal{C} = \arg\min_{\mathcal{E}^s \subset \mathcal{E}_\mathcal{C}} \sum_{(p,q) \in \mathcal{E}^s} D_\mathcal{C}(p, q) \tag{10}$$

In our implementation, $\hat{\mathcal{G}}_\mathcal{C}$ is found via the Chu-Liu/Edmonds optimum branching algorithm [37,38].

5. Dataset

The experiment used to assess the algorithm described in the previous section is based on 10 tests, where the most significant sequences of transformations (with respect to the tape music case) were applied to a set of 10 different tracks. For each track, a set of seven audio samples was created applying different acquisition setups and different tape editings, which will be described in the following paragraphs. We have selected the most representative case study operation in the tape music field in order to provide an accurate evaluation of the proposed method. The sequence of transformations can be characterized by a phylogenetic tree where each edge corresponds to a physical editing of the tape (a cut, the insertion of leader-tape) and each node can include several recording and reading settings. In the latter case, the tape was not modified, and therefore, it cannot be considered as a child in the phylogenetic sense. In the following, we will describe both the adopted reading/writing parameters, together with the tape editing settings.

The original digital audio track was generated recording a 2-min track on a virgin magnetic tape using a professional open reel-to-reel tape recorder: Studer A810. This machine provides four recording/replay speeds: 30 ips, 15 ips, 7.5 ips and 3.75 ips. The Studer A810 provides also a switchable knob to change the equalization. Table 1 shows the time constants of the equalizations for each speed; it is possible to notice that, at 30 ips, the only standard equalization is AES, whereas for other speeds, the CCIR or NAB standards can be applied. Furthermore, at 3.75 ips, only one equalization curve is available. During the recording phase, an external noise reduction system DBXType I was used.

Table 2 shows the original tracks from which the two-minute samples were extracted and the configuration of the machine during the recording phase.

Table 1. Equalization standards supported by the Studer A810 described by their time constants. Source: [39].

30 ips	15 ips	7.5 ips	3.75 ips
AES: 17.5/∞	CCIR: 35/∞	70/∞	90/3180
AES: 17.5/∞	NAB: 50/3180	50/3180	90/3180

Table 2. Samples of electroacoustic music recorded on experimental tapes with the related configuration.

	Samples			Recording Parameters		
#	Composer	Title	Year(s)	Speed	Equation	DBX
1	Luciano Berio	Differences	1958–1959	7.5	CCIR	yes
2	Pierre Boulez	Dialogue de l'ombre double	1985	7.5	CCIR	yes
3	Brian Ferneyhough	Mnemosyne	1986	7.5	CCIR	no
4	Brian Ferneyhough	Mnemosyne	1986	15	CCIR	yes
5	Bruno Maderna	Continuo	1958	15	CCIR	no
6	Bruno Maderna	Dimensioni II—invenzione su una voce	1960	7.5	NAB	yes
7	Bruno Maderna	Notturno	1956	7.5	NAB	no
8	Luigi Nono	...sofferte onde serene...	1976	15	NAB	yes
9	Gruppo NPS	Interferenze II	1965–1968	15	NAB	yes
10	Gruppo NPS	Ricerca 4	1965–1968	15	NAB	no

The first samples created for each tree consist of the digitization of the recorded samples, without any alteration, read with the correct parameter setting. These represent the roots for the respective trees. All the other samples differ from the roots for at least one alteration of the tape or different parameter in the configuration of the machine. The alterations tested in the experiments are:

- addition of a leader-tape within the tape;
- overdub with silence or with another track;
- addition of a splice within the tape.

The latter is obtained cutting the tape at 90° and then joining together the two sides with a splicing tape. Every alteration is chosen randomly and implicates a node in the lower levels of the tree.

Since the writing parameters are unknown to the analyst, the reading setup at the digitization must be guessed. As a matter of fact, our datasets include multiple digitizations of the same tape where the parameters were selected randomly. This choice leads to the availability of multiple digital copies of the same content, which need to be acknowledged as one and fused into one node in the reconstructed phylogenetic tree.

6. Results and Discussion

The proposed methodology was validated by considering three different metrics: (i) accuracy of the leader-tape detector; (ii) accuracy of the overdub detector; (iii) comparison of the estimated phylogenetic tree with the ground-truth.

The performance of the two detection modules are measured and presented here in terms of the probabilities of correct-detection and false-positive, as obtained from the employed dataset. Results are shown in Table 3, where $p(A|A)$ denotes the correct-detection probability for alteration A and $p(A|\neg A)$ denotes the false-positive probability.

Table 3. Correct-detection and false-positive probabilities for leader-tapes and overdubs.

Leader		Overdub	
$p(L|L)$	$p(L|\neg L)$	$p(O|O)$	$p(O|\neg O)$
90.0%	0.0%	75.0%	3.3%

Leader-tape detection turns out to be highly reliable, with a solid 90% rate of correct-detection and no false-positives at all. Overdub detection represents a more complex problem. On the one hand, in some cases, the detector is not able to correctly identify the presence of an overdub, as its spectral fingerprint does not appear sufficiently visible with respect to the background noise or the overdub interval limits are not sufficiently sharp in the difference of the spectrograms. On the other hand, it was observed that cases might occur in which different kinds of tape alterations (e.g., presence/absence of DBX or different equalizations) may produce artifacts that might be confused with those left by an overdub (false-positive).

However, the estimation of the phylogenetic tree does not strictly require 100% accuracy of the detectors. In fact, the tree reconstruction process involves the dissimilarity matrix as a whole, which means that it is usually robust to local noise and errors, as long as the algorithm has gathered enough information.

Since the validation dataset consisted of relatively small trees, results were obtained by qualitatively inspecting the estimated structures in comparison to the ground-truth. Three possible outcomes were observed.

1. In 50% of the cases, the estimated tree perfectly reproduces the ground-truth. Specifically, all the tracks sharing the same tape modifications (leader-tape and/or overdub) are collected in the same clique, and the resulting cliques are correctly ordered in the phylogeny sense.
2. In 40% of the cases, the estimated tree is not identical to the ground-truth, but still makes sense in phylogeny terms. For instance, in some cases, it is possible to observe that certain cliques result in being over-clustered: tracks that should belong to the same meta-node are split into more nodes, which can be siblings or in a parent-child relationship. However, the relative depths in the tree structure are maintained, and the overall phylogenetic sense is preserved. Figure 5 reports a couple of examples of this scenario.
3. In 10% of the cases, the estimated tree shows some wrong phylogenetic relations (ancestor-descendant swaps) with respect to the ground-truth.

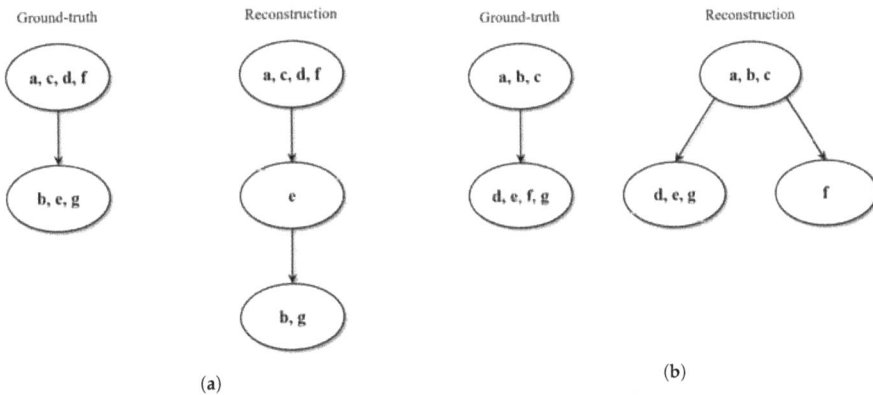

Figure 5. Examples of tree reconstruction with over-clustering errors. Datasets consist of seven audio tracks, $\{a, b, \ldots, g\}$. In (a), cluster $\{b, e, g\}$ is split into the parent-child pair $(\{e\}, \{b, g\})$; in (b), cluster $\{d, e, f, g\}$ is split into the sibling pair $(\{d, e, g\}, \{f\})$.

Finally, it is important to underline that this result assessment does not take into account alterations due to the addition of splices. The performed experiments, in fact, showed that these alterations are barely visible in the spectrogram images, or at least easily confused with other regular spectral features, making their detection problematic within a computer-vision framework. Therefore, ground-truth trees were re-designed by merging the clusters of nodes induced by a splice with their phylogenetic parents, and consequently, the algorithm was expected to reconstruct the trees accordingly.

7. Conclusions

Phylogenetic analysis of tape music is a new emerging branch of computational musicology, which requires new automatized and accurate tools to reconstruct the generation history of different copies of the same audio content. The paper has presented a phylogenetic reconstruction strategy, which relies on digitizing the analyzed tapes and then converting each audio track into a two-dimensional spectrogram. This conversion allows adopting a set of computer vision tools to align and equalize different tracks in order to infer the most likely transformation that converts one track into another. In the presented approach, overdubs, cuts and the insertion of a leader-tape were considered, as these are among the most likely transformations to be estimated in tape phylogeny. The proposed solution presents a satisfying robustness to the adoption of the wrong reading setup (i.e., with speed, equalization and filtering different from those adopted in the creation of the tape), together with a good reconstruction accuracy of the phylogenetic tree. The reconstructed dependencies proved to be correct or plausible (i.e., the temporal order of the audio content is respected in the estimated phylogenetic tree) in 90% of the experimental cases.

Future research work will be devoted to extending the proposed approach to a widened set of editing techniques, intentional and unintentional alterations, and configurations, as well as different tape recorders and syndromes. Moreover, the investigation activity has also highlighted the need for designing new objective evaluation metrics that permit measuring the accuracy of tree reconstruction in the tape music phylogeny context. Machine learning algorithms, such as the ones described in [40], could be used to enhance the phylogenetic algorithms. The same analysis can be applied to a more heterogeneous set of analogue physical support (including vinyl records, phonograph cylinders, etc.) with the final aim of a complete tool for musicological analysis of digitized analogue recordings.

Acknowledgments: The work has been supported by the Phylo4n6 project prot.BIRD165882/16, funded by the University of Padova, Italy.

Author Contributions: Sebastiano Verde designed the adopted software and performed the experimental setting. Niccolò Pretto prepared the experimental dataset, which was used in testing the proposed approach. Simone Milani supervised the software preparation and investigated the state-of-the-art in multimedia phylogeny. Sergio Canazza supervised the creation of the experimental dataset and investigated the state-of-the-art in tape computational musicology. All the authors contributed to the writing of the paper.

Conflicts of Interest: The authors declare no conflict of interest.

References

1. Pousseur, H. *Ecrits Théoriques, 1954–1967*; Editions Pierre Mardaga: Sprimont, Belgium, 2004.
2. Canazza, S. The digital curation of ethnic music audio archives: From preservation to restoration. *Int. J. Digit. Libr.* **2012**, *12*, 121–135.
3. Bressan, F.; Canazza, S. A Systemic Approach to the Preservation of Audio Documents: Methodology and Software Tools. *J. Electr. Comput. Eng.* **2013**, *2013*, 21.
4. Bressan, F.; Rodà, A.; Canazza, S.; Fontana, F.; Bertani, R. The Safeguard of Audio Collections: A Computer Science Based Approach to Quality Control—The Case of the Sound Archive of the Arena di Verona. *Adv. Multimedia* **2013**, *2013*, 14.
5. Bressan, F.; Canazza, S.; Rodá, A.; Bertani, R.; Fontana, F. Pavarotti Sings Again: A Multidisciplinary Approach to the Active Preservation of the Audio Collection at the Arena di Veronach to the Active Preservation to the Active Preservation of the Audio Collection at the Arena di Verona. *J. New Music Res.* **2013**, *42*, 364–380.
6. Canazza, S.; Fantozzi, C.; Pretto, N. Accessing tape music documents on mobile devices. *ACM Trans. Multimedia Comput. Commun. Appl.* **2015**, *12*, 20.
7. Fantozzi, C.; Bressan, F.; Pretto, N.; Canazza, S. Tape music archives: From preservation to access. *Int. J. Digit. Libr.* **2017**, *18*, 233–249.

8.   Van Huis, E. What makes a good archive? *IASA J.* **2009**, *24*, 25–28.

9.   Timpanaro, S. *The Genesis of Lachmann's Method*; University of Chicago Press: Chicago, IL, USA, 2005.

10.   Milani, S.; Fontana, M.; Bestagini, P.; Tubaro, S. Phylogenetic analysis of near-duplicate images using processing age metrics. In Proceedings of the 2016 IEEE International Conference on Acoustics, Speech and Signal Processing (ICASSP), Shanghai, China, 20–25 March 2016.

11.   Milani, S.; Bestagini, P.; Tubaro, S. Video phylogeny tree reconstruction using aging measures. In Proceedings of the 2017 European Signal Processing Conference (EUSIPCO 2017), Kos, Greece, 28 August–2 September 2017.

12.   Nucci, M.; Tagliasacchi, M.; Tubaro, S. A phylogenetic analysis of near-duplicate audio tracks. In Proceedings of the 2013 IEEE 15th International Workshop on Multimedia Signal Processing (MMSP), Pula, Italy, 30 September–2 October 2013; pp. 99–104.

13.   Kennedy, L.; Chang, S.F. Internet image archaeology: Automatically tracing the manipulation history of photographs on the web. In Proceedings of the ACM International Conference on Multimedia (ACM-MM), Vancouver, BC, Canada, 26–31 October 2008.

14.   de O. Costa, F.; Oikawa, M.A.; Dias, Z.; Goldenstein, S.; de Rocha, A.R. Image Phylogeny Forests Reconstruction. *IEEE Trans. Inf. Forensics Sec.* **2014**, *9*, 1533–1546.

15.   Dias, Z.; Goldenstein, S.; Rocha, A. Exploring heuristic and optimum branching algorithms for image phylogeny. *J. Vis. Commun. Image Represent.* **2013**, *24*, 1124–1134.

16.   Melloni, A.; Bestagini, P.; Milani, S.; Tagliasacchi, M.; Rocha, A.; Tubaro, S. Image phylogeny through dissimilarity metrics fusion. In Proceedings of the European Workshop on Visual Information Processing (EUVIP), Paris, France, 10–12 December 2014.

17.   Verde, S.; Milani, S.; Bestagini, P.; Tubaro, S. Audio phylogenetic analysis using geometric transforms. In Proceedings of the 2017 IEEE International Workshop on Information Forensics and Security (WIFS), Rennes, France, 4–7 December 2017.

18.   Zattra, L. The Assembling of Stria by John Chowning: A Philological Investigation. *Comput. Music J.* **2007**, *31*, 38–64.

19.   Sallis, F.; Bertolani, V.; Burle, J.; Zattra, L. *Live-Electronic Music. Composition, Performance and Study*; Routledge: London, UK: 2017.

20.   Orio, N.; Snidaro, L.; Canazza, S.; Foresti, G.L. Methodologies and tools for audio digital archives. *Int. J. Digit. Libr.* **2009**, *10*, 201–220.

21.   Canazza, S.; Orio, N. Digital preservation and access of audio heritage: A case study for phonographic discs. In Proceedings of the 13th Conference on Digital Libraries, Corfu, Greece, 27 September–2 October 2009; pp. 451–454.

22.   Orio, N.; Zattra, L. ACAME—Analyse Comparative Automatique de la Musique Electroacoustique; Musimediane: Paris, France, 2009.

23.   Joly, A.; Buisson, O.; Frelicot, C. Content-Based Copy Retrieval Using Distortion-Based Probabilistic Similarity Search. *IEEE Trans. Multimedia* **2007**, *9*, 293–306.

24.   De Benedictis, A.I. Scrittura e supporti nel Novecento: Alcune riflessioni e un esempio (Ausstrahlung di Bruno Maderna). In *La Scrittura Come Rappresentazione del Pensiero Musicale*; Borio, G., Ed.; ETS: Pisa, Italic, 2004; pp. 237–291.

25.   Dwyer, T. *Composing With Tape Recorders: Musique Concrete for Beginners*; Oxford University Press: London, UK, 1971.

26.   AES. *AES Recommended Practice for Audio Preservation and Restoration—Storage and Handling—Storage of Polyester-Base Magnetic Tape*; AES: New York, NY, USA, 1997 (r2012).

27.   Eilers, D.A. Splicing Tapes and Their Proper Application. *J. Audio Eng. Soc.* **1968**, *16*, 472–476.

28.   Bradley, K. *IASA TC-04 Guidelines in the Production and Preservation of Digital Audio Objects: Standards, Recommended Practices, and Strategies*, 2nd ed.; International Association of Sound and Audio Visual Archives: Aarhus, Denmark, 2009.

29.   Mallinson, J.C. Tutorial review of magnetic recording. *Proc. IEEE* **1976**, *64*, 196–208.

30.   Camras, M. *Magnetic Recording Handbook*; Van Nostrand Reinhold Co.: New York, NY, USA, 1988.

31.   National Association of Broadcaster. *Magnetic Tape Recording and Reproducing (Reel-to-Reel)*; National Association of Broadcasters: Washington, DC, USA, 1965.

32. International Electrotechnical Commission. *BS EN 60094-1:1994 BS 6288-1: 1994 IEC 94-1:1981—Magnetic Tape Sound Recording and Reproducing Systems—Part 1: Specification for General Conditions and Requirements;* International Electrotechnical Commission: Geneve, Switzerland, 1981.
33. Zanoni, M.; Lusardi, S.; Bestagini, P.; Canclini, A.; Sarti, A.; Tubaro, S. Robust music identification approach based on local spectrogram image descriptors. In Proceedings of the 142nd AES Convention, Berlin, Germany, 20–23 May 2017; p. 9763.
34. Williams, D.; Pooransingh, A.; Saitoo, J. Efficient music identification using ORB descriptors of the spectrogram image. *EURASIP J. Audio Speech Music Proc.* **2017**, *2017*, 17.
35. Bay, H.; Ess, A.; Tuytelaars, T.; Gool, L.V. Speeded-up robust features. *Comput. Vis. Image Underst.* **2008**, *110*, 346–359.
36. Fischler, M.A.; Bolles, R.C. Random sample consensus: A paradigm for model fitting with applications to image analysis and automated cartography. *Commun. ACM* **1981**, *24*, 381–395.
37. Chu, Y.J.; Liu, T.H. On the shortest arborescence of a directed graph. *Sci. Sin.* **1965**, *14*, 1396–1400.
38. Edmonds, J. Optimum branchings. *J. Res. Natl. Bur. Stand.* **1967**, *71B*, 233–240.
39. Studer. *Studer A810—Operating and Service Instruction;* Studer: Zurich, Switzerland, 2018.
40. Micheloni, E.; Pretto, N.; Canazza, S. A step toward AI tools for quality control and musicological analysis of digitized analogue recordings: Recognition of audio tape equalizations. In Proceedings of the 11th InternationalWorkshop on Artificial Intelligence for Cultural Heritage Co-Located with the 16th International Conference of the Italian Association for Artificial Intelligence (AI*IA 2017), Bari, Italic, 14–17 November 2017.

applied
sciences

MDPI

Article

SampleCNN: End-to-End Deep Convolutional Neural Networks Using Very Small Filters for Music Classification†

Jongpil Lee, Jiyoung Park, Keunhyoung Luke Kim and Juhan Nam *

Graduate School of Culture Technology, KAIST, 291 Daehak-ro, Yuseong-gu, Daejeon 34141, Korea;
richter@kaist.ac.kr (J.L.); jypark527@kaist.ac.kr (J.P.); dilu@kaist.ac.kr (K.L.K.)
* Correspondence: juhannam@kaist.ac.kr; Tel.: +82-42-350-2926
† This article is a re-written and extended version of "Sample-level Deep Convolutional Neural Networks for music auto-tagging using raw waveforms" presented at SMC 2017, Espoo, Finland on 5 July 2017.

Academic Editor: Meinard Müller
Received: 3 November 2017; Accepted: 17 January 2018; Published: 22 January 2018

Abstract: Convolutional Neural Networks (CNN) have been applied to diverse machine learning tasks for different modalities of raw data in an end-to-end fashion. In the audio domain, a raw waveform-based approach has been explored to directly learn hierarchical characteristics of audio. However, the majority of previous studies have limited their model capacity by taking a frame-level structure similar to short-time Fourier transforms. We previously proposed a CNN architecture which learns representations using sample-level filters beyond typical frame-level input representations. The architecture showed comparable performance to the spectrogram-based CNN model in music auto-tagging. In this paper, we extend the previous work in three ways. First, considering the sample-level model requires much longer training time, we progressively downsample the input signals and examine how it affects the performance. Second, we extend the model using multi-level and multi-scale feature aggregation technique and subsequently conduct transfer learning for several music classification tasks. Finally, we visualize filters learned by the sample-level CNN in each layer to identify hierarchically learned features and show that they are sensitive to log-scaled frequency.

Keywords: convolutional neural networks; music classification; raw waveforms; sample-level filters; downsampling; filter visualization; transfer learning

1. Introduction

Convolutional Neural Networks (CNN) have been applied to diverse machine learning tasks. The benefit of using CNN is that the model can learn hierarchical levels of features from high-dimensional raw data. This end-to-end hierarchical learning has been mainly explored in the image domain since the break-through in image classification [1]. However, the approach has been recently attempted in other domains as well.

In the text domain, a language model is typically built in two steps, first by embedding words into low-dimensional vectors and then by learning a model on top of the word-level vectors. While the word-level embedding plays a vital role in language processing [2], it has limitations in that the embedding space is learned separately from the word-level model. To handle this problem, character-level language models that learn from the bottom-level raw data (e.g., alphabet characters) were proposed and showed that they can yield comparable results to the word-level learning models [3,4].

In the audio domain, raw waveforms are typically converted to time-frequency representations that better capture patterns in complex sound sources. For example, spectrogram and more concise representations such as mel-filterbank are widely used. These spectral representations have served a similar role to the word embedding in the language model in that the mid-level representation are computed separately from the learning model and they are not particularly optimized for the target task. This issue has been addressed by taking raw waveforms directly as input in different audio tasks, for example, speech recognition [5–7], music classification [8–10] and acoustic scene classification [11,12].

However, the majority of previous work have focused on replacing the frame-level time-frequency transforms with a convolutional layer, expecting that the layer can learn parameters comparable to the filter banks. The limitation of this approach was pointed out by Dieleman and Schrauwen [8]. They conducted an experiment of music classification using a simple CNN that takes raw waveforms or mel-spectrogram. Unexpectedly, their CNN models with the raw waveform as input did not produce better results than those with the spectral data as input. The authors attributed this unexpected outcome to three possible causes. First, their CNN models were too simple (e.g., a small number of layers and filters) to learn the complex structure of polyphonic music. Second, the end-to-end models need an appropriate non-linearity function that can replace the log-based amplitude compression in the spectrogram. Third, the first 1D convolutional layer takes raw waveforms in a frame-level which is typically several hundred samples long. The filters in the first 1D convolutional layer should learn all possible phase variations of periodic waveforms within the length. In spectrogram, the phase variation is removed.

We recently tackled the issues by stacking 1D convolutional layers using very small filters instead of a 1D convolutional layer with the frame-level filters, inspired by the VGG networks in image classification that is built with deep stack of 3×3 convolutional layers [13,14]. The sample-level CNN model has filters with very small granularity (e.g., 3 samples) in time for all convolutional layers. The results were comparable to those using mel-spectrogram in music auto-tagging. In this paper, we term the sample-level CNN architecture as SampleCNN and extend the previous work in three ways. First, we should note that SampleCNN takes four times longer training time than a comparable CNN model that takes mel-spectrogram. In order to reduce the training time, we progressively downsample the waveforms and report the effect on performance. By reducing the band-width of music audio this way, we will be able to find the cut-off frequency where the performance starts to become degraded. Second, we extended SampleCNN using multi-level and multi-scale feature aggregation [15]. The technique proved to be highly effective in music classification tasks. We additionally evaluate the extended model in transfer learning settings where the features extracted from SampleCNN can be used for three different datasets in music genre classification and music auto-tagging. We show that the proposed model achieves state-of-the-art results. Third, we visualize learned intermediate layers of SampleCNN to observe how the filters with small granularity process music signals in a hierarchical manner. In particular, we visualize them for each of sampling rates.

2. Related Work

There are a decent number of CNN models that take raw waveforms as input. The majority of them used large-sized filters in the first convolutional layer with various size of strides to capture frequency-selective responses which were carefully designed to handle their target problems. We termed this approach as frame-level raw waveform model because the filter and stride sizes of the first convolutional layer were chosen to be comparable to the window and the hop sizes of short-time Fourier transformation, respectively [5–11].

There are a few work that used small filter and stride sizes in the first convolution layer (8 samples-sized filter [16] and 10 samples-sized filter [17,18] at 16 kHz). However, the CNN models have only two or three convolution layers, which are not sufficient to learn the complex structure of the acoustic signals. In SampleCNN, we deepen the layers even more, thereby reducing the filter and stride sizes of the first convolution layer down to two or three samples.

3. Learning Models

Figure 1 illustrates three CNN models in music auto-tagging that we compare in our experiments. Note that they are actually general architectures and so can be applied to any audio classification tasks. In this section, we describe the three models in detail.

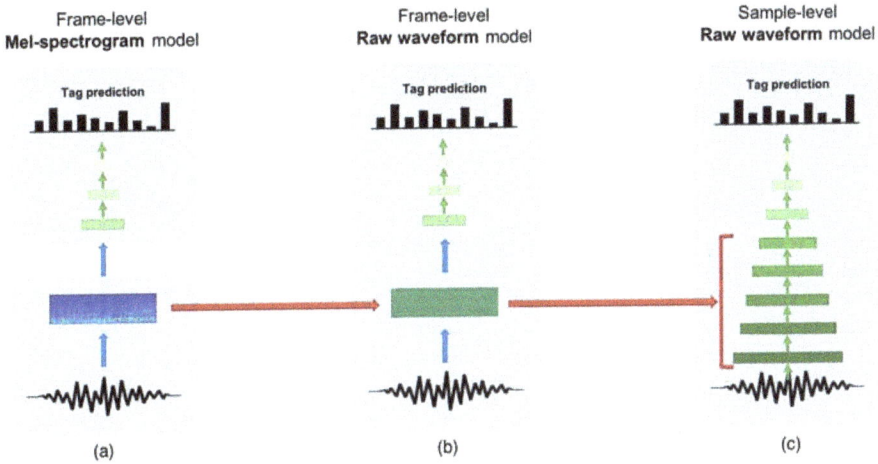

Figure 1. Comparison of (**a**) frame-level model using mel-spectrogram; (**b**) frame-level model using raw waveforms and (**c**) sample-level model using raw waveforms.

3.1. Frame-Level Mel-Spectrogram Model

This is the most common CNN model used in music classification. The time-frequency representation is usually regarded as either two-dimensional images [19,20] or one-dimensional sequence of vectors [8,21]. We only used one-dimensional(1D) CNN model for experimental comparisons because the performance gap between 1D and 2D models is not significant and the 1D model is directly comparable to models using raw waveforms.

3.2. Frame-Level Raw Waveform Model

In this model, a strided convolution layer is added beneath the bottom layer of the frame-level mel-spectrogram model. The strided convolution layer is expected to learn a filter-bank that returns a time-frequency representation. In this model, once the first strided convolution layer slides over the raw waveforms, the output feature map has the same dimensions as the mel-spectrogram. This is because the stride size, filter size, and the number of filters in the first convolution layer correspond to the hop size, window size, and the number of mel-bands in the mel-spectrogram, respectively. This configuration was used for the music auto-tagging task in [8,9] and thus we used it as a baseline model.

3.3. Sample-Level Raw Waveform Model: SampleCNN

As described in Section 1, the approach using raw waveforms should be able to address log-scale amplitude compression and phase-invariance. Simply adding a strided convolution layer is not sufficient to overcome the issues. To improve this, we add multiple layers beneath the frame-level such that the first convolution layer can handle much smaller size of samples. For example, if the stride of the first convolution layer is reduced from 729 ($=3^6$) to 243 ($=3^5$), 3-size convolution layer and max-pooling layer are added to keep the output dimensions in the subsequent convolution layers unchanged. If we repeatedly reduce the stride of the first convolution layer this way, six convolution layers (five pairs of 3-size convolution and max-pooling layer following one 3-size strided convolution layer) will be added (we assume that the temporal dimensionality reduction occurs only through max-pooling and striding while zero-padding is used in convolution to preserve the size).

We generalized the configuration as m^n-SampleCNN where m refers to the filter size (or the pooling size) of intermediate convolution layer modules and n refers to the number of the modules. The first convolutional layer is different from the intermediate convolutional layers in that the stride size is equal to the filter size. An example of m^n-SampleCNN is shown in Table 1 where m is 3 and n is 9. Note that the network is composed of convolution layers and max-pooling only, and so the input size is determined to be *stride size of the first convolutional layer* $\times m^n$. In Table 1, as the stride size of the first convolution layer is 3, the input size is set to be 59049 ($=3 \times 3^9$).

Table 1. SampleCNN configuration. In the first column (Layer), "conv 3-128" indicates that the filter size is 3 and the number of filters is 128.

3^9-SampleCNN Model			
59,049 Samples (2678 ms) as Input			
Layer	Stride	Output	# of Params
conv 3-128	3	$19,683 \times 128$	512
conv 3-128	1	$19,683 \times 128$	49,280
maxpool 3	3	6561×128	
conv 3-128	1	6561×128	49,280
maxpool 3	3	2187×128	
conv 3-256	1	2187×256	98,560
maxpool 3	3	729×256	
conv 3-256	1	729×256	196,864
maxpool 3	3	243×256	
conv 3-256	1	243×256	196,864
maxpool 3	3	81×256	
conv 3-256	1	81×256	196,864
maxpool 3	3	27×256	
conv 3-256	1	27×256	196,864
maxpool 3	3	9×256	
conv 3-512	1	9×512	393,728
maxpool 3	3	3×512	
conv 3-512	1	3×512	786,944
maxpool 3	3	1×512	
conv 1-512	1	1×512	262,656
dropout 0.5	–	1×512	
sigmoid	–	50	25,650
Total params			2.46×10^6

4. Extension of SampleCNN

4.1. Multi-Level and Multi-Scale Feature Aggregation

Music classification tasks, particularly music auto-tagging among others, have a wide variety of labels in terms of genre, mood, instruments and other song characteristics. Especially, they are positioned in different hierarchical levels and time-scales. For example, some words related to instrument ones, such as guitar and saxophone, describe objective sound sources which are usually local and repetitive within a song, whereas other labels related to genre or mood, such as rock and happy, are dependent on a larger context of music and are more complicated. In order to address this issue, we recently proposed multi-level and multi-scale feature aggregation technique [15].

The technique is conducted by combining multiple CNN models. This assumes that the hidden layers of each CNN model represent different levels of features and the models with different input sizes provide even richer feature representations by capturing both local and global characteristics of the music. In [15], they showed that different level and time-scale features have different performance sensitivity to individual tags and thus combining them all together is the best strategy to improve performance. In this work, we replace the simple CNN architectures that take mel-spectrogram as input in [15] with SampleCNNs, taking different input sizes (e.g., 700 ms to 3.5 s). Once we train the SampleCNNs as supervised feature extractors, we slide each of them over a song clip (e.g., about 30 s) and obtain features from the last three hidden layers. We then summarize them by a combination of max-pooling and average-pooling. Finally, we concatenate the multi-level and multi-scale features and feed them to a simple neural networks with two fully-connected layers to make a final prediction.

4.2. Transfer Learning

The multi-level and multi-scale feature aggregation approach can be used in a transfer learning setting by using different datasets or target tasks for the final classification after training the SampleCNNs. Especially, when the target dataset size is comparably small to the model capacity, transferred parameters can yield better performance on the target task rather than parameters trained from the innate target dataset. The applicability of transfer learning using a frame-level raw waveform model has been explored in the speech domain [17]. Here, we examine it using the sample-level raw waveform model for music genre classification and music auto-tagging with different datasets.

5. Experimental Setup

5.1. Datasets

We validate the effectiveness of the proposed method on different sizes of datasets for music genre classification and auto-tagging. All dataset splits are available on the link [22]. The details of each dataset are as follows. The numbers in the parenthesis indicate the split of training, validation and test sets.

- GTZAN [23]: 930 songs (443/197/290) (This is a fault-filtered split designed to avoid the repetition of artists across the training, validation and test sets [24]), genre classification (10 genres).
- MagnaTAgaTune (MTAT) [25]: 21,105 songs (15,244/1529/4332), auto-tagging (50 tags)
- Million Song Dataset with Tagtraum genre annotations (TAGTRAUM): 189,189 songs (141,372/10,000/37,817) (This is a stratified split with 80% training data of the CD2C version [26]), genre classification (15 genres)
- Million Song Dataset with Last.FM tag annotations (MSD) [27]: 241,889 songs (201,680/11,774/28,435), auto-tagging (50 tags)

We primarily examined the proposed model on MTAT and then verified the effectiveness of our model on MSD which is much larger than MTAT (MTAT contains 170 h long audio and MSD contains 1955 h long audio in total). We filtered out the tags and used most frequently labeled 50 tags in both datasets, following the previous work [8,19,20]. Also, all songs in the two datasets were trimmed to 29.1 s long. For transfer learning experiments, the model is first trained with the largest dataset, MSD, and the pre-trained networks are transferred to other three datasets. The evaluation is conducted with area under receiver operating characteristic (AUC) for auto-tagging datasets and accuracy for genre classification datasets.

5.2. Training Details

We used sigmoid activation for the output layer and binary cross entropy loss as the objective function to optimize. For every convolution layer, we used batch normalization [28] and ReLU activation. We should note that, in our experiments, batch normalization plays a vital role in training the deep models that take raw waveforms. We applied dropout of 0.5 to the output of the last convolution layer and minimized the objective function using stochastic gradient descent with 0.9 Nesterov momentum. The learning rate was initially set to 0.01 and decreased by a factor of 5 when the validation loss did not decrease more than 3 epochs. A total decrease of 4 times, the learning rate of the last training was 0.000016. Also, we used batch size of 23 for MTAT and 50 for MSD, respectively.

5.3. Mel-Spectrogram and Raw Waveforms

In the mel-spectrogram experiments, window sizes of 3^6, 3^5 and 3^4 are used to match up to the filter sizes in the first convolution layer of the raw waveform model as shown in Table 2. FFT size was set to 729 (=3^6) in all experiments. When the window is less than the FFT size, we zero-padded the windowed frame. The linear frequency in the magnitude spectrum is mapped to 128 mel-bands and the magnitude compression is applied with a nonlinear curve, $\log(1 + C|A|)$ where A is the magnitude and C is set to 10. Also, we conducted the input normalization simply by dividing the standard deviation after subtracting mean value of entire input data. On the other hand, we did not perform the input normalization for raw waveforms.

Table 2. Comparison of three CNN models with different window size (filter size) and hop size (stride size). n represents the number of intermediate convolution and max-pooling layer modules, thus 3^n times hop (stride) size of each model is equal to the number of input samples.

3^n Models, 59,049 Samples as Input	n	Window Size (Filter Size)	Hop Size (Stride Size)	AUC
Frame-level (mel-spectrogram)	4	729	729	0.9000
	5	729	243	0.9005
	5	243	243	0.9047
	6	243	81	0.9059
	6	81	81	0.9025
Frame-level (raw waveforms)	4	729	729	0.8655
	5	729	243	0.8742
	5	243	243	0.8823
	6	243	81	0.8906
	6	81	81	0.8936
Sample-level (raw waveforms)	7	27	27	0.9002
	8	9	9	0.9030
	9	3	3	0.9055

As described in Section 3.3, m refers to the filter size (which can be compared to a window size of FFT in the spectrogram) or pooling size (which also can be compared to a hop size of FFT in the spectrogram) of the intermediate convolution layer modules, and n refers to the number of the modules. In our previous work, we adjusted m from 2 to 5 and increased n according to the configuration of m^n-SampleCNN [13]. Among them, 3^9-SampleCNN model with 59049 samples as input worked best and thus we fix our baseline model to it. In this configuration, we can increase the filter size and stride size in the first layer by decreasing the layer depth to conduct comparison experiments between the frame-level models and the sample-level model. For example, if the hop size or the stride size of the first convolutional layer is 729 in either the frame-level mel-spectrogram model or the frame-level raw waveform model, 4 convolutional modules with 3-sized filters are added when the input size is 59,049 samples.

5.4. Downsampling

The downsampling experiments are performed using the MTAT dataset. 3^9-SampleCNN model is used with audio input sampled at 22,050 Hz. For other sampling rate experiments, we slightly modified the model configuration so that the models used for different sampling rate can have similar architecture and similar input seconds to those used in 22,050 Hz. In our previous work [13], we found that the filter size did not significantly affect performance once it reaches the sample-level (e.g., 2 to 5 samples), while the input size of the network and total layer depth are important. Thus, we configured the models as described in Table 3. For example, if the sampling rate is 2000 Hz, the first four modules use 3-sized filters and the rest 6 modules use 2-sized filters to make the total layer depth similar to the 3^9-SampleCNN. Also, 3-sized filters are used for the first four modules in all models for fairly visualizing learned filters.

Table 3. Models, input sizes and number of parameters used in the downsampling experiment. In the third column (Models), each digit from left to right stands for the filter size (or the pooling size) of the convolutional module of SampleCNN from bottom to top. Thus, the number of digits represents the layer depth of each model.

Sampling Rate	Input (in Milliseconds)	Models	# of Parameters
2000 Hz	5184 samples (2592 ms)	3-3-3-3-2-2-2-2-2	1.80×10^6
4000 Hz	10,368 samples (2592 ms)	3-3-3-2-2-2-4-2-2	1.93×10^6
8000 Hz	20,736 samples (2592 ms)	3-3-3-3-2-2-4-4-2-2	2.06×10^6
12,000 Hz	31,104 samples (2592 ms)	3-3-3-3-2-4-4-2-2	2.13×10^6
16,000 Hz	43,740 samples (2733 ms)	3-3-3-3-3-3-5-2-2	2.19×10^6
20,000 Hz	52,488 samples (2624 ms)	3-3-3-3-3-3-3-4-2	2.32×10^6
22,050 Hz	59,049 samples (2678 ms)	3-3-3-3-3-3-3-3-3	2.46×10^6

5.5. Combining Multi-Level and Multi-Scale Features

For the multi-level and multi-scale experiments described in Table 4, we used total 8 models including 2^{13}, 2^{14}, 3^8, 3^9, 4^6, 4^7, 5^5 and 5^6-SampleCNNs. Also, two fully connected layers with 4096 neurons in each layer are used as classifier.

Table 4. Comparison of various multi-scale feature combinations. Only the MTAT dataset was used.

Features from SampleCNNs Last 3 Layers (Pre-trained with MTAT)	MTAT
3^9 model	0.9046
3^8 and 3^9 models	0.9061
2^{13}, 2^{14}, 3^8 and 3^9 models	0.9061
2^{13}, 2^{14}, 3^8, 3^9, 4^6, 4^7, 5^5 and 5^6 models	0.9064

5.6. Transfer Learning

The source task for the transfer learning is fixed to music auto-tagging using MSD because the dataset contains the largest set of music. In this experiment, 3^9-SampleCNN was used. We examined the proposed model on three target datasets for genre classification and auto-tagging. We also examined the performance differences when using features from multiple levels of the pre-trained CNNs and also their combinations.

6. Results and Discussion

6.1. Mel-Spectrogram and Raw Waveforms

Table 2 shows that the sample-level raw waveform model achieves results comparable to the frame-level mel-spectrogram model. Specifically, we found that using a smaller hop size (81 samples $\approx$ 4 ms) worked better than those of conventional approaches (about 20 ms) in the frame-level mel-spectrogram model. However, if the hop size is less than 4 ms, the performance degraded. An interesting finding from the result of the frame-level raw waveform model is that when the filter length is larger than the stride, the accuracy is slightly lower than the models with the same filter length and stride. We interpret that this result is due to the learning ability of the phase variance. As the filter size decreases, the extent of phase variance that the filters should learn is reduced.

6.2. Effect of Downsampling

During the experiments, we observed that the training time of the proposed SampleCNN is about four times longer than the frame-level mel-spectrogram model because the proposed model has more network parameters with deeper layers. In order to reduce the training time, we downsampled the audio with a set of lower sampling rates including 2000, 4000, 8000, 12,000, 16,000, 20,000 Hz. This can be regarded as a time-domain counterpart of in linear-to-mel mapping in that both reduce the dimensionality of input and preserve low-frequency content. The results in Table 5 show that the performance is maintained down to 8000 Hz but it starts to be degraded from 4000 Hz. This may indicate that the relevant information to the task is concentrated below 4000 Hz (the Nyquist frequency of 8000 Hz). Also, we report the training time ratio of the models taking re-sampled audio to the model using 22,050 Hz signal as input. At the expence of the accuracy, the training time can be reduced to about half.

Table 5. Effect of downsampling on the performance and training time. MTAT is used in the experiments. We matched the depth of the models taking different sampling rate to the 3^9-SampleCNN. For example, if the sampling rate is 2000 Hz, the first four convolutional modules use 3-sized filters and the rest 6 modules use 2-sized filters to make the total layer depth similar to the 3^9-SampleCNN.

Sampling Rate	Training Time (Ratio to 22,050 Hz)	AUC
2000 Hz	0.23	0.8700
4000 Hz	0.41	0.8838
8000 Hz	0.55	0.9031
12,000 Hz	0.69	0.9033
16,000 Hz	0.79	0.9033
20,000 Hz	0.86	0.9055
22,050 Hz	1.00	0.9055

6.3. Effect of Multi-Level and Multi-Scale Features

To measure the effect of multi-level and multi-scale feature combination, we experimented with several settings in Table 4. The SampleCNN models are first trained on MTAT dataset, then this pre-trained networks are used as feature extractors for the MTAT dataset again. The results show that as more features are fusioned, the performance increases. This can be viewed similar to an ensemble method, however our approach is distinguished from it in that the feature aggregation is performed on activations of the hidden layers, not on the prediction values.

6.4. Transfer Learning and Comparison to State-of-the-Arts

In Table 6, we show the performance of the SampleCNN model and the transfer learning experiments (the bottom four lines). The results achieved state-of-the-art results on three datasets except for MSD. However, when considering that the model used in [15] utilized both multi-level and multi-scale features, the AUC score (0.8842) obtained from multi-level features only seems to be reasonable. Also, we can see that the multi-level and multi-scale aggregation technique generally improves the performance, particularly in GTZAN.

Table 6. Comparison with previous work. We report SampleCNN results on MagnaTAgaTune (MTAT) and Million Song Dataset (MSD). Furthermore, the result acquired from multi-level and multi-scale feature aggregation technique is also reported at the bottom 4 lines. "-*n* LAYER" indicates features of *n* layers below from the output are used for the transfer learning setting.

MODEL	GTZAN (Acc.)	MTAT (AUC)	TAGTRUM (Acc.)	MSD (AUC)
Bag of multi-scaled features [29]	-	0.898	-	-
End-to-end [8]	-	0.8815	-	-
Transfer learning [30]	-	0.8800	-	-
Persistent CNN [31]	-	0.9013	-	-
Time-frequency CNN [32]	-	0.9007	-	-
Timbre CNN [33]	-	0.8930	-	-
2-D CNN [19]	-	0.8940	-	0.851
CRNN [20]	-	-	-	0.862
2-D CNN [24]	0.632	-	-	-
Temporal features [34]	0.659	-	-	-
CNN using artist-labels [35]	0.7821	0.8888	-	-
multi-level and multi-scale features (pre-trained with MSD) [15]	0.720	0.9021	0.766	0.8878
SampleCNN (3^9 model) [13]	-	0.9055	-	0.8812
−3 layer (pre-trained with MSD)	0.778	0.8988	0.760	0.8831
−2 layer (pre-trained with MSD)	0.811	0.8998	0.768	0.8838
− 1 layer (pre-trained with MSD)	0.821	0.8976	0.768	0.8842
last 3 layers (pre-trained with MSD)	0.805	0.9018	0.768	0.8842

7. Visualization

In this section, we investigate two visualization techniques that can broaden our understanding of the learned hierarchical features in SampleCNN.

7.1. Learned Filters

Previous work in the music domain is limited to visualizing learned filters only on the first convolution layer [8,9,36] or visualizing responses after a filter is applied on a specific input [37,38]. The gradient ascent method has been proposed for directly seeing what is learned at a filter [39] and this technique has provided deeper understanding of what convolutional neural networks learn from images [40,41]. We applied the technique to our SampleCNN to observe how each filter in a layer processes the raw waveforms. The gradient ascent method is as follows. First, we generate random noise and back-propagate the errors in the network. The loss is set to the target filter activation. Then, we add the bottom gradients to the input with gradient normalization. By repeating this process several times, we can obtain the accumulated gradients-based waveform like signal at the input which is optimized to maximize the target filter activation. Examples of learned filters at each layer are in Figure 2. Although we can find the patterns that low-frequency filters are more visible along the layer, the estimated filters are still noisy. To show the patterns more clearly, we visualized them as spectrum in the frequency domain and sorted them by the frequency of the peak magnitude in Figure 3.

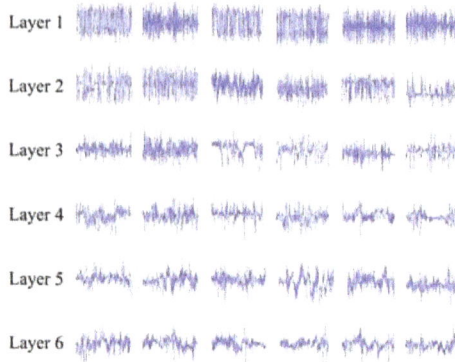

Figure 2. Examples of learned filters at each layer.

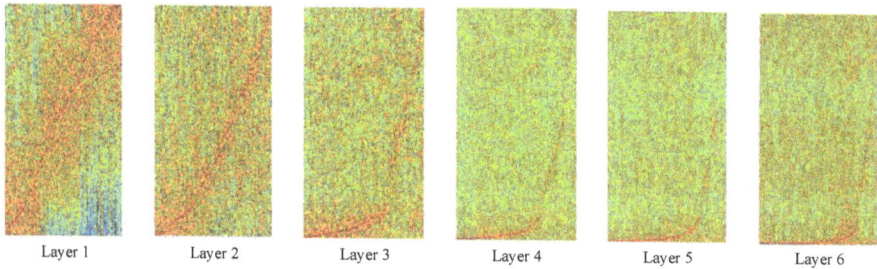

Layer 1　Layer 2　Layer 3　Layer 4　Layer 5　Layer 6

Figure 3. Spectrum of the estimated filters in the intermediate layers of SampleCNN which are sorted by the frequency of the peak magnitude. The x-axis represents the index of the filter, and the y-axis represents the frequency ranged from 0 to 11 kHz. The model used for visualization is 3^9-SampleCNN with 59,049 samples as input. Visualization was performed using the gradient ascent method to obtain the accumulated gradient-based input waveform like signal that maximizes the activation of a filter in the layers. To effectively find the filter characteristics, we set the input size to 729 samples which is close to a typical frame size.

Note that we set the input waveform estimate to 729 samples in length because, if we initialize and back-propagate to the whole input size of the networks, the estimated filters will have large dimensions such as 59,049 samples in computing spectrum. Thus, the results are equivalent to spectra from a typical frame size. The layer 1 shows the three distinctive filter bands which are possible with the filter size with 3 samples (say, a DFT size of 3). The center frequency of the filter banks increases linearly in low frequency filter banks but, as the layer goes up, it progressively becomes steeper in high frequency filter banks. This nonlinearity was found in learned filters with a frame-level end-to-end learning [8] and also in perceptual pitch scales such as mel or bark.

Finally, we visualized spectrum of the learned filter for each sampling rate up to 4th layers. In Figure 4, we can observe that all SampleCNN models focus (or zoom in) on the important low-frequency bands. We can also find that they show similar non-linear patterns to those in Figure 3.

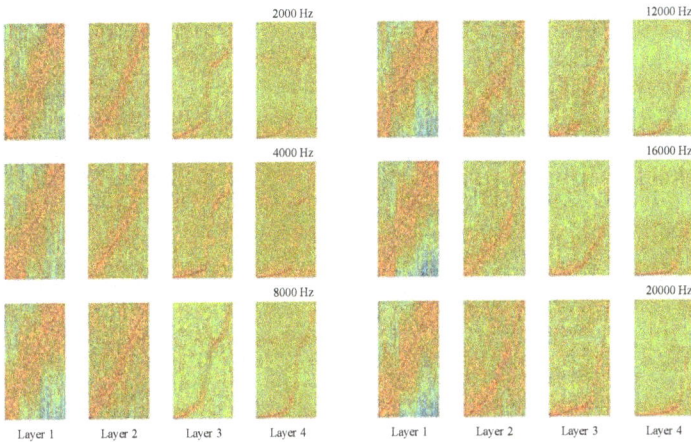

Figure 4. Spectrum visualization of learned filters for different sampling rates. The *x*-axis represents the index of the filter, and the *y*-axis represents the frequency ranged from 0 to half the sampling rate. 3-sized filters are used for the first four modules in all models for fairly visualizing learned filters.

7.2. Song-Level Similarity Using t-SNE

We extracted features from SampleCNN and aggregated them at different hierarchical levels of layer for each audio clip. We then embedded the song-level features into 2-D vectors using t-Distributed Stochastic Neighbor Embedding (t-SNE). Figure 5 visualizes the 2-D embedded features at different layer levels for selected tags to examine how multi-level feature aggregation technique enhances the performance. Songs with genre tag (*Techno*) are more closely clustered in the higher layer (-1 layer). On the other hand, songs with instrument tag (*Piano*) are more closely clustered in the lower layer (-3 layer). This may indicate that the optimal layer of feature representations can be different depending on the type of labels. Thus, combining different levels of features can improve the performance.

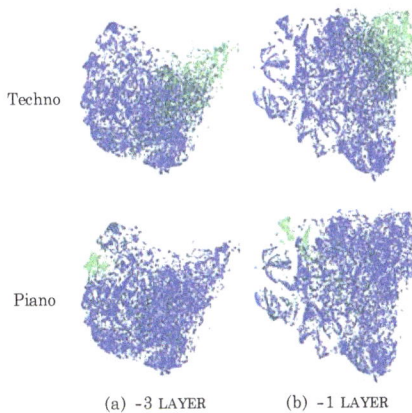

(a) -3 LAYER   (b) -1 LAYER

Figure 5. Feature visualization on songs with *Piano* tag and songs with *Techno* tag on MTAT using t-SNE. Features are extracted from (**a**) -3 LAYER and (**b**) -1 LAYER of the 3^9-SampleCNN model pre-trained with MSD.

8. Conclusions

In this article, we extend our previously proposed SampleCNN for music classification. Through the experiments, we found that downsampling music audio down to 8000 Hz does not significantly degrade performance but it saves training time. Second, transfer learning experiments with multi-level and multi-scale technique showed state-of-the-art results on most of the datasets we tested. Finally, we visualized the spectrum of the learned filters for each sampling rate and found that the SampleCNN model is actively focusing on (or zoom in on) important low-frequency bands. As future work, we will analyze why the sample-level architecture works well without input normalization and nonlinear function that compresses the amplitude, which are important when we use spectrogram as input. Also, we will investigate different filter visualization techniques to interpret the hierarchically-learned filters better.

Acknowledgments: This research was supported by Basic Science Research Program through the National Research Foundation of Korea funded by the Ministry of Science, ICT & Future Planning (2015R1C1A1A02036962) and Korea Advanced Institute of Science and Technology (G04140049).

Author Contributions: Jongpil Lee and Juhan Nam conceived and designed the experiments; Jongpil Lee and Jiyoung Park performed the experiments; Keunhyoung Luke Kim analyzed the data; Jongpil Lee and Juhan Nam wrote the paper.

Conflicts of Interest: The authors declare no conflict of interest.

References

1. Krizhevsky, A.; Sutskever, I.; Hinton, G.E. Imagenet classification with deep convolutional neural networks. In Proceedings of the 25th International Conference on Neural Information Processing Systems, Lake Tahoe, NV, USA, 3–6 December 2012; pp. 1097–1105.
2. Mikolov, T.; Sutskever, I.; Chen, K.; Corrado, G.S.; Dean, J. Distributed representations of words and phrases and their compositionality. In Proceedings of the 26th International Conference on Neural Information Processing Systems, Lake Tahoe, NV, USA, 5–10 December 2013; pp. 3111–3119.
3. Zhang, X.; Zhao, J.; Yann, L. Character-level convolutional networks for text classification. In Proceedings of the 28th International Conference on Neural Information Processing Systems, Montreal, QC, Canada, 7–12 December 2015; pp. 649–657.
4. Kim, Y.; Jernite, Y.; Sontag, D.; Rush, A.M. Character-aware neural language models. *arXiv* **2016**, arXiv:1508.06615.
5. Sainath, T.N.; Weiss, R.J.; Senior, A.W.; Wilson, K.W.; Vinyals, O. Learning the speech front-end with raw waveform CLDNNs. In Proceedings of the 16th Annual Conference of the International Speech Communication Association, Dresden, Germany, 6–10 September 2015; pp. 1–5.
6. Collobert, R.; Puhrsch, C.; Synnaeve, G. Wav2letter: An end-to-end convnet-based speech recognition system. *arXiv* **2016**, arXiv:1609.03193.
7. Zhu, Z.; Engel, J.H.; Hannun, A. Learning multiscale features directly from waveforms. In Proceedings of the Annual Conference of the International Speech Communication Association, San Francisco, CA, USA, 8–12 September 2016.
8. Dieleman, S.; Schrauwen, B. End-to-end learning for music audio. In Proceedings of the IEEE International Conference on Acoustics, Speech and Signal Processing, Florence, Italy, 4–9 May 2014; pp. 6964–6968.
9. Ardila, D.; Resnick, C.; Roberts, A.; Eck, D. Audio deepdream: Optimizing raw audio with convolutional networks. In Proceedings of the International Society for Music Information Retrieval Conference, New York, NY, USA, 7–11 August 2016.
10. Thickstun, J.; Harchaoui, Z.; Kakade, S. Learning features of music from scratch. *arXiv* **2017**, arXiv:1611.09827.
11. Dai, W.; Dai, C.; Qu, S.; Li, J.; Das, S. Very deep convolutional neural networks for raw waveforms. In Proceedings of the 2017 IEEE International Conference on Acoustics, Speech and Signal Processing, New Orleans, LA, USA, 5–9 March 2017; pp. 421–425.

Appl. Sci. **2018**, *8*, 150

12. Aytar, Y.; Vondrick, C.; Torralba, A. Soundnet: Learning sound representations from unlabeled video. In Proceedings of the International Conference on Neural Information Processing Systems, Barcelona, Spain, 5–10 December 2016; pp. 892–900.

13. Lee, J.; Park, J.; Kim, K.L.; Nam, J. Sample-level deep convolutional neural networks for music auto-tagging using raw waveforms. In Proceedings of the Sound Music Computing Conference (SMC), Espoo, Finland, 5–8 July 2017; pp. 220–226.

14. Simonyan, K.; Zisserman, A. Very deep convolutional networks for large-scale image recognition. In Proceedings of the International Conference on Learning Representations, San Diego, CA, USA, 7–9 May 2015.

15. Lee, J.; Nam, J. Multi-level and multi-scale feature aggregation using pre-trained convolutional neural networks for music auto-tagging. *IEEE Signal Process. Lett.* **2017**, *24*, 1208–1212.

16. Tokozume, Y.; Harada, T. Learning environmental sounds with end-to-end convolutional neural network. In Proceedings of the IEEE 2017 IEEE International Conference on Acoustics, Speech and Signal Processing, New Orleans, LA, USA, 5–9 March 2017; pp. 2721–2725.

17. Palaz, D.; Doss, M.M.; Collobert, R. Convolutional neural networks-based continuous speech recognition using raw speech signal. In Proceedings of the IEEE 2015 IEEE International Conference on Acoustics, Speech and Signal Processing, South Brisbane, Queensland, Australia, 19–24 April 2015; pp. 4295–4299.

18. Palaz, D.; Collobert, R.; Magimai-Doss, M. Analysis of CNN-based speech recognition system using raw speech as input. In Proceedings of the 16th Annual Conference of the International Speech Communication Association, Dresden, Germany, 6–10 September 2015; pp. 11–15.

19. Choi, K.; Fazekas, G.; Sandler, M. Automatic tagging using deep convolutional neural networks. In Proceedings of the 17th International Society of Music Information Retrieval Conference, New York, NY, USA, 7–11 August 2016; pp. 805–811.

20. Choi, K.; Fazekas, G.; Sandler, M.; Cho, K. Convolutional recurrent neural networks for music classification. In Proceedings of the IEEE 2017 IEEE International Conference on Acoustics, Speech and Signal Processing, New Orleans, LA, USA, 5–9 March 2017; pp. 2392–2396.

21. Pons, J.; Lidy, T.; Serra, X. Experimenting with musically motivated convolutional neural networks. In Proceedings of the IEEE International Workshop on Content-Based Multimedia Indexing (CBMI), Bucharest, Romania, 15–17 June 2016; pp. 1–6.

22. Lee, J. Music Dataset Split. Available online: https://github.com/jongpillee/music_dataset_split (accessed on 22 January 2017).

23. Tzanetakis, G.; Cook, P. Musical genre classification of audio signals. *IEEE Trans. Speech Audio Process.* **2002**, *10*, 293–302.

24. Kereliuk, C.; Sturm, B.L.; Larsen, J. Deep learning and music adversaries. *IEEE Trans. Multimed.* **2015**, *17*, 2059–2071.

25. Law, E.; West, K.; Mandel, M.I.; Bay, M.; Downie, J.S. Evaluation of algorithms using games: The case of music tagging. In Proceedings of the International Society for Music Information Retrieval Conference, Kobe, Japan, 26–30 October 2009; pp. 387–392.

26. Schreiber, H. Improving genre annotations for the million song dataset. In Proceedings of the International Society for Music Information Retrieval Conference, Malaga, Spain, 26–30 October 2015; pp. 241–247.

27. Bertin-Mahieux, T.; Ellis, D.P.; Whitman, B.; Lamere, P. The million song dataset. In Proceedings of the International Society for Music Information Retrieval Conference, Miami, FL, USA, 24–28 October 2011; Volume 2, pp. 591–596.

28. Ioffe, S.; Szegedy, C. Batch normalization: Accelerating deep network training by reducing internal covariate shift. In Proceedings of the The 32nd International Conference on Machine Learning, Lille, France, 6–11 July 2015.

29. Dieleman, S.; Schrauwen, B. Multiscale approaches to music audio feature learning. In Proceedings of the International Society for Music Information Retrieval Conference, Curitiba, Brazil, 4–8 November 2013; pp. 116–121.

30. Van Den Oord, A.; Dieleman, S.; Schrauwen, B. Transfer learning by supervised pre-training for audio-based music classification. In Proceedings of the International Society for Music Information Retrieval Conference, Taipei, Taiwan, 27–31 October 2014.

31. Liu, J.Y.; Jeng, S.K.; Yang, Y.H. Applying topological persistence in convolutional neural network for music audio signals. *arXiv* **2016**, arXiv:1608.07373.

32. Güçlü, U.; Thielen, J.; Hanke, M.; van Gerven, M.; van Gerven, M.A. Brains on beats. In Proceedings of the International Conference on Neural Information Processing Systems, Barcelona, Spain, 5–10 December 2016; pp. 2101–2109.

33. Pons, J.; Slizovskaia, O.; Gong, R.; Gómez, E.; Serra, X. Timbre analysis of music audio Signals with convolutional neural networks. In Proceedings of the 2017 25th European Signal Processing Conference, Kos Island, Greece, 28 August–2 September 2017.

34. Jeong, I.Y.; Lee, K. Learning temporal features using a deep neural network and its application to music genre classification. In Proceedings of the International Society for Music Information Retrieval Conference, New York, NY, USA, 7–11 August 2016; pp. 434–440.

35. Park, J.; Lee, J.; Park, J.; Ha, J.W.; Nam, J. Representation learning of music using artist labels. *arXiv* **2017**, arXiv:1710.06648.

36. Choi, K.; Fazekas, G.; Sandler, M.; Kim, J. Auralisation of deep convolutional neural networks: Listening to learned features. In Proceedings of the International Society for Music Information Retrieval Conference, Malaga, Spain, 26–30 October 2015; pp. 26–30.

37. Bittner, R.M.; McFee, B.; Salamon, J.; Li, P.; Bello, J.P. Deep salience representations for f0 estimation in polyphonic music. In Proceedings of the International Society for Music Information Retrieval Conference, Suzhou, China, 23–28 October 2017.

38. Choi, K.; Fazekas, G.; Sandler, M. Explaining deep convolutional neural networks on music classification. *arXiv* **2016**, arXiv:1607.02444.

39. Erhan, D.; Bengio, Y.; Courville, A.; Vincent, P. *Visualizing Higher-Layer Features of a Deep Network;* University of Montreal: Montreal, QC, Canada, 2009; Volume 1341, p. 3.

40. Zeiler, M.D.; Fergus, R. *Visualizing and Understanding Convolutional Networks;* Springer: Berlin, Germany, 2014; pp. 818–833.

41. Nguyen, A.; Yosinski, J.; Clune, J. Deep neural networks are easily fooled: High confidence predictions for unrecognizable images. In Proceedings of the Computer Vision and Pattern Recognition, Boston, MA, USA, 7–12 June 2015; pp. 427–436.

applied
sciences

MDPI

Article

Analyzing Free-Hand Sound-Tracings of Melodic Phrases

Tejaswinee Kelkar *and Alexander Refsum Jensenius

University of Oslo, Department of Musicology, RITMO Centre for Interdisciplinary Studies in Rhythm,
Time and Motion, 0371 Oslo, Norway; a.r.jensenius@imv.uio.no
* Correspondence: tejaswinee.kelkar@imv.uio.no; Tel.: +47-4544-8254

Academic Editor: Vesa Valimaki
Received: 31 October 2017; Accepted: 15 January 2018; Published: 18 January 2018

Abstract: In this paper, we report on a free-hand motion capture study in which 32 participants
'traced' 16 melodic vocal phrases with their hands in the air in two experimental conditions.
Melodic contours are often thought of as correlated with vertical movement (up and down) in time,
and this was also our initial expectation. We did find an arch shape for most of the tracings,
although this did not correspond directly to the melodic contours. Furthermore, representation
of pitch in the vertical dimension was but one of a diverse range of movement strategies used to
trace the melodies. Six different mapping strategies were observed, and these strategies have been
quantified and statistically tested. The conclusion is that metaphorical representation is much more
common than a 'graph-like' rendering for such a melodic sound-tracing task. Other findings include
a clear gender difference for some of the tracing strategies and an unexpected representation of
melodies in terms of a small object for some of the Hindustani music examples. The data also show
a tendency of participants moving within a shared 'social box'.

Keywords: motion; melody; shape; sound-tracing; multi-modality

1. Introduction

How do people think about melodies as shapes? This question comes out of the authors' general
interest in understanding more about how spatiotemporal elements influence the cognition of music.
When it comes to the topic of melody and shape, these terms often seem to be interwoven. In fact,
the Concise Oxford Dictionary of Music defines melody as: "A succession of notes, varying in pitch,
which has an organized and recognizable shape." [1]. Here, shape is embedded as a component in
the very definition of melody. However, what is meant by the term 'melodic shape', and how can we
study such melodic shapes and their typologies?

Some researchers have argued for thinking of free-hand movements to music (or 'air instrument
performance' [2]) as visible utterances similar to co-speech gestures [3–5]. From the first author's
experience as an improvisational singer, a critical part of learning a new singing culture was the
physical representation of melodic content. This physical representation includes bodily posture,
gestural vocabulary and the use of the body to communicate sung phrases. In improvised music,
this also includes the way in which one uses the hands to guide the music and the expectation of
a familiar audience from the performing body. These body movements may refer to spatiotemporal
metaphors, quite like the ones used in co-speech gestures.

In their theory of cognitive metaphors, Lakoff and Johnson point out how the metaphors
in everyday life represent the structure through which we conceptualize one domain with the
representation of another [6]. Zbikowski uses this theory to elaborate how words used to describe
pitches in different languages are mapped onto the metaphorical space of the 'vertical dimension' [7].
Descriptions of melodies often use words related to height, for example: a 'high'- or 'low'-pitched

voice, melodies going 'up' and 'down'. Shayan et al. suggest that this mapping might be more strongly present in Western cultures, while the use of other metaphors in other languages, such as thick and thin pitch, might explain pitch using other non-vertical one-dimensional mapping schemata [8]. The vertical metaphor, when tested with longer melodic lines, shows that we respond non-linearly to the vertical metaphors of static and dynamic pitch stimuli [9]. Research in music psychology has investigated both the richness of this vocabulary and its perceptual and metaphorical allusions [10]. However, the idea that the vertical dimension is the most important schema of melodic motion is very persistent [11]. Experimentally, pitch-height correspondences are often elicited by comparing two or three notes at a time. However, when stimuli become more complex, resembling real melodies, the persistence of pitch verticality is less clear. For 'real' melodies, shape descriptions are often used, such as arches, curves and slides [12].

In this paper, we investigate shape descriptions through a sound-tracing approach. This was done by asking people to listen to melodic excerpts and then move their body as if their movement was creating the sound. The aim is to answer the following research questions:

1. How do people present melodic contour as shape?
2. How important is vertical movement in the representation of melodic contour in sound-tracing?
3. Are there any differences in sound-tracings between genders, age groups and levels of musical experience?
4. How can we understand the metaphors in sound-tracings and quantify them from the data obtained?

The paper starts with an overview of related research, before the experiment and various analyses are presented and discussed.

2. Background

Drawing melodies as lines has seemed intuitive across different geographies and time periods, from the Medieval neumes to contemporary graphical scores. Even the description of melodies as lines enumerates some of their key properties: continuity, connectedness and appearance as a shape. Most musical cultures in the world are predominantly melodic, which means that the central melodic line is important for memorability. Melodies display several integral patterns of organization and time-scales, including melodic ornaments, motifs, repeating patterns, themes and variations. These are all devices for organizing melodic patterns and can be found in most musical cultures.

2.1. Melody, Prosody and Memory

Musical melodies may be thought of as closely related to language. For example, prosody, which can also be described as 'speech melody,' is essential for understanding affect in spoken language. Musical and linguistic experiences with melody can often influence one another [13]. Speech melodies and musical melodies are differentiated on the basis of variance of intervals, delineation and discrete pitches as scales [14]. While speech melodies show more diversity in intonation, there is lesser diversity in prosodic contours internally within a language. Analysis of these contours is used for recognition of languages, speakers and dialects in computation [15].

It has been argued that tonality makes musical melodies more memorable than speech melodies [16], while contour makes them more recognizable, especially in unfamiliar musical styles [17]. Dowling et al. suggest that adults use contour to recognize unfamiliar melodies, even when they have been transposed or when intervals are changed [16]. There is also neurological evidence supporting the idea that contour memory is independent of absolute pitch location [18]. Early research in contour memory and recognition demonstrated that acquisition of memory for melodic contour in infants and children precedes memory for intonation [17,19–22]. Melodic contour is also described as a 'coarse-grained' schema that lacks the detail from musical intervals [14].

2.2. Melodic Contour

Contour is often used to refer to sequences of up-down movement in melodies, but there are also several other terms that in different ways touch upon the same idea. Shape, for example, is more generally used for referring to overall melodic phrases. Adams uses the terms contour and shape interchangeably [23] and also adds melodic configuration and outline to the mix of descriptors. Tierney et al. discuss the biological origins and commonalities of melodic shape [24]. They also note the predominance of arch-shaped melodies in music, the long duration of the final notes of phrases, and the biases towards smooth pitch contours. The idea of shapes has also been used to analyze melodies of infants crying [25]. Motif, on the other hand, is often used to refer to a unit of melody that can be operated upon to create melodic variation, such as transposition, inversion, etc. Yet another term is that of melodic chunk, which is sometimes used to refer to the mnemonic properties of melodic units, while museme is used to indicate instantaneous perception. Of all these terms, we will stick with contour for the remainder of this paper.

2.3. Analyzing Melodic Contour

There are numerous analysis methods that can be used to study melodic contour, and they may be briefly divided into two main categories: signal-based or symbolic representations. When the contour analysis uses a signal-based representation, a recording of the audio is analyzed with computational methods to extract the melodic line, as well as other temporal and spectral features of the sound. The symbolic representations may start from notated or transcribed music scores and use the symbolic note representations for analysis. Similar to how we might whistle a short melodic excerpt to refer to a larger musical piece, melodic dictionaries have been created to index musical pieces [26]. Such indexes merit a thorough analysis of contour typologies, and several contour typologies were created to this end [23,27,28]. Contour typology methods are often developed from symbolic representations and notated as discrete pitch items. Adam's method for contour typology analysis, for example, codes the initial and final pitches of the melodies as numbers [23]. Parson's typology, on the other hand, uses note directions and their intervals as the units of analysis [26]. There are also examples of matrix comparison methods that code pitch patterns [27]. A comparison of these methods to perception and memory is carried out in [29,30], suggesting that the information-rich models do better than more simplistic ones. Perceptual responses to melodic contour changes have also been studied systematically [30–32], revealing differences between typologies and which ones come closest to resembling models of human contour perception.

The use of symbolic representations makes it easier to perform systematic analysis and modification of melodic music. While such systematic analysis works well for some types of pre-notated music, it is more challenging for non-notated or non-Western music. For such non-notated musics, the signal-based representations may be a better solution, particularly when it comes to providing representations that more accurately describe continuous contour features. Such continuous representations (as opposed to more discrete, note-based representations) allow the extraction of information from the actual sound signal, giving a generally richer dataset from which to work. The downside, of course, is that signal-based representations tend to be much more complex, and hence more difficult to generalize from.

In the field of music perception and cognition, the use of symbolic music representations, and computer-synthesized melodic stimuli, has been most common. This is the case even though the ecological validity of such stimuli may be questioned. Much of the previous research into the perception of melodic contour also suffers from a lack of representation of non-Western and also non-classical musical styles, with some notable exceptions such as [33–35].

Much of the recent research into melodic representations is found within the music information retrieval community. Here, the extraction of melodic contour and contour patterns directly from the signal is an active research topic, and efficient algorithms for extraction of the primary melody have been tested and compared in the MIREX (Music Information Retrieval Evaluation Exchange) competitions for several years. Melodic contour is also used to describe the instrumentation of music from audio signals, for example in [36,37]. It is also interesting to note that melodic contour is used as the first step to identify musical structure in styles such as in Indian Classical music [38], and Flamenco [39].

2.4. Pitch and the Vertical Dimension

As described in Section 1, the vertical dimension (up-down) is a common way to describe pitch contours. This cross-modal correspondence has been demonstrated in infants [40], showing preferences for concurrence of auditory pitch 'rising,' visuospatial height, as well as visual 'sharpness'. The association with visuospatial height is elaborated further with the SMARC (Spatial-Musical Association of Response Codes) effect [11]. Here, participants show a shorter response time for lower pitches co-occurring with left or bottom response codes, while higher pitches strongly correlated with response codes for right or top. A large body of work tries to tease apart the nuances of the suggested effect. Some of the suggestions include the general setting of the instruments and the bias of reading and writing being from left to right in most of the participants [41], as contributing factors to the manifestation of this effect.

The concepts of contour rely upon pitch height being a key feature of our melodic multimodal experience. Even the enumeration of pitch in graphical formats plays on this persistent metaphor. Eitan brings out the variety of metaphors for pitch quality descriptions, suggesting that up and down might only be one of the ways in which cross-modal correspondence with pitch appears in different languages [9,10]. Many of the tendencies suggested in the SMARC effect are less pronounced when more, and more complex, stimuli appear. These have been tested in memory or matching tasks, rather than asking people to elicit the perceived contours. The SMARC effect may here be seen in combination with the STEARC (Spatial-Temporal Association of Response Codes) effect, stating that timelines are more often perceived as horizontally-moving objects. In combination, these two effects may explain the overwhelming representation of vertical pitch on timelines. The end result is that we now tend to think of melodic representation mostly in line-graph-based terms, along the lines of what Adams ([23], p. 183) suggested:

> There is a problem of the musical referents of the terms. As metaphoric depictions, most of these terms are more closely related to the visual and graphic representations of music than to its acoustical and auditory characteristics. Indeed, word-list typologies of melodic contour are frequently accompanied by 'explanatory' graphics.

This 'problem' of visual metaphors, however, may actually be seen as an opportunity to explore multimodal perception that was not possible to understand at the time.

2.5. Embodiment and Music

The accompaniment of movement to music is understood now as an important phenomenon in music perception and cognition [42]. Research studying the close relationship between sound and movement has shed light on the mechanism to understand action as sound [43] and sound as action [44,45]. Cross-modal correspondence is a phenomenon with a tight interactive loop with the body as a mediator for perceptual, as well as performative roles [46,47]. Some of these interactions show themselves in the form of motor cortex activation when only listening to music [48]. This has further led to empirical studies of how music and body movement share a common structure that affords equivalent and universal emotional expression [49]. Mazzola et al. have also worked on a topological understanding of musical space and the topological dynamics of musical gesture [50].

Studies of Hindustani music show that singers use a wide variety of movements and gestures that accompany spontaneous improvisation [4,51,52]. These movements are culturally codified; they appear in the performance space to aid improvisation and musical thought, and they also convey this information to the listener. The performers also demonstrate a variety of imaginary 'objects' with various physical properties to illustrate their musical thought.

Some other examples of research on body movement and melody include Huron's studies of how eyebrow height accompany singing as a cue response to melodic height [53], and studies suggesting that especially arch-shaped melodies have common biological origins that are related to motor constraints [24,54].

2.6. Sound-Tracings

Sound-tracing studies aim at analyzing spontaneous rendering of melodies to movement, capturing instantaneous multimodal associations of the participants. Typically, subjects are asked to draw (or trace) a sound example or short musical excerpt as they are listening. Several of these studies have been carried out with digital tablets as the transducer or the medium [2,44,55–59]. One restriction of using tablets is that the canvas of the rendering space is very restrictive. Furthermore, the dimensionality does not evolve over time and represents a narrow bandwidth of possible movements.

An alternative to tablet-based sound-tracing setups is that of using full-body motion capture. This may be seen as a variation of 'air performance' studies, in which participants try to imitate the sound-producing actions of the music to which they listen [2]. Nymoen et al. carried out a series of sound-tracing studies focusing on movements of the hands [60,61], elaborating on several feature extraction methods to be used for sound-tracing as a methodology. The current paper is inspired by these studies, but extending the methodology to full-body motion capture.

3. Sound-Tracing of Melodic Phrases

3.1. Stimuli

Based on the above considerations and motivations, we designed a sound-tracing study of melodic phrases. We decided to use melodic phrases from vocal genres that have a tradition of singing without words. Vocal phrases without words were chosen so as to not introduce lexical meaning as a confounding variable. Leaving out instruments also avoids the problem of subjects having to choose between different musical layers in their sound-tracing.

The final stimulus set consists of four different musical genres and four stimuli for each genre. The musical genres selected are: (1) Hindustani (North Indian) music, (2) Sami joik, (3) scat singing, (4) Western classical vocalize. The melodic fragments are phrases taken from real recordings, to retain melodies within their original musical context. As can be seen in the pitch plots in Figure 1, the melodies are of varying durations with an average of 4.5 s (SD = 1.5 s). The Hindustani and joik phrases are sung by male vocalists, whereas the scat and vocalize phrases are sung by female vocalists. This is represented in the pitch range of each phrase as seen in Figure 1. The Hindustani and joik melodies are mainly sung in a strong chest voice in this stimulus set. Scat vocals are sung with a transition voice from chest to head. The Vocalizes in this set are sung by a soprano, predominantly in the head register. Hindustani and vocalize samples have one dominant vowel that is used throughout the phrase. The Scat singing examples use traditional 'shoobi-doo-wop' syllables, and joik examples in this set predominantly contain syllables such as 'la-la-lo'.

To investigate the effects of timbre, we decided to create a 'clean' melody representation of each fragment. This was done by running the sound files through an autocorrelation algorithm to create phrases that accurately resemble the pitch content, but without the vocal, timbral and vowel content of the melodic stimulus. These 16 re-synthesized sounds were added to the stimulus set, thus obtaining a total of 32 sound stimuli (summarized in Table 1).

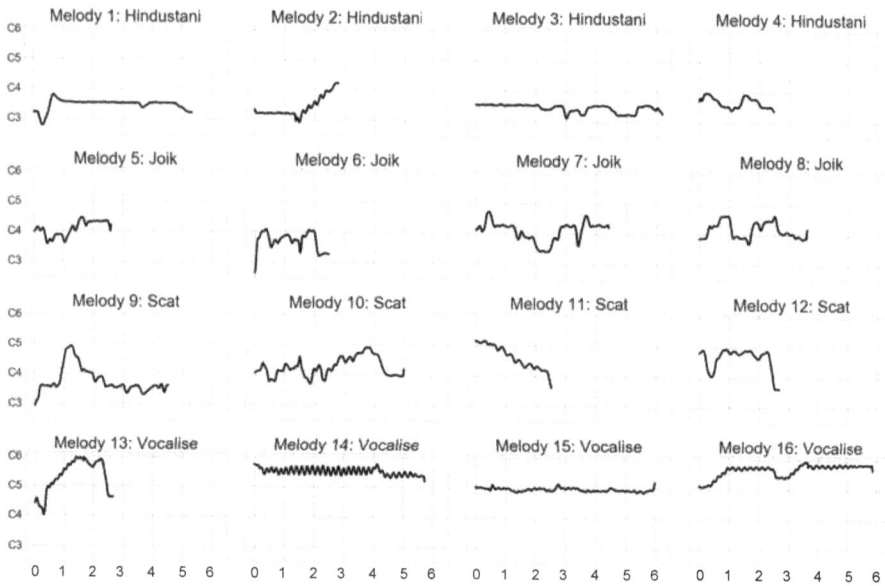

Figure 1. Pitch plots of all the 16 melodic phrases used as experiment stimuli, from each genre. The x axis represents time in seconds, and the y axis represents notes. The extracted pitches were re-synthesized to create a total of 32 melodic phrases used in the experiment.

Table 1. An overview of the 32 different stimuli: four phrases from each musical genre, all of which were presented in both normal and re-synthesized versions.

Type	Hindustani	Joik	Scat	Vocalize
Normal	4	4	4	4
Re-synthesized	4	4	4	4

3.2. Subjects

A total of 32 subjects (17 female, 15 male) was recruited, with a mean age of 31 years (SD = 9 years). The participants were mainly university students and employees, both with and without musical training. Their musical experience was quantized using the OMSI (Ollen Musical Sophistication Index) questionnaire [62], and they were also asked about the familiarity with the musical genres, and their experience with dancing. The mean of the OMSI score was 694 (SD = 292), indicating that the general musical proficiency in this dataset was on the higher side. The average familiarity with Western classical music was 4.03 out of a possible 5 points, 3.25 for jazz music, 1.87 with joik, and 1.71 with Indian classical music. Thus, two genres (vocalize and scat) were more familiar than the two others (Hindustani and joik). All participants provided their written consent for inclusion before they participated in the study, and they were free to withdraw during the experiment. The study obtained ethical approval from the Norwegian Centre for Research Data (NSD), with the project code 49258 (approved on 22 August 2016).

3.3. Procedure

Each subject performed the experiment alone, and the total duration was around 10 min. They were instructed to move their hands as if their movement was creating the melody. The use of the term creating, instead of representing, is purposeful, as in earlier studies [60,63], to avoid the act of playing or singing. The subjects could freely stand anywhere in the room and face whichever direction they liked, although nearly all of them faced the speakers and chose to stand in the center of the lab. The room lighting was dimmed to help the subjects feel more comfortable to move as they pleased.

The sounds were played at a comfortable listening level through a Genelec 8020 speaker, placed 3 m in front of the subjects. Each session consisted of an introduction, two example sequences, 32 trials and a conclusion, as sketched in Figure 2. Each melody was played twice with a 2-s pause in between. During the first presentation, the participants were asked to listen to the stimuli, while during the second presentation, they were asked to trace the melody. A long beep indicated the first presentation of a stimulus, while a short beep indicated the repetition of a stimuli. All the instructions and required guidelines were recorded and played back through the speaker to not interrupt the flow of the experiment.

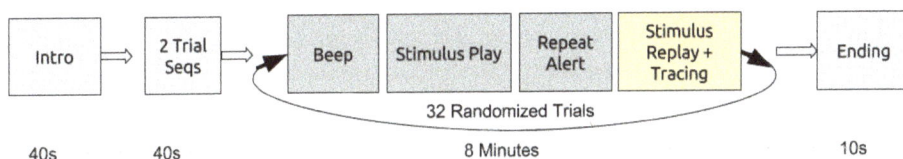

Figure 2. The experiment flow, with an approximate total duration of 10 min

The subjects' motion was recorded using an infrared marker-based motion capture system from Qualisys AB (Gothenburg, Sweden), with 8 Oqus 300 cameras surrounding the space (Figure 3a) and one regular video camera (Canon XF105 (manufactured in Tokyo, Japan)), for reference. Each subject wore a motion capture suit with 21 reflective markers on each joint (Figure 3b). The system captured at a rate of 200 Hz. The data were post-processed in the Qualisys Track Manager software (QTM, v2.16, Qualisys AB, Gothenburg, Sweden), which included labeling of markers and removal of ghost-markers (Figure 3c). We used polynomial interpolation to gap-fill the marker data, where needed. The post-processed data was exported to Python (v2.7.12 and MATLAB (R2013b, MathWorks, Natick, MA, USA) for further analysis. Here, all of the 10-min recordings were segmented using automatic windowing, and each of the segments were manually annotated for further analysis in Section 6.

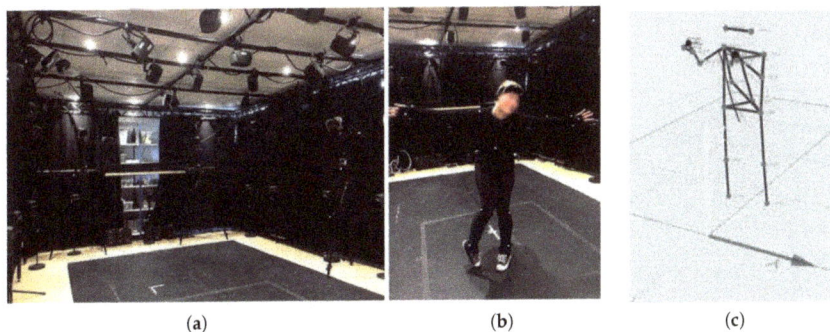

Figure 3. (a) The motion capture lab used for the experiments. (b) The subjects wore a motion capture suit with 21 reflective markers. (c) Screenshot of a stick-figure after post-processing in Qualisys Track Manager software (QTM).

4. Analysis

Even though full-body motion capture was performed, we will in the following analysis only consider data from the right and left hand markers. Marker occlusions from six of the subjects were difficult to account for in the manual annotation process, so only data from 26 participants were used in the analysis that will be presented in Section 5. This analysis is done using comparisons of means and distribution patterns. The occlusion problems were easier to tackle with the automatic analysis, so the analysis that will be presented in Section 6 was performed on data from all 32 participants.

4.1. Feature Selection from Motion Capture Data

Table 2 shows a summary of the features extracted from the motion capture data. Vertical velocity is calculated as the first derivative of the z-axis (vertical motion) for each tracing over time. 'Quantity of motion' is a dimensionless quantity representing the overall amount of motion in any direction from frame to frame. Hand distance is calculated as the euclidean distance between the x,y,z coordinates for each marker for each hand. We also calculate the sample-wise distance traveled for each hand marker.

Table 2. The features extracted from the motion capture data.

	Motion Features	Description
1	VerticalMotion	z-axis coordinates at each instant of each hand
2	Range	(Min, Max) tuple for each hand
3	HandDistance	The euclidean distance between the 2d coordinates of each hand
4	QuantityofMotion	The sum of absolute velocities of all the markers
5	DistanceTraveled	Cumulative euclidean distance traveled by each hand per sample
6	AbsoluteVelocity	Uniform linear velocity of all dimensions
7	AbsoluteAcceleration	The derivative of the absolute velocity
8	Smoothness	The number of knots of a quadratic spline interpolation fitted to each tracing
9	VerticalVelocity	The first derivative of the z-axis in each participant's tracing
10	CubicSpline10Knots	10 knots fitted to a quadratic spline for each tracing

4.2. Feature Selection from Melodic Phrases.

Pitch curves from the melodic phrases were calculated using the autocorrelation algorithm in Praat (v6.0.30, University of Amsterdam, The Netherlands), eliminating octave jumps. These pitch curves were then exported for further analysis together with the motion capture features in Python. Based on contour analysis from the literature [23,30,64], we extracted three different melodic features, as summarized in Table 3.

Table 3. The features extracted from the melodic phrases.

	Melodic Features	Description
1	SignedIntervalDirection	Interval directions (up/down) calculated for each note
2	InitialFinalHighestLowest	Four essential notes of a melody: initial, final, highest, lowest
3	SignedRelativeDistances	Feature 1 combined with relative distances of each successive note from the next, only considering the number of semitones for each successive change.

5. Analysis of Overall Trends

5.1. General Motion Contours

One global finding from the sound-tracing data is that of a clear arch shape when looking at the vertical motion component over time. Figure 4 shows the average contours calculated from the z-values of the motion capture data of all subjects for each of the melodic phrases. It is interesting to note a clear arch-like shape for all of the graphs. This fits with suggestions of a motor constraint hypothesis suggesting that melodic phrases in general have arch-like shapes [24,54]. In our study, however, the melodies have several different types of shapes (Figure 1), so this may not be the best explanation for the arch-like motion shapes. A better explanation may be that the subjects would start and end their tracing from a resting position in which the hands would be lower than during the tracing, thus leading to the arch shapes seen in Figure 4.

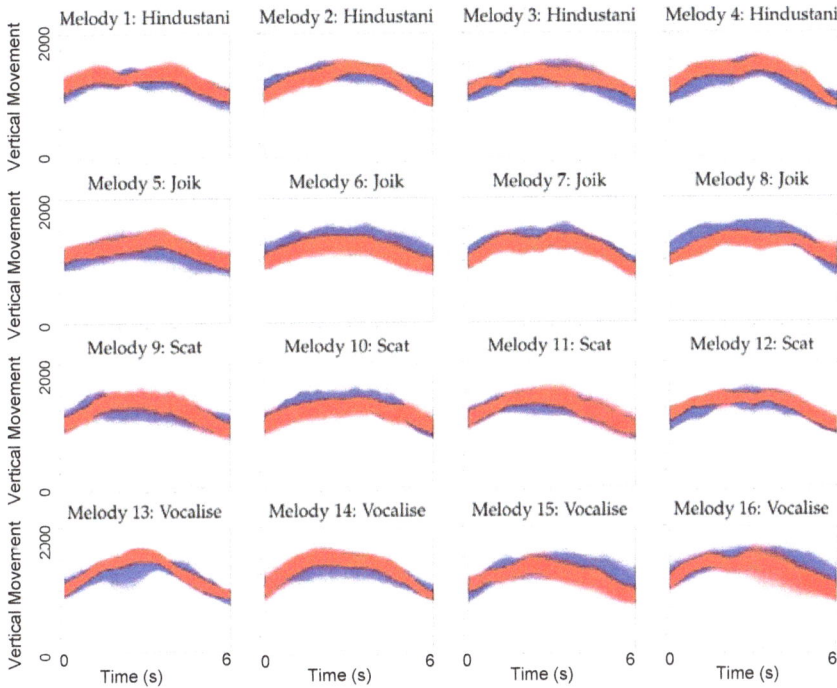

Figure 4. Average contours plotted from the vertical motion capture data in mm (z-axis) for each of the melodies (red for the original and blue for the re-synthesized versions of the melodies). The x-axis represents normalized time, and the y-axis represents aggregated tracing height for all participants.

5.2. Relationship between Vertical Movement and Melodic Pitch Distribution

To investigate more closely the relationship between vertical movement and the distribution of the pitches in the melodic fragments, we may start by considering the genre differences. Figure 5 presents the distribution of pitches in each genre in the stimulus set. These are plotted on a logarithmic frequency scale to represent the perceptual relationships between them. In the plot, each of the four genres are represented by their individual distributions. The color distinction is on the basis of whether the melodic phrase has one direction or many. Phrases closer to being ascending, descending, or stationary are coded as not changing direction. We see that in all of these conditions, the vocalize phrases in the dataset have the highest pitch profiles and the Hindustani phrases have the lowest.

If we then turn to look at the vertical dimension of the tracing data, we would expect to see a similar distribution between genres as that for the pitch heights of the melodies. Figure 6 shows the distribution of motion capture data for all tracings, sorted in the four genres. Here, the distribution of maximum z-values of the motion capture data shows a quite even distribution between genres.

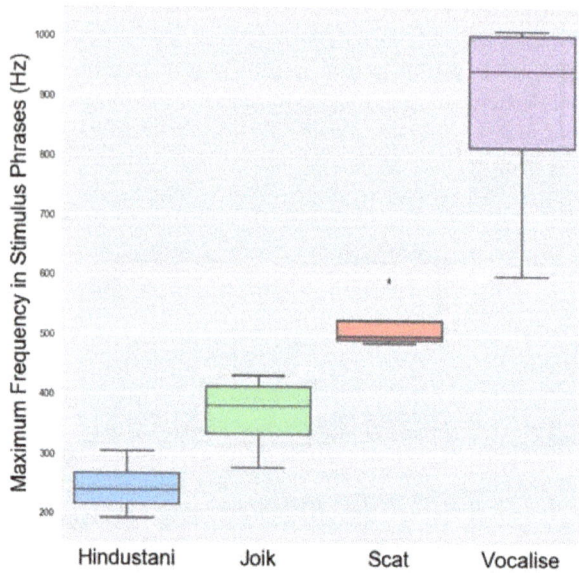

Figure 5. Pitch distribution for each genre based on mean pitches in each phrase. If movement tracings were an accurate representation of absolute pitch height, movement plots should resemble this plot.

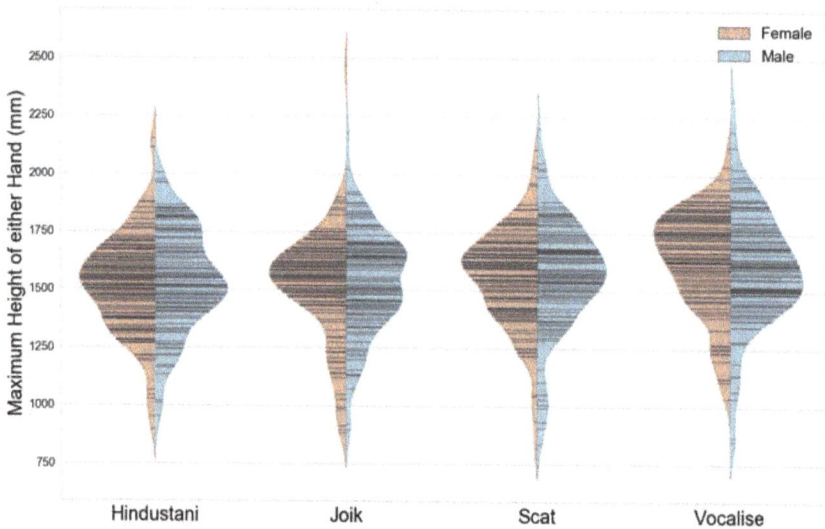

Figure 6. Plots of the maximum height of either hand for each genre. The left/pink distributions are from female subjects, while the right/blue distributions are from the male subjects. Each half of each section of the violin plot represents the probability distribution of the samples. The black lines represent each individual data point.

5.3. Direction Differences

The direction differences in the tracings can be studied by calculating the coefficients of variation of movement in all three axes for both the left (LH) and right (RH) hands. These coefficients are found to be LHvar (x,y,z) = (63.7,45.7,26.6); RHvar (x,y,z) = (56.6,87.8,23.1), suggesting that the amount of dispersion on the z-axis (the vertical) is the most consistent. This suggests that a wide array of representations in the x and y-axes are used.

The average standard deviations for the different dimensions were found to be LHstd = (99 mm, 89 mm, 185 mm); RHstd = (110 mm, 102 mm, 257 mm). This means that most variation is found in the vertical movement of the right hand, indicating an effect of right-handedness among the subjects.

5.4. Individual Subject Differences

Plots of the distributions of the quantitiy of motion (QoM) for each subject for all stimuli show a large degree of variation (Figure 7). Subjects 4 and 12, for example, have very small diversity in the average QoM for all of their tracings, whereas Subjects 2 and 5 show a large spread. There are also other participants, such as 22, who move very little on average for all their tracings. Out of the two types of representations (original and re-synthesized stimuli), we see that there is in general a larger diversity of movement for the regular melodies as opposed to the synthesized ones. However, the statistical difference between synthesized and original melodies was not significant.

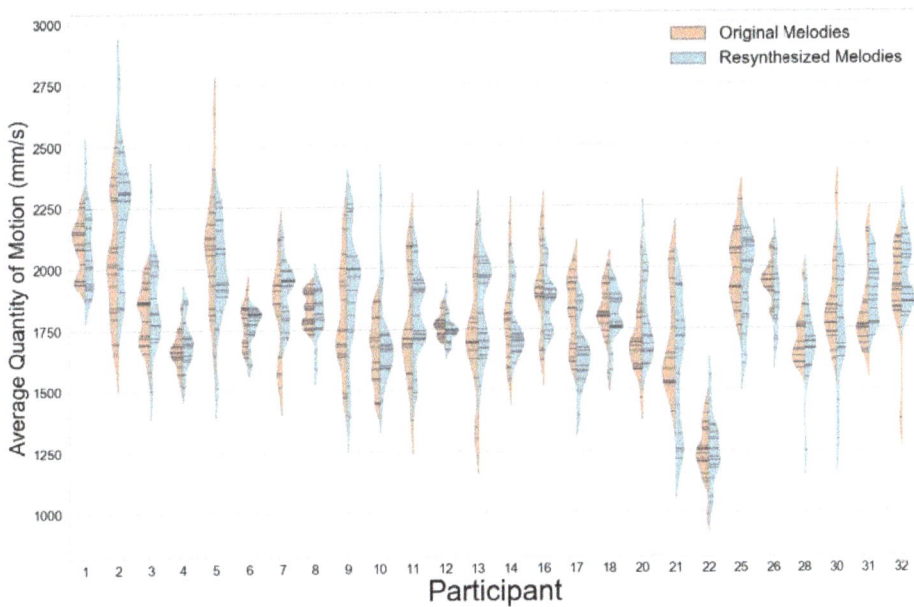

Figure 7. Distribution of the average quantity of motion for each participant. Left/red distributions are of the synthesized stimuli, while the right/green distributions are of the normal recordings.

5.5. Social Box

Another general finding from the data that is not directly related to the question at hand, but that is still relevant for understanding the distribution of data, is what we call a shared 'social box' among the subjects. Figure 6 shows that the maximum tracing height of the male subjects were higher than those of the female subjects. This is as expected, but a plot of the 'tracing volume' (the spatial distribution in all dimensions) reveals that a comparably small volume was used to represent most of the melodies (Figure 8). Qualitative observation of the recordings reveal that shorter subjects were more comfortable about stretching their hands out, while taller participants tended to use a more restrictive space relative to their height. This happened without there being any physical constraints of their movements, and no instructions that had pointed in the direction of the volume to be covered by the tracings.

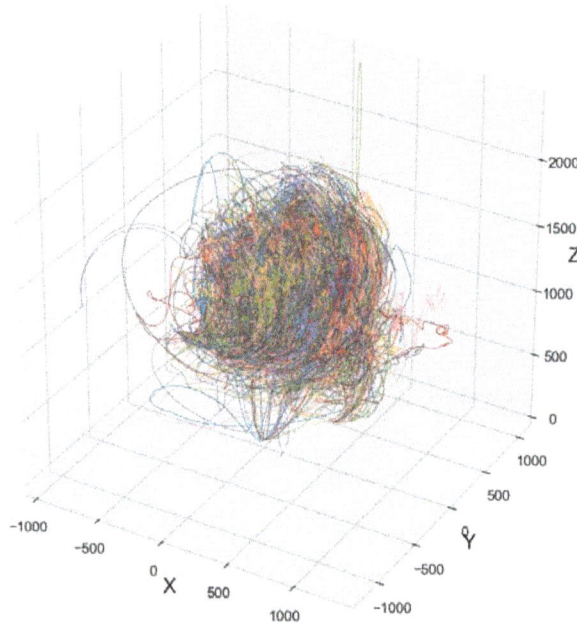

Figure 8. A three-dimensional plot of all sound-tracings for all participants reveal a fairly constrained tracing volume, a kind of 'social box' defined by the subjects. Each color represents the tracings of a single participant, and numbers along each axis are milimetres.

It is almost as if the participants wanted to fill up an invisible 'social box,' serving as the collective canvas on which everything can be represented. This may be explained by the fact that we share a world together that has the same dimensions: doors, cars, ceilings, and walls are common to us all, making us understand our body as a part of the world in a particular way. In the data from this experiment, we explore this by analyzing the range of motion relative to the heights of the participants through linear regression. The scaled movement range in the horizontal plane is represented in Figure 9. and shows that the scaled range reduces steadily over time as the height of the participants increases. Shorter participants occupy a larger area in the horizontal plane, while taller participants occupy a relatively smaller area. The R^2 coefficient of regression is found to be 0.993, meaning that this effect is significant.

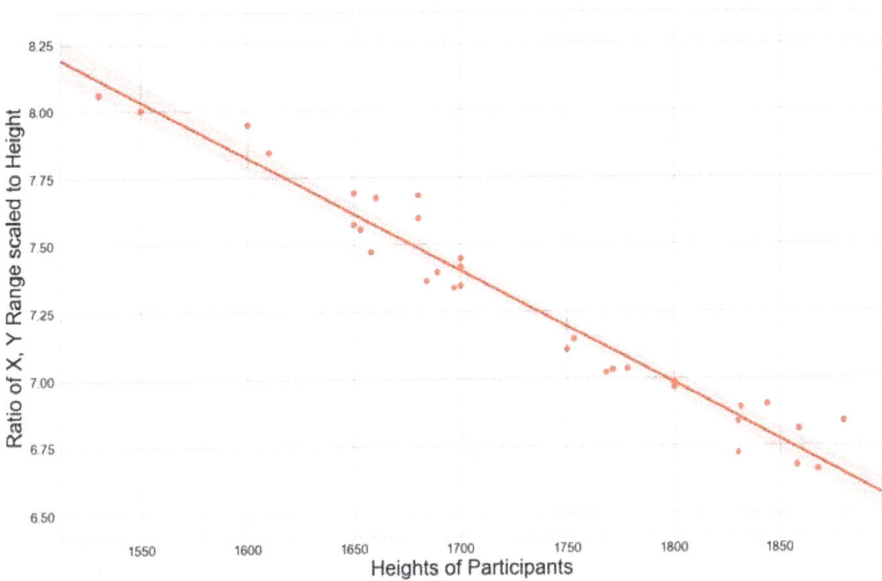

Figure 9. Regression plot of the heights of the participants against scaled (x, y) ranges. There is a clearly decreasing trend for the scaled range of movements in the horizontal plane. The taller a participant, the lower is their scaled range.

5.6. An Imagined Canvas

In a two-dimensional tracing task, such as with pen on paper, the 'canvas' of the tracing is both finite and visible all the time. Such a canvas is experienced also for tasks performed with a graphical tablet, even if the line of the tracing is not visible. In the current experiment, however, the canvas is three-dimensional and invisible, and it has to be imagined by the participant. Participants who trace by just moving one hand at a time seem to be using the metaphor of a needle sketching on a moving paper, much like an analogue ECG (Electro CardioGram) machine. Depending on the size of the tracing, the participants would have to rotate or translate their bodies to move within this imagined canvas. We observe different strategies when it comes to how they reach beyond the constraints of their kinesphere, the maximum volume you can reach without moving to a new location. They may step sideways, representing a flat canvas placed before them, or may rotate, representing a cylindrical canvas around their bodies.

6. Mapping Strategies

Through visual inspection of the recordings, we identify a limited number of strategies used in the sound-tracings. We therefore propose six schemes of representation that encompass most of the variation seen in the hands' movement, as illustrated in Figure 10 and summarized as:

1. One outstretched hand, changing the height of the palm
2. Two hands stretching or compressing an "object"
3. Two hands symmetrically moving away from the center of the body in the horizontal plane
4. Two hands moving together to represent holding and manipulating an object
5. Two hands drawing arcs along an imaginary circle
6. Two hands following each other in a percussive pattern

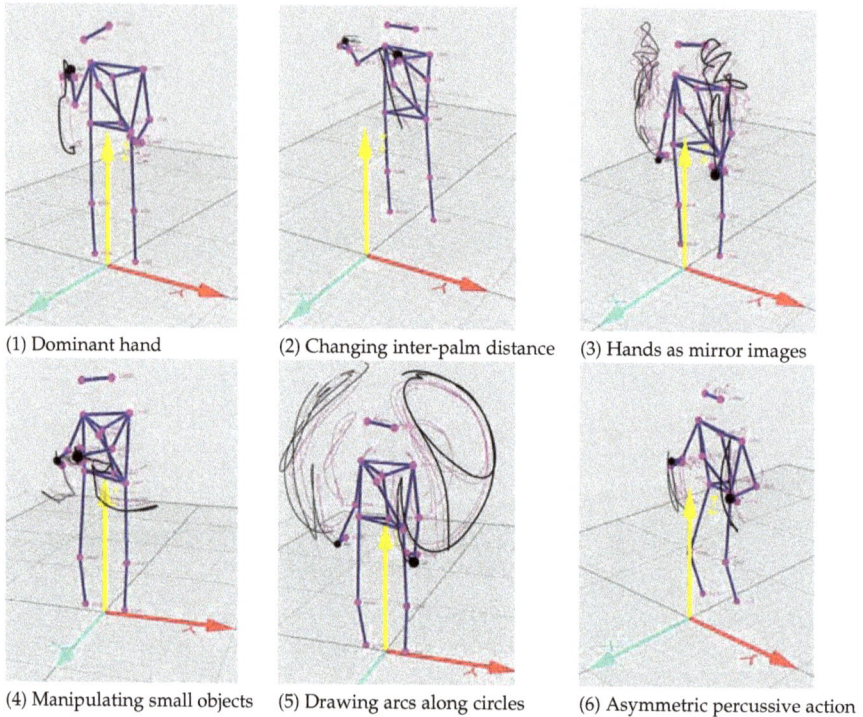

(1) Dominant hand   (2) Changing inter-palm distance   (3) Hands as mirror images

(4) Manipulating small objects   (5) Drawing arcs along circles   (6) Asymmetric percussive action

Figure 10. Motion history images exemplifying the six dominant sound-tracing strategies. The black lines from the hands of the stick figures indicate the motion traces of each tracing.

These qualitatively derived strategies were the starting point for setting up an automatic extraction of features from the motion capture data. The pipeline for this quantitative analysis consists of the following steps:

1. Feature selection: Segment the motion capture data into a six-column feature vector containing the (x,y,z) coordinates of the right palm and the left palm, respectively.
2. Calculate quantity of motion (QoM): Calculate the average of the vector magnitude for each sample.
3. Segmentation: Trim data using a sliding window of 1100 samples in size. This corresponds to 5.5 s, to accommodate the average duration of 4.5 s of the melodic phrases. The hop size for the windows is 10 samples, to obtain a large set of windowed segments. The segments that have the maximum mean values are then separated out to get a set of sound-tracings.
4. Feature analysis: Calculate features from Table 4 for each segment.
5. Thresholding: Minimize the six numerical criteria by thresholding the segments based on two-times the standard deviation for each of the computed features.
6. Labeling and separation: Obtain tracings that can be classified as dominantly belonging to one of the six strategy types.

Table 4. Quantitative motion capture features that match the qualitatively selected strategies. QoM, quantities of motion.

#	Strategy	Distinguishing Features	Description	Mean	SD
1	Dominant hand as needle	Right hand QoM much greater than left QoM	$QoM(LHY) \gg \vee \ll QoM(RHY)$	0.50	0.06
2	Changing inter-palm distance	Root mean squared difference of left, right hands in x	$RMS(LHX) - RMS(RHX)$	0.64	0.12
3	Lateral symmetry between hands	Nearly constant difference between left and right hands	$RHX - LHX = C$	0.34	0.11
4	Manipulating a small object	Right and left hands follow similar trajectories in x	$RH(x,y,z) = LH(x,y,z) + C$	0.72	0.07
5	Drawing arcs along circles	Fit of (x,y,z) for left and right hands to a sphere	$x^2 + y^2 + z^2$	0.17	0.04
6	Percussive asymmetry	Dynamic time warp of (x,y,z) of Left, Right Hands	$dtw(RH(xyz), LH(xyz))$	0.56	0.07

After running the segmentation and labeling on the complete data set, we performed a *t*-test to determine whether there is a significant difference between the labeled samples and the other samples. The results, summarized in Table 5 show that the selected features demonstrate the dominance of one particular strategy for many tracings. All except Feature 4 (manipulating a small 'object') show significant results compared to all other tracing samples for automatic annotation of hand strategies. While this feature cannot be extracted from the aforementioned heuristic, the simple feature for euclidean distance between two hands proves effective to be able to explain this strategy.

Table 5. Significance testing for each feature against the rest of the features.

Strategy #	*p*-Value
Strategy 1 vs. rest	0.003
Strategy 2 vs. rest	0.011
Strategy 3 vs. rest	0.005
Strategy 4 vs. rest	0.487
Strategy 5 vs. rest	0.003
Strategy 6 vs. rest	0.006

In Figure 11, we see that hand distance might be an effective way to compare different hand strategies. Strategy 2 performs the best on testing for separability. The hand distance for Strategy 4, for example is significantly lower than the rest. This is because this tracing style represents people who use the metaphor of an imaginary object to represent music. This imaginary object seldom changes its physical properties—its length and breadth and general size is usually maintained.

Taking demographics into account, we see that the distribution of the female subjects' tracings for vocalizes have a much wider peak than the rest of the genres. In the use of hand strategies, we observe that women use a wider range of hand strategies as compared to men (Figure 11). Furthermore, Strategy 5 (drawing arcs) is done entirely by women. The representation of music as objects is also seen to be more prominent in women, as is the use of asymmetrical percussive motion. Comparing the same distribution of genders for genres, we do not find a difference in overall movement or general body use between the genders. If anything, the 'social box effect' makes the height differences of genres smaller than they are.

In Figure 12, we visualize the use of these hand strategies for every melody by all the participants. Strategy 2 is used in 206 tracings, whereas Strategy 5 is used for only 8 tracings. Strategies 1, 3, 4 and 5 are used 182, 180, 161 and 57 times, respectively. Through this heat map in 12, we also can find some outliers for the strategies that are more infrequently used. For example, we see that Melodies 4, 13 and 16 show specially dominant use of some hand strategies.

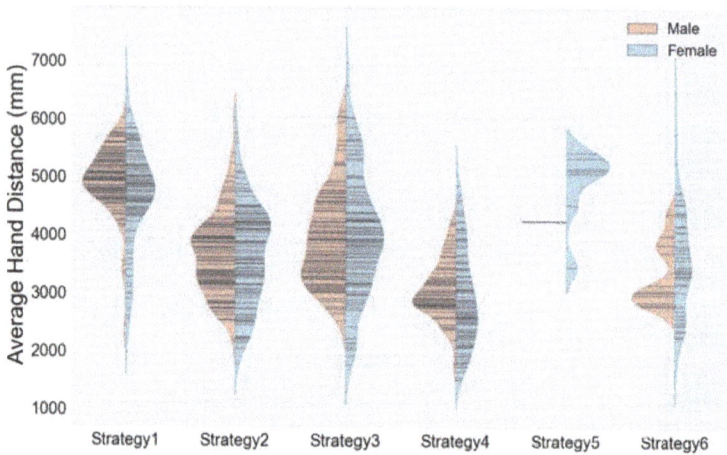

Figure 11. Hand distance as a feature to discriminate between tracing strategies.

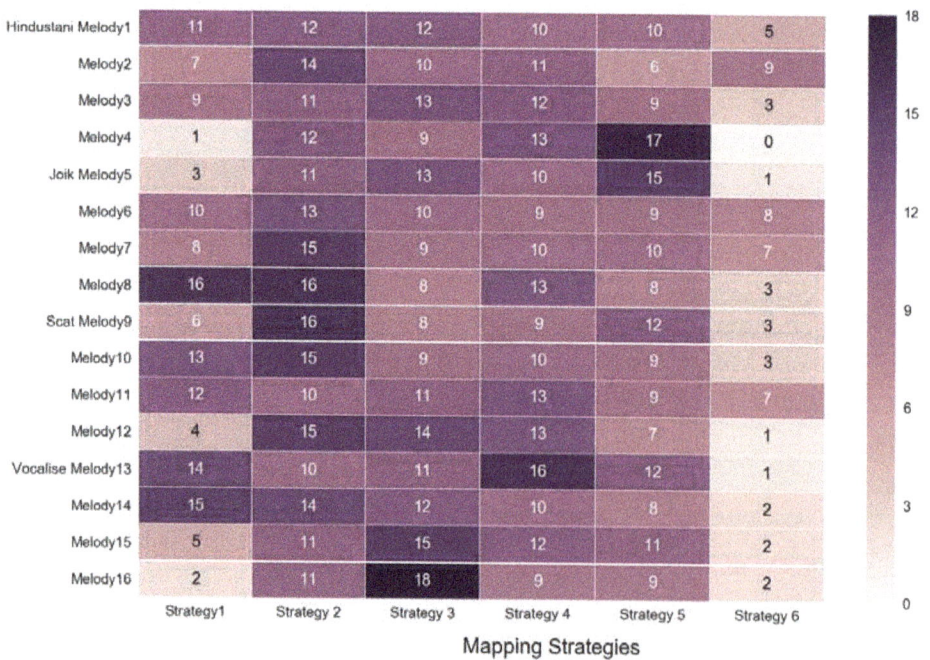

	Strategy1	Strategy 2	Strategy 3	Strategy 4	Strategy 5	Strategy 6
Hindustani Melody1	11	12	12	10	10	5
Melody2	7	14	10	11	6	9
Melody3	9	11	13	12	9	3
Melody4	1	12	9	13	17	0
Joik Melody5	3	11	13	10	15	1
Melody6	10	13	10	9	9	8
Melody7	8	15	9	10	10	7
Melody8	16	16	8	13	8	3
Scat Melody9	6	16	8	9	12	3
Melody10	13	15	9	10	9	3
Melody11	12	10	11	13	9	7
Melody12	4	15	14	13	7	1
Vocalise Melody13	14	10	11	16	12	1
Melody14	15	14	12	10	8	2
Melody15	5	11	15	12	11	2
Melody16	2	11	18	9	9	2

Mapping Strategies

Figure 12. Heat map of representation of hand strategy per melody.

7. Discussion

In this study, we have analyzed people's tracings to melodies from four musical genres. Although much of the literature points to correlations between melodic pitch and vertical movement, our findings show a much more complex picture. For example, relative pitch height appears to be much more important than absolute pitch height. People seem to think about vocal melodies as actions, rather than interpreting the pitches purely in one dimension (vertical) over time. The analysis of contour features from the literature shows that while tracing melodies through an allocentric representation of the listening body, the notions of pitch height representations matter much less than previously thought. Therefore contour features cannot be extracted merely by cross-modal comparisons of two data sets. We propose that other strategies can be used for contour representations, but this is something that will have to be developed more in future research.

According to the gestural affordances of musical sound theory [65], several gestural representations can exist for the same sound, but there is a limit to how much they can be manipulated. Our data support this idea of a number of possible and overlapping action strategies. Several spatial and visual metaphors are used by a wide range of people. One interesting finding is that there are gender differences between the representations of the different sound-tracing strategies. Women seem to show a greater diversity of strategies in general, and they also use object-like representations more often than men.

We expected musical genre to have some impact on the results. For example, given that Western vocalizes are sung with a pitch range higher than the rest of the genres in this dataset (Figure 5), it is interesting to note that, on average, people do represent vocalize tracings spatially higher than the rest of the genres. We also found that the melodies with the maximum amount of vibrato (melodies 14 and 16 in Figure 5) are represented with the largest changes of acceleration in the motion capture data. This implies that although the pitch deviation in this case is not so significant, the perception of a moving melody is much stronger by comparison to other melodies that have larger changes in pitch. It could be argued that both melody 4 and 16 contain motivic repetition that cause this pattern. However, repeating motifs are as much parts of melodies 6 and 8 (joik). The values represented by these melodies are applicable to their tracings as original as well as synthesized phrases. The effect of the vowels used in these melodies can also thus be negated. As seen in Figure 12, there are some melodies that stand out for some hand strategies. Melody 4 (Hindustani) is curiously highly represented as a small object. Melody 12 is overwhelmingly represented by symmetrical movements of both hands, while Melodies 8 and 9 are overwhelmingly represented by using 1 hand as the tracing needle.

We find it particularly interesting that subjects picked up on the idea of using small objects as a representation technique in their tracings. The use of small objects to represent melodies is well documented in Hindustani music [52,66–68]. However, the subjects' familiarity score with Hindustani music was quite low, so familiarity can not explain this interesting choice of representation in our study. Looking at the melodic features of melody 4, for example, it is steadily descending in intervals until it ascends again and comes down the same intervals. This may be argued to resemble an object that smoothly slips on a slope, and could be a probable reason for the overwhelming object representation of this particular melody. In future studies, it would be interesting to see whether we can recreate this mapping in other melodies, or model melodies in terms of naturally occurring melodic shapes born out of physical forces interacting with each other.

It is worth noting that there are several limitations with the current experimental methodology and analysis. Any laboratory study of this kind would present subjects with an unfamiliar and non-ecological environment. The results would also to a large extent be influenced by the habitus of body use in general. Experience in dance, sign language, or musical traditions with simultaneous musical gestures (such as conducting), all play a part in the interpretation of music as motion. Despite these limitations, we do see a considerable amount of consistency between subjects.

8. Conclusions

The present study shows that there are consistencies in people's sound-tracing to the melodic phrases used in the experiment. Our main findings can be summarized as:

- There is a clear arch shape when looking at the averages of the motion capture data, regardless of the general shape of the melody itself. This may support the idea of a motor constraint hypothesis that has been used to explain the similar arch-like shape of sung melodies.
- The subjects chose between different strategies in their sound-tracings. We have qualitatively identified six such strategies and have created a set of heuristics to quantify and test their reliability.

- There is a clear gender difference for some of the strategies. This was most evident for Strategy 4 (representing small objects), which women performed more than men.
- The 'obscure' strategy of representing melodies in terms of a small object, as is typical in Hindustani music, was also found in participants who had no or little exposure to this musical genre.
- The data show a tendency of moving within a shared 'social box'. This may be thought of as an invisible space that people constrain their movements to, even without any exposure to the other participants' tracings. In future studies, it would be interesting to explore how constant such a space is, for example by comparing multiple recordings of the same participants over a period of time.

In future studies, we want to investigate all of these findings in greater detail. We are particularly interested in taking the rest of the body's motion into account. It would also be relevant to use the results from such studies in the creation of interactive systems, 'reversing' the process, that is, using tracing in the air as a method to retrieve melodies from a database. This could open up some exciting end-user applications and also be used as a tool for music performance.

Supplementary Materials: The following are available online at Available online: http://www.mdpi.com/2076-3417/8/1/135/s1, supplementary archive consists of data files of the following nature: segmented motion tracings of 26 participants annotated with participant number, melody traced, and hand strategy used. The melodies are from 1 to 16, and the pitch data and sound stimuli are separately provided as well. More information about the same and code can be provided upon request.

Acknowledgments: This work was partially supported by the Research Council of Norway through its Centres of Excellence scheme, Project Number 262762.

Author Contributions: T.K. and A.R.J. both conceived and designed the experiments; T.K. performed the experiments and analyzed the data; both authors contributed analysis tools, and wrote the paper.

Conflicts of Interest: The authors declare no conflict of interest.

References

1. Kennedy, M. Rutherford-Johnson, T.; Kennedy, J. *The Oxford Dictionary of Music*, 6th ed.; Oxford University Press: Oxford, UK, 2013. Available online: http://www.oxfordreference.com/view/10.1093/acref/9780199578108.001.0001/acref-9780199578108 (accessed on 16 January 2018).
2. Godøy, R.I.; Haga, E.; Jensenius, A.R. *Playing "Air Instruments": Mimicry of Sound-Producing Gestures by Novices and Experts*; International Gesture Workshop; Springer: Berlin/Heidelberg, Germany, 2005; pp. 256–267.
3. Kendon, A. *Gesture: Visible Action as Utterance*; Cambridge University Press: Cambridge, UK, 2004.
4. Clayton, M.; Sager, R.; Will, U. In time with the music: The concept of entrainment and its significance for ethnomusicology. In *European Meetings in Ethnomusicology*; ESEM Counterpoint 1; Romanian Society for Ethnomusicology: Bucharest, Romania, 2005; Volume 11, pp. 1–82.
5. Zbikowski, L.M. Musical gesture and musical grammar: A cognitive approach. In *New Perspectives on Music and Gesture*; Ashgate Publishing Ltd.: Farnham, UK, 2011; pp. 83–98.
6. Lakoff, G.; Johnson, M. Conceptual metaphor in everyday language. *J. Philos.* **1980**, *77*, 453–486.
7. Zbikowski, L.M. Metaphor and music theory: Reflections from cognitive science. *Music Theory Online* **1998**, *4*, 1–8.
8. Shayan, S.; Ozturk, O.; Sicoli, M.A. The thickness of pitch: Crossmodal metaphors in Farsi, Turkish, and Zapotec. *Sens. Soc.* **2011**, *6*, 96–105.
9. Eitan, Z.; Schupak, A.; Gotler, A.; Marks, L.E. Lower pitch is larger, yet falling pitches shrink. *Exp. Psychol.* **2014**, *61*, 273–284.
10. Eitan, Z.; Timmers, R. Beethoven's last piano sonata and those who follow crocodiles: Cross-domain mappings of auditory pitch in a musical context. *Cognition* **2010**, *114*, 405–422.
11. Rusconi, E.; Kwan, B.; Giordano, B.L.; Umilta, C.; Butterworth, B. Spatial representation of pitch height: The SMARC effect. *Cognition* **2006**, *99*, 113–129.
12. Huron, D. The melodic arch in Western folksongs. *Comput. Musicol.* **1996**, *10*, 3–23.

13. Fedorenko, E.; Patel, A.; Casasanto, D.; Winawer, J.; Gibson, E. Structural integration in language and music: Evidence for a shared system. *Mem. Cognit.* **2009**, *37*, 1–9.

14. Patel, A.D. *Music, Language, and the Brain*; Oxford University Press: New York, NY, USA, 2010.

15. Adami, A.G.; Mihaescu, R.; Reynolds, D.A.; Godfrey, J.J. Modeling prosodic dynamics for speaker recognition. In Proceedings of the 2003 IEEE International Conference on Acoustics, Speech, and Signal Processing (ICASSP'03), Hong Kong, China, 6–10 April 2003; Volume 4.

16. Dowling, W.J. Scale and contour: Two components of a theory of memory for melodies. *Psychol. Rev.* **1978**, *85*, 341–354.

17. Dowling, W.J. Recognition of melodic transformations: Inversion, retrograde, and retrograde inversion. *Percept. Psychophys.* **1972**, *12*, 417–421.

18. Schindler, A.; Herdener, M.; Bartels, A. Coding of melodic gestalt in human auditory cortex. *Cereb. Cortex* **2012**, *23*, 2987–2993.

19. Trehub, S.E.; Becker, J.; Morley, I. Cross-cultural perspectives on music and musicality. *Philos. Trans. R. Soc. Lond. B Biol. Sci.* **2015**, *370*, 20140096.

20. Trehub, S.E.; Bull, D.; Thorpe, L.A. Infants' perception of melodies: The role of melodic contour. *Child Dev.* **1984**, *55*, 821–830.

21. Trehub, S.E.; Thorpe, L.A.; Morrongiello, B.A. Infants' perception of melodies: Changes in a single tone. *Infant Behav. Dev.* **1985**, *8*, 213–223.

22. Morrongiello, B.A.; Trehub, S.E.; Thorpe, L.A.; Capodilupo, S. Children's perception of melodies: The role of contour, frequency, and rate of presentation. *J. Exp. Child Psychol.* **1985**, *40*, 279–292.

23. Adams, C.R. Melodic contour typology. *Ethnomusicology* **1976**, *20*, 179–215.

24. Tierney, A.T.; Russo, F.A.; Patel, A.D. The motor origins of human and avian song structure. *Proc. Natl. Acad. Sci. USA* **2011**, *108*, 15510–15515.

25. Díaz, M.A.R.; García, C.A.R.; Robles, L.C.A.; Altamirano, J.E.X.; Mendoza, A.V. Automatic infant cry analysis for the identification of qualitative features to help opportune diagnosis. *Biomed. Signal Process. Control* **2012**, *7*, 43–49.

26. Parsons, D. *The Directory of Tunes and Musical Themes*; S. Brown: Cambridge, UK, 1975.

27. Quinn, I. The combinatorial model of pitch contour. *Music Percept. Interdiscip. J.* **1999**, *16*, 439–456.

28. Narmour, E. *The Analysis and Cognition of Melodic Complexity: The Implication-Realization Model*; University of Chicago Press: Chicago, IL, USA, 1992.

29. Marvin, E.W. A Generalized Theory of Musical Contour: Its Application to Melodic and Rhythmic Analysis of Non-Tonal Music and Its Perceptual and Pedagogical Implications. Ph.D. Thesis, University of Rochester, Rocheste, NY, USA, 1988.

30. Schmuckler, M.A. Testing models of melodic contour similarity. *Music Percept. Interdiscip. J.* **1999**, *16*, 295–326.

31. Schmuckler, M.A. Melodic contour similarity using folk melodies. *Music Percept. Interdiscip. J.* **2010**, *28*, 169–194.

32. Schmuckler, M.A. Expectation in music: Investigation of melodic and harmonic processes. *Music Percept. Interdiscip. J.* **1989**, *7*, 109–149.

33. Eerola, T.; Himberg, T.; Toiviainen, P.; Louhivuori, J. Perceived complexity of western and African folk melodies by western and African listeners. *Psychol. Music* **2006**, *34*, 337–371.

34. Eerola, T.; Bregman, M. Melodic and contextual similarity of folk song phrases. *Musicae Sci.* **2007**, *11*, 211–233.

35. Eerola, T. Are the emotions expressed in music genre-specific? An audio-based evaluation of datasets spanning classical, film, pop and mixed genres. *J. New Music Res.* **2011**, *40*, 349–366.

36. Salamon, J.; Peeters, G.; Röbel, A. Statistical Characterisation of Melodic Pitch Contours and its Application for Melody Extraction. In Proceedings of the ISMIR, Porto, Portugal, 8–12 October 2012; pp. 187–192.

37. Bittner, R.M.; Salamon, J.; Bosch, J.J.; Bello, J.P. Pitch Contours as a Mid-Level Representation for Music Informatics. In Proceedings of the Audio Engineering Society Conference: 2017 AES International Conference on Semantic Audio, Erlangen, Germany, 22–24 June 2017.

38. Rao, P.; Ross, J.C.; Ganguli, K.K.; Pandit, V.; Ishwar, V.; Bellur, A.; Murthy, H.A. Classification of melodic motifs in raga music with time-series matching. *J. New Music Res.* **2014**, *43*, 115–131.

39. Gómez, E.; Bonada, J. Towards computer-assisted flamenco transcription: An experimental comparison of automatic transcription algorithms as applied to a cappella singing. *Comput. Music J.* **2013**, *37*, 73–90.

40. Walker, P.; Bremner, J.G.; Mason, U.; Spring, J.; Mattock, K.; Slater, A.; Johnson, S.P. Preverbal infants are sensitive to cross-sensory correspondences much ado about the null results of Lewkowicz and Minar (2014). *Psychol. Sci.* **2014**, *25*, 835–836.
41. Timmers, R.; Li, S. Representation of pitch in horizontal space and its dependence on musical and instrumental experience. *Psychomusicol. Music Mind Brain* **2016**, *26*, 139–148.
42. Leman, M. *Embodied Music Cognition and Mediation Technology*; MIT Press: Cambridge, MA, USA, 2008.
43. Godøy, R.I. Motor-mimetic music cognition. *Leonardo* **2003**, *36*, 317–319.
44. Jensenius, A.R. Action-Sound: Developing Methods and Tools to Study Music-Related Body Movement. Ph.D. Thesis, University of Oslo, Oslo, Norway, 2007.
45. Jensenius, A.R.; Kvifte, T.; Godøy, R.I. Towards a gesture description interchange format. In Proceedings of the 2006 Conference on New Interfaces for Musical Expression, IRCAM—Centre Pompidou, Paris, France, 4–8 June 2006; pp. 176–179.
46. Gritten, A.; King, E. *Music and Gesture*; Ashgate Publishing, Ltd.: Hampshire, UK, 2006.
47. Gritten, A.; King, E. *New Perspectives on Music and Gesture*; Ashgate Publishing, Ltd.: Surrey, UK, 2011.
48. Molnar-Szakacs, I.; Overy, K. Music and mirror neurons: From motion to 'e'motion. *Soc. Cognit. Affect. Neurosci.* **2006**, *1*, 235–241.
49. Sievers, B.; Polansky, L.; Casey, M.; Wheatley, T. Music and movement share a dynamic structure that supports universal expressions of emotion. *Proc. Natl. Acad. Sci. USA* **2013**, *110*, 70–75.
50. Buteau, C.; Mazzola, G. From contour similarity to motivic topologies. *Musicae Sci.* **2000**, *4*, 125–149.
51. Clayton, M.; Leante, L. Embodiment in Music Performance. In *Experience and Meaning in Music Performance*; Oxford University Press: New York, NY, USA, 2013; pp. 188–207.
52. Rahaim, M. *Musicking Bodies: Gesture and Voice in Hindustani Music*; Wesleyan University Press: Middletown, CT, USA, 2012.
53. Huron, D.; Shanahan, D. Eyebrow movements and vocal pitch height: Evidence consistent with an ethological signal. *J. Acoust. Soc. Am.* **2013**, *133*, 2947–2952.
54. Savage, P.E.; Tierney, A.T.; Patel, A.D. Global music recordings support the motor constraint hypothesis for human and avian song contour. *Music Percept. Interdiscip. J.* **2017**, *34*, 327–334.
55. Godøy, R.I.; Haga, E.; Jensenius, A.R. Exploring Music-Related Gestures by Sound-Tracing: A Preliminary Study. 2006. Available online: https://www.duo.uio.no/bitstream/handle/10852/26899/Godxy_2006b.pdf?sequence=1&isAllowed=y (accessed on 16 January 2018).
56. Glette, K.H.; Jensenius, A.R.; Godøy, R.I. Extracting Action-Sound Features From a Sound-Tracing Study. 2010. Available online: https://www.duo.uio.no/bitstream/handle/10852/8848/Glette_2010.pdf?sequence=1&isAllowed=y (accessed on 16 January 2018).
57. Küssner, M.B. Music and shape. *Lit. Linguist. Comput.* **2013**, *28*, 472–479.
58. Roy, U.; Kelkar, T.; Indurkhya, B. TrAP: An Interactive System to Generate Valid Raga Phrases from Sound-Tracings. In Proceedings of the NIME, London, UK, 30 June–3 July 2014; pp. 243–246.
59. Kelkar, T. Applications of Gesture and Spatial Cognition in Hindustani Vocal Music. Ph.D. Thesis, International Institute of Information Technology, Hyderabad, India, 2015.
60. Nymoen, K.; Godøy, R.I.; Jensenius, A.R.; Torresen, J. Analyzing Correspondence Between Sound Objects and Body Motion. *ACM Trans. Appl. Percept.* **2013**, *10*, 9.
61. Nymoen, K.; Caramiaux, B.; Kozak, M.; Torresen, J. Analyzing Sound-Tracings: A Multimodal Approach to Music Information Retrieval. In Proceedings of the 1st International ACM Workshop on Music Information Retrieval with User-centered and Multimodal Strategies (MIRUM '11), Scottsdale, AZ, USA, 30 November 2011; ACM: New York, NY, USA, 2011; pp. 39–44.
62. Ollen, J.E. A Criterion-Related Validity Test of Selected Indicators of Musical Sophistication Using Expert Ratings. Ph.D. Thesis, The Ohio State University, Columbus, OH, USA, 2006.
63. Nymoen, K.; Torresen, J.; Godøy, R.; Jensenius, A. A statistical approach to analyzing sound-tracings. In *Speech, Sound and Music Processing: Embracing Research in India*; Springer: Berlin/Heidelberg, Germany, 2012; pp. 120–145.
64. Schmuckler, M.A. Components of melodic processing. In *Oxford Handbook of Music Psychology*; Oxford Uniersity Press: Oxford, UK, 2009; p. 93.
65. Godøy, R.I. Gestural affordances of musical sound. In *Musical Gestures: Sound, Movement, and Meaning*; Routledge: New York, NY, USA, 2010; pp. 103–125.

66. Paschalidou, S.; Clayton, M.; Eerola, T. Effort in Interactions with Imaginary Objects in Hindustani Vocal Music—Towards Enhancing Gesture-Controlled Virtual Instruments. In Proceedings of the 2017 International Symposium on Musical Acoust, Montreal, QC, Canada, 18–22 June 2017.

67. Paschalidou, S.; Clayton, M. Towards a sound-gesture analysis in Hindustani Dhrupad vocal music: Effort and raga space. In Proceedings of the ICMEM, University of Sheffield, Sheffield, UK, 23–25 March 2015; Volume 23, p. 25.

68. Pearson, L.; others. Gesture in Karnatak Music: Pedagogy and Musical Structure in South India. Ph.D. Thesis, Durham University, Durham, UK, 2016.

applied
sciences

MDPI

Article

Desert and Sonic Ecosystems: Incorporating Environmental Factors within Site-Responsive Sonic Art

Lauren Hayes [1,*,†] **and Julian Stein** [2]

[1] Arts, Media + Engineering, Arizona State University, Tempe, AZ 85287-5802, USA
[2] Design Media Arts, University of California, Los Angeles, CA 90095-1456, USA; julianstein@ucla.edu
* Correspondence: lauren.s.hayes@asu.edu; Tel.: +1-480-727-9408
† Current address: ASU Arts, Media + Engineering, 950 S Forest Mall, Tempe, AZ 85287, USA.

Academic Editor: Stefania Serafin
Received: 3 November 2017; Accepted: 9 January 2018; Published: 14 January 2018

Abstract: Advancements in embedded computer platforms have allowed data to be collected and shared between objects—or smart devices—in a network. While this has resulted in highly functional outcomes in fields such as automation and monitoring, there are also implications for artistic and expressive systems. In this paper we present a pluralistic approach to incorporating environmental factors within the field of site-responsive sonic art using embedded audio and data processing techniques. In particular, we focus on the role of such systems within an ecosystemic framework, both in terms of incorporating systems of living organisms, as well as sonic interaction design. We describe the implementation of such a system within a large-scale site-responsive sonic art installation that took place in the subtropical desert climate of Arizona in 2017.

Keywords: sonic interaction design; audio signal processing; music technology; ecosystems

---

1. Introduction

In this paper we discuss an approach to incorporating both acoustic and environmental factors within a large-scale multichannel sound installation. This research makes use of developments in portable, embeddable sensor, and microcomputer technologies. *Sounding Out Spaces* is an ongoing project that concerns context-based live electronic music and sonic art. Specifically, it aims to develop performances and installations that occur in response to a particular location or space. The sites involved range from retired industrial structures and visually stimulating landscapes, to architecture with unique acoustic properties, natural environments, and places of cultural or historical significance. Common to the collection of practices that has been developed is the theme that sound is produced in response to certain perceived or measured attributes of a particular site. These features may be acoustic, environmental, historic, and, perhaps, even imagined.

Sounding Out Spaces explores how we can use digital means to participate directly in our environment, leading to novel and inclusive experiences. These ideas are developed through iterative practice. The series began in 2014 with a focus on guerrilla performance using portable analogue technologies, and has evolved to involve, most recently, large-scale public installations. While the theoretical framework of this creative practice has been thoroughly discussed elsewhere (see [1]), the most recent work in this series, which will be addressed in what follows, focuses on developing transferable techniques and methodologies for both spontaneous and planned works. In this latest iteration we explore the implications of an ecosystemic approach. We describe a methodology where the resulting sound is contingent not only on all agents involved and their organizational relationships, but also the environment and the effects of its perturbations [2] on such a system.

Overview of Garden Ecologies

In April 2017, the latest iteration of the series was developed as a mutlichannel sonic art installation. *Sounding Out Spaces: Garden Ecologies* involved numerous audio and sensor processing platforms that were embedded throughout a community garden. These programmable digital–physical interfaces took readings from plants, vegetation, and the environment. Several new systems were developed in order to analyze and process biophysical information from the plants, soil, wind, water, and light. Along with the acoustic information from the site picked up by microphones, this data was used to affect various musical parameters of the digital signal processing (DSP) on the embedded microcomputers. The audio output of these was sent to multiple loudspeakers placed throughout the garden (see Figures 1 and 2) bringing forth an emergent, continuously evolving musical world.

Figure 1. A bird's eye view of the installation site.

Figure 2. A loudspeaker embedded in a raised bed in which lettuce is being grown.

Various community groups were invited to take part in the project through a series of sound-based workshops based at the Clark Park Community Garden, Tempe, and the Children

First Leadership Academy, Phoenix, Arizona. In these workshops, children from one of our partner organizations, Free Arts for Abused Children of Arizona, learned about listening to the environment, sound recording, and using biophysical properties of organic matter, such as plants and vegetation, along with simple electronic circuits to create new musical instruments. This culminated in a public site-responsive performance. The material produced by the children was also incorporated into the sound installation, weaving together the acoustic characteristics of the site with environmental and biophysical information from the garden and plants into an evolving and dynamic sonic ecosystem. The installation ran over two consecutive days and included an evening preview, in addition to the daytime presentations.

2. Background

In this section we outline several areas of research within the field of sound and music computing that have informed this work.

2.1. Sound and the Environment

Over the last few decades there has been a growing body of work being undertaken within the areas of soundscape composition [3], environmental sound art [4,5], and acoustic ecology [6,7]. The latter of these associated areas of research is an interdisciplinary field that "studies the social, cultural, and ecological aspects of our environment through sound" [8]. Currently, the majority of this research makes extensive use of field recordings, and collects, analyzes, and archives large amounts of sound material from the environment. Using the concept of *acoustic indices* [9], researchers can evaluate the long-term changes in the biodiversity of both terrestrial and marine ecosystems. Such estimations, which may include amplitude or intensity data, are important because of the volume of audio recordings that can accumulate over time, this often being more than can be reviewed and interpreted by human labor alone. On the other hand, much can be garnered about the conditions of a place simply by listening. Through its unfolding, sound can reveal complex, dynamic, and often otherwise hidden activity. The ambiances of a place can meander through the mundane repetitive processes of movement and growth, interrupted by sudden punctuations which disrupt the passing of time. With portable recording technology, we are able to make high-quality audio recordings of the *soundscapes* (see [10] for a critique of the notion of soundscape in its objectification of sound) of natural environments. Later in the studio, these sounds can be "humanly organized" [11] into elaborate arrangements for reception via loudspeakers within a concert setting, or through headphones when on the move. However, by bringing the outside in, through the careful capturing of the sounds that are sampled, we may lose a crucial element within the process of sensemaking [12] as these recordings are severed from their original contexts. Sound is always situated socially, culturally, and perceived within a particular space.

2.2. Ecological Systems in Sound

Concurrently, the environment has featured heavily within discussions of live electronic music in its role within the co-constitution of the body–instrument world that takes place during musical performance (see [13] for a detailed discussion of this). Out of this has grown a steadily increasing collection of theoretical developments and practices examining the ways in which ecological systems can be used to understand how musicians and audiences make sense of musical activity (see, for example, [14,15]). Additionally, this paradigm can be used to design new technological frameworks. Of these, Agostino Di Scipio's work is notable in his proposition that emergent and dynamic behavior can arise out of the structural coupling of digital processes with the space in which they are situated [16]. For Di Scipio, "Eco-systems are systems whose structure and development cannot exist (let alone be observed or modeled) except in its permanent contact with a medium" [16] (p. 271). Ecological systems—or ecosystems—typically comprise biological communities, yet the paradigm can be extended to include human society or cultural networks. It can also be used to describe the very nature of the structural relationships within certain computational compositional domains. This particular application is the focus of project that will be discussed in this paper.

Di Scipio's work challenges the often unquestioning adherence to the commonly used model of interactivity, where a musician can create, manipulate, or organize sound in time. In the digital–physical domain, this is facilitated through a digital musical instrument (DMI). DMIs are technologies which, in general, comprise a control interface (hardware) and a sound generator (software) [17]. In the ubiquitous interactivity model, which often views engagement with DMIs through the lens of human–computer interaction (HCI), the performer is in fact *necessary* in order to initiate—through some sort of physical or gestural manipulation—some changes of state within the instrument's DSP. Rather, Di Scipio proposes an *ecosystemic* model in which an evolving, self-regulating sonic entity is brought forth by "composing interactions" [16] (p. 270). In this approach, the role of the environment is integral to—and inseparable from—this process. He suggests that such ecosystems "are *autonomous* (i.e., literally, self-regulating) as their process reflects their own peculiar internal structure. Yet they cannot be isolated from the external world, and cannot achieve their own autonomous function except in close conjunction with a source of information (or energy). To isolate them from the medium is to kill them" [16] (p. 271). While this does not preclude a performer from being part of the ongoing negotiations and adjustments that take place within the system, in this model their presence is not required for musical activity.

One key approach behind Di Scipio's work involves setting up the potential for acoustic feedback to manifest from within systems of microphones and loudspeakers, which are often digitally mediated. This establishes sonic entities which self-regulate rather than ones which are the result of human organization of sound into particular aesthetic or systematic arrangements [18]. In collaboration with Dario Sanfilippo, some of the most recent work in this area focuses on techniques for incorporating the environment within these systems [19]. The environment in this work refers to the sonic ambiance of the space in which sound is being produced, along with any significant structural characteristics, and the effect that these may have on the audible feedback loops. Explicitly, the environment here is the "performance space" [19] (p. 23).

In their work, feedback occurs due to resonances both within instruments—which may comprise computers, loudspeakers, software, and other technology—as well as spaces. The instrument is said to be coupled with the environment because "they operate in a condition where they mutually affect each other" [20] (p. 31). However, in an open space it can be more difficult to find resonances, such as those formed by a the tube of an acoustic instrument such as a flute, or the four walls of a small room. An example of a solution to this problem can be found in Chris Kallmyer's piece *This Distance Makes Us Feel Closer* (2013), which uses sound to create echo and resonance where there are neither any structures nor land features to do this naturally. These are orchestrated by the artist who creates autonomous desert fog horns, which suggests the presence of reflected sound waves—despite being positioned one mile apart—in the expanse of the New Mexico desert [21]. Nevertheless, in both these examples, the sonic result is uniquely dependent upon the situatedness of the work itself—it is necessarily site-specific.

2.3. Responsiveness to Site

The term site-specific—its origins grounded within the visual arts and minimalist sculpture [22]— has become ambiguous through its usage as an umbrella term for describing art works that are either held outdoors (hosted outside of cultural institutions) [23], or works that would no longer make sense if moved to a new location [24]. In this broad view, site-specificity could be said to describe works which are constructed through a relational process, where the artist works with certain properties of a particular site. However, Miwon Kwon has noted that even the definition of *site* can be interpreted in multiple ways: as "phenomenological" [24] (p. 3), concerning the experienced physical qualities of a space; as addressing "social/institutional" [24] (p. 3) power structures; or as a site for "discursive" [24] (p. 3) activity surrounding political or economic issues.

A site-*responsive* practice [1] might be a more helpful way of describing the types of creative activities that engage with the environmental, sonic, cultural, or historical identities of a particular location—or the communities connected to that site—with careful reflection and sensitivity. It sidesteps the contradictions that arise from notions of site-specificity and focuses instead on developing a new set of methodologies to address its goals. These types of works connect "people with the space they occupy by bringing awareness to how their presence affects their environment" [25] (p. 231). Previous iterations of *Sounding Out Spaces* have addressed notions of multimodal participation that a site-responsive practice can foster [1]. In such cases, the audience is not directed to experience the work in a particular way, but is invited to explore their own trajectory through a variety of access points. Such approaches can result is unexpected behaviors. Daichi Misawa, while reflecting on his piece *Transparent Sculpture* (2012) which followed this embodied approach and premiered at Ars Electronica in 2012, observed that this can be affective "to such an extent that [the] audience started to walk, dance and scream" [26] (p. 390). This has implications not only for the accessibility of the work, but also moves away from musical experiences that Bennett Hogg describes as relying on a "framing of sounds within a stereo frame to be listened to under concert conditions [which] seems something of a betrayal of sound's potentially ecosystemic properties" [27].

3. Methodology

In this section we describe our methodology for realizing this project. We do this both by outlining the design of the technological and biological ecosystems, as well as detailing the artistic and compositional decisions that informed the use of the site. We also discuss the choice of sonic material, sensors, and the relationships between these elements. While the resulting installation was very much site-responsive, it should be noted that the aim of this work was to develop a set of techniques that could be implemented in different locations, and respond effectively and uniquely in each scenario.

3.1. Surveying the Site

The selected community garden was favorable as the location for the installation both due to the authors' familiarity with the site—a local farmers' market is hosted there each week—and its close proximity to the university area. After establishing initial links with the grassroots organization that stewards and develops the site, we discovered that the Tempe Community Action Agency (TCAA) not only facilitates opportunities for locals to build community through their gardens, but also works to serve the underprivileged by providing food and shelter [28]. In addition to pragmatic factors, such as the site being a desirable size for the scale of this iteration, the diverse range of communities involved with the garden would enable us to utilize multiple perspectives as input and feedback during development. As Christabel Stirling has observed, this is crucial in order to determine the actual social impact of public sonic art beyond the implications reported by the artists themselves [29]. Furthermore, having regular interactions with volunteers and those who were growing produce at the site provided us with knowledge about the types of vegetation and wildlife that inhabited the garden.

The project began with a survey of the location through repeated visits. This involved spending a significant amount of time listening at different parts of the site to learn about, and become familiar with, its sonic characteristics. In addition to focused, selective listening techniques [30], we also recorded on-site audio from static points around the site, and as moving trajectories. The microphone was employed as a way to amplify, or examine in detail, potentially interesting sonorities that might remain otherwise unheard. As Diane Willow—a media artist and researcher working with technology and materials—asks, "What happens when we amplify and make tangible our perception of the subtle, the ephemeral and the seemingly silent?" [31] (p. 333). This initial sampled sound was not used in the final installation itself, but was analyzed and inspected off-site. This provided a quantitative measurement of the audible frequencies and characteristics of sounds within the various regions of the site, which complemented the observations made through repeated visits. We considered:

- which aspects of the environment could be used structurally or compositionally?

- what in the environment is moving, could make sound through movement, or is already audible?
- which elements of the environment are static, and which are dynamic/changing?
- which aspects are consistent/inconsistent?

After this process of surveying, observation, listening, and analysis, the site was divided up into four regions that would each be used to experiment with a different approach to site-responsiveness.

3.2. Re-Sounding Material

While the main thrust of this project involved autonomous, self-organizing systems, we were keen to incorporate some established HCI models within the site, particularly to engage the children that we were hosting through our workshops at the garden. Building on prior research examining embodied approaches to sound and technology education [32–34], these workshops involved developing listening practices and exploring the garden in order to discover potentially interesting sounds through field recording techniques using hand-held audio recorders (https://www.zoom-na.com/products/field-video-recording/field-recording/zoom-h1-handy-recorder). We appropriated an existing art installation comprising five colorfully-painted bicycles mounted vertically on one of the walls of a structure in the garden (see Figure 3).

Figure 3. Children from Free Arts recording sounds from the site using portable sound recorders.

Our work in the garden utilized several Bela [35]-embedded microcomputing systems. Bela is a low-latency audio and sensor platform developed by Augmented Instruments Laboratory at the Centre for Digital Music (C4DM) at Queen Mary University of London. Its ability to run software built in Pure Data (Pd), a commonly used programming environment for sound synthesis and design, allowed for the rapid prototyping, development, and composition of sonic interactions to be done on-site and in the surrounding environment. Comprising low-power processors, these systems were powered using portable Universal Serial Bus (USB) battery packs. Given the the subtropical desert climate, we also experimented with solar power banks. These certainly would have been a sensible option had we run the installation over a longer period of time.

By attaching reed sensors to the bicycle forks, and a magnet on one of the spokes of each bicycle wheel, the children were able to perform the sounds that they had recorded from the garden back into the site itself. For this section of the installation, a single Bela was used to run Pd. Within the software, the speed of the rotation of each of the wheels was calculated from the time intervals detected by the reed sensors. The speed of the freewheeling was mapped in a nonlinear manner to the speed of playback of the samples. The recorded audio samples were stored in lookup tables and played back using both the digital sample position and low-frequency sawtooth waveforms. In the first case, which was audible when the wheels were spun quickly—taken as when the sensor was triggered at a frequency >1 Hz—a timbral effect was produced. The latter case—spin rate <1 Hz—allowed the samples to loop rhythmically, and then eventually slow down, and cease to sound. The children explored ways to play and manipulate their samples through the physical interfaces of the bicycles, using gestures similar to *scratching* records with the wheels, as well as using physical objects pushed against the spokes to give audible added resistance to the spin.

We also worked with MakeyMakey [36]—a low-cost electronics invention kit—connected to various plants and vegetation in the garden. By building simple connections, the plants and humans became part of a circuit. The MakeyMakey was connected to a laptop running software which, again, triggered various samples recorded by the children when the human–plant circuit was closed (see Figure 4). Through an exploratory approach, the children often connected different areas of vegetation in the garden and played these new instruments collaboratively. In some cases, more than one person had to be involved in order to complete the circuit. Sounds from the site were used by participants as improvisational material, and they were able to work sounds taken from the garden back into the space. In this way, the garden became a reconfigurable and constantly evolving digital–physical instrument.

Figure 4. Children from Free Arts creating human–plant instruments using MakeyMakeys.

3.3. Ecological Garden-Raised Beds

3.3.1. Environmental Sensing

As discussed in Section 2.2, within Di Scipio and Sanfilippo's approach, the environment is generally considered in terms of its ability to disturb and affect actual—or potential—acoustic phenomena. We build on their work by expanding the role of the environment in its ability to perturb aspects of the various autonomous systems that function within it by collecting data that is not restricted to the sonic domain. This was explored in the most technically challenging part of the installation: an ecosystemic network of 18 loudspeakers, each positioned in 18 raised garden beds among vegetables and plants. The beds were separated into three groups, each outfitted with its own system of microphones, speakers, and environmental sensors. In total, three Belas were used to run a variety of sensors in this area of the garden (see Table 1 and Figure 5 for details).

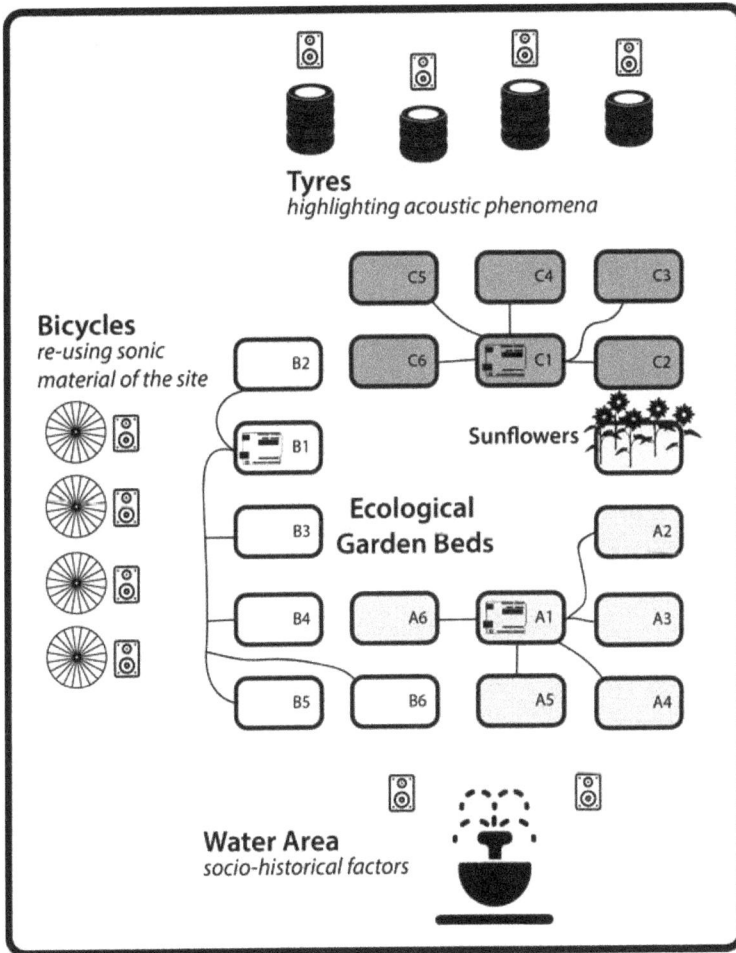

Figure 5. Layout of installations, Clark Park Community Garden, Tempe, AZ.

In addition to some high-quality portable loudspeakers (http://www.mightydwarf.com/product/blueii) (https://minirigs.co.uk/portable-speaker), it was possible to use low-cost loudspeakers (https://www.cnet.com/products/x-mini-uno-speaker-for-portable-use-xam14pu/specs/) for the majority of the beds due to the fact that they were battery powered and powerful enough to create a significant sonic impact when sounding simultaneously. In fact, it was crucial that the electronic sound did not overpower the sonic activity of the space, which was different depending on the time of day and what was happening at the site. Instead, the system was designed to fold any detectable sound into itself, and weave its own output back into the present sonic state.

Table 1. Placement of sensors, microphones, and loudspeakers within the raised beds.

Equipment	Description	Bed Location
Mighty Dwarf	portable 10 W speaker	A1, B1
Minirig	portable 15 W speaker	C1, B5
X-Mini	low-cost 2.5 W portable speaker	all remaining beds
Zoom H1 audio recorder	used for powered X-Y stereo microphones	A1, B1, C1
LOLLETTE anemometer	wind speed sensor	B1
Adafruit accelerometer	used to detect movement	B1
cadmium sulphide (CdS) photoresistor	used to detect movement via fluctuations in light	A1
Sonbest temperature/humidity sensor	used to detect slow changes in soil state	C1

Using custom software written in Pd and loosely based on Di Scipio's work on autonomous feedback systems and emergent sonic structures [18], self-adaptive feedback loops were created between the microphones and nearby loudspeakers. Essentially, these comprised a dynamic bandpass filter which would allow more or less frequencies to pass through depending on the amplitude of sound being received through the microphones. Importantly, no sound other than what was picked up by the microphones was used as source material.

The Larsen effect [20] was induced by placing the microphones proximally to the loudspeakers. Tamed by the adaptive filter, this sound was further processed using delay lines, granular synthesis, comb filters, and bandpass filters. In additional to the audible processed feedback, sounds from both humans and animals entered into the network, including significant activity from birds and crickets. The environmental sensors, which included measures for soil, wind, movement, and light, were used to inject additional disturbances into the feedback computation (see Figure 6). The resulting sound was the result of several self-organizing systems coexisting within the same environment.

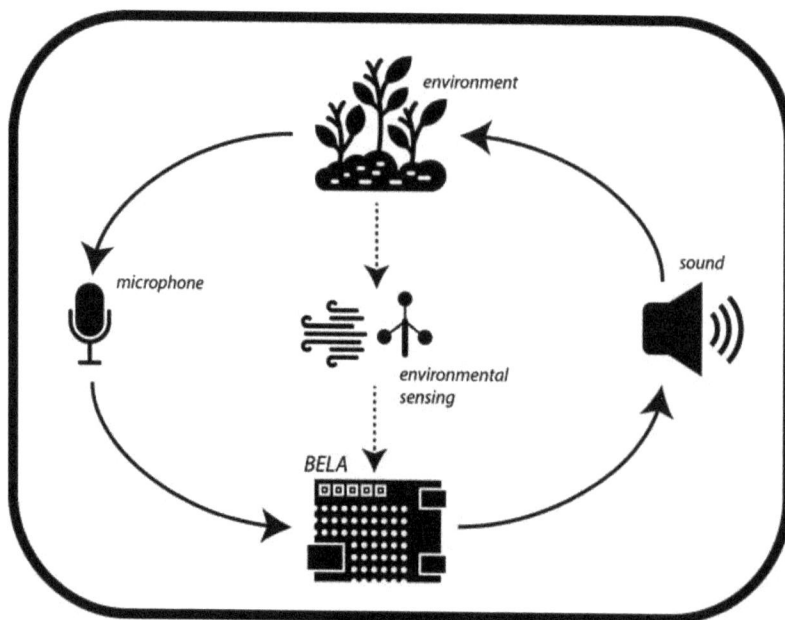

Figure 6. Signal flow of the garden bed installation.

3.3.2. Composed Relations

Being the result of the coupling of autonomous systems, the work itself was not composed, though compositional decisions could be said to reside in the mappings and relationships between sensor input and computational processes. Following several visits to and observations in the gardens, sensors were placed within the beds primarily on the basis of which sensor was the best fit for the bed's conditions. This might depend on the location of the bed within the site, or the density and type of vegetation within it. The data from the environmental sensors was then analyzed in real-time and normalized, scaled, and mapped to control various parameters of DSP on the Belas. The result was a curated collection of simple relationships between organic matter, computational processes, and the environment, combined to produce complex emergent behaviors. The aesthetic and experiential considerations were determined through repeated and ongoing listening and observing within the site on multiples days, and at different times. Below are a few examples of such composed relations:

Wind: Along with the sea, wind is arguably one of the only naturally continuous excitations [37]. Its continuous nature made it a useful resource to harness within this work. Wind speed readings from an anemometer (https://www.adafruit.com/product/1733) were used to control how quickly sounds were spatialized throughout the garden, creating a poetic connection between the movement of the wind, and the diffusion of sound within space. Minimum and maximum wind speeds were calculated continuously over the duration of the installation. When a new input value rose above a previous maximum (or below a minimum), the respective input parameter was adjusted, and the wind speed was calibrated accordingly. Due to the design of the anemometer—comprising three hemispherical cups mounted via arms onto a vertical shaft—wind speed is always measured as an average. This provided sufficiently useful wind-related data, which did not require further smoothing.

The relationship between the wind speed and sound spatialization was directly correlated: the faster the speed of the wind, the faster and more chaotic the movement of sounds. Slower wind speeds resulted in a stillness in sound content and diffusion. The spatialization was implemented using constant power panning, which was controlled by the wind speed. Timbral changes were achieved by mapping the rate of change of the scaled wind speed value to the density parameter of the granulator. This varied the number of grains of sound being played back within the software, leading to changes in perceived sound density.

Movement: In the same bed, an accelerometer (https://www.adafruit.com/product/163) was used to measure the movement of a swaying burlap sunshade. The three-dimensional data was normalized based on the minimum and maximum of values received over a sliding time-window. The method of rescaling the data was again used to account for the changing environmental conditions. In this case, the movement data was used to vary timbral aspects of the sound by modulating the speed of certain control waveforms within the granulator. The audible result was almost gestural, visually linked to the movements of the cloth, which allowed spectators to gain some insight into the processes at work.

Temperature: In another portion of the system, a soil sensor (https://cdn-shop.adafruit.com/datasheets/SLHT5.pdf) was used to measure the temperature and humidity of the soil. The initial soil state was read at the launch of the system, and subsequent fluctuations in these values modulated the overall tuning of several bandpass filters within the feedback system. The center frequencies of these bandpass filters were initially tuned to another audible source within the garden (see Section 3.4). The soil sensor values were scaled and mapping to these center frequencies. As they slowly changed over time, the resulting shifts in harmonicity led to an almost indiscernible process of detuning and retuning. Of course, in a climate more susceptible to rain—or if the bed had been watered—these changes would become more immediately perceptible to observers.

Ambient Light: Located in number 19 of the raised garden beds, a further system was developed, which involved several sunflowers. These were embedded with photoresistors (https://www.adafruit.com/product/161) to measure levels of ambient light (see Figure 7). The photoresistors were used as an affordable, lightweight solution to measure the amount of movement of individual flowers.

This was accomplished by measuring the change in light over each subsequent frame. The sensor inputs of the Bela are sampled at audio rates, which allows for smooth alignment with audio data [38]. To avoid encumbering the movement of the plants, the photoresistors were strewn together using conductive thread.

Figure 7. Sunflower embedded with photoresistor and conductive thread.

Following the movement of each flower, fluctuations in light gathered by the photoresistors were mapped to the amplitude of several subtractive synthesis engines. These comprised white noise passed through eight resonant bandpass audio filters. Each filter was tuned to a different harmonic (positive integer multiple) of the audible drone of a nearby dairy factory (see Section 3.4 for details). As with the nonlinear nature of audible feedback systems, the movement of each flower is also complex, being influenced by environmental forces and its physical structure, as well the movements of adjacent branches. By giving several flowers their own voice, this *imitation* [39] allows for further insight into the complexity of their movements, bringing attention to subtleties of the environment by emphasizing the collective rhythm of moving plants.

3.4. Highlighting Acoustic Phenomena-Tires

As mentioned above, one feature of the site that we wanted to work with was the audible drone from a local dairy factory that could be heard for several kilometers within the surrounding neighborhood. The drone was always present in the garden, although its intensity varied sporadically at different times of day.

We mimicked this drone by using three sinewave oscillators tuned to 237 Hz, 284 Hz, and 363 Hz, at amplitudes of 0.5, 0.25, and 0.4 respectively. We played this synthesized chord back through large scrap tires that had been abandoned in the space (see Figure 8). In this way, the sound of the site was reinforced, and a found acoustic phenomenon was appropriated as a key sonic element within the sound design. We were able to play on the disparity between what was included in the site—both sonically and materially—and what was omitted.

As with the more complex ecological systems described above, this part of the installation was also semi-autonomous and emerged over time through a very simple negative feedback system. Minirig speakers were placed inside the tires, along with a single omnidirectional condenser lavalier microphone (http://www.audio-technica.com/cms/wired_mics/9c6eca17168eef6f/index.html), and a Bela microcomputer. In this case, the amplitude values of the sound picked up by the microphone were fed into an accumulator with a leak value. This was fixed at a rate that would allow the most audible dynamic behavior to arise from the system, based on both self-generating feedback within the tires, as well as sounds picked up from the garden. Again, this was controlled using an adaptive filter.

Figure 8. Four abandoned tires being used as resonators.

3.5. Historical Narratives

While visiting the garden, we met many volunteers and community members who rented the raised beds in order to grow produce. Through these relationships we learned more about the history of the site. The garden was created in 2014 at the location of a former municipal pool. While this pool had recently been filled with soil and was used as a growing area, parts of its existence were still visible. We amplified the sound of water filtration systems in the garden using two omnidirectional microphones set up as boundary microphones (http://www.dpamicrophones.com/microphones/dscreet/stereo-microphone-kit-with-sc4060), and spatialized this over four high-quality loudspeakers (https://www.genelec.com/8010) to transform the seated gathering area into an auditory water bath.

This was the only part of the installation that did not use battery-powered equipment, although all elements could have been powered by a small stand-alone power source, such as a generator. As with the other self-adapting parts of the installation, this system changed over time, responding and adjusting to the level of measured sound. The sound of water picked up by the microphones was processed through a spectral filter, two different live sampling audio engines, and delay effects. These processes would become more or less audible depending on the density of the captured sound—this was again calculated using a leaky accumulator.

4. Conclusions and Future Work

As a site-responsive installation, *Garden Ecologies* aimed to engage with the sonic, cultural, historical, and environmental factors of the Clark Park Community Garden. The work involved the development of an ecosystemic framework, which included several autonomous systems coexisting in the same place, unique in output, yet responding to perturbations from the same environmental cues, and each other. We built upon prior research into musical systems that are structurally coupled to the environment by incorporating environmental information that is extra-sonic. We have developed some techniques for working with organic vegetation and changes in weather states. While this post-digital approach successfully bridges the physical and digital worlds, in future iterations we hope to explore in more depth techniques that have been established within the bio art communities, such as Laura Cinti's radical work on plant neurobiology, which engages with the perceptual and cognitive capacities of plants [40].

Technologically, this work has made significant progress in the design and creation of a portable, low-cost toolkit for site-responsive projects that can be implemented at and will be responsive to a range of locations. For example, at the time of writing, the technology had just been used in the sub-arctic tundra during the 2017 Ars Bioarctica Residency in Kilpisjärvi, Finland, a location with a drastically different environmental profile to the Sonoran Desert of Arizona.

The synchronicity between audio and sensor input afforded by the Bela platform, along with its ease of use, has certainly opened up the possibilities of working out "in the field" rather than in a studio, or gallery space. Nevertheless, despite the calibration and adaptation techniques described in this paper, significant time at the site was required to evaluate the efficacy of a particular approach or algorithm. Differences in hardware used, such as microphone type, would have a drastic impact on the sonic outcomes, as with the placement of such devices. Care was also required to select sensors that would provide meaningful data in various conditions. For example, we had to test photocells with different resistance ranges to determine which would be most useful both in bright daylight, as well as under the floodlights which lit the garden in the evenings.

We have designed the software systems as a modular framework, including tools for the scaling and mapping of data, as well as instruments for sound synthesis and feedback control. We plan to adopt a similar modular approach to the hardware, with the design of custom, robust boards that can facilitate the addition of a variety of sensors quickly and easily, and which can also powered by solar energy. Through further iterations at multiple sites we hope to begin to understand the efficacy of such an approach as an alternative to the sampling and analysis techniques that are used within acoustic ecology, as well as offering a more dynamic approach to sonification through nonlinear techniques.

Acknowledgments: We would like to thank our partner organization, Free Arts for Abused Children of Arizona, and all the volunteers that gave their time to help facilitate this project. We would also like to thank our other partners, the Clark Park Community Garden and the Tempe Community Action Agency, who were hugely instrumental in facilitating this project. We are indebted to Tobias Feltus for his thorough audio-visual documentation of the installation. Finally, the authors gratefully acknowledge the Community Partnership Grant from the City of Tempe.

Author Contributions: L.H. conceived the project; L.H. and J.S. designed and performed the experiments; L.H. and J.S. wrote the paper; J.S. created the figures; L.H. revised the paper.

Conflicts of Interest: The authors declare no conflict of interest.

Abbreviations

The following abbreviations are used in this manuscript:

DSP	digital signal processing
DMI	digital musical instrument
HCI	human–computer interaction
TCAA	Tempe Community Action Agency
USB	Universal Serial Bus
C4DM	Centre for Digital Music
Pd	Pure Data
CdS	cadmium sulphide

References

1. Hayes, L. From Site-Specific to Site-Responsive: Sound Art Performances as Participatory Milieu. In *Organised Sound*; Cambridge University Press: Cambridge, UK, 2017; Volume 22, pp. 83–92.
2. Maturana, H.R.; Varela, F.J. *Autopoiesis and Cognition: The Realization of the Living*; Reidel, D., Ed.; Boston Studies in the Philosophy of Science; Springer: Dordecht, The Netherlands, 1980; Volume 42.
3. Westerkamp, H. Linking Soundscape Composition and Acoustic Ecology. In *Organised Sound*; Cambridge University Press: Cambridge, UK, 2002; Volume 7, pp. 51–56.
4. López, F. Profound Listening and Environmental Sound Matter. In *Audio Culture: Readings of Modern Music*; Continuum International Publishing Group: New York, NY, USA, 2004; pp. 82–87.
5. Koutsomichalis, M. On Soundscapes, Phonography and Environmental Sound Art. *J. Sonic Stud.* **2013**, *4*. Available online: http://journal.sonicstudies.org/vol04/nr01/a05 (accessed on 11 January 2018).
6. Truax, B. *Handbook for Acoustic Ecology*; World Soundscape Project; Aesthetic Research Centre: Vancouver, BC, Canada, 1978.
7. Barclay, L. Sonic Ecologies: Exploring the Agency of Soundscapes in Ecological Crisis. In *Soundscape: The Journal of Acoustic Ecology*; WFAE: Charlotte, NC, USA, 2013; Volume 12.
8. Barclay, L. Acoustic Ecology in UNESCO Biosphere Reserves–Barclay & Gifford. In *International Journal of UNESCO Biosphere Reserves*; VIU Press: Nanaimo, BC, Canada, 2017; Volume 1.
9. Sueur, J.; Farina, A.; Gasc, A.; Pieretti, N.; Pavoine, S. Acoustic Indices for Biodiversity Assessment and Landscape Investigation. In *Acta Acustica United with Acustica*; Hirzel, S., Ed.; Verlag: Stuttgart, Germany, 2014; Volume 100, pp. 772–781.
10. Ingold, T. Against Soundscape. In *Autumn LEaves: Sound and the Environment in Artistic Practice*; Carlyle, A., Ed.; Double Entendre: Paris, France, 2007; pp. 10–13.
11. Blacking, J. *How Musical Is Man?* University of Washington Press: Seattle,WA, USA; London, UK, 1973.
12. Klein, G.; Moon, B.; Hoffman, R.R. Making Sense of Sensemaking 1: Alternative Perspectives. *IEEE Intell. Syst.* **2006**, *21*, 70–73.
13. Waters, S. Performance Ecosystems: Ecological Approaches to Musical Interaction. In Proceedings of the 2007 Electroacoustic Music Studies Network, Leicester, UK, 12–15 June 2007; EMS: Leicester, UK, 2007.
14. Green, O. Pondering Value in the Performance Ecosystem. *eContact!* **2008**, *10*, 197–213.
15. Davis, T. Towards a Relational Understanding of the Performance Ecosystem. In *Organised Sound*; Cambridge University Press: Cambridge, UK, 2011; Volume 16, pp. 120–124.
16. Di Scipio, A. Sound is the Interface: From Interactive to Ecosystemic Signal Processing. In *Organised Sound*; Cambridge University Press: Cambridge, UK, 2003; Volume 8.
17. Malloch, J.; Birnbaum, D.; Sinyor, E.; Wanderley, M.M. Towards a New Conceptual Framework for Digital Musical Instruments. In Proceedings of the 9th International Conference on Digital Audio Effects, Montreal, QC, Canada, 18–20 September 2006; pp. 49–52.
18. Di Scipio, A. Using PD for Live Interactions in Sound. An Exploratory Approach. In Proceedings of the 4th International Linux Audio Conference, Karlsruhe, Germany, 27–30 April 2006.
19. Sanfilippo, D.; Di Scipio, A. Environment-Mediated Coupling of Autonomous Sound-Generating Systems in Live Performance: An Overview of the *Machine Milieu* Project. In Proceedings of the 14th Sound and Music Computing Conference, Espoo, Finland, 5–8 July 2017; pp. 21–27.
20. Sanfilippo, D.; Valle, A. Feedback Systems: An Analytical Framework. In *Computer Music Journal*; MIT Press: Cambridge, MA, USA, 2013; Volume 37, pp. 12–27.
21. Gottschalk, J. *Experimental Music Since 1970*; Bloomsbury: London, UK, 2016.
22. Kaye, N. *Site-Specific Art: Performance, Place and Documentation*; Routledge: London, UK, 2000.
23. LaBelle, B. *Background Noise, Second Edition: Perspectives on Sound Art*; Bloomsbury Academic: London, UK, 2015.
24. Kwon, M. One Place after Another: Notes on Site Specificity. In *October*; MIT Press: Cambridge, MA, USA, 1997; Volume 80, pp. 85–110.
25. Rydarowski, A.; Samanci, O.; Mazalek, A. Murmur: Kinetic Relief Sculpture, Multi-Sensory Display, Listening Machine. In Proceedings of the 2nd International Conference on Tangible, Embedded and Embodied Interaction, Bonn, Germany, 18–20 February 2008; ACM: New York, NY, USA, 2008; pp. 231–238.

26. Misawa, D. Transparent Sculpture: An Embodied Auditory Interface for Sound Sculpture. In Proceedings of the 7th International Conference on Tangible, Embedded and Embodied Interaction, Barcelona, Spain, 10–13 February 2013; ACM: New York, NY, USA, 2013; pp. 389–390.

27. Hogg, B. The Violin, The River, and Me: Artistic Research and Environmental Epistemology in *balancing string* and *Devil's Water 1*, Two Recent Environmental Sound Art Projects. *HZ J.* **2013**, *18*. Available online: http://www.hz-journal.org/n18/hogg.html (accessed on 11 January 2018).

28. Tempe Comunity Action Agency. 2017. Available online: http://tempeaction.org/about-us/ (accessed on 18 December 2017).

29. Stirling, C. Sound Art/Street Life: Tracing the Social and Political Effects of Sound Installations in London. *J. Sonic Stud.* **2015**, *11*. Available online: https://www.researchcatalogue.net/view/234018/234019 (accessed on 11 January 2018).

30. Oliveros, P. *Deep Listening: A Composer's Sound Practice*; iUniverse: Lincoln, NE, USA, 2005.

31. Willow, D. Ambient Sites: Making Tangible the Subtle, Ephemeral and Seemingly Silent. In Proceedings of the Fourth International Conference on Tangible, Embedded, and Embodied Interaction, Cambridge, MA, USA, 25–27 January 2010; ACM: New York, NY, USA, 2010; pp. 333–336.

32. Droumeva, M.; Antle, A.; Wakkary, R. Exploring Ambient Sound Techniques in the Design of Responsive Environments for Children. In Proceedings of the 1st International Conference on Tangible, Embedded and Embodied Interaction, Baton Rouge, LA, USA, 15–17 February 2007; ACM: New York, NY, USA, 2007; pp. 171–178.

33. Schiettecatte, B.; Vanderdonckt, J. AudioCubes: a Distributed Cube Tangible Interface based on Interaction Range for Sound Design. In Proceedings of the 2nd International Conference on Tangible, Embedded and Embodied Interaction, Bonn, Germany, 18–20 February 2008; ACM: New York, NY, USA, 2008; pp. 3–10.

34. Hayes, L. Sound, Electronics, and Music: A Radical and Hopeful Experiment in Early Music Education. In *Computer Music Journal*; MIT Press: Cambridge, MA, USA, 2017; Voluem 41, pp. 36–49.

35. McPherson, A. Bela: An Embedded Platform for Low-Latency Feedback Control of Sound. *J. Acoust. Soc. Am.* **2017**, *141*, 3618.

36. Collective, B.M.; Shaw, D. Makey Makey: Improvising Tangible and Nature-Based User Interfaces. In Proceedings of the 6th International Conference on Tangible, Embedded and Embodied Interaction, Kingston, ON, Canada, 19–22 February 2012; ACM: New York, NY, USA, 2012; pp. 367–370.

37. Wishart, T. *On Sonic Art*; Routledge: Abingdon, UK, 1996.

38. Moro, G.; Bin, A.; Jack, R.H.; Heinrichs, C.; McPherson, A.P. Making High-Performance Embedded Instruments with Bela and Pure Data. In Proceedings of the International Conference on Live Interfaces, Brighton, UK, 29 June–3 July 2016.

39. Augoyard, J.F. *Sonic Experience: A Guide to Everyday Sounds*; McGill-Queens Press: Montreal, QC, Canada, 2006.

40. Cinti, L. The Sensorial Invisibility of Plants: An Interdisciplinary Inquiry through Bio Art and Plant Neurobiology. Ph.D. Thesis, University College London (University of London), London, UK, 2011.

applied
sciences

MDPI

Article

Application of Machine Learning for the Spatial Analysis of Binaural Room Impulse Responses

Michael Lovedee-Turner [*,†,‡] and Damian Murphy [†]

Communication Technologies Research Group, Department of Electronic Engineering, University of York, York YO10 5DD, UK; damian.murphy@york.ac.uk
* Correspondence: mjlt500@york.ac.uk; Tel.: +44-1904-324227
† Current address: Audio Lab, Department of Electronic Engineering, University of York, York YO10 5DD, UK.
‡ Binaural model code, neural network code, and direct sound and reflection dataset will be made available at: 10.5281/zenodo.1038021.

Academic Editor: Tapio Lokki
Received: 30 October 2017; Accepted: 26 December 2017; Published: 12 January 2018

Abstract: Spatial impulse response analysis techniques are commonly used in the field of acoustics, as they help to characterise the interaction of sound with an enclosed environment. This paper presents a novel approach for spatial analyses of binaural impulse responses, using a binaural model fronted neural network. The proposed method uses binaural cues utilised by the human auditory system, which are mapped by the neural network to the azimuth direction of arrival classes. A cascade-correlation neural network was trained using a multi-conditional training dataset of head-related impulse responses with added noise. The neural network is tested using a set of binaural impulse responses captured using two dummy head microphones in an anechoic chamber, with a reflective boundary positioned to produce a reflection with a known direction of arrival. Results showed that the neural network was generalisable for the direct sound of the binaural room impulse responses for both dummy head microphones. However, it was found to be less accurate at predicting the direction of arrival of the reflections. The work indicates the potential of using such an algorithm for the spatial analysis of binaural impulse responses, while indicating where the method applied needs to be made more robust for more general application.

Keywords: machine-hearing; machine-learning; binaural room impulse response; spatial analysis; direction of arrival

1. Introduction

A Binaural room impulse response (BRIR) is a measurement of the response of a room to an excitation from an (ideally) impulsive sound. The BRIR is comprised of the superposition of the direct source-to-receiver sound component, discrete reflections produced from interactions with a limited number of boundary surfaces, together with the densely-distributed, exponentially-decaying reverberant tail that results from repeated surface interactions. In particular, a BRIR is characterised by the receiver having the properties of a typical human head, that is two independent channels of information separated appropriately, and subject to spatial variation imparted by the pinnae and head. Therefore, the BRIR is uniquely defined by the location, shape and acoustic properties of reflective surfaces, together with the source and receiver position and orientation.

The BRIR is therefore a representation of the reverberant characteristics of an environment and is commonly used throughout the fields of acoustics and signal processing. Through the use of convolution, the reverberant characteristics of the room, as captured within the BRIR, can be imparted onto other audio signals, giving the perception of listening to that audio signal as if it were recorded in the BRIR measurement position. This technique for producing artificial reverberation has numerous applications, including: music production, game sound design, alongside other audio-visual media. In acoustics, the spatiotemporal characteristics of reflections arising from sound propagation and interaction within a given bounded space can be captured through measuring the room impulse response for a given source/receiver pair. One problem associated with this form of analysis is obtaining a prediction for the direction of arrival (DoA) of these reflections. Understanding the DoA of reflections can allow for the formulation of reflection backpropagation and geometric inference algorithms, amongst other features, that reveal the properties of the given acoustic environment for which the impulse response was obtained. This has applications in robot audition, sound source localisation tasks, as well as room acoustic analysis, treatment and simulation. These algorithms can be used to develop an understanding of signal propagation in a room, allowing the point of origin for acoustic events arriving at the receiver to be found. This knowledge of the signal propagation in the environment can then be used to acoustically treat the environment, improving the perceptibility of signals produced within the environment. Conversely, the inferred geometry can be used to simulate the acoustic response of the room to a different source and receiver through the use of computational acoustic simulation techniques.

Existing methods [1–3] have approached reflection DoA estimation using four or more channels, while methods looking at localising the components in two-channel BRIRs have generally shown poor accuracy for predicting the DoA of the reflections in these BRIRs [4]. This paper investigates a novel approach to using neural networks for DoA estimation for the direct and reflected sound components in BRIRs. The reduction in the number of channels available for analyses significantly adds to the complexity of extracting highly accurate direction of arrival predictions.

The human auditory system is a complex, but robust system, capable of undertaking sound localisation tasks under varying conditions with relative ease [5]. The binaural nature of the auditory system leads to two main interaural localisation cues: interaural time difference (ITD), the time of arrival difference between the signals arriving at the two ears, and interaural level difference (ILD), the frequency-dependent difference in signal loudness at the two ears due to the difference in propagation path and acoustic shadowing produced by the head [5,6]. In addition to these interaural cues, it has been shown that the auditory system makes use of self-motion [7] and the spectral filtering produced by the pinnae to improve localisation accuracy, particularly with regards to elevation and front-back confusion [5,8].

Given the robustness of the auditory system at performing localisation tasks [5], it should be possible to produce a computational approach using the same auditory cues. Due to the nature of the human auditory system, machine-hearing approaches are often implemented in binaural localisation algorithms, typically using either Gaussian mixture models (GMMs) [9–11] or neural networks (NNs) [12–15]. In most cases, the data presented to the machine-hearing algorithm fit into one of two categories: binaural cues (ITD and ILD) or spectral cues. Previous machine-hearing approaches to binaural localisation have shown good results across the training data and, in some cases, good generalisability across unknown data from different datasets [9–15].

In [14], a cochlear model was used to pre-process head-related impulse responses (HRIRs), the output of which was then used to calculate the ITD and ILD. Two different cochlear models for ITD and ILD calculation were used, as well as feeding the cochlear model output to the NN. The results presented showed that the NN was able to build up a spatial map from raw output of the cochlear model, which performed better under test conditions than using the binaural cues calculated from the output of the cochlea model.

Backman et al. [13] used a feature vector comprised of the cross-correlation function and ILD to train their NNs, which were able to produce highly accurate results within the training data. However, upon presenting the NN with unknown data, it was found to have poor generalisation.

In [12], Palomäki et al. presented approaches using a self-organising map and a multi-layer perceptron trained using the ITD and ILD values calculated from a binaural model. They found that both were capable of producing accurate results within the training data, with the self-organising map requiring the addition of head rotation to help disambiguate cue similarity between the front and back hemispheres [12]. Their findings suggested that a much larger dataset is required to achieve generalisation with the multi-layer perceptron.

In [9–11], GMMs trained using the ITD and ILD were used to classify the DoA. In both cases, the GMMs were found to produce accurate azimuthal DoA estimates. Their findings showed that the GMM's ability to accurately predict azimuth DoA was affected by the source and receiver distance and the reverberation time, with larger source-receiver distances and reverberation times generally reducing the accuracy of the model [9,10]. The results presented in [9] showed that a GMM trained with a multi-conditional training (MCT) dataset was able to localise a signal using two different binaural dummy heads with high accuracy.

Ding et al. [16] used the supervised binaural mapping technique, to map binaural features to 2D directions, which were then used to localise a sound source's azimuth and elevation position. They presented results displaying the effect of reverberation on prediction accuracy, showing that prediction accuracy decreased as reverberation times increased. They additionally showed that the use of a binaural dereverberation technique improved prediction accuracy across all reverberation times [16].

Recent work by Ma et al. [15] compared the use of GMM and deep NNs (DNNs) for the azimuthal DoA estimation task. The DNN made use of head rotation produced by a KEMAR unit (KEMAR: Knowles Electronics Manikin for Acoustic Research) is a head and torso simulator designed specifically for, and commonly used in, binaural acoustic research) [17] fitted with a motorised head. It was found that the addition of head rotation reduced the ambiguity between front and back and that DNNs outperformed GMMs, with DNNs proving better at discerning between the front and back hemispheres.

Work presented by Vesa et al. [4] investigated the problem of DoA analysis of the component parts of a BRIR. They used the continuous-wavelet transform to create a frequency domain representation of the signal, which is used to compute the ILD and ITD across frequency bands. The DoA is then computed by iterating over a database of reference HRIRs and finding the reference HRIR with the closest matching ILD and ITD values to the component of the BRIR being analysed; the DoA is then assumed to be the same as the reference HRIR. They reported mean angular errors between $28.7°$ and $54.4°$ for the component parts of the measured BRIRs.

This paper presents a novel approach for the spatial analysis of two-channel BRIRs, using a binaural model fronted NN to estimate the azimuthal direction of arrival for the direct sound and reflected components (direct sound is used to refer to the signal emitted by a loudspeaker arriving at the receiver, and the reflected component refers to a reflected copy of the emitted signal arriving at the receiver after incidence with a reflective surface) of the BRIRs. It develops and extends the approach adopted in [15] in terms of the processing used by the binaural model to extract the interaural cues, the use of a cascade-correlation neural network as opposed to the multi-layer perceptron to map the binaural cues to the direction of arrival classes, the nature of the sound components being analysed—short pulses relating to the direct sound and reflected components of a BRIR as opposed to continuous speech signals—and the method by which measurement orientations are implemented and analysed by the NN. In this paper, multiple measurement orientations are presented simultaneously to the NN, whereas in [15], multiple orientations are presented as rotations produced by a motorised head with the signals being analysed separately by the NN, which allowed for active sound source localisation in an environment.

The following sections are organised as follows; in Section 2, the implementation of the binaural model and NN, the data model used and the methodology used to generate a test dataset are discussed; Section 3 presents the test results; Section 4 discusses the findings; and Section 5 concludes the paper.

2. Materials and Methods

The proposed method uses a binaural model to produce representations of the time of arrival and frequency-dependent level differences between the signals arriving at the left and right ear of a dummy head microphone. This binaural model is used to produce a set of interaural cues for the direct sound and each detectable reflection within a BRIR. These cues alone are not sufficient to provide accurate localisation of sound sources, due to interaural cue similarities observed at mirrored source positions in the front/rear hemispheres. To distinguish between sounds arriving from either the front or rear of the head, an additional set of binaural cues is generated for the corresponding direct sound and reflected component of a BRIR captured with the dummy head having been rotated by ±90°. Presenting the NN with both the original measurement and one captured after rotating the receiver helps reduce front-back confusions, arising due to similarities in binaural cues for positions mirrored in the front and back hemispheres. The use of a rotation of ±90° was used in this study based on tests run with different rotation angles, which are presented in Section 2.2. These sets of interaural cues are then interpreted by a cascade-correlation NN, producing a prediction of the DoA for the direct sound and each detected reflection in the BRIR. The NN is trained with an MCT dataset of interaural cues extracted from HRIRs measured with a KEMAR 45BC binaural dummy head microphone, with added simulated spatially white noise at different signal-to-noise ratios. The NN is trained using mini-batches of the training dataset, and optimised using the adaptive moment (ADAM) optimiser; with the order of the training data randomised at the end of the training iteration.

2.1. Binaural Model

A binaural model inspired by the work presented in [18,19] is used to compute the temporal and frequency-dependent level differences between the signals arriving at the left and right ears of a listener. Both the temporal and spectral feature spaces provide directionally-dependent cues, produced by path differences between ears and acoustic shadowing produced by the presence of the head, which allow the human auditory system to localise a sound source in an environment [6,20]. These directionally-dependent feature spaces are used in this study to produce a feature vector that can be analysed by an NN to estimate the direction of arrival.

Prior to running the analysis of the binaural signals, the signal vectors being analysed are zero-padded by 2000 samples accounting for signal delay introduced by the application of a gammatone filter bank. This ensures that no part of the signal is lost when dealing with small windows of sound, where the filter delay would push the signal outside of the represented sample range. The zero-padded signals are then passed through a bank of 64 gammatone filters spaced equally from 80 Hz to 22 kHz using the equivalent rectangular bandwidth scale. The gammatone filter implementation in Malcolm Slaney's 'Auditory Toolbox' [21] was used in this study. The output of the cochlea is then approximated using the cochleagram function in [22] with a window size of six samples and an overlap of one sample; this produces an $F \times N$ map of auditory nerve firing rates across time-frequency units, where N is the number of time samples and F is the number of gammatone filters. The cochleagram is calculated as:

$$x_l(f,n) = y_l(f,\tau) * y_l(f,\tau)^\top \tag{1}$$

where $x_l(f,n)$ is the cochleagram output for the left channel for gammatone filter f at frame number n, $y_l(f,\tau)$ is the filtered left channel of audio at gammatone filter f and time frame τ, which is six samples in length and $(.)^\top$ signifies vector transposition [22]. The cochleagram was used to extract the features as opposed to extracting directly from the gammatone filters, as it was found to produce more accurate results when passed to the NN.

The interaural cues are then computed across the whole cochleagram producing a single set of interaural cues for each binaural signal being analysed. The first of these interaural cues is the interaural cross-correlation (IACC) function, which is computed for each frequency band as the cross-correlation between the whole approximated cochlea output x_l and x_r for the left and right channel, respectively, with a maximum lag of ± 1.1 ms. The maximum lag of ± 1.1 ms was chosen based on the maximum time delays suggested by Pulkki et al. for their binaural model proposed in [18]. The cross-correlation function is then normalised by,

$$IACC = \frac{xc_f}{x_{l,f}x_{l,f}^{\top}x_{r,f}x_{r,f}^{\top}} \tag{2}$$

where xc_f is the cross-correlation between the left and right approximated cochlea outputs for gammatone filter f. The IACC is then averaged across the 64 gammatone filters, producing the temporal feature space for the analysed signal. The maximum peak in the IACC function represents the signal delay between the left and right ear. The decision to use the entire IACC function as opposed to the ITD was based on the findings presented in [15], which suggested that features within the IACC function, such as the relationship between the main peak and any side bands, varied with azimuthal direction of arrival.

The ILD is then calculated from the cochleagram output in decibels as the loudness ratio between the two ears for each gammatone filter f such that,

$$ILD_f = 10 * log_{10}\left(\frac{\sum_{t=1}^{T} xl_{f,t}}{\sum_{t=1}^{T} xr_{f,t}}\right) dB \tag{3}$$

where $xl_{f,t}$ and $xr_{f,t}$ are the approximated cochlea output for gammatone filter f for signal x, for the left (l) and right (r) ear at time window t, and T is the total number of time windows. An example of the IACC and ILD feature vector for a HRIR at azimuth = 90° and elevation = 0° can be seen in Figure 1.

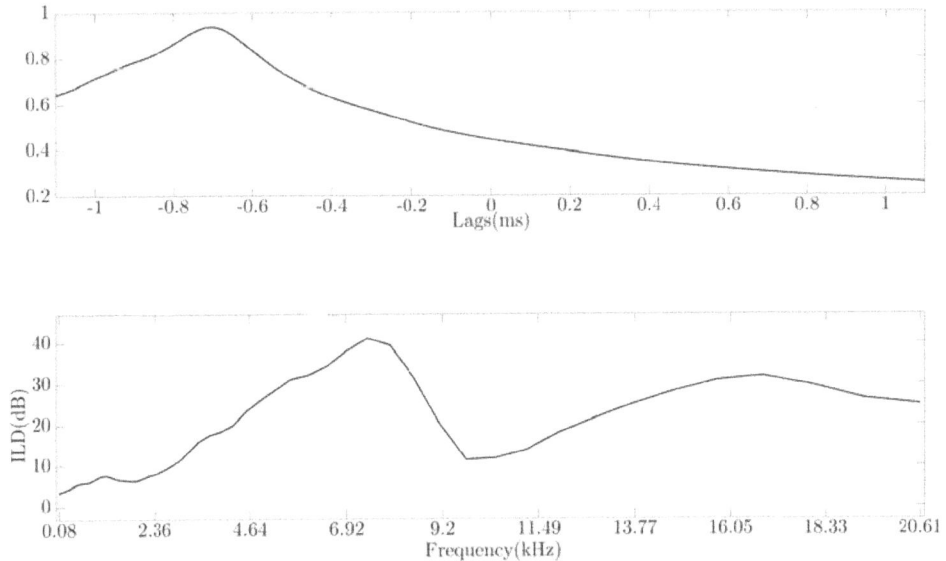

Figure 1. Example of the interaural cross-correlation function (**top**) and interaural level difference (**bottom**) for a HRIR with a source positioned at azimuth = 90 ° and elevation = 0°.

In this study, the binaural model is used to analyse binaural signals with a sampling rate of 44.1 kHz; the output of the binaural model is then an IACC function vector of length 99 and an ILD vector of length 64. This produces a feature space for a single binaural signal of length 163.

2.2. Neural Network Data Model

The binaural model presented in Section 2.1 is used to generate a training feature matrix using the un-compensated 'raw' SADIE KEMAR dataset [23]. This dataset contains an HRIR grid of 1550 points: 5° increments across the azimuth in steps of 10° elevation. To train the NN, only the HRIRs relating to 0° elevation were used, providing a dataset of 104 HRIRs. A multi-conditional training (MCT) dataset is created by adding spatially white noise to the HRIRs at 0 dB, 10 dB and 20 dB signal-to-noise ratios. This spatially white noise is generated by convolving Gaussian white noise with all 1550 HRIRs in the SADIE KEMAR dataset and averaging the resulting localised noise across the 1550 positions; producing a spatially white noise signal matrix [15]. This addition of spatially white noise is based on the findings in [9,10,15], which found that training the NN with data under different noise conditions improved generalisation. These HRIRs with added spatially white noise are then analysed by the binaural model and the output used to create the feature vector. The neural network is only trained using these HRIRs with noise mixtures, and no reflected components of BRIRs are included as part of the training data

Two training matrices are created by concatenating the feature vector of one HRIR with the feature vector produced by an HRIR corresponding to either a +90° or −90° rotation of KEMAR with the same signal-to-noise ratio. This produces two 416 × 326 feature matrices with which two neural networks can be trained with, one for each rotation. The use of an NN for each fixed rotation angle was found to produce more accurate results than having one NN trained for both.

The use of 'head rotation' has a biological precedence, in that humans use head rotation to focus on the location of a sound source; disambiguating front-back confusions that occur due to interaural cue similarities between signals arriving from opposing locations in the front and back hemispheres [6,20]. In this study, the equivalent effect of implementing a head rotation is realised by taking the impulse response measurements at two additional fixed measurement orientations (at ±90°). The use of fixed rotations reduces the number of additional signals needed to train the NN and reduces the number of additional measurements that need to be recorded. The use of additional measurement positions corresponding to receiver rotations of ±90° was found to produce lower maximum errors when compared to rotations of ±15°, ±30° and ±60° (Table 1). The two training matrices are used to train two NN, one for each rotation, and the network trained with the −90° rotation dataset is used to predict the DoA for signals that originate on the left hemisphere, while the +90° NN is used to predict the DoA for signals on the right hemisphere. Each of these NNs is trained with the full azimuth range to allow the NNs to predict the DoA for signals with ambiguous feature vectors that may be classified as originating from the wrong hemisphere. When testing the NN, the additional measurement positions are assigned to the signals based on the location of the maximum peak in the *IACC* feature vector. If the peak index in the *IACC* is less than 50 (signal originated in the left hemisphere), a receiver rotation of −90° is applied; otherwise, a receiver rotation of +90° is used. To normalise the numeric values, the training data were Gaussian-normalised to ensure each feature had zero mean and unit variance. The processing workflow for the training data can be seen in Figure 2.

Table 1. Direction of arrival accuracy comparison for the reflected component measured with the KEMAR 45BC for different fixed receiver rotation angles.

Rotation	Within ±5°	Front-Back Confusions	Max Error
KEMAR Reflections			
±15°	29.86%	15.28%	173
±30°	34.03%	6.25%	54
±60°	29.17%	9.72%	50
±90°	32.64%	9.03%	30

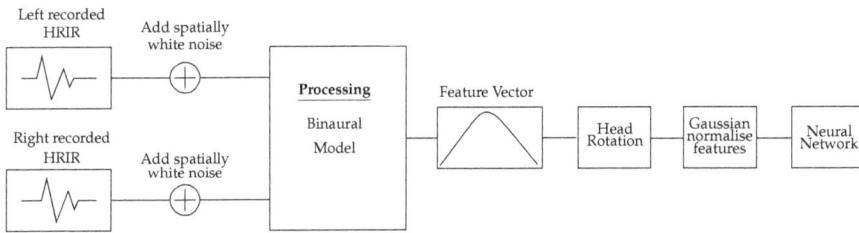

Figure 2. Signal processing chain used to generate the training data used to train the neural network.

2.3. Neural Network

TensorFlow [24], a commonly-used python library designed for the development and execution of machine learning algorithms, is used to implement a cascade-correlation NN, the topology of which connects the input feature vector to every layer within the NN. Additionally, all layers' outputs are connected to subsequent layers in the NN, as in Figure 3 [25]. The use of NN over GMM was chosen based on findings in [15], which suggested that DNN outperformed GMM for binaural localisation tasks. The decision to use the cascade-correlation NN was based on comparisons between the cascade-correlation NN architecture and the MLP, which showed that the cascade-correlation NN arrived at a more accurate solution with less training required compared to the MLP (Table 2).

Table 2. Comparison of prediction accuracy for the reflected component measured with the KEMAR 45BC using additional measurements at receiver rotations of ±90° using a multi-layer perceptron and cascade-correlation neural network. Both the multi-layer perceptron and the cascade-correlation neural network had one hidden layer with 128 neurons and an output layer with 360 neurons and were trained using the procedure discussed in Section 2.3.

Neural Network	Within ±5°	Run Time
KEMAR Reflections (Test Data)		
multi-layer perceptron	26.39%	390 Epochs 40 s
cascade-correlation	32.64%	244 Epochs 28 s

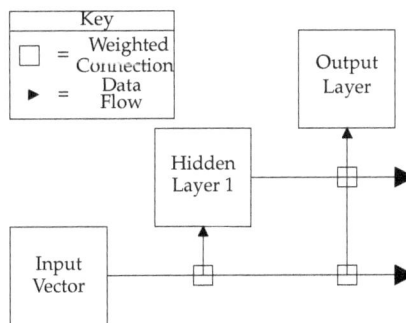

Figure 3. Cascade-correlation neural network topology used, where triangles signify the data flow and squares are weighted connections between the hidden layers and the incoming data.

The NN consists of an input layer, one hidden layer and an output layer. The input layer contains one node for each feature in the training data; the hidden layer contains 128 neurons each with a hyperbolic tangent activation function; and the output layer contains 360 neurons, one for each azimuth direction from 0° to 359°. Using 360 output neurons as opposed to 104 (one for each angle of the training dataset) allows the NN to make attempts at predicting the DoA for both known and unknown source positions. A softmax activation function is then applied to the output layer of the NN, producing a probability vector predicting the likelihood of the analysed signal having arrived from each of the 360 possible DoAs.

Each data point, whether it be a feature in the input feature vector or the output of a previous layer, is connected to a neuron via a weighted connection. The summed response of all the weighted connections linked to a neuron defines that neuron's level of activation when presented with a specific data configuration, a bias is then applied to this activation level. These weights and biases for each layer of the NN are initialised with random values, with the weights distributed such that they will be zero mean and have a standard deviation (σ) defined as:

$$\sigma_i = m^{-1/2} \tag{4}$$

where m is the number of inputs to hidden layer i [26].

The NN is trained over a maximum of 600 epochs, with the training terminating once the NN reached 100% accuracy or improvement saturation. Improvement saturation is defined as no improvement over a training period equal to 5% of the total number of epochs. Mini-batches are used to train the NN with sizes equal to 25% of the training data. The order of the training data is randomised after each epoch, so the NN never receives the same batch of data twice. The adaptive moment estimation (ADAM) optimiser [27] is used for training, using a learning rate of 0.001, a β_1 value of 0.9, a β_2 value of 0.99 and an ϵ value of 1^{-7}. The β values define the exponential decay for the moment estimates, and ϵ is the numerical stability constant [27].

The NN's targets are defined as a vector of size 360, with a one in the index relating to the DoA and all other entries equal to zero. The DoA is therefore extracted from the probability vector produced by the NN as the angle with the highest probability such that,

$$\theta = argmax \, \mathcal{P}(\theta|x) \tag{5}$$

where $\mathcal{P}(\theta|x)$ represents the probability of azimuth angle θ given the feature vector x. The probability is calculated as,

$$\mathcal{P}(\theta|x) = softmax(((x \times w_{out1}) + (\tilde{x}_1 \times w_{out2})) + b_{out}) \tag{6}$$

where w denotes a set of weights, b_{out} is the output biases and $\tilde{x}_1$ is the output from the hidden layer calculated as,

$$\tilde{x}_1 = tanh((x \times w_1) + b_1) \tag{7}$$

2.4. Testing Methodology

A key measure of the success of an NN is its ability to generalise across different datasets other than that with which it was trained. To test the generalisability of the proposed NN, a dataset was produced in an anechoic chamber for both a KEMAR 45BC [17] and Neumann KU100 [28] binaural dummy head, using an Equator D5 coaxial loudspeaker [29]. The exponential sine sweep method [30] was used to generate the BRIRs, with a swept frequency range of 20 Hz to 22 kHz over ten seconds. To be able to test the NN's performance at predicting the DoA of reflections, a flat wooden reflective surface mounted on a stand was placed in the anechoic chamber, such that a reflection with a known DoA would be produced (Figure 4). This allows for the accuracy of the NN at predicting the DoA for a reflected signal to be tested, without the presence of overlapping reflections that could occur in non-controlled environments. To approximate an omnidirectional sound source, the BRIRs were averaged over four speaker rotations (0°, 90°, 180° and 270°); omnidirectional sources are often desired in impulse response measurements for acoustic analysis [31], as they produce approximately equal acoustic excitation throughout the room. The extent to which this averaged loudspeaker response will be omnidirectional will vary across different loudspeakers, particularly at higher frequencies where loudspeakers tend to be more directional. Averaging the response of the room over speaker rotations does result in some spectral variation, particularly with noisier signals; however, this workflow is similar to that employed when measuring the impulse response of a room.

Figure 4. Measurement setup showing the reflective surface (A), KEMAR 45BC (B) and Equator D5 Coaxial Loudspeaker (C).

To calculate the required location of the reflective surface such that a known DoA would be produced, a simple MATLAB image source model based on [32] was used to calculate a point of incidence on a wall that would produce a first order reflection in a 3 m × 3 m × 3 m room with the receiver positioned in the centre of the room. The reflective surface was then placed in the anechoic chamber based on the angle of arrival and distance between the receiver and calculated point of incidence. Although care was taken to ensure accurate positioning of the individual parts of the system, it is prone to misalignments due to the floating floor in the anechoic chamber, which can lead to DoAs that differ from that which is expected.

With these BRIRs only having two sources of impulsive sounds, the direct sound and first reflection, a simple method for separating these signals was employed. Firstly, the maximum absolute peak in the signal is detected and assumed to belong to the direct sound. A 170 sample frame around the peak location indexed at $[peakIndex - 45 : peakIndex + 124]$ was used to separate the direct sound from the signal. It was ensured that all segmented audio samples only contained audio pertaining to the direct sound. The process was then run again to detect the location of the reflected component, and each segment was checked to ensure only audio pertaining to the reflected component was present (see Figure 5 for an example BRIR with window locations). When dealing with BRIRs measured in less controlled environments, a method for systematically detecting discrete reflections in the BRIR is required, and various methods have been proposed in the literature to detect reflections in impulse responses, including [4,33–35].

The separated signals were then analysed using the binaural model and a test data matrix generated by combining the segmented direct or reflected component with the corresponding rotated signal (as described in Section 2.2). The positively and negatively rotated test feature vectors were stored in separate matrices and used to test the NN trained with the corresponding rotation dataset (as described in Section 2.2). The data was then Gaussian normalised across each feature in the feature vector, using the mean and standard deviations calculated from the training data.

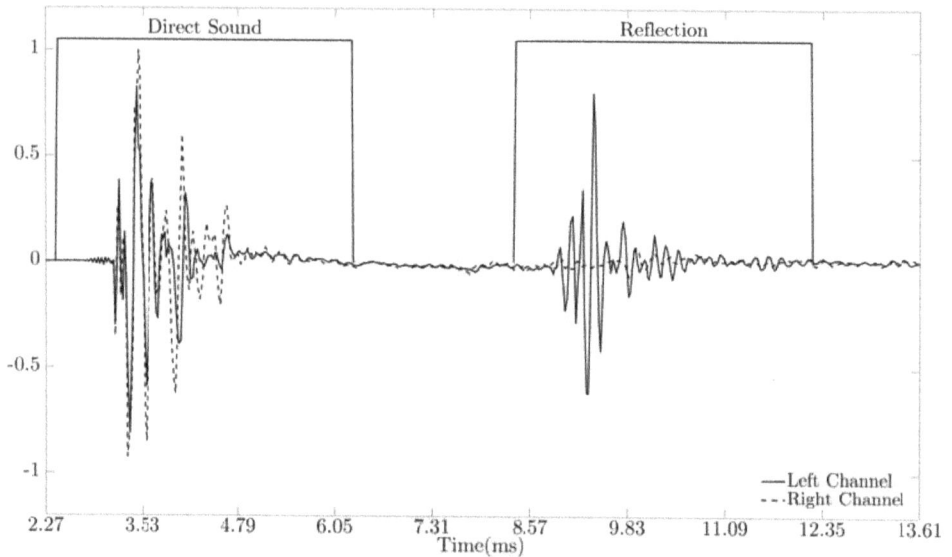

Figure 5. Example binaural room impulse response generated with source at azimuth = 0° and reflector at azimuth = 71°; the solid line is the left channel of the impulse response; the dotted line is the right channel of the impulse response; and the windowed area denotes the segmented regions using the technique discussed in Section 2.4.

The generated test data consisted of 144 of these BRIRs, with source positions from 0° to 357.5° and reflections from 1° to 358.5° using a turntable to rotate the binaural dummy head in steps of 2.5° (with the angles rounded for comparison with the NN's output). This provided 288 angles with which to test the NN: 144 direct sounds and 144 reflections. The turntable was covered in acoustic foam to attempt to eliminate any reflections that it would produce.

3. Results

The two NNs trained with the SADIE HRIR dataset (as described in Sections 2.1 and 2.2) were tested with the components of the measured test BRIRs (as described in Section 2.4), with the outputs concatenated to produce the resulting direction of arrival for the direct and reflected components. The angular error was then computed as the difference between the NN predictions and the target values. The training of the neural network generally terminated due to saturation in output performance within 122 epochs, with an accuracy of 95% and a maximum error of 5°. Statistical analysis of the prediction errors was performed using MATLAB's one-way analysis of variance (ANOVA) function [36] and is reported in the format: ANOVA(F(between group degrees of freedom, within groups degree of freedom) = F value, p = significance), all of these values are returned by the anova1 function [36].

A baseline method used as a reference to compare results obtained from the NN can be derived from the ITD equation (Equation (8) taken from [37]) rearranged for calculating the DoA,

$$ITD = \frac{d\sin(\theta_{ref})}{c} \tag{8}$$

where d is the distance between the two ears, θ_{ref} is the DoA, and c is the speed of sound [37]. The *ITD* value used for the baseline DoA predictions was measured by locating the maximum peak in the *IACC* feature vector, as calculated using the binaural model proposed in Section 2.1. The index for this peak in the *IACC* feature vector relates to one of 99 *ITD* values linearly spaced from −1.1 ms to 1.1 ms.

In Table 3, the neural network accuracy across the test data is presented. The results show that for the direct sound, the neural network predicted 64.58% and 68.06% of the DoAs within 5° for the KEMAR and KU100 dummy head, respectively. Although when analysing direct sound captured with the KU100, a greater percentage of predictions are within ±5° of the target value, the neural network makes a greater number of exact predictions and has lower relative error for KEMAR. This observation is expected given the different morpho-acoustic properties of each head and their ears, which could lead to differences in the observed interaural cues, particularly those dependent on spectral information. The results show that the neural network performs worse when analysing the reflected components. In this case, the reflected component measured with the KU100 is more accurately localised, with lower maximum error, relative error, root mean squared error and number of front-back confusions. Comparisons between the accuracy of the proposed method with the baseline shows that the NN is capable of reaching a higher degree of accuracy, with lower angular error and fewer front-back confusions.

Table 3. Direction of arrival accuracy comparison for the direct sound and reflected components measured with the KEMAR and KU100 binaural dummy heads, for both the cascade-correlation neural network and the baseline method.

Head	Exact	Within ±1°	Within ±5°	Front-Back Confusions	Average Relative Error	Root Mean Squared Error
Cascade-Correlation Neural Network						
Direct Component						
KEMAR	17.36%	21.53%	64.58%	1.39%	7.10%	5.18°
KU100	13.19%	17.36%	68.06%	0%	6.90%	6.86°
Reflected Component						
KEMAR	2.08%	11.11%	32.64%	9.03%	23.61%	13.59°
KU100	0%	9.03%	37.50%	2.78%	15.43%	8.85°
Baseline Method						
Direct Component						
KEMAR	1.39%	2.78%	11.81%	49.31%	38.78%	66.37°
KU100	1.39%	3.47%	13.19%	50%	36.01%	65.66°
Reflected Component						
KEMAR	0%	2.78%	11.11%	49.31%	38.85%	67.31°
KU100	0%	4.86%	21.53%	49.31%	36.81%	70.23°

In Figure 6, comparisons between the direct sound and reflected component for BRIRs captured with the KEMAR 45BC are presented. The boxplots show that for the direct sound, a maximum error of 12° and median error of 5° (mean error of 4.20°) were observed, while the reflected component has a maximum error of 30° and median of 8.5° (mean error of 10.87°). There is a significant difference in the neural network performance between the direct sound and reflected component, ANOVA($F(1,286) = 83.99$, $p < 0.01$). This observed difference could result from the difference in signal path distance, which was found to reduce prediction accuracy in [9,10]. May et al. reported that as source-receiver distances increased, and therefore the signal level relative to the noise floor or room reverberation decreased, the accuracy of the GMM predictions decreased. They reported that, averaged over seven reverb times, the number of anomalous predictions made by the GMM increased by ~9% between a source-receiver distance of 2 m compared to a source-receiver distance of 1 m. Further causes of error could be due to system misalignment at point of measurement or lower signal-to-noise ratios (SNR) occurring due to signal absorption at the reflector and larger propagation path (source-reflector-receiver); an average SNR of approximately 22.40 dB and 13.14 dB was observed across the direct and reflected component, respectively.

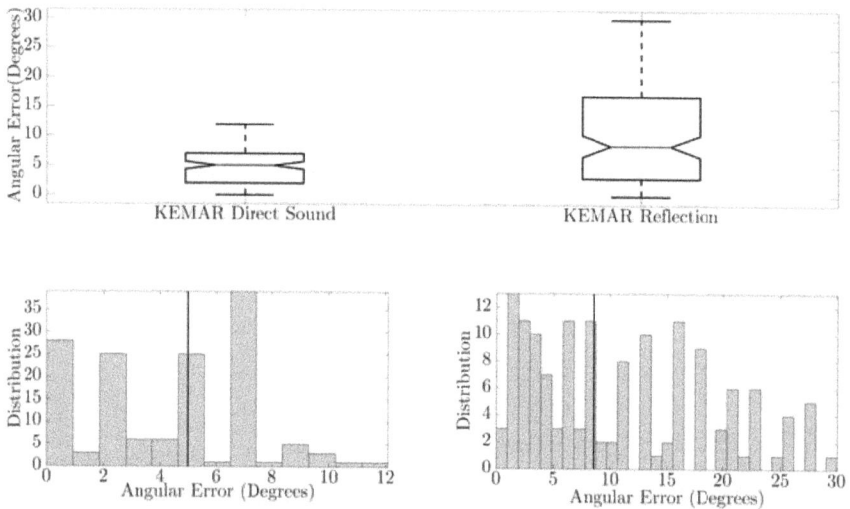

Figure 6. Comparison of angular errors in the neural network direction of arrival predictions for measurements with the KEMAR 45BC. The top image is a boxplot comparison of the angular error in the neural network predictions for the direct sound and reflected components. The bottom left is a histogram showing the error distribution for the direction of arrival predictions of the direct sound, and the bottom right is the error distribution for the direction of arrival predictions of the reflected components. The black line on the histograms depicts the median angular error.

In Figure 7, the comparison between direct sound and reflected component for BRIRs captured using the KU100 are presented. The boxplots show that for the direct sound, a maximum error of 23° is observed and a median error of 5° (mean error of 5.15°), and the reflected component had a maximum error of 19° and median of 7° (mean error of 7.51°). Although the maximum and median errors are not too dissimilar between the predictions for the direct sound and reflected component, there is a significant difference in the distribution of the angular errors, ANOVA($F(1,286) = 18.85, p < 0.01$). The direct sound DoA predictions are generally more accurate than those for the reflected component. As with the findings for the KEMAR, this could be due to the difference in signal paths between the direct sound and reflected component, system misalignment or lower SNR; an average SNR of approximately 22.41 dB and 10.91 dB was observed across direct sound and reflected components, respectively.

In Figure 8, the comparison between the two binaural dummy heads is presented for both the direct sound and reflected components of the BRIRs. The box plots show that there is no significant difference between the medians for the direct sound, and while the maximum error observed for DoA predictions with the KU100 is higher than that of the KEMAR, there is no significant difference in the angular errors between the two binaural dummy heads, ANOVA($F(1,286) = 4.29, p = 0.04$). This would suggest that for at least the direct sound, the NN is generalisable to new data, including those which are produced using a different binaural dummy head microphone from those which were used to train the NN. However, comparing the angular errors observed in the output of the NN for the reflected component shows that the KU100 has a significantly lower median angular error and performs significantly better overall when analysing the reflected components captured with the KU100, ANOVA($F(1,286) = 18.23, p < 0.01$). This observation does not match what would be expected given that the NN was trained with HRIRs captured using a KEMAR unit, suggesting that the NN should perform better or comparably when predicting the DoA for reflected signals captured using another KEMAR over the results obtained with the KU100.

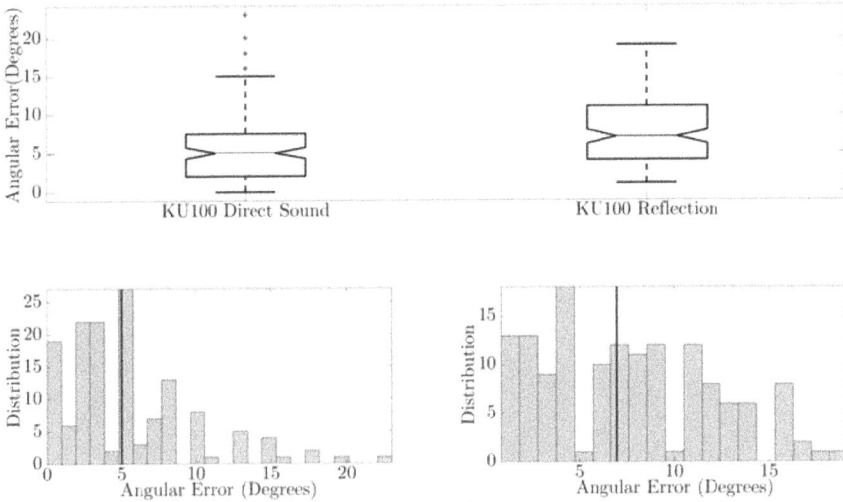

Figure 7. Comparison of angular errors in the neural network direction of arrival predictions for measurements with the KU100. The top image is a boxplot comparison of the angular error in the neural network predictions for the direct sound and reflected components; the bottom left is a histogram showing the error distribution for the direction of arrival predictions of the direct sound; and the bottom right is the error distribution for the direction of arrival predictions of the reflected components. The black line on the histograms depicts the median angular error.

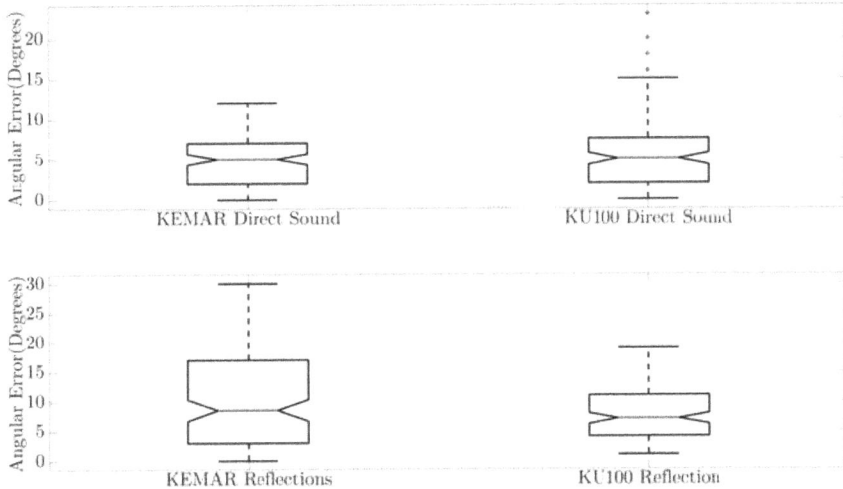

Figure 8. Boxplot comparison of angular errors in the neural network direction of arrival predictions between the KEMAR and KU100 dummy heads for direct sound (**top**) and reflected (**bottom**) components.

Figures 6 and 7 compare the accuracy of the NN predictions for direct and reflected components for each head. The difference between the direct sound and reflected component is more dissimilar for BRIRs captured with the KEMAR than the KU100, possibly suggesting the presence of an external factor that is creating ambiguity in the measured binaural cues for the reflected components captured using the KEMAR. Furthermore, comparing the interaural cues (Figures 9 and 10) between the direct sound and reflected components of the BRIR for the KEMAR and KU100 measurements shows a more distinct blurring for the reflected components measured with the KEMAR when compared to those measured with the KU100. This could suggest that a source of interference is present in the KEMAR measurements that is producing ambiguity in the measured signals' interaural cues. This could be due to noise present within the system and environment or misalignment in the measurement system for the KEMAR measurements; leading to the production of erroneous reflected signals.

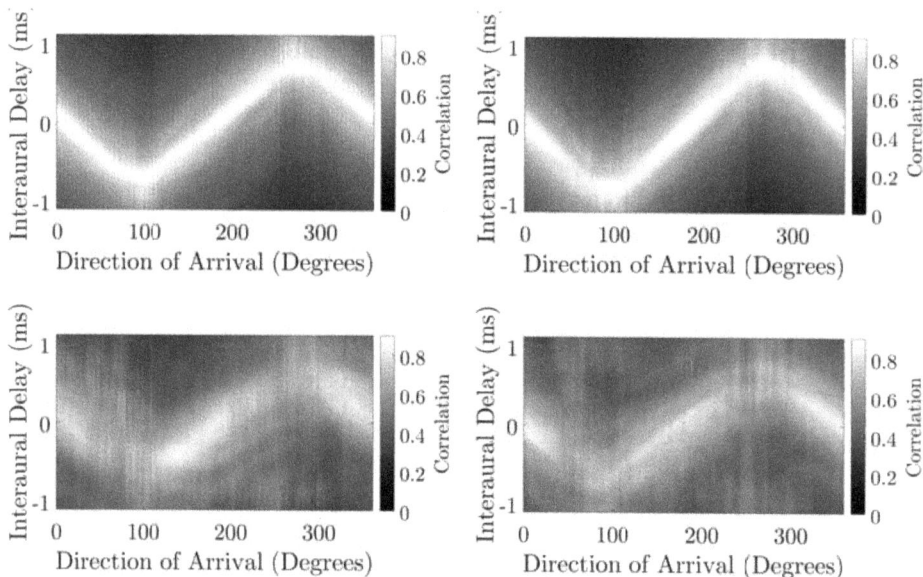

Figure 9. Comparison of interaural cross correlation across the direction of arrival for the KEMAR measured direct sound (**top left**), KEMAR measured reflection (**bottom left**), KU100 measured direct sound (**top right**) and KU100 measured reflection (**bottom right**).

By investigating the neural networks' predicted direction of arrival compared against the expected, insight can be gained into any patterns occurring in the NN output predictions. Additionally, it will show how capable the NN is at predicting the DoA for signals with a DoA not represented within the training data. In Figure 11, the predicted direction of arrival by the neural network (dashed line) is compared against the expected direction of arrival (solid line), and the plot shows the comparison for the KEMAR direct sound measurement predictions (top left), KEMAR reflection measurement predictions (bottom left), KU100 direct sound measurement predictions (top right) and KU100 reflection measurement predictions (bottom right). Generally, the direct sound measurement predictions are mapped to the closest matching DoA represented in the training database, suggesting that the NN is incapable of making predictions for untrained directions of arrival. In the case of the reflections, the NN predictions tend to plateau over a larger range of expected azimuth DoA. This observation further shows the impact of the blurring of the interaural cues (Figures 9 and 10) producing regions of ambiguous cues in the reflection measurements, causing the NN to produces regions of the same DoA prediction.

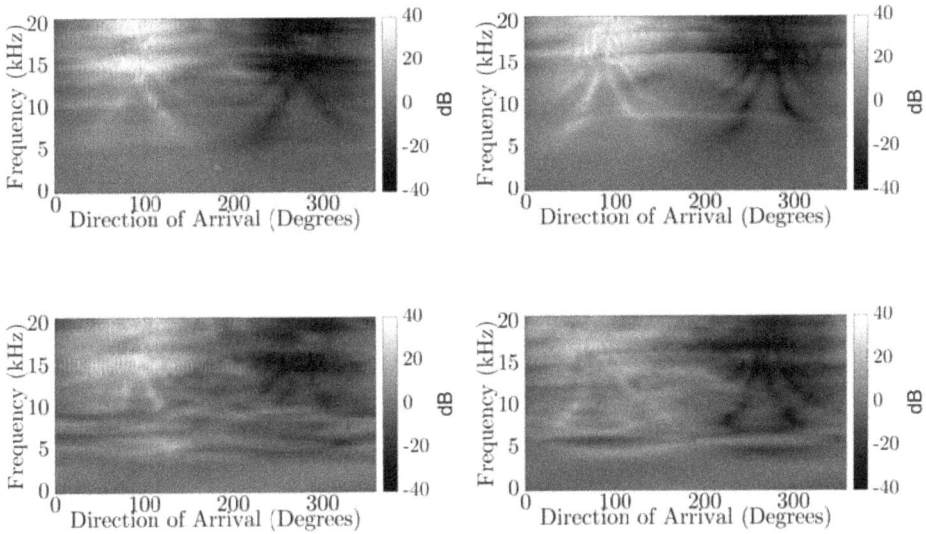

Figure 10. Comparison of interaural level difference across the direction of arrival for the KEMAR measured direct sound (**top left**), KEMAR measured reflection (**bottom left**), KU100 measured direct sound (**top right**) and KU100 measured reflection (**bottom right**).

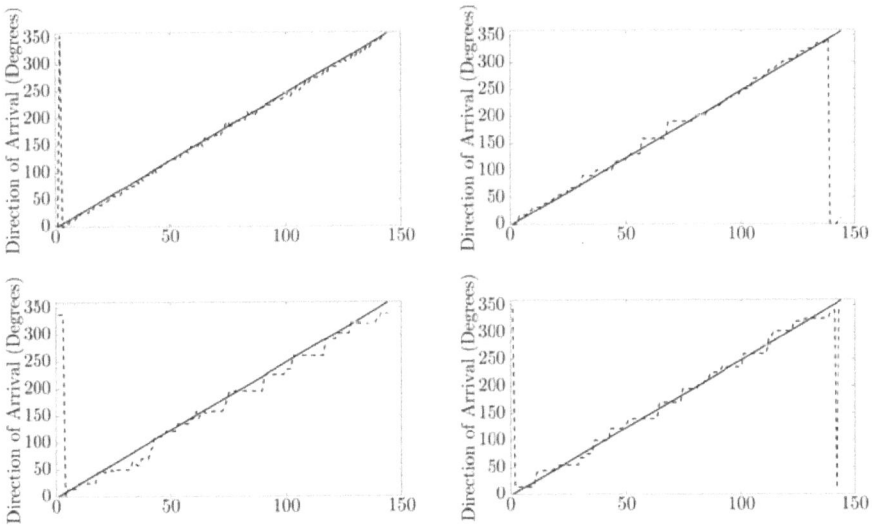

Figure 11. Plots of neural network predicted direction of arrival (dotted black line) vs. expected direction of arrival (solid line). The top left plot is for the KEMAR direct sound; the top right plot is for the KU100 direct sound; the bottom left is for the KEMAR reflection; and the bottom right is for the KU100 reflections.

4. Discussion

The results presented in Section 3 show that there is no significant difference in the accuracy of the NN when analysing the direct sound of BRIRs captured with both the KEMAR 45BC and the KU100. However, the accuracy of the NN is significantly reduced when analysing the reflected component of the BRIRs, with the NN performing better at predicting the DoA of reflected components measured with the KU100. A reduction in performance would be expected between the direct sound and reflected component, due to the lower signal-to-noise ratio that would be observed for the reflected component. It is of interest that reflections measured with the KU100 are more accurately localised than those measured with the KEMAR 45BC; this could be due to a greater degree of system misalignment in the KEMAR 45BC measurements that was not present in the KU100 measurements. An additional difference that could lead to more accurate predictions being made for the KU100 could be the diffuse-field flat frequency response of the KU100, which could produce more consistent spectral cues for the reflected component (as seen in Figure 10), leading to more accurate direction of arrival predictions by the neural network.

Analysis over different degrees of measurement orientation rotations (Table 1) showed that while the number of predictions within ±5° varies little between degrees of rotation, the maximum error in the neural networks' prediction decreases as the angle of rotation increases. Larger degrees of rotation would produce greater differences in interaural cues between the rotated and original signal, allowing the neural network to produce more accurate predictions under noisier conditions where the interaural cues become blurred. The use of additional measurement orientations decreases the number of front-back confusions, with generally larger degrees of receiver rotations producing fewer front-back hemisphere errors, except when using ±30°. Using larger degrees of rotation has the additional benefit of reducing the maximum predictions errors made by the neural network; this could be due to the greater rotational mobility allowing signals at the rear of the listener to be focused more in the frontal hemisphere; producing more accurate direction of arrival predictions. It is interesting that there is a greater percentage of front-back confusions for the KEMAR 45BC compared to the KU100; this could be due to differences in system alignment causing positions close to 90° and 270° (source facing the left or right ear) to originate from the opposite hemisphere.

The lack of significant difference between the direct sounds measured with the two binaural dummy heads agrees with the findings of May et al. [11], who found that a GMM trained with an MCT dataset was able to localise sounds captured with two different binaural dummy heads. The notable difference between the KEMAR 45BC and KU100 include: morphological differences of the head and ears between binaural dummy head microphones; the KEMAR 45BC has a torso; the KU100's microphones have a flat diffuse-field frequency response; and the material used for the dummy head microphones.

The overall accuracy of the method presented in this paper is, however, lower than that found in [11]. This could be a result of the type of signals being analysed, which, in this study, are 3.8 ms-long impulsive signals as opposed to longer speech samples. Compared to more recent NN-based algorithms [15], the proposed algorithm under performs compared to reported findings of 83.8% to 100% accuracy across different test scenarios. However, their analyses only considered signals in the frontal hemisphere around the head and considered longer audio samples for the localisation problem.

Comparing the proposed method to that presented in [12] shows that the proposed method achieves lower relative errors for the direct sound and reflections measured with both binaural dummy head microphones, compared to the 24.0% reported for real test sources using a multi-layered perceptron in [12].

The average errors reported in this paper are lower than that presented in [4], which reported average errors in the range of 28.7° and 54.4° when analysing the components of measured BRIRs. However, the results presented in [4] considered reflections with reflection orders greater than first, and therefore, further analyses of the proposed NNs' performance with full BRIRs is required for more direct comparisons to be made.

Future work will focus on improving the accuracy of the model for azimuth DoA estimation, using measured binaural room impulse responses to assess the accuracy of the neural network as reflection order and propagation path distance increases. The proposed model will then be extended on to consider estimation of elevation DoA, providing complete directional analysis of the binaural room impulse responses; the aim being for the final method to be integrated within a geometry inference and reflection backpropagation algorithm, allowing for in-depth analysis of the acoustics of a room. However, this will require higher accuracy in the DoA predictions for the reflections. Further avenues of research to improve the robustness of the algorithm could include: the use of noise reduction techniques to ideally reduce the ambiguity in the binaural cues, increasing the size of the training database used to train the neural network, investigation into using different representations of interaural cues and how they are extracted from the signals, using reflections to train the NN in addition to the HRIRs or the use of a different machine learning classifier.

5. Conclusions

The aim of this study was to investigate the application of neural networks in the spatial analysis of binaural room impulse responses. The neural network was tested using binaural room impulse responses captured using two different binaural dummy heads. The neural network was shown to have no significant difference in accuracy when analysing the direct sound of the binaural room impulse response across the two binaural dummy heads, with 64.58% and 68.06% of the predictions being within $\pm 5°$ of the expected values for KEMAR and the KU100, respectively. However, upon presenting the NN with reflected components for analysis, the accuracy of the predictions was significantly reduced. The NN also generally produces more accurate results for reflected components of the binaural room impulse response captured with the KU100. Comparisons of the interaural cues for the direct sound and reflected components show a distinct blurring in the cues for the reflected components measured with KEMAR, which is present to a lesser extent for the KU100. This blurring could be a product of lower signal-to-noise ratios or misalignment in the measurement systems, leading to greater ambiguity in the measurements. The results presented in this paper show the potential of using this technique as a tool for analysing binaural room impulse responses, while indicating that further work is required to improve the robustness of the algorithm for analysing reflections and signals with lower signal-to-noise ratios. Further development of this algorithm will investigate the application of the neural network for elevation direction of arrival analysis and integration of the method with geometry inference and reflection back-propagation algorithms, allowing for analysis of a room's geometry and its affect on sounds played within it.

Acknowledgments: Funding was provided by a UK Engineering and Physical Sciences Research Council (EPSRC) Doctoral Training Award, the Department of Electronic Engineering at the University of York, and in part by Digital Creativity Labs, EPSRC Grant Number: EP/M023265/1.

Author Contributions: Michael Lovedee-Turner developed the concepts, algorithms and experiments and wrote the paper. Damian Murphy supervised the project and paper writing, providing input throughout the development process.

Conflicts of Interest: The authors declare no conflict of interest

Abbreviations

The following abbreviations are used in this manuscript:

DoA	Direction of arrival
ITD	Interaural time difference
ILD	Interaural level difference
HRIR	Head-related impulse responses
NN	Neural network
DNN	Deep neural networks
GMM	Gaussian mixture model
IACC	Interaural cross-correlation
FFT	Fast Fourier transform

MCT   Multi-conditional training
ADAM   Adaptive moment estimation
BRIR   Binaural room impulse responses
SNR   Signal-to-noise ratio
ANOVA   Analysis of variance

References

1. Pulkki, V.; Merimaa, J. Spatial impulse response rendering II: Reproduction of diffuse sound and listening tests. *J. Audio Eng. Soc.* **2006**, *54*, 3–20.
2. Pulkki, V. Spatial sound reproduction with directional audio coding. *J. Audio Eng. Soc.* **2007**, *55*, 503–516.
3. Tervo, S.; Pätynen, J.; Lokki, T. Acoustic reflection path tracing using a highly directional loudspeaker. In Proceedings of the IEEE Workshop on Applications of Signal Processing to Audio and Acoustics, New Paltz, NY, USA, 18–21 October 2009; pp. 245–248.
4. Vesa, S.; Lokki, T. *Segmentation and Analysis of Early Reflections From a Binaural Room Impulse Response*; Technical report; Helsinki University of Technology: Helsinki, Finland, 2009.
5. Kohlrausch, A.; Braasch, J.; Kolosssa, D.; Blauert, J. An introduction to binaural processing. In *The Technology of Binaural Listening*; Blauert, J., Ed.; Springer-Verlag: Berlin/Heidelberg, Germany, 2013; pp. 1–32.
6. Howard, D.; Angus, J. *Acoustics and Psychoacoustics*, 4th ed.; Elsevier Science: Cambridge, UK, 2009.
7. Zhong, X. Localize a sound source in self motion with ITD Cues. In *Dynamic Spatial Hearing by Human and Robot Listeners*; Arizona State University: Tempe, AS, USA, 2015; pp. 52–68.
8. Musicant, A.D.; Butler, R.A. The influence of pinnae-based spectral cues on sound localization. *J. Acoust. Soc. Am.* **1984**, *75*, 1195–1200.
9. May, T.; Van De Par, S.; Kohlrausch, A. A probabilistic model for robust localization based on a binaural auditory front-end. *IEEE Trans. Audio Speech Lang. Process.* **2011**, *19*, 1–13.
10. Woodruff, J.; Wang, D. Binaural localization of multiple sources in reverberant and noisy environments. *IEEE Trans. Acoust. Speech Signal Process.* **2012**, *20*, 1503–1512.
11. May, T.; Ma, N.; Brown, G.J. Robust localisation of multiple speakers exploiting head movements and multi-conditional training of binaural cues. In Proceedings of the IEEE International Conference on Acoustics, Speech and Signal Processing (ICASSP), Brisbane, Australia, 19–24 April 2015; pp. 2679–2683.
12. Palomäki, K.; Pulkki, V.; Karjalainen, M. Neural network approach to analyze spatial sound. In Proceedings of the AES 16th International Conference: Spatial Sound Reproduction, Rovaniemi, Finland, 10–12 April 1999; pp. 233–245.
13. Backman, J.; Karjalainen, M. Modelling of human directional and spatial hearing using neural networks. In Proceedings of the IEEE International Conference on Acoustics, Speech, and Signal Processing, Minneapolis, MN, USA, 27–30 April 1993; pp. I-125–I-128.
14. Yuhas, B.P. Automated Sound Localization Through Adaptation. In Proceedings of the International Joint Conference on Neural Networks (IJCNN), Baltimore, MD, USA, 7–11 June 1992; pp. II-907–II-912.
15. Ma, N.; Brown, G.J.; May, T. Exploiting deep neural networks and head movements for binaural localisation of multiple speakers in reverberant conditions. In *Interspeech*; International Speech Communication Association: Baixas, France, 2015; pp. 1–5.
16. Ding, J.; Wang, J.; Zheng, C.; Peng, R.; Li, X. Analysis of Binaural Features for Supervised Localization in Reverberant Environments. In *Proceedings of Audio Engineering Society Convention 141*; Audio Engineering Society: Los Angeles, CA, USA, 2016; pp. 1–9.
17. GRAS. KEMAR Model 45BC. 2016. Available online: http://www.gras.dk/45bc.html (accessed on 25 October 2016).
18. Pulkki, V.; Karjalainen, M.; Huopaniemi, J. Analyzing Virtual Sound Source Attributes Using Binaural Auditory Model. *J. Audio Eng. Soc.* **1999**, *47*, 203–217.
19. Woodruff, J.; Wang, D. Sequential organization of speech in reverberant environments by integrating monaural grouping and binaural localization. *IEEE Trans. Audio Speech Lang. Process.* **2010**, *18*, 1856–1866.
20. Middlebrooks, J.C.; Green, D.M. Sound Localization By Human Listeners. *Ann. Rev. Pyschol.* **1991**, *42*, 135–159.
21. Slaney, M. Auditory Toolbox. 1998. Available online: https://engineering.purdue.edu/~malcolm/interval/1998-010/ (accessed on 28 December 2017).

22. Gao, B. Cochleagram and IS-NMF2D for Blind Source Separation. 2014. Available online: http://uk.mathworks.com/matlabcentral/fileexchange/48622-cochleagram-and-is-nmf2d-for-blind-source-separation?focused=3855900&tab=function (accessed on 28 December 2017).

23. Kearney, G. SADIE Binaural Measurements. 2016. Available online: http://www.york.ac.uk/sadie-project/binaural.html (accessed on 28 December 2017).

24. Google. TensorFlow. Available online: https://www.tensorflow.org/ (accessed on 28 December 2017).

25. Fahlman, S.E.; Lebiere, C. The Cascade-Correlation Learning Architecture. In *Advances in Neural Information Processing Systems 2*; Morgan Kaufmann Publishers Inc.: San Francisco, CA, USA, 1990; pp. 524–532.

26. LeCun, Y.A.; Bottou, L.; Orr, G.B.; Müller, K.R. Efficient BackProp. In *Neural Networks: Tricks of the Trade*; Springer: Berlin/Heidelberg, Germany, 1998; pp. 9–50.

27. Kingma, D.; Ba, J. Adam: A method for stochastic optimization. In Proceedings of the International Conference on Learning Representations, San Diego, CA, USA, 7–9 May 2015; pp. 1–15.

28. Neumann. Dummy Head KU100. Available online: https://www.neumann.com/?lang=en&id=current_microphones&cid=ku100_description (accessed on 28 December 2017).

29. Equator Audio. Equator D5 Coaxial Loudspeakers. Available online: http://www.equatoraudio.com/New-Improved-D5-Studio-Monitors-Pair-p/d5.htm (accessed on 28 November 2017).

30. Farina, A. Simultaneous measurement of impulse response and distortion with a swept-sine technique. In Proceedings of the108th Convention, Paris, France, 19–22 February 2000; pp. 1–15.

31. British Standards Institution. BSI Standard ISO 3382-1: Acoustics—Measurements of Room Acoustic Parameters Part 1: Performance Spaces (ISO 3382-1:2009). 2009. Available online: https://www.iso.org/standard/40979.html (accessed on 5 January 2018)

32. Allen, J.B. Image method for efficiently simulating small-room acoustics. *J. Acoust. Soc. Am.* **1979**, *65*, 943.

33. Kelly, I.; Boland, F. Detecting Arrivals in Room Impulse Responses with Dynamic Time Warping. *IEEE/ACM Trans. Audio Speech Lang. Process.* **2013**, *22*, 1139–1147.

34. Remaggi, L.; Jackson, P.J.B.; Coleman, P.; Wang, W. Acoustic Reflector Localization: Novel Image Source Reversion and Direct Localization Methods. *IEEE/ACM Trans. Audio Speech Lang. Process.* **2017**, *25*, 296–309.

35. Defiance, G.; Daudet, L.; Polack, J.D. Detecting arrivals within room impulse responses using matching pursuit. In Proceedings of the 11th International Conference on Digital Audio Effects (DAFx-08), Espoo, Finland, 1–4 September, 2008; pp. 1–4.

36. MATLAB. Anova1. 2017. Available online: https://uk.mathworks.com/help/stats/anova1.html (accessed on 28 December 2017).

37. Howard, D.; Angus, J. Interaural time difference. In *Acoustics and Psychoacoustics*, 2nd ed.; Focal Press: Oxford, UK, 2001.

applied
sciences

MDPI

Article

Live Convolution with Time-Varying Filters

Øyvind Brandtsegg [1,*], Sigurd Saue [1]and Victor Lazzarini [2]

[1] Norwegian University of Science and Technology, 7491 Trondheim, Norway; sigurd.saue@ntnu.no
[2] Department of Music, Maynooth University, Maynooth, W23 X021 Co. Kildare, Ireland;
    victor.lazzarini@mu.ie
*   Correspondence: oyvind.brandtsegg@ntnu.no; Tel.: +47-92-203-205

Received: 31 October 2017; Accepted: 3 January 2018; Published: 12 January 2018

Abstract: The paper presents two new approaches to artefact-free real-time updates of the impulse response in convolution. Both approaches are based on incremental updates of the filter. This can be useful for several applications within digital audio processing: parametrisation of convolution reverbs, dynamic filters, and live convolution. The development of these techniques has been done within the framework of a research project on crossadaptive audio processing methods for live performance. Our main motivation has thus been live convolution, where the signals from two music performers are convolved with each other, allowing the musicians to "play through each other's sound".

Keywords: convolution; reverberation; audio effect; live processing; morphing; cross synthesis

1. Introduction

Convolution has been used for filtering, reverberation, spatialisation, and as a creative tool for cross-synthesis in a variety of contexts [1–5]. Common to most of them is that one of the inputs is a time-invariant impulse response (characterising a filter, an acoustic space or similar), allocated, and preprocessed prior to the convolution operation. Although developments have been made to make the process latency free (using a combination of partitioned and direct convolution [6]), the time-invariant nature of the impulse response (IR) has inhibited a parametric modulation of the process. Modifying the IR traditionally has implied the need to stop the audio processing, load the new IR, and then re-start processing using the updated IR. The methods presented in this paper represents two new approaches to allow real-time modifications of the IR without interrupting the audio processing. The IR updates can be done without introducing artefacts in the audio output. Research on these methods were initiated within the project "Cross-adaptive processing as musical intervention" [7] investigating various kinds of signal interaction between audio signals from two or more live music performers. Convolution as a method for signal interaction was found desirable in the context of this research, but methods of facilitating the convolution of two live signals were needed to enable the performers to flexibly interact with the process. One of the methods presented here have been discussed by two of the authors in an earlier conference paper [8], where artefact-free updates of the filter allowed live sampling of the impulse response. The current paper expands on this work by adding another method for time-varying filter coefficients. Both methods for filter updates are based on a similar concept, where coefficients are replaced at audio rate, enabling artefact free transition from old to new filter coefficients, even with arbitrary different coefficients. Software tools from the conference paper have also been expanded to include both filter methods. A number of practical experiments in studio and live sessions have been done during the time since the conference paper was written. Reflections on these artistic explorations have been included in the present article.

1.1. Time-Varying Filters

Time-varying convolution has been explored in continuous-time systems [9], and in discrete-time systems, both finite impulse response (FIR) and infinite impulse response (IIR) coefficient-modulated filters have been extensively discussed in [10]. Applications are numerous and include, for instance, speech processing [11,12], equalization of audio signals [13–15], binaural processing [16,17], and reverberation [6,18].

Digital time-varying filters in music are typically composed of recursive filter structures of lower order and proposed approaches are particularly concerned with stabilization and transient suppression [13,19]. Strategies reported include intermediate sets of filter coefficients [15,20,21], state variable updates [11,19], and input-switching [12].

The present paper is concerned with impulse responses live sampled from audio signals. The filter order is also substantially higher than in the approaches listed above, since the audio samples involved may possibly last several seconds. Hence, we are specifically looking at time-varying FIR filters that can be dynamically updated without perceptual artefacts. Virtual acoustic reality is another application area where convolution with time-varying FIR filters find use [16,17,22]. One possible approach to avoid perceptual artefacts is to cross-fade between the outputs of several simultaneous convolution processes [23]. However, this gets prohibitively expensive for audio-rate filter updates. Jot, Larchel, and Warusfel [17] suggests incremental filter switching at high update rates (referred to as commutation) for binaural processing, while Lee et al. [18] suggests a similar approach for artificial late-field reverberation where the rate of change may adapt to characteristics of the input. These strategies minimize the artefacts, but at the cost of computing intermediate filters. Vickers [24] presents a number of frequency-domain strategies for time-varying FIR filters. In general, the solutions either violate the constraints of linear convolution and produce artefacts, or they demand extra processing power: typically more than twice the number of real multiplications per output sample compared to the common convolution implementation overlap-add short-time Fourier Transform (OLA STFT).

Instead, we will present two different techniques that perform dynamic, low-latency convolution of live-sampled impulse responses with negligible computational overhead.

1.2. Convolution and Other Sound Transformations, Live Use

Techniques for signal interaction in creative sound design have been widely used. Among these, we find Ring modulation (an early example of artistic use is Stockhausen's "Mixtur" from 1964, and an example from popular music is Black Sabbath's "Paranoid" from 1970), Vocoder (popularized by Wendy Carlos in the music for Stanley Kubrick's "A Clockwork Orange" from 1971, another popular example is Laurie Anderson's "O Superman" from 1981), Talk box (an early use by Joe Walsh in "Rocky Mountain Way", and popularized by Peter Frampton in various contexts), and Auto-Wah (popularized by Stevie Wonder on songs like "Superstition" and "Higher Ground" from the early 1970s). Contemporary electronic dance music also make extensive use of signal interaction by means of sidechain compression to create "pumping" effects (a classic example is Eric Prydz' "Call On Me" from 2004). Common to all of these methods is the immediate use of a feature from one sound to control some modulation of another (or the same) sound. The same can be said about other feature-based modulation methods used in our crossadaptive research project. Convolution has the added feature that preserves the full spectrotemporal structure of the modulation sound. Not only is it using the spectrum of one sound to filter another sound, but the temporal evolution of the modulating sound's spectrum is preserved in the filtering. Realtime spectral transformations and cross synthesis (Dynamic filtering of one signal, using the spectral envelope of another signal) has been explored during the last 30 years or so together with convolution [1,25]. Creative uses of convolution as a timbral transformative device in composition has been explored by [3–5,26,27], and the more performative aspects of real-time convolution by [28–30]. Common to all of these has been that the impulse response of the convolution process was static, and that any updates or dynamic replacement of the impulse response required some variation of a crossfading scheme between parallel convolution processes. With our currently described methods, we can update the impulse response with audio content captured from a live performance. This can be seen as a form of *live sampling*:

"Live sampling during performance.... uses the Now as its subject" [31].

Live sampling has been used as a method for creating temporal dynamism and a heightened sense of *the Now*. Perhaps the earliest occurence is in Mauricio Kagel's composition "Transición II" from 1958 [32], and the idea was further developed during the following decades amongst others by Kaffe Mathews and by Michael Waisvisz [33]. Live looping by means of improvisation instruments with long delay lines has been explored by Lawrence Casserley [34] in the Evan Parker Electroacoustic Ensemble, while utilisation in pop music have been done by artists Ed Sheeran, Boxwood, and others. In our current research project, we wanted to investigate the potential of *live sampling the impulse response*, to enable an even more intimate interaction between live sources than previously had been possible within real-time convolution and live sampling as separate domains. The recontextualisation of an audio sample recorded during the same performance, used as the acoustic environment in which the other musician can perform.

2. Time-Varying Finite Impulse Response Filters

A digital finite impulse response filter (FIR) of length N is defined by the following difference equation [35]:

$$y(n) = a_0x(n) + a_1x(n-1) + ... + a_{N-1}x(n-[N-1])$$
$$= \sum_{k=0}^{N-1} a_k x(n-k), \tag{1}$$

where $x(n)$ and $y(n)$ are the input and output signals, respectively, at time n, and a_0 to a_{N-1} are the scaling coefficients of each copy of the input signal delayed by 0 to $N-1$ samples (in this text, we will use the convention that a filter with length N has order $N-1$). When these coefficients are unchanging, the filter is a linear time-invariant filter.

For an FIR filter, its set of coefficients also make up the filter *impulse* response $h(n)$, which is the output of the filter when fed with a unit sample signal $u(n)$, which is 1 for $n = 0$ and 0 elsewhere:

$$h(n) = \sum_{k=0}^{N-1} a_k u(n-k) = a_n. \tag{2}$$

The output signal $y(n)$ can then be expressed as the *convolution* of input signal $x(n)$ and impulse response $h(n)$ (convolution is a commutative operation):

$$y(n) = \sum_{k=0}^{N-1} h(n-k)x(k) = \sum_{k=0}^{N-1} h(k)x(n-k). \tag{3}$$

The spectrum of the filter impulse response defines its *frequency* response, which determines how the filter modifies the input signal amplitudes and phases at different frequencies. A generalised form of this, called the filter *transfer function* can be obtained via the *z-transform*,

$$H(z) = \sum_{n=-\infty}^{\infty} h(n)z^{-n}, \tag{4}$$

which is a function of the complex variable z. In the usual case that the filter is of finite length N, we have:

$$H(z) = \sum_{n=0}^{N-1} h(n)z^{-n} = \sum_{n=0}^{N-1} a_n z^{-n}. \tag{5}$$

By setting $z = e^{j\omega}$ and $\omega = 2\pi k / N$, we can compute the filter spectrum via the discrete Fourier transform (DFT). This is called the filter *frequency response*. The effect of a filter in the spectral domain is defined by the following expression:

$$Y(z) = H(z)X(z), \tag{6}$$

and thus it is possible to implement the filter either in time domain as a convolution operation (Equation (3)) or in the frequency domain as a product of two spectra.

We would like to examine the cases where these coefficients in Equation (1) are not fixed, which characterizes the filter as time-varying (TV). The most general expression for a TVFIR is defined as follows [10]:

$$
\begin{aligned}
y(n) &= a_0(n)x(n) + a_1(n)x(n-1) + \ldots + a_{N-1}(n)x(n-[N-1]) \\
&= \sum_{k=0}^{N-1} a_k(n)x(n-k),
\end{aligned}
\tag{7}
$$

where we assume that the filter coefficients are drawn each from a digital signal $a_k(n)$. In this case, the impulse response will vary over time and thus becomes a function of two time variables, m and n, representing time of input signal application and time of observation, respectively:

$$h(m, n) = \sum_{k=0}^{N-1} a_k(n)u(n - m - k). \tag{8}$$

The output of this filter is defined by the expression

$$y(n) = \sum_{m=-\infty}^{\infty} h(m, n)x(m). \tag{9}$$

Since the output can only appear after the input is applied ($x(n) = 0$ for $n < 0$), the impulse response $h(m, n) = 0$ for $n < 0$ and $n < m$. If we then substitute $l = n - m$, we get the convolution

$$y(n) = \sum_{l=0}^{n} h(n - l, n)x(n - l). \tag{10}$$

The transfer function also becomes a function of two variables, z and n, and assuming causality we get:

$$H(z, n) = \sum_{m=0}^{n} h(m, n)z^{m-n} = \sum_{k=0}^{N-1} a_k(n)z^{-k}. \tag{11}$$

The next two sections present two different approaches to time-varying filtering: Dynamic replacement of impulse responses and convolution with continuously varying filters.

3. Dynamic Replacement of Impulse Responses

We first examine filter impulse responses where the coefficients are not continuously changing, but are replaced at given points in time. This approach was first presented in [8], but is given a more detailed explanation here. It is related to the superposition strategy suggested by Verhelst and Nilens [12]: when a filter change is called for, the signal input is disconnected from the currently running filter and applied to the new one. The old filter is left free-running, i.e., the old input samples propagate through the filter, and will eventually die out (for FIR filters after a time period equivalent to the filter length). The outputs of the two filters are added together. The next change adds another filter, and so on. To make this work, a number of filter processes must run in parallel.

Our approach differs from Verhelst and Nilens in that we propose to run all filters in the same filter process, simply by switching the new filter, coefficient by coefficient, into the same filter buffer as the old one. We claim that, for FIR filters, the results produced by the two approaches are equal, but that our approach avoids the added computational cost of computing two or more parallel convolution processes. By exploiting inherent properties of convolution, we also show that we may start convolving with the new filter impulse response in parallel with the generation/recording of it.

As an initial step consider a simple filter of length N that is switched on at a given time index n_s:

$$h_E(m, n) = \begin{cases} 0, & m < n_s, \\ e_{n-m}, & m \geq n_s, \end{cases} \tag{12}$$

It should be obvious that inputs $x(n)$ for $n < n_s$ will not contribute to the output. If we apply Equation (10) and once again let $l = n - m$, the first N samples of convolution output starting at time index n_s can be written as:

$$y(n) = \sum_{l=0}^{n-n_s} e_l x(n - l) \qquad (n_s \leq n \leq n_s + N - 1). \tag{13}$$

After $n = n_s + N - 1$ the sum will be upper limited by $l = N - 1$. The first output sample, $y(n_s)$, depends on the first coefficient, e_0 only. The second sample, $y(n_s + 1)$, depends on the first two, e_0 and e_1. In general, $y(n_s + k)$ depends on coefficient e_l only if $l \leq k$. Hence, the filter coefficients can be switched in one by one in parallel with the running convolution process.

Now consider the counterpart, a filter of length N that is switched *off* at the same time index n_s (and for simplicity assume that $n_s > N$, the filter length):

$$h_C(m, n) = \begin{cases} c_{n-m}, & m < n_s, \\ 0, & m \geq n_s. \end{cases} \tag{14}$$

Inputs $x(n)$ for $n \geq n_s$ will not contribute to the output. The first N samples of convolution output starting at n_s can be written as:

$$y(n) = \sum_{l=n-n_s+1}^{N-1} c_l x(n - l) \qquad (n_s \leq n \leq n_s + N - 1). \tag{15}$$

The first output sample, $y(n_s)$, does *not* depend on the first coefficient, c_0. The second sample, $y(n_s + 1)$, does not depend on the first two, c_0 and c_1. In general, $y(n_s + k)$ does not depend on coefficient c_l if $l \leq k$.

The natural extension of these two filters is the combined filter h_{C+E} where the coefficients are replaced at time index n_s:

$$h_{C+E}(m, n) = \begin{cases} c_{n-m}, & m < n_s, \\ e_{n-m}, & m \geq n_s. \end{cases} \tag{16}$$

The first N samples of convolution output starting at n_s must be equal to the sum of the two filters discussed above:

$$y(n) = \sum_{l=0}^{n-n_s} e_l x(n - l) + \sum_{l=n-n_s+1}^{N-1} c_l x(n - l) \qquad (n_s \leq n \leq n_s + N - 1). \tag{17}$$

A closer scrutiny reveals that coefficients e_l and c_l do not contribute to the same output sample $y(n)$ for any l, n. In fact, the filter coefficients c_l may be replaced with e_l one by one during the transition interval $[n_s, n_s + N - 1]$, while the convolution is running. The replacement itself will not introduce output artefacts as long as the coefficients are replaced just in time and in the correct order.

For completeness, we will examine a third filter of length N where the coefficients are replaced at two different times, $n_{s1} < n_{s2}$:

$$h_{C+D+E}(\tau, t) = \begin{cases} c_{m-n}, & m < n_{s1}, \\ d_{m-n}, & n_{s1} \leq m < n_{s2}, \\ e_{m-n}, & m \geq n_{s2}. \end{cases} \qquad (18)$$

If the time interval $n_{s2} - n_{s1} > N$, then the two transition regions $[n_{s1}, n_{s1} + N - 1]$ and $[n_{s2}, n_{s2} + N - 1]$ each behave as in Equation (17). If not, we get a subinterval $[n_{s2}, n_{s1} + N - 1]$ of the transition region where all three filters contribute to the output:

$$y(n) = \sum_{l=0}^{n-n_{s2}} e_l x(n-l) + \sum_{l=n-n_{s2}+1}^{n-n_{s1}} d_l x(n-l) + \sum_{l=n-n_{s1}+1}^{N-1} c_l x(n-l) \qquad (n_{s2} \leq n \leq n_{s1}+N-1). \quad (19)$$

Similar to the above, the coefficients e_l, d_l, and c_l do not contribute to the same output sample $y(n)$ for any l, n. Hence, several filter replacements may be carried out simultaneously without artefacts. However, the number of coefficients involved in the convolution sum will be limited by the interval $[n - n_{s2} + 1, n - n_{s1}]$ as seen in the middle term above, for a total of $(n_{s2} - n_{s1})$ summands. An alternative view of this filter replacement scheme is that the input signal $x(n)$ is split in segments $[0, n_{s1} - 1]$, $[n_{s1}, n_{s2} - 1]$, and $[n_{s2}, \infty)$ each convolved with just one set of coefficients, c_l, d_l, and e_l, respectively. The output is the sum of contributions from all three filters.

Convolution has by nature a ramp characteristic as exemplified in Figure 1 where a simple sinusoidal signal time-limited by a rectangular window is convolved with itself ($x(n) = h(n)$). When doing a gradual replacement of filter coefficients, this inherent ramp characteristic ensures a smoothly overlapping transition region between the filters h_C and h_E.

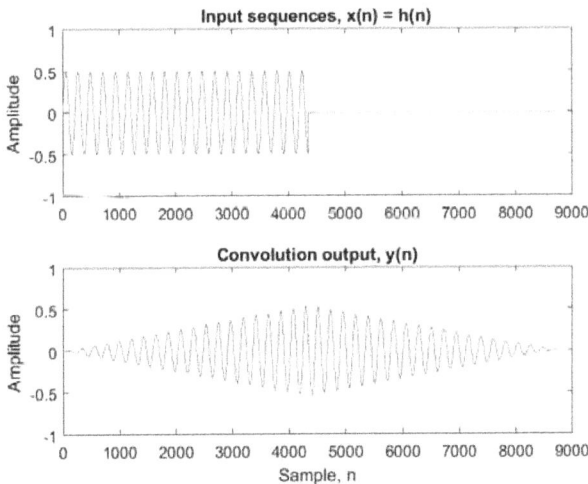

Figure 1. Convolving a time-limited sine sequence with itself.

This technique of stepwise filter replacement outperforms other methods of dynamic filter updates:

- It avoids the discontinuities and artefacts caused by instantaneous switching between filters. It makes no assumptions on similarity between filters as in the commutation [17] or on input signal properties as for the switched convolution reverberator [18].
- It saves the computational effort necessary to superpose or cross-fade parallel convolution processes [12,22].
- It provides the minimum possible latency for a filter update and even allows convolution with a filter to start in parallel with the generation/recording of the filter impulse response itself.

When filter length increases direct convolution in the time-domain is normally replaced with computationally more efficient methods in the frequency domain. A popular approach is *partitioned convolution* [6,36] where the filter impulse response and the input signal are broken into partitions and the convolution computed as multiplication of these partitions in the frequency domain (more on implementation in Section 5). The method of stepwise filter replacement still works, but now the *partition* is the unit of replacement. Replacement must be initiated at a time index n_s equal to an integer multiple of the partition length N_P: $n_s = kN_P$. A latency equal to the partition length N_P is also introduced.

Example

In order to illustrate the behavior of dynamic replacement of filter impulse responses, we present a simple example. Figure 2 shows the input signals to a convolution process. On top are the two impulse responses $h_A(n)$ and $h_B(n)$. They are both sinusoidal signals of duration 1.5 s sampled at 44,100 Hz and with a frequency of 60 and 10 Hz, respectively. At the bottom is the input signal $x(n)$, which is an impulse train with pulses at 0.3 s interval, also sampled at 44,100 Hz.

These signals are convolved using partitioned convolution (overlap-add STFT) with partition length $N_P = 256$ and FFT block size $N_B = 512$. At time index $n_s = 1$ s, a switch between the two impulse response filters is performed.

Figure 3 shows the results. At the top is the output from convolving the input signal $x(n)$ with impulse response $h_A(n)$, but only up to the time index n_s. Notice the convolution tail between 1.0 and 2.5 s, equal to the filter length. In the middle is the output from convolving the input signal $x(n)$ with impulse response $h_B(n)$ starting after the time index n_s. Finally, at the bottom, we show the output from convolving the input signal $x(n)$ with both filters, such that $h_A(n)$ is stepwise replaced by $h_B(n)$ starting at time index n_s, following the scheme introduced in the previous section. The important thing to note here is that an identical result is achieved if we instead add together the two partial outputs (top and middle).

The convolution process is working on a single impulse response buffer. During the filter transition, the buffer will contain parts of both the filters $h_A(n)$ and $h_B(n)$. Figure 4 illustrates the content of the buffer at four different points in time. Just before transition ($n_s = 1$ s), the buffer is entirely filled with impulse response $h_A(n)$. During the transition (1.5 and 2.0 s), we notice how the buffer is gradually filled with impulse response $h_B(n)$, starting from sample 0. After the transition ($n_s + N - 1 = 2.5$ s), the buffer is entirely filled with impulse response $h_B(n)$.

There are noticeable discontinuities in the buffer content during transition, but it is important to realize that each of these buffer snapshots will be multiplied with the input $x(n)$ to produce a single output sample. They are not temporal objects per se. As mentioned earlier, the method is equivalent to segmenting the input signal at the time n_s: any input $x(n)$, $n < n_s$ convolves with filter $h_A(n)$ only. Any input $x(n)$, $n \geq n_s$ convolves with filter $h_B(n)$ only.

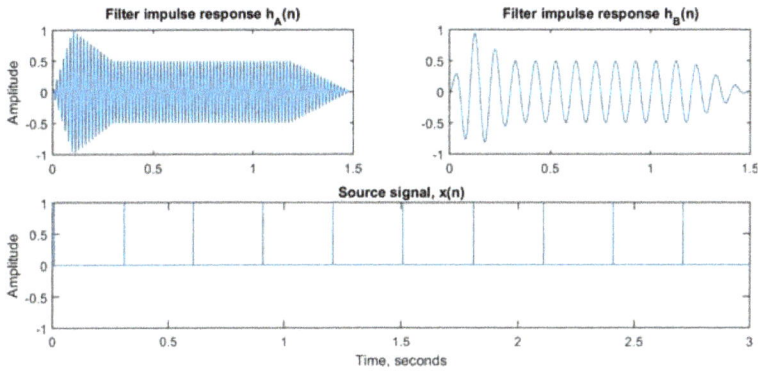

Figure 2. Demonstration of dynamic filter replacement: **Top**: The two filter impulse responses $h_A(n)$ and $h_B(n)$. **Bottom**: The input signal $x(n)$.

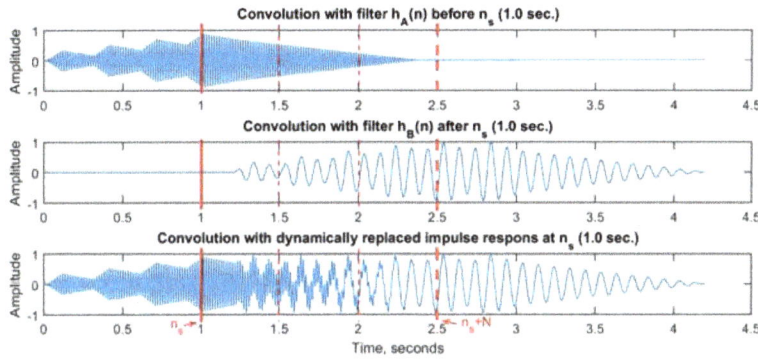

Figure 3. Demonstration of dynamic filter replacement. **Top**: The output from convolving the input $x(n)$ with impulse response $h_A(n)$ before $n_s = 1$ s **Middle**: The output from convolving the input $x(n)$ with impulse response $h_B(n)$ after $n_s = 1$ s **Bottom**: the output from convolving the input $x(n)$ with $h_A(n)$ and its stepwise with $h_B(n)$ starting at $n_s = 1$ s The vertical lines mark the time indices n_s, $n_s + N/3$, $n_s + 2N/3$, and $n_s + N$ in the transition region.

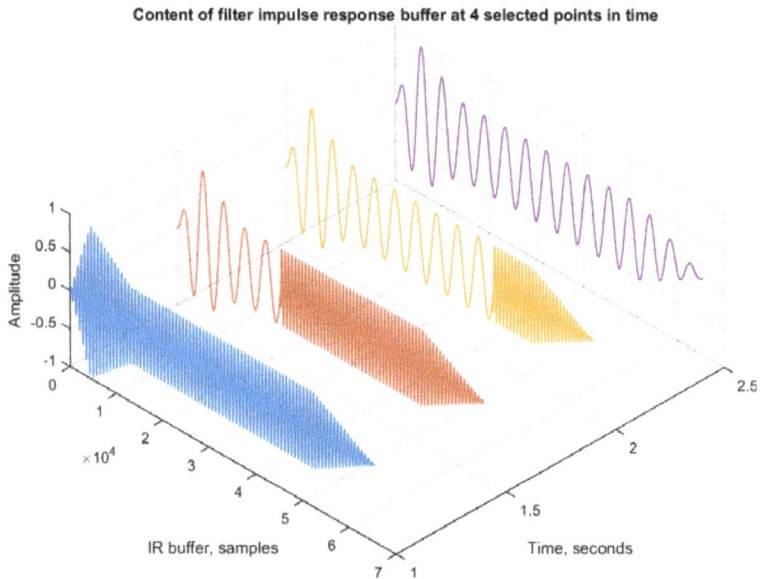

Figure 4. Demonstration of dynamic filter replacement: The content of the filter impulse response buffer at four different points in time: Before transition (1.0 s), 1/3 into the transition (1.5 s), 2/3 into the transition (2.0 s) and after the transition (2.5 s). These time indices are marked with vertical lines in Figure 3.

4. Time-Varying Convolution

Time-varying filters whose coefficients are changing at audio rates, as is the case here, have many applications in music [13,37,38]. In particular, some significant attention has been dedicated to infinite impulse response types that whose coefficients are modulated by a periodic signals [39–42]. These filters tend to be of lower order (first or second order), which have equivalent longer forms made up of two sections arranged in series, a TVFIR and a fixed-coefficient IIR [41]. It has also been shown that the most significant part of the effect of these filters is contained in the TVFIR component.

The case we will examine here presents a TVFIR whose coefficients are taken from an arbitrary input waveform, segmented by the filter length N. This is formally defined by employing the sequence $w(n)$ as the set of N filter coefficients starting at a given time index n:

$$a_k(n) = w(n+k) \qquad (0 < k < N-1). \tag{20}$$

This expression involves future values of $w(n)$ and needs to be adapted so that we can calculate the signal $y(n)$ based solely on the current and past values of the inputs. Furthermore, if we are to calculate successive values of $a_k(n)$, the following expression should apply:

$$a_k(n+1) = a_{k+1}(n), \tag{21}$$

from which we should note that the complete set of N filter coefficients will differ only by one value as the current time index n increments by one. If we compare the first N samples of $w(n)$ and $w(n+1)$, we will observe that they share $N-1$ values. To get the N coefficients $a_k(n+1)$, we only need to discard $a_0(n)$, shift all samples by one position to the left, and replace the last sample of the sequence by a new value $w(n+N)$. We can do this efficiently by applying a circular shift in the coefficients based on the filter size and current time. In this scenario, a set of N coefficient functions $c_k(n)$ can be defined as

$$c_k(n) = \begin{cases} w(n), & k = n \bmod N, \\ c_k(n-1), & \text{otherwise,} \end{cases} \tag{22}$$

where each one of the coefficients will change only once every N samples, as the signal $w(n)$ is taken in by the filter. They will hold their values until a their update time is due. Using this set of coefficients $c_k(n)$ as defined in Equation (22), the TVFIR filter expression becomes (assuming that $c_k(n) = 0$, $w(n) = 0$, and $x(n) = 0$ for $n < 0$)

$$
\begin{aligned}
y(n) &= c_0(n)x(n) + c_1(n)x(n-1) + \dots + c_{N-1}(n)x(n-N-1) \\
&= \sum_{k=0}^{N-1} c_k(n)x(n-k),
\end{aligned}
\tag{23}
$$

which defines what in this paper we call *time-varying convolution* (to distinguish it from the more general forms of TVFIR filters, even though in the other cases the convolution is actually also time varying). In this scenario, we are interpreting an arbitrary input signal of length N as an impulse response, and allow it to vary on a sample-by-sample basis. The resulting set of coefficients c_k, derived from $w(n)$, is completely replaced every N samples. There is, in fact, no particular distinction between the two inputs to the system ($x(n)$ and $w(n)$), and we may wish to view either as the "filtered" signal or the "impulse response". Taking one of these, e.g., $w(n)$, as the filter coefficients, we can determine the filter frequency response for an input $x(n)$ from the filter transfer function, which has to be determined at every sample. This becomes then a function of two variables, frequency k and time n, and can be evaluated by taking its DFT at every N samples:

$$W(n, k) - \sum_{m=0}^{2N-1} w_n(m)e^{-j\omega k}, \tag{24}$$

where $w_n(m)$ is the result of applying a rectangular window of length N to $w(n)$, localised at time index $n = lN$, with l a non-negative integer, and $\omega = 2\pi m/N$. Equally, we can take the input signal DFT,

$$X(n, k) = \sum_{m=0}^{2N-1} x_n(m)e^{-j\omega k}, \tag{25}$$

with $x_n(m)$ similarly defined. The time-varying spectrum of the output of this filter is defined by

$$Y(n, k) = W(n, k)X(n, k). \tag{26}$$

The resulting spectrum is therefore a sample-by-sample multiplication of the short-time input spectra. The convolution waveform can be obtained by applying an inverse DFT of size $2N$. Therefore, as in the time invariant case, the filter can be implemented either in the time domain with a tapped delay line or in the frequency domain using the fast Fourier transform (FFT).

The time-varying convolution effect is a type of cross-synthesis of two input signals. It tends to emphasise their common components and suppress the ones that are absent in one of them. The size of the filter will have an important role in the extent of this cross-synthesis effect and the amount of time smearing that results. As with this class of spectral processes, there is a trade-off between precise localisation in time and frequency. With shorter filter sizes, the filtering effect is not as distinct, but there is a better time definition in the output. With longer lengths, we observe more of the typical cross-synthesis aspects, but the filter will react more slowly to changes in the input signals.

4.1. Fixing Coefficients

A variation on the time-varying convolution method can be developed by allowing coefficients to be fixed for a certain amount of time. Since there is no particular distinction between the two input signals, it is possible for either of these two to be kept static at any given time. More formally, under these conditions, a given $c_k(n)$ coefficient update can be described by

$$c_{n \bmod N} = w(n - dN), \tag{27}$$

where d is a non-negative integer that depends on how long $w(n)$ is being kept static. If the input signal is varying, we can take that $d = 0$. Generally, we can assume that the if signal is held for one full filter period (defined as N samples), then $d = 1$. Under these conditions, d would be determined by how many periods the set of coefficients has been static. However, in practice, we can have a more complex sample-by-sample switching that could hold certain coefficients static, while allow others to be updated. In this case, the analysis is not so simple. In an extreme modulation example, we can have coefficients that alternate between delays of cN and dN samples (c and d integer and >0). Again, since there is no distinction between $w(n)$ and $x(n)$ in terms of their function in the cross-synthesis process, similar observations apply to the updating of the input signal samples. If we implement sample-by-sample update switches to each input, then we allow a whole range of signal "freezing" effects. Of course, depending on the size of the filter, different results might apply. For example, if we have a long filter, freezing one of the signals will create a short loop that will be applied over and over again to the other input. If the frozen signal is a genuine impulse response, say of a given space, this will work as an ordinary linear time-invariant (LTI) convolution operation. Thus, the TVFIR principles might be applied as a means of switching between different impulse responses. Smooth cross-fading can be implemented as a way of moving from one fixed FIR filter to another using the ideas developed here.

4.2. Test Signals

To illustrate some characteristics of time-varying convolution, we look at some cases of time-varying convolution using test signals as inputs. In the first example, we use a pulse train with a frequency of $f_s/1024$ Hz and a sine wave with frequency of 100 Hz using a filter size equal to 1024 samples. This is shown in Figure 5, where we can see that with a pulse at the start of each new filter, the waveform is reconstructed perfectly. This is of course equivalent to having a fixed IR consisting of a unit sample at the start.

In the second example (Figure 6), we decrease the pulse train frequency to $f_s/1124$ Hz and now the impulses are spaced by 100 more samples than one filter length. The output then contains zeros at these samples, and the sine wave is shifted in time by 100 samples, as the impulse localises the start of the sinusoid at its corresponding time. We can see how the result is that a gap is inserted in the signal.

The final example in Figure 7 shows the converse of this. If we increase the frequency to $f_s/924$ Hz, now a new sinusoid is added to the signal before the previous one has completed its N samples. The effect is to distort the waveform shape at the points where there is an overlap. The distortion appears because the impulses do not coincide with the beginning of the waveform and the waveform itself does not complete full cycles in one filter length.

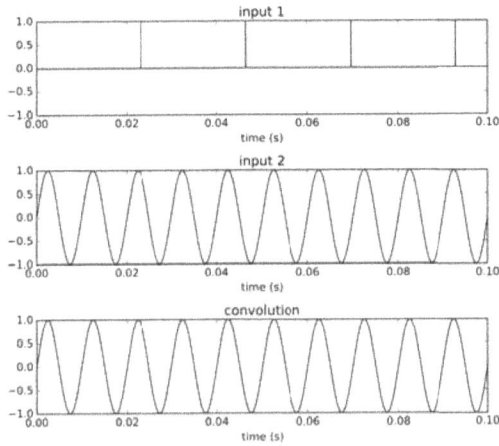

Figure 5. Time-varying convolution using a pulse train with frequency $f_s/1024$ Hz and a sine wave of 100 Hz as inputs, with filter size $N = 1024$ and sampling rate $f_s = 44{,}100$.

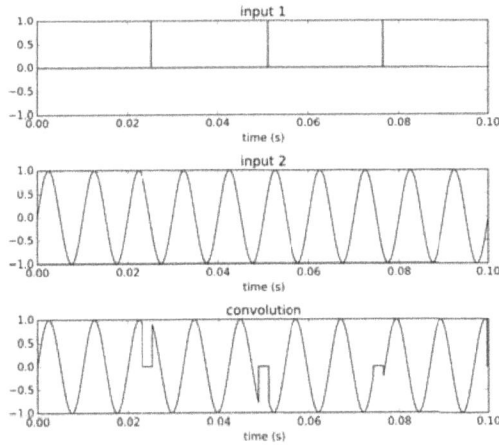

Figure 6. Time-varying convolution using a pulse train with frequency $f_s/1124$ Hz and a sine wave of 100 Hz as inputs, with filter size $N = 1024$ and sampling rate $f_s = 44{,}100$.

Note that these examples are somewhat contrived. They are used to illustrate the process of time-varying convolution in a simple way, and are unlikely to arise in practical musical applications of the process. Thus, we need not infer that the quality of cross-synthesis results will be impaired due to the obvious discontinuities seen in the some of the outputs in these examples. However, they are indicative of the fact that the result of the process is very much dependent on the types of inputs as well as the filter lengths employed.

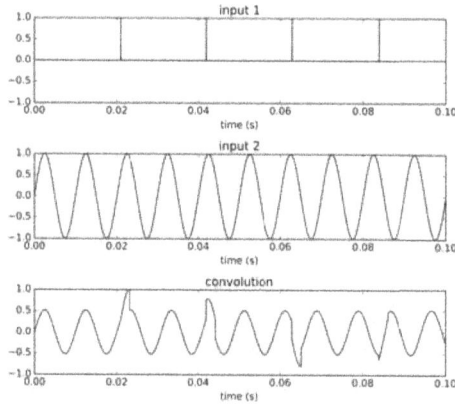

Figure 7. Time-varying convolution using a pulse train with frequency $f_s/924$ Hz and a sine wave of 100 Hz as inputs, with filter size $N = 1024$ and sampling rate $f_s = 44{,}100$.

5. Implementation

As noted above, convolution in general can be programmed employing time-domain and/or frequency-domain operations. The trade-off between the two methods are latency between input and output, and efficiency of computation. With an implementation solely using time-domain processing, which we call *direct* convolution, there is no extra latency imposed by the operation, but we have $O(N^2)$ complexity. If we implement it in the spectral domain, we can use a radix-2 FFT and reduce the computation demands to $O(N \log N)$, but since we need to wait for all of the N samples of input to be received, we will impose a latency between input and output. This can be compensated for in offline processing but not in real time. However, this latency can be reduced by moving some of the spectral operations to the time domain and reducing the transform size to a fraction of the filter size. This approach is known as *partitioned* convolution. These ideas are discussed in the remainder of this section.

5.1. Direct Convolution

A time-domain implementation of TVFIR convolution follows closely Equation (23). It employs two delay lines, one for each signal. The rotation in one of the input signals (e.g., $w(n)$ in Figure 8) to obtain the various coefficients can be simply implemented by looking up the corresponding delay line from end to beginning as if it were a static impulse response. Note that this means that the samples of this signal will always go into the delay line in reverse order. This is conceptually straightforward and the implementation very similar to a standard FIR, except for the fact that we are replacing each sample of the impulse response (as well as feeding a new sample to the delay line holding the other input signal).

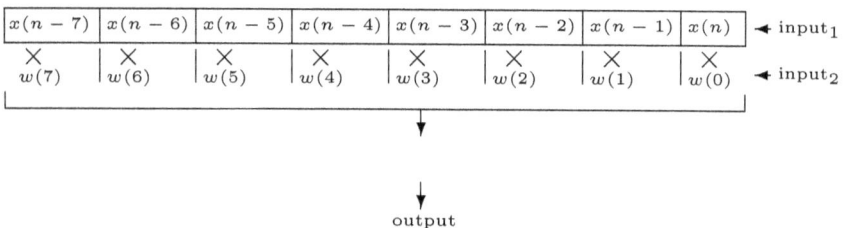

Figure 8. Direct convolution.

5.2. Fast Convolution

Fast convolution employs the fast Fourier transform (FFT) to calculate the DFT, working on blocks of N input samples, with N generally a highly composite number (such as a power-of-two). Two algorithms are more commonly used in frequency domain implementations:

- Overlap-add algorithm (OLA): N samples of each input are collected padded with zeros to make a $2N$ block to which the transforms are applied and their product taken. The output is obtained by taking the inverse FFT of the convolution spectra every N samples. Since this is a $2N$ block of samples, we will need to overlap each output block by N samples (the convolution size is actually $2N - 1$ samples, but we expect the last sample of the block to be zero). In a streaming process, this can be achieved by saving the last N samples of the previous output and mixing these with the first N samples from the current one. In this case, we will save the final N samples of the current output as we produce the final overlapped mix. This process is demonstrated in Figure 9.

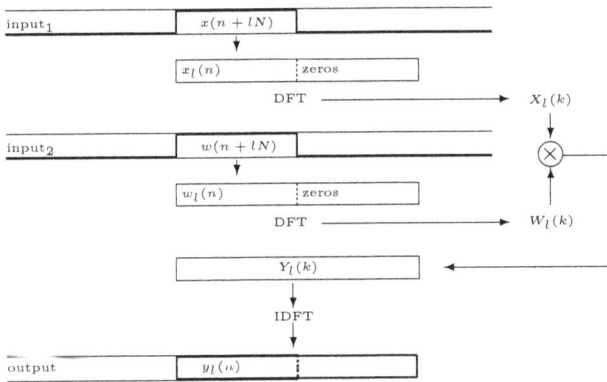

Figure 9. Fast overlap-add convolution, in this case using two arbitrary time-varying signals.

- Overlap-save algorithm (OLS): $2N$ samples are collected from one of the inputs, and N samples are collected from the other, padded to the filter length. The signals are aligned in such a way that the second half of the first input block corresponds to the start of the second. The products of their spectra is taken and then converted back to the time-domain. The first N samples of this block are discarded, and the second half is output. In a streaming implementation, each iteration will have saved the second half of the last input block (N samples) to use as the first block of the next input to the DFT. A flowchart for this algorithm is shown in Figure 10.

Since this algorithm depends on the circular property of the DFT, which cannot be guaranteed with a fully time-varying impulse response, it cannot be used in a practical implementation of the TVFIR described by Equations (24)–(26), This is because the OLS algorithm expects that the impulse response data will not vary over the duration of the convolution, which is not the case if both signals are continuously varying. Even in the more restricted scheme of stepwise replacement of impulse responses, the OLS algorithm does not appear to be applicable. With overlapping input blocks, we can no longer assume that the coefficients of the old and new filter are convolved with separate segments of the input signal.

input$_1$ | $x(n + lN)$

$x_l(n)$

DFT $\longrightarrow$ $X_l(k)$

$\otimes$

input$_2$ (impulse) | $w_l(n)$: zeros

DFT $\longrightarrow$ $W(k)$

IDFT $\longleftarrow$

discard

output | $y_l(n)$

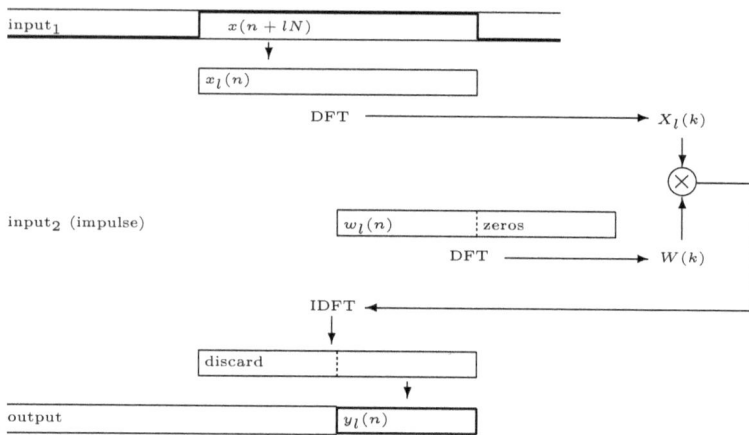

Figure 10. Fast overlap-save convolution.

As we noted before, although for longer filter lengths this is the fastest method, we will introduce a latency between input and output that might be objectionable in real-time processing. To mitigate this, the practical solution is to partition the filter size in shorter blocks and apply the process to these, as discussed in the next section.

5.3. Partitioned Convolution

Partitioned convolution attempts to balance computational load and input-output latency. The principle is based on the idea of breaking the filter down into a set of partitions and working in the frequency domain. In the case of time-varying convolution, each input signal is thus segmented and the partitions are converted to the spectral domain and placed in a separate delay line. The products of pairs of partitions (from each input) in the delay line are summed together to produce the spectrum of the convolution (Figure 11). This is converted back to the time domain and overlapped into the output stream, if we are using an overlap-add algorithm, or just output, if using overlap-save.

Conceptually, we can think that direct convolution uses a partition size of one sample, since the DFT of a single sample is an identity operation. As we saw in Section 5.1, this also involves two delay lines, with one of the signals taken in reverse order. The difference here is that to employ partitions bigger than one sample, the operations are performed in the spectral domain. Equally, we can think of fast convolution having partitions of N samples, the filter size, i.e., a single-partition.

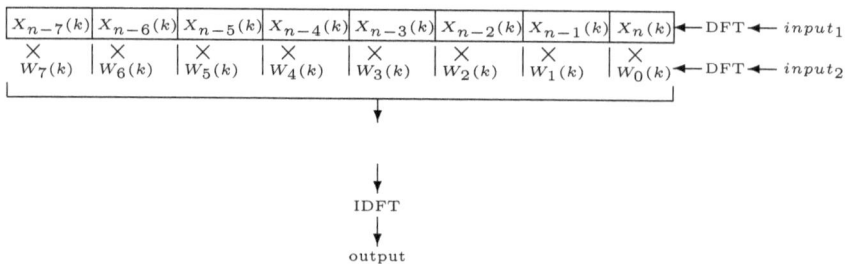

| $X_{n-7}(k)$ | $X_{n-6}(k)$ | $X_{n-5}(k)$ | $X_{n-4}(k)$ | $X_{n-3}(k)$ | $X_{n-2}(k)$ | $X_{n-1}(k)$ | $X_n(k)$ | $\longleftarrow$DFT$\longleftarrow input_1$ |

$\times$ $W_7(k)$ | $\times$ $W_6(k)$ | $\times$ $W_5(k)$ | $\times$ $W_4(k)$ | $\times$ $W_3(k)$ | $\times$ $W_2(k)$ | $\times$ $W_1(k)$ | $\times$ $W_0(k)$ $\longleftarrow$DFT$\longleftarrow input_2$

IDFT

output

Figure 11. Partitioned time-varying convolution.

5.4. Csound Opcodes

Time-varying convolution has been implemented in Csound [43] in two separate ways, which nevertheless are based on the approaches described in Section 5. The first of these is liveconv, which is an extensive modification of an existing ftconv unit generator. It implements partitioned convolution employing an external function table as a means of sourcing one of the two input signals (nominally the impulse response). The second is tvconv, which takes two audio signal inputs and applies the process for a given filter and partitioned length. In this section, we examine these two implementations in some detail.

5.4.1. liveconv

The liveconv opcode implements dynamic replacement of impulse responses (see Section 3). It employs partitioned convolution with the overlap-add (OLA) scheme. The opcode takes one input signal and a table for holding the impulse response (IR) data:

```
ares liveconv ain, ift, iplen, kupdate, kclear,
```

where its parameters are as follows:

- ares: Output signal.
- ain: Input signal.
- ift: Table number for storing the impulse response (IR) for convolution. The table may be filled with new data at any time while the convolution is running.
- iplen: Length of impulse response partition in samples; must be an integer power of two. Lower settings allow for shorter output delay but will increase CPU usage.
- kupdate: Flag indicating whether the IR table should be updated. If kupdate = 1, the IR table ift is loaded partition by partition, starting with the next partition. If kupdate = −1, the IR table ift is unloaded (cleared to zero) partition by partition, starting with the next partition. Other values have no effect.
- kclear: Flag for clearing all internal buffers. If kclear has any value ! = zero, the internal buffers are cleared immediately. This operation is not free of artefacts.

The opcode makes a clear distinction between the input signal $x_1(n)$ and the IR table. However, if the IR table is updated regularly at filter length: $t_{Rn} = n * N$ and the IR table continuously filled with data from another audio stream $x_2(n)$, then the inputs $x_1(n)$ and $x_2(n)$ will be treated on equal terms, and the opcode behave similar to tvconv without freezing.

The opcode is programmed in C and is available as part of Csound source code [44]. See Algorithm 1 below for pseudocode that highlights the most important parts. For more details, the reader is encouraged to inspect the source code at Github.

Algorithm 1: Liveconv opcode implementation. IR loading marked with blue color.

Input : Audio input, IR table, partition size and update flag

Output: Convolved audio output

```
/* Check if the IR buffer should be updated                    */
```
if *update flag is set* **then**
| Initialize process to load IR from table position 0
end
forall *samples in input audio buffer* **do**
| Read sample into an internal ring buffer;
| **if** *One complete partition is read* **then**
| | `/* This is where the stepwise loading of an IR is handled   */`
| | **forall** *loading IR processes* **do**
| | | `/* Read from IR table into internal IR buffer            */`
| | | `/* The internal buffers are accessed in reverse order    */`
| | | **forall** *samples in current IR table partition* **do**
| | | | Read into first half of internal IR buffer;
| | | **end**
| | | Pad second half of internal IR buffer with zeros;
| | | Compute FFT on internal IR buffer (replace in buffer);
| | | IR table position += partition length;
| | | **if** *IR table position $\geq$ filter length* **then**   `// The entire filter is loaded`
| | | | Terminate this loading process;
| | | **end**
| | **end**
| | `/* Start processing the audio input partition              */`
| | Pad second half of internal ring buffer with zeros;
| | Compute FFT on internal ring buffer (replace in buffer);
| | Update ring buffer position (wrap around if necessary);
| | **forall** *partitions of the filter* **do**
| | | Pairwise multiply internal IR and ring buffers;
| | | Accumulate sum in temporary output buffer;
| | **end**
| | Compute inverse FFT of temporary buffer;
| | `/* Overlap Add (OLA) current and previous output block      */`
| | **for** $i = 0$ **to** *Partition length* -1 **do**
| | | output[i] = first part of temporary[i] + saved output[i];
| | | saved output[i] = second part of temporary[i];
| | **end**
| **end**
end

5.4.2. tvconv

The tvconv opcode takes two input signals and implements time-varying convolution. We can nominally take one of these signals as the impulse response and the other as the input signal, but, in practice, no such distinction is made. The opcode takes the length of the filter and its partitions as parameters, and includes switches to optionally fix coefficients instead of updating them continuously:

```
asig tvconv  ain1, ain2, xupdate1, xupdate2, ipartsize, ifilsize,
```

where its parameters are as follows:

- `ain1, ain2`: input signals.
- `xupdate1, xupdate2`: update switches u for each input signal. If $u = 0$, there is no update from the respective input signal, thus fixing the filter coefficients. If $u > 0$, the input signal updates the filter as normal. This parameter can be driven from an audio signal, which would work on a sample-by-sample basis, from a control signal, which would work on a block of samples at a time (depending on the `ksmps` system parameter, the block size), or it can be a constant. Each input signal can be independently *frozen* using this parameter.
- `ipartsize`: partition size, an integer P, $0 < P \leq N$, where N is the filter size. For values $P > 1$, the actual partition size will be quantised to $Q = 2^k$, $k \in \mathbb{Z}$, $Q \leq P$.
- `ifilsize`: filter size, an integer N, $N \geq P$, where P is the partition size. For partition size values $P > 1$, the actual filter size will be quantised to $O = 2^k$, $k \in \mathbb{Z}$, $O \leq N$.

This opcode is programmed in C++ using the Csound Plugin Opcode Framework (CPOF) [45,46], as the `TVConv` class. In this code, there are, in fact, two implementations of the process, which are employed according to the partition size:

1. For partition size = 1: direct convolution in the time domain is used, and any filter size is allowed. The following method in `TVConv` implements this (listing 1. The vectors `in` and `ir` hold the two delay lines, which take their inputs from the signals in `inp` and `irp`. The variables `frz1` and `frz2` are signals that control the freezing/updating operation for each input.

Listing 1: Direct convolution implementation.

```
int dconv() {
    csnd::AudioSig insig(this, inargs(0));
    csnd::AudioSig irsig(this, inargs(1));
    csnd::AudioSig outsig(this, outargs(0));
    auto irp = irsig.begin();
    auto inp = insig.begin();
    auto frz1 = inargs(2);
    auto frz2 = inargs(3);
    auto inc1 = csound->is_asig(frz1);
    auto inc2 = csound->is_asig(frz2);

    for (auto &s : outsig) {
        if(*frz1 > 0) *itn = *inp;
        if(*frz2 > 0) *itr = *irp;
        itn++, itr++;
        if(itn == in.end()) {
            itn = in.begin();
            itr = ir.begin();
        }
        s = 0.;
        for (csnd::AuxMem<MYFLT>::iterator it1 = itn,
                it2 = ir.end() - 1; it2 >= ir.begin();
                it1++, it2--) {
            if(it1 == in.end()) it1 = in.begin();
            s += *it1 * *it2;
        }
        frz1 += inc1, frz2 += inc2;
        inp++, irp++;
    }
}
```

```
      return OK;
  }
```

2. For partition size > 1, partitioned convolution is used (listing 2), through an overlap-add algorithm. In this case, the process is implemented in the spectral domain, and in order to make it as efficient as possible, power-of-two partition and filter sizes are enforced internally.

Listing 2: Partitioned convolution implementation.

```
int pconv() {
  csnd::AudioSig insig(this, inargs(0));
  csnd::AudioSig irsig(this, inargs(1));
  csnd::AudioSig outsig(this, outargs(0));
  auto irp = irsig.begin();
  auto inp = insig.begin();
  auto *frz1 = inargs(2);
  auto *frz2 = inargs(3);
  auto inc1 = csound->is_asig(frz1);
  auto inc2 = csound->is_asig(frz2);

  for (auto &s : outsig) {
    if(*frz1 > 0) itn[n] = *inp;
    if(*frz2 > 0) itr[n] = *irp;

    s = out[n] + saved[n];
    saved[n] = out[n + pars];
    if (++n == pars) {
      cmplx *ins, *irs, *ous = to_cmplx(out.data());
      std::copy(itn, itn + ffts, itnsp);
      std::copy(itr, itr + ffts, itrsp);
      std::fill(out.begin(), out.end(), 0.);
      // FFT
      csound->rfft(fwd, itnsp);
      csound->rfft(fwd, itrsp);
      // increment iterators
      itnsp += ffts, itrsp += ffts;
      itn += ffts, itr += ffts;
      if (itnsp == insp.end()) {
        itnsp = insp.begin();
        itrsp = irsp.begin();
        itn = in.begin();
        itr = ir.begin();
      }
      // spectral delay line
      for (csnd::AuxMem<MYFLT>::iterator it1 = itnsp,
           it2 = irsp.end() - ffts; it2 >= irsp.begin();
           it1 += ffts, it2 -= ffts) {
        if (it1 == insp.end()) it1 = insp.begin();
        ins = to_cmplx(it1);
        irs = to_cmplx(it2);
        // spectral product
        for (uint32_t i = 1; i < pars; i++)
```

```
        ous[i] += ins[i] * irs[i];
         ous[0] += real_prod(ins[0], irs[0]);
    }
    // IFFT
    csound->rfft(inv, out.data());
    n = 0;
  }
  frz1 += inc1, frz2 += inc2;
  irp++, inp++;
 }
 return OK;
}
```

6. Applications and Use Cases

The two different convolution techniques described above has been used as the basis for two live processing instruments. The instruments are software based and designed to be used with live performers, convolving the sound produced by one musician with the sound produced by another. The instruments are packaged in the form of software plugins in the VST (Virtual Studio Technology (VST) is a software interface that integrates software audio synthesizer and effect plugins with audio editors and recording systems.) plugin format, compiled using the Cabbage [47] framework. Source code for the plugins is available at [48]). This work has been done in the context of a larger research project on crossadaptive audio processing [7], wherein these convolution techniques can be said to form a subset of a larger collection of techniques investigated. The aims of the crossadaptive research forms the motivation for the design of the instruments, and thus it is appropriate to describe these briefly before we move on. It is noteworthy that the work with the application has been done in the context of artistic research, following methods of artistic exploration rather than scientific methods. This means that there have been more of a focus on exploring the unknown potential for creative use of these techniques than there has been on testing explicit hypotheses. Thus, there is not a quantifiable verification of test results, but rather an investigation of some application examples expected to prove useful in our artistic work. Intervention into regular and habitual interaction patterns between musicians is one of the core aspects of the research. More specifically, the project aims to develop and explore various techniques of signal interaction between musical performers, such that *the actions of one* can influence *the sound of the other*. This is expected to stir up the habitual interaction patterns between experienced musicians, and thus facilitating novel ways of playing together, enabling new forms of music to emerge. Some of the crossadaptive methods will be intuitive and immediately engaging, others might pose significant challenges for the performers. In many cases, the musical output will not be immediately appealing, as the performers are given an unfamiliar context and unfamiliar tools and instruments. We do still very much rely on their performative experience to solve these challenges, and the instruments are designed to be as playable as possible within the scope and aims outlined. The crossadaptive project includes a wide range of signal interaction models, many based on feature extraction (signal analysis) and mapping the extracted features onto parametric controls of sonic manipulation by means of digital effects processing [49]. Those models of signal interaction has the advantage that any feature can be mapped to any sonic modulation; however, all mappings between features and modulations are also artificial. It is not straightforward to create a set of mappings that is as rich and intuitive to the performer as, for example, the relationship between physical gestures and sonic output of an acoustic instrument.

The convolution techniques that are the focus of this article provide a special case in this regard, as the features of the sound itself contains all aspects of modulation over the other sound. The nuances of the sound to be convolved can be multidimensional and immensely expressive in its features. Still, the performer can use his or her experience in relating the instrumental gestures to the nuances of this sound, which in turn constitute the filter coefficients. Thus, there is a high degree of control intimacy, it is intuitive for the performer to relate performative gestures to sonic output, and, with convolution, the sonic output itself is what modulates the other sound. Relating to the recording of an impulse response is as easy and intuitive for the musician as it is to relate to other kinds of live sampling. These are features of the convolution technique that makes it especially interesting in the context of the crossadaptive performance research. At the same time, the direct and performative live use of convolution techniques also opens the way for a faster and more intuitive exploration of the creative potential of convolution as an expressive sound design tool.

6.1. Liveconvolver

This instrument is designed for convolution of two live signals, but we have retained the concept of viewing one of the signals as an impulse response. This also allows us to retain an analogy with live sampling techniques, as the recording of the impulse response is indeed a form of live capture. The instrument is based on the *liveconv* opcode as described in Section 5.4.1. The overall signal flow is shown in Figure 12. The audio on the IR record input channel is continuously written to a circular buffer. When we want to replace the current IR, we read from this buffer and replace the IR partition by partition as described under Section 3. The main reason for the circular buffer is to enable transformation (for example time reversal or pitch modification) of the audio before making the IR. If the pitch and direction of the input is not modified, then the extra buffering layer does not impose any extra latency to the process.

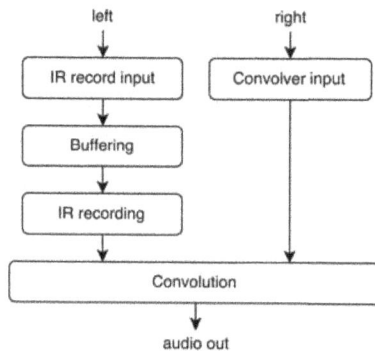

Figure 12. Liveconvolver instrument flowchart.

The single most important control of this processing instrument is the trigger for *when to replace the IR*. This can be done manually via the GUI (Graphical User Interface), see Figure 13, but, for practical reasons, we commonly use an external physical controller or pedal trigger mapped to this switch. One of the musicians then has control over *when* the IR is updated and *how long the new IR will be*. In addition to the manual control, this function can also be automatically controlled by means of transient detection or simply set to a periodic update triggered by a metronome. The manual trigger method controlled by an external pedal has been by far the most common use in our experiments.

In addition to the IR record trigger, we have added controls for highpass and lowpass filtering to enable quick adjustments to the spectral profile. When convolving two live signals, the output sound will have a tremendous dynamic range (due to the multiplication of the dynamics of each input sound) and also it will often have a quite unbalanced spectrum (spectral overlap in the two input sounds will be amplified while everything else will be highly attenuated). For this reason, some very coarse and effective controls are needed to shape the output spectrum in a live setting. The described filter controls allows such containment in a convenient manner. Due to the inherent danger of excessive resonances when recording the IR in the same room as the convolver effect is being used, we utilize a simple audio feedback prevention technique (shifting the output spectrum by a few Hz [50] by means of single sideband modulation). This helps in avoiding feedback, but can also be perceived as a subtle detuning. The amount of frequency shift can be adjusted, and, as such, there is a way to minimize the detuning amount for a given performance situation.

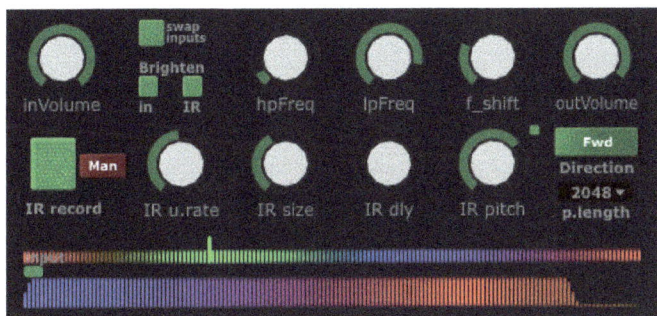

Figure 13. Liveconvolver instrument user interface. As an attempt to visualize *when* the impulse response is taken from, we use a circular colouring scheme to display the circular input buffer (thin coloured band labeled "input" in the image). We also represent the IR (broader coloured band at the bottom of the image) using the same colours. Time (of the input buffer) is thus represented by colour.

Performative Roles with Liveconv

One can distinguish two different performative roles in this kind of interplay: One performer is recording an impulse response (IR), and the other musician plays through the filter created by this IR. It is noteworthy that the two signals are mathematically equivalent in convolution, it does not matter if one convolves sound A with sound B or vice versa, the output sound will be the same (A*B). Still, the two roles facilitates quite different modes of performance, different ways of having control over the output sound. The performer recording the IR will have significant control over the spectral and temporal texture generated, while the musician playing through the filter will have control over the timing and energy flow.

Before starting the practical experimentation, the clarity and implications of these roles were not clear to us. Even so, it was clear that the two musicians would have different roles. The practical sessions would always start with a description of the convolution process and an explanation of how it would affect the sound. Then, each of the musicians would be allowed to perform in both roles, with discussions and reflections in between each take. In the vast majority of experiments, the musical performance was freely improvised, and the musicians was told to just play on the sound coming back to them. The aim was simply to see how they would respond to this new sonic environment and the interaction possibilities presented. The spontaneous and intuitive reaction of skilled performers to this unfamiliar performance scenario was expected to produce something interesting. As artistic exploration, we wanted to keep the possible outcomes as open as possible. There are many variables not explicitly controlled in these experiments. First of all, each musician has a different vocabulary of sounds, phrases, musical reaction patterns and such. Then, the sonic potential and performative affordances of the acoustic instrument being played. Since the resulting sound is a combination of two instruments (and usually two performers), the combinations of above variables affects the output greatly—likewise, the degree to which the musicians knew each other musically before the experiment. The performance conditions in terms of acoustic space, microphone selection and placement, how the processed sound is monitored (speakers, headphones, balance between acoustic and processed sound, etc.), and other external conditions would also potentially affect the output in different ways. If the aim had been to do a scientific investigation of particular aspects (of performance, interaction, timbral modulation, other), these variables would need to be controlled. However, as an artistic exploration of a hitherto unavailable technique, we have opted to keep these things open. The explorative process would thus be open to allow building on preliminary results iteratively, and allow creative input from the participants to affect the course of exploration. The most interesting products of the experiments are the audio recording themselves, as non-exhaustive examples of possible musical uses.

Even if we here identify two distinct performative roles, these were not preconceived but became apparent from noticing similarities between experiments with different musicians. Some musicians in our experiments would prefer the role of playing through the filter, perhaps since it can closely resemble the situation where one is playing through any electronic effect (e.g., an artificial reverb). Others would prefer the role of recording the IR, since this allows the real-time design of the spectrotemporal environment for the other musician to play in. This divide has been apparent in all our practical experiments thus far—for more detail, see, for example, project blog notes about sessions in San Diego [51]). If one, for example, records a single long note as the IR, there is not much to do for the performer playing through the effect other than recreating this tone at different amplitudes. Then, if the impulse response is a series of broadband clicks, this will create a delay pattern and the other musician will have to perform within the musical setting thus constituted.

In spite of the slight imbalance of power, with carefully chosen audio material, both performers can have mutual effect on the output, and the techniques facilitates a closely knitted interplay. In addition, varying the duration of the IR is also an effective way of creating distinctly different sections in an improvisation, and thus can be used to shape formal aspect of the music. Varying the length of the IR also directly affect the power balance between the performative roles: a short IR generally creates a more transparent environment for the other performer to act within. Changing between sustained and percussive sonic material in the IR also has a similar effect, while also retaining a richer image of the temporal activity in the IR recording signal.

We have experimented with the live convolver technique in studio sessions and concerts since early 2017. There have been sessions in San Diego and Los Angeles (see link in [51] and Figure 14), Oslo (see link in [52], and Trondheim (see link in [53] and Figure 15).

In addition to the creative potential of spectrotemporal morphing, there are also some clear pitfalls. As mentioned, convolution is a multiplicative process, and thus the dynamic range is very large. In addition, it naturally amplifies common frequencies between the two signals, and attenuates everything else. This leads to an unnaturally loud output when the two musicians play melodic phrases where they happen to use common tones. Many musicians will gravitate towards each other's sound, as a natural aesthetic impulse, trying to create a sonic weave that unites the two sounds. Doing this with live convolution is often counterproductive, as it creates a spectrum with a few extremely high peaks and little other information.

Figure 14. Kjell Nordeson and Øyvind Brandtsegg discussing live convolver performance.

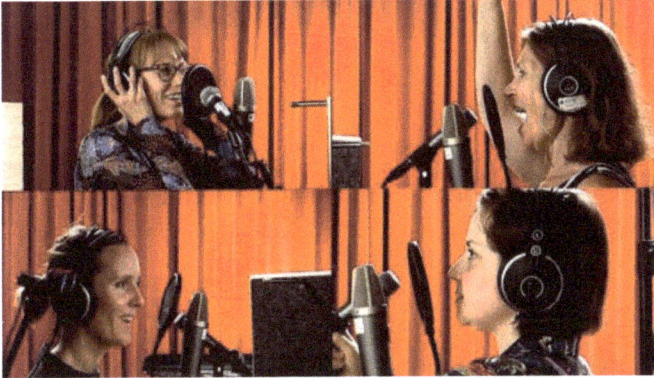

Figure 15. Session with singers from Trondheim Voices.

Another danger with live convolution when played over a P.A. in concert is the feedback potential. In this situation, the room (and the sound system) where the performance happens is convolved with itself, effectively multiplying the feedback potential of the system. The performers can also affect the feedback potential, by developing a sensitivity and understanding of which spectral material is most likely to cause feedback problems. Producing material that plays on the resonances of the room generally will increase the danger of feedback. This is another example of a counterintuitive measure for a music performer, as one would normally try to utilize the affordances of the acoustic environment and play things that resonate well in the performance space.

6.2. TV Convolver

This instrument is designed for convolution of two live signals in a streaming fashion, that is, both signals are treated equally and are continually updated. It is based on the *tvconv* opcode, as described in Section 5.4.2. It has controls for freezing each of the signals (see Figure 16), that is, bypassing the continuous update of the input buffers. Like the liveconvolver instrument, it also has controls for frequency shift amount (for reduction of feedback potential), highpass and lowpass filtering (for quick adjustment to the output spectrum, see Section 6.1 for details). In addition, it has controls for adjusting the filter length and fade time. The fade time is used in connection with freezing the filter. Even though *tvconv* allows freezing the filter at any time, we have opted to only allow freezing at filter boundaries, and also to facilitate a fade out of the coefficients towards the boundary when freezing is activated. These measures are done to avoid artefacts due to discontinuities in the filter coefficients. When used with relatively long filter lengths (typically 65,536 samples in our experiments, 1.4 s at a sampling rate of 44.1 kHz), discontinuities due to freezing could be clearly audible as clicks in the audio output. Furthermore, freezing the filter at any arbitrary point would potentially reorder the temporal structure of the live signal used for filter coefficients (see Figures 17 and 18).

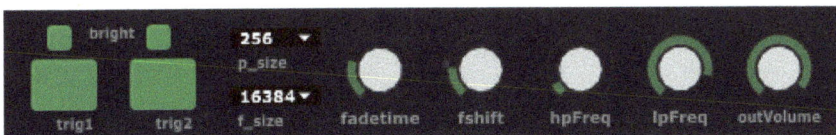

Figure 16. TV convolver GUI.

Figure 17. Example of tvconv buffer content when freezing is allowed only at filter boundaries. One contiguous block of audio remains in the filter when frozen.

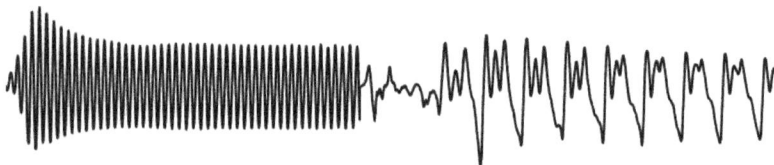

Figure 18. Example of tvconv buffer content if freezing is allowed at an arbitrary point. Old content remains in the latter part of the buffer while the first part has been written with new content.

Practical Experiments with Tvconv

The streaming nature of this variant of live convolution allows for a constantly changing effect where both of the signals are treated equally. None of them is considered the impulse response for the other to be filtered through; they are simply convolved with each other, where the time window of the convolution is set by the filter length. This instrument has been used in a studio session in Trondheim in August 2017 ([53] and Figure 15), and in several live performances in the weeks and months following the initial studio exploration. Since the two signals are treated equally, there are no technical grounds for assuming different roles for the two performers playing into the effect. This also conforms with our experience.

The effect of *tvconv* is usually harder to grasp for the performers, compared with the liveconvolver instrument. The interaction between the two signals can be musically counterintuitive. In an improvised musical dialogue, performers will sometimes use a *call and response* strategy, making a musical statement followed by a pause where other musicians are allowed to respond. With streaming convolution, this is not so productive, since there will be no output unless both input signals are sounding at the same time. Thus, it requires a special performative approach where phrases are interwoven and contrapuntal, with quicker interactions. This manner of simultaneous initiatives could be perceived as aggressive and almost disrespectful in some musical settings, as it might appear as one is trampling all over the other musician's statements. Negotiating this space of common phrasing very clearly affects the manner in which the musicians can interact, which was one of the objectives of the crossadaptive research project.

Due to the manner in which the filter is updated (see Section 5.4.2), one may perceive a variable latency when performing with tvconv. The technical latency is not variable, but the perceived latency is. This occurs when the first part of the filter contains low amplitude sounds (or sound that is otherwise low in perceptually distinct features), such that the perceived "attack" of the sound is positioned later in the filter (see Figure 19). In this case, one might perceive the filter's response as having a longer latency than the strictly technical latency of the signal processing. As the filter coefficients are continuously updated in a circular buffer, there is no explicit way of ensuring that perceptually prominent features of the input sound is written to the beginning of the filter. Admittedly, one could argue that a circular buffer does not have a *beginning* as such, but the way it is used in the convolution process means that it matters *where in this buffer* salient features of the input sound is positioned. Thus, the perceptual latency can vary within a maximum period set by the size of the filter.

Figure 19. Impulse response with initial section low on perceptual features, transient occuring later in the filter.

In addition to the immediate interaction of the streaming convolution, another manner of interaction is enabled by giving each of the performers a physical controller (midi pedal) to freeze each of the two filter buffers. If one buffer is frozen, the other performer can play in a more uninterrupted manner, being convolved through the (now static) filter. They can freely switch between freezing one or the other of the inputs and this enables a playful interaction. If both input buffers are frozen, the convolver will output a steady audio loop of length equal to the filter length. In that case, no input is updated, so the filter output is maintained even if both performers fall silent.

We also have a perceptual latency issue when freezing the filter. Since we in our instrument design have allowed freezing only at filter boundaries (see Section 6.2), a signal to freeze the filter does not always freeze the last N seconds (where N is the filter length). Rather, it will continue updating until the write index reaches the filter boundary and then freeze. The time relation between input sounds and the filter buffer is determinate, but since the filter boundaries are not visible or audible to the performers, the relationship will be perceived as somewhat random. In spite of these slight inconveniences, we have focused on the musical use of the filter in performance.

6.3. Demo Sounds

The main application of our convolution techniques has been for live performance. To allow for additional insight into how the processing techniques affect a given source material, we have made some simple demo sounds. These are available online at [54]. The same source have been used for both *liveconv* and *tvconv*, and one example output created for each of these two convolution methods. Due to the parametric nature of the methods, a large number of variations on output sounds could be conceived. These two examples simply aim at showing the techniques in their simplest form.

6.4. Future Work

There are some typical problems in using convolution for creative sound design that are even more emphasized when using it for live performance. These relate to the fact that convolution is a multiplicative process, and as such has a tremendous dynamic range. Moreover, if spectral peaks in the two signals overlap, the output sound will tend to have extreme peaks at corresponding frequencies. One possible solution to this might be utilizing a spectral whitening process as suggested by Donahue [55], although this has not yet been fully solved for the case of real-time convolution with time-varying impulse responses. Another approach might be to analyze the two input spectra and selectively reduce the amplitude of overlapping peaks, allowing control over resonant peaks without other spectral modifications. Furthermore, the impulse response update methods described in this article could be applied in more traditional manners, like parametric control of convolution reverbs and so on.

7. Conclusions

The paper has described two new approaches and implementations of time-varying convolution filters. These have been developed in the context of live convolution within improvised electroacoustic performance, as a method of crossmodulation between two audio signals. The implementations has been done in the form of opcodes for the audio programming language Csound. Software instruments for live performance with these opcodes has been built in the form of VST plugins. We have shown some usage examples from studio sessions done within the artistic research project "Cross-adaptive processing as musical intervention". These practical sessions have also revealed some performative issues and a significant creative potential for further exploration.

Acknowledgments: The research presented here is part of the project "Cross-adaptive processing as musical intervention", financed by the Norwegian Artistic Research Programme and the Norwegian University for Science and Technology. The cost for publishing in open access is covered by the Norwegian University of Science and Technology.

Author Contributions: Øyvind Brandtsegg initially formulated the need for live convolution for improvised performance and did initial investigations using existing Csound opcodes as a preparation for the common work presented here. Sigurd Saue conceived the method for incremental updates of the live convolver and implemented the liveconv opcode. Victor Lazzarini conceived the method of using two circular buffers and implemented the tvconv opcode. Finally, Øyvind Brandtsegg implemented the techniques as musical instruments and conducted the practical experiments on application. Each author has written the sections about his own contributions in this paper.

Conflicts of Interest: The authors declare no conflict of interest.

References

1. Dolson, M. Recent Advances in Musique Concrète at CARL. In Proceedings of the 1985 International Computer Music Conference, ICMC, Burnaby, BC, Canada, 19–22 August 1985.
2. Roads, C. Musical Sound Transformation by Convolution. In Proceedings of the 1993 International Computer Music Conference Opening a New Horizon ICMC, Tokyo, Japan, 10–15 September 1993.
3. Truax, B. Convolution Techniques. Available online: https://www.sfu.ca/~truax/Convolution%20Techniques.pdf (accessed on 31 October 2017).
4. Truax, B. Sound, Listening and Place: The aesthetic dilemma. *Organ. Sound* **2012**, *17*, 193–201.
5. Moore, A. *Sonic Art: An Introduction to Electroacoustic Music Composition*; Routledge: London, UK, 2016.
6. Gardner, W.G. Efficient Convolution without Input-Output Delay. *J. Audio Eng. Soc.* **1995**, *43*, 127–136.
7. Brandtsegg, Ø. Cross Adaptive Processing as Musical Intervention; Exploring Radically New Modes of Musical Interaction in Live Performance. Available online: http://crossadaptive.hf.ntnu.no/ (accessed on 7 January 2018).
8. Brandtsegg, Ø.; Saue, S. Live Convolution with Time-Variant Impulse Response. In Proceedings of the 20th International Conference on Digital Audio Effects (DAFx-17), Edinburgh, UK, 5–9 September 2017; pp. 239–246.
9. Shmaliy, Y. *Continuous-Time Systems*; Springer: Heidelberg, Germany, 2007.
10. Cherniakov, M. *An Introduction to Parametric Digital Filters and Oscillators*; John Wiley & Sons: New York, NY, USA, 2003.
11. Zetterberg, L.H.; Zhang, Q. Elimination of transients in adaptive filters with application to speech coding. *Signal Process.* **1988**, *15*, 419–428.
12. Verhelst, W.; Nilens, P. A modified-superposition speech synthesizer and its applications. In Proceedings of the IEEE International Conference on Acoustics, Speech, and Signal Processing (ICASSP'86), Tokyo, Japan, 7–11 April 1986; Volume 11, pp. 2007–2010.
13. Wishnick, A. Time-Varying Filters for Musical Applications. In Proceedings of the 17th International Conference on Digital Audio Effects (DAFx-14), Erlangen, Germany, 1–5 September 2014; pp. 69–76.
14. Ding, Y.; Rossum, D. Filter morphing for audio signal processing. In Proceedings of the IEEE ASSP Workshop on Applications of Signal Processing to Audio and Acoustics, New Paltz, NY, USA, 15–18 October 1995; pp. 217–221.
15. Zoelzer, U.; Redmer, B.; Bucholtz, J. *Strategies for Switching Digital Audio Filters*; Audio Engineering Society Convention 95; Audio Engineering Society: New York, NY, USA, 1993.
16. Carty, B. *Movements in Binaural Space: Issues in HRTF Interpolation and Reverberation, with Applications to Computer Music*; Lambert Academic Publishing: Duesseldorf, Germany, 2012.
17. Jot, J.M.; Larcher, V.; Warusfel, O. *Digital Signal Processing Issues in the Context of Binaural and Transaural Stereophony*; Audio Engineering Society Convention 98; Audio Engineering Society: New York, NY, USA, 1995.

18. Lee, K.S.; Abel, J.S.; Välimäki, V.; Stilson, T.; Berners, D.P. The switched convolution reverberator. *J. Audio Eng. Soc.* **2012**, *60*, 227–236.

19. Välimäki, V.; Laakso, T.I. Suppression of transients in variable recursive digital filters with a novel and efficient cancellation method. *IEEE Trans. Signal Process.* **1998**, *46*, 3408–3414.

20. Mourjopoulos, J.N.; Kyriakis-Bitzaros, E.D.; Goutis, C.E. Theory and real-time implementation of time-varying digital audio filters. *J. Audio Eng. Soc.* **1990**, *38*, 523–536.

21. Abel, J.S.; Berners, D. *The Time-Varying Bilinear Transform*; Audio Engineering Society Convention 141; Audio Engineering Society: New York, NY, USA, 2016.

22. Wefers, F.; Vorländer, M. Efficient time-varying FIR filtering using crossfading implemented in the DFT domain. In Proceedings of the 2014 7th Medical and Physics Conference Forum Acusticum, Cracow, Poland, 7–12 September 2014.

23. Wefers, F. *Partitioned Convolution Algorithms for Real-Time Auralization*; Logos Verlag Berlin GmbH: Berlin, Germany, 2015; Volume 20.

24. Vickers, E. *Frequency-Domain Implementation of Time-Varying FIR Filters*; Audio Engineering Society Convention 133; Audio Engineering Society: New York, NY, USA, 2012.

25. Settel, Z.; Lippe, C. Real-time timbral transformation: FFT-based resynthesis. *Contemp. Music Rev.* **1994**, *10*, 171–179, doi:10.1080/07494469400640401.

26. Wishart, T.; Emmerson, S. *On Sonic Art*; Contemporary Music Studies; Harwood Academic Publishers: Berkshire, UK, 1996.

27. Roads, C. *Composing Electronic Music: A New Aesthetic*; Oxford University Press: Oxford, UK, 2015.

28. Engum, T. Real-time Control and Creative Convolution. In Proceedings of the International Conference on New Interfaces for Musical Expression, Oslo, Norway, 30 May–1 June 2011; pp. 519–522.

29. Aimi, R.M. Hybrid Percussion: Extending Physical Instruments Using Sampled Acoustics. Ph.D. Thesis, Massachusetts Institute of Technology, Department of Architecture, Program In Media Arts and Sciences, Cambridge, MA, USA, 2007.

30. Schwarz, D.; Tremblay, P.A.; Harker, A. Rich Contacts: Corpus-Based Convolution of Contact Interaction Sound for Enhanced Musical Expression. In Proceedings of the International Conference on New Interfaces for Musical Expression, London, UK, 30 June–3 July 2014; pp. 247–250.

31. Morris, J.M. Ontological Substance and Meaning in Live Electroacoustic Music. In *Genesis of Meaning In Sound and Music, Proceedings of the 5th International Symposium on Computer Music Modeling and Retrieval, Copenhagen, Denmark, 19–23 May 2008*; Revised Papers; Ystad, S., Kronland-Martinet, R., Jensen, K., Eds.; Springer: Berlin/Heidelberg, Germany, 2009; pp. 216–226.

32. Kagel, M. Transición II; Phonophonie Liner Notes. Available online: http://www.moderecords.com/catalog/127kagel.html (accessed on 4 December 2017).

33. Emmerson, S. *Living Electronic Music*; Ashgate: Farnham, UK, 2007.

34. Casserley, L. A Digital Signal Processing Instrument for Improvised Music. *J. Electroacoust. Music* **1998**, *11*, 25–29.

35. Oppenheim, A.V.; Schafer, R.W.; Buck, J.R. *Discrete-Time Signal Processing*, 2nd ed.; Prentice-Hall, Inc.: Upper Saddle River, NJ, USA, 1999.

36. Stockham, T.G., Jr. High-speed convolution and correlation. In Proceedings of the Spring Joint Computer Conference ACM, Boston, MA, USA, 26–28 April 1966; pp. 229–233.

37. Laroche, J. On the stability of time-varying recursive filters. *J. Audio Eng. Soc.* **2007**, *55*, 460–471.

38. Lazzarini, V. *Computer Music Instruments*; Springer: Heildeberg, Germany, 2017.

39. Lazzarini, V.; Kleimola, J.; Timoney, J.; Välimäki, V. Five Variations on a Feedback Theme. In Proceedings of the 12th International Conference on Digital Audio Effects, Como, Italy, 1–4 September 2009; pp. 139–145.

40. Lazzarini, V.; Kleimola, J.; Timoney, J.; Välimäki, V. Aspects of Second-order Feedback AM synthesis. In Proceedings of the International Computer Music Conference, Huddersfield, UK, 31 July–5 August 2011; pp. 92–98.

41. Kleimola, J.; Lazzarini, V.; Välimäki, V.; Timoney, J. Feedback amplitude modulation synthesis. *EURASIP J. Adv. Signal Process.* **2011**, *2011*, doi:10.1155/2011/434378.

42. Timoney, J.; Pekonen, J.; Lazzarini, V.; Välimäki, V. Dynamic Signal Phase Distortion Using Coefficient-Modulated Allpass Filters. *J. Audio Eng. Soc.* **2014**, *62*, 596–610.

43. Lazzarini, V.; Ffitch, J.; Yi, S.; Heintz, J.; Brandtsegg, Ø.; McCurdy, I. *Csound: A Sound and Music Computing System*; Springer: Heidelberg, Germany, 2016.

44. Saue, S. Liveconv Source Code. Available online: https://github.com/csound/csound/blob/develop/Opcodes/liveconv.c (accessed on 7 January 2018).

45. Lazzarini, V. The Csound Plugin Opcode Framework. In Proceedings of the 14th Sound and Music Computing Conference 2017, Aalto University, Espoo, Finland, 5–8 July 2017; pp. 267–274.

46. Lazzarini, V. Supporting an Object-Oriented Approach to Unit Generator Development: The Csound Plugin Opcode Framework. *Appl. Sci.* **2017**, *7*, 970.

47. Walsh, R. Cabbage; A Framework for Audio Software Development. Available online: http://cabbageaudio.com/ (accessed on 7 January 2018).

48. Brandtsegg, Ø. Liveconvolver; Csound Instrument Based around the Liveconvolver Opcode. Available online: https://github.com/Oeyvind/liveconvolver (accessed on 7 January 2018).

49. Brandtsegg, Ø. A Toolkit for Experimentation with Signal Interaction. In Proceedings of the 18th International Conference on Digital Audio Effects (DAFx-15), Trondheim, Norway, 30 November–3 December 2015; pp. 42–48.

50. Wigan, E.R.; Alkin, E.G. The BBC Research Labs Frequency Shift PA Stabiliser: A Field Report. BBC Internal Memoranda. 1960. Available online: http://works.bepress.com/edmund-wigan/17/ (accessed on 7 January 2018).

51. Brandtsegg, Ø. Liveconvolver Experiences, San Diego. Available online: http://crossadaptive.hf.ntnu.no/index.php/2017/06/07/liveconvolver-experiences-san-diego/ (accessed on 7 January 2018).

52. Wærstad, B.I. Live Convolution Session in Oslo, March 2017. Available online: http://crossadaptive.hf.ntnu.no/index.php/2017/06/07/live-convolution-session-in-oslo-march-2017/ (accessed on 7 January 2018).

53. Brandtsegg, Ø. Session with 4 Singers, Trondheim, August 2017. Available online: http://crossadaptive.hf.ntnu.no/index.php/2017/10/09/session-with-4-singers-trondheim-august-2017/ (accessed on 7 January 2018).

54. Brandtsegg, Ø. Convolution Demo Sounds. Available online: http://crossadaptive.hf.ntnu.no/index.php/2017/12/07/convolution-demo-sounds/ (accessed on 7 January 2018).

55. Donahue, C.; Erbe, T.; Puckette, M. Extended Convolution Techniques for Cross-Synthesis. In Proceedings of the International Computer Music Conference 2016, Utrecht, The Netherlands, 12–16 September 2016; pp. 249–252.

applied sciences

MDPI

Article

Audlet Filter Banks: A Versatile Analysis/Synthesis Framework Using Auditory Frequency Scales

Thibaud Necciari [1,*], Nicki Holighaus [1], Peter Balazs [1], Zdeněk Průša [1], Piotr Majdak [1] and Olivier Derrien [2]

[1] Acoustics Research Institute, Austrian Academy of Sciences, Wohllebengasse 12–14, 1040 Vienna, Austria; nicki.holighaus@oeaw.ac.at (N.H.); peter.balazs@oeaw.ac.at (P.B.); zdenek.prusa@oeaw.ac.at (Z.P.); piotr@majdak.com (P.M.)

[2] Universite de Toulon, Aix-Marseille Universite, CNRS-PRISM, 31 Chemin Joseph Aiguier, 13402 Marseille CEDEX 20, France; derrien@prism.cnrs.fr

[*] Correspondence: thibaud.necciari@oeaw.ac.at; Tel.: +43-1-51581-2538

Academic Editor: Vesa Valimaki
Received: 3 November 2017; Accepted: 3 January 2018; Published: 11 January 2018

Featured Application: The proposed framework is highly suitable for audio applications that require analysis–synthesis systems with the following properties: stability, perfect reconstruction, and a flexible choice of redundancy.

Abstract: Many audio applications rely on filter banks (FBs) to analyze, process, and re-synthesize sounds. For these applications, an important property of the analysis–synthesis system is the reconstruction error; it has to be minimized to avoid audible artifacts. Other advantageous properties include stability and low redundancy. To exploit some aspects of auditory perception in the signal chain, some applications rely on FBs that approximate the frequency analysis performed in the auditory periphery, the gammatone FB being a popular example. However, current gammatone FBs only allow partial reconstruction and stability at high redundancies. In this article, we construct an analysis–synthesis system for audio applications. The proposed system, referred to as *Audlet*, is an oversampled FB with filters distributed on auditory frequency scales. It allows perfect reconstruction for a wide range of FB settings (e.g., the shape and density of filters), efficient FB design, and adaptable redundancy. In particular, we show how to construct a gammatone FB with perfect reconstruction. Experiments demonstrate performance improvements of the proposed gammatone FB when compared to current gammatone FBs in terms of reconstruction error and stability, especially at low redundancies. An application of the framework to audio source separation illustrates its utility for audio processing.

Keywords: audio signal processing; analysis–synthesis; filter bank; time-frequency transform; frames; hearing; gammatone; equivalent rectangular bandwidth (ERB); Bark scale; Mel scale

---

1. Introduction

Time-frequency (TF) transforms like the short-time Fourier or wavelet transforms play a major role in audio signal processing. They allow any signal to be decomposed into a set of elementary functions with good TF localization and perfect reconstruction is achieved if the transform parameters are chosen appropriately (e.g., [1,2]). The result of a signal analysis is a set of TF coefficients, sometimes called sub-band components, that quantifies the degree of similarity between the input signal and the elementary functions. In applications, TF transforms are used to perform sub-band processing, that is, to modify the sub-band components and synthesize an output signal. De-noising techniques [3,4], for instance, analyze the noisy signal, estimate the TF coefficients associated with noise, delete them from the set of TF coefficients, and synthesize a clean signal from the set of remaining TF coefficients. Lossy audio codecs like MPEG-2 Layer III, known as MP3 [5], or advanced audio coding (AAC) [6,7] quantize the sub-bands with a variable precision in order to reduce the digital size of audio files. In audio transformations like time-stretching or pitch-shifting [8,9], the phases of sub-band components are processed to ensure a proper phase coherence. As a last example, applications of audio source separation [10–12] or polyphonic transcriptions of music [13] rely on the non-negative matrix factorization scheme: the set of TF coefficients is factorized into several matrices that correspond to various sources present in the original signal. Each source can then be synthesized from its matrix representation. In these applications, the short-time Fourier transform (STFT) is mostly used, although modified discrete cosine transforms (MDCTs) are usually preferred in audio codecs.

Because sub-band processing may introduce audible distortions in the reconstructed signal, important properties of the analysis–synthesis system include stability (i.e., the coefficients are bounded if and only if the input signal is bounded), perfect reconstruction (i.e., the reconstruction error is only limited by numerical precision when no sub-channel processing is performed), resistance to noise, and aliasing suppression in each sub-band (e.g., [14,15] Chap. 10). Furthermore, in all applications, a low redundancy (i.e., a redundancy between 1 and 2) lowers the computational costs.

TF transforms are usually implemented as filter banks (FBs) where the set of analysis filters defines the elementary functions and the set of synthesis filters allows signal reconstruction. The TF concentration of the filters together with the downsampling factors in the sub-bands define the TF resolution and redundancy of the transform. FBs come in various flavors and have been extensively treated in the literature (e.g., [16–19]). The mathematical theory of frames constitutes an interesting alternative background for the interpretation and implementation of FBs (e.g., [20–22]). Gabor frames (sampled STFT [2,23]), for instance, are widespread in audio signal processing.

For certain applications, such as audio coding [5–7], audio equalizers [24], speech processing [25], perceptual sparsity [26,27], or source separation [11,12,28,29], exploiting some aspects of human auditory perception in the signal chain constitutes an advantage. One of the most exploited aspects of the auditory system is the auditory frequency scale, which is a simple means to approximate the frequency analysis performed in the auditory system [30]. Generally, the auditory system is a complex and in many aspects nonlinear system (for a review see, e.g., [31]). Its description ranges from simple collections of linear symmetric bandpass filters [32] through collections of asymmetric and compressive filters [33] to sophisticated models of nonlinear wave propagation in the cochlea [34]. Because nonlinear systems may complicate the inversion of the signal processing chain (e.g., [35,36]), linear approximations of the auditory system are often preferred in audio applications. In particular, gammatone filters approximate well the auditory periphery at low to moderate sound pressure levels [37,38] and are easy to implement as FIR or IIR filters [32,39–43].

Various analysis–synthesis systems based on gammatone FBs have been proposed for the purpose of audio applications (e.g., [35,39,40,44]). However, these systems do not satisfy all requirements of audio applications as, even at high redundancies, they only achieve a reconstruction error described as "barely audible". This error becomes clearly audible at low redundancies. In other words, these systems do not achieve perfect reconstruction. To our knowledge, a general recipe for constructing a gammatone FB with perfect reconstruction at redundancies close to and higher than one has not been published yet.

In this article, we describe a general recipe for constructing an analysis–synthesis system using a non-uniform oversampled FB with filters distributed on an arbitrary auditory frequency scale, enabling perfect reconstruction at arbitrary redundancies. The resulting framework is named "*Aud*let" for *aud*io processing and *aud*itory motivation. The proposed approach follows the theoretical foundation of non-stationary Gabor frames [20,45] and their application to TF transforms with a variable TF resolution [46–48]. This report extends the work reported in [20] (Section 5.1) by providing a full theoretical and practical development of the Audlet.

The manuscript is organized as follows. The next section briefly recalls the basics of non-uniform FBs, frames, and auditory frequency scales. Section 3 describes the theoretical construction of the Audlet framework. The practical implementation issues are discussed in Section 4 and Section 5 evaluates important properties and capabilities of the framework.

2. Preliminaries

2.1. Notations and Definition

In the following, we consider signals in $\ell_2(\mathbb{Z})$ as samples of a continuous signal with sampling frequency f_s, with the Nyquist frequency of $f_N = f_s/2$. We denote the normalized frequency by $\xi = f/f_s$, i.e., the interval $[0, f_N]$ corresponds to $[0, 1/2]$. The inner product of two signals x, y is $\langle x, y \rangle = \sum_n x[n] \cdot y[n]$ and the energy of a signal is defined from the inner product as $||x|| = \langle x, x \rangle$. The floor, ceiling, and rounding operators are $\lfloor \cdot \rfloor, \lceil \cdot \rceil$, and $\lfloor \cdot \rceil$, respectively. We denote the z-transform by $\mathcal{Z}: x[n] \mapsto X(z)$. By setting $z = e^{2i\pi\xi}$ for $\xi \in (-1/2, 1/2]$, the z-transform equals the discrete-time Fourier transform (DTFT). Note that the frequency domain associated to the DTFT is circular and therefore, the interval $(-1/2, 1/2]$ is considered circularly, i.e., $\xi \in \mathbb{R}$ is identified with $\xi - \lfloor \xi \rceil \in (-1/2, 1/2]$. The same applies for $(-f_N, f_N]$. Since we exclusively consider real-valued signals we deal with symmetric DTFTs, which allows us to process only the positive-frequency range. Finally, we denote the complex conjugation by an overbar, e.g., $\overline{H}$.

2.2. Filter Banks and Frames

The general structure of a non-uniform analysis FB is presented in Figure 1 (e.g., [17]). It is a collection of $K + 1$ analysis filters $H_k(z)$, where $H_k(z)$ is the z-transform of the impulse response $h_k[n]$ of the filter, and downsampling factors d_k, $k \in \{0 \dots K\}$, that divides a signal x into a set of $K + 1$ sub-band components y_k, where

$$y_k[n] = \downarrow_{d_k} \{h_k * x\}[n] \quad . \tag{1}$$

The special case where all downsampling factors are identical, i.e., $d_k = D \forall k \subset \{0 \dots K\}$, is referred to as a uniform FB.

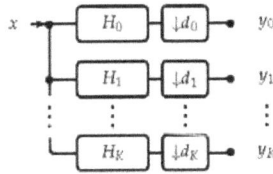

Figure 1. General structure of a non-uniform analysis filter bank (FB) $(H_k, d_k)_k$ with H_k being the z-transform of the impulse response $h_k[n]$ of the filter, also denoted as $\mathcal{A}(\cdot, (H_k, d_k)_k)$.

By analogy, a synthesis FB is a collection of $K + 1$ upsampling factors d_k and synthesis filters $G_k(z)$ (see Figure 2) that recombines the sub-band components y_k into an output signal $\tilde{x}$ according to

$$\tilde{x}[n] = 2\Re \left(\sum_{k=0}^{K} (g_k * \uparrow_{d_k} \{y_k\})[n] \right), \tag{2}$$

where $\Re$, denoting the real part, and the factor of 2 are a consequence of considering the positive frequency range only.

A synthesis FB can be generalized to a *synthesis system* (shown in Figure 3), which is a linear operator $\mathcal{S}$ that takes as an input sub-band components y_k and yields an output sequence $\tilde{x}$. For the synthesis operation, we use the notation $\widetilde{\mathcal{S}}(\cdot, (G_k, d_k)_k)$, where $(G_k, d_k)_k$ is the synthesis FB. An analysis

FB is *invertible* or *allows for perfect reconstruction* if there exists a synthesis system S that recovers x from the sub-band components y_k *without error*, i.e., $\tilde{x} = x$ for all $x \in \ell_2(\mathbb{Z})$. In other terms, the analysis–synthesis system $((H_k, d_k)_k, S)$ has the *perfect reconstruction property*. In practice, the implementation of that operation might introduce errors of the order of numerical precision.

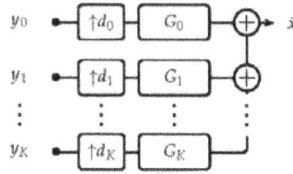

Figure 2. General structure of a non-uniform synthesis FB $(G_k, d_k)_k$, also denoted by $\tilde{S}(\cdot, (G_k, d_k)_k)$.

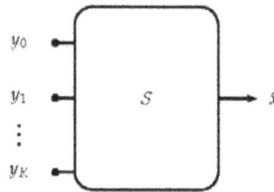

Figure 3. General structure of a synthesis system. S is a linear operator that maps the sub-band components y_k to an output signal $\tilde{x}$.

We use the mathematical theory of frames in order to analyze and design perfect reconstruction FBs (e.g., [20–22]). A *frame* over the space of finite energy signals $\ell_2(\mathbb{Z})$ is a set of functions spanning the space in a stable fashion. Consider a signal x and an analysis FB $(H_k, d_k)_k$ yielding y_k. Then, an FB constitutes a frame if and only if $0 < A \leq B < \infty$ exist such that

$$A\|x\|^2 \leq \sum_k \|y_k\|^2 \leq B\|x\|^2, \forall x \in \ell^2(\mathbb{Z}) \tag{3}$$

where A and B are called the lower and upper frame bounds of the system, respectively. The existence of A and B guarantees the invertibility of the FB. Several numerical properties of an FB can be derived from the frame bounds. In particular, the ratio $\sqrt{B/A}$ corresponds to the *condition number* [49] of the FB, i.e., it determines the stability and reconstruction error of the system. Furthermore, the ratio B/A characterizes the overall frequency response of the FB. A ratio $B/A = 1$, for instance, means a perfectly flat frequency response. This is often desired in signal processing because, in that particular case, the analysis and synthesis FB are the same. Specifically, the synthesis filters are obtained by time-reversing the analysis filters, i.e., $G_k(z) = \overline{H}_k(z)$.

The frame bounds A and B correspond to the infimum and supremum, respectively, of the eigenvalues of the operator $\tilde{S}(\mathcal{A}(\cdot, (H_k, d_k)_k), (H_k, d_k)_k)$ associated with the system $(H_k, d_k)_k$. In practice, these eigenvalues can be computed using iterative methods (see Sections 3.2 and 3.3).

2.3. Auditory Frequency Scales

An important aspect of the auditory system to consider in auditory-motivated analysis is the frequency-to-place transformation that occurs in the cochlea. Briefly, when a sound reaches the ear it produces a vibration pattern on the basilar membrane. The position and width of this pattern along the membrane depend on the spectral content of the sound; high-frequency sounds produce maximum excitation at the base of the membrane, while low-frequency sounds produce maximum excitation at the apex of the membrane. This property of the auditory system can be modeled in a first approximation as a bank of bandpass filters, named "critical bands" or "auditory filters", whose center frequencies and bandwidths respectively approximate the place and width of excitation on the basilar membrane. The frequency and bandwidth of the auditory filters are nonlinear functions of frequency. These functions, called auditory frequency scales, are derived from psychoacoustic experiments (see e.g., [50], Chapter 3 for a review). The Bark, the equivalent rectangular bandwidth (ERB), and Mel scales are commonly used in hearing science and audio signal processing [30]. To refer to the different frequency mappings we introduce the function $F: f \to$ Scale where f is frequency in Hz and Scale is an auditory unit that depends on the scale. The ERB rate, for instance, is [30]

$$F_{\text{ERB}}(f) = 9.265 \ln \left(1 + \frac{f}{228.8455} \right) \tag{4}$$

and its inverse is

$$f = F_{\text{ERB}}^{-1}(F_{\text{ERB}}) = 228.8455 \left(e^{F_{\text{ERB}}/9.265} - 1 \right). \tag{5}$$

The ERB (in Hz) of the auditory filter centered at frequency f is

$$B_{\text{ERB}}(f) = 24.7 + \frac{f}{9.265}. \tag{6}$$

Expressions for the Bark and Mel scales are respectively provided in [51,52]. For scales that do not specify a bandwidth function, like the Mel scale, we propose the following function: $B_{\text{scale}}(f) = \frac{\partial(F_{\text{scale}}^{-1})}{\partial f}(F_{\text{scale}}(f))$. This ensures a proper overlap between the filters' passband.

3. The Proposed Approach

This section describes the analysis FB and synthesis stage of the Audlet FB. The FB is entirely designed in the frequency domain, which simplifies the assessment of properties such as invertibility and the amount of aliasing. Note that the purpose of this section is to provide a mathematical framework for general FB regardless of the practical implications. The implementation of the Audlet framework is addressed separately in Section 4.

3.1. Analysis Filter Bank

The analysis FB consists of Audlet filters H_k, $k \in \{1, \ldots, K-1\}$, a low-pass filter H_0, and a high-pass filter H_K. In total, it consists of $K+1$ filters. The Audlet filters are defined by

$$H_k(e^{2i\pi\xi}) = d_k^{\frac{1}{2}} w \left(\frac{f_s \cdot \xi - f_k}{\Gamma_k} \right) \quad k \in \{1, \ldots, K-1\} \tag{7}$$

where $w(\xi)$ is a prototype filter's shape centered at frequency 0. Any symmetric or asymmetric window is an eligible w. The main condition on w is that its frequency response must decay away from 0 on both sides. The parameters $\Gamma_k = \beta B_{scale}(f_k)$ and f_k control the bandwidth and center frequency, respectively, of the filter H_k. The parameter β allows for the filter bandwidths to be compressed/expanded. Note that when $\beta \neq 1$, the bandwidth of the filters H_k deviates from the human auditory filters' bandwidth.

To determine K and construct the sets $\{f_k\}$ and $\{\Gamma_k\}$, the first step consists in choosing an essential frequency range $[f_{\min}, f_{\max}] \subseteq [0, f_N]$, a frequency mapping F_{Scale}, and a filter density $V \in \mathbb{R}^+$ of filters per Scale unit. The set $\{d_k\}$ is considered arbitrary for now. An optimal choice of downsampling factors d_k is provided in Section 3.1.3.

3.1.1. Construction of the Set $\{f_k\}$

The center frequency f_1 is given by

$$f_1 = \max\{f_{\min}, F_{\text{Scale}}^{-1}(1/V)\} \tag{8}$$

and the subsequent f_k's are obtained iteratively by

$$f_k = F_{\text{Scale}}^{-1}(F_{\text{Scale}}(f_1) + (k-1)/V). \tag{9}$$

The iteration is processed as long as $f_k \leq f_{\max}$ and $f_k < f_N$, resulting in $K-1$ filters, with K determined by

$$K = \min \left\{ \underset{k \in \mathbb{N}}{\text{argmax}} \left(\frac{k-1}{V} \leq F_{\text{Scale}}(f_{\max}) - F_{\text{Scale}}(f_1) \right), \right.$$
$$\left. \underset{k \in \mathbb{N}}{\text{argmax}} \left(\frac{k-1}{V} < F_{\text{Scale}}(f_N) - F_{\text{Scale}}(f_1) \right) \right\}.$$

Note that $f_{\max}$ should be slightly higher than the highest frequency of interest in the analyzed signals. Finally, $f_0 = 0$ and $f_K = f_N$. At this stage, the "restricted" frequency response of the FB (i.e, restricted to the filters $H_1, \ldots, H_{K-1}$) is given by

$$\mathcal{H}_0^{(r)}(\xi) = \widetilde{\mathcal{H}_0}^{(r)}(\xi) + \overline{\widetilde{\mathcal{H}_0}^{(r)}(-\xi)}, \quad \text{with}$$

$$\widetilde{\mathcal{H}_0}^{(r)}(\xi) := \sum_{k=1}^{K-1} d_k^{-1/2} |H_k(e^{2\pi i \xi})|^2, \text{ for all } \xi \in (-1/2, 1/2].$$

To obtain a perfect reconstruction system, the frequency response of the system should optimally cover the frequency range $[0, f_N]$. However, this may not be the case for $\widetilde{\mathcal{H}_0}^{(r)}(\xi)$ because the amplitude of the filter H_1 (and/or H_{K-1}) may vanish at frequencies between 0 and f_1 (resp., between f_{K-1} and f_N). To circumvent this problem, a low-pass filter H_0 and high-pass filter H_K are included.

3.1.2. Construction of H_0 and H_K

The purpose of the filters H_0 and H_K is to stabilize the FB response $\mathcal{H}_0$ by compensating for the potentially low amplitude of $\mathcal{H}_0^{(r)}(\xi)$ in the range $[0, f_1[\cup]f_{K-1}, f_N]$. While the content in the frequency bands 0 and K might carry some perceptually relevant information, most applications will not modify the corresponding coefficients. Consequently, it is crucial that H_0 and H_K are mostly concentrated outside $[f_1, f_{K-1}]$, but their time domain behavior is only of secondary importance. Nonetheless, we propose a construction that retains some smoothness in frequency and thus, by Fourier duality, h_0 and h_K have appropriate decay.

There is no canonical method that provides optimal compensation and time localization for any valid set of Audlet parameters. In [46], for instance, plateau functions with raised cosine flanks were proposed. This method might result in additional ripples in $\mathcal{H}_0$ if w is not a raised-cosine window. Alternatively, in [47], H_0 and H_K were constructed from a set of virtual filters extending the FB beyond $[f_1, f_{K-1}]$. An adaptation of this method to the Audlet framework is unnecessarily complex and unintuitive. Instead, we propose the following. We define

$$M = \max_{\xi \in [0, 1/2]} \mathcal{H}_0^{(r)}(\xi) \quad \text{and} \quad \mathcal{H}_{\text{inv}}^{(r)} = \sqrt{(M - \mathcal{H}_0^{(r)})_+}.$$

The function $\mathcal{H}_{\text{inv}}^{(r)}$ is nonnegative and has at least the same differentiability as w (taking the positive part $(\cdot)_+$ is only necessary in the special cases considered in the remark below). However, any ripples in $\mathcal{H}_0^{(r)}$ replicate in $\mathcal{H}_{\text{inv}}^{(r)}$. To reduce this rippling effect and introduce strict band-limitation of H_0 and H_K, we multiply $\mathcal{H}_{\text{inv}}^{(r)}$ with appropriately localized plateau functions P_0 and P_K. Assume that $f_{p,s}^-, f_{p,e}^-, f_{p,s}^+, f_{p,e}^+ \in (0, f_N)$ are chosen such that $f_1 < f_{p,s}^- < f_{p,e}^- < f_{p,s}^+ < f_N$ and $f_1 < f_{p,s}^- < f_{p,e}^- < f_{p,s}^+ < f_N$ and let

$$P_0(\xi) = \begin{cases} 1/\sqrt{2} & \text{if } \xi f_s \in (-f_{p,s}^-, f_{p,s}^-) \\ \cos\left(\pi \frac{|\xi| f_s - f_{p,s}^-}{f_{p,e}^- - f_{p,s}^-}\right)/\sqrt{2} & \text{if } |\xi| f_s \in [f_{p,s}^-, f_{p,e}^-] \\ 0 & \text{elsewhere,} \end{cases}$$

and

$$P_K(\xi) = \begin{cases} 1/\sqrt{2} & \text{if } \xi f_s \in (-f_N, -f_{p,s}^+] \cup (f_{p,s}^+, f_N] \\ 1/\sqrt{2} - \cos\left(\pi \frac{|\xi| f_s - f_{p,s}^+}{f_{p,s}^+ - f_{p,e}^+}\right)/\sqrt{2} & \text{if } |\xi| f_s \in [f_{p,e}^+, f_{p,s}^+] \\ 0 & \text{elsewhere.} \end{cases}$$

The frequency $f_{p,s}^-$ (resp. $f_{p,s}^+$) defines the width of the plateau in P_0 (resp. P_K). The region $[f_{p,s}^-, f_{p,e}^-]$ ($[f_{p,e}^+, f_{p,s}^+]$) defines the transition area of P_0 (P_K) (see Figure 4). The filters H_0 and H_K are finally defined by their DTFTs as

$$H_0(e^{2\pi i(\cdot)}) = P_0 \cdot \mathcal{H}_{\text{inv}}^{(r)}, \quad \text{and} \quad H_K(e^{2\pi i(\cdot)}) = P_K \cdot \mathcal{H}_{\text{inv}}^{(r)}. \tag{10}$$

We propose selecting $0 < \kappa_1 < \kappa_2$, such that $F_{\text{Scale}}(f_{K-1}) - F_{\text{Scale}}(f_1) \geq \kappa_1 + \kappa_2$ and fix

$$f_{p,s}^- = F_{\text{Scale}}^{-1}(F_{\text{Scale}}(f_1) + \kappa_1), \quad f_{p,e}^- = F_{\text{Scale}}^{-1}(F_{\text{Scale}}(f_1) + \kappa_2),$$

$$f_{p,s}^+ = F_{\text{Scale}}^{-1}(F_{\text{Scale}}(f_{K-1}) - \kappa_1), \quad f_{p,e}^+ = F_{\text{Scale}}^{-1}(F_{\text{Scale}}(f_{K-1}) - \kappa_2).$$

This choice ensures that $P_0^2 + P_K^2 \leq 1$, preventing overcompensation, and is properly adapted to the scale used. By default, we set $\kappa_1 = 3/V, \kappa_2 = 4/V$, such that $f_{p,s}^- = f_4, f_{p,e}^- = f_5, f_{p,s}^+ = f_{K-4}, f_{p,e}^+ = f_{K-5}$. The intuition here is that from f_4 (resp. f_{K-4}) onward, the restricted FB response $\mathcal{H}_0^{(r)}$ is expected to be stable already, and that the size of the transition area ensures a sufficiently smooth roll-off. It should be noted that, although the filters proposed above are chosen to be strictly band-limited, a similar construction with time-limited, but only approximately band-limited, filters is also conceivable, by smoothly truncating h_0, h_K instead of H_0, H_K.

Remark 1. *The choice of raised cosine transition areas provides continuously differentiable P_0, P_K. If additional decay of h_0, h_K is desired, the construction of a compactly supported plateau function of arbitrary differentiability is standard, e.g., through convolution of a characteristic function with a smooth function. There are some corner cases in which one or both of the compensation filters h_0, h_K are unnecessary, namely if f_1 is very close to 0 (resp. f_{K-1} to f_N). In that case the maximum M should be computed over the interval $[f_1/f_s, 1/2]$ (resp. $[0, f_{K-1}/f_s]$) and we set $H_0 = 0$ ($H_K = 0$). A rule of thumb is if $\min_{\xi \in [0, f_1/f_s]} \mathcal{H}_0^{(r)}(\xi) \geq (1-\epsilon) \min_{\xi \in [f_1/f_s, f_{K-1}/f_s]} \mathcal{H}_0^{(r)}(\xi)$, for some $\epsilon \ll 1$, then the low-pass filter H_0 is not required. An analogous argument is valid for H_K.*

The total frequency response of the analysis FB (i.e., including the $K+1$ filters) is then

$$\mathcal{H}_0(\xi) := \widetilde{\mathcal{H}}_0(\xi) + \widetilde{\mathcal{H}}_0(-\xi), \quad \text{with} \tag{11}$$

$$\widetilde{\mathcal{H}}_0(\xi) := \sum_{k=0}^{K} d_k^{-1} |H_k(e^{2\pi i \xi})|^2, \text{ for all } \xi \in (-1/2, 1/2].$$

and the redundancy of the FB is

$$R = d_0^{-1} + 2\sum_{k=1}^{K-1} d_k^{-1} + d_K^{-1}. \tag{12}$$

The factor of 2 stems from the fact that coefficients in the 1-st to $(K-1)$-th sub-bands may be complex valued.

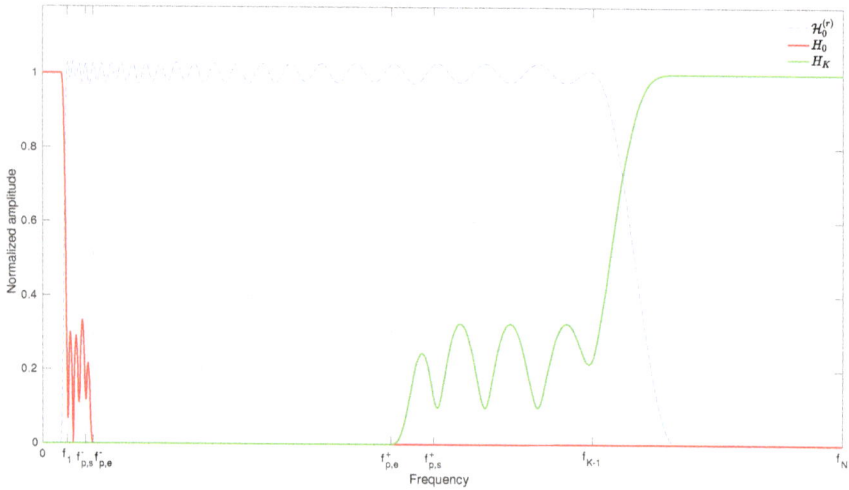

Figure 4. Illustration of the frequency allocations of the filters H_0 (red line) and H_K (green line) given the restricted frequency response $\mathcal{H}_0^{(r)}(\xi)$ (dashed line) of an FB.

3.1.3. Construction of the Set $\{d_k\}$

Downsampling the filters' outputs, i.e., using $d_k > 1$ for some or all $k \in \{0, \ldots, K\}$, has the advantage of reducing R but introduces aliasing. The amount of aliasing can be determined from the frequency domain representation of the output signal $\tilde{X}(z) = \sum_k \mathcal{Z}\left(g_k * \uparrow_{d_k} \{y_k\}\right)[n]$, also called the *alias domain* representation [16,17]. For $\xi \in (-1/2; 1/2]$, $\tilde{X}(z)$ reduces to the following ([20] Section 4)

$$\tilde{X}(e^{2i\pi\xi}) = \frac{1}{D}\left[X(e^{2i\pi(\xi+0/D)}) \cdots X(e^{2i\pi(\xi+(D-1)/D)})\right]\mathcal{H}_j(\xi) \tag{13}$$

where $D = \text{lcm}(\{d_k\}_k)$ and

$$\mathcal{H}_j(\xi) := \widetilde{\mathcal{H}}_j(\xi) + \overline{\widetilde{\mathcal{H}}_j(-\xi)} \quad \text{with} \tag{14}$$

$$\widetilde{\mathcal{H}}_j(\xi) = \sum_{\substack{k \in \{0,\ldots,K\}, \\ \text{s.t. } j \in \frac{D}{d_k}\mathbb{Z}}} d_k^{-1} H_k(e^{2\pi i\xi}) \overline{H_k(e^{2\pi i(\xi+j/D)})},$$

for all $j \in \{0, \ldots, D-1\}$. The term $\mathcal{H}_0$ in (14) represents the frequency response of the FB, while the terms $\mathcal{H}_j$, $j \neq 0$, represent the alias components. Thus, an alias-free system is obtained when $\mathcal{H}_0 = C > 0$ and $\mathcal{H}_j = 0$, $\forall j \neq 0$. While this is not always achievable, choosing d_k's to be *inversely proportional to the filters' bandwidth* yields a close-to-optimal solution [19], i.e.,

$$d_k = \left\lfloor \frac{c_{bw}f_s}{\Gamma_k} \right\rfloor \quad \text{for } k = 1, \ldots, K-1. \tag{15}$$

For a targeted redundancy R_t, combining (15) and (12) while disregarding the floor operator $\lfloor \cdot \rfloor$ leads to

$$c_{bw} = \frac{2}{R_t f_s} \sum_{k=1}^{K-1} \beta B_{\text{scale}}(f_k). \tag{16}$$

Since the H_k values are strictly decaying away from f_k with a bandwidth of Γ_k, choosing d_k's according to (15) ensures an even distribution of the overall aliasing across channels.

Using (15) to derive d_0 and d_K may result in a large amount of aliasing because H_0 and H_K may feature large plateaus depending on f_{min} and f_{max}. We propose instead choosing d_0 and d_K according to

$$d_0 = \left\lfloor \frac{f_s}{2f_{p,s}^- + \frac{\beta B_{scale}(f_{p,s}^-)}{c_{bw}}} \right\rfloor \qquad (17)$$

$$d_K = \left\lfloor \frac{f_s}{2(f_N - f_{p,s}^+) + \frac{\beta B_{scale}(f_{p,s}^+)}{c_{bw}}} \right\rfloor . \qquad (18)$$

Note that R_t controls the d_k only for $k = 1, \ldots, K - 1$, while the actual redundancy R depends on all d_k, i.e., including d_0 and d_K. As a result, the value of R is slightly larger than R_t.

3.2. Invertibility Test

Overall, the design of an Audlet analysis FB involves a set of seven parameters: the perceptual scale, frequency range $[f_{min}, f_{max}]$, filter shape w, filter density V and bandwidth factor β, and a target redundancy R_t. To check that a given parameter set results in a stable and invertible system, three methods exist:

1. An eigenvalue analysis of the linear operator corresponding to analysis with $(H_k, d_k)_k$ followed by FB synthesis with $(H_k, d_k)_k$. The frame bounds A and B correspond to the smallest (infimum) and largest (supremum) eigenvalues of the resulting operator, respectively. The largest eigenvalue can be estimated by numerical methods with reasonable efficiency but estimating the smallest eigenvalue directly is highly computationally expensive. In the next section we discuss an alternative method that consists in approximating the inverse operator and estimating its largest eigenvalue, the reciprocal of which is the desired lower frame bound A (see also Section 5 for an example frame bounds analysis).

2. Computation of A and B directly from the overall FB response, i.e., verification that $0 < A \leq \mathcal{H}_0(\xi) \leq B < \infty$ for some constants A, B and almost every $\xi \in (-1/2, 1/2]$.

3. Checking of whether the overall aliasing is dominated by $\mathcal{H}_0$, i.e., if there exist $0 < A_0 \leq B_0 < \infty$ that satisfy

$$A_0 \leq \mathcal{H}_0(\xi) \pm \sum_{j=1}^{D-1} |\mathcal{H}_j(\xi)| \leq B_0, \qquad (19)$$

for almost every $\xi \in [-1/2, 1/2]$. This method is a straightforward application of [20] (Proposition 5). The inner term in (19) can be computed or, at least, estimated by direct computation.

While method 1 above can always be applied, the applicability of methods 2 and 3 depends on w. If w is compactly supported in the interval $[a, b]$ and $0 < \frac{b-a}{\Gamma_{K-1}} \leq f_s$, $d_k \leq \frac{f_s}{(b-a)\Gamma_k}$ $\forall k \in \{1, \ldots, K-1\}$ (i.e., $c_{bw} \leq (b-a)^{-1}$), $d_0 \leq \frac{f_s}{2f_{p,e}^-}$, and $d_K^{-1} \leq \frac{f_s}{f_s - 2f_{p,e}^+}$, then the alias terms $\mathcal{H}_j$, $j \in \{1, \ldots, D-1\} = 0$. This setting corresponds to the *painless* case [53]. This is the only case when method 2 can be applied. If w has no compact support but is mostly concentrated on $[a, b]$ and decays outside, the alias terms $\mathcal{H}_j$, $j \in \{1, \ldots, D-1\}$ exist but may be small compared to $\mathcal{H}_0$. In that case, method 3 can be applied.

In terms of computational costs, method 1 is by far the most demanding of the three. Still, if a certain parameters set is used over a large number of analyses, this one-time investment to determine invertibility easily pays off. However, the user must still be aware of the potential inaccuracies induced by numerical eigenvalue computation.

3.3. Synthesis Stage

The synthesis stage consists of a linear operator $\mathcal{S}((y_k)_k)$ mapping the sub-band signals y_k to the output signal $\tilde{x}$ (see Figure 3) such that the input signal x is recovered. For uniform analysis FBs, i.e., $d_k = D \forall k$, the operator S can be structured as in Figure 2. In that case, *exact* dual filter G_k's can be computed [54] with a factorization algorithm that generalizes [23]. The synthesis is then performed by computing $\widetilde{S}((y_k)_k, (G_k, D)_k)$.

For non-uniform analysis FBs, we implement S using a *conjugate gradient* (CG) iteration [49,55,56]. This is a very efficient iterative algorithm that is guaranteed to converge when $(H_k, d_k)_k$ forms a frame, i.e., whenever stable perfect reconstruction is possible. Given the Hermitian operator $\widetilde{S}(A(x, (H_k, d_k)_k), (H_k, d_k)_k)$, the CG approximates the action of the inverse operator. For Hermitian operators, the CG converges monotonously to 0. In addition, for problems of size P, the CG is guaranteed to converge within P steps. In practice, convergence speed depends solely on the (potentially unknown) condition number of the linear problem at hand, which, in this case, equals $\sqrt{B/A}$. Often, it is beneficial to use a preconditioning step to improve the condition number. We propose the operator $\mathcal{F}^{-1} \mathrm{diag}(1/\mathcal{H}_0) \mathcal{F}$ as preconditioner (see also [48,57,58]). A robust implementation of the appropriate preconditioned CG (PCG) algorithm is conceptually straightforward and was provided in [48].

In the following, we describe a heuristic variant of this PCG algorithm with asymmetric preconditioning, that enables efficient implementation even if the intermediate solutions (denoted by x_j below) are only given in the time domain. Experimentally, Algorithm 1 was observed to converge in the same number of iterations as the robust implementation from [48] (divergence was observed only if the filters H_k were set to uniformly distributed random noise). We denote the analysis of x with respect to the analysis FB $(H_k, d_k)_k$ by $(y_k)_k = A(x, (H_k, d_k)_k)$. We denote the synthesis from $(y_k)_k$ with respect to the synthesis FB $(G_k, d_k)_k$ by $\tilde{x} = \widetilde{S}((y_k)_k, (G_k, d_k)_k)$. The composition $\tilde{x} = \widetilde{S}(A(x, (H_k, d_k)_k), (G_k, d_k)_k)$ thus represents analysis followed by synthesis.

Algorithm 1 Synthesis by means of conjugate gradients

Initialize $(H_k, d_k)_k$, $(y_k)_k$
$\qquad\qquad x_0 \in \ell_2(\mathbb{Z})$ (arbitrary)
$\qquad\qquad j = 0$ and $\varepsilon > 0$ (error tolerance)
$\mathcal{H}_0 \leftarrow \sum_k d_k^{-1} H_k$
for $k = 0, \dots, K+1$ **do**
$\qquad G_k \leftarrow H_k / \mathcal{H}_0$
end for
$b \leftarrow \widetilde{S}((y_k)_k, (G_k, d_k)_k)$
$r_0 \leftarrow b - \widetilde{S}(A(x_0, (H_k, d_k)_k), (G_k, d_k)_k)$
$p_0 \leftarrow r_0$
while $r_j > \varepsilon$ **do**
$\qquad q_j \leftarrow \widetilde{S}(A(p_j, (H_k, d_k)_k), (G_k, d_k)_k)$
$\qquad a_j \leftarrow |r_j|^2 / \langle p_j, q_j \rangle$
$\qquad x_{j+1} \leftarrow x_j + a_j p_j$
$\qquad r_{j+1} \leftarrow r_j - a_j q_j$
$\qquad b_j \leftarrow |r_{j+1}/r_j|^2$
$\qquad p_{j+1} \leftarrow r_{j+1} + b_j p_j$
$\qquad j \leftarrow j+1$
end while

To speed up convergence we use *approximate dual filters* as an initial choice for G_k's,

$$G_k(e^{2i\pi\xi}) := \frac{H_k(e^{2i\pi\xi})}{\mathcal{H}_0(\xi)}. \tag{20}$$

We interpret G_k's as approximate dual filters because in the absence of aliasing (i.e., if $\mathcal{H}_j = 0$, $\forall j \neq 0$), the application of G_k exactly cancels all ripples in the frequency response $\mathcal{H}_0$.

Hence, the analysis-synthesis system $\widetilde{S}(A(x, (H_k, d_k)_k), (G_k, d_k)_k)$ can be interpreted as a preconditioned variant of $\widetilde{S}(A(x, (H_k, d_k)_k), (H_k, d_k)_k)$ [48,57,58].

Note that in the painless case, evoked in Section 3.2, the operator $\widetilde{S}(A(x, (H_k, d_k)_k), (G_k, d_k)_k)$ equals the identity and thus, synthesis is performed simply by applying $\widetilde{S}((y_k)_k, (G_k, d_k)_k)$ once.

Although this is not apparent from the iterative inversion scheme described above, the proposed synthesis stage acts in a similar fashion to an FB. More specifically, if $D = \text{lcm}(\{d_k\}_k)$ and $(\tilde{H}_j, D)_j$ is the equivalent uniform FB associated with $(H_k, d_k)_k$ [16,20], then iterating the CG algorithm until convergence is equivalent to computing the FB synthesis with respect to the canonical dual FB of $(\tilde{H}_j, D)_j$, which is of the form $(\tilde{G}_j, D)_j$, for some sequences of filters $(\tilde{G}_j)_j$ (see [21]). Since convergence is achieved within numerical precision in a small number of CG steps we can assume that the proposed synthesis system is characterized by the properties of the filters $(\tilde{G}_j)_j$. We cannot easily compute those filters, but it is well known that the ratio of the optimal frame bounds B/A (see Section 3.2) is closely related to the similarity of a system and its canonical dual [59]. If $B/A \approx 1$, then we can expect $\tilde{G}_j \approx \tilde{H}_j$, for all j. Since each $\tilde{H}_j$ is just a delayed version of some H_k, the time- and frequency-domain localization of the synthesis system matches that of the analysis system.

For larger values of B/A, the duality of $(\tilde{H}_j, D)_j$ and $(\tilde{G}_j, D)_j$ implies that $(\tilde{G}_j, D)_j$ has to account for the discrepancies of $(\tilde{H}_j, D)_j$ [59]. These considerations apply to any dual FB pair, Audlet or not. The Audlet FB is constructed in such a way that, given the prototype filter w and filter density V, the frequency response of $(H_k, d_k)_k$ is as flat as possible, such that the B/A depends mostly on the presence of aliasing. The required aliasing compensation often implies a widening of the dual filters' essential support and essential passband, proportional to the amount of aliasing present.

4. Implementation

4.1. Practical Issues

The general mathematical framework described in the previous section is valid for band-limited filters and more classical FIR filters. Although the impulse responses of band-limited filters are theoretically infinite, their decay can be controlled by design such that they can be truncated with a minor loss of precision. In our implementation, we instead choose an alternative approach similar to "fast Fourier transform (FFT) filter banks" proposed by Smith [60]. We start by considering the input signal as a finite-length vector in $\mathbb{R}^L$, $L \in \mathbb{N}$. In an overlap-add block-processing scheme like the one proposed in [46,60], such a sequence would be a single windowed block possibly zero-padded on both ends. In the offline setting assumed in this paper, the sequence represents the entire input signal. We discretize the continuous frequency ξ by assuming the sequence is one period of an L-periodic signal. This introduces circular boundary effects that can be diminished by zero padding (increasing L), provided the filters' impulse responses decay rapidly. Increasing L preserves the perfect reconstruction property. Such assumptions allow implementing the filtering, downsampling, and upsampling directly in the frequency domain using sampled frequency responses of analysis and synthesis filters H_k and G_k, respectively. The filtering with an analysis filter followed by downsampling is done using the standard point-wise product of the L-point FFT of the signal with a sampled frequency response, while the downsampling is achieved by folding the result to a sequence of length L/d_k (manual aliasing) and performing L/d_k-point inverse FFT (IFFT). Performing downsampling this way is exactly equivalent to time-domain downsampling by a factor of d_k. Upsampling and filtering is achieved by taking a L/d_k-point FFT of the sub-band, periodizing the result to length L followed by a point-wise product with the sampled frequency response of a synthesis filter. A final L-point IFFT brings the result back to the time domain. In this framework, working with strictly band-limited filters is even advantageous for two reasons. First, the frequency domain point-wise product can be restricted to the filter bandwidth and second, for band-limited filters, the parameters can be chosen such that the system is painless [53] (no aliasing is introduced by downsampling), for which the approximate dual filters from (20) are *exact* and thus achieve perfect reconstruction.

4.2. Code

We provide code for performing an Audlet analysis/synthesis as part of the Matlab/Octave "large time-frequency analysis toolbox (LTFAT)" toolbox [61,62] available at http://ltfat.github.io/. The analysis filters are generated by the function audfilters. The function allows to construct at will uniform or non-uniform Audlet FBs with integer or rational downsampling factors, thus offering flexibility in FB design. Rational downsampling factors can be achieved in the time domain by properly combining upsamplers and downsamplers (e.g., [19]). In LTFAT the sampling rate changes are directly performed in the frequency domain by periodizing and folding the $Y_k(z)$'s, then performing an inverse DFT [63]. This technique allows to achieve rational downsampling factors at low computational costs. The desired number of channels in the frequency range $[f_{min}, f_{max}]$ can be set by specifying either K or V. The function audfilters also accepts parameters Scale, β, w, and R_t. Currently, three scales (ERB—the default—as well as Bark and Mel) are available. Possible choices of w include (but are not limited to) Hann (default), Blackman, Nuttall, gammatone, or Gaussian. If R_t is specified, c_{bw} is inferred from R_t according to (15)–(18). Otherwise $c_{bw} = 1$. The analysis of a signal is performed by filterbank. The synthesis is performed by ifilterbankiter that implements Algorithm 1. In the painless case, the more computationally efficient synthesis can be achieved by first computing the *exact* synthesis FB with filterbankdual and then synthesizing the signal with ifilterbank. The function filterbankdual can also be used to check whether a given analysis FB qualifies for the painless case.

Example scripts to perform Audlet analyses/syntheses in various FB settings are provided as Supplementary Material (see Archive S1). The supplementary material also demonstrates the realization of iterative reconstruction.

Note that for real-time implementations using macro blocks like in [46], the overall redundancy depends also on the overlap between the blocks. For analysis or processing purposes, the sub-bands can be combined in an overlap-add manner closely approximating the true non-blocked sub-bands. The perfect reconstruction property within the blocks is preserved.

4.3. Computational Complexity

In [45,64] it was shown that the frequency-domain computation of an FB analysis (H_k, d_k) is $\mathcal{O}(L \log L)$, obtained as the sum of: (1) an L-point FFT ($\mathcal{O}(L \log L)$); (2) point-wise multiplication with the filter frequency responses ($\sum_k L_k$); and (3) an L/d_k-point IFFT ($\mathcal{O}(\sum_k L/d_k \log L/d_k)$) for each filter, and similarly for FB synthesis with respect to (H_k, d_k). Here, L_k denotes the bandwidth of H_k in samples.

In the painless case, the same analysis applies to the dual FB $(G_k, d_k)_k$. In general, every iteration of the CG has the complexity of FB analysis with (H_k, d_k) followed by FB synthesis with (G_k, d_k). For a given analysis system (H_k, d_k), the number of iterations required for numerical convergence relies only on the frame bound ratio B/A and is completely independent of the signal under scrutiny (see also [48] for a visualization of convergence in various settings).

5. Evaluation

In this section we evaluate three important properties of the Audlet, namely its simple and versatile FB design, perfect reconstruction, and utility for audio applications that perform sub-channel processing. This evaluation comprises two parts:

1. The construction of uniform and non-uniform gammatone FBs and examination of their stability and reconstruction property at low and high redundancies. For this purpose we replicated the simulations described in [44] (Section IV), which we consider as state of the art.

2.   The construction of various analysis–synthesis systems and use to perform sub-band processing. For this purpose we considered the example application of audio source separation because it is intuitive, clear, and it easily demonstrates the behavior of the system when attempting modification of an audio signal. In this application we assess the effects of perfect reconstruction, bandwidth and shape of the filters, and auditory scale on the quality of sub-channel processing.

Scripts to reproduce the results of these evaluations are provided as Supplementary Material (see Archive S1).

5.1. Construction of Perfect-Reconstruction Gammatone FBs

5.1.1. Method

To construct a gammatone FB we use the prototype filter shape in the frequency domain of a complex gammatone filter of order γ centered at zero [42,43]

$$H_{GT,\gamma,\alpha}(e^{2i\pi\xi}) = \left(1 + i\alpha^{-1}\xi\right)^{-\gamma}. \tag{21}$$

An order $\gamma = 4$ and bandwidth factor $\alpha = 1.019$ are usually chosen for emulating the human auditory filters [38]. Because $H_{GT,\gamma,\alpha}$ has an infinite support in the frequency domain, it can be truncated to become a compactly-supported gammatone filter shape by

$$w_{csGT,\gamma,\alpha}(\xi) = \begin{cases} H_{GT,\gamma,\alpha}(e^{2i\pi\xi}) & \text{if } |H_{GT,\gamma,\alpha}(e^{2i\pi\xi})| \geq \epsilon, \\ 0 & \text{otherwise.} \end{cases} \tag{22}$$

where ϵ is a threshold that allows to trade accuracy for computational efficiency. Once an essential frequency range and a filter density are chosen, the set of gammatone filters is generated according to (7) using $w(\xi) = H_{GT,\gamma,\alpha}(e^{2i\pi\xi})$ (or $w = w_{csGT,\gamma,\alpha}$ if a painless system is desired) and $\beta = 1$. In Figure S2 in supplementary material, the frequency response and impulse response of two gammatone filters computed using (7) and (21) with center frequencies $f_k = 258$ and 4000 Hz are displayed.

To examine the stability and reconstruction property of the proposed gammatone construction, we replicated the two simulations described in [44] (Section IV). The first simulation considers uniform FBs and the second simulation considers non-uniform FBs. The uniform FBs were evaluated by two measures: the ratio B/A and reconstruction error in terms of signal-to-noise ratio (SNR). The non-uniform FBs were evaluated only by the SNR. We compared our results to those from Strahl and Mertins (S–M) [44] where available.

The FB settings were as follows. The sampling rate was $f_s = 44.1$ kHz, the essential frequency range was $[f_{min} = 20$ Hz, $f_{max} = 20000$ Hz$]$, and the scale was ERB. The gammatone filters in [44] were implemented as FIR filters, that is, the H_k's had an infinite frequency response. Thus, in the following simulations we used $w(\xi) = H_{GT,4,1.019}(e^{2i\pi\xi})$. In the uniform case, the downsampling factors d_k's, $k \in \{1, \ldots, K-1\}$, were set to a constant D; d_0 and d_K were chosen according to (17) and (18), respectively. The evaluation was performed for all combinations of $D \in \{1, 2, 4, 6, 8\}$ and $K \in \{51, 76, 101, 151\}$ (our K corresponds to $M + 1$ in [44]). For the synthesis stage, Algorithm 1 was used with an error tolerance $\varepsilon = 10^{-9}$. The ratio B/A was calculated for the *full* frequency range (i.e., from 0 to f_N) by iteratively computing the eigenvalues of the operator S associated with the system $(H_k, d_k)_k$ [65]. The SNR was calculated as $||x||^2/||x - \tilde{x}||^2$ in dB for x being a Gaussian white noise with a length of 30,000 samples.

In the non-uniform case, K was fixed to 51 and the FBs were evaluated for various values of R. We considered the oversampling factors $O \in \{1, 2, 4, 6, 8\}$ used in [44]. The relationship between O and R is $O = R/2 - \frac{1}{2}(d_0^{-1} + d_K^{-1})$ because in [44], O was $\sum_{k=1}^{K-1} d_k^{-1}$, which considers only the real part of the coefficients, and h_0 and h_K were not included. For simplicity, our FBs were designed for

$R_t \in \{2, 4, 8, 12, 16\}$. Similar to [44], two sets of d_k were used to achieve the various R_t's. The first set consisted of d_k's that were inversely proportional to the filters' bandwidth according to (15). The second set was exactly that mentioned in [44] (Appendix B). For each set, d_0 and d_k were chosen according to (17) and (18), respectively. All other FB and signal parameters were as in the uniform case.

5.1.2. Results and Discussion

The ratios B/A computed for the uniform gammatone FBs for various combinations of D and K and those reported in [44] (Figure 5) are listed in Table 1. For $K = 51$–101, our ratios B/A decreased with increasing K. This is a consequence of the increasing overlap between filters with increasing K, which in turn yields a flatter FB response. Increasing K to 151 did not result in smaller ratios. This can be attributed to the steep flank of H_K in that setting. This can be counteracted by increasing the values of κ_1 and κ_2 when very small filter spacing V (equivalently, large K) is used. Our framework generally achieved comparable or smaller ratios than those from [44]. Note that in [44], B/A was calculated for the frequency range from 0.06 to 17 kHz. These ratios, when calculated for the full frequency range, might have been larger than those listed in Table 1. Consequently, the actual difference between Audlet and S–M ratios might be larger than that reflected in Table 1.

Table 1. Ratios B/A for various combinations of D and K obtained for the proposed Audlet framework and reported in [44] (S–M).

K	Framework	$D = 1$	$D = 2$	$D = 4$	$D = 6$	$D = 8$
51	Audlet	1.124	1.124	1.125	1.134	1.157
	S–M	1.100	> 10	> 10	> 10	> 10
76	Audlet	1.007	1.007	1.009	1.021	1.073
	S–M	1.100	2	2	3	6
101	Audlet	1.003	1.003	1.005	1.017	1.068
	S–M	1.003	1.003	1.003	2	4
151	Audlet	1.015	1.015	1.016	1.025	1.066
	S–M	1.003	1.003	1.003	1.100	2

The SNRs achieved with our framework were 180 dB (or larger) for all tested combinations of D and K. The limit of 180 dB is the consequence of the error tolerance of 10^{-9} in the PCG algorithm. In comparison, SNRs reported in [44] for $D = 1$ ranged between 30 and 72 dB and increased with increasing K (SNRs for other D's were not reported).

The SNRs computed for the non-uniform gammatone FBs for various R are listed in Table 2 together with those reported in [44]. In all conditions, our framework achieved SNRs of at least 170 dB. In contrast, the system from [44] offered decent reconstruction (SNR $\geq$ 15 dB) only in configurations involving small downsampling factors (i.e., at large R).

Table 2. Signal-to-noise ratios (SNRs; in dB) obtained for the Audlet framework and reported in [44] (Figure 10) (S–M).

	d_k Based on (15)–(18)			d_k from [44]		
R_t	R	Audlet	S–M	R	Audlet	S–M
2	2.40	> 180	5	2.38	> 170	10
4	4.46	> 180	7	4.38	> 190	13
8	8.60	> 180	10	8.38	> 200	17
12	12.73	> 220	9	12.38	> 210	18
16	16.87	> 260	15	16.38	> 200	19

Overall, we conclude that the reconstruction quality of currently available gammatone FB implementations deteriorates at low redundancies. This may hinder the quality of sub-channel processing in audio applications but, as it seems, the reconstruction quality can be improved by using the Audlet framework.

It might appear intriguing that we obtained larger SNRs than in [44] even in conditions with similar ratios B/A (compare the condition with $K = 151$ and $D = 1$ in Table 1). The good performance achieved by our framework can mostly be explained by the design of our synthesis stage. In contrast, most analysis–synthesis systems based on gammatone filters, such as [44], use synthesis filters that are time-reversed versions of the analysis filters, i.e., $G_k(e^{2i\pi\xi}) = \overline{H_k}(e^{2i\pi\xi})$ that translates to $g_k[n] = \overline{h_k}[-n]$ in the discrete-time domain (e.g., [35,39,40]). Such a synthesis stage provides perfect reconstruction if and only if the frame bound ratio is equal to one [20].

5.2. Utility for Audio Applications

5.2.1. Method

This experiment is an example application of the Audlet framework to audio source separation. Given a mixture of instrumental music and voice, we constructed various analysis–synthesis systems and separated the voice from the music. The systems were designed so as to assess the effects of perfect reconstruction, shape and bandwidth of the filter, and auditory scale on the quality of sub-channel processing at low, mid, and high redundancies. Four systems were implemented:

trev_gfb: a state-of-the-art gammatone FB with approximate reconstruction (the acronym **trev** stands for "time reversal"). The H_k's followed (7) with $w(\xi) = w_{csGT,4,1.019}(\xi)$ (22) with a threshold $\epsilon = 10^{-5}$. The synthesis filters $G_k(e^{2i\pi\xi}) = \overline{H_k}(e^{2i\pi\xi})$. This corresponds to the baseline system used in audio applications like [11,28,29].

Audlet_gfb: an Audlet FB with a gammatone prototype. The H_k's were computed as in **trev_gfb** but the synthesis stage was Algorithm 1. This system aims to compare to the baseline system and assess the effect of perfect reconstruction.

Audlet_hann: an Audlet FB with a Hann prototype. This system aims to assess the effect of filter shape.

STFT_hann: an STFT using a 1024-point Hann window. Synthesis was achieved by the dual window [2]. The time step was then adapted to match the desired redundancy R_t. This corresponds to the baseline system used in most audio applications (e.g., [10,66]). This system aims to assess the use of an auditory frequency scale.

The effect of filter bandwidth was assessed by varying parameter β. Specifically, two values were tested: $\beta \in \{1, 1/6\}$. Using a value of $\beta \neq 1$ means a clear departure from auditory perception but may help better resolve spectral components, particularly at high frequencies where the auditory filters become really broad (see (6)). Accordingly, many audio applications that rely on constant-Q or wavelet transforms use 12 or more bins per octave (e.g., [46,63]).

The performance of all systems were evaluated at three redundancies: $R_t \in \{1.1, 1.5, 4\}$. To this end, (15) was used with c_{bw} adjusted such that R_t was achieved. The quality of the separation was assessed by computing energy ratio- and perceptually-based objective measures according to [67]. Energy ratio measures include the signal-to-distortion ratio (SDR) and signal-to-artifact ratio (SAR). Perceptual measures include the overall perceptual score (OPS) and target perceptual score (TPS). OPS assesses the general audio quality of the separation, while TPS assesses the preservation of the target. All measures were computed using the PEASS toolbox [67].

The following parameters were fixed for systems trev_gfb, Audlet_gfb and Audlet_hann: $f_s = 22.05$ kHz, $[f_{min}, f_{max}] = [20, 10,000]$, Scale = ERB, and $K = 209$ filters corresponding to $V = 6$ filters/ERB.

The signal mixture, shown in Figure 5a, was created by adding an instrumental music signal to a singing voice signal (target), shown in Figure 5b. The separation was performed by analyzing the mixture with the analysis FB, applying a binary TF mask to the sub-band components by point-wise multiplication, and computing the output signal from the modified sub-band components using the synthesis stage. This operation corresponds to the application of a frame multiplier in signal processing [9,68]. In order to create the binary masks, the target signal was analyzed by the FB and the magnitude of the coefficients was hard thresholded with a threshold of –25 dB. The threshold value was varied between −40 and −20 dB in 5-dB steps. While the threshold value did affect the separation performance, all configurations were affected equally. The value of –25 dB was selected because it yielded good separation results for both the gammatone and Hann prototypes. Four masks were created in total, one for each analysis filter's shape and each β. The two masks for $\beta = 1/6$ are displayed in Figure 5c,d. Because the frequency resolution of the STFT does not match those of other FBs, an additional mask was computed for the STFT.

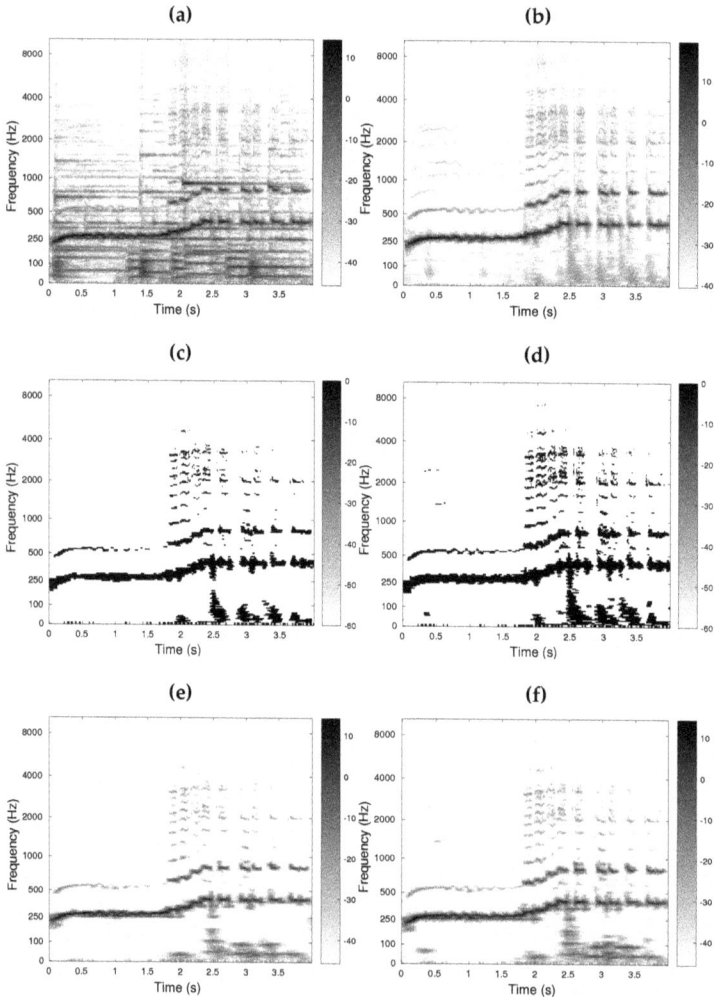

Figure 5. Source separation for $R_t = 4$ and $\beta = 1/6$ displayed as time-frequency (TF) plots: the magnitude of each sub-band component (in dB) as a function of time (in s). (**a**) Shows the mixture analyzed by a gammatone FB; (**b**) Shows the target (voice) analyzed by a gammatone FB; (**c**) Shows the binary mask obtained for Audlet_hann; (**d**) Shows the binary mask obtained for trev_gfb and Audlet_gfb—the black and white dots in the masks represent '1' and '0' entries, respectively; (**e**,**f**) Show the target separated by Audlet_hann and Audlet_gfb, respectively.

5.2.2. Results and Discussion

Figure 5e,f show the voice signal separated using Audlet_hann and Audlet_gfb, respectively, for $R_t = 4$ and $\beta = 1/6$. The objective quality measures are listed in Table 3. Audio files are available on the companion webpage: http://www.kfs.oeaw.ac.at/audletFB. The following observations can be made.

First, system Audlet_gfb outperformed trev_gfb in most conditions. This demonstrates the role of perfect reconstruction in the quality of sub-channel processing. In other words, using the Audlet framework can improve the reconstruction quality. Note that for $\beta = 1/6$, the performance of trev_gfb improved with increasing R_t and tended towards the performance of Audlet_gfb. This is due to the decrease in the amount of aliasing with increasing R_t. For trev_gfb and $R_t = 4$, very little aliasing was present and a good performance was achieved despite the approximate reconstruction of trev_gfb.

Second, the performance of Audlet_hann was comparable to that of Audlet_gfb in almost every measure. Although the filter shape did not play a major role in this particular example, it may have a larger impact in other applications.

Third, for all configurations, reducing β from 1 to 1/6 generally improved all quality measures. This suggests that, depending on the application, a departure from the human auditory perception may improve signal processing performance. In the present application, for instance, finely tuned filters are required to resolve all harmonics and therefore properly separate the signals.

Finally, while STFT_hann performed comparably to Audlet_hann at the highest R, the performance of STFT_hann dropped at mid and low redundancies. This suggests that using an auditory frequency scale may improve signal processing performance at low redundancies.

Table 3. Objective quality measures for the separated voice signal. The signal-to-distortion ratio (SDR) and signal-to-artifact ratio (SAR) are in dB; the larger the ratio, the better the separation result. Overall perceptual score (OPS) and target perceptual score (TPS) are without unit; they indicate scores between 0 (bad quality) and 1 (excellent quality). The corresponding audio files are available on the companion webpage. STFT: short-time Fourier transform.

System	R_t	SDR		SAR		OPS		TPS	
		$\beta = 1$	1/6	1	1/6	1	1/6	1	1/6
trev_gfb		0.1	5.8	3.2	9.2	0.26	0.26	0.06	0.12
Audlet_gfb	1.1	4.7	10.7	8.5	19.0	0.25	0.31	0.11	0.20
Audlet_hann		4.7	11.8	7.6	18.3	0.26	0.34	0.05	0.26
STFT_hann		−1.7		0.5		0.46		0.02	
trev_gfb		2.4	8.5	5.7	13.5	0.24	0.30	0.11	0.17
Audlet_gfb	1.5	6.9	11.1	12.5	20.5	0.24	0.35	0.13	0.29
Audlet_hann		7.0	12.8	11.1	20.1	0.22	0.36	0.07	0.35
STFT_hann		2.4		9.2		0.22		0.04	
trev_gfb		7.0	10.7	12.0	18.9	0.24	0.37	0.24	0.34
Audlet_gfb	4	9.0	11.4	18.3	21.6	0.27	0.38	0.32	0.39
Audlet_hann		11.1	13.1	19.4	21.7	0.25	0.37	0.21	0.32
STFT_hann		11.4		20.5		0.38		0.34	

6. Conclusions

A framework for the construction of oversampled perfect-reconstruction FBs with filters distributed on auditory frequency scales has been presented. This framework was motivated by auditory perception and targeted at audio signal processing; it has thus been named "Audlet". The proposed approach has its foundation in the mathematical theory of frames. The analysis FB design is directly performed in the frequency domain and allows for various filter shapes, and uniform or non-uniform settings with low redundancies. The synthesis is achieved using a (heuristic) preconditioned conjugate-gradient iterative algorithm. The convergence of the algorithm has been observed for Audlet FBs that constitute a frame. This is possible even for redundancies close to 1. For higher redundancies and filters with a compact support in the frequency domain, a so-called "painless" system can be achieved. In this case the exact dual FB can be calculated, which in turn results in a computationally more efficient synthesis.

We showed how to construct a gammatone FB with perfect reconstruction. The proposed gammatone FB was compared to widely used state-of-the-art implementations of gammatone FB with approximate reconstruction. The results showed the better performance of the proposed approach in terms of reconstruction error and stability, especially at low redundancies. An example application of the framework to the task of audio source separation demonstrated its utility for audio processing.

Overall, the Audlet framework provides a versatile and efficient FB design that is highly suitable for audio applications requiring stability, perfect reconstruction, and a flexible choice of redundancy. The framework is implemented in the free Matlab/Octave toolbox LTFAT [61,62].

Supplementary Materials: Supplementary material available online at www.mdpi.com/2076-3417/8/1/96/s1 is provided by the authors. Archive S1: Matlab functions and test audio files to perform Audlet analyses/syntheses in various FB settings and reproduce all results presented in the manuscript. The archive, about 2.6 MB in size, also includes a brief documentation. Figure S2: Frequency response and impulse response of two gammatone filters computed using the proposed framework.

Acknowledgments: The authors would like to thank Damián Marelli for insightful discussions and help on the theoretical development on non-uniform FBs. This work was partly supported by the Austrian Science Fund (FWF) START-project FLAME ("Frames and Linear Operators for Acoustical Modeling and Parameter Estimation"; Y 551-N13), the French-Austrian ANR-FWF project POTION ("Perceptual Optimization of Time-Frequency Representations and Audio Coding; I 1362-N30"), and the Austrian-Czech FWF-GAČR project MERLIN ("Modern methods for the restoration of lost information in digital signals; I 3067-N30"). Open access publication costs were covered by FWF.

Author Contributions: T.N., P.B and N.H. conceived the study; N.H., T.N. and Z.P. wrote the software; T.N. and N.H. conceived and performed the experiments; T.N., N.H., P.B., P.M. and O.D. analyzed the data; Z.P. contributed software and analysis tools; T.N. wrote the original draft; T.N., N.H., P.B., Z.P., P.M. and O.D. reviewed and edited the manuscript.

Conflicts of Interest: The authors declare no conflict of interest.

References

1. Flandrin, P. *Time-Frequency/Time-Scale Analysis*; Wavelet Analysis and Its Application; Academic Press: San Diego, CA, USA, 1999; Volume 10.
2. Gröchenig, K. *Foundations of Time-Frequency Analysis*; Birkhäuser: Boston, MA, USA, 2001.
3. Kamath, S.; Loizou, P. A multi-band spectral subtraction method for enhancing speech corrupted by colored noise. In Proceedings of the 2002 IEEE International Conference on Acoustics, Speech, and Signal Processing, Orlando, FL, USA, 13–17 May 2002; Volume 4.
4. Majdak, P.; Balazs, P.; Kreuzer, W.; Dörfler, M. A time-frequency method for increasing the signal-to-noise ratio in system identification with exponential sweeps. In Proceedings of the 2011 IEEE International Conference on Acoustics, Speech and Signal Processing (ICASSP), Prague, Czech Republic, 22–27 May 2011.
5. International Organization for Standardization. *ISO/IEC 11172-3: Information Technology—Coding of Moving Pictures and Associated Audio for Digital Storage Media at up to About 1.5 Mbits/s, Part 3: Audio*; Technical Report; International Organization for Standardization (ISO): Geneva, Switzerland, 1993.
6. International Organization for Standardization. *ISO/IEC 13818-7: 13818-7: Generic Coding of Moving Pictures and Associated Audio: Advanced Audio Coding*; Technical Report; International Organization for Standardization (ISO): Geneva, Switzerland, 1997.
7. International Organization for Standardization. *ISO/IEC 14496-3/AMD-2: Information Technology—Coding of Audio-Visual Objects, Amendment 2: New Audio Profiles*; Technical Report; International Organization for Standardization (ISO): Geneva, Switzerland, 2006.
8. Průša, Z.; Holighaus, N. Phase vocoder done right. In Proceedings of the 25th European Signal Processing Conference (EUSIPCO–2017), Kos Island, Greece, 28 August–2 September 2017; pp. 1006–1010.
9. Sirdey, A.; Derrien, O.; Kronland-Martinet, R. Adjusting the spectral envelope evolution of transposed sounds with gabor mask prototypes. In Proceedings of the 13th International Conference on Digital Audio Effects (DAFx-10), Graz, Austria, 10 September 2010; pp. 1–7.

10. Leglaive, S.; Badeau, R.; Richard, G. Multichannel Audio Source Separation with Probabilistic Reverberation Priors. *IEEE/ACM Trans. Audio Speech Lang. Process.* **2016**, *24*, 2453–2465.

11. Gao, B.; Woo, W.L.; Khor, L.C. Cochleagram-based audio pattern separation using two-dimensional non-negative matrix factorization with automatic sparsity adaptation. *J. Acoust. Soc. Am.* **2014**, *135*, 1171–1185.

12. Unoki, M.; Akagi, M. A method of signal extraction from noisy signal based on auditory scene analysis. *Speech Commun.* **1999**, *27*, 261 – 279.

13. Bertin, N.; Badeau, R.; Vincent, E. Enforcing Harmonicity and Smoothness in Bayesian Non-Negative Matrix Factorization Applied to Polyphonic Music Transcription. *IEEE Trans. Audio Speech Lang. Process.* **2010**, *18*, 538–549.

14. Cvetković, Z.; Johnston, J.D. Nonuniform oversampled filter banks for audio signal processing. *IEEE Speech Audio Process.* **2003**, *11*, 393–399.

15. Smith, J.O. Spectral Audio Signal Processing. Online Book. 2011. Available online: http://ccrma.stanford. edu/~jos/sasp/ (accessed on 9 January 2018).

16. Akkarakaran, S.; Vaidyanathan, P. Nonuniform filter banks: New results and open problems. In *Beyond Wavelets*; Studies in Computational Mathematics; Elsevier: Amsterdam, The Netherlands, 2003; Volume 10, pp. 259–301.

17. Vaidyanathan, P. *Multirate Systems And Filter Banks*; Electrical Engineering, Electronic and Digital Design; Prentice Hall: Englewood Cliffs, NJ, USA, 1993.

18. Vetterli, M.; Kovačević, J. *Wavelets and Subband Coding*; Prentice Hall PTR: Englewood Cliffs, NJ, USA, 1995.

19. Kovačević, J.; Vetterli, M. Perfect reconstruction filter banks with rational sampling factors. *IEEE Trans. Signal Process.* **1993**, *41*, 2047–2066.

20. Balazs, P.; Holighaus, N.; Necciari, T.; Stoeva, D. Frame theory for signal processing in psychoacoustics. In *Excursions in Harmonic Analysis*; Applied and Numerical Harmonic Analysis; Birkäuser: Basel, Switzerland, 2017; Volume 5, pp. 225–268.

21. Bölcskei, H.; Hlawatsch, F.; Feichtinger, H. Frame-theoretic analysis of oversampled filter banks. *IEEE Trans. Signal Process.* **1998**, *46*, 3256–3268.

22. Cvetković, Z.; Vetterli, M. Oversampled filter banks. *IEEE Trans. Signal Process.* **1998**, *46*, 1245–1255.

23. Strohmer, T. Numerical algorithms for discrete Gabor expansions. In *Gabor Analysis and Algorithms: Theory and Applications*; Feichtinger, H.G., Strohmer, T., Eds.; Birkäuser: Boston, MA, USA, 1998; pp. 267–294.

24. Härmä, A.; Karjalainen, M.; Savioja, L.; Välimäki, V.; Laine, U.K.; Huopaniemi, J. Frequency-Warped Signal Processing for Audio Applications. *J. Audio Eng. Soc.* **2000**, *48*, 1011–1031.

25. Gunawan, T.S.; Ambikairajah, E.; Epps, J. Perceptual speech enhancement exploiting temporal masking properties of human auditory system. *Speech Commun.* **2010**, *52*, 381 – 393.

26. Balazs, P.; Laback, B.; Eckel, G.; Deutsch, W.A. Time-Frequency Sparsity by Removing Perceptually Irrelevant Components Using a Simple Model of Simultaneous Masking. *IEEE Trans. Audio Speech Lang. Process.* **2010**, *18*, 34–49.

27. Chardon, G.; Necciari, T.; Balazs, P. Perceptual matching pursuit with Gabor dictionaries and time-frequency masking. In Proceedings of the 39th International Conference on Acoustics, Speech, and Signal Processing (ICASSP 2014), Florence, Italy, 4–9 May 2014.

28. Wang, D.; Brown, G.J. *Computational Auditory Scene Analysis: Principles, Algorithms, and Applications*; Wiley-IEEE Press: Hoboken, NJ, USA, 2006.

29. Li, P.; Guan, Y.; Xu, B.; Liu, W. Monaural Speech Separation Based on Computational Auditory Scene Analysis and Objective Quality Assessment of Speech. *IEEE Trans. Audio Speech Lang. Process.* **2006**, *14*, 2014–2023.

30. Glasberg, B.R.; Moore, B.C.J. Derivation of auditory filter shapes from notched-noise data. *Hear. Res.* **1990**, *47*, 103–138.

31. Rosen, S.; Baker, R.J. Characterising auditory filter nonlinearity. *Hear. Res.* **1994**, *73*, 231–243.

32. Lyon, R. All-pole models of auditory filtering. *Divers. Audit. Mech.* **1997**, pp. 205–211.

33. Irino, T.; Patterson, R.D. A Dynamic Compressive Gammachirp Auditory Filterbank. *Audio Speech Lang. Process.* **2006**, *14*, 2222–2232.

34. Verhulst, S.; Dau, T.; Shera, C.A. Nonlinear time-domain cochlear model for transient stimulation and human otoacoustic emission. *J. Acoust. Soc. Am.* **2012**, *132*, 3842–3848.

35. Feldbauer, C.; Kubin, G.; Kleijn, W.B. Anthropomorphic coding of speech and audio: A model inversion approach. *EURASIP J. Adv. Signal Process.* **2005**, *2005*, 1334–1349.

36. Decorsière, R.; Søndergaard, P.L.; MacDonald, E.N.; Dau, T. Inversion of Auditory Spectrograms, Traditional Spectrograms, and Other Envelope Representations. *IEEE Trans. Audio Speech Lang. Process.* **2015**, *23*, 46–56.
37. Lyon, R.; Katsiamis, A.; Drakakis, E. History and future of auditory filter models. In Proceedings of the 2010 IEEE International Symposium on Circuits and Systems (ISCAS), Paris, France, 30 May–2 June 2010; pp. 3809–3812.
38. Patterson, R.D.; Robinson, K.; Holdsworth, J.; McKeown, D.; Zhang, C.; Allerhand, M.H. Complex sounds and auditory images. In Proceedings of the Auditory Physiology and Perception: 9th International Symposium on Hearing, Carcens, France, 9–14 June 1991; pp. 429–446.
39. Hohmann, V. Frequency analysis and synthesis using a Gammatone filterbank. *Acta Acust. United Acust.* **2002**, *88*, 433–442.
40. Lin, L.; Holmes, W.; Ambikairajah, E. Auditory filter bank inversion. In Proceedings of the 2001 IEEE International Symposium on Circuits and Systems (ISCAS 2001), Sydney, Australia, 6–9 May 2001; Volume 2, pp. 537–540.
41. Slaney, M. *An Efficient Implementation of the Patterson-Holdsworth Auditory Filter Bank*; Apple Computer Technical Report No. 35; Apple Computer, Inc.: Cupertino, CA, USA; 1993; pp. 1–42.
42. Holdsworth, J.; Nimmo-Smith, I.; Patterson, R.D.; Rice, P. *Implementing a Gammatone Filter Bank*; Annex c of the Svos Final Report (Part A: The Auditory Filterbank); MRC Applied Psychology Unit: Cambridge, UK, 1988.
43. Darling, A. *Properties and Implementation of the Gammatone Filter: A Tutorial*; Technical Report; University College London, Department of Phonetics and Linguistics: London, UK, **1991**, pp. 43–61.
44. Strahl, S.; Mertins, A. Analysis and design of gammatone signal models. *J. Acoust. Soc. Am.* **2009**, *126*, 2379–2389.
45. Balazs, P.; Dörfler, M.; Holighaus, N.; Jaillet, F.; Velasco, G. Theory, Implementation and Applications of Nonstationary Gabor Frames. *J. Comput. Appl. Math.* **2011**, *236*, 1481–1496.
46. Holighaus, N.; Dörfler, M.; Velasco, G.; Grill, T. A framework for invertible, real-time constant-Q transforms. *Audio Speech Lang. Process.* **2013**, *21*, 775–785.
47. Holighaus, N.; Wiesmeyr, C.; Průša, Z. A class of warped filter bank frames tailored to non-linear frequency scales. *arXiv* **2016**, arXiv:1409.7203.
48. Necciari, T.; Balazs, P.; Holighaus, N.; Søndergaard, P. The ERBlet transform: An auditory-based time-frequency representation with perfect reconstruction. In Proceedings of the 2013 IEEE International Conference on Acoustics, Speech and Signal Processing (ICASSP), Vancouver, BC, Canada, 26–31 May 2013; pp. 498–502.
49. Trefethen, L.N.; Bau, D.,III. *Numerical Linear Algebra*; SIAM: Philadelphia, PA, USA, 1997.
50. Moore, B.C.J. *An Introduction to the Psychology of Hearing*, 6th ed.; Emerald Group Publishing: Bingley, UK, 2012.
51. Zwicker, E.; Terhardt, E. Analytical expressions for critical-band rate and critical bandwidth as a function of frequency. *J. Acoust. Soc. Am.* **1980**, *68*, 1523–1525.
52. O'shaughnessy, D. *Speech Communication: Human and Machine*; Addison-Wesley: Boston, MA, USA, 1987.
53. Daubechies, I.; Grossmann, A.; Meyer, Y. Painless nonorthogonal expansions. *J. Math. Phys.* **1986**, *27*, 1271–1283.
54. Průša, Z.; Søndergaard, P.L.; Rajmic, P. Discrete Wavelet Transforms in the Large Time-Frequency Analysis Toolbox for Matlab/GNU Octave. *ACM Trans. Math. Softw.* **2016**, *42*, 32:1–32:23.
55. Hestenes, M.R.; Stiefel, E. Methods of conjugate gradients for solving linear systems. *J. NBS* **1952**, *49*, 409–436.
56. Gröchenig, K. Acceleration of the frame algorithm. *IEEE Trans. Signal Process.* **1993**, *41*, 3331–3340.
57. Eisenstat, S.C. Efficient implementation of a class of preconditioned conjugate gradient methods. *SIAM J. Sci. Stat. Comput.* **1981**, *2*, 1–4.
58. Balazs, P.; Feichtinger, H.G.; Hampejs, M.; Kracher, G. Double preconditioning for Gabor frames. *IEEE Trans. Signal Process.* **2006**, *54*, 4597–4610.
59. Christensen, O. *An Introduction to Frames and Riesz Bases*; Applied and Numerical Harmonic Analysis; Birkhäuser: Boston, MA, USA, 2016.
60. Smith, J.O. Audio FFT filter banks. In Proceedings of the 12th International Conference on Digital Audio Effects (DAFx-09), Como, Italy, 1–4 September 2009; pp. 1–8.
61. Søndergaard, P.L.; Torrésani, B.; Balazs, P. The Linear Time Frequency Analysis Toolbox. *Int. J. Wavelets Multiresolut. Inf. Process.* **2012**, *10*, 1250032.
62. Průša, Z.; Søndergaard, P.L.; Holighaus, N.; Wiesmeyr, C.; Balazs, P. The large time-frequency analysis toolbox 2.0. In *Sound, Music, and Motion*; Springer: Berlin, Germany, 2014; pp. 419–442.

63. Schörkhuber, C.; Klapuri, A.; Holighaus, N.; Dörfler, M. A matlab toolbox for efficient perfect reconstruction time-frequency transforms with log-frequency resolution. In Proceedings of the Audio Engineering Society 53rd International Conference on Semantic Audio, London, UK, 27–29 January 2014.

64. Velasco, G.A.; Holighaus, N.; Dörfler, M.; Grill, T. Constructing an invertible constant-Q transform with nonstationary Gabor frames. In Proceedings of the 14th International Conference on Digital Audio Effects (DAFx-11), Paris, France, 19–23 September 2011; pp. 93–99.

65. Lehoucq, R.; Sorensen, D.C. Deflation Techniques for an Implicitly Re-Started Arnoldi Iteration. *SIAM J. Matrix Anal. Appl.* **1996**, *17*, 789–821.

66. Le Roux, J.; Vincent, E. Consistent Wiener Filtering for Audio Source Separation. *Signal Process. Lett. IEEE* **2013**, *20*, 217–220.

67. Emiya, V.; Vincent, E.; Harlander, N.; Hohmann, V. Subjective and Objective Quality Assessment of Audio Source Separation. *IEEE Trans. Audio Speech Lang. Process.* **2011**, *19*, 2046–2057.

68. Balazs, P. Basic Definition and Properties of Bessel Multipliers. *J. Math. Anal. Appl.* **2007**, *325*, 571–585.

applied
sciences

MDPI

Article

A Real-Time Sound Field Rendering Processor

Tan Yiyu [1,2,*], Yasushi Inoguchi [2], Makoto Otani [3], Yukio Iwaya [4] and Takao Tsuchiya [5]

[1] RIKEN Advanced Institute for Computational Science, Kobe, Hyogo 650-0047, Japan

[2] Research Center for Advanced Computing Infrastructure, Japan Advanced Institute of Science & Technology, Nomi, Ishikawa 923-1292, Japan; inoguchi@jaist.ac.jp

[3] Department of Architecture and Architectural Engineering, Kyoto University, Kyoto 615-8540, Japan; otani@archi.kyoto-u.ac.jp

[4] Department of Electrical Engineering and Information Technology, Tohoku Gakuin University, Sendai, Miyagi 980-8511, Japan; iwaya.yukio@mail.tohoku-gakuin.ac.jp

[5] Faculty of Science and Engineering, Doshisha University, Kyotanabe, Kyoto 610-0321, Japan; ttsuchiy@mail.doshisha.ac.jp

* Correspondence: tan.yiyu@riken.jp; Tel.: +81-78-940-5833

Academic Editor: Vesa Valimaki

Received: 3 November 2017; Accepted: 18 December 2017; Published: 28 December 2017

Abstract: Real-time sound field renderings are computationally intensive and memory-intensive. Traditional rendering systems based on computer simulations suffer from memory bandwidth and arithmetic units. The computation is time-consuming, and the sample rate of the output sound is low because of the long computation time at each time step. In this work, a processor with a hybrid architecture is proposed to speed up computation and improve the sample rate of the output sound, and an interface is developed for system scalability through simply cascading many chips to enlarge the simulated area. To render a three-minute Beethoven wave sound in a small shoe-box room with dimensions of 1.28 m × 1.28 m × 0.64 m, the field programming gate array (FPGA)-based prototype machine with the proposed architecture carries out the sound rendering at run-time while the software simulation with the OpenMP parallelization takes about 12.70 min on a personal computer (PC) with 32 GB random access memory (RAM) and an Intel i7-6800K six-core processor running at 3.4 GHz. The throughput in the software simulation is about 194 M grids/s while it is 51.2 G grids/s in the prototype machine even if the clock frequency of the prototype machine is much lower than that of the PC. The rendering processor with a processing element (PE) and interfaces consumes about 238,515 gates after fabricated by the 0.18 μm processing technology from the ROHM semiconductor Co., Ltd. (Kyoto Japan), and the power consumption is about 143.8 mW.

Keywords: sound field rendering; FPGA; FDTD

1. Introduction

Sound field rendering exhibits numerical methods to model sound propagation behavior in spatial and time domains, and is fundamental to numerous scientific and engineering applications, which vary widely from interactive computer games and virtual reality to highly accurate computations for offline applications like architecture design. To date, many analysis algorithms, including geometric methods and wave-based methods, have already been proposed to analyze sound wave propagations. In particular, wave-based methods are popularly applied because of their high accuracy, in which a sound space is discretized into small grids, and an analysis algorithm is applied on each grid to model sound behavior at discrete time steps. Among wave-based methods, the finite difference time domain (FDTD) method has been widely applied and has become an essential algorithm in room acoustics owing to its ease of implementation and parallelization. FDTD was introduced to analyze acoustical behavior by O. Chiba et al., and D. Botteldooren et al. [1–3]. However, numerical dispersion is an inherent problem constraining the valid bandwidth in the FDTD method. To reduce numerical dispersion, L. Savioja et al., G.R. Campos et al., and D.T. Murphy et al. applied digital waveguide mesh topologies [4–6]; K. Kowalczyk and M. van Walstijn proposed a second-order accurate FDTD scheme, and the 27-point compact explicit FDTD scheme was introduced [7]. J. van Mourik and D. Murphy investigated a set of high-order explicit "large-star" stencils, which could obtain less dispersiveness at low frequencies and provide high valid bandwidth [8]. H. Brian and B. Stefan proposed the fourth-order accurate explicit and implicit FDTD schemes for 2D and 3D wave equations [9,10], respectively. They recently presented a set of two-step explicit FDTD schemes with high-order accuracy in both space and time for 3D wave equations [11].

On the other hand, since spatial grids are usually oversampled to suppress the numerical dispersion errors in the FDTD method [10], the memory usage and computing power required is significant. Generally, the computing power of solving such wave equations increases as the fourth power of frequency [12], and it is increased proportionally with the volume of sound spaces. For example, every doubling of the frequency band induces a 16-fold increase in the computational load [2]. Given the auditory range of humans (20 Hz–20 kHz), analyzing sound wave propagation in a space corresponding to a concert hall or a cathedral (e.g., volume of 10,000–15,000 m^3) for the maximum simulation frequency of 20 kHz requires petaflops of computing power and terabytes of memory. As a result, the traditional sound rendering systems based on computer simulations demand huge computation power, especially for broadband simulations extending into the kilohertz range. They require a PC cluster or supercomputer for computations because of constraints of memory bandwidth and the performance of arithmetic units in a single PC. Although the performance of the arithmetic units can be enhanced through increasing the clock frequency of processors or using multicores, it is constrained by the power wall and dark silicon problems.

In recent years, general-purpose graphic processing units (GPGPUs) and FPGAs have been applied to speed up the arithmetic operations in sound rendering systems [13–22]. Although GPGPU-based solutions achieve high computation performance through increasing system threads, the input and output interfaces are more difficult to customize according to applications. Therefore, it is very hard to directly input the live signals and output the rendered results in such sound rendering systems in interactive and real-time applications. In some solutions, to apply GPGPUs in real-time sound rendering, the rendered results are sent to a buffer, and then the audio cards in the host machine are driven to output the rendered results through calling their application program interfaces (APIs). The constraint of these solutions is system scalability because the number of audio cards is limited by the number of peripheral component interconnect (PCI) slots inside the host machine. Especially in multi-channel applications, such as 128 channels, it is impossible to output using the audio cards inserted in the host machine, instead, the external professional equipment is required. Then, how to output the run-time rendered results from the GPGPU to the output equipment is a problem because no interfaces are provided to the external devices. In contrast, the input/output (I/O) interfaces can be customized in accordance to applications in FPGA. For multi-channel applications, the I/O interfaces may be designed according to applications to output the rendered results directly or output them to the external professional devices at run-time.

Different than the software-based solutions in computer simulations and GPGPUs, FPGA-based sound rendering solutions implement sound analysis equations by the configurable logic blocks directly, and hundreds of arithmetic units are coordinated to work in parallel to improve computation performance [17–22]. Furthermore, the input and output interfaces are easily tailored according to applications in FPGA. From the point of view of real-time processing, FPGA seems a promising solution to real-time sound rendering applications. In our previous work, a FPGA-based accelerator for real-time sound rendering was developed to enlarge the simulated space at the expense of the computation speed [19]. However, the sample rate of the output sound was low and sound quality was not good.

On the other hand, multiple chips are generally needed to perform rendering tasks for a large sound space because of the limited hardware resources inside a single chip. The connection interfaces between chips, therefore, become important, which affect data exchange and system reliability. In this research, a real-time sound rendering processor based on the hardware-oriented FDTD (HO-FDTD) was investigated to address the problems we met in previous work. To verify our proposal, a prototype machine was implemented using FPGA, and a trial chip including a PE and interfaces was fabricated by using the 0.18 μm processing technology from the ROHM semiconductor Co., Ltd. The processor has the hybrid architecture to improve the sampling rate of the output sound, and it provides simple interfaces for system scalability. The main contributions of this work are shown as follows.

(1) The hybrid architecture to speed up computation and improve the sampling rate of the output sound. The system architecture and function modules are introduced.
(2) Simple interface for system scalability. The data transceiver, receiver, and decoder are introduced, and the related operation flows are described.
(3) Design and implementation of the FPGA-based prototype machine and application specific integrated circuit (ASIC), which achieve significant performance gain over multi-core based software simulation.
(4) Evaluation and analysis of system performance based on the prototype machine and ASIC, including rendering time, sample rate of the output sound, and throughput.

The rest of this paper is organized as follows. The rendering algorithm is introduced in Section 2, including the updated equations for general grids and grids on a reflective boundary. In Section 3, the system architecture and design are described, as well as the design issues and the functions of modules in hardware systems. System performance of the FPGA-based prototype machine and ASIC are estimated in Section 4, followed by conclusions drawn in Section 5.

2. HO-FDTD Algorithm

The HO-FDTD algorithm, a hardware-oriented FDTD algorithm proposed for real-time sound field rendering in our previous work [18], was applied as the rendering algorithm in this research. In the HO-FDTD algorithm, different formulas are applied to calculate the sound pressures of grids.

2.1. General Grids

Within an enclosure, the dynamics of an acoustic field is governed by the following two basic equations [23].

$$\nabla P + \rho \frac{\partial u}{\partial t} = 0 \tag{1}$$

$$\frac{\partial P}{\partial t} + \rho c^2 \nabla \bullet u = 0 \tag{2}$$

Here, P and u are the pressure and particle velocity, respectively; both are functions of time and a spatial coordinate. The physical constants ρ and c are the air density and wave speed in air, respectively, ∇ and $\nabla \bullet$ are the three-dimensional gradient and divergence operations. The differential wave equation (Equation (3)) may be derived by inserting Equation (2) into Equation (1) and eliminating the particle velocity.

$$\frac{\partial^2 P}{\partial t^2} + c^2 \nabla^2 P = 0 \tag{3}$$

where $\nabla^2 = \frac{\partial^2}{\partial x^2} + \frac{\partial^2}{\partial y^2} + \frac{\partial^2}{\partial z^2}$ is the Laplacian operator in 3D sound spaces. Then, the wave equation in a 3D sound space can be described by the time domain formulation in Equation (4)

$$\frac{\partial^2 P}{\partial t^2} = c^2 \left(\frac{\partial^2 P}{\partial x^2} + \frac{\partial^2 P}{\partial y^2} + \frac{\partial^2 P}{\partial z^2} \right) \tag{4}$$

By applying the center differential method in Equation (4), and letting $\Delta x = \Delta y = \Delta z = \Delta l$, the discretion of Equation (4) yields Equation (5).

$$
\begin{aligned}
P^{n+1}(i,j,k) = \chi^2[&P^n(i+1,j,k) + P^n(i-1,j,k) + P^n(i,j+1,k) + P^n(i,j-1,k) \\
+&P^n(i,j,k+1) + P^n(i,j,k-1)] + (2-6\chi^2)P^n(i,j,k) - P^{n-1}(i,j,k)
\end{aligned}
\tag{5}
$$

where $\chi = c\Delta t/\Delta l$ represents the Courant number, and n is a discretized time step. In general, $\chi \leq 1/\sqrt{3}$ for a three-dimensional sound space. Equation (5) indicates that three multiplications, six additions, and one subtraction are needed to calculate sound pressure of a grid. When it is implemented by hardware, at least two multipliers, six adders, and one subtractor are required. In order to reduce the multiplication operations, which need more clock cycles and hardware resources, χ is assumed to be $1/2$, and Equation (5) is rewritten as [18]

$$
\begin{aligned}
P^{n+1}(i,j,k) = \tfrac{1}{4}[&P^n(i+1,j,k) + P^n(i-1,j,k) + P^n(i,j+1,k) + P^n(i,j-1,k) \\
+&P^n(i,j,k+1) + P^n(i,j,k-1) + 2P^n(i,j,k)] - P^{n-1}(i,j,k)
\end{aligned}
\tag{6}
$$

In Equation (6), multipliers are replaced by right and left shifters in hardware to save hardware resources and improve system timing performance. In principle, Equation (6) is a seven-point stencil wave equation in which to calculate sound pressure of a grid needs the sound pressures of its six neighbors at previous time step.

2.2. Boundary Condition

A reflective boundary can be modeled as a locally reacting surface by assuming that a wave does not propagate along with the boundary surface, and the acoustical behavior is affected by the sound pressure and particle velocity perpendicular to the boundary surface. If a sound wave travels in a positive axis (x, y, z) direction, the boundary impedance Z is represented by the sound pressure and the particle vibration through Equation (7) [23,24].

$$
Z = \frac{P}{U}
\tag{7}
$$

Here, U is the particle velocity component perpendicular to the boundary. Differentiating both sides of Equation (7) and substituting U by the momentum conservation equation of wave propagation, the boundary conditions are obtained in terms of sound pressure [25].

$$
\frac{\partial P}{\partial t} = -c\zeta \nabla P
\tag{8}
$$

where $\zeta = Z/\rho c$ is the normalized boundary impedance. For a rectangular sound space, boundary grids are classified into interior grids of a boundary, edges, and corners according to their position. Different formulas are applied to update sound pressures of different types of boundary grids because their conditions are different. For example, for the interior grids of right boundary, Equation (9) is derived by applying the centered finite difference method on Equation (8) and assuming the normalized boundary impedances of all boundaries are ζ.

$$
\frac{P^{n+1}(i,j,k) - P^{n-1}(i,j,k)}{2\Delta t} = -c\zeta \frac{P^n(i+1,j,k) - P^n(i-1,j,k)}{2\Delta x}
\tag{9}
$$

By rearranging the terms in Equation (9) and introducing the parameter χ, Equation (10) is derived to express a virtual point $P^n(i+1,j,k)$, which lies outside of the sound space [25].

$$
P^n(i+1,j,k) = P^n(i-1,j,k) + \frac{1}{\chi\zeta}(P^{n-1}(i,j,k) - P^{n+1}(i,j,k))
\tag{10}
$$

Substituting the related items in Equation (5) by Equation (10), then

$$P^{n+1}(i,j,k) = [\chi^2(2P^n_{i-1,j,k}(i-1,j,k) + P^n(i,j+1,k) + P^n(i,j-1,k) + P^n(i,j,k+1)$$
$$+P^n(i,j,k-1)) + 2(1-3\chi^2)P^n(i,j,k) + (\tfrac{\chi}{\xi}-1)P^{n-1}_{i,j,k}(i,j,k)]/(\tfrac{\chi}{\xi}+1) \tag{11}$$

If the reflection factor R is defined as $(\xi-1)/(\xi+1)$ and χ is assumed to be 1/2, Equation (11) is changed to

$$P^{n+1}(i,j,k) = \tfrac{1+R}{2(3+R)}[2P^n(i-1,j,k) + P^n(i,j+1,k) + P^n(i,j-1,k)$$
$$+P^n(i,j,k+1) + P^n(i,j,k-1) + 2P^n(i,j,k)] - \tfrac{3R+1}{3+R}P^{n-1}(i,j,k) \tag{12}$$

Equation (12) consists of two parts, one is the sum associated with the sound pressures of a grid and its neighbor grids at the time step n, and another corresponds to the sound pressure of a grid at the time step $n-1$. Compared with Equation (6), except the multiplicands, Equation (12) only replaces the sound pressure of the virtual grid $P^n(i+1,j,k)$ by the sound pressure of the neighbor grid $P^n(i-1,j,k)$ in the sum part. Moreover, for the interior grids on other boundaries, the multiplicands of two parts are the same while just substituting the sound pressure of the virtual grid by the sound pressure of the related neighbor grid in the summation. For example, when a grid is on the interior of the left boundary, the updated equation is achieved through substituting the sound pressure of the virtual grid $P^n(i-1,j,k)$ with the sound pressure of the neighbor grid $P^n(i+1,j,k)$ in Equation (12). The sum part is, therefore, changed from $(2P^n(i-1,j,k) + P^n(i,j+1,k) + P^n(i,j-1,k) + P^n(i,j,k+1) + P^n(i,j,k-1) + 2P^n(i,j,k))$ to $(2P^n(i+1,j,k) + P^n(i,j+1,k) + P^n(i,j-1,k) + P^n(i,j,k+1) + P^n(i,j,k-1) + 2P^n(i,j,k))$. The similar derivation procedure can be applied to edges and corners by using different boundary conditions. For example, when grids are on edges, which are intersections of two boundary planes, two boundary conditions are satisfied simultaneously. Expressions for two virtual points are consequently required to be derived.

Equations (6) and (12) indicate that the sound pressures of grids are calculated by the sound pressure of their neighbors at previous time steps, and no data dependency exists during computation. Hence, the equation may be implemented through pipelining to improve performance in hardware. We observe that Equations (6) and (12) consist of the sum of the sound pressures of a grid and its neighbors at the time step n, and the sound pressure of a grid at the time step $n-1$. For different types of grids, the updated equations have similar formats except for the multiplicands for the sum and $P^{n-1}(i,j,k)$. Thus, a uniform updated Equation (13) can be derived.

$$P^{(n+1)}(i,j,k) = D1 * [P^n(i-1,j,k) + P^n(i+1,j,k) + P^n(i,j-1,k) + P^n(i,j+1,k)$$
$$+P^n(i,j,k-1) + P^n(i,j,k+1) + 2P^n(i,j,k)] - D2 * P^{(n-1)}(i,j,k) \tag{13}$$

The D1 and D2 are shown in Table 1. It is worth noting that the part of summing in Equation (13) is changed according to grid positions. For grids on boundaries, the sound pressures of the virtual grids are replaced by those of the related neighbor grids.

Table 1. Parameters.

Grid Position	D1	D2
General	$\frac{1}{4}$	1
Interior	$\frac{R+1}{2(R+3)}$	$\frac{3R+1}{R+3}$
Edge	$\frac{R+1}{8}$	R
Corner	$\frac{R+1}{2(5-R)}$	$\frac{5R-1}{5-R}$

3. System Architecture

Since a sound space is divided into small grids, the simple architecture is to apply a computing unit at each grid to analyze the sound behavior. At a time step, computing units read data from their neighbors, carry out a computation, and output the calculation results to their neighbors. The temporary data are kept by computing units for further calculation. The whole system is fully parallel architecture, and the sample rate of the output sound is affected by the clock frequency of the computing units and the cycle count taken by the computing unit to complete computation in a time step. The computing units usually run at more than 100 MHz and complete computation in a time step less than 10 cycles. The rendering systems with the parallel architecture, therefore, achieve high sample rate in the output sound. However, the consumed hardware resources are increased exponentially as the number of grids is increased. The simulated area by a chip is hence small. To extend the simulated area, the rendering system with the time-sharing architecture was proposed [19], in which all data were stored in the block random access memories (RAMs) inside FPGA and sound pressures were calculated grid by grid through a computing unit. Although the simulated sound area by a single FPGA was enlarged by 37 times, the sample rate of the output sound was just 12.5 kHz, which would be further reduced as the number of grids was increased. To address these problems, a real-time sound rendering processor with the hybrid architecture was proposed in this research. The whole system is shown in Figure 1, and consists of the Computing Engine, in/out buffers, and the interfaces at six directions for data exchange when multiple chips are cascaded to perform sound rendering in a large sound space. The functions of components are shown as follows in detail.

- Computing Engine. The Computing Engine calculates the sound pressures of grids, and it is the core of the system. In the current solution, a sound space is divided into small sub-spaces, for example, a sound space with $8 \times 8 \times 8$ grids can be divided into 8 small sub-spaces with each having $4 \times 4 \times 4$ grids. A PE is used to analyze sound behavior in each small sub-space, and all PEs are cascaded to work in parallel to perform rendering in the whole sound space. The sound pressures of the grids on the boundaries between neighboring small sound spaces are written into the relevant buffers for further use by the neighbor PEs. Therefore, two buffers are required between two neighboring PEs. They are applied to keep the input data from and output data for the neighbor PE.

- INTERFACEs. The INTERFACEs provide the possibility for system scalability to extend the simulated sound area. From Equation (6), when multiple processors are cascaded, one chip exchanges data with neighbor chips in six directions, namely top, down, right, left, front, and back. Thus, a processor provides six interfaces for data communication.

- IN_BUFs and OUT_BUFs. The IN_BUFs and OUT_BUFs are utilized to store the sound pressures of grids on the boundaries between the neighboring small sound spaces when multiple processors are cascaded to perform rendering in a much larger sound space. The data from the neighbor processors are stored in the IN_BUFS and the data for the neighbor chips are kept by the OUT_BUFs.

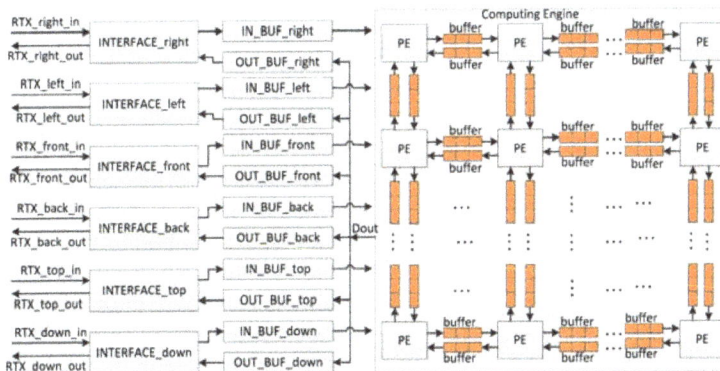

Figure 1. System diagram. Processing element (PE), receiving/transmission (RTX).

3.1. Processing Elements

The PE should be as simple as possible to reduce hardware resource consumption and improve system clock frequency. As shown in Figure 2, a PE consists of the computing unit, the grid position controller, system controller, five buffers, three multiplexers, and two block RAMs: RAM_1 and RAM_2. The PE is based on the time-sharing architecture [19] to extend the rendered area. At a time step, the computing unit reads data from the relevant buffers (buffer 1–5) or the IN_BUFs in accordance with the grid position, carries out computation, and writes the results to the RAM or the OUT_BUFs. After the calculation at a grid is completed, the computation is then shifted to the next grid until sound pressures of all grids are calculated. The modules in a PE are introduced as follows:

- Computing Unit. The computing unit calculates the sound pressure of a grid according to the input sound pressures at previous time steps, location indicators, and incident data. Based on Equation (13), a uniform computing unit was designed, which consists of a 7-input adder, a subtractor, two 32-bit fixed-point multipliers, and four multiplexers [19]. In Figure 3, the multipliers are for boundary grids while they are replaced by the right and left shifters for grids outside boundaries. The multiplicands are selected by the multiplexers according to the location indicator of a grid.

- Grid Position Controller. The grid position controller generates the grid position by using a counter, which is updated at every clock cycle.

- System Controller. The system controller maintains the computation flow and generates control signals, such as grid location flag (loc_indicator), read/write enable signal (we) of the RAMs, reading/writing addresses of the RAMs (raddr_RAM and waddr_RAM), and RAM selection signal (ram_we_sel).

- RAM_1 and RAM_2. The sound pressures of grids at previous one and two time steps, namely $P^{n-1}(i, j, k)$ and $P^{n-2}(i, j, k)$, are stored in the RAM_1 and RAM_2. During computation, sound pressures at different time steps are stored in and read out from the RAM_1 and RAM_2 alternatively, and the calculation results at current time step are kept by the same RAM as that in which the sound pressures of grids at the previous two time steps are stored. For example, at a time step, $P^{n-1}(i, j, k)$ and $P^{n-2}(i, j, k)$ are stored in RAM_1 and RAM_2, respectively. Then, the calculation results at the current time step are written into the RAM_2. At the next time step, the reading and writing operations for the RAMs are switched; $P^{n-2}(i, j, k)$ is read out from the RAM_1 while others are taken from the RAM_2. In addition, the calculation results are stored in the RAM_2. Such switching for RAM operations is repeated until all calculated time steps are over. The writing-enable signals of the RAMs are controlled by the signals ram_we_sel and data_dvld output by the system controller and the computing unit, respectively. When computations at a time step are finished, the signal ram_we_sel is reversed to invert the writing-enable signals of the RAMs. The size of RAMs is determined by the number of grids and data width. If data are 32-bit, and a sound space has N × M × L grids, each RAM is 4 NML bytes.

- Buffer 1–5. Data are read out from the RAMs and written in the five buffers in advance to reduce data access latency during calculation. The buffers are updated along with the computation. If data width is 32-bit, and a sound space has N × M × L grids, each buffer is 4 NM bytes in size.

- Multiplexers. Three multiplexers are used to select data for the computing unit. In the system, the input data of the computing unit may be from the local buffers within the same PE or the external input buffer, in which the data from the neighbor PE are stored.

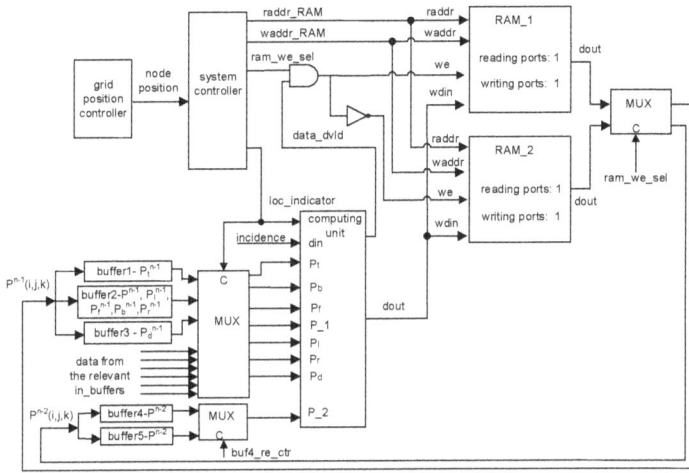

Figure 2. System diagram of the PE.

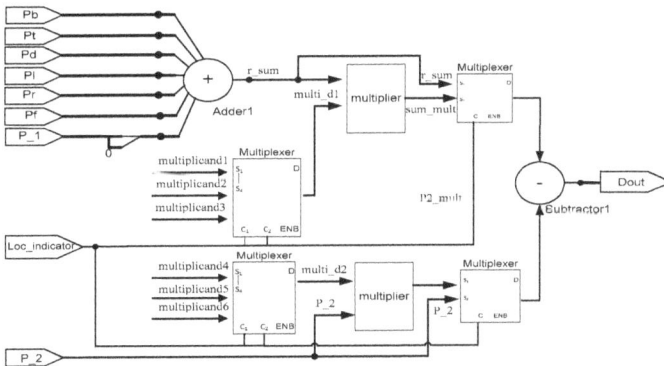

Figure 3. Computing unit.

3.2. INTERFACEs

The INTERFACEs are provided for system scalability. When multiple processors are cascaded, control instructions and temporary data during computation need to exchange with the neighbor processors. Through the INTERFACEs, multiple processors are easily connected to each other to extend the rendered sound space. As shown in Figure 4, the INTERFACE consists of receiver, decoder, and transmitter. And they are introduced more detail as follows.

- Receiver. The receiver receives serial data from the neighbor processors, checks the data type (data or control instructions), and stores data to the first in first out (FIFO) buffer or buffer according to the data type. The receiver is composed of a data type detector (RX_FHD) and a state machine (RX_STATE) to control the receiving flow. As shown in Figure 5, the system firstly detects the 8-bit type flag, which is AB and 54 in hexadecimal for pure data and the control instructions, respectively. After the type flag is received, system then receives the data length, which is 8-bit and denotes the data length in byte. Finally, data are received and stored in the RX_FIFO for the control instructions or IN_BUF for the pure data.

- Decoder. The DECODER decodes and executes the control instructions. Eleven control instructions are defined for system configuration and data communication. Furthermore, an automatic instruction forwarding mechanism is provided to forward the control instructions to other processors when multiple processors are cascaded.
- Transmitter. The transmitter (TX_STATE) transmits the control instructions or pure data to the neighbor processors. Data are transmitted in the manner of data frame, which consists of type flag, data length, chip ID, and the relevant pure data or instructions. Before data transmission, the bus is checked. If it is free, transmission is started; otherwise, the system waits for some clock cycles and checks again until the bus is free. Three types of data are transmitted, which are pure data, forwarded instructions, and the acknowledgement instruction. To transmit the pure data or the forwarded instructions, the data valid instruction is firstly sent out. Then, the system waits for the acknowledgment signal from the receiver. After the signal is received, the related data frame is transmitted serially. When the system receives the data valid instruction from the neighbor processors, it responds the data communication request by sending back the acknowledgement instruction. The whole procedure is shown in Figure 6 and is controlled by a state machine.

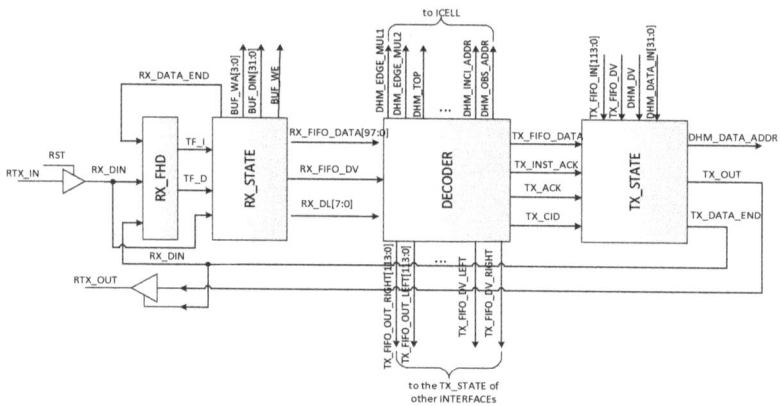

Figure 4. Diagram of the INTERFACE.

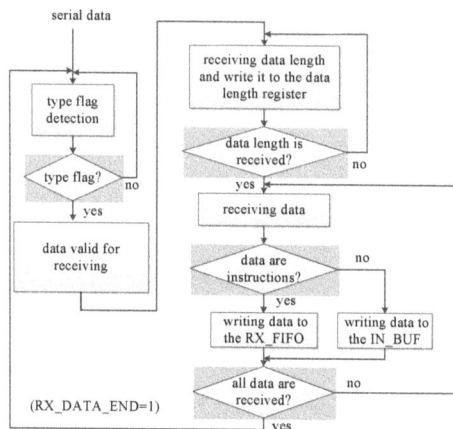

Figure 5. Data receiving.

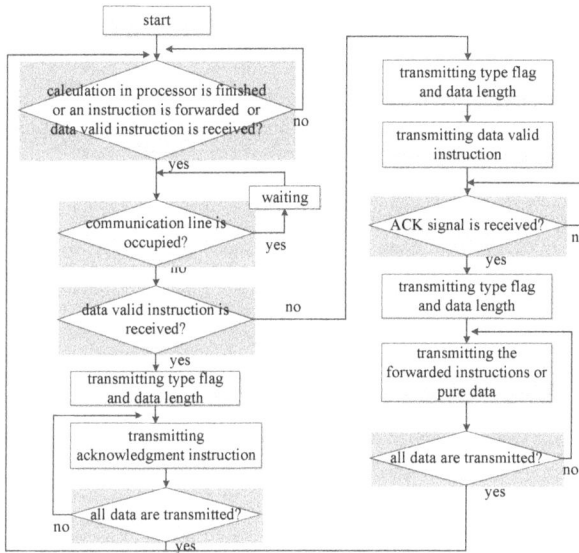

Figure 6. Data transmission.

4. Performance Estimation

To verify and estimate system performance, the register transfer level (RTL) model and cycle-accurate simulator of the processor were developed using VHDL and C programming language. Furthermore, the prototype machine was investigated and implemented using a processor-based FPGA machine TD-SPP3000 from Tokyo Electron Device Ltd., and sound propagation in a small shoe-box type room was examined. In the defined room, both length and width were 1.28 m, and height was 0.64 m; the cube grid size Δl was 4×10^{-2} m; the reflection factor of the boundaries was 0.95; and the incident and observation points were at the middle of the room. Therefore, the small room was discretized into a mesh with $32 \times 32 \times 16$ grids. The hardware development environment was a 64-bit Windows 7 platform with FPGA tools Xilinx ISE 14.3 and ModelSim SE 10.1d. For comparison, the counterpart system was also developed by C++ programming language, and executed on a personal computer (PC) with 32 GB RAM and an Intel i7-6800K six-core processor running at 3.4 GHz. The software environment of the PC was 64-bit CentOS 7.0 with gcc 4.9.4. The reference C++ codes were compiled and optimized by using the command g++ with the option -O3, mcmodel = large, and -fopenmp to use all six cores in the PC. Data were 32-bit fixed-point in the prototype machine while they were integer in the software simulations.

The prototype machine consisted of two FPGA boards, and each board contained two XC5VLX330T-FF1738 FPGA chips from Xilinx. A high-speed A/D board (ADS5474) was attached to the FPGA1 on the board 1 to sample the incident signals. Then, the sampled data were processed by the rendering processor implemented by the FPGA1 on the board 1. The sound pressures at the observation point were transferred to the D/A board (DAC5682Z) on the board 2 through the advanced telecommunication computing architecture (ATCA) bus and output to drive the speakers directly. The rendering processor and the A/D and D/A boards all ran at 200 MHz. In the rendering processor, each PE processed the sound rendering in a sound space with $4 \times 4 \times 4$ grids, and all PEs worked in parallel to carry out rendering at the whole sound space. Thus, computation at each time step took 64 cycles, and each cycle was 5 $(1/(200 \times 10^6))$ ns.

4.1. Performance of the Prototype Machine

4.1.1. Rendering Time

Table 2 shows the rendering time taken by the prototype machine and the software simulation on the PC using six cores to render a three-minute Beethoven wave sound in the defined small shoe-box type room.

Table 2. Rendering time (s).

Grid	Prototype	PC
32 × 32 × 16	run-time	762.25

PC: personal computer.

As shown in Table 2, the rendering task consumes about 12.70 min (762.25/60) in the offline software simulation on the PC, while it is handled at run-time on the prototype machine. Figure 7 shows the computation flow in the software simulation. In the software simulation, all the incident data were firstly read from an incident wave file and stored in a buffer. At each time step, an incident datum is read and sound pressures of all grids are calculated, the sound pressure of the observation point is updated, and the sound pressures of grids at current time step and previous time steps are swapped. This procedure is iterated until all time steps are completed. Since the swap operation must be operated after the sound pressures of all grids are obtained, the outer loop cannot be parallelized while the computation and data swap modules (shown in yellow in Figure 7) are parallelized by using OpenMP. Figures 8 and 9 present parts of the source code of the computation and data swap modules in which the directives of OpenMP are shown by the bold words. The three loops in the computation and data swap modules are parallelized through being collapsed into one large iteration space and then divided according to the valid threads. Figure 10 depicts the computation time in the case of different numbers of cores being applied in the software simulation. As the number of cores is increased, the computation time is decreased due to system parallelization using OpenMP. When six cores are all used for computation, the computation time is 762.25 s, which is about 18% of the computation time when a single core is applied in calculation.

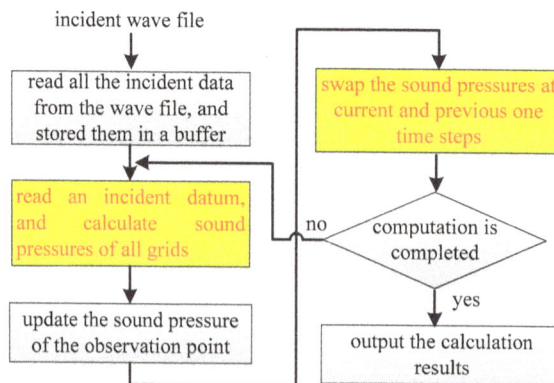

Figure 7. Computation flow of the software simulation.

In the prototype machine, the incident wave sound is played by a media player and sampled by the A/D board as incidences at each time step; then rendering is carried out and, finally, the rendered results are output through the D/A board to drive the speakers directly. Therefore, the rendering is carried out at run-time, and the rendered results at each time step are output by the D/A board directly. When the input incident wave is finished, the rendering is also completed after the operations when the final time step is over. Consequently, the whole rendering procedure lasts the same period as the length of the incident sound plus the time taken by the computation at the final time step, namely 3 min plus 64 cycles (3 s + 320 ns). In the computer-based software simulation, data are stored in the external main memory, the rendered results are temporarily kept by an array at each time step, and finally written to a wave file. During computation, main memory is accessed frequently to read data out or write data back, which is time-consuming. In contrast, data are stored in the on-chip memory (block RAMs inside FPGA) in the prototype machine, and they are accessed in one or two cycles. Furthermore, five buffers are applied to read data out in advance to reduce data access overhead in the PE. On the other hand, the rendering processor is the hybrid architecture, and many PEs work in parallel to speed up computation. Because each PE is applied to analyze sound behavior at a sound space with $4 \times 4 \times 4$ grids, the prototype machine contains 256 (($32 \times 32 \times 16)/(4 \times 4 \times 4)$) PEs to work in parallel, and speeds up computation by 256 times in comparison with the rendering system with the time-sharing architecture, in which only one PE is applied to carry out rendering.

```
static void wave_pressure(int it) {
   // elm_k[ix][iy][iz] : sound pressures at current time step
   // elm_k1[ix][iy][iz] : sound pressures at previous one time step
   // elm_k2[ix][iy][iz] : sound pressures at previous two time step
   ......
#pragma omp parallel for collapse(3)
for ( int ix =1; ix < nx+1; ix++) {
 for ( int iy =1; iy < ny+1; iy++) {
  for( int iz = 1; iz < nz+1; iz++) {
   .....
   elm_k[ix][iy][iz]= int((elm_k1[ix-1][iy][iz]+elm_k1[ix+1][iy][iz]+elm_k1[ix][iy-1][iz] +
    elm_k1[ix][iy+1][iz]+elm_k1[ix][iy][iz-1]+elm_k1[ix][iy][iz+1]+2*elm_k1[ix][iy][iz] ) * d1)
    + datain_dhm - int(d2 * elm_k2[ix][iy][iz] ); }
      }
     }
    }
```

Figure 8. Snapshot of the computation module.

```
static void wave_pressure_updated(){
   #pragma omp parallel for collapse(3)
   for (int ix = 0; ix <=nx+1 ; ix++)
    for (int iy =0; iy <= ny+1; iy++)
     for (int iz =0; iz <= nz+1; iz++)
      {
       elm_k2[ix][iy][iz] = elm_k1[ix][iy][iz];//store the old value of K1 to K2;
       elm_k1[ix][iy][iz] = elm_k[ix][iy][iz]; // assign the old value of K to K1
      }
   }
```

Figure 9. Snapshot of the data swap module.

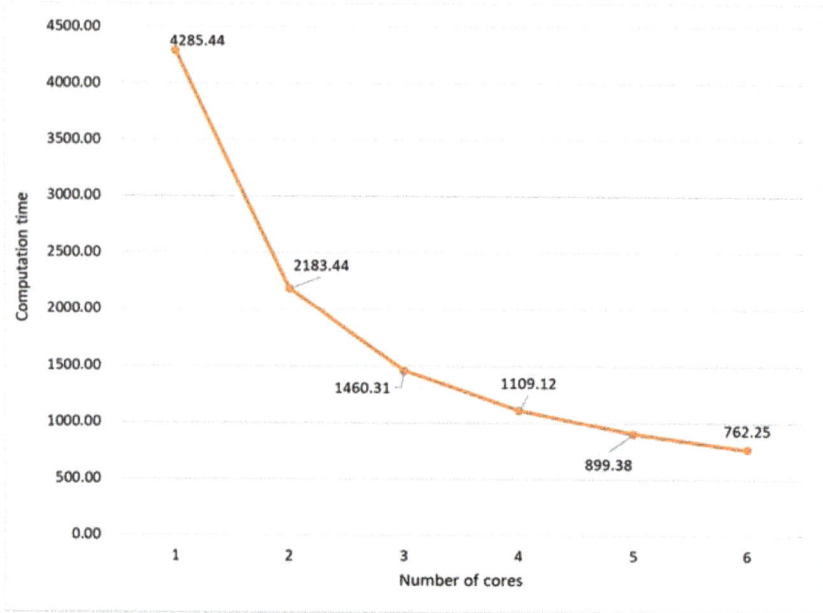

Figure 10. Computation time by using the OpenMP in software simulation.

The rendering processor is designed through pipelining, and the rendered results are consecutively output by the D/A board after a one-cycle delay. To investigate the effect of such a small delay on the output and system stability, a pulse with amplitude of 16,384 Pa was launched into the prototype system, and the impulse response of the defined shoe-box room is presented in Figure 11. As shown in Figure 11, the output was just delayed one-cycle, and the system became stable after 400 time steps. Thus, the small delay almost has no effect on the system.

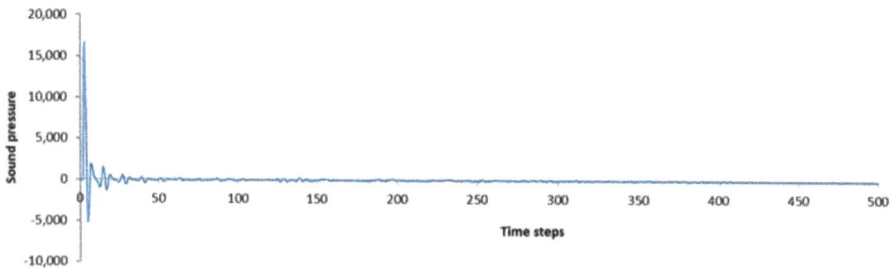

Figure 11. Impulse response of the defined room.

4.1.2. Sample Rate of the Output Sound

Although the prototype machine carries out sound rendering at run-time, the output sound quality is worse than that output by the offline computer-based software simulation. In the software simulation, incident data are read from a sound wave file directly at each time step, and the rendering results are written into another sound wave file after the computation is finished. The output sound wave, therefore, has same sample rate and bitrate as the incidence. However, in the prototype machine, the sample rate of the output sound is calculated through Equation (14).

$$f_{sampe} = \frac{f_{clk}}{M} \tag{14}$$

f_{clk} is the system clock frequency, and M denotes the number of grids processed by a PE. In the current prototype machine, since each PE performs sound rendering in a sound space with $4 \times 4 \times 4$ grids, and the system clock frequency is 200 MHz, the sample rate of the output sound is about 3.125 MHz (200 MHz/64). Compared with the rendering system with the time-sharing architecture, in which the sample rate of the output sound is about 12.5 kHz [19], the prototype achieves about 256 times gain in sample rate of the output sound. In the rendering system with the time-sharing architecture, only a computing unit is applied to calculate the sound pressure grid by grid. Hence, as the number of grids is increased, the computation time at a time step will become longer, which results in a low sample rate of the output sound. However, the proposed rendering processor is the hybrid architecture, in which the top level is the parallel architecture, and the PE is based on the time-sharing architecture. Therefore, the computation is speeded up, and high sample rate is obtained in the output sound.

On the other hand, the incidence is input through an A/D converter and the rendered results are output through a D/A converter in the prototype machine. Because the A/D converter is 14-bit, each incident datum is 14-bit, but it is 16-bit in the software simulation. This long data width of the incidence results in high accuracy in computation in the software simulation. Furthermore, each PE performs sound rendering in a sound space with $4 \times 4 \times 4$ grids, if the computation at a grid is completed at one cycle through a pipelining technique, calculation at a time step takes 64 cycles. In other words, the incidence will be input in the system every 64 cycles from the A/D converter. This may result in the information loss in the incident wave sound, and reduce the computation accuracy. Another factor to affect the output sound quality comes from the electronic noise of the A/D and D/A boards. When the prototype machine has no input at the A/D converter, the electronic noise can be heard at the output of the D/A converter. Compared with professional audio devices, current A/D and D/A converters provides worse performance in suppressing electronic noise.

4.1.3. Throughput

The throughput denotes the number of grids updated per second, and is calculated by Equation (15).

$$D_{throughput} = \frac{N_{grid}}{T_{total}} * N_{time\_step} \tag{15}$$

Here, N_{grid} is the number of grids, N_{time_step} is the number of time steps, and T_{total} is the calculation time. In the software simulation, the incident sound wave has 9,022,848 data, and each datum will be input into the system as an incidence. Nevertheless, the time steps are 9,022,848. From Table 2 and Figure 10, when six processor cores are applied, the throughput in the software simulation is about 194 ($32 \times 32 \times 16/(762.25/9,022,848)$) M grids/s. In the prototype machine, the computations at each time step are completed in 64 ($4 \times 4 \times 4$) cycles. Thus, the throughput is about 51.2 ($32 \times 32 \times 16/(64 \times 1/0.2)$) G grids/s. Even if the clock frequency of the prototype is much lower than the PC, the throughput is much higher. Compared with the PC-based simulation, the prototype machine achieves about 263.9 times gain in throughput.

4.2. System Implementation by ASIC

A trial processor with a PE and interfaces was developed and taped out by using the 0.18 μm processing technology from the ROHM semiconductor Co. Ltd. through the fabrication service provided by the VLSI design and education center at University of Tokyo. When the PE is utilized to process $4 \times 4 \times 4$ grids, the layouts of the processor and the whole chip are shown in Figure 12a,b, respectively. The chip is 2.5 mm $\times$ 5.0 mm in size, and contains 89 pins [26]. The whole system consumes 238,515 gates, and the power consumption is about 143.8 mW. The system clock frequency is 200 MHz.

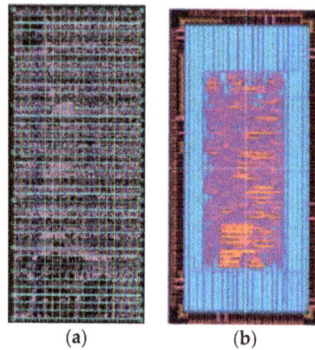

(a) (b)

Figure 12. (**a**) Layout of the processor (**b**) Layout of the chip.

5. Conclusions

Real-time sound rendering is computation-intensive. The output sound quality and system scalability are two concerning issues in the design of a real-time sound rendering system by hardware. In this research, a real-time sound rendering processor with the hybrid architecture was investigated and implemented to speed up computation and improve the sample rate of the output sound. While rendering sound in a small shoe-box room with dimensions of 1.28 m $\times$ 1.28 m $\times$ 0.64 m, the proposed processor performs sound rendering at real-time, while the offline computer-based software simulation takes about 12.70 min. Compared with the FPGA-based sound rendering system with the time-sharing architecture, our processor achieves 256 times increase in computation speed and improvement in the sample rate of the output sound.

Furthermore, owing to the limited hardware resources inside a single chip, multiple chips are usually required to carry out rendering for a large sound space. To make the system easily extendable, interfaces were provided in the proposed sound rendering processor for system scalability. Through the interfaces, multiple processors are easily connected to each other to extend the rendered sound space. Although the current interface achieves good performance in data transmission and system scalability, the data transmission speed is limited due to data width. In future work, the serial advanced technology attachment (SATA) interface will be investigated and applied in the processor to enhance the data transfer speed.

Acknowledgments: This work was supported in part by the Strategic Information and Communications R&D Promotion Programme (SCOPE), Ministry of Internal Affairs and Communications. Thanks for Professor Imamura Toshiyuki's valuable discussion on the system parallelization by using OpenMP.

Author Contributions: Tan Yiyu and Yasushi Inoguchi specified the system architecture and designed the system. Tan Yiyu and Takao Tsuchiya discussed and derived the algorithm, Makoto Otani, Yukio Iwaya, Takao Tsuchiya, and Yasushi Inoguchi conceived the system solution and helped debug prototype system. Tan Yiyu and Yasushi Inoguchi wrote and revised the paper.

Conflicts of Interest: The authors declare no conflict of interest.

References

1. Botteldooren, D. Acoustical finite-difference time-domain simulation in a quasi-Cartesian grid. *J. Acoust. Soc. Am.* **1994**, *95*, 2313–2319. [CrossRef]
2. Botteldooren, D. Finite-difference time-domain simulation of low-frequency room acoustic problems. *J. Acoust. Soc. Am.* **1995**, *98*, 3302–3308. [CrossRef]

3. Chiba, O.; Kashiwa, T.; Shimoda, H.; Kagami, S.; Fukai, I. Analysis of sound fields in three dimensional space by the time-dependent finite-difference method based on the leap frog algorithm. *J. Acoust. Soc. Jpn.* **1993**, *49*, 551–562.

4. Savioja, L.; Valimaki, V. Interpolated rectangular 3-D digital waveguide mesh algorithms with frequency warping. *IEEE Trans. Speech Audio Process.* **2003**, *11*, 783–790. [CrossRef]

5. Campos, G.R.; Howard, D.M. On the computational efficiency of different waveguide mesh topologies for room acoustic simulation. *IEEE Trans. Speech Audio Process.* **2005**, *13*, 1063–1072. [CrossRef]

6. Murphy, D.T.; Kelloniemi, A.; Mullen, J.; Shelley, S. Acoustic modeling using the digital waveguide mesh. *IEEE Signal Process. Mag.* **2007**, *24*, 55–66. [CrossRef]

7. Kowalczyk, K.; van Walstijn, M. Room acoustics simulation using 3-D compact explicit FDTD schemes. *IEEE Trans. Audio Speech Lang. Process.* **2011**, *19*, 34–46. [CrossRef]

8. Van Mourik, J.; Murphy, D. Explicit higher-order FDTD schemes for 3D room acoustic simulation. *IEEE/ACM Trans. Audio Speech Lang. Process.* **2014**, *22*, 2003–2011. [CrossRef]

9. Hamilton, B.; Bilbao, S. Fourth-order and optimised finite difference schemes for the 2-D wave equation. In Proceedings of the 16th Conference on Digital Audio Effects (DAFx-13), Maynooth, Ireland, 2–6 September 2013.

10. Hamilton, B.; Bilbao, S.; Webb, C.J. Revisiting implicit finite difference schemes for 3D room acoustics simulations on GPU. In Proceedings of the International Conference on Digital Audio Effects (DAFx-14), Erlangen, Germany, 1–5 September 2014; pp. 41–48.

11. Brian, H.; Stefan, B. FDTD methods for 3-D room acoustics simulation with high-order accuracy in space and time. *IEEE/ACM Trans. Audio Speech Lang. Process.* **2017**, *25*, 2112–2124.

12. Valimäki, V.; Parker, J.D.; Savioja, L.; Smith, J.O.; Abel, J.S. Fifty years of artificial reverberation. *IEEE Trans. Audio Speech Lang. Process.* **2012**, *20*, 1421–1448. [CrossRef]

13. Ishii, T.; Tsuchiya, T.; Okubo, K. Three-dimensional sound field analysis using compact explicit-finite difference time domain method with Graphics Processing Unit Cluster System. *Jpn. J. Appl. Phys.* **2013**, *52*, 07HC11. [CrossRef]

14. Tsuchiya, T. Three-dimensional sound field rendering with digital boundary condition using graphics processing unit. *Jpn. J. Appl. Phys.* **2010**, *49*, 07HC10. [CrossRef]

15. Savioja, L. Real-time 3D finite difference time domain simulation of low and mid-frequency room acoustics. In Proceedings of the International Conference on DAFx, Graz, Austria, 6–10 September 2010; pp. 77–84.

16. Tanaka, M.; Tsuchiya, T.; Okubo, K. Two-dimensional numerical analysis of nonlinear sound wave propagation using constrained interpolation profile method including nonlinear effect in advection equation. *Jpn. J. Appl. Phys.* **2011**, *50*, 07HE17. [CrossRef]

17. Tan, Y.Y.; Inoguchi, Y.; Sugawara, E.; Otani, M.; Iwaya, Y.; Sato, Y.; Matsuoka, H.; Tsuchiya, T. A real-time sound field renderer based on digital Huygens' model. *J. Sound Vib.* **2011**, *330*, 4302–4312.

18. Tan, Y.Y.; Inoguchi, Y.; Sato, Y.; Otani, M.; Iwaya, Y.; Matsuoka, H.; Tsuchiya, T. A hardware-oriented finite-difference time-domain algorithm for sound field rendering. *Jpn. J. Appl. Phys.* **2013**, *52*, 07HC03.

19. Tan, Y.Y.; Inoguchi, Y.; Sato, Y.; Otani, M.; Iwaya, Y.; Matsuoka, H.; Tsuchiya, T. A real-time sound rendering system based on the finite-difference time-domain algorithm. *Jpn. J. Appl. Phys.* **2014**, *53*, 07KC14.

20. Tan, Y.Y.; Inoguchi, Y.; Sato, Y.; Otani, M.; Iwaya, Y.; Matsuoka, H.; Tsuchiya, T. Analysis of sound field distribution for room acoustics: from the point of view of hardware implementation. In Proceedings of the International Conference on DAFx, York, UK, 17–21 September 2012; pp. 93–96.

21. Tan, Y.Y.; Inoguchi, Y.; Sugawara, E.; Sato, Y.; Otani, M.; Iwaya, Y.; Matsuoka, H.; Tsuchiya, T. A FPGA implementation of the two-dimensional digital huygens model. In Proceedings of the International Conference on Field Programmable Technology (FPT), Beijing, China, 8–10 December 2010; pp. 304–307.

22. Inoguchi, Y.; Tan, Y.Y.; Sato, Y.; Otani, M.; Iwaya, Y.; Matsuoka, H.; Tsuchiya, T. DHM and FDTD based hardware sound field simulation acceleration. In Proceedings of the International Conference on DAFx, Paris, France, 19–23 September 2011; pp. 69–72.

23. Kuttruff, H. *Room Acoustics*; Taylor & Francis: New York, NY, USA, 2009.

24. Maxwell, J.C. *A Treatise on Electricity and Magnetism*, 3rd ed.; Clarendon: Oxford, UK, 1892; Volume 2, pp. 68–73.

25. Kowalczyk, K.; Walstijn, M.V. Formulation of locally reacting surfaces in FDTD/K-DWM modelling of acoustic spaces. *Acta Acust. United Acust.* **2008**, *94*, 891–906. [CrossRef]
26. Yiyu, T.; Inoguchi, Y.; Otani, M.; Iwaya, Y.; Tsuchiya, T. Design of a Real-time Sound Field Rendering Processor. In Proceedings of the RISP International Workshop on Nonlinear Circuits, Communications and Signal Processing, Honolulu, HI, USA, 6–9 March 2016; pp. 173–176.

applied
sciences

MDPI

Article

Populating the Mix Space: Parametric Methods for Generating Multitrack Audio Mixtures

Alex Wilson * and Bruno M. Fazenda

Acoustics Research Centre, School of Computing, Science and Engineering, University of Salford, Greater Manchester, Salford M5 4WT, UK; b.m.fazenda@salford.ac.uk
* Correspondence: a.d.wilson2@salford.ac.uk

Academic Editor: Tapio Lokki
Received: 31 October 2017; Accepted: 4 December 2017; Published: 20 December 2017

Featured Application: The numerical methods described in this paper can be used in the automatic creation of artificial datasets of audio mixes, as real-world mixes are both scarce and costly to produce. Such datasets can be used for a variety of applications, such as material for signal analysis, audio stimuli in psychoacoustic testing or as a population of solutions to be optimised, thus forming the basis of an automatic mixing system. Within this paper, the application of interest is testing the robustness of tempo estimation to re-mixing.

Abstract: The creation of multitrack mixes by audio engineers is a time-consuming activity and creating high-quality mixes requires a great deal of knowledge and experience. Previous studies on the perception of music mixes have been limited by the relatively small number of human-made mixes analysed. This paper describes a novel "mix-space", a parameter space which contains all possible mixes using a finite set of tools, as well as methods for the parametric generation of artificial mixes in this space. Mixes that use track gain, panning and equalisation are considered. This allows statistical methods to be used in the study of music mixing practice, such as Monte Carlo simulations or population-based optimisation methods. Two applications are described: an investigation into the robustness and accuracy of tempo-estimation algorithms and an experiment to estimate distributions of spectral centroid values within sets of mixes. The potential for further work is also described.

Keywords: intelligent music production; music information retrieval; multitrack mixing; stereo panning; audio equalisation; tempo estimation; spectral centroid

---

1. Introduction

The mixing of audio signals is a complicated optimisation problem, in which an audio engineer must consider a vast number of technical and aesthetic considerations in order to achieve the desired result. Traditionally, many tasks in audio mixing are performed on a mixing console. Typically, such a device consists of a series of channel strips, one representing each audio track, on which various operations can be performed such as adjustments in equalisation, panning and overall level. While this format is useful for allowing a hands-on interaction with the audio content, it is not the most direct or efficient way of exploring these parameters and discovering mixes in the process.

One legacy of this console design philosophy is that, in the literature, it has become commonplace to define a mix as the sum of the input tracks, subject to control vectors for gain, panning, equalisation etc., [1–3]. Subsequently, a number of publications [4–6] have referred to a mix of n tracks as a point in an n-dimensional vector space, with each axis as the gain of a given track. While effective in certain cases, and certainly straightforward to visualise, this definition produces a solution space which is sub-optimal when searching for mixes.

The following are equations used to define a mix, according to various previous works. Note that the nomenclature has not been changed from the original texts. Equation (1) was used by [1], stating simply that a mix is the sum of all individual channels.

$$\text{mix} = \sum_{n=1}^{N} \text{Ch}_n[t] \tag{1}$$

This definition seems logical and even trivial, if inspired by a summing mixer, and has become the foundation for a series of more elaborate definitions, such as adding a gain vector, a to each track, allowing for time-dependent changes to the track gains, simulating the movement of individual faders [2].

$$y[n] = \sum_{k=1}^{K} a_k[n] \times x_k[n] \tag{2}$$

In a review paper from 2011 [3], Equation (3) was used, adding generic control vectors c which modulate the input signals x. These control vectors allow for a variety of results, such as polarity correction, delay correction, panning and source separation, depending on their implementation.

$$\text{mix}_l(n) = \sum_{m=0}^{M-1} \sum_{k=0}^{K-1} c_{k,m,l}(n) \times x_m(n) \tag{3}$$

Each of these equations considers the mix as the sum of the input tracks, although there is little agreement on terminology or nomenclature in this general definition. What is important to realise here is that these expressions characterise not strictly the mix itself but the output of a summing mixer, or conventional fader-based mixing console. As will be shown in Section 2, the set of unique mixes is a subset of this set, as illustrated by Equation (4). We refer to this subset as the mix-space, introduced in [7]. It is this space that a mixing console should directly explore, rather than the gain-space. Section 2 presents an updated definition of the term *mix*, which produces concise solution spaces by exploring only the parameter space ϕ, avoiding the redundancies in g, which represents the gain vector of the system.

$$\left(\underbrace{g_1, g_2, g_3, \dots, g_n}_{\text{gain-space}} \right) = \left(\underbrace{r}_{\text{master volume}}, \underbrace{\phi_1, \phi_2, \dots, \phi_{n-1}}_{\text{mix-space}} \right) \tag{4}$$

The primary contributions of this work are as follows: (a) the mix-space as a theoretical framework in which existing audio mixes can be examined, in contrast to the gain-space, and (b) methods for the generation of audio mixes in the mix-space. These contributions are described in Section 2.

The creation of artificial datasets relating to music mixing practice helps to overcome one of the main obstacles in the field of mix analysis, which is the lack of available data and the cost associated with gathering new data from mix engineers. Thus far, it has been difficult to make statistical inference about music mixing practice as available studies have only had access to small datasets of user-generated audio mixes, with few exceptions [8].

Thus far, the numerical methods in this paper have been applied in creating an initial population for evolutionary algorithms [9,10]. Further applications are explored in Section 3 and discussed in Section 4.

2. Theoretical Framework

Adjustment of track level, pan position and equalisation are common in audio processing. While level and pan are fundamental operations in multichannel mixing, equalisation is one of the most commonly used processors. Together, these three operations form a basic channel strip. As such, the scope of this paper considers these three operations.

2.1. Track Gains

Consider the trivial case where two audio signals are to be mixed, where only the absolute levels of each signal can be adjusted. In Figure 1, the gains of two signals are represented by x and y, where both are positive-bound. Consider the point p as a configuration of the signal gains, i.e., (p_x, p_y). From this point, the values of x and y are both increased in equal proportion, arriving at the point p'. The magnitude of p is less than that of p' ($\|p\| < \|p'\|$) yet since the ratio of x to y is identical, the angles subtended by the vectors from the y-axis are equal ($\angle p = \angle p'$). In the context of a mix of two tracks, what this means is that the volume of p' is greater than p, yet the blend of input tracks is the same.

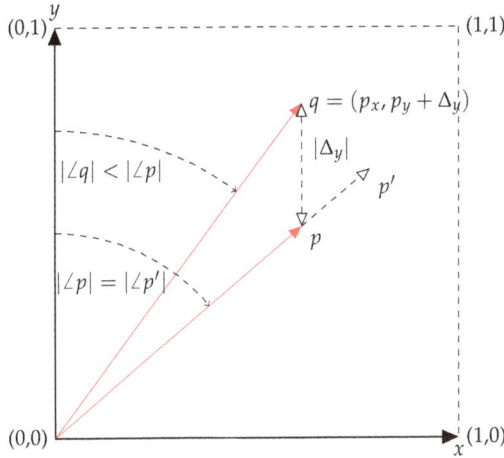

Figure 1. Points p, p' and r, in 2-track gain space. Note that the audio output at points p and p' is the same 'mix'.

As an alternate to Equation (1), a mix can be thought of as the relative balance of audio signals. From this definition, the points p and p' are the same mix, only p' is being presented at a greater volume. If the listener has control over the master volume of the system, then any difference between p and p' becomes ambiguous.

Definition 1. *Mix: an audio stream constructed by the superposition of others in accordance with a specific blend, balance or ratio.*

From p, the level of fader y can be increased by Δ_y, arriving at q. In this particular example, the value of Δ_y was chosen such that $\|q\| = \|p'\|$. However, for any $|\Delta_y| > 0$, $\angle q \neq \angle p'$. Therefore, q clearly represents a different mix to either p or p'. Consequently, the definition of a mix is clarified by what it is not: when two audio streams contain the same blend of input tracks but the result is at different overall amplitude levels, these two outputs can be considered the same mix. For this mixing example, where there are $n = 2$ signals, represented by n gain values, the mix is dependant on $n - 1$ variables; in this case, the angle to the vector. The ℓ_2 norm of the vector is simply proportional to the overall loudness of the mix.

Figure 2a shows a similar structure, with $n = 3$. Here, the point p' is also an extension of p. As in Figure 1, q is located by increasing the value of y from the point p and $\|q\| = \|p'\|$. Here, the values of each angle are explicitly determined and displayed. All three vectors share the equatorial angle of $60°$. The polar angle of p and p' is $50°$, while the polar angle of q is less than this, at $\approx 37°$. As in the two-dimensional case, it is the angles which determine the parameters of the mix and the norm of the vector is related to the overall loudness.

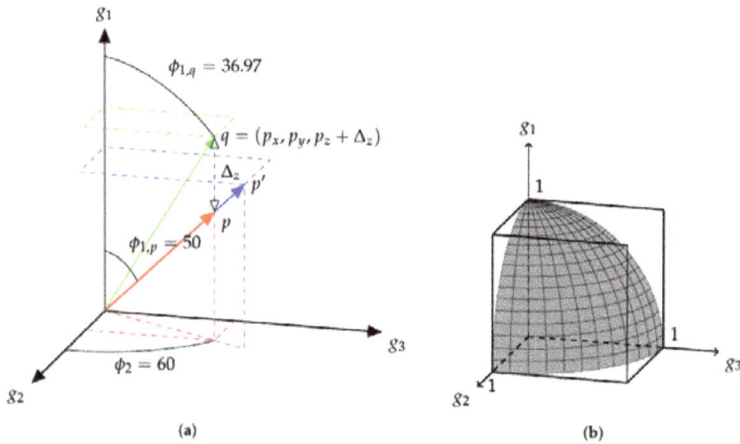

Figure 2. Graphical representation of three mixes in mix-space. While shown for three tracks, this is generalisable to any number of tracks n, using hyperspherical coordinates. (a) Mix at a point in 3-track gain space. Note that the audio output at points p and p' is the same 'mix', despite the vectors having different lengths in this space; (b) For a 3-track mixture, while the cube ($\mathbb{R}^3$) represents all outputs of a summing mixer, the surface of the sphere ($\mathbb{S}^2$) represents all possible mixes.

While Figures 1 and 2a show a space of track gains, there is clearly a redundancy of mixes in this space. What is ultimately desired is a space of mixes.

Definition 2. *Mix-space: a parameter space containing all the possible audio mixes that can be achieved using a defined set of processes.*

It becomes apparent that a Euclidean space with track gains as basis vectors is not an efficient way to represent a space of mixes, according to Definition 2. This explains why Equation (1) would not be appropriate when searching for mixes. If, in Figure 2a, a set of m points randomly selected on $\mathbb{R}^3$ were chosen, the number of mixes could be less than m, as the same mix could be chosen multiple times at different overall volumes. A set of m randomly selected points on a sphere of any radius ($\mathbb{S}^2$) would result in a number of mixes equal to m. This surface is represented in Figure 2b, which shows the portion of a unit-sphere in positively-unbounded $\mathbb{R}^3$, upon which exist all possible mixes of three tracks.

While both the 2-content of $\mathbb{S}^2$ (surface area) and the 3-content of the enclosing $\mathbb{R}^3$, (volume) both, strictly, contain an infinite amount of points, the reduced dimensionality of $\mathbb{S}^2$ makes it a more attractive content to use in optimisation, as $\mathbb{S}^2$ is a subset of $\mathbb{R}^3$ (in this context, *content* can be considered as "hypervolume". See http://mathworld.wolfram.com/Content.html). As a consequence, the *mix-space*, ϕ, is a more compact representation of audio mixes than the gain-space, g.

While the examples so far have used polar and spherical coordinates, for $n = 2$ and $n = 3$ respectively, to extend the concept to any n dimensions, hyperspherical coordinates are used. The conversion from Cartesian to hyperspherical coordinates is given below in Equation (5). The inverse operation, from hyperspherical to Cartesian, is provided in Equation (6), based on [11]. Here, g_j is the gain of the jth track out of a total of n tracks. The angles are represented by ϕ_i. By convention, ϕ_{n-1} is the equatorial angle, over the range $[0, 2\pi)$ radians, while all other angles range over $[0, \pi]$ radians.

$$r = \sqrt{g_n{}^2 + g_{n-1}{}^2 + \cdots + g_2{}^2 + g_1{}^2}$$
$$\phi_i = \arccos \frac{g_i}{\sqrt{g_n{}^2 + g_{n-1}{}^2 + \cdots + g_i{}^2}} \quad , \text{where } i = [1, 2, \ldots, n-3], i \in \mathbb{Z}$$

$$\vdots$$

$$\phi_{n-2} = \arccos \frac{g_{n-2}}{\sqrt{g_n^2 + g_{n-1}{}^2 + g_{n-2}{}^2}} \tag{5}$$

$$\phi_{n-1} = \begin{cases} \arccos \frac{g_{n-1}}{\sqrt{g_n^2 + g_{n-1}{}^2}} & g_n \geq 0 \\ 2\pi - \arccos \frac{g_{n-1}}{\sqrt{g_n^2 + g_{n-1}{}^2}} & g_n < 0 \end{cases}$$

$$g_1 = r \cos \phi_1$$
$$g_j = r \cos \phi_j \prod_{i=1}^{j-1} \sin \phi_i \, , \text{where } j = [2, 3, \ldots n-2], j \in \mathbb{Z} \tag{6}$$
$$g_n = r \prod_{i=1}^{n-1} \sin \phi_i$$

Figure 3 represents a comparable 4-track mixing exercise, as described in [7]. The four audio sources were specifically chosen for this example (vocals, guitar, bass and drums) and assigned to g_1, g_2, g_3 and g_4 respectively. Consequently, the set of mixes is represented by a 3-sphere of radius r. Due to the deliberate assignment of tracks in this example, the parameters ϕ_1, ϕ_2 and ϕ_3 represent a set of inter-channel balances which, due to the specific relationships of instruments, have importance to musicians and audio engineers: ϕ_3 determines the balance of bass to drums, the rhythm section in this case; ϕ_2 describes the projection of this balance onto the g_2 axis, i.e., the blend of guitar to rhythm section, and finally, ϕ_1 describes the balance of the vocal to this backing track.

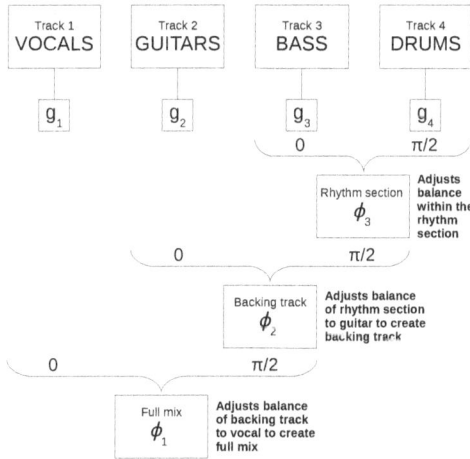

Figure 3. Schematic representation of a four-track mixing task, with track gains g_1, g_2, g_3, g_4, and the semantic description of the three ϕ terms, when adjusted from 0 to $\pi/2$. Figure taken from [7].

From here, the parameter space comprising the $n-1$ angular components of the hyperspherical coordinates of a $(n-1)$-sphere in a n-dimensional gain-space, is referred to as a $(n-1)$-dimensional mix-space. More simply, this can be stated by saying the mix-space is the surface of a hypersphere in gain-space. In the case of music mixing, only the positive values of g are of interest. Subsequently, the interesting region of the mix-space is only a small proportion of the total hypersurface. This fraction is $1/2^n$.

As each point in ϕ represents a unique mix, the process of mixing can be represented as a path through the space. In Figure 4a, a random walk begins at the point marked 'o' in the 2D mix-space (the origin [0,0], which corresponds to a gain vector of [1,0,0]). The model for the walk is a simple Brownian motion (http://people.sc.fsu.edu/~jburkardt/m_src/brownian_motion_simulation/brownian_motion_simulation.html). After 30 s, the walk is stopped and the final point reached is marked '×'. The gain values for each of the three tracks are shown in Figure 4b and it is clear that the random walk is on a 2-sphere, as anticipated. The time-series of gain values is shown in Figure 4c. Note that $g \in [-1, 1]$, so for positive g the region explored is as represented in Figure 2b.

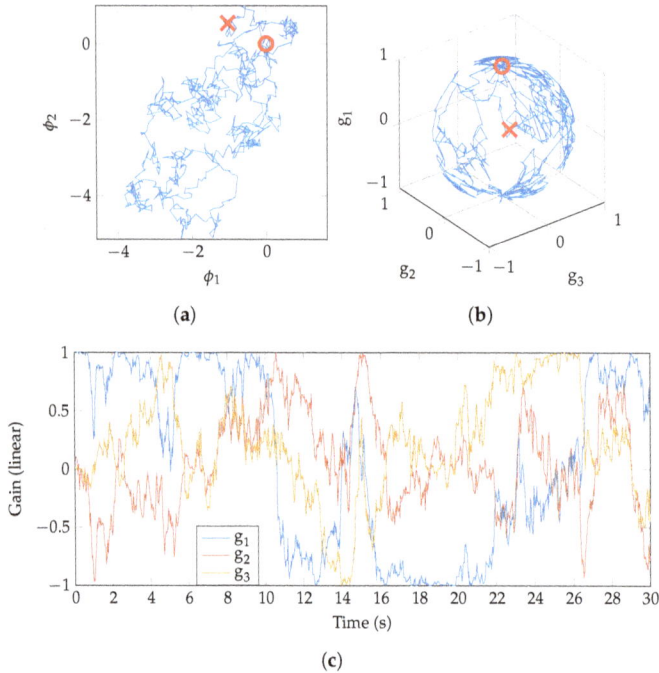

Figure 4. A time-varying mix can be considered as a path in the mix-space. Here, a random time-varying mix is generated by means of a random walk. (**a**) Random walk in mix-space; Brownian motion, halted after 30 s; (**b**) Random walk from Figure 4a converted to gain-space; (**c**) Time series of gain values for each of the three tracks.

When presented in isolation, such a random mix, whether static or time-varying, may be unrealistic. It is hypothesised that real mix engineers do not carry out a random walk but a guided and informed walk, from some starting point ("source") to their ideal final mix ("sink"). For further discussion of these terms, see [7], which uses the mix-space as a framework for the analysis of a simple 4-track mixing experiment. The power in these methods comes from generating a large number of mixes, more so than realistically could be obtained from real-world examples, and estimating parameters using statistical methods. Further generation and statistical analysis of time-varying mixes is left to further work.

2.2. Generating Gain Vectors by Sampling the Mix-Space

A set of mixes can be generated by choosing points in the mix-space. In selecting a suitable parametric distribution, it is important to note that linear distributions, such as the normal distribution, are not appropriate as the domain in question is not linear but a spherical surface. The statistics of such distributions are described by a number of equivalent terms in the literature, such as circular, spherical or directional statistics. In order to generate points close to a desired position on the

$(n-1)$-sphere, points are generated from a von-Mises–Fisher (vMF) distribution. The probability density function of the vMF distribution for a random n-dimensional unit vector $\mathbf{x}$ is given by

$$f_n(\mathbf{x}; \mu, \kappa) = C_n(\kappa) e^{\kappa \mu^T \mathbf{x}}$$

where $\kappa \geq 0, ||\mu|| = 1, n \geq 2$ and the normalisation constant $C_n(\kappa)$ is given by

$$C_n(\kappa) = \frac{\kappa^{n/2-1}}{(2\pi)^{n/2} I_{n/2-1}(\kappa)}.$$

Here, I_v is the modified Bessel function of the first kind at order v. The parameters μ and κ are called the mean direction and concentration parameter, respectively. The greater the value of κ, the higher the concentration of the distribution around the mean direction μ, resulting in lower variance. The distribution is unimodal for $\kappa > 0$ and is uniform on $\mathbb{S}^{n-1}$ for $\kappa = 0$. Further details can be found in [12,13]. The `SphericalDistributionsRand` (https://github.com/yuhuichen1015/SphericalDistributionsRand) code, based on the work of [14], was used to generate points according to a vMF distribution. In the context of audio mixes, μ (where $|\mu| = 1$) represents the mix about which others are distributed, akin to the mean in a normal distribution. The κ term represents the diversity of mixes generated, analogous (but inversely proportional) to variance. An example is shown in Figure 5, where three distributions are drawn from a 2-sphere.

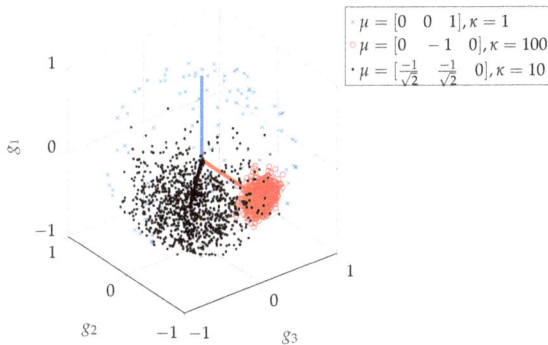

Figure 5. Three sets of mixes, drawn from the mix-space. This shows the effect of varying the concentration parameter κ, that a larger value results in less diversity.

2.2.1. Simple Mixing Model

From here, the example mixing session described is an 8-track session, containing vocals, guitars, bass and drums [15]. For $n = 8$ tracks, the gains required for the equal-loudness mix (once all audio tracks have been normalised in perceived loudness) are distributed around the following μ—each track gain is equal to n^{-2}, such that $|\mu| = 1$.

$$\mu = [0.3536 \quad 0.3536 \quad 0.3536 \quad 0.3536 \quad 0.3536 \quad 0.3536 \quad 0.3536 \quad 0.3536]$$

Previous studies have indicated that, while a good initial guess, presenting each track at equal loudness is not an ideal final mix. As suggested by three recent PhD theses on the topic [15–17], vocals are often the loudest element in a mix. To this equal loudness configuration, a vocal boost is added according to p.157 of [16], i.e., a boost of 6.54 dB. This addition of 6.54 dB to the vocal track produces the following vector, where track 8 is vocals.

$$\mu = [0.3536 \quad 0.3536 \quad 0.3536 \quad 0.3536 \quad 0.3536 \quad 0.3536 \quad 0.3536 \quad 0.7507]$$

If the previous vector was, then it is clear that this point is no longer on the unit 7-sphere. To project the point back onto the unit 7-sphere, the vector is normalised by dividing by the ℓ_2 (Euclidean) norm, resulting in the following.

$$\mu = [0.2948 \quad 0.2948 \quad 0.2948 \quad 0.2948 \quad 0.2948 \quad 0.2948 \quad 0.2948 \quad 0.6259] \qquad (7)$$

This vector is the new μ on the unit 7-sphere about which a set of mixes will be generated. The result is shown in Figure 6a. Each mix generated draws a gain value for each track such that the ℓ_2 norm is equal to 1. Note that the median values closely match the vector μ, as expected. Of course, there may not exist a mix which has these median values. This specific value of κ was chosen to avoid generating negative gains, achieved through trial and error. For a distribution which produces negative gains, the absolute value could be taken to avoid inverting the phase of the tracks. Ignoring phase, a gain of g is perceptually equal to $-g$, meaning that the shape of the distribution would be altered if negative gains were included.

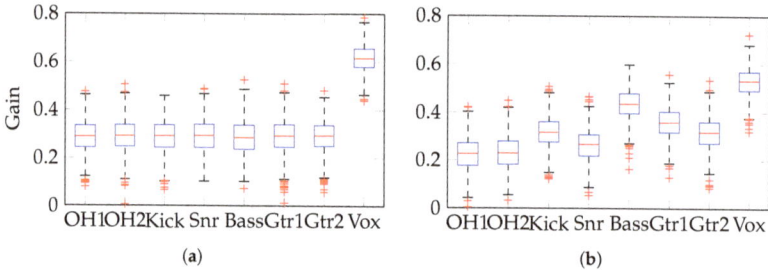

Figure 6. Boxplots of track gains for two generated datasets of mixes, drawn from separate distributions. (a) μ = Equation (7), $\kappa = 200$; (b) μ = Equation (8), $\kappa = 200$

2.2.2. Perceptual Mixing Model

Rather than a simple vocal boost, what is required is a more informed choice of instrument levels. In [7], a simple 4-track mixing exercise was reported, where participants created mixes of vocals, guitars, bass and drums using only volume faders. This experiment was expanded to an 8-track format, as in this paper, and is reported in [15]. Participants were asked the same task, only this time stereo-panning and a basic 3-band EQ was added. The median instrument levels obtained from this experiment are shown in Equation (8). Since participants had the ability to pan sources; the median levels were available for left and right channels separately, which are shown in Equations (10) and (11). Figure 6b shows the mixes obtained when the target vector is based on these median track levels, known as $\mu_{informed}$. It can be seen that the levels of bass guitar and kick drum are higher than average, while drum overheads have been attenuated. Vocals are set high in the mix, as seen in the mono experiment [7,15] and other previous studies [16,17]. Matlab code for generating sets of mixes, as in Figure 6, is available for download (https://github.com/alexwilson101/PopulateMixSpace).

$$\mu_{informed} = [0.2254 \quad 0.2282 \quad 0.3221 \quad 0.2679 \quad 0.4437 \quad 0.3616 \quad 0.3221 \quad 0.5387] \qquad (8)$$

2.3. Track Panning

Thus far, only mono mixes have been considered, where all audio tracks are summed to one channel. In creative music production, it is rare that mono mixes are encountered. The same mathematical formulations of the mix-space can be used to represent panning. Consider Figure 4, which shows track gains in the range $[-1, 1]$. Should these be replaced with track pan positions p_n (with -1 and 1 corresponding to extreme left and right pan positions, for example) then the mix-space (or "pan-space") can be used to generate a position for each track in the stereo field. To avoid confusion with the earlier use of ϕ, the pan-space is denoted by θ, although the formalism is identical.

$$\left(\underbrace{p_1, p_2, p_3, \ldots, p_n}_{\text{absolute panning}} \right) = \left(\underbrace{r_{\text{pan}}, \underbrace{\theta_1, \theta_2, \ldots, \theta_{n-1}}_{\text{width-scaling}}}_{\text{pan-space}} \right) \tag{9}$$

However, the mix-space for gains (ϕ) takes advantage of the fact that a mix (in terms of track gains only) is comprised of a series of inter-channel gain ratios, meaning that the radius r is arbitrary and represents a master volume. In terms of track panning, one obtains a series of inter-channel panning ratios, the precise meaning of which is not intuitive. Additionally, the radius r_{pan} would still be required to determine the exact pan position of the individual tracks. Therefore, the pan-space describes the relative pan positions of audio tracks to one another.

For a simple example with only two tracks, the meaning of r_{pan} and θ is relatively simple to understand. Consider the unit circle in a plane where the Cartesian coordinates (x, y) represent the pan positions of two tracks, as shown in Figure 7. Mix A is at the point $(\frac{1}{\sqrt{2}}, \frac{1}{\sqrt{2}})$: both tracks are panned at the same position. As this is a circle with arbitrary radius, r_{pan}, then the radius controls how far positive (right) the two tracks are panned, from 0 (centre) to $+1$ (far right). Mix B does the same but towards the left channel. One may ask whether A and B are identical "panning-mixes", as p and p' in Figure 1 were identical "level-mixes"?

Now consider mix C, where one track is panned left and the other right. Mix D is simply the mirror image of this. Are *these* to be considered as the same mix, or as different mixes? Here, r_{pan} adjusts the distance between the two tracks, from both centre when $r_{\text{pan}} = 0$, to $(-1, 1)$ when $r_{\text{pan}} = \sqrt{2}$ (as indicated by mix C'). Does a change in r_{pan} change the mix, or is it simply the same mix only wider/narrower? Overall, the angle θ adjusts the panning mix and r_{pan} is used to obtain absolute positions in the stereo field, at a particular width-scale (i.e., to zoom in or zoom out).

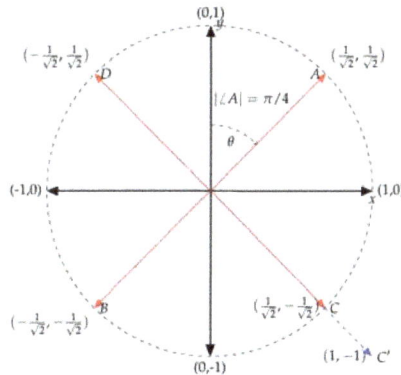

Figure 7. Panning of two tracks, represented as a 1-sphere. The panning mix is determined by the angle θ with r_{pan} acting as a scaling variable, adjusting the overall width of the mix. For example, C' is a wider version of C.

2.3.1. Method 1—Separate Left and Right Gain Vectors

The method for random gains (see Section 2.2) was used to create separate mixes for the left and right channels of a stereo mix. In absolute terms, hard-panning only exists when the gain in one channel is 0 (*perceptually*, the impression of hard-panning can be achieved when the difference between one channel and the other is sufficiently large [18]). Since the vocal boost prevents any vocal gain of zero, the panning of the vocals is much less wide than the other tracks. Additionally, since $\kappa = 200$ was chosen to prevent any negative gains, there are few zero-gain instances; therefore, there is a lack of hard-panning. Figure 8a,b show the gain settings produced and a boxplot of the resulting pan positions is shown in Figure 8c, where the inter-quartile range extends to ± 0.4 for the seven instrument tracks and about ± 0.2 for the vocals. The estimated density of pan positions for each track is shown, illustrating the relatively narrow vocal panning. As expected, these estimated density functions are Gaussian, to a good approximation.

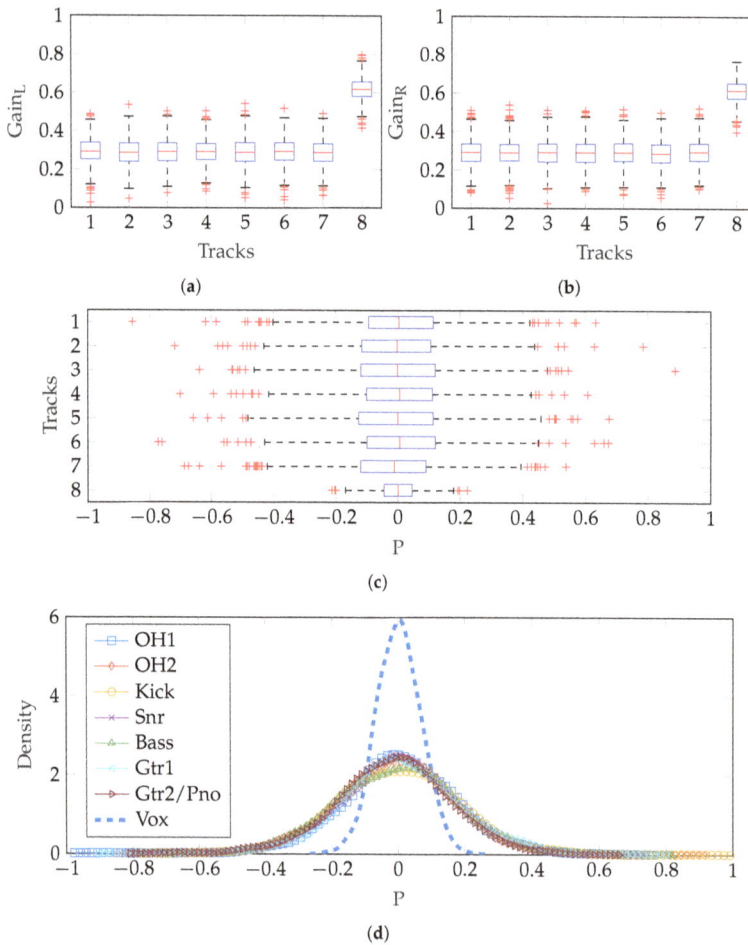

Figure 8. Panning method 1—separate vMF distributions for gain$_L$ and gain$_R$, both using Equation (7). (**a**) Boxplot of track gains for left channel, using Equation (7); (**b**) Boxplot of track gains for right channel, using Equation (7); (**c**) Boxplot of pan positions for each track; (**d**) Probability density of pan positions for each track.

Rather than using the same μ for both left and right channels, a unique choice of μ_L and μ_R can be made, as described in Section 2.2.2. The vectors used are shown in Equations (10) and (11). When summed to mono, this is equivalent to Equation (8).

$$\mu_L = [0.2741 \quad 0.1354 \quad 0.3361 \quad 0.2657 \quad 0.4401 \quad 0.3796 \quad 0.2566 \quad 0.5651] \tag{10}$$

$$\mu_R = [0.1189 \quad 0.2597 \quad 0.3162 \quad 0.2612 \quad 0.4683 \quad 0.2935 \quad 0.3727 \quad 0.5531] \tag{11}$$

Figure 9 shows the difference in gains produced for left and right channels. There were some negative track gains produced: when generating audio mixes, the absolute magnitude of the gain was used to avoid phase inversions which would alter spatial perception of the stereo overhead pair. It is clear that the similarity of vocals gains in left and right channels produces a limited variety of pan positions close to the central position, as shown in Figure 9c,d. Other instruments are panned with mean position and variance in accordance with the experimental results [15].

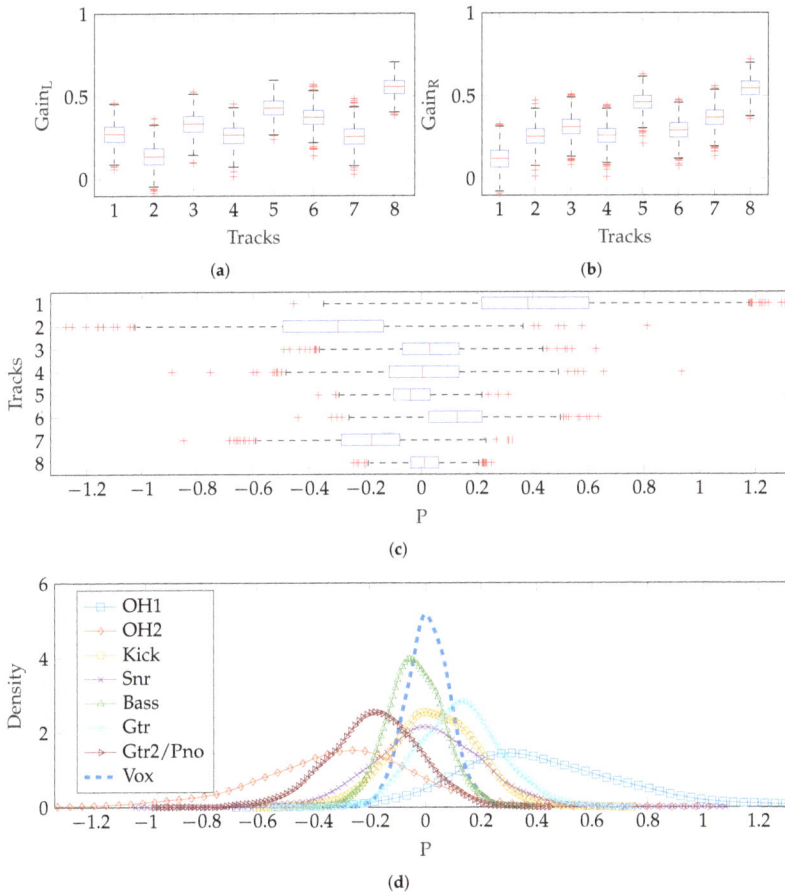

Figure 9. Panning method 1b—separate vMF distributions for left and right channels but using unique μ vectors, shown in Equations (10) and (11). (**a**) Boxplot of track gains for left channel, using Equation (10); (**b**) Boxplot of track gains for right channel, using Equation (11); (**c**) Boxplot of pan positions for each track. Where $|P| > 1$, this is caused by negative track gains; (**d**) Probability density of pan positions for each track.

2.3.2. Method 2—Separate Gain and Panning

This method involved generating random mono mixes as Section 2.2 (using Equation (7) and then generating pan positions separately. A μ_{pan} was created for a vMF distribution. This vector was based on experimental results reported in [15], which showed that, generally, overheads and guitars were widely panned while kick, snare, bass and vocals were positioned centrally.

$$\mu_{pan} = [-0.5 \quad 0.5 \quad 0 \quad 0 \quad 0 \quad -0.4 \quad 0.4 \quad 0] \tag{12}$$

This then needs to be a unit vector for it to be used in creating vMF-distributed points. Consequently, the precise values are not critically important, as it is the relative pan positions that are reflected in the normalised vector and r_{pan} which would be used to adjust the scaling of these relative positions.

$$\mu_{pan} = [-0.5522 \quad 0.5522 \quad 0 \quad 0 \quad 0 \quad -0.4417 \quad 0.4417 \quad 0] \tag{13}$$

Three different values for κ were used, which illustrates how this parameter controls the distribution of panning. The results are shown in Figure 10, where the influence of κ is clear. When $\kappa \to 0$, the distribution of pan positions approaches uniform over the sphere, and so the median pan positions are close to 0 (central position in the stereo field) for all tracks, regardless of μ_{pan}. As κ increases, the distribution of pan positions is narrower, more concentrated on the specific pan positions specified in μ_{pan}.

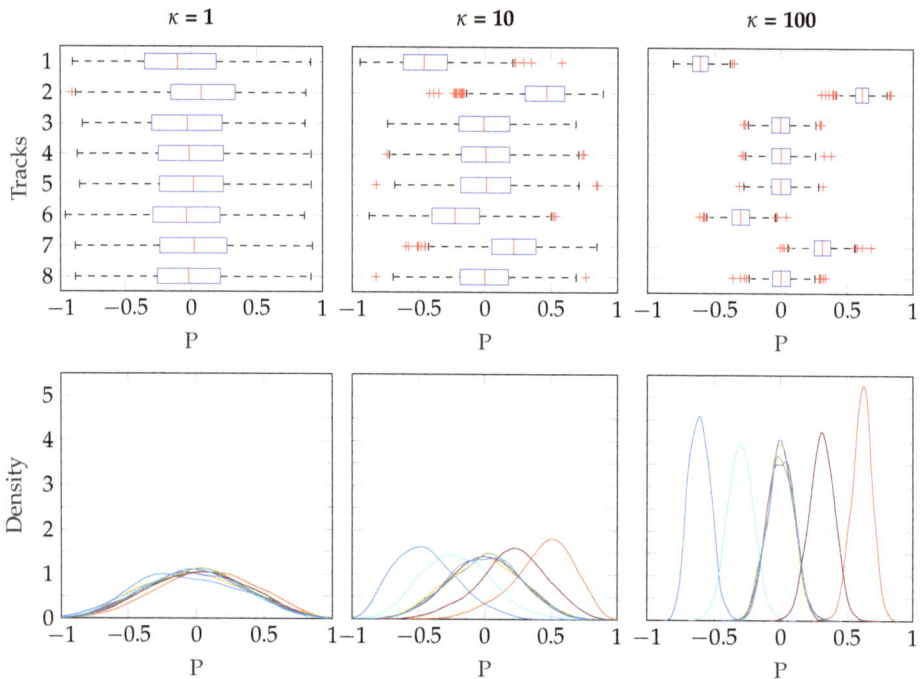

Figure 10. Panning method 2—generating vMF distributions in panning space. As expected, increasing κ (concentration parameter) results in a narrower range of pan positions for each track, around the target vector Equation (13).

Figure 11 shows an example of two mixes created using this method. The gains and pan positions of each track are displayed. It is clear that the instruments are typically panned close to the positions specified in the pan vector (Equation (13). In this example, $r_{pan} = 1$; increasing this parameter would produce wider mixes, while a decrease would produce a less wide mix.

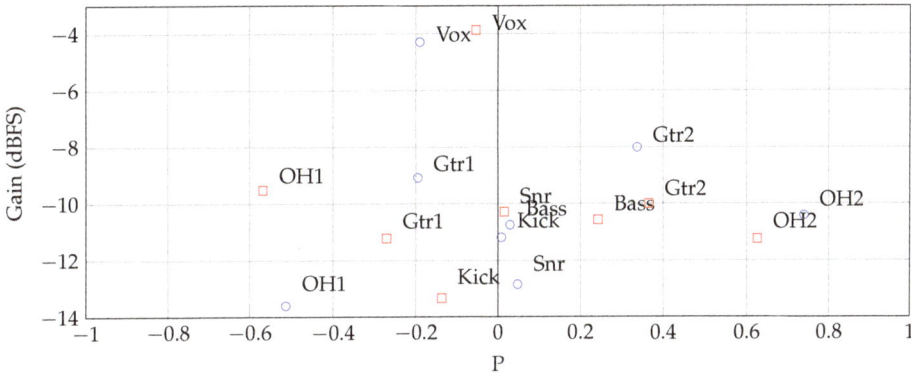

Figure 11. Two random mixes generated using panning method 2, shown as squares and circles. Each mix has a different gain vector (based on Equation (7) and different pan vector (based on Equation (13).

2.4. Track Equalisation

Similarly to how the mix can be considered as a series of *inter-channel* gain ratios, when the frequency-response of a single audio track is split into a fixed number of bands, the *inter-band* gain ratios can be used to construct a *tone-space* using the same formulae. For three bands, with gain of low, middle and high bands in the filter being g_{low}, g_{mid} and g_{high} respectively, the problem is comparable to the 3-track mixing problem shown in Figure 2a. Again, one can convert this to spherical coordinates (by Equations (5) and obtain $[r_{EQ}, \psi_1, \psi_2]$, yet, in this case, the values of ψ_n control the EQ filter applied, and r_{EQ} is the total amplitude change produced by equalisation (to avoid confusion, ψ is used in place of ϕ when referring to equalisation). As before, if all three bands are increased or decreased by the same proportion, then the tone of the instrument does not change apart from an overall change in presented amplitude, r_{EQ}. Analogous to its use in track gains, the value of ψ_2 adjusts the balance between g_{mid} and g_{high}, while ψ_1 adjusts the balance of g_{low} to the previous balance.

$$\left(\underbrace{g_1, g_2, g_3, \ldots, g_{n_{bands}}}_{\text{gains of filter bands}} \right) = \left(\underbrace{r_{EQ}}_{\text{scaling}}, \underbrace{\psi_1, \psi_2, \ldots, \psi_{n-1}}_{\text{tone-space}} \right) \tag{14}$$

In Figure 12, five points are randomly chosen in the *tone-space*. These co-ordinates are converted to three band gains as before, except that, in order to centre on a gain vector of $[1, 1, 1]$, $r_{EQ} = \sqrt{n_{bands}}$, which is $\sqrt{3}$ in this example. Of course, this method can be used for any number of bands.

With this method, one must assume that an audio track has equal amplitude in each band, which is rarely the case. When g_L is increased on a hi-hat track, there may be little effect, compared to a bass guitar. Therefore, the loudness change is a function of r_{EQ} and the spectral envelope of the track, prior to equalisation. This is not considered here and is left to further work.

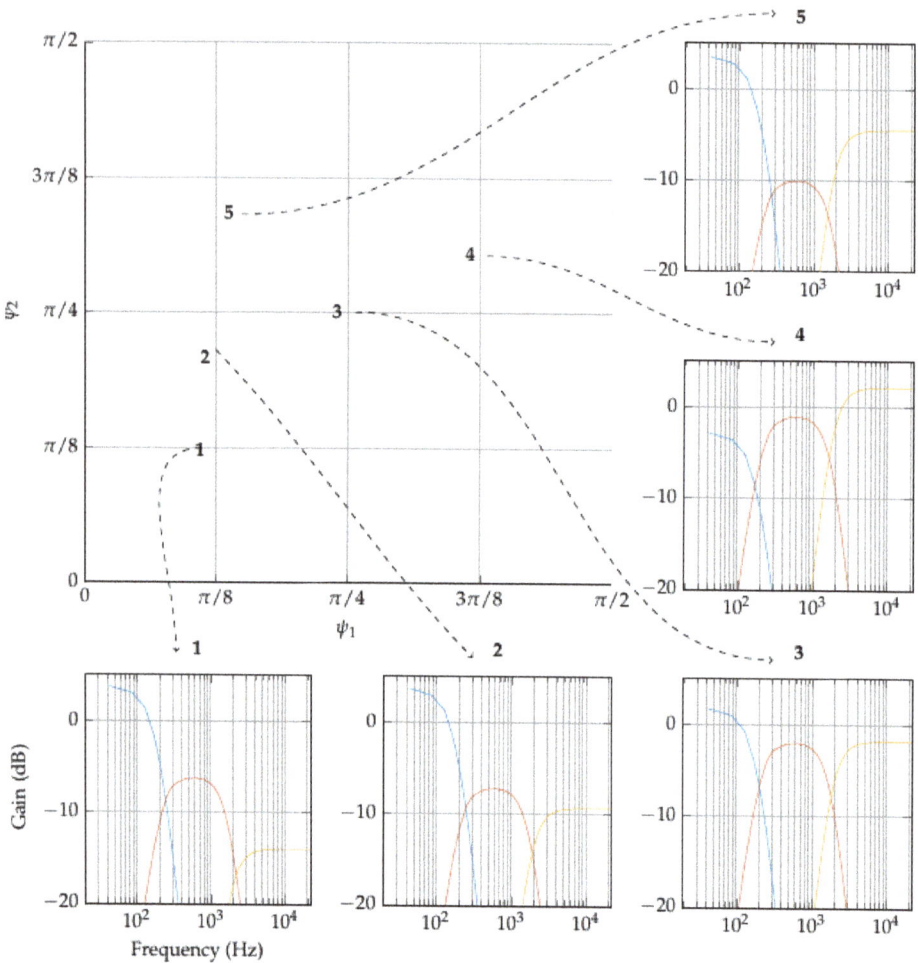

Figure 12. Five randomly-chosen examples of 3-band equalisation, chosen from the tone-space. As $\psi_2 \to 0$, the gain of the high band decreases. As $\psi_1 \to 0$, the gain of the low band increases at the expense of the other two bands; their balance is determined by ψ_2.

3. Applications

Being able to generate artificial datasets of audio mixtures in the mix-space has a variety of applications. Two such applications are described here. The procedure is similar for both experiments: an audio mix is created using a generated gain vector and raw multitrack audio, resulting in a generated mix from which audio signal features may be determined. Feature extraction used the MIRtoolbox [19], version 1.6.1. Equations (7) and (8) were used to create two sets of mixes. These experiments use sets of 500 mixes, rather than 1000 as outlined in earlier sections. It can be shown that the distributions of audio signal features do not change much beyond 500 mixes [15]. The reduced computation time is advantageous in these examples.

The test audio in these experiments is 30-second segments of the songs "Burning Bridges", "I'm Alright" and "What I Want" as used in previous studies [8,15], available from the Mixing Secrets free multitrack download library (http://www.cambridge-mt.com/ms-mtk.htm). The raw multitrack audio was reduced to the required eight tracks and each track was normalised in perceived loudness according to a modified form of ITU BS.1770 [20]. The songs "I'm Alright" and "What I Want" feature a track of piano as track #7, in place of 'Gtr 2'.

3.1. Testing the Robustness of Tempo Estimation Algorithms to Changes in the Mix

In the absence of any time-stretching processes, the tempo of each mix should be identical for a given song. As a result, if the tempo of alternate mixes is estimated and any disagreement is found, this suggests limitations in the tempo-estimation algorithm. In this section, the process of estimating tempo across a large set of artificial mixes is presented as a means of assessing the performance of tempo-estimation algorithms. Two such algorithms are tested herein: the classic and metre-based [21] implementations of `mirtempo` in the MIRtoolbox. In short, the classic tempo estimation algorithm performs onset detection based on the amplitude envelope of the audio. Periodicities in the detected onsets are determined by finding peaks in the autocorrelation function. The metre method additionally takes into account the metrical hierarchy of the audio, allowing for a more consistent tempo-tracking. Whichever tempo-estimation is used, the resultant tempo is the mean value over the 30-second audio segment. Panning and equalisation were not considered here as tempo was estimated from a mono signal.

Figures 13a and 14a show the results for "Burning Bridges", where it is clear that the classic method performs poorly. The correct tempo of 100 bpm is estimated for only a small percentage of the mixes while all others are estimated close to 133 bpm (see Figure 14a). This leads to a high mean squared error (MSE) as shown in Table 1. A similar flaw is evident for "I'm Alright" where the tempo is again overestimated by roughly 33% for both mix distributions (see Figures 13b and 14b). This indicates a consistent error in the tempo-estimation routine, which is being revealed by these mix distributions. The metre-based method performs much better, estimating the correct tempo in almost all cases and exhibiting a lower MSE, with only a small amount of absolute error (0.1–0.2 bpm). The performance of the classic tempo-estimation method is improved for "What I Want", where both methods are found to have a high level of accuracy, as shown in Figures 13c and 14c. For both distributions, the metre-based version produces clusters of solutions for "What I Want", although the tempo represented by largest cluster is consistent.

It is conceivable that no tempo-estimation algorithm is able to obtain the correct result in all cases. What this experiment reveals is that there is also variation within the mixes of a given song, with some mixes providing the correct tempo and other mixes yielding error, with different estimation methods showing varying levels of robustness to mixing practice.

Table 1. Summary of tempo estimation accuracy results. Shown is the mean squared error (MSE) in each set of 500 mixes.

Audio	BPM	Mixes	Mirtempo (Classic)	Mirtempo (Metre)
Burning Bridges	100	Equation (7)	998.54	13.05
		Equation (8)	1082	0.0147
I'm Alright	96	Equation (7)	738.4297	16.4854
		Equation (8)	742.37	16.7442
What I Want	99	Equation (7)	1.5110	0.3803
		Equation (8)	0.8394	0.4251

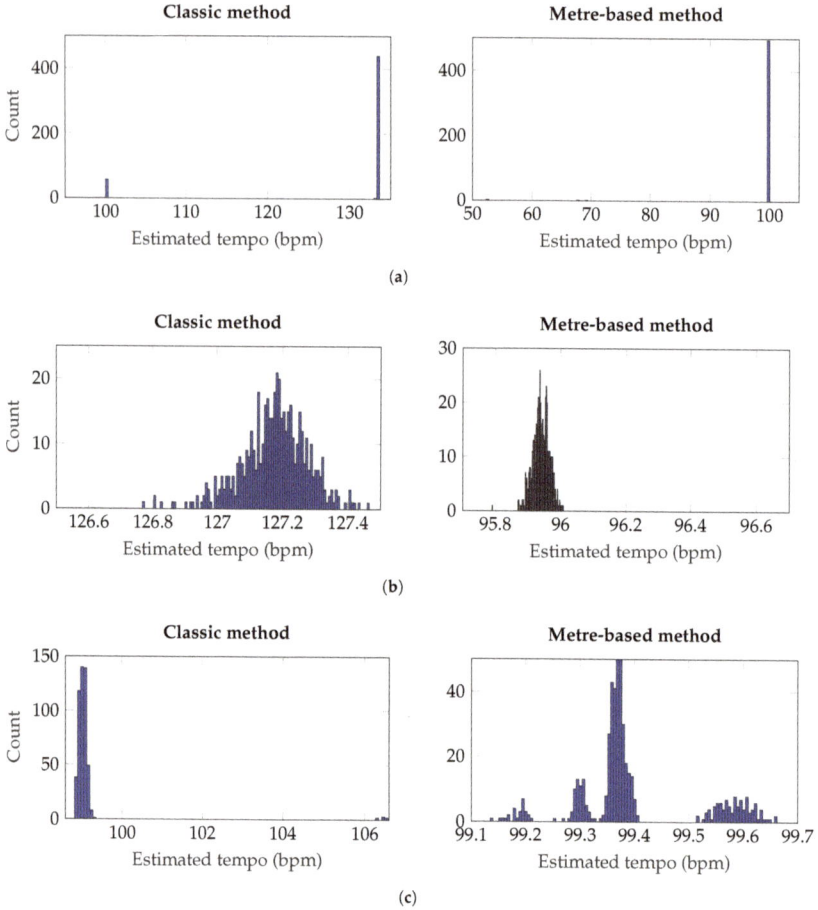

Figure 13. Estimated tempo for three songs, 500 mixes each using Equation (7). In each histogram, the data is split into 100 bins. Overall, performance is better for the metre-based method, as it demonstrates greater accuracy and improved robustness to changes in the mix. (**a**) "Burning Bridges"—The correct tempo is ≈100 bpm; (**b**) "I'm Alright"—The correct tempo is ≈96 bpm; (**c**) "What I Want"—The correct tempo is ≈99 bpm.

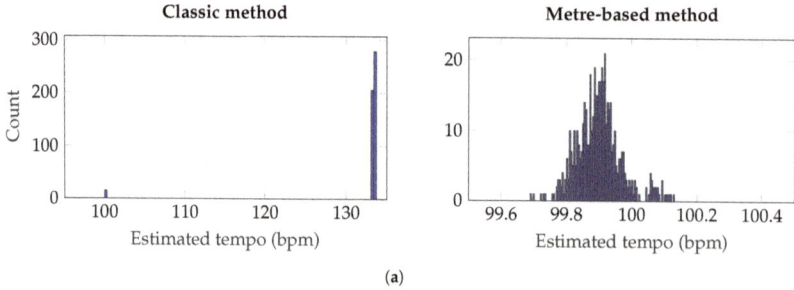

Figure 14. *Cont.*

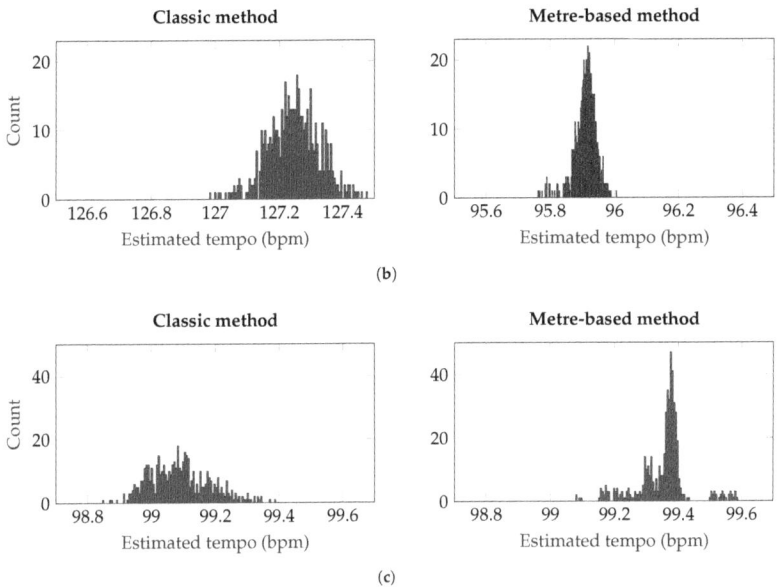

Figure 14. Estimated tempo for three songs, 500 mixes each using Equation (8). In each histogram, the data is split into 100 bins. Overall, performance is better for the metre-based method, as it demonstrates greater accuracy and improved robustness to changes in the mix; (**a**) "Burning Bridges"—The correct tempo is ≈100 bpm; (**b**) "I'm Alright"—The correct tempo is ≈96 bpm; (**c**) "What I Want"—The correct tempo is ≈99 bpm.

3.2. Estimation of Spectral Centroid in Sets of Mixes

It is common to use the spectral centroid as a feature to describe the timbre of an audio signal, specifically as an approximation to perceptual brightness [22–24]. However, where the spectral centroid of a mixed recording is evaluated, it is not clear that the value obtained is typical of the recording as a whole, or if it simply relates to that specific mix of the recording. This is especially problematic in an object-based audio broadcast, where no reference mix exists. This applies to any signal feature, not just the spectral centroid. As studies of features across multiple alternate mixes are still rare in the literature [8,25,26], this issue has not been adequately investigated.

A previous work by the authors [8] reports on the spectral centroid of 1501 user-generated mixes of 10 songs. The number of mixes per song ranges from 97 to 373. The estimated probability distributions of spectral centroid are shown (among other signal features relating to amplitude, timbre and spatial properties), indicating that the median spectral centroid can vary by song, although it is still possible for significant overlap in distributions to exist.

The work in this section investigates the distributions of the spectral centroid that occur for artificial mixes drawn from different mix-space distributions. Equation (7) describes a simple model for mixes while Equation (8) shows the result of a perceptual level-balancing experiment. What is it about the mix that changes when these levels are adjusted? In this section, an estimation of the median spectral centroid produced by these two sets of mixes is made using Monte Carlo methods.

The experiment was conducted as follows. Using $\mu =$ Equation (7) and $\kappa = 200$, a set of 500 gain vectors was generated. For each of these vectors, a mix was created and the spectral centroid was measured. This resulted in 500 measurements of spectral centroid, the density of which was estimated using Kernel Density Estimation (KDE). This procedure was repeated for a second set of 500 mixes, generated using $\mu =$ Equation (8) and $\kappa = 200$. The estimated density distribution of both is plotted in

Figure 15. These distributions were compared using a Wilcoxon rank sum test, which tests the null hypothesis that the distributions of both samples are equal. This null hypothesis was rejected in each case, as shown by the *p*-values in each subplot of Figure 15 ($p < 0.05$ in each case).

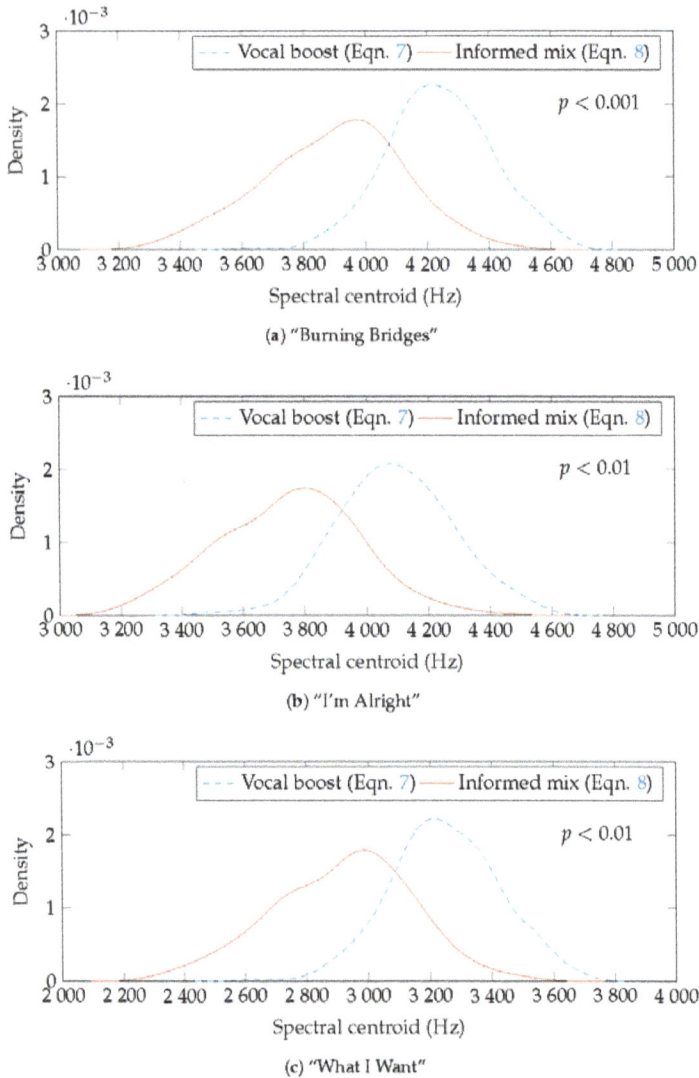

(a) "Burning Bridges"

(b) "I'm Alright"

(c) "What I Want"

Figure 15. Probability distribution of spectral centroid as a function of mix-space parameters; (a) "Burning Bridges"; (b) "I'm Alright"; (c) "What I Want".

The significant difference between the medians of the two groups illustrates that there is a coarse perceptual difference in timbre, generally, between mixes drawn from the two distributions. This is true for all three songs considered. Of course, whether or not there is a significant difference between the medians of the two groups depends on the chosen parameters: the μ vectors must be perceptually different but if κ is low enough, then the distributions will overlap, regardless of the choice of μ (recall that as $\kappa \to 0$, the distribution approaches uniformity). The choice of κ depends on the application.

The higher spectral centroid in the simple equal-gain-with-vocal-boost approach (Equation (7)) is caused by an overestimation in the level of the drum overheads and vocal, and an underestimation of the level of bass and kick drum, when compared to the results of the perceptual test (Equation (8)). The distributions of the spectral centroid for these artificially-generated mixes were compared to the distributions of the spectral centroid for user-generated mixes, as were reported in previous work by the authors [8]. For "Burning Bridges" and "What I Want", the peak of the $\mu_{informed}$ distribution compares well to the user-generated mixes (approx 3.8 kHz and 3.2 kHz respectively). In the case of "I'm Alright", $\mu =$ Equation (7) yields a better match to real mixes (approx 4.2 kHz); however, the 373 user-generated mixes of this song from [8] did contain a large proportion of highly amateur, potentially low-quality, mixes. For further comparison of artificial mixes and user-generated mixes, see [15].

This experiment shows that a set of mixes can be obtained by sampling the mix-space but that perceptually-relevant mixes are more likely to be obtained if some level of human guidance is fed into the system. The parametric mixing model for this experiment did not feature panning or equalisation. It has been shown that the addition of equalisation broadens the distribution of spectral centroid values, as would be expected given the wider variety of instrument tone [15].

4. Discussion

4.1. Artificial Datasets for Testing of Processes

The theoretical framework presented in this paper provides for a space of mixes that can be explored, using evolutionary computing, machine learning or similar computational methods. Applications of this include the creation of an initial population of solutions to be used in the search of balance-mixes [9] and electric guitar tones [10], both using interactive genetic algorithms. These approaches have yielded positive results, as the user is able to search the space effectively and find the desired solution.

For subjective testing, the methods presented in this paper have the advantage that each mix is generated at a constant perceived loudness, as the magnitude of the gain vector can be set to a constant (such as $r = 1$ in Equation (4). In both [9,10], which used an *interactive* genetic algorithm, test participants were asked to rate subjectively the solutions presented. Being generated at a consistent loudness level allowed for fair evaluations, while avoiding the additional computational time required for specific loudness-normalisation to be applied to each generated mix. This allows a more free exploration of the solution space, since audio stimuli can be generated in real-time using this method.

Currently, newly-developed algorithms for tempo estimation, key estimation etc., are evaluated during specific challenges, such as the MIREX audio tempo estimation challenge (http://www.music-ir.org/mirex/wiki/2017:Audio_Tempo_Estimation), using standard datasets of audio recordings. We propose that sets of artificially generated mixes be considered as a standard test, in order to examine the level of robustness to mixing practice, as in Section 3.1.

4.2. Signal Analysis of Audio Mixing Practices

Of course, more conventional experiments can be analysed in this framework. In a level-balancing task, where participants were asked to set track gains to their desired levels, the resulting gains can be converted to the mix-space and analysed therein [7]. This allows differences in cohorts to be investigated: thus far, the different mixes produced by headphone or loudspeaker users has been investigated [15] in addition to checking if changing the initially presented rough mix influences the mixing-decisions [7], a hypothesis also supported by later work [6].

A recent work analysed the audio mixes of broadcast audio stems (dialogue, foreground sound effects, background sound effects and music) as produced by hearing-impaired listeners [27]. This 4-track mixing scenario is equivalent to that represented by Figure 3. The changes in level made to each mix stem were reported in a bar chart, showing an increase in dialogue level and a decrease in the level of the other three stems. From a mix-space perspective, we know that these two strategies are equivalent. The mixes created from such an experiment can be more effectively analysed in a 3-dimensional mix-space, in which it could be more clear how different cohorts (such as hearing-impaired listeners) would balance the four tracks in different ways. If the needs of the user demand a change to the audio mix, as in the case of increasing speech intelligibility, then the path from the current mix to the desired mix may be more easily determined in the mix-space.

As object-based audio broadcast becomes commonplace, audio signal feature extraction algorithms will need to be robust to changes in the audio object, be it changes in amplitude, panning, equalisation, or other parameters. It has been shown that the measured value of pulse clarity (a measure of how easy it is to pick out the underlying rhythm of a mix [28]) varies with object loudness, typically decreasing as the mix moves into regions of the mix-space where the relative level of vocals is increased [15].

5. Conclusions

A method for the creation of artificial audio mixes has been presented. This has been achieved by the parametric generation of points in a novel "mix-space", a concise representation of three audio processing activities: level-balancing, stereo-panning and equalisation.

This method has been used for a number of application thus far: in creating an initial population for evolutionary algorithms [9,10] and two simple experiments estimating the values of audio signal features using Monte Carlo techniques. This has revealed limitations in tempo-estimation algorithms. This paper suggests that, in the future, such algorithms need to be robust to changes in instrument level and other mixing practices. This will allow such routines to be applied to an object-based paradigm of audio broadcast, where no reference mix may exist on which to determine the value of the feature.

Future work is required to further generalise the presented models to audio mixing practices, such as dynamic range processing, as well as implementing a fully-parametric model of time-varying mixes and the related statistical analysis.

Author Contributions: Portions of the work described in this paper are from the PhD thesis of A.W., under the supervision of B.M.F. A.W. conceived and designed the experiments; A.W. performed the experiments; A.W. and B.M.F. analyzed the data; A.W. contributed materials/analysis tools; A.W. and B.M.F. wrote the paper.

Conflicts of Interest: The authors declare no conflict of interest. No funding sponsors had a role in the design of the study; in the collection, analyses, or interpretation of data; in the writing of the manuscript, or in the decision to publish the results.

References

1. Gonzalez, E.; Reiss, J. Improved control for selective minimization of masking using Inter-Channel dependancy effects. In Proceedings of the 11th International Conference on Digital Audio Effects (DAFx-08), Espoo, Finland, 1–4 September 2008.
2. Tsilfidis, A.; Papadakos, C.; Mourjopoulos, J. Hierarchical perceptual mixing. In Proceedings of the 126th AES Convention, Munich, Germany, 7–10 May 2009.
3. Reiss, J.D. Intelligent systems for mixing multichannel audio. In Proceedings of the IEEE 17th International Conference on Digital Signal Processing, Corfu, Greece, 6–8 July 2011.
4. Cartwright, M.; Pardo, B.; Reiss, J. Mixploration: Rethinking the audio mixer interface. In Proceedings of the ACM 19th International Conference on Intelligent User Interfaces, Haifa, Israel, 24–27 February 2014.
5. Terrell, M.; Simpson, A.; Sandler, M. The mathematics of mixing. *J. Audio Eng. Soc.* **2014**, *62*, 4–13.
6. Jillings, N.; Stables, R. A semantically powered digital audio workstation in the browser. In Proceedings of the Audio Engineering Society International Conference on Semantic Audio, Erlangen, Germany, 22–24 June 2017.
7. Wilson, A.; Fazenda, B.M. Navigating the Mix-Space: Theoretical and practical level-balancing technique in multitrack music mixtures. In Proceedings of the 12th Sound and Music Computing Conference, Maynooth, Ireland, 24–26 October 2015.

8. Wilson, A.; Fazenda, B. Variation in Multitrack Mixes: Analysis of Low-level Audio Signal Features. *J. Audio Eng. Soc.* **2016**, *64*, 466–473.

9. Wilson, A.; Fazenda, B. An evolutionary computation approach to intelligent music production, informed by experimentally gathered domain knowledge. In Proceedings of the 2nd AES Workshop on Intelligent Music Production, London, UK, 13 September 2016.

10. Wilson, A. Perceptually-motivated generation of electric guitar timbres using an interactive genetic algorithm. In Proceedings of the 3rd Workshop on Intelligent Music Production, Salford, UK, 14 September 2017.

11. Blumenson, L.E. A Derivation of n-Dimensional Spherical Coordinates. *Am. Math. Mon.* **1960**, *67*, 63–66.

12. Fisher, N.I. *Statistical Analysis of Circular Data*; Cambridge University Press: Cambridge, UK, 1995.

13. Mardia, K.V.; Jupp, P.E. *Directional Statistics*; John Wiley & Sons: Hoboken, NJ, USA, 2009; Volume 494.

14. Chen, Y.H.; Wei, D.; Newstadt, G.; DeGraef, M.; Simmons, J.; Hero, A. Statistical estimation and clustering of group-invariant orientation parameters. In Proceedings of the IEEE 18th International Conference on Information Fusion, Washington, DC, USA, 6–9 July 2015.

15. Wilson, A. Evaluation and Modelling of Perceived Audio Quality in Popular Music, towards Intelligent Music Production. Ph.D. Thesis, University of Salford, Salford, UK, 2017.

16. Pestana, P. Automatic Mixing Systems Using Adaptive Audio Effects. Ph.D. Thesis, Universidade Catolica Portuguesa, Lisbon, Portugal, 2013.

17. De Man, B. Towards a Better Understanding of Mix Engineering. Ph.D. Thesis, Queen Mary, University of London, London, UK, 2017.

18. Lee, H.; Rumsey, F. Level and time panning of phantom images for musical sources. *J. Audio Eng. Soc.* **2013**, *61*, 978–988.

19. Lartillot, O.; Toiviainen, P. A matlab toolbox for musical feature extraction from audio. In Proceedings of the 10th International Conference on Digital Audio Effects (DAFx-07), Bordeaux, France, 10–15 September 2007.

20. Pestana, P.D.; Reiss, J.D.; Barbosa, A. Loudness measurement of multitrack audio content using modifications of ITU-R BS.1770. In Proceedings of the 34th AES Convention; Audio Engineering Society, Rome, Italy, 4 May 2013.

21. Lartillot, O.; Cereghetti, D.; Eliard, K.; Trost, W.J.; Rappaz, M.A.; Grandjean, D. Estimating Tempo and metrical features by tracking the whole metrical hierarchy. In Proceedings of the 3rd International Conference on Music & Emotion (ICME3), Jyväskylä, Finland, 11–15 June 2013.

22. von Bismarck, G. Timbre of steady sounds: A factorial investigation of its verbal attributes. *Acta Acust. United Acust.* **1974**, *30*, 146–159.

23. Grey, J.M.; Gordon, J.W. Perceptual effects of spectral modifications on musical timbres. *J. Acoust. Soc. Am.* **1978**, *63*, 1493–1500.

24. McAdams, S.; Winsberg, S.; Donnadieu, S.; De Soete, G.; Krimphoff, J. Perceptual scaling of synthesized musical timbres: Common dimensions, specificities, and latent subject classes. *Psychol. Res.* **1995**, *58*, 177–192.

25. De Man, B.; Leonard, B.; King, R.; Reiss, J. An analysis and evaluation of audio features for multitrack music mixtures. In Proceedings of the 15th International Society for Music Information Retrieval Conference, Taipei, Taiwan, 27–31 October 2014.

26. Wilson, A.; Fazenda, B.M. 101 Mixes: A statistical analysis of mix-variation in a dataset of multitrack music mixes. In Proceedings of the 139th AES Convention, Audio Engineering Society, New York, NY, USA, 29 October–1 November 2015

27. Shirley, B.G.; Meadows, M.; Malak, F.; Woodcock, J.S.; Tidball, A. Personalized object-based audio for hearing impaired TV viewers. *J. Audio Eng. Soc.* **2017**, *65*, 293–303.

28. Lartillot, O.; Eerola, T.; Toiviainen, P.; Fornari, J. Multi-feature modeling of pulse clarity: Design, validation and optimization. In Proceedings of the 9th International Society for Music Information Retrieval Conference, Philadelphia, PA, USA, 14–18 September 2008.

applied
sciences

MDPI

Article

Virtual Analog Models of the Lockhart and Serge Wavefolders [†]

Fabián Esqueda [1,*], Henri Pöntynen [1], Julian D. Parker [2] and Stefan Bilbao [3]

[1] Aalto Acoustics Lab, Aalto University, 02150 Espoo, Finland; henri.pontynen@aalto.fi
[2] Native Instruments GmbH, D-10997 Berlin, Germany; julian.parker@native-instruments.de
[3] Acoustics and Audio Group, University of Edinburgh, Edinburgh EH9 3FD, Scotland, UK; s.bilbao@ed.ac.uk
[*] Correspondence: fabian.esqueda@aalto.fi; Tel.: +358-50-4646-041
[†] This paper is an extended version of our paper published in the 14th Sound and Music Computing Conference (SMC-17), Espoo, Finland, 5–8 July 2017.

Academic Editor: Stefania Serafin
Received: 12 October 2017; Accepted: 13 December 2017; Published: 20 December 2017

Abstract: Wavefolders are a particular class of nonlinear waveshaping circuits, and a staple of the "West Coast" tradition of analog sound synthesis. In this paper, we present analyses of two popular wavefolding circuits—the Lockhart and Serge wavefolders—and show that they achieve a very similar audio effect. We digitally model the input–output relationship of both circuits using the Lambert-W function, and examine their time- and frequency-domain behavior. To ameliorate the issue of aliasing distortion introduced by the nonlinear nature of wavefolding, we propose the use of the first-order antiderivative method. This method allows us to implement the proposed digital models in real-time without having to resort to high oversampling factors. The practical synthesis usage of both circuits is discussed by considering the case of multiple wavefolder stages arranged in series.

Keywords: acoustic signal processing; circuit modeling; nonlinear waveshaping; antialiasing; synthesis; music

1. Introduction

Nonlinear waveshaping is a technique used in sound synthesis to generate complex harmonic spectra. It consists of processing a signal with low harmonic content (typically a sinusoid) using a nonlinear mapping function designed to introduce harmonic overtones to the output signal [1]. The first documented use of waveshaping in the digital domain can be traced back to 1969, when Jean-Claude Risset emulated the sound of a clarinet by distorting a sinusoid with a clipping function [2]. Waveshaping techniques were extensively researched within the context of computer music in the 1970s, with several authors exploring the use of Chebyshev polynomials in particular, as an accurate and computationally cheap alternative to additive synthesis [1,3–5]. The underlying principles behind waveshaping synthesis are closely related to other well-known synthesis techniques, such as frequency modulation (FM) and phase distortion (PD) synthesis [6,7]. These two techniques rely on distorting the frequency and phase, respectively, of sinusoidal oscillators. Recent research on the topic of distortion-based synthesis has explored the use of logic operators in lieu of traditional polynomial waveshapers [8], and proposed extensions to both FM and PD synthesis [9,10].

The use of waveshaping in the analog domain began in the 1950s, when guitar players started deliberately overdriving their tube amplifiers to alter the timbre of their instrument [11]. In 1961, Gibson released the "Maestro FZ-1 Fuzz Tone", the first commercially available fuzz distortion pedal, which exploited the saturating behavior of transistors to introduce harmonic distortion [12]. Most guitar distortion pedals, including popular designs such as the Ibanez Tube Screamer and Electro-Harmonix Big Muff Pi, operate under this same basic principle [13,14].

In analog synthesizers, the use of distortion-based methods is one of the cornerstones of "West Coast" synthesis, a paradigm pioneered by California-native Don Buchla during the 1960s. Buchla's instruments focused on timbre manipulation at oscillator level by employing a variety of techniques such as nonlinear waveshaping, oscillator synchronization and pitch modulation [15–17]. This approach to sound synthesis contrasts that of traditional subtractive synthesis, where timbre is typically controlled by filtering harmonically-rich oscillator waveforms, like sawtooth and square waves, using resonant filters [18]. In recent years, West Coast synthesis has become increasingly popular, with contemporary manufacturers such as Make Noise and Verbos Electronics releasing their own takes on classic Buchla circuits.

This study presents virtual analog (VA) models for two analog synthesizer circuits: the Lockhart wavefolder and the wavefolder used in the middle section of the Serge Wave Multipliers. Wavefolding is a type of nonlinear waveshaping common in West Coast synthesis where portions of the input signal that exceed certain threshold are inverted or "folded back", hence the name of the effect. The two circuits considered in this study were chosen because of the strong similarities between their general behavior. In a similar way to guitar distortion pedals, both wavefolders exploit the saturating behavior of semiconductor p–n junctions (i.e., transistors/diodes) to implement a folding function.

Wavefolders are amongst the most emblematic building blocks of West Coast synthesis. In spite of that, they have been mostly overlooked by both VA and digital waveshaping research. We have recently begun to fill this research gap in [17], which presents a VA model of the wavefolder circuit in the seminal Buchla 259 module. Previous work on circuit-based VA modeling has researched the filters found in vintage synthesizers such as those produced by Moog [19–22], Electronic Music Studios (EMS) [23,24], Korg [25,26] and Buchla [16]. Extensive work has also been done on modeling guitar distortion pedals [13,27], tube amplifiers [11,28,29], modulation effects [30–33] and the Roland TR-808 drum machine [34,35]. Measurement-based VA modeling, commonly known as "black-box modeling", has also been thoroughly studied within the context of guitar amplifiers and pedals [36–38]. This approach is particularly useful when the original circuit schematics are not available.

A major challenge in VA modeling of nonlinear circuits, and digital waveshaping in general, is aliasing suppression. Early research on waveshaping synthesis addressed this issue by using low-order polynomial transfer functions, which not only allowed full parametric control of the produced spectrum but also ensured that the output waveform was bandlimited [4]. In VA modeling, high oversampling factors are usually neccesary to prevent harmonics introduced by nonlinearities from reflecting into the baseband as aliases [13]. Oversampling increases the computational requirements of the model, by introducing additional filtering stages and scaling the number operations required to compute each output sample. For VA models that require evaluating transcendental functions, as is the case with the proposed Lockhart and Serge models, these added costs could compromise the integration of the system within a larger, real-time computer music system.

A sizable portion of VA research has concentrated on designing efficient algorithms to generate alias-free geometric waveforms like those used in analog subtractive synthesizers, the so-called classic analog waveforms. Well-known techniques include the bandlimited impulse train (BLIT) family of methods, which involves the use of bandlimited basis functions and their integrated forms [39–41], and the use of differentiated polynomial waveforms (DPW) [42–44]. Moreover, Välimäki and Franck have applied the antialiasing principle behind the DPW algorithm to tackle aliasing in wavetable oscillators [45]. Recent work on antialiasing techniques has extended the use of the bandlimited ramp (BLAMP) method, originally proposed to antialias triangular oscillators in [25,41], to special cases of linear piecewise nonlinearities such as signal rectification, and inverse/hard clipping [17,46,47].

In this work, we propose the use of the antiderivative antialiasing method introduced in [48,49]. This approach can be used to reduce the aliasing caused by arbitrary nonlinear waveshaping functions and is applicable to the proposed wavefolder models. In its first-order form, the method can be derived by analytically convolving a linear continuous-time representation of the input signal with a rectangular lowpass kernel [48]. As shown in this work, the use of the antiderivative method reduces the oversampling requirements of the proposed wavefolder models.

A VA model of the Lockhart wavefolder was originally presented in [50]. This paper extends that work by introducing a second wavefolding circuit and studying the similarities between both systems. Additionally, we present a different treatment of the required Lambert-W function and an extended evaluation of the proposed antialiasing method in terms of computational costs.

This paper is organized as follows. Sections 2 and 3 describe the model derivation of the Lockhart and Serge wavefolders, respectively. Time-domain simulations of the circuits are also presented in these two sections. Section 4 deals with two implications of VA wavefolding in the digital domain, namely aliasing suppression and evaluation of the Lambert-W function. Section 5 presents frequency-domain results of the Lockhart and Serge wavefolders, as well as an evaluation of the proposed antialiasing method in terms of perceived sound quality and computational costs. Section 6 discusses the practical synthesis usage of both circuits and compares the behavior of the middle Serge Wave Multiplier with a recommended four-stage topology built around the Lockhart wavefolder. Concluding remarks and perspectives appear in Section 7.

2. The Lockhart Wavefolder

Figure 1a shows a simplified circuit diagram of the Lockhart wavefolder. This circuit was designed in 1973 by R. Lockhart Jr., who intended it to be used as a general-purpose frequency tripler [51]. Following its publication in Bernie Hutchin's Electronotes [52], Lockhart's design was repurposed as a wavefolder by Ken Stone, who realized its potential as a simple yet interesting waveshaper [53,54]. The Lockhart wavefolder has become ubiquitous in the music synthesizer do-it-yourself (DIY) community. For example, it is the core processor in Yves Usson's "Metalizer" module [55].

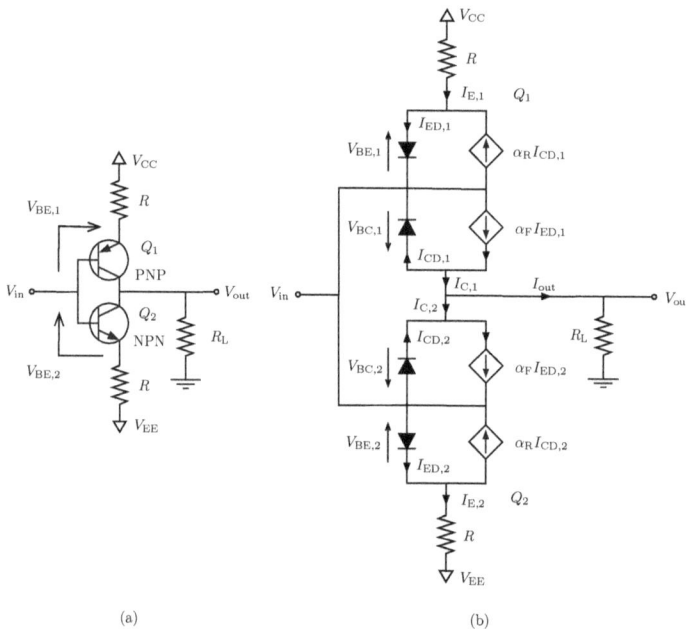

(a)   (b)

Figure 1. (**a**) Simplified schematic of the Lockhart wavefolder circuit (adapted from [53]); and (**b**) its Ebers–Moll large-signal equivalent model.

The main modifications made by Stone to Lockhart's original design were the addition of an input potentiometer to attenuate the amplitude of the input waveform, and an inverting amplifier at the output of the circuit [54]. For the sake of simplifying the analysis, these are not shown in Figure 1a. The inverting stage at the output is reintroduced in Section 2.3. In our treatment of the circuit, we introduce the load resistance R_L as an additional parameter which can be used to further control the timbre of the folded waveform.

2.1. Circuit Analysis

The Lockhart wavefolder consists of an NPN and a PNP bipolar junction transistors connected at their base and collector terminals. In order to model the large-signal behavior of the circuit, we replace transistors Q_1 and Q_2 with their corresponding Ebers-Moll large-signal models [23,24]. Figure 1b shows the large-signal equivalent circuit of the Lockhart wavefolder. We use a double subscript notation to distinguish the voltages and currents in transistor Q_1 from those in Q_2. For example, $I_{ED,2}$ denotes the current through the base-emitter diode in Q_2. Component values for the circuit have been compiled in Table 1.

Table 1. Component values for the Lockhart wavefolder circuit.

Component	Value (kΩ)	Component	Value (V)
R	15	V_{CC}	15
R_L	1–50	V_{EE}	−15

We begin the analysis of the circuit by assuming that the supply voltages are equal but opposite in sign (i.e., $V_{CC} = -V_{EE}$), and that $|V_{in}| \ll V_{CC}$. This assumption means that the base-emitter junctions of both Q_1 and Q_2 will always be forward-biased and their voltage drops will remain approximately constant for all expected values of V_{in}. In Ken Stone's version of the circuit, V_{in} is assumed to be bounded between approximately ± 1.2 V [53]. Applying KVL around both input–emitter loops gives us

$$V_{in} = V_{CC} - RI_{E,1} - V_{BE,1} \tag{1}$$

$$V_{in} = V_{BE,2} + RI_{E,2} + V_{EE}, \tag{2}$$

where $I_{E,1}$ and $I_{E,2}$ are the emitter currents, and $V_{BE,1}$ and $V_{BE,2}$ are the voltages across the base–emitter junctions of Q_1 and Q_2, respectively. Solving Equations (1) and (2) for $I_{E,1}$ and $I_{E,2}$ gives us:

$$I_{E,1} = \frac{V_{CC} - V_{BE,1} - V_{in}}{R} \tag{3}$$

$$I_{E,2} = \frac{V_{in} - V_{BE,2} - V_{EE}}{R}. \tag{4}$$

Next, we apply KCL at the collector node, which gives us

$$I_{out} = I_{C,1} - I_{C,2}, \tag{5}$$

where

$$I_{C,1} = \alpha_F I_{ED,1} - I_{CD,1} \tag{6}$$

$$I_{C,2} = \alpha_F I_{ED,2} - I_{CD,2}. \tag{7}$$

If we then assume that the contributions of the reverse currents $\alpha_R I_{CD_1}$ and $\alpha_R I_{CD_2}$ to the total currents associated with the emitter nodes are negligible (i.e., $\alpha_R \approx 0$), we can establish that

$$I_{ED,1} \approx I_{E,1} \tag{8}$$

$$I_{ED,2} \approx I_{E,2}. \tag{9}$$

Assuming $\alpha_F = 1$, as suggested in [23], and inserting Equations (8) and (9) into Equations (6) and (7), respectively, yields a new expression for the total output current of the circuit:

$$I_{out} = I_{E,1} - I_{E,2} - I_{CD,1} + I_{CD,2}. \tag{10}$$

We then combine Equations (3) and (4) to derive an expression for the difference between emitter currents:

$$I_{E,1} - I_{E,2} = \frac{V_{CC} + V_{EE} - 2V_{in} + V_{BE,2} - V_{BE,1}}{R}. \tag{11}$$

Since $V_{CC} + V_{EE} = 0$, and voltage drops $V_{BE,1}$ and $V_{BE,2}$ are assumed to be constant and equal, their contribution to this expression disappears. Therefore, we can simplify this result as:

$$I_{E,1} - I_{E,2} = -\frac{2}{R} V_{in}. \tag{12}$$

Substituting Equation (12) into Equation (10) produces an expression for the total output current I_{out} in terms of the input voltage and the currents through the collector diodes:

$$I_{out} = -\frac{2}{R} V_{in} - I_{CD,1} + I_{CD,2}. \tag{13}$$

The current–voltage (I–V) relationship of diodes can be modeled using Shockley's ideal diode equation, defined as

$$I_D = I_s \left(\exp\left(\frac{V_D}{\eta V_T} \right) - 1 \right), \tag{14}$$

where I_D is the current through the diode, I_s is the reverse bias saturation current, V_D is the voltage across the diode, V_T is thermal voltage and η is the ideality factor of the diode [56]. For the p–n junctions inside transistors we can assume a reverse saturation current value $I_s = 10^{-17}$ A and an ideality factor $\eta = 1$. A thermal voltage value $V_T = 25.864$ mV is used throughout this study.

Applying Shockley's diode equation to the collector diodes and substituting into Equation (13) gives us:

$$I_{out} = -\frac{2V_{in}}{R} - \underbrace{I_s \left(\exp\left(\frac{V_{CD,1}}{\eta V_T} \right) - 1 \right)}_{I_{CD,1}} + \underbrace{I_s \left(\exp\left(\frac{V_{CD,2}}{\eta V_T} \right) - 1 \right)}_{I_{CD,2}}. \tag{15}$$

Next, we use KVL to derive expressions for $V_{CD,1}$ and $V_{CD,2}$ in terms of V_{in} and V_{out}:

$$V_{CD,1} = V_{out} - V_{in} \tag{16}$$

$$V_{CD,2} = V_{in} - V_{out}. \tag{17}$$

Now, the collector diodes in the large-signal model are antiparallel. Therefore, we can make the further assumption that only one of them will conduct at a time depending on the polarity of V_{in}. A similar treatment is presented in [57] for the case of diode pairs in guitar distortion circuits. This means that

$$I_{CD,1} \approx 0 \quad \text{for} \quad V_{in} \geq 0 \quad \text{and} \quad I_{CD,2} \approx 0 \quad \text{for} \quad V_{in} < 0.$$

By combining these new assumptions with Equations (15)–(17) we arrive at the piecewise expression

$$I_{out} = -\frac{2}{R} V_{in} + \lambda I_s \left(\exp\left(\frac{\lambda (V_{in} - V_{out})}{\eta V_T} \right) - 1 \right), \tag{18}$$

where $\lambda = \text{sgn}\,(V_{\text{in}})$ and sgn () is the signum function

$$\text{sgn}(x) := \begin{cases} -1 & \text{if} \quad x < 0 \\ 0 & \text{if} \quad x = 0 \\ 1 & \text{if} \quad x > 0. \end{cases} \tag{19}$$

Equation (18) can be further simplified if we consider that the independent constant factor λI_s that results from its expansion will be very small ($\pm 10^{-17}$ A) and can therefore be neglected. This gives us:

$$I_{\text{out}} = -\frac{2}{R} V_{\text{in}} + \lambda I_s \exp\left(\frac{\lambda\,(V_{\text{in}} - V_{\text{out}})}{\eta V_{\text{T}}}\right). \tag{20}$$

Finally, we multiply both sides of this expression by R_{L} to derive an input–output voltage relationship for the Lockhart wavefolder:

$$V_{\text{out}} = -\frac{2R_{\text{L}}}{R} V_{\text{in}} + \lambda R_{\text{L}} I_s \exp\left(\frac{\lambda\,(V_{\text{in}} - V_{\text{out}})}{\eta V_{\text{T}}}\right). \tag{21}$$

2.2. Explicit Formulation

Equation (21) describes an implicit relationship between the input and output voltages of the circuit; it cannot be solved algebraically. Instead, a closed-form solution for V_{out} can be derived with the help of the Lambert-W function. The use of the Lambert-W function $W()$ has been previously researched within the context of VA modeling. Several authors have used it to solve the implicit I–V relationship of diodes [56,58,59]. Parker and D'Angelo used $W()$ to model the Buchla Lowpass-Gate, a synthesizer circuit that employs a resistive opto-isolator (also known as a vactrol) in its control path [16]. Strictly speaking, $W()$ is multivalued; however, in this work, we only utilize the upper branch of the function. This branch is known as $W_0()$ in the literature [56,60].

The Lambert-W function is used to solve equations of the form

$$(A + Bx)\exp\,(Cx) = D, \tag{22}$$

which have the explicit solution

$$x = \frac{1}{C} W\left(\frac{CD}{B}\exp\left(\frac{AC}{B}\right)\right) - \frac{A}{B}, \tag{23}$$

where A, B, C and $D \in \mathbb{R}$ [58].

Equation (21) can be arranged in the form described by Equation (22) by first rewriting it as

$$V_{\text{out}} + \frac{2R_{\text{L}}}{R} V_{\text{in}} = \lambda R_{\text{L}} I_s \exp\left(\frac{\lambda V_{\text{in}}}{\eta V_{\text{T}}}\right) \exp\left(\frac{-\lambda V_{\text{out}}}{\eta V_{\text{T}}}\right), \tag{24}$$

and dividing both sides by $\exp\,(-\lambda V_{\text{out}}/\eta V_{\text{T}})$, which gives us

$$\left(V_{\text{out}} + \frac{2R_{\text{L}}}{R} V_{\text{in}}\right)\exp\left(\frac{\lambda V_{\text{out}}}{\eta V_{\text{T}}}\right) = \lambda R_{\text{L}} I_s \exp\left(\frac{\lambda V_{\text{in}}}{\eta V_{\text{T}}}\right). \tag{25}$$

Solving for V_{out} as defined in Equation (23) yields an explicit model for the Lockhart wavefolder

$$V_{\text{out}} = \lambda \eta V_{\text{T}} W\,(\Delta \exp\,(\lambda \beta V_{\text{in}})) - \alpha V_{\text{in}}, \tag{26}$$

where

$$\alpha = \frac{2R_{\text{L}}}{R}, \quad \beta = \frac{2R_{\text{L}} + R}{\eta V_{\text{T}} R} \quad \text{and} \quad \Delta = \frac{R_{\text{L}} I_s}{\eta V_{\text{T}}}.$$

Table 2 summarizes all parameter values for the proposed Lockhart wavefolder model.

Table 2. Parameter values for the Lockhart wavefolder described by Equation (26).

Name	Value	Name	Value	Name	Value
R	$15\,\text{k}\Omega$	I_s	$10^{-17}\,\text{A}$	V_T	$25.864\,\text{mV}$
R_L	1–50 $\text{k}\Omega$	η	1	–	–

2.3. Model Discretization and Evaluation

The voltages inside the Lockhart wavefolder are time-dependent. Therefore, we can describe the continuous-time model defined by Equation (26) as being of the form

$$V_{\text{out}}(t) = f(V_{\text{in}}(t)), \tag{27}$$

where $f()$ is the transfer function of the system and t is time. In the synthesis literature, the term "transfer function" is commonly used to denote the waveshaping function [4]. It should not be confused with the s- and z-domain transfer functions used in linear system analysis.

As previously mentioned, Ken Stone's circuit features an inverting stage before the output which can be modeled by inverting the polarity of the right-hand side of Equation (26):

$$V_{\text{out}} = \alpha V_{\text{in}} - \lambda \eta V_T W \left(\Delta \exp \left(\lambda \beta V_{\text{in}} \right) \right). \tag{28}$$

While including this step is not strictly necessary, we have chosen to do so, as it will facilitate the evaluation of the model. Now, given the form described by Equation (27), the Lockhart model can be discretized trivially by replacing all continuous-time signals with their discrete-time equivalents, i.e.,

$$V_{\text{out}}[n] = f(V_{\text{in}}[n]), \tag{29}$$

where n is the sample index.

The time-domain behavior of the proposed circuit model was validated by comparing it against a reference simulation obtained using the SPICE (Simulation Program with Integrated Circuit Emphasis) software LTspice (Version IV, Linear Technology, Milpitas, CA, USA, 2016) [61]. The results of this simulation are shown in Figure 2a for values of V_{in} between -1.5 and 1.5 V. Figure 2b shows the transfer function of the proposed model implemented in MATLAB (Version R2017a, MathWorks, Natick, MA, USA, 2017) using Equation (28) and MATLAB's native "lambertw" function. In both cases, four different values of R_L were simulated: 1, 5, 10 and $50\,\text{k}\Omega$. From these figures, we can observe the general behavior of the Lockhart wavefolder. At low input values, the system behaves linearly, whereas for high input values the circuit inverts the slope of the driving signal. The transition between non-folded and folded portions of the signal is gradual, which responds to the characteristic soft saturating behavior of p–n junctions. The region where the transfer function folds the input signal is indicated with a blue arrow in Figure 2b for the case when $R_L = 50\,\text{k}\Omega$. As shown in these Figures, increasing the value of R_L sharpens the shape of the transfer function.

The curves shown in Figure 2a,b indicate a good match between the SPICE simulations and the proposed digital model. Figure 3 shows the absolute value (in mV) of the difference between both simulations. From this plot, we can observe that the difference between the curves is indeed very small, below 1 mV for all values of V_{in} measured. These small differences are perceptually insignificant and can be attributed to the simplifications made during the analysis of the circuit and to the way in which SPICE computes currents flowing through semiconductor devices. For example, SPICE introduces a small fictitious conductance in parallel with each p–n junction in order to aid the convergence of its iterative solvers. Additionally, the SPICE diode model will account for the small reverse current that flows when the voltage across the diode is negative [61].

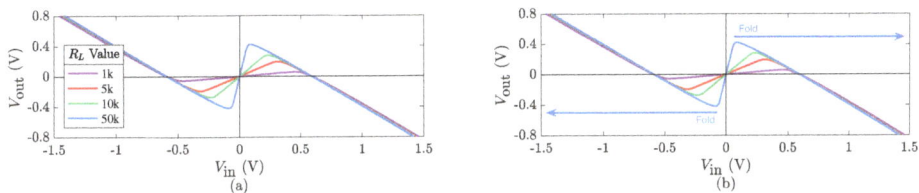

Figure 2. Transfer function of the Lockhart wavefolder simulated using: (**a**) SPICE (Simulation Program with Integrated Circuit Emphasis); and (**b**) the proposed virtual analog (VA) model. Different colors indicate different values of R_L.

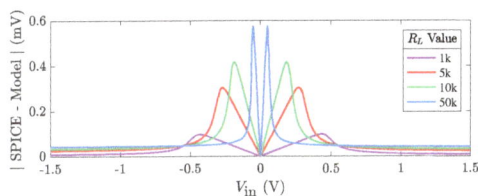

Figure 3. Absolute value of the difference between a SPICE simulation of the Lockhart wavefolder and its proposed VA model.

Figure 4 shows a time-domain view of the output of the proposed model when driven by a 500-Hz sinusoidal waveform with a peak amplitude of 1 V for two different load resistance values, $R_L = 10$ and 50 kΩ. A sampling rate $F_s = 44.1$ kHz was used to generate these figures, which are plotted against their corresponding SPICE simulations. From these results, we can once again observe the effect of wavefolding and the impact of R_L on the overall shape of the output. For high values of R_L the transition region between folded and non-folded values becomes very small and the resulting waveform is almost discontinuous (see Figure 4b). In the frequency domain, this will translate to higher harmonic content, similar to that of a square wave oscillator. A more detailed frequency-domain analysis of the Lockhart circuit is presented in Section 5 of this study.

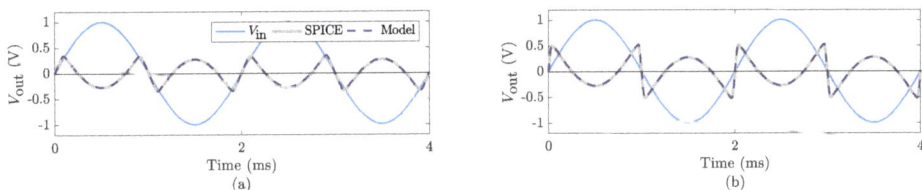

Figure 4. Time-domain view of the proposed Lockhart wavefolder model plotted against its SPICE simulation for a 500-Hz sinusoidal input (peak amplitude 1 V) with load resistance: (**a**) $R_L = 10$ kΩ; and (**b**) $R_L = 50$ kΩ.

3. The Serge Middle Wave Multiplier

The second circuit considered in this study is the middle section of the Serge Wave Multipliers (often abbreviated as the Serge VCM). The Serge VCM is a synthesizer module designed in 1977 by West Coast designer Serge Tcherepnin, founder of Serge Modular Music Systems. It offered three separate and independent analog sound processors, namely the "top", "middle" and "bottom" sections. As described in an original 1980 Serge product catalog, "The middle section generates a sweep of the odd harmonics (1, 3, 5, 7, 9, 11 and 13th) when a triangle wave is applied to its input... This module can be used to explore timbral areas beyond the range of ring modulation because there are more varied harmonics than the sum and difference tones" [62].

The middle Serge VCM is essentially a waveshaping circuit consisting of six identical wavefolding stages arranged in series. An amplifier at the input of the circuit is used to modulate the gain of the input waveform and control the amount of folds introduced [63]. In this section we focus on the analysis of a single folding stage. The transfer function and frequency-domain behavior of the complete system are presented in Section 6. Figure 5 shows the schematic of a single wavefolding stage in the circuit. Component information is given in Table 3.

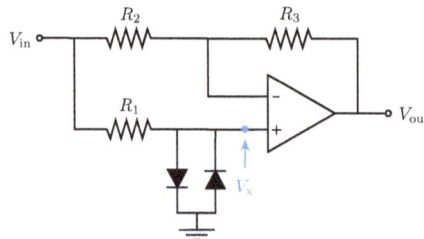

Figure 5. Schematic of a single folding cell in the middle section of the Serge Wave Multipliers (VCM). Figure adapted from [63].

Table 3. Component information for the Serge wavefolder circuit shown in Figure 5.

Component	Value (kΩ)	Component	Description
R_1	33	Diodes	1N4148 or similar
R_2	100	Op-Amp	TL072 or similar
R_3	100	–	–

To derive the transfer function for the Serge wavefolding circuit, we first assume ideal op-amp behavior and derive an expression for V_{out} in terms of V_{in} and V_x, the voltage at the non-inverting input of the amplifier. This gives us:

$$V_{\text{out}} = V_x - \frac{R_3}{R_2}(V_{\text{in}} - V_x). \tag{30}$$

Since in this case $R_3 = R_2$, we can further simplify this result as:

$$V_{\text{out}} = 2V_x - V_{\text{in}}. \tag{31}$$

Next, we derive an expression for V_x by considering the subcircuit shown in Figure 6, which is essentially a diode pair similar to those found in guitar distortion circuits [13,57,58].

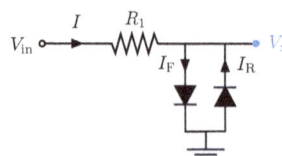

Figure 6. Equivalent view of the diode saturator at the non-inverting input of the op-amp in Figure 5.

Applying KVL around the outer loop of the circuit yields the relation

$$V_{\text{in}} = R_1 I + V_x, \tag{32}$$

where I is the current through resistor R_1. Then, we apply KCL at the output node of the circuit, which gives us

$$I = I_F - I_R. \tag{33}$$

Combining Equation (32) with Equation (33) and applying Shockley's diode equation gives us

$$\frac{V_{in} - V_x}{R_1} = \underbrace{I_s \left(\exp \left(\frac{V_x}{\eta V_T} \right) - 1 \right)}_{I_F} - \underbrace{I_s \left(\exp \left(\frac{-V_x}{\eta V_T} \right) - 1 \right)}_{I_R}. \tag{34}$$

As before, we assume the diodes will not conduct simultaneously and arrive at the piecewise relationship

$$V_{in} - V_x = \lambda R_1 I_s \left(\exp \left(\frac{\lambda V_x}{\eta V_T} \right) - 1 \right), \tag{35}$$

where once again $\lambda = \mathrm{sgn}(V_{in})$. To further simplify this expression we neglect the constant factor $\lambda R_1 I_s$ that results from its expansion. This gives us:

$$V_{in} - V_x = \lambda R_1 I_s \exp \left(\frac{\lambda V_x}{\eta V_T} \right). \tag{36}$$

Next, we rearrange this equation in the Lambert-W form described by Equation (22) by dividing both sides by $\exp \left(\lambda V_x / \eta V_T \right)$. This yields

$$(V_{in} - V_x) \exp \left(-\frac{\lambda V_x}{\eta V_T} \right) = \lambda R_1 I_s, \tag{37}$$

which can be solved for V_x as:

$$V_x = V_{in} - \lambda \eta V_T W \left(\frac{R_1 I_s}{\eta V_T} \exp \left(\frac{\lambda V_{in}}{\eta V_T} \right) \right). \tag{38}$$

As a final step, we insert Equation (38) into Equation (31) to derive a complete expression for the transfer function of a single wavefolding stage in the Serge middle VCM:

$$V_{out} = V_{in} - 2\lambda \eta V_T W \left(\frac{R_1 I_s}{\eta V_T} \exp \left(\frac{\lambda V_{in}}{\eta V_T} \right) \right). \tag{39}$$

Figure 7 shows the transfer function of the circuit, evaluated in MATLAB for values of V_{in} between -1.5 and 1.5 V. As before, the model was discretized trivially and is presented against its corresponding SPICE simulation. Parameter values used in this simulation are compiled in Table 4. The value of parameters I_s and η for the 1N4148 diode were matched to those of its corresponding SPICE model [61]. Figure 7b shows the absolute difference between both simulations. These results indicate a good match between the models, as the maximum difference was once again found to be below 1 mV.

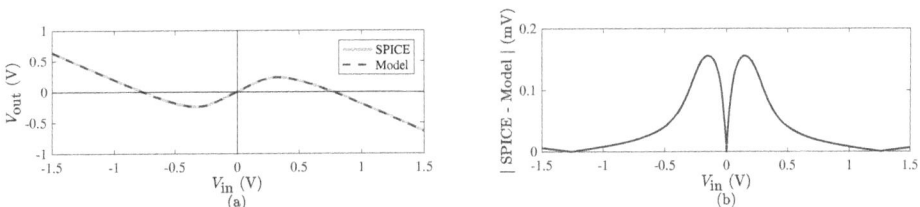

Figure 7. (a) Transfer function of a single wavefolding stage in the Serge middle VCM measured using SPICE and the proposed model; and (b) the absolute difference between these two curves.

Table 4. Simulation parameters for a single folding stage in the middle Serge Wave Multiplier.

Name	Value	Name	Value
I_s	2.52 nA	R_1	33 kΩ
η	1.752	V_T	25.864 mV

Finally, Figure 8 shows the output of the Serge wavefolder for a 500-Hz sinusoidal input. As expected, the circuit behaves as a wavefolder, folding portions of the input waveform whose absolute value exceeds approximately 0.3 V. This behavior is similar to that of the Lockhart wavefolder (cf. Figure 4a).

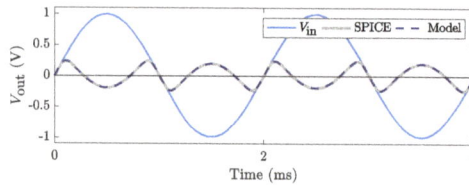

Figure 8. Time-domain view of the Serge wavefolder model plotted against its SPICE simulation for a 500-Hz sinusoidal input with peak amplitude if 1 V.

3.1. Model Equivalence

Equation (39) shares a close resemblance with Equation (28), the proposed Lockhart wavefolder model. In fact, both expressions have the same form, which consists of the difference between a portion of the input signal and an input-dependent nonlinear element. In the case of the Lockhart wavefolder, when the $R_L = R/2$ Equation (28) simplifies to

$$V_{\text{out}} = V_{\text{in}} - \lambda \eta V_T W \left(\frac{R_L I_s}{\eta V_T} \exp \left(\frac{\lambda V_{\text{in}}}{\eta V_T} \right) \right),$$

(40)

which is remarkably close to Equation (39), with the only difference being the missing factor of two outside the Lambert-W function. This factor accounts for the difference between physical parameters I_s and η in each circuit. Figure 9 shows a comparison of the transfer functions for the Lockhart ($R_L = 7.5$ kΩ) and Serge wavefolders implemented using the parameter values in Tables 2 and 4, respectively. From this figure, we can observe that the only significant difference between both transfer functions is in their sharpness at the folding points. This means the Lockhart wavefolder will introduce sharper folds which will translate into brighter sounds at the output. From this analysis, it is clear that both circuits result in a similar audio effect, even though they are produced using different architectures.

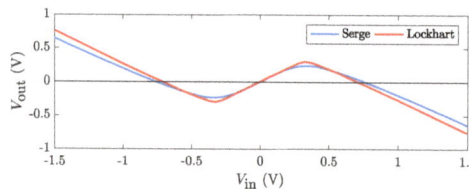

Figure 9. Transfer functions for the proposed Serge and Lockhart ($R_L = 7.5$ kΩ) wavefolder models.

4. Wavefolding in the Digital Domain

In the previous sections, the time-domain behavior of the Lockhart and Serge wavefolder models was examined via trivial discretization of their characteristic transfer functions. In this section, we move on to consider two implications of virtual analog wavefolding: evaluation of the Lambert-W function and aliasing.

4.1. Evaluating the Lambert-W Function

A particular challenge of using the Lambert-W function in VA modeling, where real-time operation is paramount, is that of computational efficiency. Optimizing the evaluation of $W()$ is an active research topic (see, e.g., [64]). For the case of guitar distortion circuits, Paiva et al. proposed the use of a simplified iterative method which relies on a lookup table for its initial guess [57]. In this work, we propose approximating the value of $W()$ directly using Fritsch's iteration, as suggested in [60]. In order to compute w_m, an approximation to $W(x)$, where $x \in \mathbb{R}_{>0}$, we iterate over

$$w_{m+1} = w_m(1 + \varepsilon_m), \tag{41}$$

where

$$\varepsilon_m = \left(\frac{r_m}{1 + w_m}\right)\left(\frac{q_m - r_m}{q_m - 2r_m}\right) \tag{42}$$

$$r_m = \ln\left(\frac{x}{w_m}\right) - w_m \tag{43}$$

$$q_m = 2\left(1 + w_m\right)\left(1 + w_m + \frac{2}{3}r_m\right), \tag{44}$$

and $m = 0, 1, 2, \ldots, M - 1$. The value w_0 is an initial guess and M is the number of iterations required for ε_m to approximate zero within machine-size floating point precision. The special case $W(0) = 0$ is defined separately.

The efficiency of Fritsch's iteration will depend on the choice of initial guess. As explained in [60], an initial guess within 10^{-4} of the solution will yield an approximation to $W()$ accurate to within 10^{-16} in just one iteration. Figure 10 shows the approximate times required to compute $W(x)$ for a set of values of x between 10^{-24} and 10^{300} using Fritsch's iteration and the previously-proposed Halley's method [50]. This range was chosen as it covers all values of interest. For instance, when $V_{\text{in}} = 5\,\text{V}$, the argument of $W()$ in the Serge wavefolder model will be approximately 1.52×10^{44}. All times were computed by averaging the result of 30 iterations implemented under identical circumstances. A piecewise approximation was used to compute the initial guess, as described in [60]. From this plot, we can observe that Fritsch's iteration outperforms Halley's method by up to approximately 11 times. A MATLAB implementation of the Lambert-W function used to perform these measurements can be found in the accompanying website for this article.

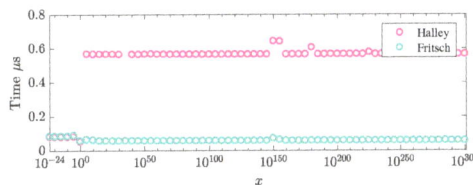

Figure 10. Averaged processing times required to compute $W(x)$ using Halley's method and Fritsch's iteration.

4.2. Aliasing Considerations

As discussed in Section 1, nonlinear waveshaping in the digital domain is susceptible to aliasing distortion due to its frequency-expanding nature. Wavefolding is no exception to this problem. As an arbitrary input waveform is folded, new harmonic overtones will be added to the frequency spectrum. Harmonics at frequencies exceeding half of the sampling rate, or the Nyquist limit, will be reflected into the audio baseband as aliases. Aliasing is known to cause unpleasant artifacts—such as beating and inharmonicity—that cannot be tolerated in a music computing scenario. Oversampling is commonly employed to mitigate this issue; however, this approach increases the computational requirements of the system by introducing additional operations.

We propose the use of the first-order antialiasing method presented in [48,49]. This method is designed to reduce aliasing caused by memoryless waveshaping functions with the form described by Equation (29). The antialiased output of the waveshaping function is defined as

$$V_{\text{out}}[n] = \frac{F(V_{\text{in}}[n]) - F(V_{\text{in}}[n-1])}{V_{\text{in}}[n] - V_{\text{in}}[n-1]},$$ (45)

where $F()$ is the antiderivative of $f()$, the original transfer function. For the case of the Lockhart wavefolder defined by Equation (28), the integrated transfer function is defined as

$$F(V_{\text{in}}) = \frac{\alpha}{2} V_{\text{in}}^2 - \frac{\eta V_{\text{T}}}{2\beta} \left[\Psi_1(\Psi_1 + 2) \right],$$ (46)

where

$$\Psi_1 = W\left(\Delta \exp\left(\lambda \beta V_{\text{in}}\right)\right),$$ (47)

and α, β and Δ remain as before. This result showcases an advantageous property of the Lambert-W function $W()$; its antiderivative is defined in terms of $W()$ itself. Therefore, computing $F()$ does not pose a major increase in computational costs with respect to evaluating simply $f()$. For the case of the Serge wavefolder defined by Equation (39), the required antiderivative is given by

$$F(V_{\text{in}}) = \frac{V_{\text{in}}^2}{2} - (\eta V_{\text{T}})^2 \left[\Psi_2(\Psi_2 + 2) \right],$$ (48)

where

$$\Psi_2 = W\left(\frac{R_1 I_{\text{s}}}{\eta V_{\text{T}}} \exp\left(\frac{\lambda V_{\text{in}}}{\eta V_{\text{T}}}\right)\right).$$ (49)

When $V_{\text{in}}[n] \approx V_{\text{in}}[n-1]$, Equation (45) can become ill-conditioned. This is avoided by defining the special case

$$V_{\text{out}}[n] = f\left(\frac{V_{\text{in}}[n] + V_{\text{in}}[n-1]}{2}\right),$$ (50)

when $|V_{\text{in}}[n] - V_{\text{in}}[n-1]|$ is smaller than a predetermined threshold [48]. This special case simply bypasses the antialiased form while compensating for the half-sample latency of the method.

5. Results

This section examines the frequency-domain behavior of the Lockhart and Serge wavefolders and their proposed antialiased forms. Next, we evaluate the computational costs of the antiderivative method with respect to oversampling for the case of the Lockhart model.

5.1. Frequency-Domain Behavior

The spectrogram in Figure 11 shows the effect of increasing the value of R_{L} in the Lockhart wavefolder model for a constant 150-Hz sinusoidal input. As expected, the level of harmonic distortion introduced by the circuit is proportional to the value of this resistance. Therefore, this parameter can be used for additional timbral control. It should be noted that due to the antisymmetric nature of the folding function, the system introduces odd harmonics only for input signals centered around zero. Since the level of harmonics introduced by the Lockhart wavefolder depends on the choice of R_{L}, we consider the highest recommended case $R_{\text{L}} = 50\,\text{k}\Omega$ as a worst-case scenario in terms of aliasing distortion.

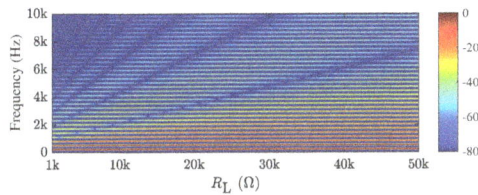

Figure 11. Spectrogram of the Lockhart wavefolder under 150-Hz sinusoidal input for values of R_L between 1 and 50 kΩ.

Figure 12 shows the spectrograms of a linear sweep from 20 Hz–5 kHz with peak amplitude of 1 V processed by the proposed Lockhart and Serge wavefolder models. This frequency range was chosen as it covers all fundamental frequencies of musical interest. A sample rate $F_s = 1$ MHz was used to generate these figures in order to simulate an ideal alias-free continuous-time behavior. These results will be used as a reference when evaluating the performance of the proposed antialiased forms. From these spectrograms we can observe how the Lockhart wavefolder is capable of generating brighter sounds. This perceptual attribute can be varied by changing the value of R_L.

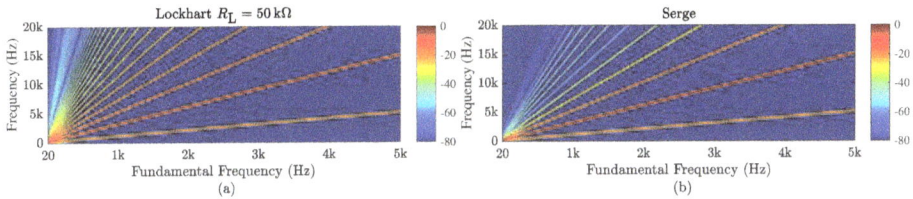

Figure 12. Spectogram for a linear sweep from 20 Hz to 5 kHz processed using: (**a**) the proposed Lockhart wavefolder model ($R_L = 50$ kΩ); and (**b**) the proposed Serge wavefolder model. A sample rate $F_s = 1$ MHz was used to simulate analog behavior.

Figure 13a shows the result of processing the same linear sweep using the trivial (i.e., non-antialiased) Lockhart model at a standard audio rate of $F_s = 44.1$ kHz. When compared to Figure 12a, we can clearly observe the high levels of aliasing distortion introduced by the model. This is somewhat ameliorated in Figure 13b, where the sweep has been processed at a sample rate of $F_s = 88.2$ kHz (i.e., two-times oversampling). Figure 13c,d shows the result of processing the sweep using the proposed antialiasing method at audio rate and with two-times oversampling, respectively. As shown in these spectrograms, there is a significant reduction in aliasing, particularly below the fundamental frequency. This behavior is advantageous in music applications because at low frequencies the audibility of aliasing distortion is only limited by the hearing threshold. On the other end of the spectrum, the masking effects of harmonics will help suppress the audible effects of high-frequency aliases [65].

The spectrograms in Figure 14a,b show the outcome of processing the 1 V linear sweep with the proposed Serge wavefolder model at audio rate and with two-times oversampling, respectively. When compared with Figure 13a,b it is evident that the Serge wavefolder model is less susceptible to aliasing distortion. This can be attributed to the fact that its transfer function is not as sharp as that of the Lockhart, particularly when $R_L = 50$ kΩ. Figure 14c,d shows the result of processing the linear sweep using the antiderivative method at audio rate and with oversampling by two. In this case, operating at audio rate yields very effective results as there are very few visible aliases left below the fundamental. When combined with oversampling by two the antiderivative method produces a nearly alias-free spectrum for the measured frequency range.

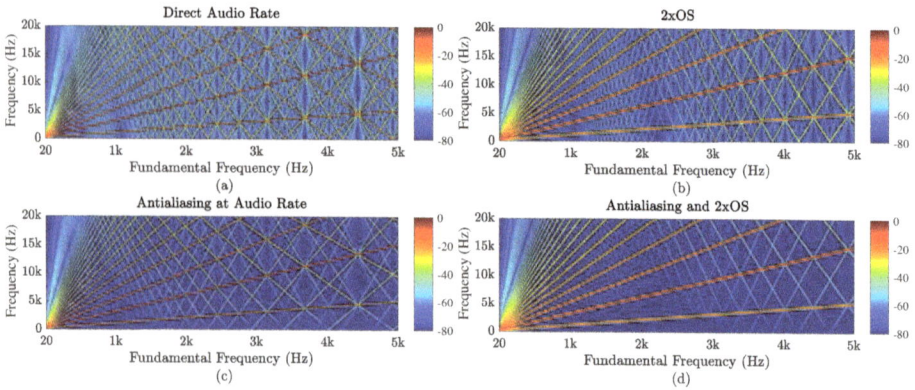

Figure 13. Spectrogram for a 1 V linear sweep from 20 Hz–5 kHz processed with the proposed Lockhart wavefolder ($R_L = 50\,k\Omega$) model: (**a**) at audio rate; (**b**) using two times oversampling; (**c**) with antialiasing at audio rate; and (**d**) with antialiasing and oversampling by two.

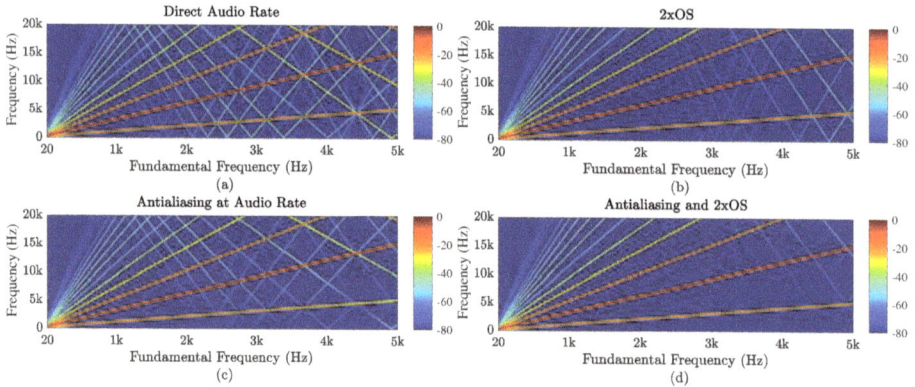

Figure 14. Spectrogram for a 1 V linear sweep from 20 Hz–5 kHz processed with the proposed Serge wavefolder model: (**a**) at audio rate; (**b**) using two times oversampling; (**c**) with antialiasing at audio rate; and (**d**) with antialiasing and oversampling by two.

The performance of the proposed antialiasing method was further evaluated by computing the A-weighted noise-to-mask ratio (ANMR) for a set of sinusoidal input signals processed by both wavefolder models. The ANMR has been previously researched as a perceptually-informed measure to evaluate the audibility of aliasing distortion [41,65]. The algorithm computes the power ratio in decibels between the wanted harmonics and aliased components, but takes into account the masking effects of the former. An A-weighting filter is applied to all signals prior to the evaluation of the ANMR in order to account for the frequency-dependent sensitivity of hearing for low-level sounds. Signals with an ANMR value below $-10\,dB$ are considered to be completely free from perceivable aliasing. A detailed account of this method can be found in [65].

Figure 15a compares the measured ANMRs for a set of sinusoidal inputs with fundamental frequencies between 1 and 5 kHz processed by the Lockhart model at different sampling rates. The ideal alias-free signals required to compute these values were synthesized using Fourier analysis and additive synthesis, as suggested in [41]. All signals were downsampled back to audio rate (i.e., 44.1 kHz) prior to evaluation. A dashed horizontal line has been used to indicate the $-10\,dB$ hearing threshold for aliasing distortion. In Figure 15a we can observe the significant increase in signal quality obtained by the proposed antialiasing method when applied to the Lockhart wavefolder, even when operating at audio rate. Moreover, these measurements show that the performance of the proposed method, when combined with two-times oversampling, is on par with oversampling by a factor of 8. For all fundamental frequencies below approximately 4.2 kHz, the ANMR lies below the $-10\,\text{dB}$ line. This range can be regarded as sufficient for musical applications if we consider that the highest fundamental frequency on a standard grand piano is 4186.01 Hz (MIDI note C8).

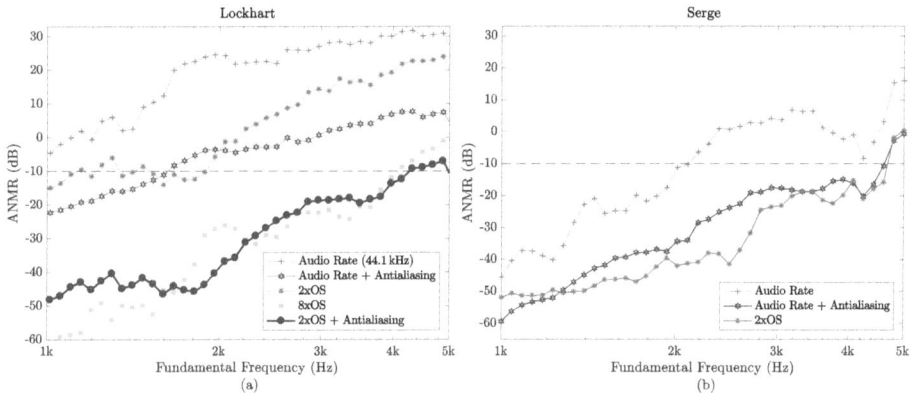

Figure 15. Measured A-weighted noise-to-mask ratios (ANMRs) for a range of sinusoidal waveforms processed: (**a**) using the Lockhart wavefolder model ($R_L = 50\,\text{k}\Omega$) under six different sampling rates; and (**b**) using the Serge wavefolder model under two different sampling rates, with and without the proposed antialiasing method. Values below the $-10\,\text{dB}$ threshold indicate lack of perceivable aliasing.

Figure 15b shows the measured ANMRs for the Serge wavefolder. When implemented at audio rate, the output is free from perceivable aliasing for fundamental frequencies up to approximately 2 kHz. These measurements go in accordance with the Spectrogram in Figure 14a, which shows aliasing is significantly more evident above this frequency. The use of the antiderivative method yields results comparable to those of oversampling by a factor of two, with all measured fundamental frequencies below approximately 4.6 kHz lying below the $-10\,\text{dB}$ aliasing threshold. Overall, these results indicate the proposed Serge wavefolder model can operate at audio rate with the help of the antiderivative method, therefore avoiding the need for oversampling.

5.2. Computational Costs

The computational costs of the antialiased Lockhart wavefolder model were measured by porting the algorithms into C code using the 128-bit *long double* data type. Table 5 shows the computation times for a 1-s 100-Hz sinusoidal input processed using the proposed model for different peak amplitude values. These results were computed by averaging the processing times of one hundred implementations. All tests were performed under identical circumstances, using a fixed resistance value of $R_L = 50\,\text{k}\Omega$, the highest recommended value. From these results we can observe that the complexity of the model does not depend on the input and that the overhead of implementing the antialiasing method is minimal. When operating at audio rate, the added computation time is approximately 1 ms for a 1-second simulation. Moreover, these time measurements show that the antialiased Lockhart model, implemented at a sample rate $F_s = 88.2\,\text{kHz}$, is approximately 3.5 times faster than oversampling by a factor of 8 (i.e., $F_s = 352.8\,\text{kHz}$) and nearly twice as fast as oversampling by a factor of 4. Changing the value of R_L did not affect the execution times of the algorithms.

Table 5. Averaged computation times (in milliseconds) for the proposed Lockhart wavefolder model ($R_L = 50\,k\Omega$) implemented in C for a 1-s 100-Hz sinusoidal input sampled at different oversampling (OS) rates and with different peak amplitude levels.

Amplitude (V)	Audio Rate (ms)	Audio Rate w/Antialiasing (ms)	OSx2 (ms)	OSx4 (ms)	OSx8 (ms)	OSx2 w/ Antialiasing (ms)
1	11.5	12.5	23.4	46.6	92.9	25.4
5	11.6	12.6	23.3	46.8	92.7	25.3
10	11.5	12.6	23.7	46.7	92.9	25.5
15	11.5	12.7	23.5	46.5	92.9	25.5

For high values of R_L, long double representation is necessary to account for the large values that will result at the argument of the Lambert-W function. For smaller values, 64-bit precision will be sufficient to accommodate most input levels of interest. For instance, when $R_L = 7.5\,k\Omega$ the signal at the input of the proposed Lockhart wavefolder model can have a peak amplitude of up to 9 V.

The measurements in Table 5 were conducted by synthesizing all input signals at the target rates. In practical implementations, oversampling will require additional pre- and post-filtering stages that will further increase the computational costs of the system. The complexity and costs of these filtering stages will be directly proportional to the required oversampling factor. This constitutes another advantage of the proposed antialiasing method.

6. Practical Synthesis Usage

In practical sound synthesis applications, a single folding stage is rarely used, as the timbral variety it can produce is quite limited. Most analog designs, for example the Intellijel μFold and the aforementioned Yusynth Metalizer, employ several wavefolding stages arranged in series. The number of stages varies according to the design, but typically cascades of two to six stages are used. As mentioned in Section 3, the Serge middle VCM utilizes six identical folding stages. Figure 16 shows a simplified block diagram representation of the Serge middle VCM based on the original design [63]. Blocks labeled "SWF" correspond to the proposed Serge wavefolder model. An ad hoc gain factor of four, not present in the original circuit, has been added to compensate for the scaling of the signal introduced by the cascade of wavefolders.

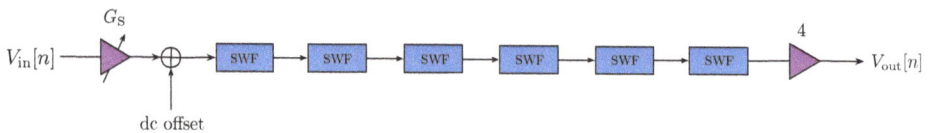

Figure 16. Block diagram representation of the Serge middle VCM. Blocks labeled "SWF" indicate the Serge wavefolder model.

In cascaded wavefolder structures like the one shown in Figure 16, timbral control can be achieved in two manners. The first is by adjusting the gain of the input waveform (using G_S in this case). This parameter controls the amount of folds introduced, allowing the overall brightness of the sound to be varied. It can be modulated in real-time to provide articulation to a sound similar to filtering in subtractive synthesis or modulation index in FM synthesis. The second way to control timbre is by adding a dc offset to the input of the wavefolder. This breaks the aforementioned symmetry of the folding function and introduces even harmonics. When modulated by using, for example, a low-frequency oscillator, this parameter provides an effect reminiscent of pulse-width modulation. Figure 17a shows the transfer function of the Serge middle VCM model for the case of zero dc offset. This plot was generated by defining V_{in} to have a constant value of 1 V and sweeping through values of G_S between -8 and 8. Figure 17b shows the output of the Serge middle VCM when driven by a

100-Hz sinusoid with $G_S = 6$. For simplicity, in this section, we assume the range of V_{in} to be fixed at ± 1 V; therefore, all gain modulation is done using G_S only.

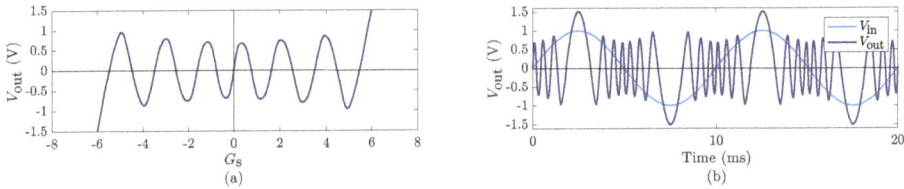

Figure 17. (**a**) Transfer function of the proposed Serge middle VCM; and (**b**) its output when driven by a 100-Hz sinewave for $G_S = 6$ and zero dc offset.

The spectrogram in Figure 18a shows the effect of increasing G_S from 0 to 6 for a 150-Hz sinusoidal input. This plot effectively depicts the rich harmonic patterns introduced by the system, which are far more complex than those introduce by traditional waveshaping methods and offer a wide timbral palette for sound synthesis. The fluctuations in energy at the fundamental and first few harmonics indicate the gain values at which each new fold is introduced. Figure 17b shows the effect of introducing a dc offset at the input of the system for a constant 200-Hz sinusoidal input. This result shows how the use of a dc offset can extend the timbral possibilities of the system even further, by introducing complex patterns consisting of both even and odd harmonics.

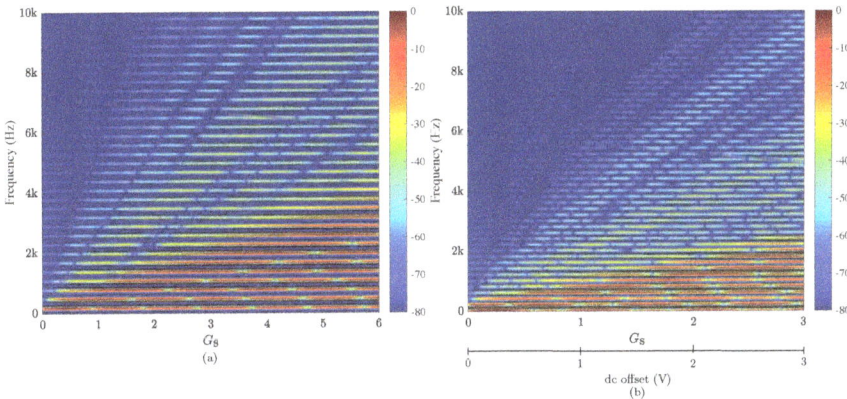

Figure 18. Spectrogram of: (**a**) a 150-Hz sinewave with peak amplitude 1 V processed by the proposed Serge middle VCM with varying gain G_S from 0–6; and (**b**) a 200-Hz sinewave processed with varying gain G_S from 0–3 and dc offset from 0–3 V.

Now, although the Lockhart wavefolder was originally designed to operate as a standalone unit, it can be adapted into a series topology with relative ease. Here, we propose using the wavefolding structure shown in Figure 19 to expand the synthesis capabilities of the Lockhart wavefolder. This design, while not based on any existing circuit, is comparable to that of the Yusynth Metalizer which also utilizes four Lockhart circuits in series [55]. The following paragraphs describe the sections of this proposed topology. Its frequency-domain behavior is then examined and compared with that of the Serge middle VCM.

The blocks labeled "LWF" in Figure 19 correspond to the proposed Lockhart wavefolder model. In order for this cascade of Lockhart wavefolders to behave as expected, we need to make sure that the individual folding stages satisfy two criteria. Firstly, the individual folders must provide approximately unity gain for small input values, i.e., below the folding point, and approximately negative unity gain beyond the folding point. Secondly, each stage should start folding at the same point with respect to its individual input.

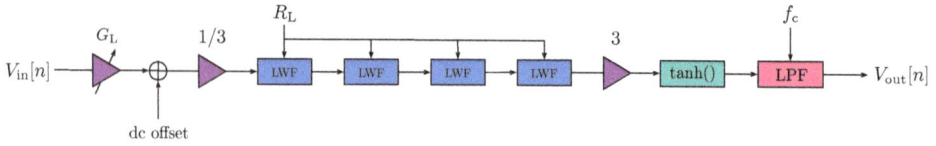

Figure 19. Block diagram for the proposed VA cascaded Lockhart wavefolder topology. Blocks labeled "LWF" and "LPF" indicate the Lockart wavefolder model and lowpass filtering, respectively.

We can meet these criteria with the proposed Lockhart model by selecting an appropriate value for R_L and adding static gain stages before and after the folding stages. These gain blocks will also help compensate for the attenuation introduced by the folding operation. First, we choose a value of R_L for which the Lockhart wavefolder exhibits unity gain for small input values. Having found this resistance value, the pre- and post-gain stages can be determined by measuring the value of $|V_{out}|$ at exactly the folding point. The pre-gain is taken to be approximately this value, and the post-gain is taken to be its inverse. In Section 3.1, it was shown that for $R_L = 7.5\,\text{k}\Omega$ the Lockhart wavefolder exhibits approximately unity gain below the folding point. This value leads to pre- and post-gains of approximately 1/3 and 3, respectively.

Figure 20a shows the transfer function of the proposed structure measured at the output of the post-gain block. We can observe how the folds introduced by this structure are evenly distributed, unlike those in Figure 17a. As with the Serge middle VCM, timbral control is achieved by modulating the value of G_L and by adding a dc offset. The static gain blocks ensure the amplitude of the folded output is bounded between approximately $\pm 1\,\text{V}$ for values of G_L between -10 and 10 (assuming once more that V_{in} has a peak amplitude of 1 V). Figure 20b shows the time-domain result of processing a 100-Hz sinusoidal input with the proposed structure for $G_L = 10$ and zero dc offset. In this particular design, additional timbral control can be achieved by modulating the value of R_L.

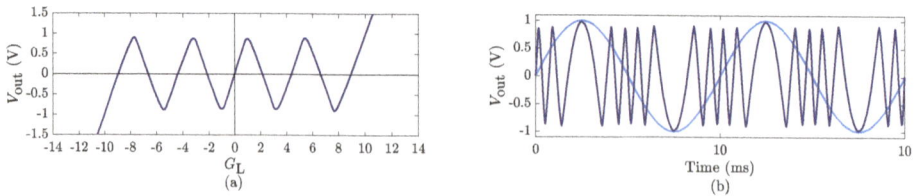

Figure 20. (a) Transfer function of the proposed cascaded Lockhart wavefolder structure measured after the post-gain block; and (b) its output when driven by a 100-Hz sinusoidal input for $G_L = 10$ and zero dc offset.

Lastly, we add two optional blocks. The first is a tanh() function after the post-gain block to model the behavior of an output buffering stage and to limit the range of the output waveform. This tanh() block can also be antialiased using the antiderivative method described by Equation (45). The antiderivative of the tanh() function is given by log(cosh()) [48]. The second optional block is a static one-pole lowpass filter with a cutoff at $f_c = 1.3\,\text{kHz}$ whose purpose is to act as a simple tone control. A similar static filtering stage can be found at the output of the Buchla 259 wavefolder [17]. The s-domain transfer function of this filtering stage is defined as

$$H(s) = \frac{w_c}{s + w_c},\tag{51}$$

where $w_c = 2\pi f_c$.

Finally, we examine the time-varying behavior of the proposed structure by considering the case of a 150-Hz input sinewave with variable gain G_L and dc offset. Figure 21 shows the spectrogram that results from varying G_L from 0 to 15. As expected, the system introduces complex harmonic patterns similar to those shown in Figure 18a. Likewise, Figure 18b demonstrates the effect of varying G_L from 0 to 10 while simultaneously increasing the level of dc offset from 0 to 5 V. This response is comparable to that of the Serge middle VCM (see Figure 18b).

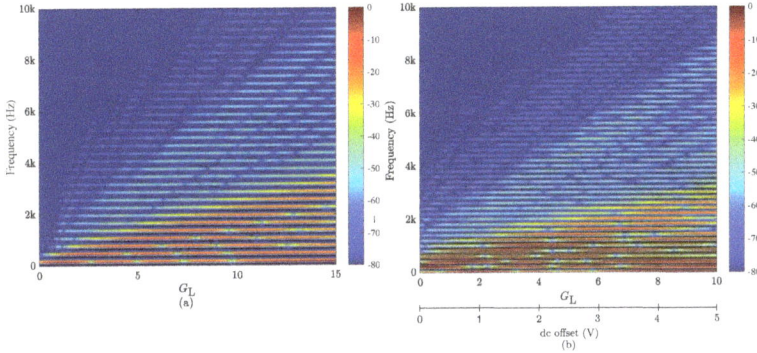

Figure 21. Spectrogram of: (**a**) a 150-Hz sinewave with peak amplitude 1 V processed by the proposed cascaded Lockhart topology with varying gain G_L from 0 to15; and (**b**) a 200-Hz sinewave processed with varying gain G_S from 0 to 0 and dc offset from 0 to 5 V.

A real-time demo of the proposed Lockhart wavefolder topology implemented using Max/MSP and Gen is available at http://research.spa.aalto.fi/publications/papers/smc17-wavefolder.

7. Conclusions

In this work, we have explored the behavior of two West Coast synthesizer circuits: the Lockhart and Serge wavefolders. By means of circuit analysis, we have derived closed-form expressions for the characteristic transfer functions of both systems. These transfer functions were validated against SPICE simulations implemented using LTspice. The results obtained indicate a good match between the proposed models and their corresponding SPICE simulations. In addition to this, we observed that the behavior of both circuits is very similar, despite the fact that their designs are fundamentally different.

The issue of aliasing caused by wavefolding in the digital domain was treated by incorporating the first-order antiderivative method. Within the context of the Lockhart wavefolder, it was shown that the proposed antialiased model is perceptually free from the effects of aliasing distortion when implemented at a sampling rate of $F_s = 88.2$ kHz. A thorough evaluation of the proposed Lockhart model indicates that this configuration yields a signal quality equivalent to that of oversampling by a factor of eight (i.e., $F_s = 352.8$ kHz) at nearly a fourth of the computational expenses. For the case of the Serge wavefolder, the use of the antiderivative method produces an increase in signal quality equivalent to that of oversampling by a factor of two (i.e., $F_s = 88.2$ kHz).

Furthermore, a recommended synthesis topology built around the Lockhart model consisting of four cascaded wavefolding stages, a saturator and a lowpass filter was presented. This topology was compared against a model of the Serge middle VCM built using six wavefolding stages. These structures illustrate the capabilities of wavefolding in a synthesis environment. However, it should be noted that the discussed topologies are not unique, as they can be modified according to the needs of the particular application. This effectively showcases the flexibility of VA models.

Supplementary Materials: The following are available online at http://research.spa.aalto.fi/publications/papers/smc17-wavefolder: a real-time Max/MSP demo of the proposed Lockhart wavefolder topology implemented using Gen~and a MATLAB implementation of the Lambert-W function using Fritsch's iteration.

Acknowledgments: The authors would like to thank Ken Stone for the valuable correspondence about the origins of the Lockhart circuit and Geoffrey Gormond for fruitful discussions on the behavior of the Serge VCM. Special thanks go to Prof. Vesa Välimäki for his assistance generating the ANMR curves. The work of Fabián Esqueda is supported by the Aalto ELEC Doctoral School.

Author Contributions: Fabián Esqueda and Henri Pöntynen conceived the idea for the paper and oversaw its development. Henri Pöntynen and Julian D. Parker wrote Sections 2.1 and 6, respectively, with additional work by Fabián Esqueda. Fabián Esqueda wrote the rest of the paper, conducted all measurements and generated all figures, with additional comments and corrections from the other three co-authors.

Conflicts of Interest: The authors declare no conflict of interest.

References

1. Le Brun, M. Digital waveshaping synthesis. *J. Audio Eng. Soc.* **1979**, *27*, 250–266.
2. Risset, J.C. *An Introductory Catalog of Computer Synthesized Sounds*; Bell Laboratories: Murray Hill, NJ, USA, 1969.
3. Schaefer, R.A. Electronic musical tone production by nonlinear waveshaping. *J. Audio Eng. Soc.* **1970**, *18*, 413–417.
4. Arfib, D. Digital synthesis of complex spectra by means of multiplication of nonlinear distorted sine waves. In Proceedings of the 59th Convention of the Audio Engineering Society, Hamburg, Germany, 28 February–3 March 1978.
5. Roads, C. A tutorial on non-linear distortion or waveshaping synthesis. *Comput. Music J.* **1979**, *3*, 29–34.
6. Chowning, J.M. The synthesis of complex audio spectra by means of frequency modulation. *J. Audio Eng. Soc.* **1973**, *21*, 526–534.
7. Ishibashi, M. Electronic Musical Instrument. Patent No. 4,658,691, 21 April 1987.
8. Kleimola, J. Audio synthesis by bitwise logical modulation. In Proceedings of the International Conference on Digital Audio Effects (DAFx), Espoo, Finland, 1–4 September 2008.
9. Timoney, J.; Lazzarini, V.; Lysaght, T. A modified FM synthesis approach to bandlimited signal generation. In Proceedings of the International Conference on Digital Audio Effects (DAFx), Espoo, Finland, 1–4 September 2008.
10. Kleimola, J.; Lazzarini, V.; Timoney, J.; Välimäki, V. Vector phaseshaping synthesis. In Proceedings of the International Conference on Digital Audio Effects (DAFx), Paris, France, 19–23 September 2011.
11. Pakarinen, J.; Yeh, D.T. A review of digital techniques for modeling vacuum-tube guitar amplifiers. *Comput. Music J.* **2009**, *33*, 85–100.
12. Hobbs, R.V.; Snoddy, G.T. Tone Modifier for Electrically Amplified Electro-Mechanically Produced Musical Tones. Patent No. 3,213,181, 19 October 1965.
13. Yeh, D.T.; Abel, J.S.; Smith, J.O., III. Simplified, physically-informed models of distortion and overdrive guitar effects pedals. In Proceedings of the International Conference on Digital Audio Effects (DAFx), Bordeaux, France, 10–15 September 2007.
14. Werner, K.J.; Nangia, V.; Smith, J.O., III; Abel, J.S. Resolving wave digital filters with multiple/multiport nonlinearities. In Proceedings of the International Conference on Digital Audio Effects (DAFx), Trondheim, Norway, 30 November–3 December 2015.
15. Buchla Electronic Musical Instruments. The History of Buchla. Available online: https://buchla.com/history/ (accessed on 3 October 2017).
16. Parker, J.; D'Angelo, S. A digital model of the Buchla lowpass-gate. In Proceedings of the International Conference on Digital Audio Effects (DAFx), Maynooth, Ireland, 2–5 September 2013.
17. Esqueda, F.; Pöntynen, H.; Välimäki, V.; Parker, J.D. Virtual analog Buchla 259 wavefolder. In Proceedings of the International Conference on Digital Audio Effects (DAFx), Edinburgh, UK, 5–9 September 2017.
18. Moog, R.A. A voltage-controlled low-pass high-pass filter for audio signal processing. In Proceedings of the 7th Convention of the Audio Engineering Society, New York, NY, USA, 11–15 October 1965.
19. Stilson, T.; Smith, J.O., III. Analyzing the Moog VCF with considerations for digital implementation. In Proceedings of the International Computer Music Conference, Hong Kong, China, 19–24 August 1996.
20. Huovilainen, A. Non-linear digital implementation of the Moog ladder filter. In Proceedings of the International Conference on Digital Audio Effects (DAFx), Naples, Italy, 5–8 October 2004.

21. Hélie, T. On the use of Volterra series for real-time simulations of weakly nonlinear analog audio devices: application to the Moog ladder filter. In Proceedings of the International Conference on Digital Audio Effects (DAFx), Montreal, QC, Canada, 18–20 September 2006.

22. D'Angelo, S.; Välimäki, V. Generalized Moog ladder filter: Part II—Explicit nonlinear model through a novel delay-free loop implementation method. *IEEE/ACM Trans. Audio Speech Lang. Process.* **2014**, *22*, 1873–1883.

23. Civolani, M.; Fontana, F. A nonlinear digital model of the EMS VCS3 voltage-controlled filter. In Proceedings of the International Conference on Digital Audio Effects (DAFx), Espoo, Finland, 1–4 September 2008.

24. Fontana, F.; Civolani, M. Modeling of the EMS VCS3 voltage-controlled filter as a nonlinear filter network. *IEEE Trans. Audio Speech Lang. Process.* **2010**, *18*, 760–772.

25. Huovilainen, A. Design of a Scalable Polyphony-MIDI Synthesizer for a Low Cost DSP. Master's Thesis, Aalto University, Espoo, Finland, 2010.

26. Rest, M.; Parker, J.; Werner, K.J. WDF modeling of a Korg MS-50 based non-linear diode bridge VCF. In Proceedings of the International Conference on Digital Audio Effects (DAFx), Edinburgh, UK, 5–7 September 2017.

27. Yeh, D.T.; Abel, J.S.; Smith, J.O., III. Simulation of the diode limiter in guitar distortion circuits by numerical solution of ordinary differential equations. In Proceedings of the International Conference on Digital Audio Effects (DAFx), Bordeaux, France, 10–15 September 2007.

28. De Paiva, R.C.D.; Pakarinen, J.; Välimäki, V.; Tikander, M. Real-time audio transformer emulation for virtual tube amplifiers. *EURASIP J. Adv. Signal Process.* **2011**, *2011*, 1–15.

29. Dunkel, W.R.; Rest, M.; Werner, K.J.; Olsen, M.J.; Smith, J.O., III. The Fender Bassman 5F6-A family of preamplifier circuits—A wave digital filter case study. In Proceedings of the International Conference on Digital Audio Effects (DAFx), Brno, Czech Republic, 5–9 September 2016.

30. Huovilainen, A. Enhanced digital models for analog modulation effects. In Proceedings of the International Conference on Digital Audio Effects (DAFx), Madrid, Spain, 20–22 September 2005.

31. Holters, M.; Zölzer, U. Physical modelling of a wah-wah effect pedal as a case study for application of the nodal DK method to circuits with variable parts. In Proceedings of the International Conference on Digital Audio Effects (DAFx), Paris, France, 19–23 September 2011.

32. Eichas, F.; Fink, M.; Holters, M.; Zölzer, U. Physical modeling of the MXR Phase 90 guitar effect pedal. In Proceedings of the International Conference on Digital Audio Effects (DAFx), Erlangen, Germany, 1–5 September 2014.

33. Parker, J. A simple digital model of the diode-based ring modulator. In Proceedings of the International Conference on Digital Audio Effects (DAFx), Paris, France, 19–23 September 2011.

34. Werner, K.J.; Abel, J.S.; Smith, J.O. A physically-informed, circuit-bendable, digital model of the Roland TR-808 bass drum circuit. In Proceedings of the International Conference on Digital Audio Effects (DAFx), Erlangen, Germany, 1–5 September 2014.

35. Werner, K.J.; Abel, J.S.; Smith, J.O. The TR-808 cymbal: A physically-informed, circuit-bendable, digital model. In Proceedings of the International Computer Music Conference (ICMC)/Sound and Music Computing Conference (SMC), Athens, Greece, 14–20 September 2014.

36. De Paiva, R.C.D.; Pakarinen, J.; Välimäki, V. Reduced-complexity modeling of high-order nonlinear audio systems using swept-sine and principal component analysis. In Proceedings of the 45th Audio Engineering Society Conference: Applications of Time-Frequency Processing in Audio, Helsinki, Finland, 1–4 March 2012.

37. Eichas, F.; Zölzer, U. Black-box modeling of distortion circuits with block-oriented models. In Proceedings of the International Conference on Digital Audio Effects (DAFx), Brno, Czech Republic, 5–9 September 2016.

38. Eichas, F.; Möller, S.; Zölzer, U. Block-oriented gray box modeling of guitar amplifiers. In Proceedings of the International Conference on Digital Audio Effects (DAFx), Edinburgh, UK, 5–9 September 2017.

39. Stilson, T.; Smith, J.O., III. Alias-free digital synthesis of classic analog waveforms. In Proceedings of the International Computer Music Conference, Hong Kong, China, 19–24 August 1996.

40. Brandt, E. Hard sync without aliasing. In Proceedings of the International Computer Music Conference, Havana, Cuba, 17–23 September 2001.

41. Välimäki, V.; Pekonen, J.; Nam, J. Perceptually informed synthesis of bandlimited classical waveforms using integrated polynomial interpolation. *J. Acoust. Soc. Am.* **2012**, *131*, 974–986.

42. Välimäki, V. Discrete-time synthesis of the sawtooth waveform with reduced aliasing. *IEEE Signal Process. Lett.* **2005**, *12*, 214–217.

43. Välimäki, V.; Nam, J.; Smith, J.O., III; Abel, J.S. Alias-suppressed oscillators based on differentiated polynomial waveforms. *IEEE Trans. Audio Speech Lang. Process.* **2010**, *18*, 786–798.

44. Kleimola, J.; Välimäki, V. Reducing aliasing from synthetic audio signals using polynomial transition regions. *IEEE Signal Process. Lett.* **2012**, *19*, 67–70.

45. Franck, A.; Välimäki, V. Higher-order integrated wavetable synthesis. In Proceedings of the International Conference on Digital Audio Effects (DAFx), York, UK, 17–21 September 2012.

46. Esqueda, F.; Bilbao, S.; Välimäki, V. Aliasing reduction in clipped signals. *IEEE Trans. Signal Process.* **2016**, *60*, 5255–5267.

47. Esqueda, F.; Välimäki, V.; Bilbao, S. Rounding corners with BLAMP. In Proceedings of the International Conference on Digital Audio Effects (DAFx), Brno, Czech Republic, 5–9 September 2016.

48. Parker, J.; Zavalishin, V.; Le Bivic, E. Reducing the aliasing of nonlinear waveshaping using continuous-time convolution. In Proceedings of the International Conference on Digital Audio Effects (DAFx), Brno, Czech Republic, 5–9 September 2016.

49. Bilbao, S.; Esqueda, F.; Parker, J.D.; Välimäki, V. Antiderivative antialiasing for memoryless nonlinearities. *IEEE Signal Process. Lett.* **2017**, *24*, 1049–1053.

50. Esqueda, F.; Pöntynen, H.; Parker, J.D.; Bilbao, S. Virtual analog model of the Lockhart wavefolder. In Proceedings of the 14th Sound and Music Computing Conference (SMC), Espoo, Finland, 5–8 July 2017.

51. Lockhart, R., Jr. Non-selective frequency tripler uses transistor saturation characteristics. *Electron. Des.* **1973**, *17*.

52. Hutchins, B.A. Frequency multiplication and division methods. *Electronotes* **1976**, *70*, 3–17.

53. Stone, K. Simple wave folder for music synthesizers. Available online: https://www.cgs.synth.net/modules/cgs52_folder.html (accessed on 19 September 2017).

54. Stone, K. (CGS, Melbourne, Australia). Private communication, 2017.

55. Usson, Y. Yusynth Metalizer. Available online: http://yusynth.net/modular/en/metalizer/ (accessed on 18 September 2017).

56. Banwell, T.; Jayakumar, A. Exact analytical solution for current flow through diode with series resistance. *Electron. Lett.* **2000**, *36*, 291–292.

57. De Paiva, R.C.D.; D'Angelo, S.; Pakarinen, J.; Välimäki, V. Emulation of operational amplifiers and diodes in audio distortion circuits. *IEEE Trans. Circuits Syst. II Exp. Briefs* **2012**, *59*, 688–692.

58. Werner, K.J.; Nangia, V.; Bernardini, A.; Smith, J.O., III; Sarti, A. An improved and generalized diode clipper model for wave digital filters. In Proceedings of the 139th Convention of the Audio Engineering Society, New York, NY, USA, 29 October–1 November 2015.

59. Bernardini, A.; Werner, K.J.; Sarti, A.; Smith, J.O., III. Modeling nonlinear wave digital elements using the Lambert function. *IEEE Trans. Circuits Syst. I Regul. Pap.* **2016**, *63*, 1231–1242.

60. Veberič, D. Lambert W function for applications in physics. *Comput. Phys. Commun.* **2012**, *183*, 2622–2628.

61. Linear Technology. LTspice. Available online: http://www.linear.com (accessed on 2 October 2017).

62. Serge Modular Music Systems, San Francisco, CA, USA. New Products from Serge Modular: Series 79 Synthesizer Systems, 1980. Available online: http://www.serge.synth.net/documents/catalog/SergeModular-Series79%20.pdf (accessed on 21 September 2017).

63. Stone, K. Serge Wave Multipliers for music synthesizers. Available online: https://www.cgs.synth.net/modules/cgs113_vcm.html (accessed on 21 September 2017).

64. Fukushima, T. Precise and fast computation of Lambert W-functions without transcendental function evaluations. *J. Comput. Appl. Math.* **2013**, *244*, 77–89.

65. Lehtonen, H.M.; Pekonen, J.; Välimäki, V. Audibility of aliasing distortion in sawtooth signals and its implications for oscillator algorithm design. *J. Acoust. Soc. Am.* **2012**, *132*, 2721–2733.

applied
sciences

MDPI

Article

Playing for a Virtual Audience: The Impact of a Social Factor on Gestures, Sounds and Expressive Intents

Simon Schaerlaeken *, Didier Grandjean † and Donald Glowinski †

Neuroscience of Emotion and Affective Dynamics Lab, Faculty of Psychology and Educational Sciences and Swiss Center for Affective Sciences, University of Geneva, 1205 Geneva, Switzerland; didier.grandjean@unige.ch (D.G.); donald.glowinski@unige.ch (D.G.)
* Correspondence: simon.schaerlaeken@unige.ch; Tel.: +41-78-943-7215
† Co-senior authors.

Academic Editor: Stefania Serafin
Received: 30 October 2017; Accepted: 13 December 2017; Published: 19 December 2017

Abstract: Can we measure the impact of the presence of an audience on musicians' performances? By exploring both acoustic and motion features for performances in Immersive Virtual Environments (IVEs), this study highlights the impact of the presence of a virtual audience on both the performance and the perception of authenticity and emotional intensity by listeners. Gestures and sounds produced were impacted differently when musicians performed at different expressive intents. The social factor made features converge towards values related to a habitual way of playing regardless of the expressive intent. This could be due to musicians' habits to perform in a certain way in front of a crowd. On the listeners' side, when comparing different expressive conditions, only one congruent condition (projected expressive intent in front of an audience) boosted the participants' ratings for both authenticity and emotional intensity. At different values for kinetic energy and metrical centroid, stimuli recorded with an audience showed a different distribution of ratings, challenging the ecological validity of artificially created expressive intents. Finally, this study highlights the use of IVEs as a research tool and a training assistant for musicians who are eager to learn how to cope with their anxiety in front of an audience.

Keywords: immersive virtual environment; music; performance; expressiveness; authenticity; emotions; social

1. Introduction

Musical performances are the result of a complex interactive phenomenon between the musicians and the audience who attends and appreciates them. However, the understanding of this complex phenomenon could benefit from a scientific approach. First, experiments on musical performances could bring insights into how to deal with musicians' performance anxiety, as musicians' training involves learning to cope with stress during performances. However, opportunities to train musicians on how to regulate their emotions during concerts are limited. The most effective method seems to combine relaxation training with exposure to stressful events (to build up realistic expectations of what will be felt during performances) and cognitive restructuring (to counteract on self-handicapping habitual thoughts and attitudes) [1]. In fact, musicians' experience and ability to resist stress mainly depends on the opportunities they have during their career to perform during live performances and having repeated peer sessions. Second, researchers interested in the effect of audiences on musicians and their body language are usually left with uncontrolled and highly variable situations when they study concerts. On a practical side, recording physiological and motion capture data during concert can hinder musicians during concerts, thus impacting the quality of their performance, making concerts a challenging environment for scientific research.

Immersive Virtual Environments (IVEs) have been used by researchers to control for these complex parameters [2]. IVEs allow researchers to create realistic virtual environments with unlimited configurations that can adapt in real time to users' behavior. It allows for researchers to control for environmental parameters such as the audience, the space between different objects, and the lighting. IVEs offer researchers the opportunity to compute perceptual analyses and create new roads for computational development. This environment combined with a motion capture set-up was used in this study to precisely record expressive musical gestures and explore the possible underlying behavioral mechanisms impacted by the presence of an audience in the context of different levels of expressive intents.

1.1. Controlling Environmental Variables in Virtual Reality

Musicians develop unique abilities allowing them to adapt their behavior to different social and environmental contexts [2]. Consequently, it is necessary to control for many parameters when recording musical performances, e.g., sounds, lights, the presence of other musicians or audience. Thus IVEs represent a key methodological tool for psychological research as it can provide greater experimental control, more precise measurements, ease of replication across participants, and high ecological validity, making it extremely attractive for researchers [3,4]. They can also provide live feedbacks to participants. Virtual reality and IVEs have been used in research for patients suffering from post-traumatic stress disorder [5–7] and for treating phobias such as fear of flying [8] and arachnophobia [9]. The use of such technology has also been proven to be efficient for treating social phobia and reducing the fear of the public speaking [10,11]. Few studies have considered using virtual environments in music to study performance anxiety [12,13]. In a recent study by Williamon et al. (2014), musicians were invited to enter IVEs to train their ability to cope with the pressure of performing live and rated such tool useful for developing their performance skills and very realistic. It demonstrated that simulated environments are able to offer a realistic experience of performance contexts.

1.2. Music Performance: From Sound to Gesture

Communicating and expressing emotions through music is the main reason why people engage in this activity [14]. Evidence points at a general ability to accurately recognize emotions expressed with music (e.g., happiness, sadness, and nostalgia) [15–21]. Regardless of cultural background or musical training, people are generally able to name the intended emotion, providing evidence for an universal recognition not only of the expression of basic emotion but also of more complex feelings [22–24]. Moreover, many studies have tried to capture the acoustic cues that musicians use to convey specific emotions (e.g., [19,20,25,26]). These cues involve changes in tempo, sound level, articulation, timbre, timing, tone attack and decay, intonation, vibrato extent and frequency, accents on particular notes, etc. On the listener's side, judgments of intended emotions have been related to specific musical features, including tempo, articulation, intensity, and timbre [17,27–29]. In comparison, the same observations can be made for vocal expression and emotional prosody, as specific acoustic cues are predominant in accurately recognizing the intended emotions [30].

While auditory information plays a crucial role in music communication models, Finnäs (2001) noticed an increasing interest in the visual component influencing the perception of the musical performance [31]. While the auditory stream convey emotions in music, its associated movements also contain significant information. Many musical traditions have included a combination of both audio and visual stimulations during the experience of music performance [32,33]. Such practice still remains in our mediatized society [34]. The visual component of the live music performance contributes significantly to the appreciation of music performance [35,36]. Malin (2008) even concludes that a "variety of musical properties and types of evaluations can be affected by the visual information" [37]. Although both auditory and visual kinematic cues contribute significantly to the perception of overall expressiveness, the effect of visual kinematic cues appears to be somewhat stronger [38]. It also provides preliminary evidence of cross-modal interactions in the perception of auditory and visual expressiveness in music performance. The visual component should not be categorized as a marginal phenomenon in music perception, but as an important factor in the communication of meaning. This process of cross-modal integration exists for many genre of music, from classical to pop and rock music [33,36]. All in all, visual kinematic cues have been found to influence the perception of phrasing and musical tension [33], felt emotion [39] the perception of emotional expression [40], and the overall appreciation of the performance (for a meta-analysis, see [38]).

A crucial visual cue is related to musicians' gestures, and how their bodies move during performances. Two types of movements can be here distinguished: instrumental actions and ancillary/expressive movements [41,42]. The former are creating sound while the latter have an intrinsic relationship with the music, representing a link between the music and the expressive intention of the musician [43]. Musical gestures are mainly made to produce sounds but are also used by the musician as means to convey or express emotions (see review Expressive Gesture [44,45]). Musicians' expressive gestures fall into two categories: (i) communicating their expressive intentions; and (ii) expressing their feelings without intending to communicate them [46]. Gestures contribute to communicate information to the audience as well as the other musicians. Expressive movements occur frequently in musical performances, even though these movements are not mandatory for musical performance such as during training [47]. Furthermore, across performers, these idiosyncratic expressive movements appear to have some consistencies [42,48]. Finally, Vines, Krumhansl, Wanderley, and Levitin (2006) concluded that these movements are not randomly performed, but rather are used to communicate a holistic, musical, expressive unit [33]. Understanding how this unit works is the primary goal of researchers interested in musical gestures.

1.2.1. Emotional Intensity and Expressive Intents in Musical Performances

With the recognition of emotions in music, the emotional intensity and expressive intents have been shown to be dependent on auditory cues. In their study on such information, researchers asked participants to rate the emotional expressiveness of music performances in which timing and intensity were parametrically manipulated [49]. Emotion judgments monotonically increased with performance variability, and timing changes were reported to explain more variance in emotional expressiveness than sound intensity. Changes in tempo and sound intensity in a music performance were also shown to be correlated with one another, and with real-time ratings of emotional arousal [50]. A systematic relationship between emotionality ratings, timing, and loudness was highlighted when listeners rated their moment-to-moment level of perceived emotionality while listening to music performances. Therefore, the variation of acoustic features associated with the expressive intent of the musician during a performance appears to have a crucial impact on the emotionality perceived. On the visual side, Davidson (2005) demonstrated that certain perceptual elements of a musician's gestures are sufficient for the audience to identify a musician's expressive intent [43]. She suggested the use of three level of expressiveness to be able to study the link between expressive gestures and musical performance: (1) without expression, labeled as "deadpan"; (2) with normal expression/concert-like, labeled as "projected"; and (3) with exaggerated expression, labeled as "exaggerated". Some body parts have been reported to convey more expressive information, specifically head, shoulders, arms, and torso [46,48,51–53]. Some motion features have also been associated with expressive motion such as the quantity of motion [54]. The use of motion features helps understand broad, unrefined body reaction and gives a first glimpse of the behavioral components of expressive gestures. It might help understand how musicians cope with the audience [28,55,56]. For example, musician facing an audience and playing in exaggerated expressive manners could be affected by the amount of supplementary stress caused by the difficulty of the task. All in all, as mentioned by Shaffer (1992) [57], "a performer can be faithful to its structure and at the same time have the freedom to shape its moods" (p. 265). This corresponds to a phenomenon called *performance expression*. It refers to "the small and large variations in timing, dynamics, timbre, and pitch that form the micro-structure of a performance and differentiate it from another performance of the same music" ([58], p. 118).

1.2.2. Authenticity in Musical Performances

The importance of authenticity is undervalued in emotion research and musical performances. Authenticity could be an underlying factor of emotion communication through music. For example, in popular music culture, audio-visual performances convey markers of authenticity, which are essential for the creation of credibility and emotions [34]. In popular music, as the saying goes, "seeing is believing" ([34], p. 85). In everyday life, the anthropologist Erving Goffman, one of the great pioneers of social science research on emotions, affirmed that "We all play emotion theater most of the time". Goffman (1982) demonstrated that human beings mostly try to present themselves in the best light and always stage their daily lives to protect themselves [59]. Faked emotions or the modulation of the expression of emotions play a central role in self-preservation by keeping inappropriate emotional expressions to damage self-presentation. Scherer et al. (2013) argued that one should abandon the idea that, for the sake of complete authenticity, actors should live through "real emotions" on the stage [60]. Specific emotional expressions are only credible, i.e., appear authentic, when they can be perceived as generated by appraisals that fit the respective circumstances. This means that in order to succeed in appearing credible, the artist must: (1) pick the most appropriate set of vocal, facial and body expression elements for the respective emotions and combine them dynamically, in a psycho-biologically valid fashion; (2) achieve precise synchronization of the respective processes, letting the expression unfold in an appropriate fashion; and (3) handle the situational development appropriately in terms of its dynamic flow [60]. These requirements demand the highest amount of professionalism when trying to voluntary display a certain dynamic forms of expression. In music performance, one could also argue that self-awareness is a key feature in the perception of authentic emotion. Musicians exhibit this awareness at different times with both their technique and their emotional expressiveness. For example, after sight-reading a new music for some time, as the musician builds up the motor repertoire required, he or she can focus more on putting more expressive intent into his/her movements. Once these abilities become automatized, i.e., habits for a specific performance, they are no longer at the forefront of the individual's consciousness, the musician will then begin to bring components of their own personal performance style to the music. This, in turn, contributes to the perceived authenticity of the ultimate performance.

Even though the view of what is an authentic musical performance is subjective and based on individual bias, listeners tend to agree that authentic musical performance styles all have a sense of uniqueness. For example, celebrated pianist Glenn Gould is often noted as an exemplary expressive musician with an extremely particular performance style. This individuality is one of the hallmarks of authenticity in performance. Wöllner (2013) suggests that individual artistic expression can be quantified as such when the performance matches the listener's "mental prototype" of what a unique and authentic performance would look and sound like [61]. Overall, it is important to note that, as implied in the BRECVEMA model regarding "appreciation emotions" in aesthetic judgment [62], while the "mental prototype" we all use when making judgments about the authenticity of a performance are socially and culturally driven, the gestures that characterize "authenticity" in music are extremely useful in analyzing how skilled musicians play "emotion theater" to create moving and expressively credible performances.

1.3. Goal of This Study

This study aims to investigate the impact of the audience presence on both aspects of a music performance, from both performers and observers' views. By studying the difference in acoustic and motion features at different levels of expressive intent, we want to demonstrate the impact of the audience presence on the link between the expressive intents performed on the musician's side and the emotionality perceived on the observer's side. We hypothesized that the presence of a virtual audience would hinder the movements of the musicians due to the stress generated by the act of performing live. More specifically, it would reduce the differences in acoustic and motion features between the different expressive conditions. Consequently, this would also impact the emotional intensity and authenticity perceived by the audience. The participants should therefore report similar values across expressive conditions.

To understand such complex phenomenon, we recorded musicians playing with different expressive manners in front of a virtual crowd or an empty room. We analyzed motion and acoustic features and measured the impact of our social factor, i.e., the audience. Afterwards, we presented video clips of musicians playing and asked participants to rate both the emotional intensity and authenticity perceived. We performed a series of analysis to link these values to the audio and visual cues explored.

2. Materials and Methods

This experiment was divided into two phases: the recording sessions and the rating experiment. Both phases were approved by the local ethical committee of the department of Psychology, University of Geneva. These two phases aim to emphasize, respectively, on the proximal and distal cues of a Brunswik lens model [63]. This type of model has been shown to be highly representative in the case of emotional prosody [64] and in music [27].

2.1. Recording Session

Four violinists (3 females, $M_{age} = 22$) took part in the recordings. They were paid according to the ethical protocol. They agreed with the use of the material recorded as stimuli for this study. Musicians performed inside an Immersive Virtual Environment (IVE) with the use of a system of three screens, seven TITAN QUAD 3D projectors (Digital Projection Limited, Manchester, UK), and stereo glasses presenting seamless and perceptively coherent 3D images. Two different virtual environments were created for this experiment: a room filled with an audience behaving naturally and attending the concert, and the same room without the audience (Appendix Figure A1). The virtual audience was composed of high quality agents with realistic facial expressions and behaviors. An agent is created using four different components: (1) realistic body from a 3D scan of real actors; (2) realistic body animations created from motion capture footage; (3) accurate and controllable facial expressions based on FACSGen [65,66]; and (4) expressive behavior and social interaction modeling. The audience behavior could smoothly change from engaged to disengaged behaviors. Specifically, in the engaged behavior, the audience attention increases through the convergence of individual gazes focusing on the musician [67]. On the other hand, a disengaged audience is rendered by allowing the gaze of avatar to wander around as generally observed in distracted people. Furthermore, for realism purposes, each avatar had random idle animations of their body while looking at the musician in order to approximate usually seen fluent behaviors at concerts. To avoid unnatural uniformity, part of the audience (5–10%) is always modeled in the disengaged condition when the majority is engaged, and vice versa [68].

Each musician was instructed to play 30-second-long interpretations of Bach's Partita No. 2 in D minor, BWV 1004: Sarabande. The part of the musical score to be played was carefully selected to correspond to complete musical phrases. The excerpts were interpreted according to three selected expressive intents: deadpan, projected, and exaggerated [43]. The different combinations of conditions were performed in a pseudo-randomized order by each musician. Six excerpts were recorded for each of the 4 musicians, adding up to a total of 24 excerpts. The sound was captured using an Olympus LS-10 (Olympus, Tokyo, Japan). Motion capture data was also recorded for every piece using a VICON optical motion tracking system (Vicon Motion Systems Ltd., Oxford, UK) composed of eight Bonita 3 cameras (Vicon Motion Systems Ltd., Oxford, UK). A total of 26 markers were used, covering selected body parts based on recent literature on the analysis of music performance, i.e., the head, arms, and torso [48,51–53].

2.2. Rating Experiment

Forty participants took part in the rating study (19 females). All of them spoke French as first language. The average age was 23 years ($SD = 7.19$) and most participants were psychology students. Participants completed the experiment on a computer. The experiment itself was programmed with Limesurvey [69] and ran on computers with a screen resolution of 1280 × 1024 pixels. Loudness was set to 50% and could be adjusted by the participant. Headphones were provided. The experiment lasted ~30 min. The participants had to complete a musical habits questionnaire before starting the experiment. Stimuli were fully mixed together and presented in a unique random order for each participant. The participant listened to each stimulus while watching the point light display (PLD) of the musician's movements and then answered multiple questions. They rated the emotional intensity of the stimuli, the authenticity, each of the 9 emotions from the Geneva Emotional Music Scales (GEMS) (Appendix Figure A3) [24]. All ratings were done on sliders from 0 to 100. The participants were also asked to rate the importance of each body part in evaluating the general emotional intensity. This process was repeated for the 24 stimuli per participant.

2.3. Multi-Modal Expressive Behavior Analysis

Drawing upon the recent studies [70,71], we considered two types of expressive body features: the kinetic energy and the Body Twist Index (BTI). The former helps understand broad, unrefined body reactions and gives a glimpse of the behavioral components of expressive gestures. The latter captures body shape related information, i.e., the relative displacement of body parts with respect to other ones. In the case of violin player, this second feature is critical since the upper and lower parts of the body are dissociated. Violinists tend to twist their body more while playing compared to cello players for example.

Listening to music does not involved watching or performing gesture in around 80% of the time [72]. We therefore computed acoustic features and focused essentially on the metrical centroid [73]. This feature offers a very detailed description of the metrical structure of a musical piece. Time-related aspect of music is thought to have an impact on the emotional arousal and linked to the notion of musical entrainment [62]. It is also the preferred acoustic cue used by listener to perceived different emotions in a piece of music [27]. The metrical centroid is expressed in beat per minutes (BPM). Low BPM values indicate a prevalence of high metrical level (i.e., slow pulsations corresponding to whole notes, bars, etc.). High BPM values indicate on the contrary that more elementary metrical levels predominate (i.e., very fast levels corresponding to very fast rhythmical values). We hypothesized that these three features could help modeling critical changes in musician's expressive responses to the presence of an audience.

Both motion features were calculated using authors' MATLAB (v2016b, MathWorks, Inc., Natick, MA, USA, 2016) toolbox (build upon the MoCap Toolbox [74]). The kinetic energy was computed for every marker of the motion capture data and then averaged across markers and over time. The Body Twist Index was represented by the average angle between the pelvis and a perpendicular line to the shoulders, considering only the top quantile (above quantile 75%) of the data for each excerpt. Both features were z-scored per musician (Figure 1). The acoustic feature was computed with the MIRToolbox 1.6.1 [75]. It was computed on overlapping frames of the musical excerpts (duration: 1 s, hop: 0.25 s). The average value of the dynamic acoustic feature was also z-scored per musician (Figure 1).

Linear models were used in this study for modeling the performance of the musicians while linear mixed models were used to estimate the participants' ratings. Mixed models offer two advantages: they incorporate random effects and they allow handling correlated data and unequal variance [76]. When using features as fixed effect in the modeling of participants' perception of authenticity and emotional intensity, we divided them into four bins ("0–25","25–50","50–75", and "75–100"). This allowed contrasts to be computed between bins. Comparing model was done using Chi-squared testing. All *p*-values were corrected using False Discovery Rate (FDR).

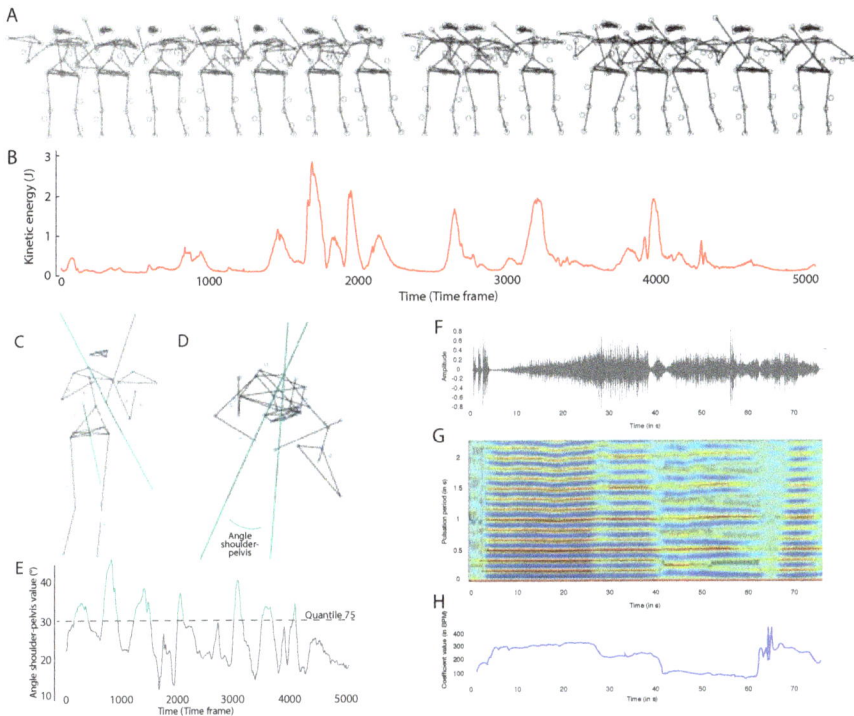

Figure 1. Motion and acoustic features computed: (**A**) Chronograph of the motion capture data for one excerpt played in front of an audience while exaggerating expressive intent. (**B**) Kinetic energy associated with movements of all the markers for the excerpt depicted in (**A**). (**C**) 3D view of the motion capture and the line associated with the pelvis and shoulder. (**D**) Transversal view of the motion capture data and the angle computed between the line from the pelvis and shoulders. (**E**) Computation of Body Twist Index. The angle between the aforementioned lines is computed over the duration of the excerpt. The Body Twist Index consists of the average of all values comprised in the top quantile (above the quantile 75 value). (**F**) Sound profile of the performance of the Scherzo of L. van Beethoven's Symphony No.9 in D minor, op.125. (**G**) Corresponding autocorrelogram with tracking of the metrical structure. (**H**) Corresponding metrical centroid curve (Copyright Grandjean, D. et al., 2013 [73]).

3. Results

In this section, we present the analysis of both the proximal and distal components of the performance, i.e., from both performers' and observers' side.

3.1. Proximal Performance Data Analysis

In order to characterize the musical gestures and motion, we computed the z-scored values of all features (kinetic energy, body twist index, and metrical centroid) recorded during the performance at different expressive intents with the presence or not of a virtual audience. Based on the marginality principle, we considered only a linear model containing main effects of both the expressive intents and the social factor as well as the interaction between these factors ($F_{kineticenergy}(5, 18) = 39.78, p = 4.1 \times 10^{-9}$, $R^2_{adjusted} = 0.89$ & $F_{BTI}(5, 18) = 6.77, p = 0.001, R^2_{adjusted} = 0.55, F_{metricalcentroid}(5, 17) = 16.58$, $p = 5.17 \times 10^{-6}, R^2_{adjusted} = 0.78$).

Both motion features increased with stronger expressive intents in the absence of the social factor (Figure 2). When the audience was present, while the kinetic energy still increased with the expressive intent, the difference with the other social condition was not significant. When comparing the presence and absence of the audience, the impact of the social factor only appeared for the deadpan expressive condition for the BTI. In this case, the presence of the audience was characterized by a significant increase in the "twist" angle ($F_{BTI,DP,Social}(1,18) = 5.41, p = 0.032$). Noteworthy was the significant difference for the BTI in the deadpan and exaggerated condition when the audience was present or not. The BTI value increased in the deadpan condition with the presence of an audience while it was diminished in the exaggerated condition ($F_{BTI,DP/EXAG,Social}(1,18) = 5.64, p = 0.028$).

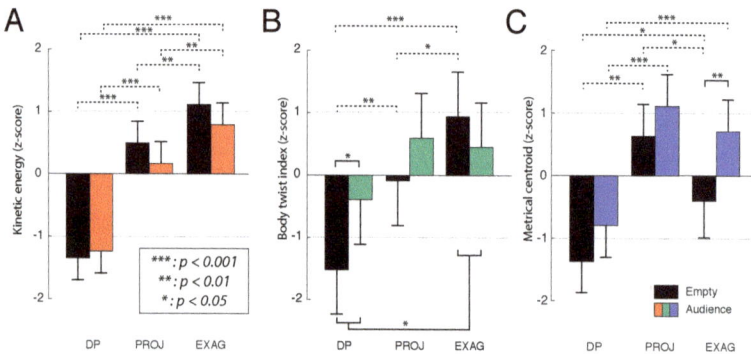

Figure 2. Impact of the interaction of the expressiveness and the presence of an audience on body features (deadpan: DP, projected: PROJ, and exaggerated: EXAG): (**A**) Kinetic energy (red); (**B**) Body Twist Index (green); and (**C**) Metrical centroid (blue). (* $p < 0.05$, ** $p < 0.01$, *** $p < 0.001$).

In the case of the acoustic feature, metrical centroid, we observed a different pattern when the virtual audience was absent (Figure 2). When asked to play with a "projected" expressive intent, the metrical centroid of the performance was significantly greater, meaning that the more elementary metrical levels predominated (i.e., very fast levels corresponding to very fast rhythmical values). It was however lower for both the "deadpan" and "exaggerated" conditions, contrarily to the linear increase observed in motion features. The effect of the social factor was highlighted in the increase of the beats per minute of the metrical centroid for every expressive intent, especially for the "exaggerated" condition where this increase was significant ($F_{MetCent,EXAG,Social}(1,17) = 9.24, p = 0.007$).

3.2. Distal Participant Data Analysis

The second part of our analyses focused on the ratings given by the participants on both authenticity and emotional intensity. Across all stimuli, the reliability of our participants' ratings was high, $\alpha = 0.92$ for authenticity and $\alpha = 0.93$ for intensity. Participants were grouped together into two categories, music-lovers and musicians, based on their responses for the musical habits questionnaires [77]. No significant difference in means was observed between groups for both authenticity ($M_{musicians} = 43.155, SD_{musicians} = 25.28, M_{music-lovers} = 46.217, SD_{music-lovers} = 23.98$, $t(819) = -1.829, p = 0.067, d = -0.12$) and intensity ($M_{musicians} = 42.308, SD_{musicians} = 24.598$, $M_{music-lovers} = 45.314, SD_{music-lovers} = 22.059, t(819) = -1.89, p = 0.059, d = -0.13$). The difference between both groups is marginal. Moreover, this difference is trivial due to the very small effect size associated with the p-value being influenced by the large number of trials (see Cohen's guideline [78]). We thus concluded that an analysis could be performed on the dataset as a whole. Responses for authenticity and emotional intensity were also highly correlated ($r = 0.727$). This is noticeable with the relatively similar outcomes of the evaluated models. This next section focuses on separate model estimations for both authenticity and emotional intensity. Two different models were implemented. First, we estimated both dependent variables using both the expressive intent and the social factor. Second, we explored the impact of the variation of both motion and acoustic features on the perceived emotional intensity and authenticity.

3.2.1. Interaction Effect of the Expressive Intents and the Presence of an Audience on Perceived Authenticity and Emotional Intensity

Two models were computed to estimate the influence of the expressive intents and the presence of an audience, respectively, on the perceived authenticity and the emotional intensity. The first model estimated the perceived emotional intensity using the different categories of expressive intent and the presence of an audience, as well as the interaction, as fixed effects, and with the participants and the musician at play as random effects (a model computing only the main effect and not the interaction is not presented in this article based on the principle of marginality; however, graphs related to such model can be found in the Supplementary Materials (Appendix Figure A2)). This model was significantly better than a model using only the main fixed effects, no interaction, and the same random effects (intensity: $\chi^2(3, N_{trials} = 875) = 12.036$, $p = 0.01$, $AIC = 7640.6$, $BIC = 7688.4$, $R_m^2 = 0.08$, $R_c^2 = 0.42$). The second model estimated the perceived authenticity using the same fixed and random effects. This model was significantly better than a model using only the main fixed effects, no interaction, and the same random effects (authenticity: $\chi^2(3, N_{trials} = 875) = 18.026$, $p = 8.68 \times 10^{-4}$, $AIC = 7738.7$, $BIC = 7786.4$, $R_m^2 = 0.11$, $R_c^2 = 0.42$).

Both perceived authenticity and emotional intensity increased significantly with every increment of the musicians' expressiveness in the case of the absence of audience (intensity: $\chi^2_{EmoInt,DP/Proj,Absence}(1, N_{trials} = 875) = 22.909$, $p = 1.69 \times 10^{-6}$, $\chi^2_{EmoInt,Proj/Exag,Absence}(1, N_{trials} = 875) = 13.765$, $p = 0.0002$, and authenticity: $\chi^2_{Auth,DP/Proj,Absence}(1, N_{trials} = 875) = 33.624$, $p = 6.6 \times 10^{-9}$, $\chi^2_{Auth,Proj/Exag,Absence}(1, N_{trials} = 875) = 11.009$, $p = 0.0009$) (Figure 3). When the audience was present, the ratings associated with such stimuli were only significantly different between the deadpan and projected condition for both dependent variables (intensity: $\chi^2_{EmoInt,DP/Proj,Absence}(1, N_{trials} = 875) = 40.608$, $p = 1.8e^{-10}$, and authenticity: $\chi^2_{Auth,DP/Proj,Presence}(1, N_{trials} = 875) = 74.007$, $p < 2.2 \times 10^{-16}$). The comparison between the presence and the absence of a virtual audience highlighted significantly different results only for the projected condition ($\chi^2_{EmoInt,Proj,Social}(1, N_{trials} = 875) = 9.327$, $p = 0.002$ and ($\chi^2_{Auth,Proj,Social}(1, N_{trials} = 875) = 15.896$, $p = 6.69 \times 10^{-5}$).

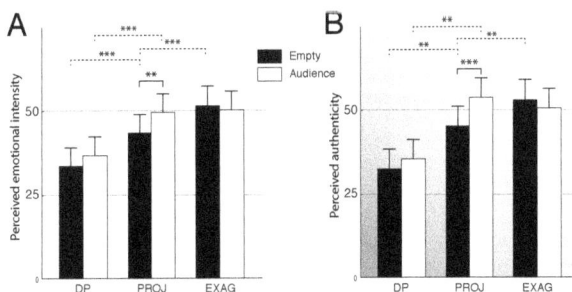

Figure 3. Interaction of the expressiveness and the presence of an audience on: (**A**) the perceived emotional intensity; and (**B**) the perceived authenticity (deadpan: DP, projected: PROJ, and exaggerated: EXAG) (** $p < 0.01$, *** $p < 0.001$).

3.2.2. Effect of the Motion and Acoustic Features and the Presence of an Audience

Perceived authenticity and emotional intensity were also modeled using the motion and acoustic features computed on the material recorded. Three models were computed using, respectively, one of the three features, energy kinetic, body twist index, and metrical centroid. The interaction of the features, used here as continuous predictors, and the fixed effect representing the presence of an audience were used in these models. The participants and the musicians were used as random effects. When comparing such model with models with no interaction, only the models for kinetic energy and metrical centroid were significantly improving the model accuracy for both emotional intensity (kinetic energy: $\chi^2(4, N_{trials} = 875) = 13.572$, $p = 0.011$, $AIC = 7649.9$, $BIC = 7707.2$, $R_m^2 = 0.8$, $R_c^2 = 0.42$, metrical centroid: $\chi^2(4, N_{trials} = 875) = 32.133$, $p = 1.8 \times 10^{-6}$, $AIC = 7653.7$, $BIC = 7711$, $R_m^2 = 0.08$, $R_c^2 = 0.43$) and authenticity (kinetic energy: $\chi^2(4, N_{trials} = 875) = 12.339$, $p = 0.018$, $AIC = 7750.3$, $BIC = 7817.6$, $R_m^2 = 0.105$, $R_c^2 = 0.42$, metrical centroid: $\chi^2(4, N_{trials} = 875) = 35.714$, $p = 3.31 \times 10^{-7}$, $AIC = 7760.5$, $BIC = 7817.8$, $R_m^2 = 0.10$, $R_c^2 = 0.42$).

When the musician was playing in front of an empty room, the recorded material was rated as more emotionally intense and authentic as the kinetic energy was increasing. The stimuli associated with higher value for kinetic energy were significantly rated higher for both dependent variables (intensity: $\chi^2_{EmoInt,Low/MidKinEn,Absence}(1, N_{trials} = 875) = 4.368$, $p = 0.036$, $\chi^2_{EmoInt,Mid/HighKinEn,Absence}(1, N_{trials} = 875) = 4.948$, $p = 0.026$; and authenticity: $\chi^2_{Auth,Low/MidKinEn,Absence}(1, N_{trials} = 875) = 6.66$, $p = 0.009$, $\chi^2_{Auth,Mid/HighKinEn,Absence}(1, N_{trials} = 875) = 4.47$, $p = 0.034$) In the case of metrical centroid, mid-range values were rated as more intense and authentic compared to extreme values (Figure 4). The rating associated with lower values were significantly different from the high values, while the high values were significantly (and marginally in the case of authenticity) different from the mid values (intensity: $\chi^2_{EmoInt,Low/HighMetCent,Absence}(1, N_{trials} = 875) = 16.473$, $p = 4.9 \times 10^{-5}$, $\chi^2_{EmoInt,Mid/HighMetCent,Absence}(1, N_{trials} = 875) = 5.288$, $p = 0.021$; and authenticity: $\chi^2_{Auth,Low/HighMetCent,Absence}(1, N_{trials} = 875) = 21.277$, $p = 3.9 \times 10^{-6}$, $\chi^2_{Auth,Mid/HighMetCent,Absence}(1, N_{trials} = 875) = 3.773$, $p = 0.052$).

The introduction of an audience in front of the musicians influenced how the musicians performed (Figure 2) but also brought significant changes in how intense and authentic the performance was perceived. Those changes highlighted the emergence of a bipartite distribution of our data instead of the tripartite grouping of the expressive intents (Figure 4). For both the authenticity and the emotional intensity, the two levels of the bipartite distribution were significantly different from each another (intensity: $\chi^2_{EmoInt,KinEn,Low/HighKinEn,Audience}(1, N_{trials} = 875) = 59.625$, $p = 1.14 \times 10^{-14}$, $\chi^2_{EmoInt,Low/HighMetCent,Audience}(1, N_{trials} = 875) = 36.056$, $p = 1.91 \times 10^{-9}$, authenticity: $\chi^2_{Auth,Low/HighFeatKinEn,Audience}(1, N_{trials} = 875) = 81.819$, $p < 2.2 \times 10^{-16}$, $\chi^2_{Auth,Low/HighMetCent,Audience}(1, N_{trials} = 875) = 48.918$, $p = 2.6 \times 10^{-12}$). The values of the features at which the separation occurred was calculated. It corresponds at 6.73×10^{-3} W for the kinetic energy ($kineticenergy_{min} = 2.207 \times 10^{-3}$ W and $kineticenergy_{max} = 0.0304$ W) and 308.18 BPM for the metrical centroid ($metricalcentroid_{min} = 260.48$ BPM and $metricalcentroid_{max} = 335.57$ BPM).

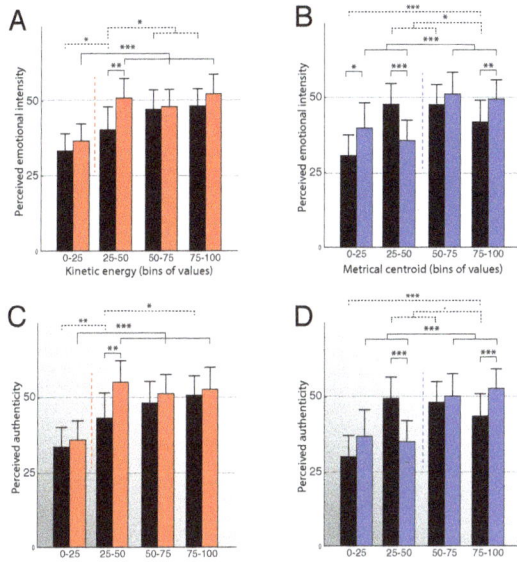

Figure 4. Impact of the interaction of the computed features: (**A,C**) kinetic energy (red); and (**B,D**) metrical centroid (blue); and the presence of an audience on: (**A,B**) the perceived emotional intensity; and (**C,D**) the perceived authenticity (deadpan: DP, projected: PROJ, and exaggerated: EXAG) ($p < 0.01$, * $p < 0.05$, ** $p < 0.01$, *** $p < 0.001$).

4. Discussion

In this study, we highlighted the changes in motion and acoustic features associated with different expressive manners. We measured the impact of the presence of virtual audience on those features. We also modeled the perceived emotional intensity and authenticity associated with different expressive manners and feature values. We provided scientific evidence about how emotional and authentic musical performances could be perceived.

4.1. Absence of an audience

We observed that different expressive intents were characterized by different quantities of body movements, as already demonstrated in the literature, in the absence of an audience [43,54]. Specifically, this study explored two features, the kinetic energy and newly developed body twist index, highlighting that energetic movements and wider body twists were associated with the magnitude of the expressive intent. Both features were increasing with the different expressive intents. Generally, it was revealed that musicians tended to make more movements when playing expressively [54]. This absence of the audience allowed them to fully twist and put more energy into their gestures. When analyzing the sound produced, the metrical centroid representing the metrical structure of the piece was also impacted by the different expressive manners. However, its value did not linearly increase between deadpan, projected and exaggerated expressive intents. The projected expressive intent was characterized by the highest value for the metrical centroid feature. This highlighted the predominance of faster notes within the musical piece. The other two expressive intents could be symbolized by longer notes or slow pulsations. In the case of the deadpan style, this suggested a more controlled way of playing, while emphasizing on a regular and slower beat. When musicians were exaggerating, the excessive expressive intent was marked on both slower and faster pulsations at different timing driving the decrease in the centroid value. As musicians dedicated more attention to their expressive intent, they change the way they played a certain piece. This confirms the phenomenon of performance expression [57,58].

When studying participants' perception of emotional intensity and authenticity, both ratings were affected significantly by the different expressive intents in the absence of an audience. Ratings were significantly increasing with the expressive conditions showing greater emotional intensity and authenticity in the exaggerated condition. The study of the impact of the physical attributes of the performance on the perception of emotional intensity and authenticity was conducted using feature values instead of the well-documented expressive conditions. The perception associated to stimuli with no virtual audience displayed was in line with the feature values obtained for different expressive intent. The deadpan (low kinetic energy and low metrical centroid), projected (medium kinetic energy and high metrical centroid), and exaggerated (high kinetic energy and medium metrical centroid) conditions are perceived, respectively, as relatively low, medium and high emotional intensity and authenticity. They showed a tripartite distribution of values which would fit with the three expressive conditions [43].

Before addressing the impact of the social factor on participants' ratings, we also noted the high correlation between authenticity and emotional intensity. Two conclusions can be drawn from such results. Firstly, the difficulty to discriminate between both ratings might be due to an underlying link between them. In music, authenticity represents an important part of the performance to make the listener feel the desired emotions [60]. Secondly, such link should be further untangled with a different experimental setup. We propose the use of "fake" stimuli, where musicians would fake the emotion felt and performed. This could for example be conducted by using mood induction procedures for "real" emotion stimuli while asking musicians to fake the rest of the stimuli. The efficiency of mood induction procedures have been already proven, especially for negative emotions [79].

4.2. Impact of the Presence of a Virtual Audience

When considering the impact of the presence of an audience on the body features, the data showed an interaction effect between the social factor and the expressive conditions on the features values. The only non-significant model is associated to kinetic energy. This expressive cue is well-known and extensively used by musicians to impact their communication of expressive intents. The presence of an audience does not significantly impact the well-regulated quantity of movement during the performance. The use of two other features, body twist index and metrical centroid, brings complementary information on the modification of such expressive performances when playing in front of an audience. In both features, the differences between the three expressive conditions tend to fade away with the presence of an audience. Values for each features converge towards ones for the projected expressive intent. This phenomenon could be linked to habits [80]. Musicians that are used to play in front of an audience could be tuning their movements in a certain way to make them comfortable yet expressive. In this study, the presence of an audience seems to push musicians to use this set of usual movements. When playing with no expressive intent, the musician could feel the need to still express some emotions for the audience to enjoy the performance more. Deadpan expressive intents are usually performed in a controlled environment, e.g., at home, alone while rehearsing, when the musician's locus on control is self-oriented. Internals' performance and stress are proven to be better controlled, for instance in a work environment [81]. The presence of an audience produces considerable effect, shifting the locus of control, disrupting the habits associated with this expressive condition and putting musicians in a more complex, stressful and less controlled situation. To counterbalance this effect, musicians tend to reach back to a controlled and habitual situation mimicking projected expressive movements. Similarly, exaggerating expressive intents in front of a crowd could put musicians in an uncomfortable position where their play seems a bit less authentic. To counterbalance this effect, musicians will then naturally try to be less expressive and go back to a projected expressive condition. Consequently, our first hypothesis is therefore validated showing that gestures become less differentiable across conditions.

Similar effects could be observed on the emotional intensity and authenticity. Both deadpan and exaggerated stimuli produce responses similar to the projected conditions when an audience is present during the musicians' performances. As the values of the features tend to converge towards the likes of the projected expressive intent, perception converges towards a uniform value represented by the projected scenario. This phenomenon is most likely due to the unease felt by the musician when playing with an unfamiliar expressive style in a habitual situation. The difference in perceptual intensity and authenticity is the strongest in the case of musicians playing as they would in a concert and facing a virtual audience. Contrarily to deadpan and exaggerated conditions, the ease coming with the congruence of the common context—playing in a concert-like fashion and facing an audience—allows musicians to appear more authentic and communicate their expressive intent better in the process.

When playing in front of an audience, the previously observed tripartite distribution of ratings disappears. For both emotional intensity and authenticity, the distribution based on the both kinetic energy and metrical centroid becomes dichotomist. The emergence of such threshold between perception of high and low emotional intensity/authenticity is a repercussion of the convergence of exaggerated performances toward a concert-like situation. In front of an audience, both projected and exaggerated context are evaluated as highly authentic and emotionally intense. The second hypothesis stating that different expressive conditions would be rated similarly for emotional intensity and authenticity in the social condition is therefore verified. Furthermore, the calculated threshold could consequently be used for automatic detection of emotional intensity.

Both models using the expressive conditions and the features values are converging towards one conclusion: the use of the methodological framework designed by Davidson, 2005, is here showing its limits in its ecological validity [43]. While the three expressive intents can be communicated accurately by the musicians to the listeners, the presence of an audience, even a virtual one, during the recordings reduces the ability to differentiate between such categories of expressive intents. The distinction between the projected and exaggerated conditions is blurred and we suggest to take into account such modifications when recording musicians in front of a public, virtual or real. This audience effect on the expressive intents might here be due to an emergence of non-explicit and involuntary regulation processes in social context, driving a strong impact of such conditions. Such processes should be further explored.

4.3. Interactive Virtual Environments as a Tool to Study Musical Performances

Our findings supports the use of tools such as IVEs in music research . Previous studies in the domain of social phobias already recommended the use of IVE for coping with stress generated by the fear of public speaking [10,11]. Similarly, IVEs could become crucial tools to train musicians to cope with stress related to live performances and aid learning [2]. As shown in this study, the presence of an audience impacts both the movements and sounds related to the performance and, consequently, the perception of listeners. The system developed in this study could also be easily adapted to context-sensitivity in real-time and could provide feedbacks to the musicians, e.g., when their movements are radically diverging from a previous recorded performance. It could help musicians understand the impact of stress on their performance allowing them to develop coping mechanisms for musical performance anxiety. Finally, research on communication of emotions, alongside with this study and the IVE, might be used by music teachers to enhance performers' expressiveness [82,83].

5. Conclusions

To conclude, the presence of an audience generated important variations in both acoustic and motion features related to music performance. This influence is to be taken into account when approaching music research during concerts. Immersive Virtual Environments could therefore be utilized both for research and as a tool for training musicians to cope with audience anxiety.

Acknowledgments: We thank both the National Centers of Competence in Research (NCCRs) and the Swiss National Fund (SNF) for funding this study.

Author Contributions: Glowinski, Donald and Grandjean, Didier conceived and designed the experiments; Glowinski, Donald performed the experiments; Schaerlaeken, Simon and Glowinski, Donald analyzed the data; Grandjean, Didier contributed reagents/materials/analysis tools; and Schaerlaeken, Simon wrote the first draft of the paper, which was revised by Glowinski, Donald and Grandjean, Didier.

Conflicts of Interest: The authors declare no conflict of interest.

Appendix A

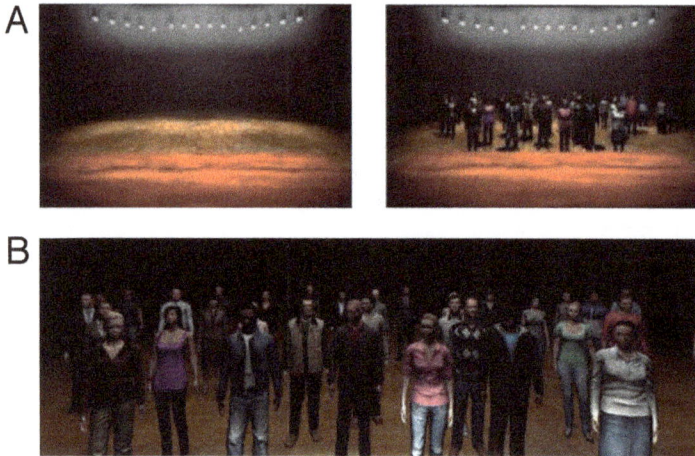

Figure A1. Interactive Virtual Environment: (**A**) Example views of both social conditions (left, empty; right, audience); (**B**) details of a disengaged audience (deadpan: DP, projected: PROJ, and exaggerated: EXAG).

Appendix B

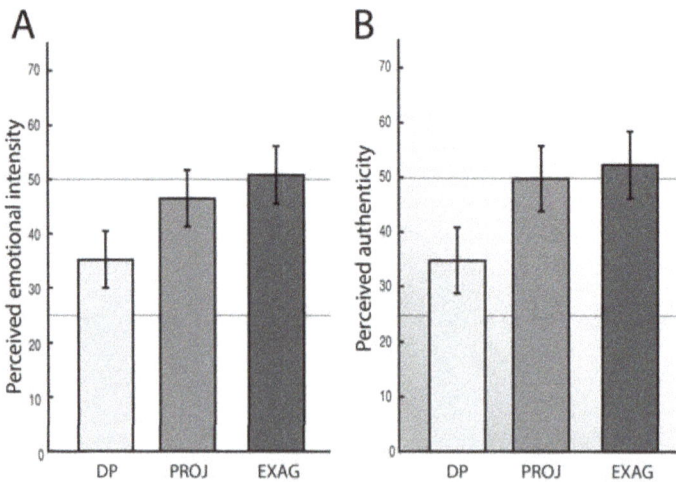

Figure A2. Impact of the expressiveness on (**A**) the perceived emotional intensity; (**B**) the perceived authenticity (deadpan: DP, projected: PROJ, and exaggerated: EXAG).

Appendix C

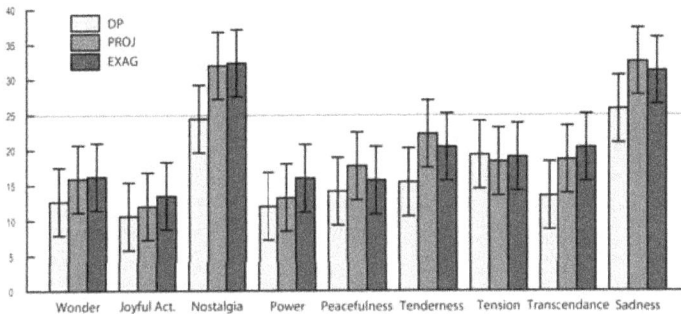

Figure A3. Impact of the interaction between the Geneva emotional scale and the expressiveness (deadpan: DP, projected: PROJ, and exaggerated: EXAG).

References

1. Wilson, G.D.; Roland, D. Performance anxiety. In *The Science and Psychology of Music Performance: Creative Strategies for Teaching and Learning*; Oxford University Press: Oxford, UK, 2002; pp. 47–61.
2. Williamon, A.; Aufegger, L.; Eiholzer, H. Simulating and stimulating performance: Introducing distributed simulation to enhance musical learning and performance. *Front. Psychol.* **2014**, *5*, 1–9.
3. Blascovich, J.; Loomis, J.; Beall, A.C.; Swinth, K.R.; Hoyt, C.L.; Bailenson, N.; Bailenson, J.N. Immersive virtual environment technology as a methodological tool for social psychology. *Psychol. Inq.* **2002**, *13*, 103–124.
4. Sanchez-Vives, M.V.; Slater, M. From presence to consciousness through virtual reality. *Nat. Rev. Neurosci.* **2005**, *6*, 332–339.
5. Difede, J.; Cukor, J.; Hoffman, H.G. Virtual Reality Exposure Therapy for the Treatment of Posttraumatic Stress Disorder Following September 11, 2001. *J. Clin. Psychiatry* **2007**, *68*, 1639–1647.
6. Rizzo, A.; Reger, G.; Gahm, G.; Difede, J.; Rothbaum, B.O. Virtual reality exposure therapy for combat-related PTSD. In *Post-Traumatic Stress Disorder*; Springer: New York, NY, USA, 2009; pp. 375–399.
7. Rizzo, A.S.; Buckwalter, J.G.; Forbell, E.; Reist, C.; Difede, J.; Rothbaum, B.O.; Lange, B.; Koenig, S.; Talbot, T. Virtual Reality Applications to Address the Wounds of War. *Psychiatr. Ann.* **2013**, *43*, 123–138.
8. Rothbaum, B.O.; Hodges, L.; Smith, S.; Lee, J.H.; Price, L. A controlled study of virtual reality exposure therapy for the fear of flying. *J. Consult. Clin. Psychol.* **2000**, *68*, 1020–1026.
9. Bouchard, S.; Côté, S.; St-Jacques, J.; Robillard, G.E.; Renaud, P. Effectiveness of virtual reality exposure in the treatment of arachnophobia using 3D games. *Technol. Health Care* **2006**, *14*, 19–27.
10. Pertaub, D.; Slater, M.; Barker, C. An experiment on fear of public speaking in virtual reality. In *Studies in Health Technology and Informatics*; IOS Press: Amsterdam, The Netherlands, 2001; pp. 372–378.
11. North, M.M.; North, S.M.; Coble, J.R. Virtual reality therapy: An effective treatment for the fear of public speaking. *Int. J. Virtual Real. IJVR* **2015**, *3*, 1–6.
12. Orman, E.K. Effect of virtual reality graded exposure on anxiety levels of performing musicians: A case study. *J. Music Ther.* **2004**, *41*, 70–78.
13. Bissonnette, J.; Dubé, F.; Provencher, M.D.; Moreno Sala, M.T. Evolution of music performance anxiety and quality of performance during virtual reality exposure training. *Virtual Real.* **2016**, *20*, 71–81.
14. Juslin, P.N.; Laukka, P. Expression, Perception, and Induction of Musical Emotions: A Review and a Questionnaire Study of Everyday Listening. *J. New Music Res.* **2004**, *33*, 217–238.
15. Behrens, G.A.; Green, S.B. The ability to identify emotional content of solo improvisations performed vocally and on three different instruments. *Psychol. Music* **1993**, *21*, 20–33.
16. Gabrielsson, A. Expressive intention and performance. In *Music and the mind machine*; Springer: Berlin/Heidelberg, Germany, 1995; pp. 35–47.

17. Gabrielsson, A.; Juslin, P.N. Emotional Expression in Music Performance: Between the Performer's Intention and the Listener's Experience. *Psychol. Music* **1996**, *24*, 68–91.
18. Juslin, P.N. Emotional Communication in Music Performance: A Functionalist Perspective and Some Data. *Music Percept. Interdiscip. J.* **1997**, *14*, 383–418.
19. Juslin, P.N. Perceived Emotional Expression in Synthesized Performances of a Short Melody: Capturing the Listener's Judgment Policy. *Music. Sci.* **1997**, *1*, 225–256.
20. Juslin, P.N.; Madison, G. The Role of Timing Patterns in Recognition of Emotional Expression from Musical Performance. *Music Percept. Interdiscip. J.* **1999**, *17*, 197–221.
21. Laukka, P.; Juslin, P.N. Improving emotional communication in music performance through cognitive feedback. *Music. Sci. J. Eur. Soc. Cognit. Sci. Music* **2000**, *4*, 151–183.
22. Balkwill, L.l.; Thompson, W.F. A Cross-Cultural Investigation of the Perception of Emotion in Music: Psychophysical and Cultural Cues. *Music Percept. Interdiscip. J.* **1999**, *17*, 43–64.
23. Fritz, T.; Jentschke, S.; Gosselin, N.; Sammler, D.; Peretz, I.; Turner, R.; Friederici, A.D.; Koelsch, S. Universal Recognition of Three Basic Emotions in Music. *Curr. Biol.* **2009**, *19*, 573–576.
24. Zentner, M.; Grandjean, D.; Scherer, K.R. Emotions evoked by the sound of music: Characterization, classification, and measurement. *Emotion* **2008**, *8*, 494–521.
25. Jansens, S.; Bloothooft, G.; de Krom, G. Perception And Acoustics Of Emotions In Singing. In Proceedings of the Fifth European Conference on Speech Communication and Technology, Rhodes, Greece, 22–25 September 1997.
26. Mergl, R.; Piesbergen, C.; Tunner, W. *Musikalisch-Improvisatorischer Ausdruck und Erkennen von Gefühlsqualitäten*; Hogrefe: Göttingen, Germany, 2009; pp. 1–11.
27. Juslin, P.N. Cue utilization in communication of emotion in music performance: relating performance to perception. *J. Exp. Psychol. Hum. Percept. Perform.* **2000**, *26*, 1797–1813.
28. Juslin, P.N.; Laukka, P. Communication of emotions in vocal expression and music performance: Different channels, same code? *Psychol. Bull.* **2003**, *129*, 770–814.
29. Juslin, P.N.; Sloboda, J.A. *Music and Emotion: Theory and Research.*; Oxford University Press: Oxford, UK, 2001.
30. Banse, R.; Scherer, K.R. Acoustic profiles in vocal emotion expression. *J. Personal. Soc. Psychol.* **1996**, *70*, 614–636.
31. Finnäs, L. Presenting music live, audio-visually or aurally—does it affect listeners' experiences differently? *Br. J. Music Educ.* **2001**, *18*, 55–78.
32. Frith, S. *Performing Rites: On the Value of Popular Music*; Harvard University Press: Cambridge, MA, USA, 1998.
33. Vines, B.W.; Krumhansl, C.L.; Wanderley, M.M.; Levitin, D.J. Cross-modal interactions in the perception of musical performance. *Cognition* **2006**, *101*, 80–113.
34. Auslander, P. *Liveness: Performance in a Mediatized Culture*; Routledge: Abingdon, UK, 2008.
35. Bergeron, V.; Lopes, D.M. Hearing and seeing musical expression. *Philos. Phenomenol. Res.* **2009**, *78*, 1–16.
36. Cook, N. Beyond the notes. *Nature* **2008**, *453*, 1186–1187.
37. Malin, Y. Metric Analysis and the Metaphor of Energy: A Way into Selected Songs by Wolf and Schoenberg. *Music Theory Spectr.* **2008**, *30*, 61–87.
38. Platz, F.; Kopiez, R. When the eye listens: A meta-analysis of how audio-visual presentation enhances the appreciation of music performance. *Music Percept.* **2012**, *30*, 71–83.
39. Chapados, C.; Levitin, D.J. Cross-modal interactions in the experience of musical performances: Physiological correlates. *Cognition* **2008**, *108*, 639–651.
40. Vines, B.W.; Krumhansl, C.L.; Wanderley, M.M.; Dalca, I.M.; Levitin, D.J. Music to my eyes: Cross-modal interactions in the perception of emotions in musical performance. *Cognition* **2011**, *118*, 157–170.
41. Cadoz, C.; Wanderley, M.M.; Cadoz, C.; Wanderley, M.M.; Music, G.; Wanderley, M. Gesture-Music. In *Trends Gestural Control Music*; IRCAM: Paris, France, 2000.
42. Wanderley, M.M.; Vines, B.W.; Middleton, N.; McKay, C.; Hatch, W. The Musical Significance of Clarinetists' Ancillary Gestures: An Exploration of the Field. *J. New Music Res.* **2005**, *34*, 97–113.
43. Davidson, J.W. Bodily communication in musical performance. In *Musical Communication*; OUP Oxford: Oxford, UK, 2005; pp. 215–238.
44. Glowinski, D.; Mancini, M.; Cowie, R.; Camurri, A.; Chiorri, C.; Doherty, C. The movements made by performers in a skilled quartet: A distinctive pattern, and the function that it serves. *Front. Psychol.* **2013**, *4*, 1–9.

45. Palmer, C. *Music Performance: Movement and Coordination*; Elsevier: Amsterdam, The Netherlands, 2012.
46. Dahl, S.; Friberg, A. Visual Perception of Expressiveness in Musicians' Body Movements. *Music Percept. Interdiscip. J.* **2007**, *24*, 433–454.
47. Wanderley, M.M. *Quantitative Analysis of Non-Obvious Performer Gestures*; Springer: Berlin/Heidelberg, Germany, 2002; pp. 241–253.
48. Nusseck, M.; Wanderley, M.M. Music and Motion—How Music-Related Ancillary Body Movements Contribute to the Experience of Music. *Music Percept. Interdiscip. J.* **2009**, *26*, 335–353.
49. Bhatara, A.K.; Duan, L.M.; Tirovolas, A.; Levitin, D.J. Musical expression and emotion: Influences of temporal and dynamic variation. 2009, manuscript submitted for publication.
50. Sloboda, J.A.; Lehmann, A.C. Tracking Performance Correlates of Changes in Perceived Intensity of Emotion During Different Interpretations of a Chopin Piano Prelude. *Music Percept.* **2001**, *19*, 87–120.
51. Sakata, M.; Wakamiya, S.; Odaka, N.; Hachimura, K. Effect of body movement on music expressivity in jazz performances. In *International Conference on Human-Computer Interaction*; Springer: Berlin/Heidelberg, Germany, 2009; pp. 159–168.
52. Thompson, M.R.; Luck, G. Exploring relationships between pianists' body movements, their expressive intentions, and structural elements of the music. *Musicae Scientiae* **2012**, *16*, 19–40.
53. Van Zijl, A.G.W.; Luck, G. Moved through music: The effect of experienced emotions on performers' movement characteristics. *Psychol. Music* **2013**, *41*, 175–197.
54. Camurri, A.; Lagerlöf, I.; Volpe, G. Recognizing emotion from dance movement: Comparison of spectator recognition and automated techniques. *Int. J. Hum. Comput. Stud.* **2003**, *59*, 213–225.
55. Leman, M. *Embodied Music Cognition and Mediation Technology*; Mit Press: Cambridge, CA, USA, 2008.
56. Mancas, M.; Madhkour, R.B.; Beul, D.D. Kinact: A saliency-based social game. In Proceedings of the 7th International Summer Workshop on Multimodal Interfaces, Plzen, Czech Republic, 1–26 August 2011; Volume 4, pp. 65–71.
57. Shaffer, L.H. How to interpret music. In *Cognitive Bases of Musical Communication*; American Psychological Association: Washington, DC, USA, 1992; pp. 263–278.
58. Palmer, C. Music Performance. *Ann. Rev. Psychol.* **1997**, *48*, 115–138.
59. Goffman, E. *The Presentation of Self in Everyday Life*; Penguin Books: London, UK, 1959.
60. Scherer, K.R.; Keith, G.; Schaufer, L.; Taddia, B.; Pregardien, C. The singer's paradox: on authenticity in emotional expression on the opera stage. In *Emotion*; OUP Oxford: Oxford, UK, 2013; pp. 55–73.
61. Wöllner, C. How to quantify individuality in music performance? Studying artistic expression with averaging procedures. *Front. Psychol.* **2013**, *4*, 1–3.
62. Juslin, P.N. From Everyday Emotions to Aesthetic Emotions: Towards a Unified Theory of Musical Emotions. *J. Psychol.* **2013**, *10*, 235–266.
63. Brunswik, F. *Perception and the Representative Design of Experiments*; Univer: Berkeley, CA, USA, 1956.
64. Grandjean, D.; Bänziger, T.; Scherer, K.R. Intonation as an interface between language and affect. *Prog. Brain Res.* **2006**, *156*, 235–247.
65. Krumhuber, E.G.; Tamarit, L.; Roesch, E.B.; Scherer, K. FACSGen 2.0 animation software: Generating three-dimensional FACS-valid facial expressions for emotion research. *Emotion* **2012**, *12*, 351.
66. Roesch, E.; Tamarit, L.; Reveret, L.; Grandjean, D.; Sander, D.; Scherer, K. FACSGen: A tool to synthesize emotional facial expressions through systematic manipulation of facial action units. *J. Nonverbal Behav.* **2011**, *35*, 1–16.
67. Guadagno, R.E.; Blascovich, J.; Bailenson, J.N.; Mccall, C. Virtual humans and persuasion: The effects of agency and behavioral realism. *Media Psychol.* **2007**, *10*, 1–22.
68. Garau, M.; Slater, M.; Vinayagamoorthy, V.; Brogni, A.; Steed, A.; Sasse, M.A. The impact of avatar realism and eye gaze control on perceived quality of communication in a shared immersive virtual environment. In Proceedings of the SIGCHI Conference on Human Factors in Computing Systems, Ft. Lauderdale, FL, USA, 5–10 April 2003; pp. 529–536.
69. Schmitz, C. LimeSurvey: An Open Source Survey Tool. 2012. Available online: http://www.limesurvey.org (accessed on 5 November 2016).
70. Dardard, F.; Gnecco, G.; Glowinski, D. Automatic classification of leading interactions in a string quartet. *ACM Trans. Interact. Intell. Syst.* **2016**, *6*, 1–27.

71. Glowinski, D.; Baron, N.; Shirole, K.; Coll, S.Y.; Chaabi, L.; Ott, T.; Rappaz, M.A.; Grandjean, D. Evaluating music performance and context-sensitivity with Immersive Virtual Environments. *EAI Endors. Trans. Creat. Technol.* **2015**, *2*, 1–10.

72. Juslin, P.N.; Liljestrom, S.; Laukka, P.; Vastfjall, D.; Lundqvist, L.O. Emotional reactions to music in a nationally representative sample of Swedish adults: Prevalence and causal influences. *Music. Sci.* **2011**, *15*, 174–207.

73. Lartillot, O.; Cereghetti, D.; Eliard, K.; Trost, W.J.; Rappaz, M.-A.; Grandjean, D. Estimating Tempo and Metrical Features by Tracking the Whole Metrical Hierarchy. In Proceedings of the 3rd International Conference on Music & Emotion (ICME3), Jyväskylä, Finland, 11–15 June 2013; pp. 11–15.

74. Burger, P. *MoCap Toolbox—A Matlab Toolbox for Computational Analysis of Movement Data*; Logos Verlag Berlin: Berlin, Germany, 2013; pp. 172–178.

75. Lartillot, O.; Toiviainen, P. A matlab toolbox for musical feature extraction from audio. In Proceedings of the 10th Int Conference on Digital Audio Effects DAFx07, Bordeaux, France, 10–15 September 2007; pp. 1–8.

76. McLean, R.A.; Sanders, W.L.; Stroup, W.W. A unified approach to mixed linear models. *Am. Stat.* **1991**, *45*, 54.

77. Zentner, M. (University of Innsbruck, Innsbruck, Austria). Personal Communication, 2004.

78. Cohen, J. A power primer. *Psychol. Bull.* **1992**, *112*, 155–159.

79. Westermann, R.; Stahl, G.; Hesse, F. Relative effectiveness and validity of mood induction procedures: analysis. *Eur. J. Soc. Psychol.* **1996**, *26*, 557–580.

80. Graybiel, A.M. Habits, rituals, and the evaluative brain. *Annu. Rev. Neurosci.* **2008**, *31*, 359–387.

81. Anderson, C.R. Locus of control, coping behaviors, and performance in a stress setting: A longitudinal study. *J. Appl. Psychol.* **1977**, *62*, 446.

82. Juslin, P.N.; Karlsson, J.; Lindström, E.; Friberg, A.; Schoonderwaldt, E. Play it again with feeling: Computer feedback in musical communication of emotions. *J. Exp. Psychol. Appl.* **2006**, *12*, 79.

83. Juslin, P.N.; Persson, R.S. Emotional communication. In *The Science and Psychology of Music Performance: Creative Strategies for Teaching and Learning*; Oxford University Press: Oxford, UK, 2002; pp. 219–236.

applied sciences

MDPI

Article

Mobile Music, Sensors, Physical Modeling, and Digital Fabrication: Articulating the Augmented Mobile Instrument [†]

Romain Michon [1],[*], Julius Orion Smith [1], Matthew Wright [1], Chris Chafe [1], John Granzow [1],[2] and Ge Wang [1]

[1]   Center for Computer Research in Music and Acoustics (CCRMA) – Stanford University, Stanford, CA 94305-8180, USA; jos@ccrma.stanford.edu (J.O.S.); matt@ccrma.stanford.edu (M.W.); cc@ccrma.stanford.edu (C.C.); jgranzow@umich.edu (J.G.); ge@ccrma.stanford.edu (G.W.)

[2]   School of Music, Theater, and Dance – University of Michigan, Ann Arbor, MI 48109-2085, USA

[*]   Correspondence: rmichon@ccrma.stanford.edu; Tel.: +1-650-723-4971

[†]   This paper is a re-written and expanded version of "Passively Augmenting Mobile Devices Towards Hybrid Musical Instrument Design", published in the 2017 New Interfaces for Musical Expression Conference (NIME-17), Copenhagen, Denmark, 15–19 May 2017.

Academic Editor: Stefania Serafin
Received: 31 October 2017; Accepted: 13 December 2017; Published: 19 December 2017

Abstract: Two concepts are presented, extended, and unified in this paper: mobile device augmentation towards musical instruments design and the concept of hybrid instruments. The first consists of using mobile devices at the heart of novel musical instruments. Smartphones and tablets are augmented with passive and active elements that can take part in the production of sound (e.g., resonators, exciter, etc.), add new affordances to the device, or change its global aesthetics and shape. Hybrid instruments combine physical/acoustical and "physically informed" virtual/digital elements. Recent progress in physical modeling of musical instruments and digital fabrication is exploited to treat instrument parts in a multidimensional way, allowing any physical element to be substituted with a virtual one and vice versa (as long as it is physically possible). A wide range of tools to design mobile hybrid instruments is introduced and evaluated. Aesthetic and design considerations when making such instruments are also presented through a series of examples.

Keywords: mobile music; physical modeling; musical instrument design

1. Introduction

1.1. Physical Interfaces and Virtual Instruments: Remutualizing the Instrument

The concept of musical controller is not new and was perhaps invented when the first organs were made centuries ago. However, the rise of analog synthesizers in the middle of the twentieth century, followed a few decades later by digital synthesizers almost systematized the dissociation of the control-interface and sound-generation in musical instrument design. This gave birth to a new family of musical instruments known as "Digital Musical Instruments" (DMIs).

Marc Battier defines DMIs from a "human computer interaction (HCI) standpoint" as "instruments that include a separate gestural interface (or gestural controller unit) from a sound generation unit [1]." Thus, this feature that originally resulted from logical engineering decisions encouraged by the use of flexible new technologies, became one of the defining components of DMIs [2,3]. This characteristic has been extensively commented upon and studied in the New Interfaces for Musical Expression (NIME) literature. In particular, Marcello Wanderley highlights the fact that "with the separation of the DMI into two independent units, basic interaction characteristics of existing instruments may be lost and/or difficult to reproduce" [4].

Perry Cook provides an exhaustive overview of the risks associated with "abstracting the controller from the synthesizer," [5] which might sometimes result in a "loss of intimacy" between performer and instrument. More specifically, he associates the flaws of "demutualized instruments" to the lack of haptic feedback (which has been extensively studied [4,6]), the lack of "fidelity in the connections from the controller to the generator," and the fact that "no meaningful physics goes on in the controller."

This paper addresses the first and the third issues pointed out by Cook and builds upon his work by providing a framework to design remutualized instruments reconciling the haptic, the physical, and the virtual.

1.2. Augmented and Acoustically Driven Hybrid Instruments: Thinking DMIs As a Whole

Augmented and acoustically driven hybrid instruments are two special kinds of DMIs combining acoustical and virtual elements to make sound. By doing so, they often blur the interface/synthesizer boundary, making them more mutualized and unified as a whole, partly solving the issue presented in Section 1.1.

Augmented instruments are based on acoustic instruments that are "enhanced" using virtual elements. Digital technologies can be added to the existing tool-set of instrument designers. There exist dozens of examples of such instruments in the computer music literature [7–12].

Instead of being based on existing acoustic instruments, acoustically driven hybrid instruments use acoustic elements (e.g., membrane, solid surfaces, strings, etc.) to drive virtual (i.e., electronic, digital, etc.) ones. The electric guitar is a good example of this kind of instrument. Acoustically driven hybrid instruments have been extensively studied and theorized by Roberto Aimi in his PhD thesis [13]. Their goal is to play to the strengths of physical/acoustical elements (e.g., imperfection, tangibility, randomness, etc.) and combine them with the infinite possibilities of digital ones.

A specific kind of acoustically driven hybrid instruments uses digital physical models (see Section 1.5) as the virtual portion of the instrument. The Korg Wavedrum (http://www.korg.com/us/products/drums/wavedrum_global_edition/—All URLs were verified on 1 December 2017.) is probably one of the earliest examples of this type of instrument. It uses the sound excitations created on a physical drum membrane to drive a wide range of physical models. The same technique has been used as the core of a wide range of other musical instruments [13–17].

By using acoustical elements as an interface, instruments presented in this section implement a form of "passive haptic feedback," (i.e., the performer physically feels these elements while actuating them, even though they are not transmitting information from the virtual simulation via active force feedback). Moreover, they force the instrument maker to "co-design synthesis algorithms and controllers" (see Section 1.1 and [5]) reinforcing "the sense of intimacy, connectedness, and embodiment for the player and audience" [5].

1.3. Mobile Devices as Musical Instruments

Mobile devices (smart-phones, tablets, etc.) have been used as musical instruments for the past ten years both in the industry (e.g., GarageBand (http://www.apple.com/ios/garageband) for iPad, Smule's apps (https://www.smule.com), moForte's *GeoShred* (http://www.moforte.com/geoshredapp), etc.) and in the academic community [18–22]. As stand alone devices they present a promising platform for the creation of versatile instruments for live music performance. Within a single entity, sounds can be generated and controlled, differentiating them from most Digital Musical Instruments (DMIs), and allowing the creation of instruments much closer to "traditional" acoustic instruments in this respect. This resemblance is pushed even further with mobile phone orchestras such as *MoPhO* (http://mopho.stanford.edu) [23], where each performer uses a mobile phone as a independent musical instrument.

1.4. Augmenting Mobile Devices

Despite all their qualities, mobile devices were never designed to be used as musical instruments and lack some crucial elements to compete with their acoustic counterparts. This problem can be solved by adding prosthetics to the device implementing missing features or enhancing existing ones. Augmentations can be classified in two categories: passive and active.

Passive augmentations leverage built-in elements of the device (e.g., speaker, microphone, motion sensors, touchscreen, etc.) to implement new features or improve existing ones. Thus, the scope of this kind of augmentation is limited to what the host device can offer. There exist many examples of this type of augmentation ranging from passive amplifiers (e.g., Etsy Amplifiers (https://www.etsy.com/market/iphone_amplifier)) to smart-phone-based motion synthesizer controller (e.g., AAUG Motion Synth Controller (http://www.auug.com/)).

Active augmentations rely on electronic elements to add new features to mobile devices. Thus, they must be connected to it using one of its input or output ports (e.g., headphone jack, USB, etc.). Unlike passive augmentations, their scope is more or less infinite. Active augmentations range from external speakers to smart-phone-based musical instruments such as the Artiphon INSTRUMENT 1 (https://artiphon.com/).

1.5. Physical Modeling

Waveguide synthesis has been used since the second half of the 1980's to model a wide range of musical instruments [24–27]. The main advantages of this technique are its simplicity and efficiency while still sounding adequately real. It allows for the accurate modeling of a wide range of instruments (string and wind instruments, tonal percussion, etc.) just with a single "filtered delay loop." This technique was used in many commercial synthesizers in the 1990s such as the Yamaha VL1.

While any instrument part implementing a quasi harmonic series (e.g., a linear string, tube, etc.) can be modeled with a single digital waveguide, other parts must be modeled using other techniques such as modal synthesis.

Modal synthesis [28] consists of implementing each mode of a linear system as an exponentially decaying sine wave. Each mode can then be configured with its frequency, gain, and resonance duration (T60). Since each harmonic is implemented with an independent sine wave generator, this technique is a lot more computationally expensive than waveguide modeling. The parameters of a modal synthesizer (essentially a list of frequencies, gains, and T60s) can be calculated from the impulse response of a physical object [29] or by using the Finite Element Method (FEM) on a volumetric mesh [30]. This technique strengthens the link between physical elements and their virtual counterparts as it allows for the design of an instrument part on a computer using CAD software, and turn it into a physical model that could also be materialized using digital fabrication. This concept is further developed in Section 5.2.

Other methods such as *finite-difference schemes* [31] can be used to implement physical models of musical instruments and provide more flexibility and accuracy in some cases. However, most of them are computationally more expensive than waveguide models and modal synthesis. An overview of these techniques is provided in [27]. Since this paper is targeting the use of physical models on mobile platforms with a limited computational power, we're focusing on CPU-efficient techniques.

1.6. 3D Printing, Acoustics, and Musical Instrument Design/Lutherie

3D printing has been used extensively in the past few years to make novel, traditional, acoustic, digital, etc., musical instruments [32]. While high-end 3D printers can be used to make full size traditional acoustic musical instruments, cheaper printers are often utilized to augment or modify existing instruments or to make new ones from random objects. Fast prototyping and iterative design are at the heart of this new approach to lutherie and musical instrument making in general [33].

While string instruments are particularly well represented [34–38], many experiments around wind instruments have been conducted as well [39–41].

1.7. Towards the Hybrid Mobile Instrument

In a previous publication [42], we introduced the concept of "augmented mobile-device" and we presented the BLADEAXE: an acoustically driven hybrid instrument partly based on acoustic elements used to generate sound excitations and an iPad. The iPad was used both as a controller, and to implement virtual physical-model-based elements of the instrument. The BLADEAXE was the last iteration of a series of mobile-device-based instruments that we developed during the past four years.

In this paper, we generalize the various concepts introduced by the BLADEAXE and propose a framework centered around the FAUST programming language [43] to facilitate the design of "hybrid mobile instruments." Such instruments combine physical/acoustical elements and physically informed virtual/digital elements (i.e., physical models) (Throughout this paper, "physical elements" designate tangible acoustical musical instrument parts and "virtual elements" designate digital, physically informed (based on a physical model) instrument parts.). Virtual elements are implemented on the mobile device that serves as the "core" of the system. Modern digital fabrication (e.g., 3D printing, etc.) is combined to physical modeling techniques to approach musical instrument design in a multidimensional way. By being standalone, implementing passive haptic feedback, and having a unified design between the interface and the sound generation unit, we believe that hybrid mobile instruments solve some of the flaws of DMIs highlighted by Perry Cook [5].

First, we present `faust2smartkeyb`, a FAUST-based tool facilitating the design of musical apps and focusing on skill transfer. It serves as the "glue" between the different building blocks of mobile-device-based hybrid musical instruments (i.e., physically modeled parts, built-in and external sensors, touchscreen interface, connections to digitally fabricated elements, etc.). Next, we introduce MOBILE3D, an OpenScad (http://www.openscad.org) library to help design mobile device passive augmentations using DIY (*Do It Yourself*) digital fabrication techniques such as 3D printing and laser cutting. We give an exhaustive overview of the taxonomy of the various types of passive augmentations that can be implemented on mobile devices through a series of examples and we demonstrate how they leverage existing components on the device. Next, a framework based on the Teensy board (https://www.pjrc.com/teensy/) and FAUST, to design and make active mobile device augmentations is presented and evaluated through the results of a workshop. Finally, the concept of "acoustically driven hybrid mobile instrument" is studied and a framework centered around the FAUST Physical Modeling Toolkit to design virtual and physical musical instrument parts is presented.

2. `faust2smartkeyb`: Facilitating Musical Apps Design and Skill Transfer

Making musical apps for mobile devices involves the use and mastery of various technologies, standards, programming languages, and techniques ranging from low level C++ programming for real-time DSP (Digital Signal Processing) to advanced interface design. This adds up to the variety of the platforms (e.g., iOS, Android, etc.) and of their associated tools (e.g., Xcode, Android Studio, etc.), standards, and languages (e.g., JAVA, C++, Objective-C, etc.).

While there exists a few tools to facilitate the design of musical apps such as libpd, [44] Mobile CSOUND [45], and more recently JUCE (https://www.juce.com) and SuperPowered (http://superpowered.com), none of them provides a comprehensive cross-platform environment for musical touchscreen interface design, high level DSP programming, turnkey instrument physical model prototyping, built-in sensors handling and mapping, MIDI and OSC compatibility, etc.

Earlier works inspired the system presented in this section and served as its basis. `faust2ios` and `faust2android` [46] are command line tools to convert FAUST codes into fully working Android and iOS applications. The user interface of apps generated using this system corresponds to the standard UI specifications provided in the FAUST code and is made out of sliders, buttons, groups, etc. More recently, `faust2api`, a lower level tool to generate audio engines with FAUST featuring polyphony, built-in sensors mapping, MIDI and OSC (Open Sound Control) support, etc., for a wide range of platforms including Android and iOS was introduced [47].

Despite the fact that user interfaces better adapted to musical applications (e.g., piano keyboards, (x, y) controllers, etc.) can replace the standard UI of a FAUST object in apps generated by `faust2android` [48], they are far from providing a generic solution to capture musical gestures on a touchscreen and to allow for musical skill transfer.

In this section, we present `faust2smartkeyb` (faust2smartkeyb is now part of the FAUST distribution.), a tool based on `faust2api` to facilitate the creation of musical apps for Android and iOS. The use of musical instrument physical models in this context and in that of acoustically driven hybrid instrument design (see Section 5) is emphasized. Similarly, allowing the design of interfaces implementing skill transfer from existing musical instruments is one of our main focus.

2.1. Apps Generation and General Implementation

`faust2smartkeyb` works the same way than most FAUST targets/"architectures" [49] and can be called using the `faust2smartkeyb` command-line tool:

```
faust2smartkeyb [options] faustFile.dsp
```

where `faustFile.dsp` is a FAUST file declaring a SMARTKEYBOARD interface (see Section 2.2) and `[options]` is a set of options to configure general parameters of the generated app (e.g., Android vs. iOS app, internal number of polyphony voices, etc.). An exhaustive list of these options is available in the `faust2smartkeyb` documentation [50].

The only required option is the app type (`-android` or `-ios`). Unless specified otherwise (e.g., using the `-source` option), `faust2smartkeyb` will compile the app directly in the terminal and upload it on any Android device connected to the computer if the `-install` option is provided. If `-source` is used, an Xcode (https://developer.apple.com/xcode/) or an Android Studio (https://developer.android.com/studio) project is generated, depending on the selected app type (see Figure 1).

`faust2smartkeyb` is based on `faust2api` [47] and takes advantage of most of the features of this system. It provides polyphony, MIDI, and OSC support and allows for SMARTKEYBOARD interfaces to interact with the DSP portion of the app at a very high level (see Figure 1).

`faust2smartkeyb` inherits some of `faust2api`'s options. For example, an external audio effect FAUST file can be specified using `-effect`. This is very useful to save computation when implementing a polyphonic synthesizer [47]. Similarly, `-nvoices` can be used to override the default maximum number of polyphony voices (twelve) of the DSP engine generated by `faust2api` (see Figure 1).

The DSP engine generated by `faust2api` is transferred to a template Xcode or Android Studio project (see Figure 1) and contains the SMARTKEYBOARD declaration (see Section 2.2). The interface of the app, which is implemented in JAVA on Android and in Objective-C on iOS, is built from this declaration. While OSC support is built-in in the DSP engine and works both on iOS and Android, MIDI support is only available on iOS thanks to Rt-MIDI [47] (see Figure 1). On Android, raw MIDI messages are retrieved in the JAVA portion of the app and "pushed" to the DSP engine. MIDI is only supported since Android-23 so `faust2smartkeyb` apps wont have MIDI support on older Android versions.

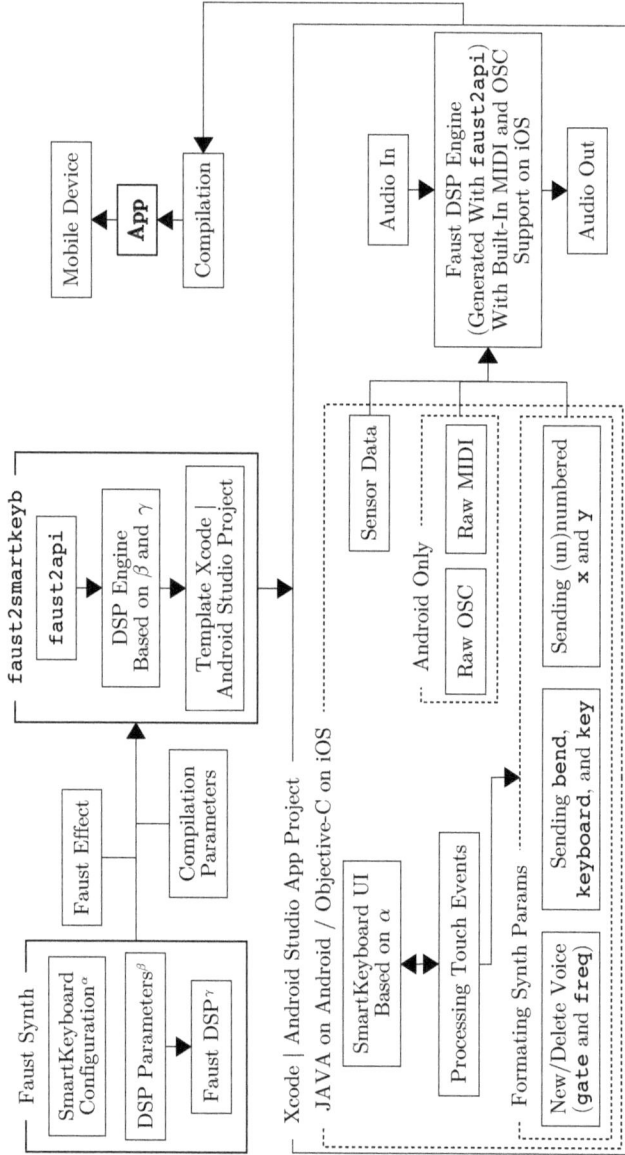

Figure 1. Overview of `faust2smartkeyb`.

2.2. Architecture of a Simple faust2smartkeyb Code

The SMARTKEYBOARD interface can be declared anywhere in a FAUST file using the SmartKeyboard{} metadata:

```
declare interface "SmartKeyboard{
  // configuration keys
}";
```

It is based on the idea that any touchscreen musical interface can be implemented as a set of keyboards with different key numbers (like a table with columns and cells, essentially). Various interfaces ranging from drum pads, isomorphic keyboards, (x, y) controllers, wind instruments fingerings, etc. can be implemented using this paradigm. The position of fingers in the interface can be continuously tracked and transmitted to the DSP engine both as high level parameters formatted by the system (e.g., frequency, note on/off, gain, etc.) or low level parameters (e.g., (x, y) position, key and keyboard ID, etc.). These parameters are declared in the FAUST code using default parameter names (see [50] for an exhaustive list).

By default, the screen interface is a polyphonic chromatic keyboard with thirteen keys whose lowest key is a C5 (MIDI note number 60). A set of key/value pairs can be used to override the default look and behavior of the interface (see [50] for an exhaustive list). Code Listing 1 presents the FAUST code of a simple app where two identical keyboards can be used to control a simple synthesizer based on a band-limited sawtooth wave oscillator and a simple exponential envelope generator. Since MIDI support is enabled by default in apps generated by `faust2smartkeyb` and that the SMARTKEYBOARD standard parameters are the same as the one used for MIDI in FAUST, this app is also controllable by any MIDI keyboard connected to the device running it. A screen-shot of the interface of the app generated from Code Listing 1 can be seen in Figure 2.

```
declare interface "SmartKeyboard{
  'Number of Keyboards':'2'
}";
import("stdfaust.lib");
f = nentry("freq",200,40,2000,0.01);
g = nentry("gain",1,0,1,0.01);
t = button("gate");
envelope = t*g : si.smoo;
process = os.sawtooth(f)*envelope <: _,_;
```

Listing 1: Simple SMARTKEYBOARD FAUST app.

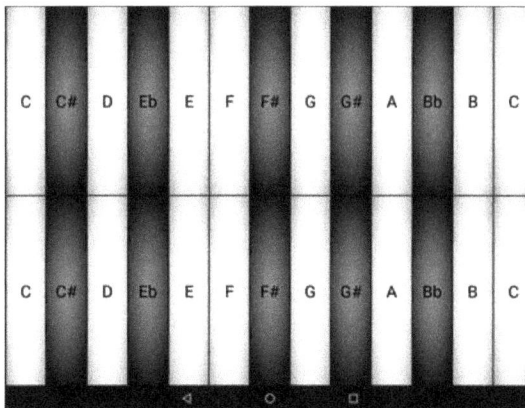

Figure 2. Simple SMARTKEYBOARD interface.

2.3. Preparing a FAUST Code for Continuous Pitch Control

In `faust2smartkeyb` programs, pitch is handled using the `freq` and `bend` standard parameters [50]. The behavior of the formatting of these parameters can be configured using specific keys.

freq gives the "reference frequency" of a note and is tied to the gate parameter. Every time gate goes from 0 to 1 (which correlates with a new note event), the value of freq is updated. freq always corresponds to an integer MIDI pitch number which implies that its value is always quantized to the nearest semitone.

Pitch can be continuously updated by using the bend standard parameter. bend is a ratio that should be multiplied to freq. E.g.,:

```
f = nentry("freq",200,40,2000,0.01);
bend = nentry("bend",1,0,10,0.01) : si.polySmooth(t,0.999,1);
freq = f*bend;
```

The state of polyphonic voices is conserved in memory until the app is ended. Thus, the value of bend might jump from one value to another when a new voice is activated. polySmooth() is used here to smooth the value of bend to prevent clicks, only after the voice started. This suppresses any potential "sweep" that might occur if the value of bend changes abruptly at the beginning of a note.

2.4. Configuring Continuous Pitch Control

The Rounding Mode configuration key has a significant impact on the behavior of freq, bend, and gate. When Rounding Mode = 0, pitch is fully "quantized," and the value of bend is always 1. Additionally, a new note is triggered every time a finger slides to a new key, impacting the value of freq and gate. When Rounding Mode = 1, continuous pitch control is activated, and the value of bend is constantly updated in function the position of the finger on the screen. New note events updating the value of freq and gate are only triggered when fingers start touching the screen. While this mode might be useful in some cases, it is hard to use when playing tonal music as any new note might be "out of tune."

When Rounding Mode = 2, "pitch rounding" is activated and the value of bend is rounded to match the nearest quantized semitone when the finger is not moving on the screen. This allows for generated sounds to be "in tune" without preventing slides, vibratos, etc. While the design of such a system has been previously studied [51], we decided to implement our own algorithm for this (see Figure 3). touchDiff is the distance on the screen between two touch events for a specific finger. This value is smoothed (sTouchDiff) using a unity-dc-gain one pole lowpass filter in a separate thread running at a rate defined by configuration key Rounding Update Speed. Rounding Smooth corresponds to the pole of the lowpass filter used for smoothing (0.9 by default). A separate thread is needed since the callback of touch events is only called when events are received. If sTouchDiff is greater than Rounding Threshold during a certain number of cycles defined by Rounding Cycles, then rounding is deactivated and the value of bend corresponds to the exact position of the finger on the screen. If rounding is activated, the value of bend is rounded to match the nearest pitch of the chromatic scale.

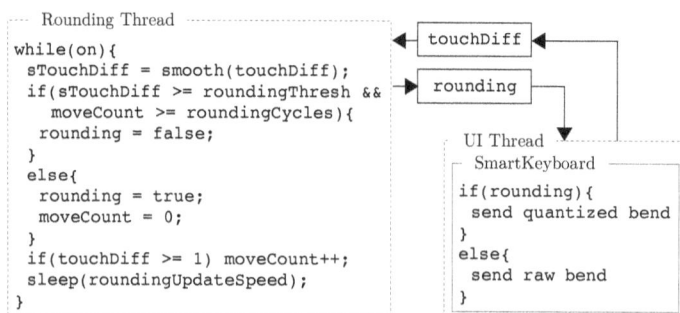

Figure 3. SMARTKEYBOARD pitch rounding "pseudo code" algorithm.

2.5. Using Specific Scales

A wide range of musical scales (see [50] for an exhaustive list), all compatible with the system described in Section 2.4, can be used with the SMARTKEYBOARD interface and configured using the `Keyboard N - Scale` key. When other scales than the chromatic scale are used, keys on the keyboard all have the same color.

Custom scales and temperaments can be implemented using the `Keyboard N - Scale` configuration key. It allows us to specify a series of intervals to be repeated along the keyboard (not necessarily at the octave). Intervals are provided as semitones and can have a decimal value. For example, the chromatic scale can be implemented as:

```
Keyboard N - Scale = {1}
```

Similarly, the standard equal-tempered major scale can be specified as:

```
Keyboard N - Scale = {2,2,1,2,2,2,1}
```

A 5-limit just intoned major scale (rounded to the nearest 0.01 cents) could be:

```
Keyboard N - Scale = {2.0391,1.8243,1.1173,2.0391,2.0391,1.8243,1.1173}
```

Equal-tempered Bohlen-Pierce (dividing 3:1 into 13 equal intervals) would be:

```
Keyboard N - Scale = {146.304230835802}
```

Alternatively, custom scales and pitch mappings can be implemented directly from the FAUST code using some of the lower level standard parameters returned by the SMARTKEYBOARD interface (e.g., x, y, `key`, `keyboard`, etc.).

2.6. Handling Polyphony and Monophony

By default, the DSP engine generated by `faust2api` has twelve polyphony voices. This parameter can be overridden using the `-nvoices` option when executing the `faust2smartkeyb` command. This system works independently from the monophonic/polyphonic configuration of the SMARTKEYBOARD interface. Indeed, even when a keyboard is monophonic, a polyphonic synthesizer might still be needed to leave time for the release of an envelope generator, for example.

The `Max Keyboard Polyphony` key defines the maximum number of voices of polyphony of a SMARTKEYBOARD interface. Polyphony is tied to fingers present on the screen, in other words, one finger corresponds to one voice. If `Max Keyboard Polyphony = 1`, then the interface becomes "monophonic." The monophonic behavior of the system is configured using the `Mono Mode` key [50].

2.7. Other Modes

In some cases, both the monophonic and the polyphonic paradigms are not adapted. For example, when implementing an instrument based on a physical model, it might be necessary to use a single voice and constantly run it. This might be the case of a virtual wind instrument where notes are "triggered" by some of the continuous parameters of the embouchure and not by discrete events such as the one created by a key. This type of system can be implemented by setting the `Max Keyboard Polyphony` key to zero. In that case, the first available voice is triggered and ran until the app is killed. Adding new fingers on the screen will have no impact on that and the `gate` parameter wont be sent to the DSP engine. `freq` will keep being sent unless the `Keyboard N - Send Freq` is set to zero. Since this parameter is keyboard specific, some keyboards in the interface might be used for pitch control while others might be used for other types of applications (e.g., X/Y controller, etc.).

It might be useful in some cases to number the standard x and y parameters in function of the fingers present on the screen. This can be easily accomplished by setting the `Keyboard N - Count Fingers` key to one. In that case, the first finger to touch the screen will send the x0 and y0 standard parameters to the DSP engine, the second finger x1 and y1, and so on.

2.8. Example: Violin App

Implementation strategies greatly varies from one instrument to another, so giving a fully representative example is impossible. Instead, we focus on a specific implementation of a violin here where strings are excited by an interface independent from the keyboards used to control their pitch. This illustrates a "typical" physical model mapping where the MIDI concept of note on/off event is not used. More examples are available on-line (*Making Faust-Based Smartphone Musical Instruments* On-Line Tutorial: https://ccrma.stanford.edu/~rmichon/faustTutorials/#making-faust-based-smartphone-musical-instruments).

Unlike plucked string instruments, bowed string instruments must be constantly excited to generate sound. Thus, parameters linked to bowing (i.e., bow pressure, bow velocity, etc.) must be continuously controlled. The faust2smartkeyb code presented in Listing 2 is a violin app where each string is represented by one keyboard in the interface. An independent surface can be used to control the bow pressure and velocity. This system is common to all strings that are activated when they are touched on the screen. This virtual touchscreen interface could be easily be substituted by a physical one using the technique presented in Section 4.

The SMARTKEYBOARD configuration declares 5 keyboards (4 strings and one control surface for bowing). "String keyboards" are tuned like on a violin (G, D, A, E) and are configured to be monophonic and implement "pitch stealing" when a higher pitch is selected. Bow velocity is computed by measuring the displacement of the finger touching the 5th keyboard (bowVel). Bow pressure just corresponds to the y position of the finger on this keyboard. Strings are activated when at least one finger is touching the corresponding keyboard (as (i)).

The app doesn't take advantage of the polyphony support of faust2smartkeyb and a single voice is constantly ran after the app is launched (Max Keyboard Polyphony = 0). Four virtual strings based on a simple violin string model (violinModel()) implemented in the FAUST Physical Modeling Library (see Section 5.2) are declared in parallel and activated in function of events happening on the screen.

```
declare interface "SmartKeyboard{
   'Number of Keyboards':'5','Max Keyboard Polyphony':'0',
   'Rounding Mode':'2','Send Fingers Count':'1',
   'Keyboard 0 - Number of Keys':'19',
   [...same for next 3 keyboards...]
   'Keyboard 4 - Number of Keys':'1',
   'Keyboard 0 - Lowest Key':'55','Keyboard 1 - Lowest Key':'62',
   'Keyboard 2 - Lowest Key':'69','Keyboard 3 - Lowest Key':'76',
   'Keyboard 0 - Send Keyboard Freq':'1',
   [...same for next 3 keyboards...]
   'Keyboard 4 - Send Freq':'0','Keyboard 4 - Send Key X':'1',
   'Keyboard 4 - Send Key Y':'1','Keyboard 4 - Static Mode':'1',
   'Keyboard 4 - Key 0 - Label':'Bow'
}";
/////////////////////////// SMARTKEYBOARD PARAMETERS ///////////////////////////
kbfreq(0) = hslider("kb0freq",220,20,10000,0.01);
kbbend(0) = hslider("kb0bend",1,0,10,0.01);
[...same for the 3 next keyboards...]
kb4k0x = hslider("kb4k0x",0,0,1,1) : si.smoo;
kb4k0y = hslider("kb4k0y",0,0,1,1) : si.smoo;
kbfingers(0) = hslider("kb0fingers",0,0,10,1) : int;
[...same for the 3 next keyboards...]
/////////////////////////// MODEL PARAMETERS ///////////////////////////
```

```
sl(i) = kbfreq(i)*kbbend(i) : pm.f2l : si.smoo; // strings lengths
as(i) = kbfingers(i)>0; // activate string
bowPress = kb4k0y; // could also be controlled by an external controller
bowVel = kb4k0x-kb4k0x' : abs : *(8000) : min(1) : si.smoo;
bowPos = 0.7; // could be controlled by an external controller
////////////////////////// ASSEMBLING MODELS /////////////////////////////////
process = par(i,4,pm.violinModel(sl(i),bowPress,bowVel*as(i),bowPos))
    :> _;
```

Listing 2: `faust2smartkeyb` app implementing a violin with an independent interface for bowing.

Alternatively, the bowing interface could be removed and the bow velocity could be calculated based on the displacement on the y axis of a finger on a keyboard, allowing one to excite the string and control its pitch with a single finger. However, concentrating so many parameters on a single gesture tends to limit the affordances of the instrument. The code presented in Listing 2 could be easily modified to implement this behavior.

Mastering a musical instrument, should it be fully acoustic, digital, or hybrid, is a time consuming process. While skill transfer can help reduce its duration, we do not claim that the instruments presented in this paper are faster to learn than any other type of instrument. Virtuosity can be afforded by the instrument, but it still depends on the musicianship of the performer.

This section just gave an overview of some of the features of `faust2smartkeyb`. More details about this tool can be found in its documentation [50] as well as on the corresponding on-line tutorials.

3. Passively Augmenting Mobile Devices

In this section, we try to generalize the concept of "passively augmented mobile device" briefly introduced in Section 1.4 and we provide a framework to design this kind of instrument. We focus on "passive augmentations" leveraging existing components of hand-held mobile devices in a very lightweight, non-invasive way (as opposed to "active augmentation" presented in Section 4 that require the use of electronic components). We introduce MOBILE3D, an OpenScad (http://www.openscad.org) library to help design mobile device augmentations using DIY digital fabrication techniques such as 3D printing and laser cutting. We give an exhaustive overview of the taxonomy of the various types of passive augmentations that can be implemented on mobile devices through a series of examples and we demonstrate how they leverage existing components on the device. Finally, we evaluate our framework and propose future directions for this type of research.

3.1. Mobile 3D

MOBILE3D is an OpenScad library facilitating the design of mobile device augmentations. OpenScad is an open-source Computer Assisted Design (CAD) software using a high level functional programming language to specify the shape of any object. It supports fully parametric parts, permitting users to rapidly adapt geometries to the variety of devices available on the market.

MOBILE3D is organized in different files that are all based on a single library containing generic standard elements (`basics.scad`) ranging from simple useful shapes to more advanced augmentations such as the ones presented in the following sections. A series of device-specific files adapt the elements of `basics.scad` and are also available for the iPhone 5, 6, and 6 Plus and for the iPod Touch. For example, a generic horn usable as a passive amplifier for the built-in speaker of a mobile device can be simply created with the following call in OpenScad:

```
include <basics.scad>
SmallPassiveAmp();
```

To generate the same object specifically for the iPhone 5, the following code can be written:

```
include <iPhone5.scad>
```

```
iPhone5_SmallPassiveAmp();
```

Finally, the shape of an object can be easily modified either by providing parameters as arguments to the corresponding function, or by overriding them globally before the function is called. If this approach is chosen, all the parts called in the OpenScad code will be updated, which can be very convenient in some cases. For example, the radius (expressed in millimeters here) of `iPhone5_SmallPassiveAmp()` can be modified locally by writing:

```
include <iPhone5.scad>
iPhone5_SmallPassiveAmp(hornRadius=40);
```

or globally by writing:

```
include <iPhone5.scad>
iPhone5_SmallPassiveAmp_HornRadius = 40;
iPhone5_SmallPassiveAmp();
```

MOBILE3D is based on two fundamental elements that can be used to quickly attach any prosthetic to the device: the top and bottom *holders* (see Figure 4). They were designed to be 3D printed using elastomeric material such as *NinjaFlex®* (https://ninjatek.com) in order to easily install and remove the device without damaging it. They also help reducing printing duration, which is often a major issue during prototyping. These two holders glued to a laser-cut plastic plate form a sturdy case, whereas completely printing this part would take much more time.

Figure 4 presents an example of an iPhone 5 augmented with a passive amplifier. The bottom holder and the horn were printed separately and glued together, but they could also have been printed as one piece. In this example, the bottom and top holders were printed with PLA (PolyLactic Acid), which is a hard plastic, and they were mounted on the plate using Velcro®. This is an alternative solution to using *NinjaFlex®* that can be useful when augmenting the mobile device with large appendixes requiring a stronger support.

The passive amplifier presented in Figure 4 was made by overriding the default parameters of the `iPhone5_SmallPassiveAmp()` function:

```
include <lib/iPhone5.scad>
iPhone5_SmallPassiveAmp_HornLength = 40;
iPhone5_SmallPassiveAmp_HornRadius = 40;
iPhone5_SmallPassiveAmp_HornDeformationFactor = 0.7;
iPhone5_SmallPassiveAmp();
```

An exhaustive list of all the elements available in MOBILE3D can be found on the project webpage (https://ccrma.stanford.edu/~rmichon/mobile3D).

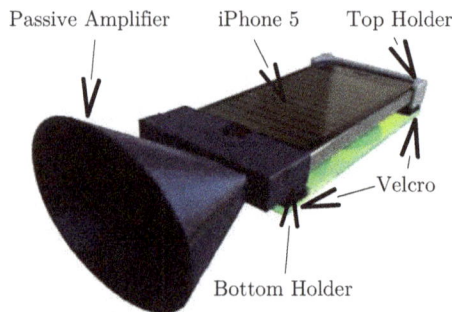

Figure 4. iPhone 5 augmented with a horn used as passive amplifier on its built-in speaker (instrument by Erin Meadows).

3.2. Leveraging Built-In Sensors and Elements

Mobile devices host a wide range of built-in sensors and elements that can be used to control sound synthesizers (see Section 2). While the variety of available sensors and elements differs from one device to another, most smart-phones have at least a touch screen, a loudspeaker, a microphone, and some type of motion sensor (accelerometer, gyroscope, etc.). In this section, we'll focus on these four elements and we'll demonstrate how they can be "augmented" for specific musical applications.

3.2.1. Microphone

While the built-in microphone of a mobile device can simply serve as a source for any kind of sound process (e.g., audio effect, physical model, etc.), it can also be used as a versatile, high rate sensor [52]. In this section, we demonstrate how it can be augmented for different kinds of uses.

One of the first concrete uses of the built-in microphone of a mobile device to control some sound synthesis process was done with Smule's Ocarina [22]. There, the microphone serves as a blow sensor by measuring the gain of the signal created when blowing on it to control the gain of an ocarina sound synthesizer.

MOBILE3D contains an object that can be used to leverage this principle when placed in front of the microphone (see Figure 5). It essentially allows for the performer to blow into a mouthpiece mounted on the device. The air-flow is directed through a small aperture inside the pipe, creating a sound that can be recorded by the microphone and analyzed in the app using standard amplitude tracking techniques. The air-flow is then sent outside of the pipe, preventing it from ever being in direct contact with the microphone.

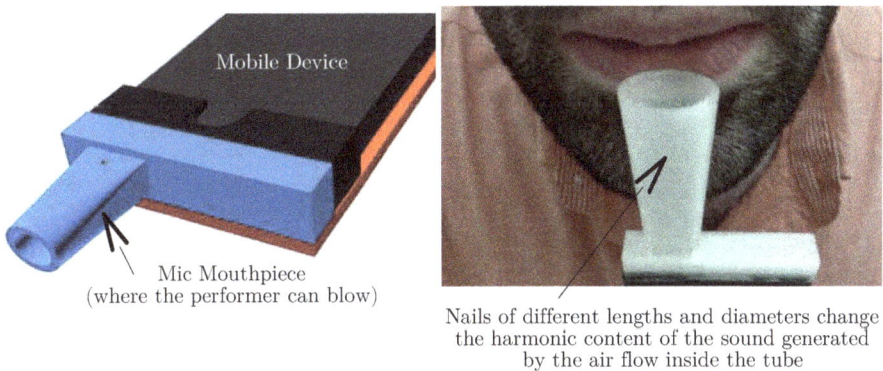

Figure 5. Mouthpiece for mobile device built-in mic (**on the left**) and frequency-based blow sensor for mobile device built-in microphone (**on the right**).

The acquired signal is much cleaner than when the performer blows directly onto the mic, allowing us to generate precise control data. Additionally, condensation never accumulates on the mic which can help extend the duration of its life, etc.

The built-in microphone of mobile devices has already been used as a data acquisition system to implement various kinds of sensors using frequency analysis techniques [53]. MOBILE3D contains an object using similar principles that can be used to control some of the parameters of a synthesizer running on a mobile device. It is based on a conical tube (see Figure 5) where dozens of small tines of different length and diameter are placed inside it. These tines get thicker towards the end of the tube and their length varies linearly around it. When the performer blows inside the tube, the resulting airflow hits the nails, creating sounds with varying harmonic content. By directing the airflow towards different locations inside the tube, the performer can generate various kind of sounds that can be recognized in the app using frequency analysis techniques. The intensity and the position of the airflow around the tube can be measured by keeping track of the spectral centroid of the generated sound, and used to control synthesis parameters.

The same approach can be used with an infinite number of augmentations with different shapes. While our basic spectral-centroid-based analysis technique only allows us to extract two continuous parameters from the generated signal, it should be possible to get more of them using more advanced techniques.

3.2.2. Speaker

Even though their quality and power has significantly increased during the last decade, mobile device built-in speakers are generally only good for speech, not music. This is mostly due to their small size and the lack of a proper resonance chamber to boost bass, resulting in a very curvy frequency response and a lack of power.

There exists a wide range of passive amplifiers on the market to boost the sound generated by the built-in speakers of mobile devices, also attempting to flatten their frequency response (see Section 1.4). These passive amplifiers can be seen as resonators driven by the speaker. In this section, we present various kinds of resonators that can be connected to the built-in speaker of mobile devices to amplify and/or modify their sound.

MOBILE3D contains multiple passive amplifiers of various kinds that can be used to boost the loudness of the built-in speaker of mobile devices (e.g., see Figure 4). Some of them were designed to maximize their effect on the generated sound [54]. Their shape can vary greatly and will usually be determined by the type of the instrument. For example, if the instrument requires the performer to make fast movements, a small passive amplifier will be preferred to a large one, etc. Similarly, the orientation of the output of the amplifier will often be determined by the way the performer holds the instrument, etc. These are design decisions that are left up to the instrument designer.

3D printed musical instrument resonators (e.g., guitar body, etc.) can be seen as a special case of passive amplifiers. MOBILE3D contains a few examples of such resonators that can be driven by the device's built-in speakers. While they don't offer any significant advantage over "standard" passive amplifiers like the one presented in the previous paragraph, they are aesthetically interesting and perfectly translate the idea of hybrid instrument developed in Section 5.

Another way to use the signal generated by the built-in speakers of mobile devices is to modify it using dynamic resonators. For example, in the instrument presented in Figure 6, the performer's hand can filter the generated sound to create a *wah* effect. This can be very expressive, especially if the signal has a dense spectral content. This instrument is featured in the teaser video (https://www.youtube.com/watch?v=dGBDrmvG4Yk) of the workshop presented in Section 3.4.

Waveguide driving the output of the built-in speaker to the mouth of the performer

Figure 6. Hand resonator (**on the left**) and mouth resonator (**on the right**) for mobile device built-in speaker.

Similarly, the sound generated by the built-in speaker is sent to the mouth of the performer in the instrument presented in Figure 6. The sound is therefore both modulated acoustically and through the embedded synthesis and touch-screen. The same result can obviously be achieved by directly applying the mouth of the performer to the speaker, but the augmentation presented in Figure 6 increases the effect of the oral cavity on the sound through a passive wave guide.

3.2.3. Motion Sensors

Most mobile devices have at least one kind of built-in motion sensor (e.g., accelerometer, gyroscope, etc.). They are perfect to continuously control the parameters of sound synthesizer and have been used as such since the beginning of mobile music (see Section 1.3).

Augmentations can be made to mobile devices to direct and optimize the use of this type of sensor. This kind of augmentation can be classified in two main categories:

- augmentations to create specific kinds of movements (spin, swing, shake, etc.),
- augmentations related to how the device is held.

Figure 7 presents a "sound toy" where a mobile device can be spun like a top. This creates a slight "Leslie effect", increased by the passive amplifier. Additionally, the accelerometer and gyroscope data are used to control the synthesizer running on the device. This instrument is featured in the teaser video of the workshop presented in Section 3.4.

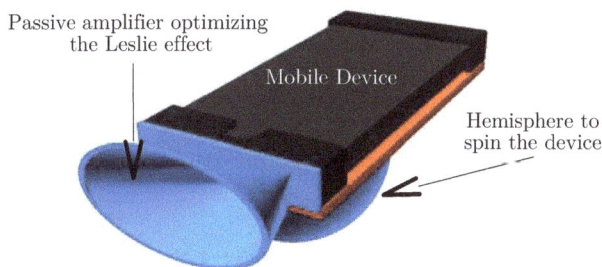

Figure 7. Mobile-device-based top creating a "Leslie" effect when spun.

Another example of motion-sensor-based augmentation is presented in Figure 8 and described with more details in Section 3.4. It features a smart-phone mounted on a bike wheel where, once again, the gyroscope and accelerometer data are used to control the parameters of a synthesizer running on the device. Similarly, a "rolling smart-phone" is presented in Figure 8 and described in Section 3.4. MOBILE3D contains a series of templates and functions to make this kind of augmentation.

Augmentations leveraging built-in sensors related to how the device is held are presented in more detail in Section 3.3.

Wheels   Embedded passive amplifier

Figure 8. Rolling mobile phone with phasing effect by Revital Hollander (**on the left**) and mobile device mounted on a bike wheel by Patricia Robinson (**on the right**).

3.2.4. Other Sensors

Most mobile devices host built-in sensors that exceed the ones presented in the previous sections and are not supported yet in MOBILE3D. For example, built-in cameras can be used as very versatile sensors [52], and a wide range of passive augmentations could be applied to them to "customize" their use for musical ends. We plan to support more sensors in MOBILE3D in the future.

3.3. Holding Mobile Devices

Mobile devices were designed to be held in a specific way, mostly so that they can be used conveniently both as a phone and to use the touch-screen (see Section 1.3). Passive augmentations can be designed to hold mobile devices in different ways to help carry out specific musical gestures, better leveraging the potential of the touch-screen and of built-in sensors.

More generally, this type of augmentation is targeted towards making mobile-device-based musical instruments more engaging and easier to play.

In this section, we give a brief overview of the different types of augmentations that can be made with MOBILE3D to hold mobile devices in different ways.

3.3.1. Wind Instrument Paradigm

One of the first attempts to hold a smart-phone as a wind instrument was Ocarina, where the screen interface was designed to be similar to a traditional ocarina. The idea of holding a smart-phone as such is quite appealing since all fingers (beside the thumbs) of both hands perfectly fit on the screen (thumbs can be placed on the other side of the device to hold it). However, this position is impractical since at least one finger has to be on the screen in order to hold the device securely. The simple augmentation presented in Figure 9 solves this problem by adding "handles" on both sides of the device so that it can be held using the palm of the two hands, leaving all fingers (including the thumbs) free to carry out any action. Several functions and templates are available in MOBILE3D to design these types of augmentations.

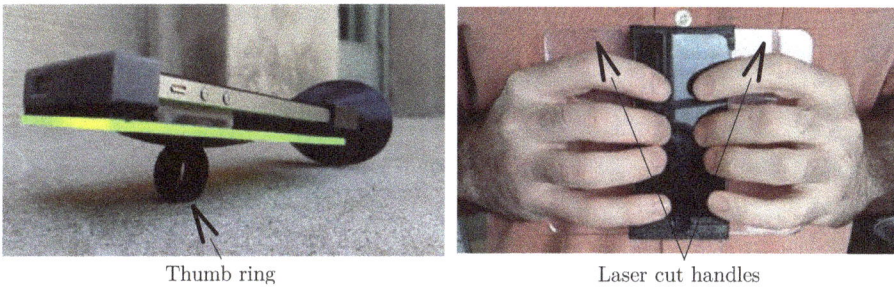

Thumb ring   Laser cut handles

Figure 9. Thumb-held mobile-device-based musical instrument (**on the left**) and smart-phone augmented to be held as a wind instrument (**on the right**).

3.3.2. Holding the Device with One Hand

MOBILE3D contains several functions and templates to hold mobile devices with one hand, leaving at least four fingers available to perform on the touch-screen. This way to hold the device opens up a wide range of options to fully take advantage of the built-in motion sensors and easily execute free movements. Additionally, the performer can decide to use two devices in this case (one for each hand).

The instrument presented in Figure 9 uses one of MOBILE3D's ring holders to hold the device with only the thumb.

3.3.3. Other Holding Options

There are obviously many other options to hold mobile-devices to carry out specific musical gestures. For example, one might hold the device in one hand and perform it with the other, etc. In any case, we believe that MOBILE3D provides enough options to cover the design needs for most musical instruments.

3.4. More Examples and Evaluation

To evaluate MOBILE3D and the framework presented in this paper, we organized a one-week workshop last summer at Stanford's *Center for Computer Research in Music and Acoustics* (CCRMA) called *The Composed Instrument Workshop: Intersections of 3D Printing and Digital Audio for Mobile Platforms* (Workshop Web-Page: https://ccrma.stanford.edu/~rmichon/composedInstrumentWorkshop/). We taught the seven participants how to make basic musical smart-phone apps using `faust2smartkeyb` (see Section 2) and how to use MOBILE3D to design mobile device augmentations. They were free to make any musical instrument or sound toy for their final project. Some examples of these instruments are presented in Figures 4 and 8.

In only one week, participants mastered all these techniques and designed and implemented very original instrument ideas. This helped us debug and improve MOBILE3D with new objects and features.

4. Actively Augmenting Mobile Devices

While the non-invasive and lightweight character of passive mobile device augmentations (see Section 3) contributes to the overall physical coherence of hybrid instruments, their simplicity can sometimes be a limitation as they remain tied to what built-in active elements of the device (e.g., touchscreen, microphone, speaker, etc.) can offer. Inversely, active augmentations can take any form and can be used to implement almost anything that mobile devices don't have. While their level of complexity can be more or less infinite, we praise for an incremental approach where instrument designers should first take advantage of elements already available on mobile devices, and then use active augmentations parsimoniously to implement what they could not have done otherwise.

In this section, we provide a framework/method to make active augmentations for mobile devices, towards mobile hybrid musical instrument design (see Section 5). Unlike passive augmentations, the scope of active augmentations is almost infinite and any musical controller could probably fit in this category. Thus, we will only consider the tools to carry out this task and let design or aesthetic considerations up to the instrument maker.

4.1. Active Augmentation Framework

In our view, mobile device augmentations should supplement existing built-in sensors (e.g., touchscreen, motion sensors, etc.) and remain as lightweight and confined as possible. Indeed, there's often not much to add to a mobile device to turn it into a truly expressive musical instrument. NUANCE [55] is a good example of that since it adds a whole new level of expressivity to the touchscreen, simply by using a few sensors. On the other hand, unlike passive augmentations, active augmentations can be used to add an infinite number of features.

In this section, we introduce a framework for designing active mobile device augmentations supplementing sensors already available on the device. This allows us to keep our augmentations lightweight and powered by the device, preserving the standalone aspect and partly the physical coherence of the instrument.

To keep our augmentations simple, we propose to use a wired solution for transmitting sensor data to the mobile device, which also allows us to power the augmentation. Augmentations requiring an external power supply (e.g., battery) are discarded and are not considered in the frame of this work.

MIDI is a standard universal way to transmit real-time musical (and non-musical) control data to mobile devices, so we opted for this solution. Teensys such the Teensy 3.2 (https://www.pjrc.com/store/teensy32.html) are micro-controllers providing built-in USB MIDI support, making them particularly well suited to be used in our framework.

Teensyduino (https://www.pjrc.com/teensy/teensyduino.html) (Teensy's IDE), comes with a high level library part of `Bounce.h` for sending MIDI over USB. The code presented in Listing 3 demonstrates how to use this library to send sensor values on a MIDI "Continuous Controller" (CC).

```
#include <Bounce.h>
void setup() {
}
void loop() {
  int sensorValue = analogRead(A0);
  int midiCC = 10; // must match the faust configuration
  int midiValue = sensorValue*127/1024; // value between 0-127
  int midiChannel = 0;
  usbMIDI.sendControlChange(midiCC,midiValue,midiChannel); // send!
  delay(30); // wait for 30ms
}
```

Listing 3: Simple Teensy code sending sensor data in MIDI format over USB.

Once uploaded to the microcontroller, the Teensy board can be connected via USB to any MIDI-compatible mobile device (iOS and Android) to control some of the parameters of a `faust2smartkeyb` app (see Section 2). This will require the use of a USB adapter, depending on the type of USB plug available on the device. MIDI is enabled by default in `faust2smartkeyb` apps and parameters in the FAUST code can be mapped to a specific MIDI CC by using a metadata (see Section 2.2):

```
frequency = nentry("frequency[midi:ctrl 10]",1000,20,2000,0.01);
```

Here, the `frequency` parameter will be controlled by MIDI messages coming from MIDI CC 10 and mapped to the minimum (20 Hz for MIDI CC 10 = 0) and maximum (2000 Hz for MIDI CC 10 = 127) values defined in the `nentry` declaration. Thus, if this parameter was controlling the frequency of an oscillator and that the Teensy board running the code presented in Listing 3 was connected to the mobile device running the corresponding `faust2smartkeyb` app, the sensor connected to the A0 pin of the Teensy would be able to control the frequency of the generated sound.

Other types of MIDI messages (e.g., `sendNoteOn()`) can be sent to a `faust2smartkeyb` app using the same technique.

Most of the parameters controlled by elements on the touchscreen or by built-in sensors of the app presented in Section 2.8 could be substituted by external sensors or custom interfaces using the technique described above.

4.2. Examples and Evaluation: CCRMA Mobile Synth Summer Workshop

The framework presented in Section 4.1 was evaluated within a two weeks workshop at CCRMA at the end of June 2017 (https://ccrma.stanford.edu/~rmichon/mobileSynth: this webpage contains more details about the different instruments presented in the following subsections.). It was done in continuity with the FAUST *Workshop* taught the previous years (https://ccrma.stanford.edu/~rmichon/faustWorkshops/2016/) and the *Composed Instrument Workshop* presented in Section 3.4. During the first week (*Mobile App Development for Sound Synthesis and Processing in Faust*), participants learned how to use FAUST through `faust2smartkeyb` and made a wide range of musical apps. During the second week (*3D Printing and Musical Interface Design for Smart-phone Augmentation*), they designed various passive (see Section 3) and active augmentations using the framework presented in Section 4.1. They were encouraged to first use elements available on the device (e.g., built-in sensors, touchscreen, etc.) and then think about what was missing to their instrument to make it more expressive and controllable.

This section presents selected works from students of the workshop.

4.2.1. *Bouncy-Phone* by Casey Kim

Casey Kim designed *Bouncy-Phone*, an instrument where a 3D printed spring is "sandwiched" between an iPhone and an acrylic plate hosting a set of photo-resistors (see Figure 10). The interface on the touchscreen implements two parallel piano keyboards controlling the pitch of a monophonic synthesizer. The instrument is played by blowing onto the built-in microphone, in a similar way than Ocarina. The x axis of the accelerometer is mapped to the frequency of a lowpass filter applied to the generated sound. The spring is used to better control the position of the device in space in order to finely tune the frequency of the filter. The shades created by the two hands of the performer between the phone and the acrylic plate are used to control the parameters of various audio effects.

4.2.2. *Something Else* by Edmond Howser

Edmond Howser designed *Something Else*, an instrument running a set of virtual strings based on physical models from the FAUST Physical Modeling Library (see Section 5.2). The touchscreen of an iPhone can be used to trigger sound excitations of different pitches. A set of three photoresistors were placed in 3D printed cavities (see Figure 10) that can be covered by the fingers of the performer to progressively block the light, allowing for a precise control of the parameters associated to them. These sensors were mapped to the parameters of a set of audio effects applied to the sounds generated by the string physical models. The instrument is meant to be held as a trumpet with three fingers on top of it (one per photoresistor) and fingers from the other hand on the side, on the touchscreen.

4.2.3. *Mobile Hang* by Marit Brademann

Mobile Hang is an instrument based on an iPhone designed by Marit Brademann. A 3D printed prosthetic is mounted on the back of the mobile device (see Figure 10). It hosts a Teensy board as well as a set of force sensitive resistors that can be used to trigger a wide range of percussion sounds based on modal physical models of the FAUST Physical Modeling Library (see Section 5.2) with different velocities. A large hole placed in the back of the tapping surface allows for the performer to hold the instrument with the thumb of his right hand. The left hand is then free to interact with the different (x, y) controllers on the touchscreen controlling the parameters of various effects applied to the generated sounds. *Mobile Hang* also takes advantage of the built-in accelerometer of the device to control additional parameters.

Photoresistors   FSRs

Figure 10. *Bouncy-Phone* by Casey Kim, *Something Else* by Edmond Howser, and *Mobile Hang* by Marit Brademann (from left to right).

5. Articulating the Hybrid Mobile Instrument

Current technologies allow one to blur the boundary between the physical/acoustical and the virtual/digital world. Transforming a physical object into its virtual approximation can be done easily using various techniques (see Section 1.5). On the other hand, recent progress in digital fabrication, with 3D printing in particular (see Section 1.6), allows us to materialize 3D virtual objects. Even though 3D printed acoustic instruments don't compete yet with "traditionally made" ones, their quality keeps increasing and they remain perfectly usable.

This section generalizes some of the concepts used by the BLADEAXE [42], where sound excitations made by physical objects are used to drive physical-model-based virtual elements. It allows for instrument designers to arbitrarily choose the nature (physical or virtual) of the different parts of their creations.

We introduce a series of tools completing the framework presented in this paper to approach musical instrument design in a multimodal way where physical acoustical parts can be "virtualized" and vice versa. First, we give an overview of our framework to design mobile hybrid instruments. We provide a set of rules to help the instrument designer to make critical decisions about the nature (acoustical or digital) of the different parts of his instrument in the context of mobile devices. Then we introduce the FAUST Physical Modeling Library (FPML), "the core" of our framework, that can be used to implement a wide range of physical models of musical instruments to be run on a mobile device (e.g., using `faust2smartkeyb`). Finally, we demonstrate how custom models can be implemented using MESH2FAUST [56] and FMPL.

5.1. Framework Overview

5.1.1. From Physical to Virtual

In Section 1.5, we gave an overview of different physical modeling techniques that can be used to make virtual versions of physical objects designed to generate sound (i.e., musical instruments). The framework presented in this section is a bit more limiting and focuses on specific modeling techniques that are flexible and computationally cheap (which is a crucial feature for mobile development).

Various linearizable acoustical physical objects can be easily turned into modal physical models using their impulse response [28]. Pierre-Amaury Grumiaux et al. implemented ir2faust [57], a command-line tool taking an impulse response in audio format and generating the corresponding FAUST physical model compatible with the FAUST Physical Modeling Library presented in Section 5.2. This technique is commonly used to make signal models of musical instrument parts (e.g., acoustic resonators such as violin and guitar bodies, etc.).

Modal physical models can also be generated by carrying out a finite element analysis (FEM) on a 3D volumetric mesh. Meshes can be made "from scratch" or using a 3D scanner, allowing musical instrument designers to make virtual parts using a CAD model. MESH2FAUST [56] can be used to carry out this type of task. Modal models generated by this tool are fully compatible with the FAUST Physical Modeling Library presented in Section 5.2. While this technique is more flexible and allows us to model elements "from scratch," generated models are usually not as accurate as the one deduced from the impulse response of a physical object that faithfully reproduce its harmonic content.

Even though it is tempting to model an instrument in its whole using its complete graphical representation, better results are usually obtained using a modular approach where each part of the instrument (e.g., strings, bridge, body, etc.) are modeled as single entities. The FAUST Physical Modeling Library introduced in Section 5.2 implements a wide range of ready-to-use musical instrument parts. Missing elements can then be easily created using MESH2FAUST or ir2faust. Various examples of such models are presented in Sections 5.2.2 and 5.2.3.

5.1.2. From Virtual to Physical

3D printing can be used to materialize virtual representation of musical instrument parts under certain conditions. Thus, most elements provided to MESH2FAUST [56] can be printed and turned into physical objects.

5.1.3. Connecting Virtual and Physical Elements

Standard hardware for digitizing mechanical acoustic waves and vice versa can be used to connect the physical and virtual elements of a hybrid instrument (see Figure 11). Piezos (contact microphones) can capture mechanical waves on solid surfaces (e.g., guitar body, string, etc.) and microphones mechanical air waves (e.g., in a tube, etc.). Captured signals can be digitized using an analog to digital converter (ADC). Inversely, digital audio signals can be converted to analog signals using a digital to analog converter (DAC) and then to mechanical waves with a transducer (for solid surfaces) or a speaker (for the air).

In some cases, a unidirectional connection is sufficient as waves travel in only one direction and are not (or almost not) reflected. This is the case of the BLADEAXE [42] where sound excitations (i.e., plucks) are picked up using piezos and transmitted to virtual strings. This type of system remains simple and works relatively well as the latency of the DAC or the ADC doesn't impact the characteristics the generated sound.

On the other hand, a bidirectional connection (see Section 5.2.1) might be necessary in other cases. Indeed, reflection waves play a crucial role in the production of sound in some musical instruments such as woodwinds. For examples, connecting a physical clarinet mouthpiece to a virtual bore will require the use of a bidirectional connection in order for the frequency of vibration of the reed to be coupled to the tube it is connected to. This type of connection extends beyond the instrument to the performer that constantly adjusts its various parameters in function of the generated sound [5]. However, implementing this type of system can be very challenging as the DAC and the ADC will add latency, which in the case of the previous example will artificially increase the length of the virtual bore. Thus, using low latency DACs and ADCs is crucial when implementing this type of systems sometimes involving the use of active control techniques [58,59].

More generally, the use of high-end components with a flat frequency response is very important when implementing any kind of hybrid instruments. Also, hardware can become very invasive in some cases, and it is the musical instrument designer's responsibility to find the right balance between all these parameters.

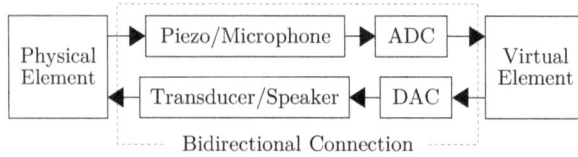

Figure 11. Bidirectional connection between virtual and physical elements of a hybrid instrument.

5.1.4. Adapting This Framework to Mobile Devices

Beyond this theoretical modularity (keeping in mind that audio latency can be a limiting factor in some cases) where any part of mobile hybrid instruments can either be physical or virtual, some design "templates" are more efficient than others. Here, we give some guidelines/rules to restrain the scope of our framework to optimize its results when making mobile hybrid instruments.

In the context of augmented mobile instruments where standalone aspects and lightness are key factors, the number of physical/acoustical elements of hybrid instruments must be scaled down compared to what is possible with a desktop-based system. Indeed, transducers are large and heavy components requiring the use of an amplifier, which itself needs a large power source other than the mobile device battery, etc. Similarly, multichannel ADCs and DACs can take a fair amount of space and will likely need to be powered with an external battery/power supply.

Even though applications generated with `faust2smartkeyb` (see Section 2) are fully compatible with external USB ADC/DACs, we believe that restraining hybrid mobile instruments to their built-in ADC/DACs helps preserve their compactness and playability.

Beyond the aesthetic and philosophical implications of hybrid instruments (which are of great interest but are not the object of this paper), their practical goal is to leverage the benefits of physical and virtual elements to combine them. In practice, the digital world is more flexible and allows us to model/approximate many physical elements. However, even with advanced sensor technologies, it often fails to capture the intimacy (see Section 1.1) between a performer and an acoustic instrument allowing us to directly interact with its sound generation unit (e.g., plucked strings, hand drum, etc.) [13].

Thus, a key factor in the success of hybrid mobile instruments lies in the use of a physical/acoustical element as the direct interface for the performer, enabling passive haptic feedback and taking advantage of the randomness and unpredictability of acoustical elements (see Section 1.2). In other words, even though it is possible to combine any acoustical element with any digital one, we encourage instrument designers to use acoustical excitations to drive virtual elements (see Figure 12), implementing the concept of "acoustically driven hybrid instruments" presented in Section 1.2. While the single analog input available on most mobile devices allows for the connection of one acoustical element, having access to more independent analog inputs would significantly expend the scope of the type of instruments implementable with our framework. This remains one of its main limitation.

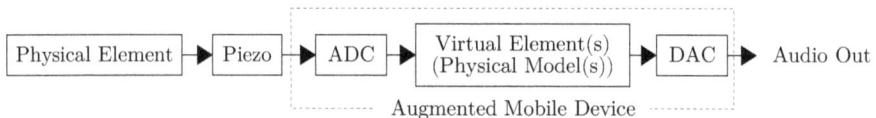

Figure 12. "Typical" acoustically driven mobile hybrid instrument model.

5.2. FAUST *Physical Modeling Library*

More than just a set of functions, the FAUST Physical Modeling Library provides a comprehensive environment to implement physical models of musical instrument parts fully compatible with the hybrid instrument paradigm described in Section 5. This section summarizes its various features.

5.2.1. Bidirectional Block-Diagram Algebra

In the physical world, waves propagate in multiple dimensions and directions across the different parts of musical instruments. Thus, coupling between the constituting elements of an instrument sometimes plays an important role in its general acoustical behavior. In Section 5.1.3, we highlighted the importance of bidirectional connections to implement coupling between the performer, the physical, and the virtual elements of a hybrid instrument. While these types of connections happen naturally between physical elements, it is necessary to implement them when connecting virtual elements together.

The block-diagram algebra of FAUST allows us to connect blocks in a unidirectional way (from left to right) and feedback signals (from right to left) can be implemented using the tilde (~) diagram composition operation:

```
process = (A : B) ~ (C : D) ;
```

where A, B, C, and D are hypothetical functions with a single argument and a single output. The resulting FAUST-generated block diagram can be seen in Figure 13.

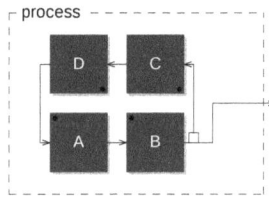

Figure 13. Bidirectional construction in FAUST using the tilde diagram composition operation.

In this case, the D/A and the C/B couples can be seen as bidirectional blocks/functions that could implement some musical instrument part. However, the FAUST semantics doesn't allow them to be specified as such from the code, preventing the implementation of "bidirectional functions." Since this feature is required to create a library of physical modeling elements, we had to implement it.

Bidirectional blocks in the FAUST Physical Modeling Library all have three inputs and outputs. Thus, an empty block can be expressed as:

```
emptyBlock = _,_,_;
```

The first input and output correspond to left-going waves (e.g., C and D in Figure 13), the second input and output to right-going waves (e.g., A and B in Figure 13), and the third input and output can be used to carry any signal to the end of the algorithm. As we'll see in Section 5.2.2, this can be useful when picking up the sound at the middle of a virtual string, for example.

Bidirectional blocks are connected to each other using the chain primitive which is part of physmodels.lib. For example, an open waveguide (no terminations) expressed as:

```
waveguide(nMax,n) = par(i,2,de.fdelay4(nMax,n)),_;
```

where nMax is the maximum length of the waveguide and n its current length, could be connected to our emptyBlock:

```
foo = chain(emptyBlock : waveguide(256,n) : emptyBlock) ;
```

Note the use of `fdelay4` in `waveguide`, which is a fourth order fractional delay line [60].

The FAUST compiler is not able yet to generate the block diagram corresponding to the previous expression in an organized bidirectional way (see Section 5.3). However, a "hand-made" diagram can be seen in Figure 14.

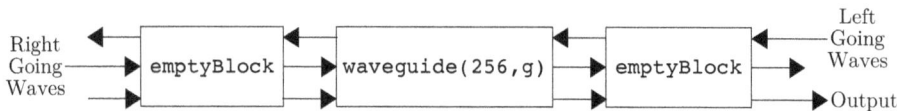

Figure 14. Bidirectional construction in FAUST using the `chain` primitive.

The placement of elements in a `chain` matters and corresponds to their order in the physical world. For example, for a set of hypothetical functions implementing the different parts of a violin, we could write:

```
violin = chain(nuts : string : bridge : body);
```

The main limitation of this system is that it introduces a one sample delay in both directions for each block in the `chain` due to the internal use of ~ [49]. This has to be taken into account when implementing certain types of elements such as a string or a tube.

Terminations can be added on both sides of a chain using `lTermination(A,B)` for a left-side termination and `rTerminations(B,C)` for a right-side termination where B can be any bidirectional block, including a `chain`, and A and C are functions that can be put between left and right-going signals (see Figure 15).

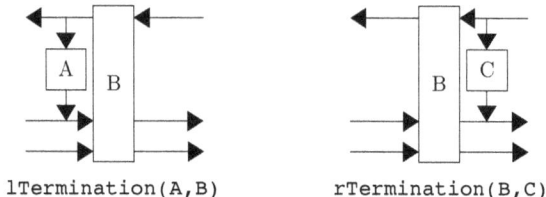

Figure 15. `lTermination(A,B)` and `rTermination(B,C)` in the FAUST Physical Modeling Library.

A signal x can be fed anywhere in a `chain` by using the `in(x)` primitive. Similarly, left and right-going waves can be summed and extracted from a chain using the `out` primitive (see Code Listing 4).

Finally, a chain of blocks A can be "terminated" using `endChain(A)` which essentially removes the three inputs and the first two outputs of A.

Assembling a simple waveguide string model with "ideal" rigid terminations is simple using this framework:

```
string(length,pluckPosition,excitation) = endChain(wg)
with{
  maxStringLength = 3; // in meters
  lengthTuning = 0.08; // adjusted "by hand"
  tunedLength = length-lengthTuning;
  nUp = tunedLength*pluckPosition; // upper string segment length
  nDown = tunedLength*(1-pluckPosition); // lower string segment length
  lTerm = lTermination(*(-1),basicBlock); // phase inversion
  rTerm = rTermination(basicBlock,*(-1)); // phase inversion
```

```
stringSegment(maxLength,length) = waveguide(nMax,n)
with{
   nMax = maxLength : 12s; // meters to samples
   n = length : 12s/2; // meters to samples
};
wg = chain(lTerm : stringSegment(maxStringLength,nUp) :
   in(excitation) : out : stringSegment(maxStringLength,nDown) :
   rTerm); // waveguide chain
};
```

Listing 4: "Ideal" string model with rigid terminations.

In this case, since in and out are placed next to each other in the chain, the position of excitation and the position of the pickup are the same as well.

5.2.2. Assembling High Level Parts: Violin Example

FPML contains a wide range of ready-to-use instrument parts and pre-assembled models. An overview of the content of the library is provided in the FAUST libraries documentation [60]. Detailing the implementation of each function of the library would be interesting, however this section focuses on one of its models: violinModel (see Code Listing 5) which implements a simple bowed string connected to a body through a bridge.

```
violinModel(stringLength,bowPressure,bowVelocity,bowPosition) =
endChain(modelChain)
with{
   stringTuning = 0.08;
      stringL = stringLength-stringTuning;
      modelChain = chain(
         violinNuts :
         violinBowedString(stringL,bowPressure,bowVelocity,bowPosition) :
         violinBridge : violinBody : out
      );
};
```

Listing 5: violinModel: a simple violin physical model from the FAUST Physical Modeling Library.

violinModel assembles various high-level functions implementing violin parts. violinNuts is a termination applying a light low-pass filter on the reflected signal. violinBowedString is made out of two open string segments allowing us to choose the bowing position. The bow nonlinearity is implemented using a table. violinBridge implements the "right termination" as well as the reflectance and the transmittance filters [27]. Finally, violinBody is a simple violin body modal model.

In addition to its various models and parts, the FPML also implements a series of ready-to-use models hosting their own user interface. The corresponding functions end with the _ui suffix. For example:

```
process = pm.violin_ui;
```

is a complete FAUST program adding a simple user interface to control the violin model presented in Code Listing 5.

While [...]_ui functions associate continuous UI elements (e.g., knobs, sliders, etc.) to the parameters of a model, functions ending with the _ui_midi prefix automatically format the parameters linked the FAUST MIDI parameters (i.e., frequency, gain, and note-on/off) using envelope generators. Thus, such functions are ready to be controlled by a MIDI keyboard.

Nonlinear behaviors play an important role in some instruments (e.g., gongs, cymbals, etc.). While waveguide models and modal synthesis are naturally linear, nonlinearities can be introduced using nonlinear allpass ladder filters [61]. `allpassNL` implements such a filter in the FAUST Physical Modeling Library.

Some of the physical models of the FAUST-STK [62] were ported to FPML and are available through various functions in the library.

5.2.3. Example: Marimba Physical Model Using FPML and MESH2FAUST

This section briefly demonstrates how a simple marimba physical model can be made using MESH2FAUST and FPML (An extended version of this example with more technical details is also available in the corresponding on-line tutorial: https://ccrma.stanford.edu/~rmichon/faustTutorials/ #making-custom-elements-using-mesh2faust). The idea is to use a 3D CAD model of a marimba bar, generate the corresponding modal model, and then connect it to a tube model implemented in FPML.

A simple marimba bar 3D model can be made by extruding a marimba bar cross section using the Inkscape to OpenSCAD tool part of MESH2FAUST [56]. The resulting CAD model is then turned into a volumetric mesh by importing it to MeshLab and by uniformly re-sampling it to have approximately 4500 vertices. The mesh produced during this step (`marimbaBar.obj` in the following code listing) can then be processed by MESH2FAUST using the following command (A complete listing of MESH2FAUST's options can be found in its on-line documentation: https://github.com/grame-cncm/faust/blob/master-dev/tools/physicalModeling/mesh2faust/README.md.):

```
mesh2faust --infile marimbaBar.obj --nsynthmodes 50 --nfemmodes 200
  --maxmode 15000 --expos 2831 3208 3624 3975 4403 --freqcontrol
  --material 1.3E9 0.33 720 --name marimbaBarModel
```

The material parameters are those of rosewood which is traditionally used to make marimba bars. The number of modes is limited to 50 and various excitation positions were selected to be uniformly spaced across the horizontal axis of the bar. `frequency control mode` is activated to be able to transpose the modes of the generated model in function of the fundamental frequency making the model more generic.

A simple marimba resonator was assembled using FPML and is presented in Code Listing 6. It is made out of an open tube where two simple lowpass filters placed at its extremities are used to model the wave reflections. The model is excited on one side of the tube and sound is picked-up on the other side.

```
marimbaResTube(tubeLength,excitation) = endChain(tubeChain)
with{
  lengthTuning = 0.04; tunedLength = tubeLength-lengthTuning;
  absorption = 0.99; lowpassPole = 0.95;
  endTubeReflexion = si.smooth(lowpassPole)*absorption;
  tubeChain = chain(
    in(excitation) : terminations(endTubeReflexion,
    openTube(maxLength,tunedLength),
    endTubeReflexion) : out
  );
};
```

Listing 6: Simple marimba resonator tube implemented with FPML.

Code Listing 7 demonstrates how the marimba bar model generated with MESH2FAUST (`marimbaBarModel`) can be simply connected to the marimba resonator. A unidirectional connection can be used in this case since waves are only transmitted from the bar to the resonator.

```
marimbaModel(freq,exPos) =
  marimbaBarModel(freq,exPos,maxT60,T60Decay,T60Slope) :
  marimbaResTube(resTubeLength)
with{
  resTubeLength = freq : f21;
  maxT60 = 0.1; T60Decay = 1; T60Slope = 5;
};
```

Listing 7: Simple marimba physical model.

This model is now part of the FAUST Physical Modeling Library. It could be easily used with `faust2smartkeyb` to implement a marimba app (see Section 2) as well as with any of the FAUST targets (e.g., Web App, Plug-In, etc.). More examples of models created using this technique can be found on-line (Faust Physical Modeling Toolkit Webpage: https://ccrma.stanford.edu/~rmichon/pmFaust/).

5.3. Discussion and Future Directions

The framework presented in this section remains limited by several factors. Audio latency induced by ADCs and DACs prevents in some cases the implementation of cohesive bidirectional chains between physical and virtual elements. Audio latency reduction has been an ongoing research topic for many years and more work has to be done in this direction. This problem is exacerbated by the use of mobile devices at the heart of these systems that are far from being specialized for this specific type of application (i.e., operating system optimizations inducing extra latency, number of analog inputs and outputs, etc.). On the other hand, we believe that despite the compromises that they entail, mobile devices remain a versatile, and yet easy to customize platform well suited to implement hybrid instruments (e.g., the BLADEAXE [42]).

The FAUST Physical Modeling Library is far from being exhaustive and many models and instruments could be added to it. We believe that MESH2FAUST will help enlarge the set of functions available in this system.

The framework presented in Section 5.2.1 allows us to assemble the different parts of instrument models in a simple way by introducing a bidirectional block diagram algebra to FAUST. While it provides a high level approach to physical modeling, FAUST is not able to generate the corresponding block diagram in a structured way. This would be a nice feature to add.

Similarly, we would like to extend the idea of being able to make multidimensional block diagrams in FAUST by adding new primitives to the language.

More generally, we hope to make more instruments using this framework and use them on stage for live performance.

6. Conclusions

By combining physical and virtual elements, hybrid instruments are "physically coherent" by nature and allow instrument designers to play to the strengths of both acoustical and digital elements. Current technologies and techniques allow us to blur the boundary between the physical and the virtual world enabling musical instrument designers to treat instrument parts in a multidimensional way. The FAUST Physical Modeling Toolkit presented in Section 5 facilitates the design of such instruments by providing a way to approach physical modeling of musical instruments at a very high level.

Mobile devices combined with physical passive or active augmentations are well suited to implement hybrid instruments. Their built-in sensors, standalone aspect, and computational capabilities are the core elements required to implement the virtual portion of hybrid instruments. `faust2smartkeyb` facilitates the design of mobile apps using elements from the FAUST Physical Modeling Library, implementing skill transfer, and serving as the glue between the various parts of the instrument. Mobile devices might limit the scope of hybrid instruments by scaling down the number of connections between acoustical and digital elements because of technical limitations. However, we demonstrated that a wide range of instruments can still be implemented using this type of system.

The framework presented in this paper is a toolkit for musical instrument designers. By facilitating skill transfer, it can help accelerate the learning process of instruments made with it. However, musical instruments remain a tool controlled by the performer. Having a well designed instrument leveraging some of the concepts presented here doesn't mean that it will systematically play beautiful music and generate pleasing sounds: this is mostly up to the performer.

We believe that mobile hybrid instruments presented in this paper help reconcile the haptic, the physical, and the virtual, partially solving some of the flaws of DMIs depicted by Perry Cook [5].

We recently finished releasing the various elements of the framework presented in this paper and we hope to see the development of more mobile-device-based hybrid instruments in the future.

Author Contributions: Romain Michon is the main author of this article and the primary developer of the various tools that it presents. Julius Orion Smith helped with the development of the FAUST Physical Modeling Toolkit. He also provided critical feedback for the different steps of this project along with Matthew Wright and Chris Chafe. John Granzow is the author of some of the functions of MOBILE3D and co-taught the various workshops presented in this article with Romain Michon. Finally, Ge Wang participated in the development of `faust2smartkeyb`.

Conflicts of Interest: The authors declare no conflict of interest.

Abbreviations

The following abbreviations are used in this manuscript:

DMI	Digital Musical Instruments
HCI	Human Computer Interaction
NIME	New Interfaces for Musical Expression
PLA	PolyLactic Acid
DIY	Do It Yourself
IDE	Integrated Development Environment
FPML	FAUST Physical Modeling Library

References

1. Battier, M. Les Musiques électroacoustiques et l'environnement informatique. Ph.D. Thesis, University of Paris X, Nanterre, France, 1981.
2. Magnusson, T. Of Epistemic Tools: musical instruments as cognitive extensions. *Organised Sound* **2009**, *14*, 168–176.
3. Jordà, S. Digital Lutherie Crafting Musical Computers for New Musics' Performance and Improvisation. Ph.D. Thesis, Universitat Pompeu Fabra, Barcelona, Spain, 2005.
4. Wanderley, M.M.; Depalle, P. Gestural Control of Sound Synthesis. *Proc. IEEE* **2004**, *92*, 632–644.
5. Cook, P. Remutualizing the Instrument: Co-Design of Synthesis Algorithms and Controllers. In Proceedings of the Stockholm Music Acoustics Conference (SMAC-03), Stockholm, Sweden, 6–9 August 2003.
6. Leonard, J.; Cadoz, C. Physical Modelling Concepts for a Collection of Multisensory Virtual Musical Instruments. In Proceedings of the Conference on New Interfaces for Musical (NIME15), Baton Rouge, LA, USA, 31 May–3 June 2015.
7. Lähdeoja, O. An Approach to Instrument Augmentation: The Electric Guitar. In Proceedings of the 2008 Conference on New Interfaces for Musical Expression (NIME-08), Genova, Italy, 5–7 June 2008.
8. Bevilacqua, F.; Rasamimanana, N.; Fléty, E.; Lemouton, S.; Baschet, F. The augmented violin project: Research, composition and performance report. In Proceedings of the International Conference on New Interfaces for Musical Expression, Paris, France, 4–8 June 2006.
9. Young, D. The Hyperbow Controller: Real-Time Dynamics Measurement of Violin Performance. In Proceedings of the 2002 Conference on New Instruments for Musical Expression (NIME-02), Dublin, Ireland, 24–26 May 2002.

10. Overholt, D. The Overtone Violin: A New Computer Music Instrument. In Proceedings of the 2005 International Computer Music Conference (ICMC-05), Barcelona, Spain, 4–10 September 2005.
11. Impett, J. A Meta-Trumpet(er). In Proccedings of the International Computer Music Conference (ICMC-94), Aarhus, Denmark, 12–17 September 1994.
12. Burtner, M. The Metasaxophone: Concept, Implementation, and Mapping Strategies for a New Computer Music Instrument. *Organised Sound* **2002**, *7*, 201–213.
13. Aimi, R.M. Hybrid Percussion: Extending Physical Instruments Using Sampled Acoustics. Ph.D. Thesis, Massachusetts Institute of Technology, Cambridge, MA, USA, 2007.
14. Puckette, M. Playing a Virtual Drum from a Real One. *J. Acoust. Soc. Am.* **2011**, *130*, 2432.
15. Momeni, A. Caress: An Enactive Electro-acoustic Percussive Instrument for Caressing Sound. In Proceedings of the International Conference on New Interfaces for Musical Expression (NIME-15), Baton Rouge, LA, USA, 31 May–3 June 2015.
16. Schlessinger, D.; Smith, J.O. The Kalichord: A Physically Modeled Electro-Acoustic Plucked String Instrument. In Proceedings of the 9th International Conference on New Interfaces for Musical Expression (NIME-09), Pittsburgh, PA, USA, 4–6 June 2009.
17. Berdahl, E.; Smith, J.O. A Tangible Virtual Vibrating String. In Proceedings of the Eighth International Conference on New Interfaces for Musical Expression (NIME-08), Genova, Italy, 5–7 June 2008.
18. Tanaka, A. Mobile Music Making. In Proceedings of the 2004 Conference on New Interfaces for Musical Expression (NIME-04), Hamamatsu, Japan, 3–5 June 2004.
19. Geiger, G. Using the Touch Screen as a Controller for Portable Computer Music Instruments. In Proceedings of the 2006 International Conference on New Interfaces for Musical Expression (NIME-06), Paris, France, 4–8 June 2006.
20. Gaye, L.; Holmquist, L.E.; Behrendt, F.; Tanaka, A. Mobile Music Technology: Report on an Emerging Community. In Proceedings of the International Conference on New Interfaces for Musical Expression (NIME-06), Paris, France, 4–8 June 2006.
21. Essl, G.; Rohs, M. Interactivity for Mobile Music-Making. *Organised Sound* **2009**, *14*, 197–207.
22. Wang, G. Ocarina: Designing the iPhone's Magic Flute. *Comput. Music J.* **2014**, *38*, 8–21.
23. Wang, G.; Essl, G.; Penttinen, H. Do Mobile Phones Dream of Electric Orchestra? In Proceedings of the International Computer Music Conference (ICMC-08), Belfast, Ireland, 24–29 August 2008.
24. Smith, J.O. Physical Modeling Using Digital Waveguides. *Comput. Music J.* **1992**, *16*, 74–91.
25. Karjalainen, M.; Välimäki, V.; Tolonen, T. Plucked-String Models: From the Karplus-Strong Algorithm to Digital Waveguides and Beyond. *Comput. Music J.* **1998**, *22*, 17–32.
26. Välimäki, V.; Takala, T. Virtual Musical Instruments—Natural Sound Using Physical Models. *Organised Sound* **1996**, *1*, 75–86.
27. Smith, J.O. *Physical Audio Signal Processing for Virtual Musical Instruments and Digital Audio Effects*; W3K Publishing: Palo Alto, CA, USA, 2010. Available online: https://ccrma.stanford.edu/~jos/pasp/ (accessed on 16 December 2017).
28. Adrien, J.M. The Missing Link: Modal Synthesis. In *Representations of Musical Signals*; MIT Press: Cambridge, MA, USA, 1991; pp. 269–298.
29. Karjalainen, M.; Smith, J.O. Body Modeling Techniques for String Instrument Synthesis. In Proceedings of the International Computer Music Conference (ICMC-96), Hong Kong, China, 19–24 August 1996.
30. Bruyns, C. Modal Synthesis for Arbitrarily Shaped Objects. *Comput. Music J.* **2006**, *30*, 22–37.
31. Bilbao, S. *Numerical Sound Synthesis: Finite Difference Schemes and Simulation in Musical Acoustics*; John Wiley and Sons: Chichester, UK, 2009.
32. Lipson, H.; Kurman, M. *Fabricated: The New World of 3D Printing*; Wiley: Indianapolis, IN, USA, 2013.
33. Granzow, J. Additive Manufacturing for Musical Applications. Ph.D. Thesis, Stanford University, Stanford, CA, USA, 2017.
34. Orrù, F. Francesco Orrù's Portfolio. Available online: https://www.myminifactory.com/users/4theswarm (accessed on 16 December 2017).
35. Diegel, O. Odd Guitars Website. Available online: http://www.oddguitars.com/ (accessed on 16 December 2017).
36. Bernadac, L. 3D Varius Website. Available online: http://www.3d-varius.com/ (accessed on 16 December 2017).
37. Summit, S. System and Method for Designing and Fabricating String Instruments. U.S. Patent 20140100825 A1, 2014.

38. Hovalabs. Hovalin Website. Available online: http://www.hovalabs.com/hova-instruments/hovalin (accessed on 16 December 2017).

39. Zoran, A. The 3D Printed Flute: Digital Fabrication and Design of Musical Instruments. *J. New Music Res.* **2011**, *40*, 379–387.

40. Bailey, N.J.; Cremel, T.; South, A. Using Acoustic Modelling to Design and Print a Microtonal Clarinet. In Proceedings of the 9th Conference on Interdisciplinary Musicology (CIM14), Berlin, Germany, 4–6 December 2014.

41. Dabin, M.; Narushima, T.; Beirne, S.T.; Ritz, C.H.; Grady, K. 3D Modelling and Printing of Microtonal Flutes. In Proceedings of the 16th International Conference on New Interfaces for Musical Expression (NIME-16), Brisbane, Australia, 11–15 July 2016.

42. Michon, R.; Smith, J.O.; Wright, M.; Chafe, C. Augmenting the iPad: the BladeAxe. In Proceedings of the International Conference on New Interfaces for Musical Expression (NIME-16), Brisbane, Australia, 11–15 July 2016.

43. Orlarey, Y.; Letz, S.; Fober, D. Faust: An Efficient Functional Approach to DSP Programming. In *New Computational Paradigms for Computer Music*; Delatour: Paris, France, 2009.

44. Brinkmann, P.; Kirn, P.; Lawler, R.; McCormick, C.; Roth, M.; Steiner, H.C. Embedding PureData with libpd. In Proceedings of the Pure Data Convention, Weimar, Germany, 8–11 August 2011.

45. Lazzarini, V.; Yi, S.; Timoney, J.; Keller, D.; Pimenta, M. The Mobile Csound platform. In Proceedings of the International Conference on Computer Music (ICMC-12), Ljubljana, Slovenia, 9–14 September 2012.

46. Michon, R. faust2android: A Faust Architecture for Android. In Proceedings of the 16th International Conference on Digital Audio Effects (DAFx-13), Maynooth, Ireland, 2–6 September 2013.

47. Michon, R.; Smith, J.; Chafe, C.; Letz, S.; Orlarey, Y. faust2api: A comprehensive API generator for Android and iOS. In Proceedings of the Linux Audio Conference (LAC-17), Saint-Etienne, France, 18–21 May 2017.

48. Michon, R.; Smith, J.O.; Orlarey, Y. MobileFaust: A Set of Tools to Make Musical Mobile Applications with the Faust Programming Language. In Proceedings of the Linux Audio Conference (LAC-15), Mainz, Germany, 18–21 May 2015.

49. GRAME. *FAUST Quick Reference*; Centre National de Création Musicale: Lyon, France, 2017.

50. Faust2smartkeyb documentation. Available online: https://ccrma.stanford.edu/~rmichon/smartKeyboard/ (accessed on 16 December 2017).

51. Perrotin, O.; d'Alessandro, C. Adaptive Mapping for Improved Pitch Accuracy on touch User Interfaces. In Proceedings of the International Conference on New Interfaces for Musical Expression, Seoul, Korea, 27–30 May 2013.

52. Misra, A.; Essl, G.; Rohs, M. Microphone as Sensor in Mobile Phone Performance. In Proceedings of the New Interfaces for Musical Expression conference (NIME-08), Genova, Italy, 5–7 June 2008.

53. Laput, G.; Brockmeyer, E.; Hudson, S.; Harrison, C. Acoustruments: Passive, Acoustically-Driven, Interactive Controls for Handheld Devices. In Proceedings of the Conference for Human-Computer Interaction (CHI), Seoul, Republic of Korea, 18–23 April 2015.

54. Fletcher, N.H.; Rossing, T.D. *The Physics of Musical Instruments*, 2nd ed.; Springer Verlag: Berlin, Germany, 1998.

55. Michon, R.; Smith, J.O.; Chafe, C.; Wright, M.; Wang, G. Nuance: Adding Multi-Touch Force Detection to the iPad. In Proceedings of the Sound and Music Computing Conference (SMC-16), Hamburg, Germany, 31 August–3 September 2016.

56. Michon, R.; Martin, S.R.; Smith, J.O. Mesh2Faust: A Modal Physical Model Generator for the Faust Programming Language—Application to Bell Modeling. In Proceedings of the International Computer Music Conference (ICMC-17), Beijing, China, 15–20 October 2017.

57. Grumiaux, P.A.; Michon, R.; Arias, E.G.; Jouvelot, P. Impulse-Response and CAD-Model-Based Physical Modeling in Faust. In Proceedings of the Linux Audio Conference (LAC-17), Saint-Etienne, France, 18–21 May 2017.

58. Fuller, C.; Elliott, S.; Nelson, P. *Active Control of Vibration*; Academic Press: Cambridge, MA, USA, 1996.

59. Meurisse, T.; Mamou-Mani, A.; Benacchio, S.; Chomette, B.; Finel, V.; Sharp, D.; Caussé, R. Experimental Demonstration of the Modification of the Resonances of a Simplified Self-Sustained Wind Instrument Through Modal Active Control. *Acta Acust. United Acust.* **2015**, *101*, 581–593.

60. Faust Libraries Documentation. Available online: http://faust.grame.fr/library.html (accessed on 16 December 2017).
61. Smith, J.O.; Michon, R. Nonlinear Allpass Ladder Filters in Faust. In Proceedings of the 14th International Conference on Digital Audio Effects, Paris, France, 19–23 September 2011.
62. Michon, R.; Smith, J.O. Faust-STK: A set of linear and nonlinear physical models for the Faust programming language. In Proceedings of the 14th International Conference on Digital Audio Effects (DAFx-11), Paris, France, 19–23 September 2011.

Article

A Neural Parametric Singing Synthesizer Modeling Timbre and Expression from Natural Songs [†]

Merlijn Blaauw [*,‡] and **Jordi Bonada** [‡]

Music Technology Group, Universitat Pompeu Fabra, 08012 Barcelona, Spain; jordi.bonada@upf.edu

* Correspondence: merlijn.blaauw@upf.edu; Tel.: +34-93-542-2199

† This paper is an extended version of our paper published in Blaauw, M.; Bonada, J. A neural parametric singing synthesizer. In Proceedings of the 18th Annual Conference of the International Speech Communication Association (Interspeech), Stockholm, Sweden, 20–24 August 2017.

‡ These authors contributed equally to this work.

Academic Editor: Vesa Valimaki

Received: 3 November 2017; Accepted: 12 December 2017; Published: 18 December 2017

Abstract: We recently presented a new model for singing synthesis based on a modified version of the WaveNet architecture. Instead of modeling raw waveform, we model features produced by a parametric vocoder that separates the influence of pitch and timbre. This allows conveniently modifying pitch to match any target melody, facilitates training on more modest dataset sizes, and significantly reduces training and generation times. Nonetheless, compared to modeling waveform directly, ways of effectively handling higher-dimensional outputs, multiple feature streams and regularization become more important with our approach. In this work, we extend our proposed system to include additional components for predicting F0 and phonetic timings from a musical score with lyrics. These expression-related features are learned together with timbral features from a single set of natural songs. We compare our method to existing statistical parametric, concatenative, and neural network-based approaches using quantitative metrics as well as listening tests.

Keywords: singing synthesis; machine learning; deep learning; conditional generative models; autoregressive models

---

1. Introduction

Many of today's more successful singing synthesizers are based on concatenative methods [1,2]. That is, they transform and concatenate short waveform units selected from an inventory of recordings of a singer. While such systems are able to achieve good sound quality and naturalness in certain settings, they tend to be limited in terms of flexibility, and can be difficult to extend or significantly improve upon. One notable limitation is that jointly sampling musical and phonetic contexts usually is not feasible, forcing timbre and expression to be modeled disjointly, from separate, specialized corpora. Machine learning approaches, such as statistical parametric methods [3,4], are much less rigid and do allow for things such as combining data from multiple speakers, model adaptation using small amounts of training data, and joint modeling of timbre and expression from a single corpus of natural songs. However, until recently, these approaches have been unable to match the sound quality of concatenative methods, in particular suffering from oversmoothing in frequency and time.

Recent advances in generative models for Text-to-Speech Synthesis (TTS) using Deep Neural Networks (DNNs), in particular the WaveNet model [5], showed that model-based approaches can achieve sound quality on-par or even beyond that of concatenative systems. This model's ability to accurately generate raw speech waveform sample-by-sample, clearly shows that oversmoothing is not an issue. Recently, we presented a model for singing synthesis based on the WaveNet model [6], with an important difference being that we model vocoder features rather than raw waveform. While a vocoder unavoidably introduces some degradation in sound quality, we consider the degradation introduced by current models to still be the dominant factor. Thus, if we can improve the quality of the generative model, we should be able to achieve a quality closer to the upper bound the vocoder can provide, i.e., round-trip vocoder analysis-synthesis without modification. Additionally, by decomposing the signal into phonetic and pitch components, we are able to conveniently synthesize any melody with any lyrics, and require less training data to sufficiently cover the entire pitch-timbre space.

Our previously presented system only generated timbral features, and did not generate features related to musical expression, such as F0 and phonetic timings. Additionally, the corpora used to train the models were specialized recordings similar to those used for building concatenative voices. In this work, we extend our previously presented system to also include F0 and phonetic timing prediction, and train the entire system from a single corpus of natural singing. We feel that this is an important step forward towards capturing all aspects of a singer's voice in a natural setting. Finally, we provide detailed quantitative and qualitative experiments and results.

2. Proposed System

2.1. Overview

The task of singing synthesis mimics the task of a singer during a studio recording, that is, interpret a musical score with lyrics to produce a singing waveform signal. The goal of our system is to model a specific singer's voice and a specific style of singing. To achieve this, we first record a singer singing a set of musical scores. From these recording, acoustic features are extracted using the analysis part of a vocoder. Additionally, the recordings are phonetically transcribed and segmented. Note level transcription and segmentation can be generally obtained from the musical scores, as long as the singer did not excessively deviate from the written score.

During training, our model learns to produce acoustic features given phonetic and musical input sequences, including the begin and end time of each segment. However, during generation, we only have access to the note begin and end times, and phoneme sequence corresponding to each note (generally a syllable). As we do not have access to the begin and end times of each phoneme, these must be predicted using a phonetic timing model. The next step is to predict F0 from the timed musical and phonetic information, using a pitch model. The predicted phonetic timings and F0 are then used by the timbre model to generate the remaining acoustic features such as the harmonic spectral envelope, aperiodicity envelope and voiced/unvoiced (V/UV) decision. Finally, the synthesis part of the vocoder is used to generate the waveform signal from the acoustic features. An overview of the entire system is depicted in Figure 1.

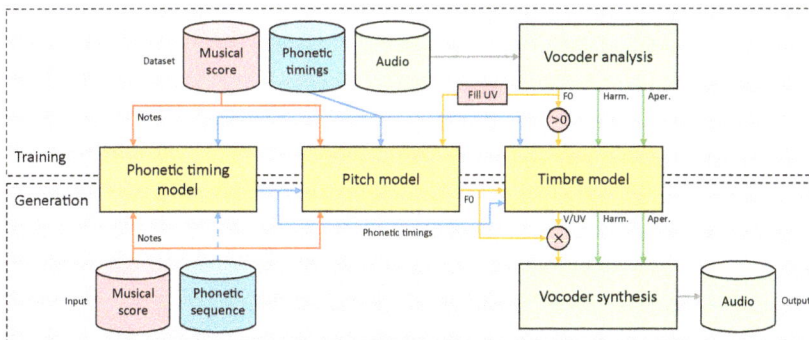

Figure 1. Diagram depicting an overview of the system with its different components. Here, V/UV is the predicted voice/unvoiced decision, and the Fill UV block fills unvoiced isegments by interpolation.

2.2. Modified WaveNet Architecture

The main building block of our system is based on the WaveNet model and architecture. A key aspect of this model is that it is autoregressive. That is, the prediction at each timestep depends on (a window of) predictions of past timesteps. In our case, a timestep corresponds to a single frame of acoustic features. Additionally, the model is probabilistic, meaning that the prediction is a probability distribution rather than a single value. In order to control the prediction, e.g., by phonetic and musical inputs, the predicted distribution is not only conditioned on past predictions, but also on control inputs. This model is implemented using a powerful, yet efficient neural network architecture.

The network we propose, depicted in Figure 2, shares most of its architecture with WaveNet. Like this model we use gated convolutional units instead of gated recurrent units, such as Long Short-Term Memory (LSTM) units, to speed up training. The input is fed through an initial causal convolution which is then followed by stacks of 2×1 dilated convolutions [7] where the dilation factor is doubled for each layer. This allows exponentially growing the model's receptive field, while linearly increasing the number of required parameters. To increase the total nonlinearity of the model without excessively growing its receptive field, the dilation factor is increased up to a limit and then the sequence is repeated. We use residual and skip connections to facilitate training deeper networks [8]. As we wish to control the synthesizer by inputting notes and lyrics, we use a conditional version of the model. At every layer, before the gated nonlinearity, feature maps derived from control inputs are summed to the feature maps from the layer's main convolution. In our case, we do the same thing at the output stack, similar to [9].

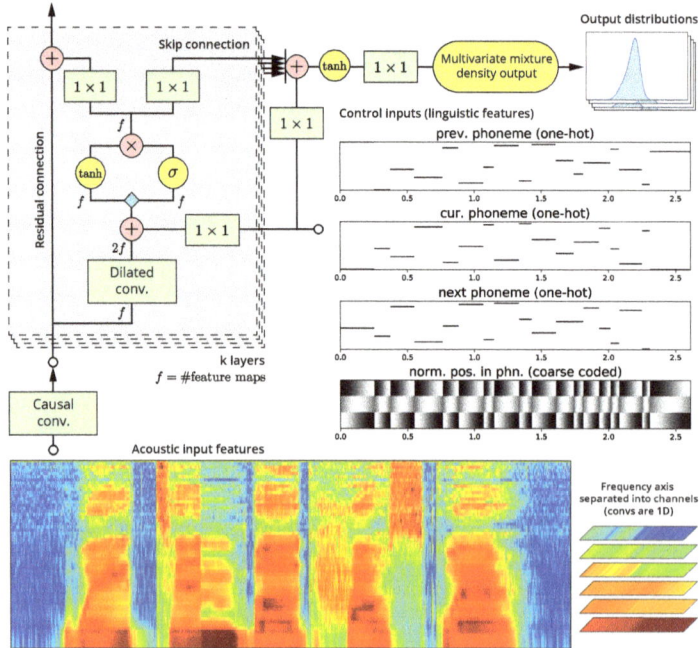

Figure 2. Overview of the modified WaveNet network architecture. In this case, the network depicted predicts harmonic spectral envelope features (top-right and bottom), given control inputs (mid-right).

The underlying idea of this model is that joint probability over all timesteps can be formulated as a product of conditional probabilities for a single timestep with some causal ordering. The conditional probability distributions are predicted by a neural network trained to maximize likelihood of a observation given past observations. To synthesize, predictions are made by sampling the predicted distribution conditioned on past predictions, that is, in a sequential, autoregressive manner. However, while models on which we base our model like WaveNet, or PixelCNN [10] and PixelRNN [11] before it, perform this factorization for univariate variables (e.g., individual waveform samples or pixel channels), we do so for multivariate vectors corresponding to a single frame,

$$p(\mathbf{x}_1, \dots, \mathbf{x}_T \mid \mathbf{c}) = \prod_{t=1}^{T} p(\mathbf{x}_t \mid \mathbf{x}_{<t}, \mathbf{c}), \tag{1}$$

where $\mathbf{x}_t$ is an N-dimensional vector of acoustic features $[x_{t,1}, \dots, x_{t,N}]$, $\mathbf{c}$ is an T-by-M-dimensional matrix of control inputs, and T is the length of the signal. In our case, we consider the variables within a frame to be conditionally independent,

$$p(\mathbf{x}_t \mid \mathbf{x}_{<t}, \mathbf{c}) = \prod_{i=1}^{N} p(x_{t,i} \mid \mathbf{x}_{<t}, \mathbf{c}). \tag{2}$$

In other words, a single neural network predicts the parameters of a multivariate conditional distribution with diagonal covariance, corresponding to the acoustic features of a single frame.

The main reason for choosing this model is that, unlike raw audio waveform, features produced by a parametric vocoder have two dimensions, similar to (single channel) images. However, unlike images, these two dimensions are not both spatial dimensions, but rather time-frequency dimensions. The translation invariance that 2D convolutions offer is an undesirable property for the frequency (or cepstral quefrency) dimension. Therefore, we model the features as 1D data with multiple channels. Note that these channels are only independent within the current frame; the prediction of each of the features in the current frame still depends on all of the features of all past frames within the receptive field (the range of input samples that affect a single output sample). This can be explained easily as all input channels of the initial causal convolution contribute to all resulting feature maps, and so on for the other convolutions.

Predicting all channels at once rather than one-by-one simplifies the models, as it avoids the need for masking channels and separating them in groups. This approach is similar to [12], where all three RGB channels of a pixel in an image are predicted at once, although in our work we do not incorporate additional linear dependencies between channel means.

2.2.1. Constrained Mixture Density Output

Many of the architectures on which we base our model predict categorical distributions, using a softmax output. The advantage of this nonparametric approach is that no a priori assumptions have to be made about the (conditional) distribution of the data, allowing things such as skewed or truncated distributions, multiple modes, and so on. Drawbacks of this approach include an increase in model parameters, values are no longer ordinal, and the need to discretize data which is not naturally discrete or has high bitdepth.

Because our model predicts an entire frame at once, the issue of increased parameter count is aggravated. Instead, we opted to use a mixture density output similar to [12]. This decision was partially motivated because in earlier versions of our model with softmax output [13], we noted the predicted distributions were generally quite close to Gaussian or skewed Gaussian. In our model we use a mixture of four continuous Gaussian components, constrained in such a way that there are only four free parameters (location, scale, skewness and a shape parameter). Figure 3 shows some of the typical distributions that the contraints imposed by this parameter mapping allow. We found such constraints to be useful to avoid certain pathological distributions, and in our case explicitly not allowing multimodal distributions was helpful to improve results. We also found this approach speeds up convergence compared to using categorical output. See Appendix A for details.

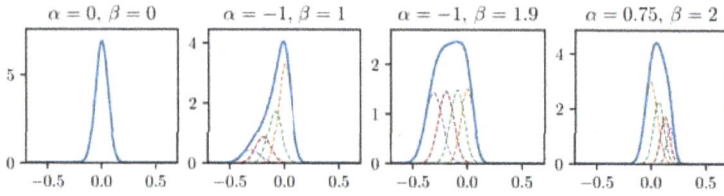

Figure 3. Example distributions of the constrained mixture density output. All subplots use location $\xi = 0$ and scale $\omega = 6 \times 10^{-2}$, but varying skewness α and shape β. The plots show the resulting mixture distributions (solid) and the four underlying Gaussian components (dashed).

2.2.2. Regularization

While the generation process is autoregressive, during training rather than using past predictions, groundtruth past samples are used. This is a practical necessity as it allows the computations to be parallelized. However this also causes a number of issues. One issue, known as exposure bias [14], results in the model becoming biased to the groundtruth data it is exposed to during training, and causing errors to accumulate at each autoregressive generation step based on its own past predictions. In our case, such errors cause a degradation in synthesis quality, e.g., unnatural timbre shifts over time. Another notable issue is that as the model's predictions are conditioned on both past timesteps and control inputs, the network may mostly only pay attention to past timesteps and ignore the control inputs [15]. In our case, this can result in the model occasionally changing certain lyrics rather than follow those dictated by its control inputs.

One way to reduce the exposure bias issue may be to increase the dataset size, so that the model is exposed to a wider range of data. However, we argue that the second problem is mostly a result of the inherent nature of the data modeled. Unlike raw waveform, vocoder features are relatively smooth over time, more so for singing where there are many sustained vowels. This means that, usually, the model will be able to make accurate predictions given the highly correlated past timesteps.

As a way around both these issues, we propose using a denoising objective to regularize the network,

$$\mathcal{L} = -\log p\left(x_t \mid \tilde{x}_{<t}, c\right) \quad \text{with} \quad \tilde{x}_{<t} \sim p\left(\tilde{x}_{<t} \mid x_{<t}\right), \tag{3}$$

where $p\left(\tilde{x} \mid x\right)$ is a Gaussian corruption distribution,

$$p\left(\tilde{x} \mid x\right) = \mathcal{N}\left(\tilde{x}; x, \lambda I\right), \tag{4}$$

with noise level $\lambda \geq 0$. That is, Gaussian noise is added to the input of the network, while the network is trained to predict the uncorrupted target.

When sufficiently large values of λ are used, this technique is very effective for solving the problems noted above. However, the generated output can also become noticeably more noisy. One way to reduce this undesirable side effect is to apply some post processing to the predicted output distribution, much in the same vein as the temperature softmax used in similar models (e.g., [9]).

We have also tried other regularization techniques, such as dropout, but found them to be ultimately inferior to simply injecting input noise.

2.3. Timbre Model

This model is responsible for generating acoustic features related to the timbre of the voice. It consists of a multistream variant of the modified WaveNet architecture. Control inputs are the sequence of timed phonemes and F0, predicted by the timing model and pitch model respectively. The predicted timbrical acoustic features can be combined with the predicted F0 to generate the final waveform using the synthesis stage of the vocoder.

2.3.1. Multistream Architecture

Most parametric vocoders separate the speech signal into several components. In our case, we use three feature streams; a harmonic spectral envelope, an aperiodicity envelope and a voiced/unvoiced decision (continuous pitch is predicted by the pitch model and given as a control input). These components are largely independent, but their coherence is important (e.g., synthesizing a harmonic component corresponding to a voiced frame as unvoiced will generally cause artifacts, and vice versa). Rather than jointly modeling all data streams with a single model, we decided to model these components using independent networks. This approach gives us more fine-grained control over each stream's architecture, and also avoids the possibility of streams with lower perceptual importance interfering with streams of higher perceptual importance. For instance, the harmonic component is by far the most important, therefore we would not want any other jointly modeled stream potentially reducing model capacity dedicated to this component.

To encourage predictions to be coherent, we concatenate the predictions of one network to the input of another, as depicted in Figure 4. In our current system, the aperiodic component depends on the harmonic component, and the voiced/unvoiced decision depends on both harmonic and aperiodic components. All the networks are similar, but have slightly different hyperparameters (see Table A1 in Appendix C for details). The voiced/unvoiced decision network has a Bernoulli output distribution rather than a mixture density (see Section 2.2.1). While we found this approach to generally work well, we did not exhaustively investigate the many other alternative approaches.

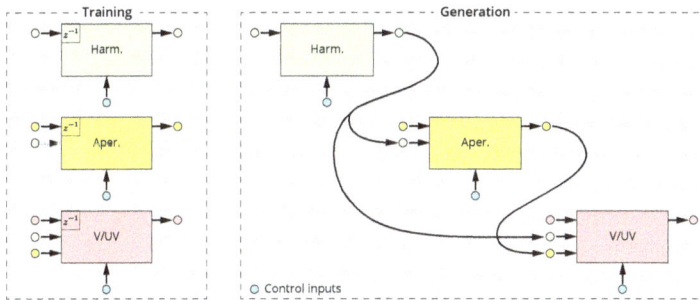

Figure 4. Diagram depicting the cascaded multistream architecture for training and generation phases. The "z^{-1}" blocks represent unit delays. The upward inputs represent control inputs, shared between all streams. Autoregressive connections in generation phase are not shown.

2.3.2. Handling Long Notes

In most datasets, not all note durations will be exhaustively covered. In particular, the case of synthesizing notes significantly longer than the notes in the dataset can be problematic. This issue manifests itself mainly as a repetition in time of some of the transitions predicted by the timbre model, causing a kind of stutter. To reduce such artifacts, we compute the control feature corresponding to the frame position within the phoneme (see Section 2.6) with a nonlinear mapping depending on the length of the phoneme. The idea behind this is that the edges of a phoneme, where the transitions are likely to be, will maintain their original rate, while the more stable center parts will be expanded more.

2.4. Pitch Model

Generating expressive F0 contours for singing voice is quite challenging. Not only is this because of its importance to the overall results, but also because in singing voice there are many factors that simultaneously affect F0. There are a number of musical factors, including melody, various types of attacks, releases and transitions, phrasing, vibratos, and so on. Additionally, phonetics can also cause inflections in F0, so-called microprosody [16]. Some approaches try to decompose these factors to various degrees, for instance by separating vibratos [4] or using source material without consonants [1,17]. In our approach, however, we model the F0 contour as-is, without any decomposition. As such, F0 is predicted from both musical and phonetic control inputs, using a modified WaveNet architecture (see Table A1 in Appendix C for details).

2.4.1. Data Augmentation

One issue with modeling pitch, is that obtaining a dataset that sufficiently covers all notes in a singer's register can be challenging. Assuming that pitch gestures are largely independent of absolute pitch, we apply data augmentation by pitch shifting the training data, similar to [18]. While training, we first draw a pitch shift in semitones from a discrete uniform random distribution, for each sample in the minibatch,

$$pshift \sim \mathcal{U}\left(pshift_{min}, pshift_{max}\right) \tag{5}$$

$$pshift_{min} = pitch_{min}^{singer} - pitch_{max}^{sample} \tag{6}$$

$$pshift_{max} = pitch_{max}^{singer} - pitch_{min}^{sample}, \tag{7}$$

where $pshift_{min}$ and $pshift_{max}$ define the maximum range of pitch shift applied to each sample. These ensure that all notes of the melody within a sample can occur at any note within the singer's register. Finally, this pitch shift is applied to both the pitch used as a control input and the target output pitch,

$$\hat{pitch}_{cond} = pitch_{cond} + pshift \tag{8}$$

$$\hat{f}_0 = f_0 \, 2^{\frac{1}{12}pshift} . \tag{9}$$

2.4.2. Tuning Postprocessing

For pitch in singing voice, one particular concern is ensuring that the predicted F0 contour is in tune. The model described above does not enforce this constraint, and in fact we observed predicted pitch to sometimes be slightly out of tune. If we define "out of tune" as simply deviating a certain amount from the note pitch, it is quite normal for F0 to be out of tune for some notes in expressive singing, without perceptually sounding out of tune. One reason why our model sometimes sounds slightly out of tune may be that such notes are reproduced in different musical context where they do sound out of tune. We speculate that one way to combat this is may be use a more extensive dataset.

We improve tuning of our system by applying a moderate postprocessing of predicted F0. For each note (or segment within a long note), the perceived pitch is estimated using F0 and its derivative. The smoothed difference between this pitch and the score note pitch is used to correct the final pitch used to generate the waveform. Appendix B discusses the algorithm in detail.

2.5. Timing Model

The timing model is used to predict the duration of each phoneme in the sequence to synthesize. Unlike with TTS systems where phoneme durations are generally predicted in a freerunning manner, in singing synthesis, the phoneme durations are heavily constrained by the musical score. In our proposed system we enforce this constraint using a multistep prediction. First, the note timing model predicts the deviations of note (and rest) onsets with respect to nominal onsets in the musical score. At the same time phoneme durations are predicted by the phoneme duration model. Finally, a simple fitting heuristic is used to ensure the predicted phoneme durations fit within the available note duration, after adjusting timing. This approach is somewhat similar to the approach taken by [19].

2.5.1. Note Timing Model

Most singers will not follow the timing of a musical score exactly. Slightly advancing or delaying notes is part of normal expressive singing, and is the result of the given musical and linguistic context and the style of the singer. Additionally, there may be a small truly random component, simply because most singers cannot sing with exact timing.

Note onset deviations are computed from a musical score and phonetic segmentation of the corresponding utterance by the singer. We define a note onset deviation as the difference between the onset of the first syllabic nucleus in a note and that note's nominal onset as written in the musical score. These deviations are also computed for rest notes, or equivalently, note offsets before a rest.

We use a neural network to predict these deviations from note-level musical and linguistic input features. These input features are designed by hand, in part because using note-level data means we have relatively few samples compared to phoneme or frame-level data. We assume that these features contain most or all contextual information relevant to computing note time deviations, therefore we can use a simple feedforward neural network, without the need for a recurrent or convolutional architecture. To avoid making any assumptions about the (conditional) probability distribution of the note onset deviations, we use a nonparametric approach by using a softmax output and discretizing the deviations to multiples of the hoptime. Details of the input features and network architecture are available in Table A2 (Appendix C).

2.5.2. Phoneme Duration Model

Phoneme durations are obtained in a similar way. They are first computed from the given phonetic segmentation, and then discretized on a log scale, similar to [20]. A neural network is used to predict the phoneme durations from phoneme-level musical and linguistic input features. Unlike the note timing model, in this case we do require some local context information, so we use a convolutional architecture. Here we assume the range of context information affecting the duration of a phoneme to be limited by the musical score and the linguistic constraints on the number of possible onset and coda consonants. Therefore, the limited receptive field of a convolutional neural net should not be a significant disadvantage over a recurrent neural net's unbound receptive field. See Table A2 in Appendix C for details.

2.5.3. Fitting Heuristic

The fitting heuristic is used to conform the total of predicted phoneme durations to the available note duration predicted by the note timing model. The basic strategy is to expand or shrink the (principal) vowel, ensuring it is always at least some given percentage of the note duration, by also shrinking consonants if needed.

First, the sequence of phonemes to fit in the note duration is obtained by "shifting" onset consonants to the preceding note. The sequences will thus always start with a vowel (or silence for rests), followed by zero or more consonants formed by the note's coda consonants and the next note's onset consonants. In cases where a note contains multiple syllables, the secondary vowels are handled as if they were consonants. Then, the sequence of N predicted durations $d_0, d_1, \ldots, d_{N-1}$ is fit into the available note duration d_a,

$$
r = \min\left(1, \frac{d_a(1 - r_0)}{\sum_{i=1}^{N-1} d_i}\right), \tag{10}
$$

where r_0 is the minimum fraction of the note's duration to be occupied by the primary syllabic nucleus.

$$
\hat{d}_i = \begin{cases} d_a - r\sum_{j=1}^{N-1} d_j & \text{for } i = 0 \\ rd_i & \text{for } i = 1, 2, \ldots, N-1. \end{cases} \tag{11}
$$

2.6. Acoustic and Control Frontend

We use an acoustic frontend based on the WORLD vocoder [21] (D4C edition [22]) with a 32 kHz sample rate and 5 ms hop time. The dimensionality of the harmonic component is reduced to 60 log Mel-Frequency Spectral Coefficients (MFSCs) by truncated frequency warping in the cepstral domain [23] with an all-pole filter with warping coefficient $\alpha = 0.45$. The dimensionality of the aperiodic component is reduced to four coefficients by exploiting WORLD's inherently bandwise aperiodic analysis. All acoustic features are min/max normalized before feeding them to the neural network.

The control frontend produces linguistic and musical features that control the synthesizer. The linguistic features we use are relatively simple compared to most TTS systems as we omit most of the features that are principally used to predict prosody. The main linguistic features we use are previous, current and next phoneme identity encoded as one-hot vectors. We assume that the lyrics input is a phonetic rather than orthographic sequence. For datasets that do not already include aligned phonetic and acoustic features, we apply a forced alignment using a speaker-dependent Hidden Semi-Markov Model (HSMM) trained using deterministic annealing [24]. The most important musical features are note pitch and duration, as one-hot and 4-state coarse coded vectors respectively. Additionally, we include the normalized position of the current frame within the current phoneme and note as a 3-state coarse coded vectors, roughly corresponding to the probability of being in the beginning, middle or end of the phoneme or note respectively. See Table A1 in Appendix C for a complete listing of the control features used.

2.7. Audio Generation Speed

One special concern with autoregressive models, especially those generating raw waveform, is that the time required to generate a sequence can exceed several times the sequence's duration. Our approach generating vocoder features has the advantage that timesteps have to be produced at a much lower rate, as well as requiring a significantly reduced network architecture to achieve a similar receptive field. However, even in this case, as autoregressive inference is inherently sequential, it cannot exploit massively parallel hardware such as modern GPUs. Therefore, naive implementations of the generation algorithm still tend to be relatively slow. By caching calculations between timesteps, we were able to implement a fast generation algorithm. While this algorithm was developed independently, it is essentially identical to those proposed in other works [25,26]. Using this algorithm, our model can achieve generation speeds of 10–15× real-time on CPU. Combined with low memory and disk footprints, these relatively fast generation speeds make the system competitive with most existing systems in terms of deployability.

3. Related Work

Our method is heavily based on a class of fully-visible probabilistic autoregressive generative models that use neural networks with similar architectures. This type of model was first proposed to model natural images (PixelCNN) [9,10,12], but was later also applied to modeling raw audio waveform (WaveNet) [5], video (Video Pixel Networks) [27] and text (ByteNet) [28].

Soon after WaveNet, there have been several other related works on text-to-speech. Deep Voice [26] obtains real-time inference by using a deeper, but narrower architecture, and heavily optimized generation algorithm. It also introduces a pipeline comprised of solely neural network-based building blocks, although these are independently trained and targets are still obtained in a traditional way, i.e., by using an F0 estimator, phonetic dictionary, and so on. Deep Voice 2 [20] improves the components of this pipeline, and explores multispeaker training which allows modeling hundreds of voices with less than half an hour of data per speaker. The SampleRNN [29] model proposes an alternative architecture for unconditional raw waveform generation based on multiscale hierarchical Recurrent Neural Networks (RNNs) rather than dilated Convolutional Neural Networks (CNNs). Char2Wav [30] uses a SampleRNN component as a neural vocoder for synthesizing predicted vocoder parameters. The vocoder parameters are predicted by an attention-based sequence-to-sequence (seq2seq) model, which allows for a fully end-to-end system, generating speech signals from unaligned orthographic or phonetic sequences. Another end-to-end system is Tacotron [31], which proposes a sophisticated seq2seq model able to predict magnitude spectrum frames from text.

More traditional neural parametric speech synthesizers tend to be based on feedforward architectures such as DNNs and Mixture Density Networks (MDNs) [32], or on recurrent architectures such as Long Short-Term Memory RNNs (LSTM-RNNs) [33]. Feedforward networks learn a framewise mapping between linguistic and acoustic features, thus potentially producing discontinuous output. This is often partly mitigated by predicting static, delta and delta-delta feature distributions combined with a parameter generation algorithm that maximizes output probability [34]. Recurrent architectures avoid this issue by propagating hidden states (and sometimes the output state) over time. In contrast, autoregressive architectures such as the one we propose make predictions based on predicted past acoustic features, allowing, among other things, to better model rapid modulations such as plosive and trill consonants.

There have been several works proposing different types of singing synthesizers. The more prominent of which are based on concatenative methods [1,2] and statistical parametric methods centered around Hidden Markov Models (HMMs) [3,4]. Similar to in this work, an important benefit of statistical models is that they allow joint modeling of timbre and musical expression from natural singing [18,35]. Many of the techniques developed for HMM-based TTS are also applicable to singing synthesis, e.g., speaker-adaptive training [36]. The main drawback of HMM-based approaches is that phonemes are modeled using a small number of discrete states and within each state statistics are constant. This causes excessive averaging, an overly static "buzzy" sound and noticeable state transitions in long sustained vowels in the case of singing. More recently, the work on HMM singing synthesis was extended to feedforward DNNs [37], albeit with a somewhat limited architecture.

4. Experiments

The goal of our experiments is mainly to compare our system against competing systems such as concatenative unit selection, HMM or DNN systems. We are also interested in having some indication of the absolute performance of our system, i.e., compared to a reference recording.

We conducted two sets of experiments; one set of experiments involve systems trained on a dataset of natural singing and the second set involve systems trained on a dataset of what we call pseudo singing. Pseudo singing are recordings of something in between speech and singing, using a constant cadence and one or more constant pitches. One limitation of pseudo singing is that it can only be used to train timbre models, as it does not contain musical expression. However, the reason for also conducting experiments with this kind of data is two-fold: first, we expect the performance of in particular unit selection systems to be notably better with pseudo singing datasets, as the more stable and coherent data is better suited for this type of system; and, second, we have access to a wider range of datasets of this kind, including more languages. As we only compare the performance of the timbre model of different systems when using pseudo singing, we generate sequences in a so-called performance driven manner, that is, F0 and phonetic timings that control the timbre model are obtained from a reference recording.

The webpage accompanying this article, http://www.dtic.upf.edu/~mblaauw/NPSS/, contains several demo songs synthesized by our system, after training on both kinds of data.

4.1. Datasets

For systems trained on natural singing, we use a public dataset published by the Nagoya Institute of Technology (Nitech), identified as NIT-SONG070-F001 (http://hts.sp.nitech.ac.jp/archives/2.3/HTS-demo_NIT-SONG070-F001.tar.bz2). This dataset consists of studio quality recordings of a female singer singing Japanese children songs. The original dataset consists of 70 songs, but the public version consists of a 31 song subset (approximately 31 min, including silences). Out of these 31 songs, we use 28 for training and 3 for testing (utterances 015, 029 and 040).

We use three proprietary datasets from training systems on pseudo singing; an English male voice (M1), an English female voice (F1) and Spanish female voice (F2). The studio quality recordings consist of short sentences which were sung at a single pitch and an approximately constant cadence. The sentences were selected to favor high diphone coverage. The Spanish dataset contains 123 sentences, while the English datasets contain 524 sentences (approximately 16 and 35 min respectively, including silences). A randomly selected 10% of sentences are used for testing.

Note that these datasets are small compared to the datasets typically used to train TTS systems. However, for natural singing, many-hour datasets would exceed the repertoire of most singers. For pseudo singing, as only timbre is captured in a very constrained setting, substantially larger datasets would likely yield diminishing returns.

4.2. Compared Systems

- **NPSS**: Our system, which we call Neural Parametric Singing Synthesizer (NPSS), as described in Section 2.
- **IS16**: A concatenative unit selection-based system [1], which was the highest rated system in the Interspeech 2016 Singing Synthesis Challenge.
- **Sinsy-HMM**: A publicly accessible implementation of the Sinsy HMM-based synthesizer (http://www.sinsy.jp/). This system is described in [4,35], although the implementation may differ to some degree from any single publication, according to one of the authors in private correspondence. While the system was trained on the same NIT-SONG070-F001 dataset, it should be noted that the full 70 song dataset was used, including the 3 songs we use for testing.
- **Sinsy-DNN**: A publicly accessible implementation of the Sinsy feedforward DNN-based synthesizer (http://www.sinsy.jp/) [37]. The same caveats as with Sinsy-HMM apply here. Additionally, the DNN voice is marked as "beta", and thus should be considered still experimental. The prediction of timing and vibrato parameters in this system seems to be identical to Sinsy-HMM at the time of writing. Thus, only timbre and "baseline" F0 is predicted by the DNN system.
- **HTS**: A HMM-based system, similar to Sinsy-HMM, but consisting of a timbre model only, and trained on pseudo singing. The standard demo recipe from the HTS toolkit (version 2.3) [38] was followed, except for a somewhat simplified context dependency (just the two previous and two following phonemes).

4.3. Methodology

We compare the different systems using a set of quantitative and qualitative tests. Finding perceptually relevant metrics to compare generative models quantitatively tends to be very challenging, as is the case with expressive singing voice. Although we pay special attention to the metrics we use, this should be kept in mind when comparing values. Qualitative tests tend to be more conclusive, but can also be challenging when evaluating multidimensional aspects such as "expression". It should be noted that the quantitative metrics for the systems trained on pseudo singing are evaluated with respect to a pseudo singing reference. Therefore, these results might not directly correspond to our end goal, expressive singing, as evaluated in the listening tests and quantitative metrics for the systems trained on natural singing.

4.3.1. Quantitative Metrics

For all metrics, we apply a simple linear time mapping to reduce misalignments due to predicted timings possibly differing from reference timings.

- **Mel-Cepstral Distortion (MCD)**: Mel-Cepstral Distortion (MCD) is a common perceptually motivated metric for the quantitative evaluation of timbre models. In our case, some moderate modifications are made to improve robustness for singing voice; Mel-cepstral parameters are extracted from WORLD spectra, rather than STFT spectra, to better handle high pitches. To reduce the effect of pitch mismatches between reference and prediction, we filter pairs of frames with a pitch difference exceeding ±200 cents. Similarly, to increase robustness to small misalignments in time, frames with a modified z-score exceeding 3.5 are not considered [39]. MCD is computed for harmonic components, using 33 (0–13.6 kHz) coefficients.

- **Band Aperiodicity Distortion (BAPD)**: Identical to MCD, except computed over linearly spaced band aperiodicity coefficients. BAPD is computed for aperiodic components, using 4 (3–12 kHz) coefficients.

- **Modulation Spectrum (MS) for Mel-Generalized Coefficients (MGC)**: One issue with framewise metrics, like MCD, is that these do not consider the behavior of the predicted parameter sequences over time. In particular, the common issue of oversmoothing is typically not reflected in these metrics. A recently proposed metric, the Modulation Spectrum (MS) [40], allows visualizing the spectral content of predicted time sequences. For instance, showing oversmoothing as a rolloff of higher modulation frequencies. We are mainly interested in the lower band of the MS (e.g., <25 Hz), because the higher band of the reference (natural singing) can be overly affected by noise in the parameter estimation. To obtain a single scalar metric, we use the Modulation Spectrum Log Spectral Distortion (MS-LSD) between the modulation spectra of a predicted parameter sequence and a reference recording.

- **Voiced/unvoiced decision metrics**: In singing voice, there is a notable imbalance between voiced and unvoiced frames due to having many long, sustained vowels. As both false positives (unvoiced frames predicted as voiced) and false negatives (voiced frames predicted as unvoiced) can result in highly noticeable artifacts, we list both False Positive Rate (FPR) and False Negative Rate (FPR) for this estimator. All silences are excluded.

- **Timing metrics**: Metrics for the timing model are relatively straight forward, e.g., Mean Absolute Error (MAE), Root Mean Squared Error (RMSE) or Pearson correlation coefficient r between onsets or durations. We list errors for note onsets, offsets and consonant durations separately to ensure the fitting heuristic affects the results only minimally.

- **F0 metrics**: Standard F0 metrics such as RMSE are given, but it should be noted that these metrics are often not very correlated to perceptual metrics in singing [41]. For instance, starting a vibrato slightly early or late compared to the reference may be equally valid musically, but can be the cause the two F0 contours to become out of phase, resulting in high distances.

- **Modulation Spectrum (MS) for log F0**: Similar to timbre, we use MS-based metrics to get a sense of how close the generated F0 contours are in terms of variability over time. The MS of F0 is computed by first segmenting the score into sequences of continuous notes, without rests. Then, for each sequence, the remaining unvoiced regions in the log F0 curve are filled using cubic spline interpolation. We apply a Tukey window corresponding to a 50 frame fade in and fade out, and subtract the per-sequence mean. Then, the modulation spectra are computed using a Discrete Fourier Transform (DFT) size 4096, and averaged over all sequences.

4.3.2. Listening Tests

For the listening tests, all stimuli were downsampled to 32 kHz, which is the lowest common denominator between the different systems.

- **Mean Opinion Score (MOS)**: For the systems trained on natural singing, we conducted a MUSHRA [42] style listening test. The 40 participants, of which 8 indicated native or good knowledge of Japanese, were asked to rate different versions of the same audio excerpt compared to a reference. The test consisted to 2 short excerpts (<10 s) for each of the 3 validation set songs, in 7 versions (reference, hidden reference, anchor and 4 systems), for a total of 42 stimuli. The scale used as 0–100, divided into 5 segments corresponding to a 5-scale MOS test. The anchor consisted of a distorted version of the NPSS synthesis, applying the following transformations: 2D Gaussian smoothing ($\sigma = 10$) of harmonic, aperiodic and F0 parameters, linearly expanding the spectral envelope by 5.2%, random pitch offset (±100 cents every 250 ms, interpolated by cubic spline), and randomly "flipping" 2% of the voiced/unvoiced decisions. We excluded 59 of the total 240 tests performed, as these had a hidden reference rated below 80 (ideally the rating should be 100). We speculate that these cases could be due to the relative difficulty of the listening test for untrained listeners.

- **Preference Test**: For the systems trained on pseudo singing, we conducted an AB preference test. The 18 participants were asked for their preference between two different stimuli, or indicate no preference. The stimuli consisted of two short excerpts (<10 s) of one song per voice/language. Versions with and without background music were presented. We perform pairwise comparisons between our system and two other systems, resulting in a total of 24 stimuli.

5. Results

5.1. Quantitative Results

For systems trained on natural singing, Tables 1–3 list quantitative metrics related to timbre, timing and pitch models respectively. Examples of different modulation spectra for timbre and pitch are shown in Figures 5 and 6. For systems trained on pseudo singing, Table 4 lists quantitative metrics related to timbre models.

Table 1. Quantitative results for the timbre models trained on natural singing. Note that for the IS16 system the Modulation Spectrum Log Spectral Distortion (MS-LSD) and Voiced/Unvoiced (V/UV) metrics are omitted as it does not use predicted harmonic features (MS-LSD is computed from predicted features, not analyzed features) or V/UV decision. The HTS system is only considered when comparing systems trained on pseudo singing, but should be roughly equivalent to Sinsy-HMM.

| System | Harmonic | | Aperiodic | V/UV | |
	MCD (dB)	MS-LSD (<25 Hz/Full, dB)	BAPD (dB)	FPR (%)	FNR (%)
IS16	6.94	-	3.84		-
Sinsy-HMM	7.01	8.09/18.50	4.09	15.90	0.68
Sinsy-DNN	**5.41**	13.76/29.87	5.02	**13.75**	0.63
NPSS	5.54	**7.60/11.65**	**3.44**	16.32	0.64

Table 2. Quantitative results for the timing models trained on natural singing. The table lists Mean Absolute Error (MAE) and Root Mean Squared Error (RMSE), both in 5 ms frames, and Pearson correlation coefficient r. Note that the Sinsy-DNN system uses the same HMM-based duration model as the Sinsy-HMM system, so it is excluded from the comparison. The IS16 system used durations predicted by the NPSS system. The HTS system is only considered when comparing systems trained on pseudo singing, but should be roughly equivalent to Sinsy-HMM.

| System | Note Onset Deviations | | | Note Offset Deviations | | | Consonant Durations | | |
	MAE	RMSE	r	MAE	RMSE	r	MAE	RMSE	r
Sinsy-HMM	7.107	9.027	0.379	13.800	**17.755**	0.699	4.022	5.262	0.589
NPSS	**6.128**	**8.383**	**0.419**	**12.100**	18.645	**0.713**	**3.719**	**4.979**	**0.632**

Table 3. Quantitative results of pitch models trained on natural singing. Table shows log F0 Modulation Spectrum Log Spectral Distortion (MS-LSD) in dB. The F0 Root Mean Squared Error (RMSE) in cents and Pearson correlation coefficient *r* are also given for reference. The IS16 and HTS systems are excluded from this comparison because they are not suitable for modeling F0 from natural singing.

System	MS-LSD (<25 Hz, dB)	RSME (Cents)	*r*
Sinsy-HMM	5.052	**81.795**	**0.977**
Sinsy-DNN	2.858	83.706	0.976
NPSS	**2.008**	105.980	0.963

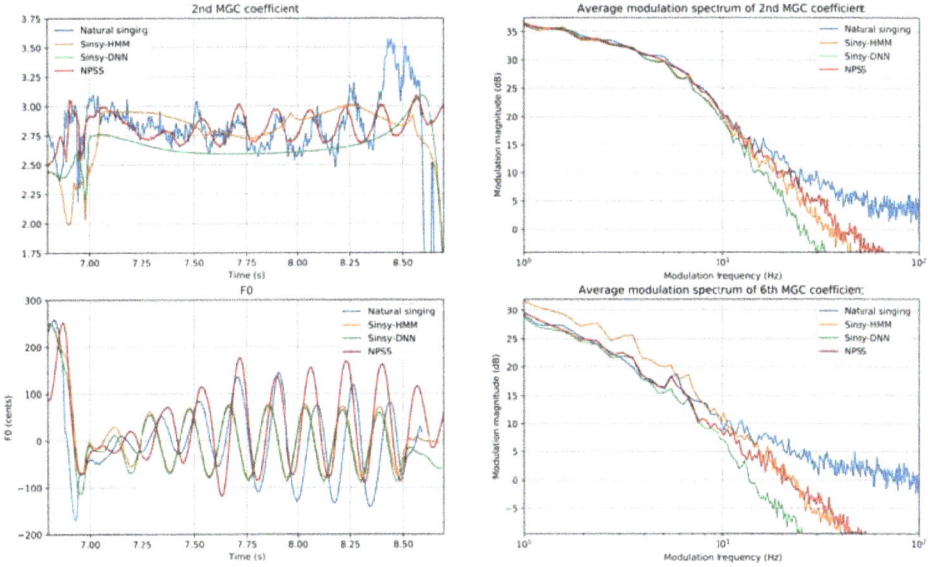

Figure 5. Comparing the average modulation spectrum of harmonic Mel-Generalized Coefficient (MGC) features. In the plotted excerpt, the relation between pitch and timbre during vibratos can be observed.

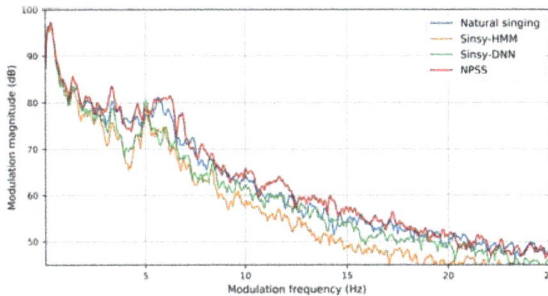

Figure 6. Comparing the average modulation spectrum of log F0 contours predicted by various systems and natural singing.

Table 4. Quantitative results for the timbre models trained on pseudo singing, separated by voice/language. The IS16 system is excluded from the quantitative metrics because removing utterances from the dataset to use for testing would mean missing diphones would have to be replaced. The Sinsy-HMM and Sinsy-DNN systems were excluded from this comparison, as the only available models are trained on natural singing. The listed metrics are Mel-Cepstral Distortion (MCD) and Modulation Spectrum Log Spectral Distortion (MS-LSD) for harmonic features, Band Aperiodicity Distortion (BAPD) for aperiodic features, and False Positive Rate (FPR) and False Negative Rate (FNR) for voiced/unvoiced (V/UV) features.

Voice (Language)	System	Harmonic		Aperiodic	V/UV	
		MCD (dB)	MS-LSD (<25 Hz/Full, dB)	BAPD (dB)	FPR (%)	FNR (%)
M1 (Eng.)	HTS	4.95	11.09/22.44	2.72	16.10	**2.46**
	NPSS	5.14	**7.79/8.18**	**2.44**	**11.22**	2.65
F1 (Eng.)	HTS	4.75	10.25/22.09	4.07	**15.60**	1.01
	NPSS	4.95	**5.68/9.04**	**3.83**	15.79	**0.56**
F2 (Spa.)	HTS	4.88	11.07/22.28	3.62	1.85	**2.21**
	NPSS	5.27	**8.02/6.59**	**3.38**	**1.40**	3.20

These metrics show that, for some of the framewise metrics, such as harmonic MCD, our system is slightly behind. For some other metrics, such as the timing errors or aperiodic BAPD, our system is slightly ahead. For systems trained on pseudo singing the differences tend to be a little bigger, we argue that this is due that predicting averages for this kind of data results in good results for these kind of metrics. However, in all metrics based on the modulation spectrum, which considers variations in time, NPSS shows an improvement over the other systems.

When we compare an example of generated harmonic parameters during a vibrato in the left two subplots of Figure 5, we notice the features predicted by NPSS having more detail than Sinsy-HMM and Sinsy-DNN. In particular the framewise conditioning of harmonic features on F0 in NPSS, causes the harmonic features to modulate along the vibrato, similar to what happens in the reference recording. In the modulation spectrum analysis on the right-hand side of Figure 5, we can see that overall NPSS tends to follow the modulation spectrum of the reference recording a little closer than Sinsy-HMM and Sinsy-DNN in lower modulation frequencies. Compared to especially Sinsy-DNN, NPSS has less rolloff in higher modulation frequencies, indicating less oversmoothing over time. However, all systems have less high frequency modulation spectrum content than the reference recording, indicating none of the systems are able to reproduce all the details of the original signal.

The analysis of the modulation spectrum of the log F0 predicted by different systems is shown in Figure 6. We can see that overall NPSS matches the modulation spectrum of the reference recording similarly or slightly better than Sinsy-HMM, but notably better than Sinsy-DNN. When we focus our attention to the range of modulation frequencies corresponding to vibratos in this voice, 5–7 Hz, we see that Sinsy-HMM and Sinsy-DNN have a sharp peak at 5 Hz, whereas for NPSS this whole range has increased energy, similar to the reference. This may indicate that NPSS produces a wider range of vibrato rates, similar to a real singer. In Sinsy-HMM and Sinsy-DNN vibrato parameters (rate and depth) are modeled separately from the base F0, which may explain their tendency to produce very controlled, regular vibratos.

5.2. Qualitative Results

Results of the listening tests comparing different systems trained on natural singing are listed in Table 5. For systems trained on pseudo singing, results of the preference listening test are shown in Figure 7.

Table 5. Mean opinion scores for systems trained on natural singing, displayed on a 1–5 scale with their respective 95% confidence intervals. The HTS system is only considered when comparing systems trained on pseudo singing, but should be roughly equivalent to Sinsy-HMM.

System	Mean Opinion Score
Hidden reference	4.76 ± 0.04
IS16	2.36 ± 0.11
Sinsy-HMM	2.98 ± 0.10
Sinsy-DNN	2.77 ± 0.10
NPSS	**3.43 ± 0.11**

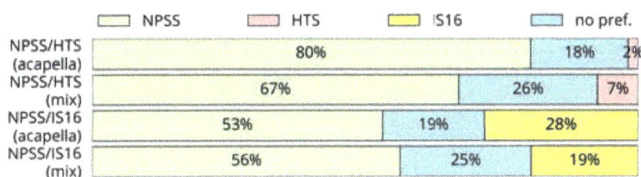

Figure 7. Results of the preference test for systems trained on pseudo singing. The Sinsy-HMM and Sinsy-DNN systems were excluded from this comparison, as the only available models are trained on natural singing.

In the listening tests, NPSS is clearly ahead of competing systems. In the MOS test for systems trained on natural singing, NPSS is around a third between the second best rated system (Sinsy-HMM) and the reference. Here, it should be noted that the concatenative system, IS16, performs worst, showing that this kind of system is poorly suited for this kind of data. In contrast, the preference test for systems trained on pseudo singing, shows a strong preference for NPSS over the HTS system, and a moderate preference over the IS16 system, which was designed for this kind of data. The correlation between the qualitative results and the quantitative metrics based on the modulation spectrum indicate that this may be a metric with higher perceptual relevance than the framewise metrics such as MCD.

In our experience NPSS, HMM and DNN systems all produce quite coherent timbres. The concatenative system in contrast tends to produce more discontinuous timbres, especially when using a dataset of natural singing, or other artifacts at concatenation boundaries, e.g., in fast singing or when phonetic segmentation is not perfect. We found NPSS to generally produce less static features over time, and less coloring of timbre. Compared to HMM and DNN systems, the autoregressive generation of NPSS seems to help in reproducing rapidly varying consonants, although these can occasionally sound better still in the concatenative system. In terms of expression, the HMM system produces very coherent behavior, which while perhaps a little less human, tends to generally sound quite pleasant. NPSS on the other hand, seems to be more varied, but this also means that results are sometimes better than other times. One notable quality of NPSS is that the framewise conditioning of timbre on pitch means that vibratos produce natural, synchronized modulations in both pitch and timbre (see, e.g., Figure 5), unlike in the other systems which condition on note pitch.

6. Conclusions

We presented a singing synthesizer based on neural networks, which can generate synthetic singing voice given a musical score with lyrics. From a single set of relatively few songs, the system is able to learn both timbre and expression. Separate, but interconnected models learn phonetic timing, pitch and timbre. The core building block of the system is a variant of the WaveNet architecture, modified to allow generating features obtained from a parametric vocoder. This autoregressive approach offers improved reproduction of consonants and a more natural variation of predicted parameters over time, compared to competing approaches such as statistical parametric systems.

Compared to concatenative approaches, our model allows for greater flexibility and is more robust to small misalignments between phonetic and acoustic features in the training data. In listening test our system was rated to reduce the gap between the second best system and the reference recording by about a third. While correlating this with quantitative metrics is challenging, metrics that take into account variations over time, such as the modulation spectrum, do seem to corroborate this. The relatively small CPU, memory and disk footprint allows for many practical applications of our system. We hope that in the near future we can evaluate our model trained on natural singing for a wider range of languages and datasets. Further exploring the flexibility offered by this neural approach, such as the area of multispeaker training is also promising, as it might help to overcome the issue of limited dataset sizes typical of singing voice.

Acknowledgments: We gratefully acknowledge the support of NVIDIA Corporation with the donation of the Titan X Pascal GPU used for this research. We thank Nagoya Institute of Technology for providing the NIT-SONG070-F001 dataset (licensed under CC BY 3.0), Zya for providing the English datasets, and Voctro Labs for providing the Spanish dataset and the implementation of the fast generation algorithm. This work is partially supported by the Spanish Ministry of Economy and Competitiveness under the CASAS project (TIN2015-70816-R).

Author Contributions: Jordi Bonada and Merlijn Blaauw designed and implemented the proposed system; Jordi Bonada implemented the fast generation algorithm; Jordi Bonada and Merlijn Blaauw designed the experiments; and Merlijn Blaauw performed the experiments and wrote the paper.

Conflicts of Interest: The authors declare no conflict of interest.

Appendix A. Details Constrained Gaussian Mixture

The output mixture density we call Constrained Gaussian Mixture (CGM), is a mixture of $K = 4$ Gaussians,

$$p(x) = \sum_{k=0}^{K-1} w_k \mathcal{N}(x; \mu_k, \sigma_k^2). \tag{A1}$$

The 12 mixture parameters w_k, μ_k, σ_k for $k = 0, 1, \dots, K-1$ are computed from four free parameters: location ξ, scale ω, skewness α and shape β (see Figure 3 for some example distributions). Assuming the network predicts four outputs with linear activations, a_0, a_1, a_2, a_3, we apply some nonlinearities to obtain the free parameters in suitable ranges,

$$\xi = 2\operatorname{sigm}(a_0) - 1 \qquad \text{range } [-1, 1] \tag{A2}$$

$$\omega = \frac{2}{255} e^{4\operatorname{sigm}(a_1)} \qquad \text{range } \left[\frac{2}{255}, \frac{2e^4}{255}\right] \tag{A3}$$

$$\alpha = 2\operatorname{sigm}(a_2) - 1 \qquad \text{range } [-1, 1] \tag{A4}$$

$$\beta = 2\operatorname{sigm}(a_3) \qquad \text{range } [0, 2] . \tag{A5}$$

Then, we map predicted location ξ, scale ω, skewness α and shape β to Gaussian mixture parameters μ_k, σ_k, w_k for $k = 0, 1, \dots, K-1$,

$$\sigma_k = \omega e^{(|\alpha|\gamma_s - 1)k} \tag{A6}$$

$$\mu_k = \xi + \sum_{i=0}^{k-1} \sigma_k \gamma_u \alpha \tag{A7}$$

$$w_k = \frac{\alpha^{2k} \beta^k \gamma_w^k}{\sum_{i=0}^{K-1} \alpha^{2i} \beta^i \gamma_w^i}, \tag{A8}$$

where γ_u, γ_s and γ_w are constants tuned by hand,

$$\gamma_u = 1.6 \tag{A9}$$

$$\gamma_s = 1.1 \tag{A10}$$

$$\gamma_w = \frac{1}{1.75} . \tag{A11}$$

A temperature control is achieved by first shifting component means towards their global weighted average,

$$\bar{\mu} = \sum_{k=0}^{K-1} \mu_k w_k \tag{A12}$$

$$\hat{\mu}_k = \mu_k + (\bar{\mu} - \mu_k)(1 - \tau) , \tag{A13}$$

where $0 < \tau \leq 1$ is the temperature. Then, the component variances are scaled by the temperature,

$$\hat{\sigma}_k = \sigma_k \sqrt{\tau} . \tag{A14}$$

Appendix B. Details Tuning Postprocessing

The principal idea behind the tuning correction postprocessing is simple; apply the difference between the perceived pitch of a note, given its predicted F0 contour, and the pitch of the corresponding note in the score. However, robustly estimating the perceived pitch of a note from the corresponding F0 contour is nontrivial. In singing voice there are many factors that affect F0, but may not influence the perceived note pitch. These factors include vibratos, scoops, releases, transitions, microprosody due to consonants and so on. Therefore, simple estimators, such as directly taking the mean of the framewise F0 over the note duration, will typically yield poor results.

To obtain a more robust estimate of the perceived note pitch, $\overline{F0}$, we compute a weighted average of the predicted F0 over the note's duration,

$$\overline{F0} = \frac{\sum_i F0_i w_i}{\sum_i w_i} , \tag{A15}$$

where $F0_i$ and w_i correspond to the i-th frame within a given note of the predicted F0 vector and weighting vector respectively. To simplify notation, throughout this section "F0" refers to log F0 in semitones. The weighting vector in Equation (A15) is composed of a number of different factors that correspond to different heuristics designed to make the estimate more robust,

$$w = w_e w_d w_p w_t . \tag{A16}$$

The first of these factors, w_e, is a weighting to reduce the influence of the edges of the note, where most of the transition effects will typically be located. We compute w_e as a Tukey window with $\alpha = 0.5$. That is, we apply a cosine-taper weighting along the first and last 25% of the note duration.

The second factor, w_d, is a weighting depending on the derivative of the F0 contour. The idea is that the portion of the note where F0 is mostly flat will contribute more to the perceived pitch than portions where F0 fluctuates due to transitions or microprosody. We first estimate the derivative by convolving the signal with a 3rd order 1st derivative Savitzky-Golay FIR filter, s_d, with a length of 11 frames (55 milliseconds),

$$dF0 = F0 \circledast s_d , \tag{A17}$$

where $\circledast$ denotes the convolution operator. Then, we compute the weighting factor, w_d, as follows,

$$w_{d,i} = \frac{1}{\min(1 + 27|dF0_i|, 15)} , \tag{A18}$$

where the constants were obtained empirically.

The third factor, w_p, is a weighting depending on the phoneme corresponding to each frame p_i,

$$w_{p,i} = \begin{cases} 2, & \text{for } p_i \in \{vowel, syllabic\ consonant\} \\ 0, & \text{for } p_i \in \{silence, pause, breath\} \\ 1, & \text{otherwise.} \end{cases} \tag{A19}$$

The idea is that frames corresponding to vowels typically contribute more to the perceived pitch than consonants, which often contain microprosody effects.

The last factor, w_t, is a weighting depending on the distance from the target pitch, based on the assumption that detuning in the perceived pitch will typically be caused by relatively small deviations. Other factors, such as scoops or microprosody, may cause relatively big deviations, but these tend not to contribute to the perceived detuning. We use a pitch deviation of ± 1 semitone as a threshold,

$$w_{t,i} = \begin{cases} 1, & \text{for } |F0_{tar} - F0_i| \leq 1 \\ 1/|F0_{tar} - F0_i|, & \text{otherwise.} \end{cases} \tag{A20}$$

Finally, the required amount of pitch correction, $cF0$, is computed for each frame in a note as follows,

$$cF0_i = F0_{tar} - \overline{F0}, \tag{A21}$$

where $F0_{tar}$ is the note's target pitch, as is written in the score. For rests, we do not apply any correction, $cF0_i = 0$. These framewise correction vectors are then concatenated for all notes and rests in the sequence. As the resulting vector may be discontinuous, we smooth it by zero-phase filtering with a Gaussian window with a length of 30 frames (150 milliseconds).

As the above method computes a notewise correction, it is based on the assumption that the detuning will be approximately constant along a note. However, this is not always the case, especially for longer notes. There can for instance be a pitch trend along a note's duration, which may sound like the singer is slowly trying to reach the correct pitch. To reduce this kind of detuning, we divide longer notes in smaller sub-note segments, and compute the per-segment correction as described above. However, prior to the final smoothing step, instead of a constant correction per segment, we obtain the framewise correction by linearly interpolating each segment's correction at its center.

Appendix C. Model Hyperparameters

Table A1 lists the hyperparameters for the timbre model and pitch model, which both use the same modified WaveNet architecture. Table A2 list the hyperparameters for timing models, which use a simpler architecture. All models are trained using the Adam optimizer [43] with standard parameters $\beta_1 = 0.9$, $\beta_2 = 0.999$, $\epsilon = 1 \times 10^{-8}$; initial learning rates and inverse time decays are listed in the tables. Training a complete system takes around 10 h on a single Titan X Pascal GPU. While we found these settings to work well experimentally, they have not been exhaustively optimized.

Table A1. Hyperparameters for networks based on WaveNet architecture.

Hyperparameter	Timbre Model			Pitch Model
	Harmonic	Aperiodic	V/UV	F0
Feature dimensionality	60	4	1	1
Additional inputs (dim.)	-	harmonic (60)	harmonic (60) aperiodic (4)	-

Table A1. *Cont.*

Hyperparameter	Timbre Model			Pitch Model
	Harmonic	**Aperiodic**	**V/UV**	**F0**
Control inputs	prev. phn. identity (one-hot) cur. phn. identity (one-hot) next phn. identity (one-hot) pos.-in-phn. (coarse) F0 (coarse)			prev. phn. class (one-hot) cur. phn. class (one-hot) next phn. class (one-hot) pos.-in-phn. (coarse) prev. note pitch (one-hot) cur. note pitch (one-hot) next note pitch (one-hot) prev. note dur. (coarse) cur. note dur. (coarse) next note dur. (coarse) pos.-in-note (coarse)
Input noise level λ	0.4	0.4	0.4	0.4
Generation temperature τ	piecewise linear (0,0.05; 3,0.05; 8,0.5; 60,0.5)	0.01	-	0.01
Initial causal convolution	10×1	10×1	10×1	20×1
Residual channels	130	20	20	100
Dilated convolutions	2×1	2×1	2×1	2×1
Num. layers	5	5	5	13
Num. layers per stage	3	3	3	7
Dilation factors	1, 2, 4, 1, 2	1, 2, 4, 1, 2	1, 2, 4, 1, 2	1, 2, 4, 8, 16, 32, 64, 1, 2, 4, 8, 16, 32
Receptive field (ms)	100	100	100	1050
Skip channels	240	16	4	100
Output stage	tanh $\to 1 \times 1$ $\to 60\times$ CGM$_{K=4}$	tanh $\to 1 \times 1$ $\to 4\times$ CGM$_{K=4}$	tanh $\to 1 \times 1$ $\to 1\times$ sigmoid	tanh $\to 1 \times 1$ $\to 1\times$ CGM$_{K=4}$
Batch size	32	32	32	64
Num. valid out timesteps	210	210	210	105
Learning rate (initial, decay, interval)	5×10^{-4}, 1×10^{-5}, 1	5×10^{-4}, 1×10^{-5}, 1	5×10^{-4}, 1×10^{-5}, 1	1×10^{-3}, -
Num. epochs (updates)	1650 (82,500)	1650 (82,500)	1650 (82,500)	235 (11,750)

Table A2. Hyperparameters for timing networks.

Hyperparameter	Note Timing	Phoneme Duration
Input features	note duration (one-hot) prev. note duration (one-hot) 1st phoneme class (one-hot) note position in bar (normalized) note is rest num. coda consonants prev. note prev. note is rest	phoneme identity (one-hot) phoneme class (one-hot) phoneme is vowel phoneme kind (onset/nucleus/coda/inner) note duration (one-hot) prev. note duration (one-hot) next note duration (one-hot)
Target range (frames)	$[-15,14]$, $[-30,29]$ for rests	$[5, 538]$
Target discretization	30 bins, linear	50 bins, log scale

Table A2. *Cont.*

Hyperparameter	Note Timing	Phoneme Duration
Architecture	input → dropout (0.81) $1 \times 1 \to 256\times$ ReLU → dropout (0.9) $1 \times 1 \to 64\times$ ReLU → dropout (0.9) $1 \times 1 \to 32\times$ ReLU → dropout (0.81) $1 \times 1 \to$ 30-way softmax	input → dropout (0.8) $3 \times 1 \to 256\times$ gated tanh → dropout (0.8) 3×1 (dilation = 2) → $64\times$ gated tanh → dropout (0.8) $1 \times 1 \to 32\times$ gated tanh → dropout (0.64) $1 \times 1 \to$ 50-way softmax
Batch size	32	16
Learning rate	2×10^{-4}	2×10^{-4}
Number of epochs	140	210

References

1. Bonada, J.; Umbert, M.; Blaauw, M. Expressive singing synthesis based on unit selection for the singing synthesis challenge 2016. In Proceedings of the 17th Annual Conference of the International Speech Communication Association (Interspeech), San Francisco, CA, USA, 8–12 September 2016; pp. 1230–1234.
2. Bonada, J.; Serra, X. Synthesis of the Singing Voice by Performance Sampling and Spectral Models. *IEEE Signal Process. Mag.* **2007**, *24*, 67–79.
3. Saino, K.; Zen, H.; Nankaku, Y.; Lee, A.; Tokuda, K. An HMM-based singing voice synthesis system. In Proceedings of the 9th International Conference on Spoken Language Processing (ICSLP—Interspeech), Pittsburgh, PA, USA, 17–21 September 2006; pp. 2274–2277.
4. Oura, K.; Mase, A.; Yamada, T.; Muto, S.; Nankaku, Y.; Tokuda, K. Recent development of the HMM-based singing voice synthesis system—Sinsy. In Proceeedings of the 7th ISCA Workshop on Speech Synthesis (SSW7), Kyoto, Japan, 22–24 September 2010; pp. 211–216.
5. Van den Oord, A.; Dieleman, S.; Zen, H.; Simonyan, K.; Vinyals, O.; Graves, A.; Kalchbrenner, N.; Senior, A.W.; Kavukcuoglu, K. WaveNet: A generative model for raw audio. *CoRR arXiv* **2016**, arXiv:1609.03499.
6. Blaauw, M.; Bonada, J. A neural parametric singing synthesizer. In Proceedings of the 18th Annual Conference of the International Speech Communication Association (Interspeech), Stockholm, Sweden, 20–24 August 2017; pp. 1230–1234.
7. Yu, F.; Koltun, V. Multi-Scale Context Aggregation by Dilated Convolutions. In Proceedings of the 4th International Conference on Learning Representations (ICLR), San Juan, Puerto Rico, 2–4 May 2016.
8. He, K.; Zhang, X.; Ren, S.; Sun, J. Deep residual learning for image recognition. In Proceedings of the 34th IEEE Conference on Computer Vision and Pattern Recognition (CVPR), Las Vegas, NV, USA, 27–30 June 2016; pp. 770–778.
9. Reed, S.; van den Oord, A.; Kalchbrenner, N.; Bapst, V.; Botvinick, M.; de Freitas, N. *Generating Interpretable Images with Controllable Structure*; Technical Report; Google DeepMind: London, UK, 2016.
10. van den Oord, A.; Kalchbrenner, N.; Vinyals, O.; Espeholt, L.; Graves, A.; Kavukcuoglu, K. Conditional image generation with PixelCNN decoders. In Proceedings of the Advances in Neural Information Processing Systems 29 (NIPS), Barcelona, Spain, 5–10 December 2016; pp. 4790–4798.
11. van den Oord, A.; Kalchbrenner, N.; Kavukcuoglu, K. Pixel recurrent neural networks. In Proceedings of the 33rd International Conference on Machine Learning (ICML), New York, NY, USA, 19–24 June 2016; Volume 48, pp. 1747–1756.
12. Salimans, T.; Karpathy, A.; Chen, X.; Kingma, D.P. PixelCNN++: Improving the PixelCNN with discretized logistic mixture likelihood and other modifications. In Proceedings of the 5th International Conference on Learning Representations (ICLR), Toulon, France, 24–26 April 2017.
13. Blaauw, M.; Bonada, J. A Singing Synthesizer Based on PixelCNN. Presented at the María de Maeztu Seminar on Music Knowledge Extraction Using Machine Learning (Collocated with NIPS). Available online: http://www.dtic.upf.edu/~mblaauw/MdM_NIPS_seminar/ (accessed 1 October 2017).
14. Ranzato, M.; Chopra, S.; Auli, M.; Zaremba, W. Sequence level training with recurrent neural networks. In Proceedings of the 4th International Conference on Learning Representations (ICLR), San Juan, Puerto Rico, 2–4 May 2016.

15. Wang, X.; Takaki, S.; Yamagishi, J. A RNN-based quantized F0 model with multi-tier feedback links for text-to-speech synthesis. In Proceedings of the 18th Annual Conference of the International Speech Communication Association (Interspeech), Stockholm, Sweden, 20–24 August 2017; pp. 1059–1063.

16. Taylor, P. *Text-to-Speech Synthesis*; Cambridge University Press: Cambridge, UK, 2009; Chapter 9.1.4, p. 229.

17. Umbert, M.; Bonada, J.; Blaauw, M. Generating singing voice expression contours based on unit selection. In Proceedings of the 4th Stockholm Music Acoustics Conference (SMAC), Stockholm, Sweden, 30 July–3 August 2013; pp. 315–320.

18. Mase, A.; Oura, K.; Nankaku, Y.; Tokuda, K. HMM-based singing voice synthesis system using pitch-shifted pseudo training data. In Proceedings of the 11th Annual Conference of the International Speech Communication Association (Interspeech), Makuhari, Chiba, Japan, 26–30 September 2010; pp. 845–848.

19. Nakamura, K.; Oura, K.; Nankaku, Y.; Tokuda, K. HMM-based singing voice synthesis and its application to Japanese and English. In Proceedings of the 39th IEEE International Conference on Acoustics, Speech and Signal Processing (ICASSP), Florence, Italy, 4–9 May 2014; pp. 265–269.

20. Arik, S.Ö.; Diamos, G.; Gibiansky, A.; Miller, J.; Peng, K.; Ping, W.; Raiman, J.; Zhou, Y. Deep voice 2: Multi-speaker neural text-to-speech. In Proceedings of the Advances in Neural Information Processing Systems 30 (NIPS), Long Beach, CA, USA, 4–9 December 2017.

21. Morise, M.; Yokomori, F.; Ozawa, K. WORLD: A vocoder-based high-quality speech synthesis system for real-time applications. *IEICE Trans. Inf. Syst.* **2016**, *99*, 1877–1884.

22. Morise, M. D4C, a band-aperiodicity estimator for high-quality speech synthesis. *Speech Commun.* **2016**, *84*, 57–65.

23. Tokuda, K.; Kobayashi, T.; Masuko, T.; Imai, S. Mel-generalized cepstral analysis—A unified approach to speech spectral estimation. In Proceedings of the 3rd International Conference on Spoken Language Processing (ICSLP), Yokohama, Japan, 18–22 September 1994.

24. Ueda, N.; Nakano, R. Deterministic annealing EM algorithm. *Neural Netw.* **1998**, *11*, 271–282.

25. Ramachandran, P.; Paine, T.L.; Khorrami, P.; Babaeizadeh, M.; Chang, S.; Zhang, Y.; Hasegawa-Johnson, M.; Campbell, R.; Huang, T. Fast generation for convolutional autoregressive models. In Proceedings of the 5th International Conference on Learning Representations (ICLR), Toulon, France, 24–26 April 2017.

26. Arik, S.Ö.; Chrzanowski, M.; Coates, A.; Diamos, G.; Gibiansky, A.; Kang, Y.; Li, X.; Miller, J.; Raiman, J.; Sengupta, S.; et al. Deep voice: Real-time neural text-to-speech. In Proceedings of the 34th International Conference on Machine Learning (ICML), Stockholm, Sweden, 10–15 July 2017; pp. 195–204.

27. Kalchbrenner, N.; van den Oord, A.; Simonyan, K.; Danihelka, I.; Vinyals, O.; Graves, A.; Kavukcuoglu, K. Video pixel networks. *CoRR arXiv* **2016**, arXiv:1610.00527.

28. Kalchbrenner, N.; Espeholt, L.; Simonyan, K.; van den Oord, A.; Graves, A.; Kavukcuoglu, K. Neural machine translation in linear time. *CoRR arXiv* **2016**, arXiv:1610.10099.

29. Mehri, S.; Kumar, K.; Gulrajani, I.; Kumar, R.; Jain, S.; Sotelo, J.; Courville, A.C.; Bengio, Y. SampleRNN: An unconditional end-to-end neural audio generation model. In Proceedings of the 5th International Conference on Learning Representations (ICLR), Toulon, France, 24–26 April 2017.

30. Sotelo, J.; Mehri, S.; Kumar, K.; Santos, J.F.; Kastner, K.; Courville, A.; Bengio, Y. Char2Wav: End-to-End speech synthesis. In Proceedings of the 5th International Conference on Learning Representations (ICLR), Toulon, France, 24–26 April 2017.

31. Wang, Y.; Skerry-Ryan, R.J.; Stanton, D.; Wu, Y.; Weiss, R.J.; Jaitly, N.; Yang, Z.; Xiao, Y.; Chen, Z.; Bengio, S.; et al. Tacotron: A fully end-to-end text-to-speech synthesis model. In Proceedings of the 18th Annual Conference of the International Speech Communication Association (Interspeech), Stockholm, Sweden, 20–24 August 2017; pp. 4006–4010.

32. Zen, H.; Senior, A. Deep mixture density networks for acoustic modeling in statistical parametric speech synthesis. In Proceedings of the 39th IEEE International Conference on Acoustics, Speech, and Signal Processing (ICASSP), Florence, Italy, 4–9 May 2014; pp. 3872–3876.

33. Zen, H.; Sak, H. Unidirectional long short-term memory recurrent neural network with recurrent output layer for low-latency speech synthesis. In Proceedings of the 40th IEEE International Conference on Acoustics, Speech, and Signal Processing (ICASSP), South Brisbane, QLD, Australia, 19–24 April 2015; pp. 4470–4474.

34. Tokuda, K.; Yoshimura, T.; Masuko, T.; Kobayashi, T.; Kitamura, T. Speech parameter generation algorithms for HMM-based speech synthesis. In Proceedings of the 25th IEEE International Conference on Acoustics, Speech, and Signal Processing (ICASSP), Istanbul, Turkey, 5–9 June 2000; Volume 3, pp. 1315–1318.

35. Oura, K.; Mase, A.; Nankaku, Y.; Tokuda, K. Pitch adaptive training for HMM-based singing voice synthesis. In Proceedings of the 37th IEEE International Conference on Acoustics, Speech and Signal Processing (ICASSP), Kyoto, Japan, 25–30 March 2012; pp. 5377–5380.

36. Shirota, K.; Nakamura, K.; Hashimoto, K.; Oura, K.; Nankaku, Y.; Tokuda, K. Integration of speaker and pitch adaptive training for HMM-based singing voice synthesis. In Proceedings of the 39th IEEE International Conference on Acoustics, Speech and Signal Processing (ICASSP), Florence, Italy, 4–9 May 2014; pp. 2559–2563.

37. Nishimura, M.; Hashimoto, K.; Oura, K.; Nankaku, Y.; Tokuda, K. Singing voice synthesis based on deep neural networks. In Proceedings of the 17th Annual Conference of the International Speech Communication Association (Interspeech), San Francisco, CA, USA, 8–12 September 2016; pp. 2478–2482.

38. Zen, H.; Nose, T.; Yamagishi, J.; Sako, S.; Masuko, T.; Black, A.W.; Tokuda, K. The HMM-based speech synthesis system (HTS) version 2.0. In Proceedings of the 6th ISCA Workshop on Speech Synthesis (SSW6), Bonn, Germany, 22–24 August 2007; pp. 294–299.

39. Iglewicz, B.; Hoaglin, D.C. *How to Detect and Handle Outliers*; ASQC Basic References in Quality Control; ASQC Quality Press: Milwaukee, WI, USA, 1993.

40. Takamichi, S.; Toda, T.; Black, A.W.; Neubig, G.; Sakti, S.; Nakamura, S. Postfilters to modify the modulation spectrum for statistical parametric speech synthesis. *IEEE/ACM Trans. Audio Speech Lang. Process.* **2016**, *24*, 755–767.

41. Umbert, M.; Bonada, J.; Goto, M.; Nakano, T.; Sundberg, J. Expression control in singing voice synthesis: Features, approaches, evaluation, and challenges. *IEEE Signal Process. Mag.* **2015**, *32*, 55–73.

42. ITU-R Recommendation BS.1534-3. *Method for the Subjective Assessment of Intermediate Quality Levels of Coding Systems*; Technical Report; International Telecommunication Union: Geneva, Switzerland, 2015.

43. Kingma, D.P.; Ba, J.L. Adam: A method for stochastic optimization. In Proceedings of the 3rd International Conference on Learning Representations (ICLR), San Diego, CA, USA, 7–9 May 2015.

Article

A Psychoacoustic-Based Multiple Audio Object Coding Approach via Intra-Object Sparsity

Maoshen Jia [1,*], Jiaming Zhang [1], Changchun Bao [1] and Xiguang Zheng [2]

[1]   Beijing Key Laboratory of Computational Intelligence and Intelligent System, Faculty of Information Technology, Beijing University of Technology, Beijing 100124, China; zjm@emails.bjut.edu.cn (J.Z.); baochch@bjut.edu.cn (C.B.)

[2]   Faculty of Engineering & Information Sciences, University of Wollongong, Wollongong NSW2522, Australia; xz725@uow.edu.au

*   Correspondence: jiamaoshen@bjut.edu.cn; Tel.: +86-150-1112-0926

Academic Editor: Vesa Valimaki
Received: 29 October 2017; Accepted: 12 December 2017; Published: 14 December 2017

Abstract: Rendering spatial sound scenes via audio objects has become popular in recent years, since it can provide more flexibility for different auditory scenarios, such as 3D movies, spatial audio communication and virtual classrooms. To facilitate high-quality bitrate-efficient distribution for spatial audio objects, an encoding scheme based on intra-object sparsity (approximate k-sparsity of the audio object itself) is proposed in this paper. The statistical analysis is presented to validate the notion that the audio object has a stronger sparseness in the Modified Discrete Cosine Transform (MDCT) domain than in the Short Time Fourier Transform (STFT) domain. By exploiting intra-object sparsity in the MDCT domain, multiple simultaneously occurring audio objects are compressed into a mono downmix signal with side information. To ensure a balanced perception quality of audio objects, a Psychoacoustic-based time-frequency instants sorting algorithm and an energy equalized Number of Preserved Time-Frequency Bins (NPTF) allocation strategy are proposed, which are employed in the underlying compression framework. The downmix signal can be further encoded via Scalar Quantized Vector Huffman Coding (SQVH) technique at a desirable bitrate, and the side information is transmitted in a lossless manner. Both objective and subjective evaluations show that the proposed encoding scheme outperforms the Sparsity Analysis (SPA) approach and Spatial Audio Object Coding (SAOC) in cases where eight objects were jointly encoded.

Keywords: audio object coding; sparsity; psychoacoustic model; multi-channel audio coding

---

1. Introduction

With the development of multimedia video/audio signal processing, multi-channel 3D audio has been widely employed for applications, such as cinemas and home theatre systems, since it can provide excellent spatial realism of the original sound field, as compared to the traditional mono/stereo audio format.

There are multiple formats for rendering 3D audio, which contain channel-based, object-based and HOA-based audio formats. In traditional spatial sound rendering approach, the channel-based format is adopted in the early stage. For example, the 5.1 surround audio format [1] provides a horizontal soundfield and it has been widely employed for applications, such as the cinema and home theater. Furthermore, typical '3D' formats include a varying number of height channels, such as 7.1 audio format (with two height channels). As the channel number increases, the audio data will raise dramatically. Due to the bandwidth constrained usage scenarios, the spatial audio coding technique has become an ongoing research topic in recent decades. In 1997, ISO /MPEG (Moving Picture Experts Group) designed the first commercially-used multi-channel audio coder MPEG-2

Advanced Audio Coding (MPEG-2 AAC) [2]. It could compress multi-channel audio by adding a number of advanced coding tools to MPEG-1 audio codecs, delivering European Broadcasting Union (EBU) broadcast quality at a bitrate of 320 kbps for a 5.1 signal. In 2006, MPEG Surround (MPS) [3,4] was created for highly transmission of multi-channel sound by downmixing the multi-channel signals into mono/stereo signal and extracting Interaural Level Differences (ILD), ITD (Interaural Time Differences) and IC (Interaural Coherence) as side information. Spatially Squeezed Surround Audio Coding (S^3AC) [5–7], as a new method instead of original "downmix plus spatial parameters" model, exploited spatial direction of virtual sound source and mapping the soundfield from 360° into 60°. At the receiver, the decoded signals can be achieved by inverse mapping the 60° stereo soundfield into 360°.

However, such channel-based audio format has its limitation on flexibility, i.e., each channel is designated to feed a loudspeaker in a known prescribed position and cannot be adjusted for different reproduction needs by the users. Alternatively, a spatial sound scene can be described by a number of sound objects, each positioned at a certain target object position in space, which can be totally independent from the locations of available loudspeakers [8]. In order to fulfill the demand of interactive audio elements, object-based (a.k.a. object-oriented) audio format enables users to control audio content or sense of direction in application scenarios where the number of sound sources varies, sources move are commonly encountered. Hence, object signals generally need to be rendered to their target positions by appropriate rendering algorithms, e.g., Vector Base Amplitude Panning (VBAP) [9]. Therefore, object-based audio format can personalize customer's listening experience and make surround sound more realistic. By now, object-based audio has been commercialized in many acoustic field, e.g., Dolby ATMOS for cinemas [10].

To facilitate high-quality bitrate-efficient distribution of audio objects, several methods have been developed, one of these techniques is MPEG Spatial Audio Object Coding (SAOC) [11,12]. SAOC encodes audio objects into a mono/stereo downmix signal plus side information via Quadrature Mirror Filter (QMF) and extract the parameters that stand for the energy relationship between different audio objects. Additionally, Directional Audio Coding (DirAC) [13,14] compress a spatial scene by calculating a direction vector representing spatial location information of the virtual sources. At the decoder side, the virtual sources are created from the downmixed signal at positions given by the direction vectors and they are panned by combining different loudspeakers through VBAP. The latest MPEG-H 3D audio coding standard incorporates the existing MPEG technology components to provide universal means for carriage of channel-based, object-based and Higher Order Ambisonics (HOA) based inputs [15]. Both MPEG-Surround (MPEG-S) and SAOC are included in MPEG-H 3D audio standard.

Recently, a Psychoacoustic-based Analysis-By-Synthesis (PABS) method [16,17] was proposed for encoding multiple speech objects, which could compress four simultaneously occurring speech sources in two downmix signals relied on inter-object sparsity [18]. However, with the number of objects increases, the inter-object sparsity becomes weakened, which leads to quality loss of decoded signal. In our previous work [19–21], a multiple audio objects encoding approach was proposed based on intra-object sparsity. Unlike the inter-object sparsity employed in PABS framework, this encoding scheme exploited the sparseness of object itself. That is, in a certain domain, an object signal can be represented by a small number of time-frequency instants. The evaluation results validated that this intra-object based approach achieved a better performance than PABS algorithm and retain the superior perceptual quality of the decoded signals. However, the aforementioned technique still has some restrictions which leads to a sub-optimum solution for object compression. Firstly, Short Time Fourier Transform (STFT) is chosen as the linear time-frequency transform to analyze audio objects. Yet the energy compaction capability of STFT is not optimal. Secondly, the above object encoding scheme concentrated on the features of object signal itself without considering the psychoacoustic, thus it is not an optimal quantization means for Human Auditory System (HAS).

This paper expands on the contributions in [19]. Based on intra-object sparsity, we propose a novel encoding scheme for multiple audio objects to further optimize our previous proposed approach and minimize the quality loss caused by compression. Firstly, by exploiting intra-object sparsity in the Modified Discrete Cosine Transform (MDCT) domain, multiple simultaneously occurring audio objects are compressed into a mono downmix signal with side information. Secondly, psychoacoustic model is utilized in the proposed codec to accomplish an optimal quantization for HAS. Hence, a Psychoacoustic-based Time-Frequency (TF) instants sorting algorithm is proposed for extracting the dominant TF instants in the MDCT domain. Furthermore, by utilizing these extracted TF instants, we propose a fast algorithm of Number of Preserved Time-Frequency Bins (*NPTF*, defined in Appendix A) allocation strategy to ensure a balanced perception quality for all object signals. Finally, the downmix signal can be further encoded via SQVH technique at desirable bitrate and the side information is transmitted in a lossless manner. In addition, a comparative study of intra-object sparsity of audio signal in the STFT domain and MDCT domain is presented via statistical analysis. The results show that audio objects have sparsity-promoting property in the MDCT domain, which means that a greater data compression ratio can be achieved.

The remainder of the paper is structured as follows: Section 2 introduces the architecture of the encoding framework in detail. Experimental results are presented and discussed in Section 3, while the conclusion is given in Section 4. Appendix A investigates the sparsity of audio objects in the STFT and MDCT domain, respectively.

2. Proposed Compression Framework

In the previous work, we adopted STFT as time-frequency transform to analyze the sparsity of audio signal and designed a codec based on the intra-object sparsity. From the statistical results of sparsity presented in Appendix A, we know that audio signals satisfy the approximate k-sparsity both in the STFT and MDCT domain, i.e., the energy of audio signal is almost concentrated in k time-frequency instants. In other words, audio signals have sparsity-promoting property in the MDCT domain in contrast to STFT, that is, $k(r_{FEPR})_{\mathrm{MDCT}} < k(r_{FEPR})_{\mathrm{STFT}}$. By using this advantage of MDCT, a multiple audio objects compression framework is proposed in this section based on intra-object sparsity. The proposed encoding scheme consists of five modules: time-frequency transform, active object detection, psychoacoustic-based TF instants sorting, *NPTF* allocation strategy and Scalar Quantized Vector Huffman Coding (SQVH).

The following process is operated in a frame-wise fashion. As is shown in Figure 1, all input audio objects (Source 1 to Source Q) are converted into time-frequency domain using MDCT. After active object detection, the TF instants of all active objects will be sorted according to Psychoacoustic model in order to extract the most perceptually important time-frequency instants. Then, a *NPTF* allocation strategy among all audio objects is proposed to counterpoise the energy of all preserved TF instants of each object. Thereafter, the extracted time-frequency instants are downmixed into a mono mixture stream plus side information via downmix processing operation. Particularly attention is that the downmix signal can be further compressed by existing audio coding methods. In this proposed method, SQVH technique is employed after de-mixing all TF instants, because it can compress audio signal at desirable bitrate. At the receiving end, Source 1 to Source Q can be decoded by exploiting the received downmix signal and the side information. The detailed contents are described below.

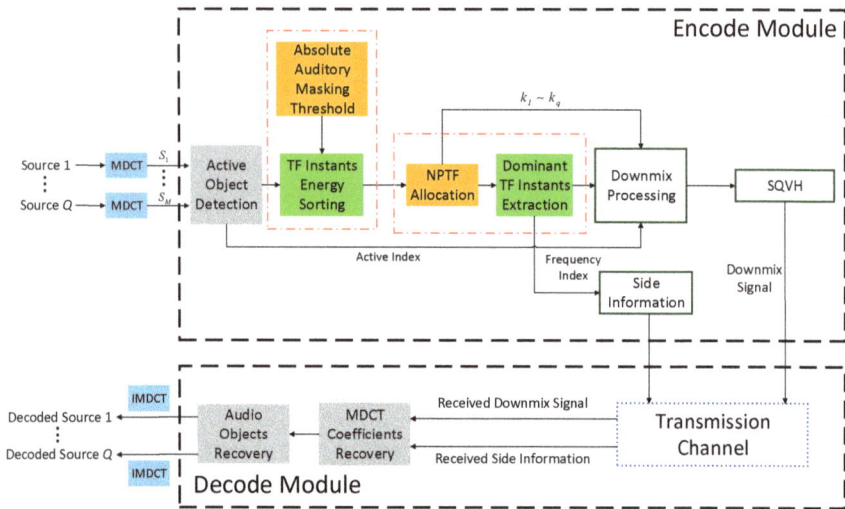

Figure 1. The block diagram for the proposed compression framework. (MDCT, Modified Discrete Cosine Transform; IMDCT, Inverse Modified Discrete Cosine Transform; NPTF, Number of Preserved Time-Frequency Bins; SQVH, Scalar Quantized Vector Huffman Coding; TF, Time-Frequency).

2.1. MDCT and Active Object Detection

In n^{th} frame, an input audio object $s_n = [s_n(1), s_n(2), \ldots, s_n(M)]$ is transformed into the MDCT domain, denoted by $S(n, l)$, where n $(1 \leq n \leq N)$ and l $(1 \leq l \leq L)$ are frame number and frequency index, respectively. $M = 1024$ is the frame length. Here, a 2048-points MDCT is applied with 50% overlapped [22]. By this overlap, discontinuity at block boundary is smoothed out without increasing the number of transform coefficients. Afterwards, MDCT of an original signal s_n can be formulated as:

$$S(n,l) = 2\left[s_n \cdot \left(\boldsymbol{\varphi}_l^1\right)^T + s_{n+1} \cdot \left(\boldsymbol{\varphi}_l^2\right)^T\right] \tag{1}$$

where $L = 1024$, $\boldsymbol{\varphi}_l^1 \triangleq \{\varphi_l^1(1), \varphi_l^1(2), \cdots, \varphi_l^1(M)\}$, $\boldsymbol{\varphi}_l^2 \triangleq \{\varphi_l^2(1), \varphi_l^2(2), \cdots, \varphi_l^2(M)\}$ are the basis functions corresponding to n^{th} frame and $(n + 1)^{th}$ frame. $\varphi_l^1(m) = \omega(m) \cdot \cos\left[\frac{\pi}{M} \cdot \left(m + \frac{M+1}{2}\right) \cdot \left(l - \frac{1}{2}\right)\right]$, $\varphi_l^2(m) = \omega(m + M) \cdot \cos\left[\frac{\pi}{M} \cdot \left(m + \frac{3M+1}{2}\right) \cdot \left(l - \frac{1}{2}\right)\right]$ and ^{T} is the transpose operation. In addition, a Kaiser–Bessel derived (KBD) short-time window slid along the time axis with 50% overlapping between frames is used as window function $\omega(m)$.

In order to ensure the encoding scheme only encodes active frames without processing the silence frames, an Active Object Detection technique is applied to check the active audio objects in the current frame. Hence, Voice Activity Detection (VAD) [23] is utilized in this work, which is based on the short-time energy of audio in the current frame and comparison with the estimated background noise level. Each source uses a flag to indicate whether it is active in current frame. i.e.,

$$flag = \begin{cases} 1, & \text{if the current object is active} \\ 0, & \text{otherwise} \end{cases} \tag{2}$$

Afterwards, only the frames which are detected as active will be sent into the next module. In contrast, the mute frames will be ignored in the proposed codec. This procedure ensures that silence frames cannot be selected.

2.2. Psychoacoustic-Based TF Instants Sorting

In Appendix A, it is proved that the majority of the frame energy concentrates in finite k time-frequency instants for each audio object. For this reason, we can extract these k dominant TF instants for compression. In our previous work [19–21], TF instants are sorted and extracted by natural ordering via the magnitude of the normalized energy. However, this approach does not take into account HAS. It is well-known that HAS is not equally sensitive to all frequencies within the audible band since it has a non-flat frequency response. This simply means that we can hear some tones better than others. Thus, tones played at the same volume (intensity) at different frequencies are perceived as if they are being played at different volumes. For the purpose of enhance perceptual quality, we design a novel method through absolute auditory masking threshold to extract the dominant TF instants.

The absolute threshold of hearing characterizes the amount of energy needed in a pure tone such that it can be detected by a listener in a noiseless environment and it is expressed in terms of dB Sound Pressure Level (SPL) [24]. The quiet threshold is well approximated by the continuous nonlinear function, which is based on a number of listeners that were generated in a National Institutes of Health (NIH) study of typical American hearing acuity [25]:

$$T(f) = 3.64 \times (f/1000)^{-0.8} - 6.5 \times e^{-0.6(f/1000-3.3)^2} + 10^{-3} \times (f/1000)^4 \tag{3}$$

where $T(f)$ reflects the auditory properties for human ear in the STFT domain. Hence, the $T(f)$ should be discretized and converted into the MDCT domain. The whole processing procedure includes two steps: inverse time-frequency transform and MDCT [26]. After these operations, absolute auditory masking threshold in the MDCT domain is denoted as $T_{mdct}(l)$ (dB expression), where $l = 1, 2, \ldots, L$. Then, an L-dimensional Absolute Auditory Masking Threshold (AAMT) vector $T \equiv [T_{mdct}(1), T_{mdct}(2), \ldots, T_{mdct}(L)]$ is generated for subsequent computing. From psychoacoustic theory, it is clear that if there exists a TF bin (n_0, l_0) that the difference between $S_{dB}(n_0, l_0)$ (dB expression of $S(n_0, l_0)$) and $T_{mdct}(l_0)$ is larger than other TF bins, which means that $S(n_0, l_0)$ can be perceived more easily than other TF components, but not vice versa. Specifically, any signals below this threshold curve (i.e., $S_{dB}(n_0, l_0) - T_{mdct}(l_0) < 0$) is imperceptible (because $T_{mdct}(l)$ is the lowest limit of HAS). Rely on this phenomenon, the AAMT vector T is used for extracting the perceptual dominant TF instants efficiently.

For q^{th} $(1 \leq q \leq Q)$ audio object $S_q(n, l)$, whose dB expression is written as $S_{q\_dB}(n, l)$. An aggregated vector can be attained by converging each $S_{q\_dB}(n, l)$ denoted as $S_{q\_dB} \equiv [S_{q\_dB}(n, 1), S_{q\_dB}(n, 2), \ldots, S_{q\_dB}(n, L)]$. Subsequently, a perceptual detection vector is designed as:

$$P_q = S_{q\_dB} - T \equiv [P_q(n, 1), P_q(n, 2), \cdots, P_q(n, L)] \tag{4}$$

where $P_q(n, l) = S_{q\_dB}(n, l) - T_{mdct}(l)$. To sort each element in P_q according to the magnitude in descending order, mathematically, a new vector can be attained as:

$$P'_q \equiv \left[P_q(n, l_1^q), \cdots, P_q(n, l_L^q) \right] \tag{5}$$

the elements in P'_q satisfy:

$$P_q(n, l_i^q) \geq P_q(n, l_j^q), \ \forall i < j, \ i, j \in \{1, 2, \cdots, L\} \tag{6}$$

where $l_1^q, \cdots, l_L^q$ is the reorder frequency index which represent the perceptual significantly TF instants in order of importance for HAS. In other words, $S_q(n, l_1^q)$ is the most considerable component with respect to HAS. In contrast, $S_q(n, l_L^q)$ is almost the least significant TF instant for HAS.

2.3. NPTF Allocation Strategy

Allocating the *NPTF* for each active object signal can be actualized with various manners according to realistic application scenarios. As a most common used means called simplified average distribution method, all active objects share the same *NPTF* has been employed in [19,21]. This allocation method balances a tradeoff between computational complexity and perceptual quality. Therefore, it is a simple and efficient way. Nonetheless, this allocation strategy cannot guarantee all decoded objects with similar perceptual quality. Especially, the uneven quality can be emerged if there exists big difference of intra-object sparseness amongst objects. To conquer the above-mentioned issue, an Analysis-by-Synthesis (ABS) framework was proposed to balance the perceptual quality for all objects through solving a minimax problem via the iterative processing [20]. The test results show that this technique yields the approximate evenly distributed Frame Energy Preservation Ratio (*FEPR*, defined in Appendix A) for all objects. Despite the harmonious perceptual quality can be maintained, the attendant problem which is the sharp increase in computational complexity cannot be neglected. Accordingly, relied on the TF sorting result obtained in Section 2.2, an *NPTF* allocation strategy for obtaining a balanced perceptual quality of all inputs is proposed in this work.

In the nth frame, we assume that the qth object will be distributed k_q *NPTF*, i.e., k_q TF instants will be extracted for coding. An Individual Object Energy Retention ratio (*IOER*) function for the qth object is defined by:

$$f_{IOER}(k,q) = \frac{\sum_{i=1}^{k} S_q\left(n, l_i^q\right)}{\sum_{l=1}^{L} S_q(n,l)} \qquad (7)$$

where l_i^q is the reorder frequency index obtained in the previous section. *IOER* function represents the energy of the k perceptual significant elements against the original signal $S_q(n, l)$. Thus, k_q will be allocated for each object with approximate *IOER*. Under the criterion of minimum mean-square error, for all $q \in \{1, 2, \ldots, Q\}$ the k_q can be attained via a constrained optimization equation as follow:

$$\min_{k_1, k_2, \cdots, k_Q} \sum_{q=1}^{Q} \left\| f_{IOER}(k_q, q) - \bar{f} \right\|^2$$
$$\text{s.t.} \sum_{q=1}^{Q} k_q = L \qquad (8)$$

where $\bar{f} = \frac{1}{Q} \sum_{q=1}^{Q} f_{IOER}(k,q)$ represents the average energy of all objects. The optimal solution k_1, k_2, $\ldots$, k_Q for each object are the desired $NPTF_1$, $NPTF_2$, $\ldots$, $NPTF_Q$, which can be searched by our proposed method elaborated in Algorithm 1.

The proposed *NPTF* allocation strategy allows different reserved TF instants (i.e., MDCT coefficients) for each object among a certain group of multi-track audio objects without iterative processing, therefore, the computational complexity decrease rapidly through the dynamic TF instants distribution algorithm. In addition, a sub-equal perception quality for each object can be maintained via our proposed *NPTF* allocation strategy rather than pursuit the quality of a particular object.

Thereafter, vector P_q' needs to be extract the $NPTF_q(k_q)$ elements to forming a new vector $\widetilde{P}_q \equiv \left[P_q(n, l_1^q), \cdots, p_q(n, l_{NPTF_q}^q) \right]$. It should be note that $l_1^q, l_2^q, \ldots, l_{NPTF_q}^q$ indicate the origin of $S_q\left(n, l_1^q\right), S_q\left(n, l_2^q\right), \ldots, S_q\left(n, l_{NPTF_q}^q\right)$, respectively. We group $l_1^q, l_2^q, \ldots, l_{NPTF_q}^q$ into a vector $I_q \equiv \left[l_1^q, l_2^q, \ldots, l_{NPTF_q}^q \right]$, in the meantime, a new vector containing all extracted TF instants $\hat{S}_q \equiv \left[S_q\left(n, l_1^q\right), S_q\left(n, l_2^q\right), \ldots, S_q\left(n, l_{NPTF_q}^q\right) \right]$ is generated. Finally, both I_q and $\hat{S}_q$ should be stored locally and sent into the Downmix Processing module.

Algorithm 1: NPTF allocation strategy based on bisection method

---

Input: Q ▶ number of audio objects

Input: $\left\{S_q(n,l)\right\}_{q=1}^{Q}$ ▶ MDCT coefficients of each audio object

Input: $\left\{l_i^q\right\}_{i=1}^{L}$ ▶ reordered frequency index by psychoacoustic model

Input: *BPA*   ▶ lower limit used in dichotomy part

Input: *BPB*   ▶ upper limit used in dichotomy part

Input: *BPM*   ▶ median used in dichotomy part

Output: K ▶ desired NPTF allocation result

1. Set $K = \varnothing$
2. **for** $q = 1$ to Q **do**
3.   **for** $k = 1$ to L **do**
4.   Calculate IOER function $f_{\text{IOER}}(k,q)$ using $\left\{S_q(n,l)\right\}_{q=1}^{Q}$ and $\left\{l_i^q\right\}_{i=1}^{L}$ in Formula (12).
5.   **end for**
6. **end for**
7. Initialize $BPA = 0$, $BPB = 1$, $BPM = 0.5 \cdot (BPA + BPB)$, $STOP = 0.01$ chosen based on a series of informal experimental results.
8. **while** $(BPB - BPA > STOP)$ **do**
9.   Find the index value corresponding to BPM value in IOER function (i.e., $f_{\text{IOER}}(k_q, q) \approx BPM$), denoted by k_q.
10.   **if** $\sum\limits_{q=1}^{Q} k_q > L$ **then**
11. $BPB = BPM$,
12. $BPM = [0.5 \cdot (BPA + BPB)]$.
13.   **else**
14. $BPA - BPM$,
15. $BPM = [0.5 \cdot (BPA + BPB)]$.
16.   **end if**
17.   **end while**
18. $K = \left\{k_q\right\}_{q=1}^{Q}$
19. **return** K

2.4. Downmix Processing

After extracting the dominant TF instants $\hat{S}_q$, source 1 to source Q only contains the perception significantly MDCT coefficients of all active audio objects. However, each source include a number of zero entries, hence, the downmix processing must be exploited which aims to redistributing the nonzero entries of the extracted TF instants from 1 to L in the frequency axis to generate the mono downmix signal.

For each active source q, a k-sparse $(k = NPTF_q)$ approximation signal of $S_q(n,l)$ can be attained by rearrange $\hat{S}_q$ in the original position, expressed as:

$$\widetilde{S}_q(n,l) = \begin{cases} S_q(n,l), & \text{if } l \in I_q \\ 0, & \text{otherwise} \end{cases} \tag{9}$$

The downmix matrix is denoted as $\boldsymbol{D}_n \equiv \left[\widetilde{S}_1, \widetilde{S}_2, \cdots, \widetilde{S}_Q\right]^{\text{T}}$, where $\widetilde{S}_q \equiv \left[\widetilde{S}_q(n,1), \widetilde{S}_q(n,2), \ldots, \widetilde{S}_q(n,L)\right]$ and ^{T} is the transpose operation. This matrix is sparse matrix containing $M \times L$ entries. Through a column-wise scanning of $\boldsymbol{D}_n$ and sequencing the nonzero entries onto the frequency axis according to the scanning order, the mono downmix signal and side information can be obtained via Algorithm 2.

Figure 2 indicates the demixing procedure in accordance with an example of eight simultaneously occurring audio objects. Each square represents a time-frequency instant. The preserved TF components for each sound source (a total of 8 audio objects in this example) are represented by various color-block and shading.

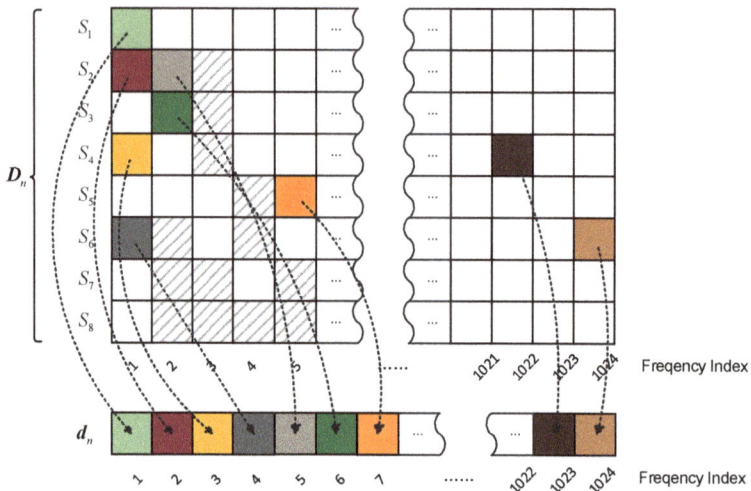

Figure 2. Example of TF (Time-Frequency) instants extraction and de-mixing procedure with eight unique simultaneously occurring sources.

Furthermore, the above-presented downmix processing guarantees the redistributed TF components locating in the nearby frequency position as their original position, which is prerequisite for subsequent Scalar Quantized Vector Huffman Coding (SQVH). Consequently, the downmix signal d_n can be further encoded by SQVH technique. Meanwhile, the side information compressed via the Run Length Coding (RLC) and the Golomb-Rice coding [19] at about 90 kbps.

2.5. Downmix Signal Compressing by SQVH

SQVH is a kind of efficient transform coding method which is used in fixed bitrate codec [26–28]. In this section, SQVH with variable bitrate for encoding downmix signal is designed and described as follows.

For the nth frame, the downmix signal d_n attained in Algorithm 2 can be expressed as:

$$d_n \equiv [d_n(1), d_n(2), \cdots, d_n(L)] \tag{10}$$

d_n need to be divided into 51 sub-bands, each sub-band contains 20 TF instants, respectively (without considering the last 4 instants). The sub-band power (spectrum energy) is determined for each of the 51 regions and it is defined as root-mean-square (*rms*) value of coterminous 20 MDCT coefficients computed as:

$$R_{rms}(r) = \sqrt{\frac{1}{20} \sum_{l=1}^{20} d_n^2(20(r-1)+l)} \tag{11}$$

where r is region index, $r = 0, 1, \ldots, 50$. The region power is then quantized with a logarithmic quantizer, $2^{(i/2+1)}$ are set to be quantization values, where i is an integer in the range $[-8, 31]$. $R_{rms}(0)$ is the lowest frequency region, which is quantized with 5 bits and transmitted directly in transmission channel. The quantization indices of the remaining 50 regions, which are differentially coded against the last highest-numbered region and then Huffman coded with variable bitrates. In each sub-band, the Quantized Index (*QI*) value can be given by:

$$QI_r(l) = \min\left\{ \left\lfloor \frac{|d_n(20 \cdot (r-1)+l)|}{R_{rms}(r) \times q_{stepsize}} + b \right\rfloor, \quad MAX \right\} \tag{12}$$

where $q_{stepsize}$ is quantization steps, b is an offset value according to different categories, $\lfloor\ \rfloor$ denotes a round-up operation, *MAX* is maximum of MDCT coefficients corresponding to that **category** and l represents the lth vector in the region r. There are several **categories** designed in SQVH coding. The **category** assigned to a region defines the quantization and coding parameters such as quantization step size, offset, vector dimension v_d and an expected total number of bits. The coding parameters for different category is given in Table 1.

Algorithm 2: Downmix processing compression algorithm

Input: Q	► number of audio objects
Input: L	► frequency index
Input: λ	► downmix signal index
Input: $\widetilde{S}_q$	► k-sparse approximation signal of S_q
Output: SI_n	► side information matrix
Output: d_n	► downmix signal

1.   Initialize $\lambda= 1$.
2.   Set $SI_n = 0$, $d_n = 0$.
3.   **for** $l = 1$ to L **do**
4.   **for** $q = 1$ to Q **do**
5.   **if** $\widetilde{S}_q(n,l) \neq 0$ **then**
6. $d_n(\lambda) = \widetilde{S}_q(n,l)$.
7. $SI_n(q, l) = 1$.
8.   Increment λ.
9.   **end if**
10.   **end for**
11. **end for**
12. **return** d_n and SI_n

Table 1. The coding parameters for different category.

Categories	$q_{stepsize}$	b	*MAX*	v_d	Bit Count
0	$2^{-1.5}$	0.3	13	2	52
1	$2^{-1.0}$	0.33	9	2	47
2	$2^{-0.5}$	0.36	6	2	43
3	$2^{0.0}$	0.39	4	4	37

As is depicted in Table 1, four categories are selected in this work. Category 0 has the smallest quantization step size and uses the most bits, but not vice-versa. The set of scalar values, $QI_r(l)$, correspond to a unique vector is identified by an index as follows:

$$v_{index}(i) = \sum_{j=0}^{v_d-1} QI_r(i \times v_d + j)(MAX + 1)^{\lfloor v_d-(j+1)\rfloor} \tag{13}$$

where i represents the ith vector in region r and j is the index to the jth value of $QI_r(l)$ in a given vector. Then, all vector indices are Huffman coded with variable bit-length code for that region. Three types of bit-stream distributions are given in the proposed method, whose performance is evaluated in next section.

2.6. Decoding Process

In decoding stage, MDCT coefficients recovery is an inverse operation of de-mixing procedure, thus it needs the received downmix signal and the side information as auxiliary information. The downmix signal is decoded by the same standard audio codec as used in the encoder and the side information is decoded by the lossless codec. Thereafter, all recovered TF instants are assigned to the corresponding audio object. Finally, all audio object signals are obtained by transforming back to the time domain using the IMDCT.

3. Performance Evaluation

In this section, a series of objective and subjective tests are presented, which aim to examine the performance of the proposed encoding framework.

3.1. Test Conditions

The QUASI audio database [29] is employed as the test database in our evaluation work, which offers a vast variety categories of audio object signals (e.g., piano, vocal, drums, vocal, etc.) sampled at 44.1 kHz. All the test audio data are selected from this database. Four test files are used for evaluate the encoding quality when multiple audio objects are active simultaneously. Each test file consists of eight audio segments which is created with the length of 15 s. In other words, eight audio segments representing eight different types of audio objects are grouped together to form a multi-track test audio file, where the notes are also different among the eight tracks. The MUltiple Stimuli with Hidden Reference and Anchor (MUSRHA) methodology [30] and Perceptual Evaluation of Audio Quality (PEAQ) are employed in subjective and objective evaluation, respectively. Moreover, there are 15 listeners who took part in each subjective listening test. A 2048-points MDCT is utilized with 50% overlapping while adopting KBD window as window function.

3.2. Objective Evaluations

The first experiment is performed in the lossless transmission case, it means that both the downmix signal and the side information are compressed using lossless techniques. The Sparsity Analysis (SPA) multiple audio objects compression technique proposed in our previous work is served as reference approach [19] (named "SPA-STFT") because of its superior performance. Meanwhile, the intermediate step given by SPA that uses the MDCT (named 'SPA-MDCT') is also compared in this test. The Objective Difference Grade (ODG) score calculated by the PEAQ of ITU-R BS.1387 is chosen as the evaluation criterion, which reflect the perceptual difference between the compressed signal and the original one. The ODG values vary from 0 to -4 with 0 being imperceptible loss in quality and -4 being a very annoying degradation in quality. What needs to be emphasized is that ODG scores cannot be treated as an absolute criterion because it only provide a relative reference value of the perceptual quality. Condition 'Pro' represents the objects encoded by our proposed encoding framework while condition 'SPA-STFT' and 'SPA-MDCT' are the reference approaches. Note that 'SPA-STFT' encoding approach exploits a 2048-points Short Time Fourier Transform (STFT) with 50% overlapping.

Statistical results are shown in Figure 3 where each subfigure corresponds to an eight-track audio file. From each subfigure, it can be observed that the decoded signals through our proposed encoding framework has the highest ODG score compared to both the SPA and the MDCT-based SPA approach, which indicates that the proposed framework can cause less damage to audio quality compared to these two reference approaches.

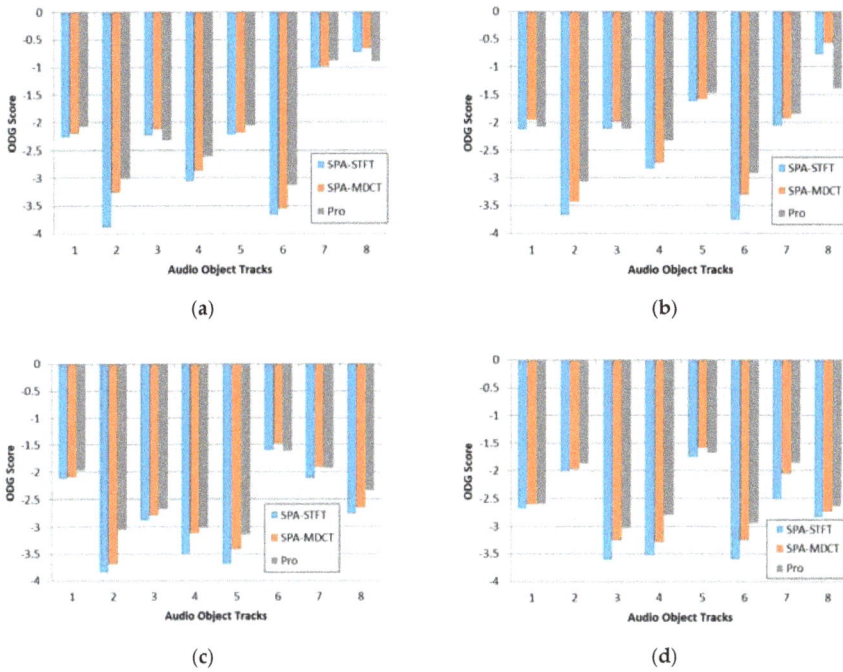

(a)

(b)

(c)

(d)

Figure 3. ODG (Objective Difference Grade) Score for the proposed audio object encoding approach and the SPA (Sparsity Analysis) framework (both in the STFT (Short Time Fourier Transform) and MDCT domain). (**a–d**) represent the results for 4 multi-track audio files.

In addition, the performance of the MDCT-based SPA approach is better than the SPA, which prove that the selection of MDCT as time-frequency transform is efficient. Furthermore, in order to observe the quality differences of decoded objects, the standard deviation of each file is given as follow:

As illustrated in Figure 4, our proposed encoding framework has a lower standard deviation than the reference algorithms for each multi-track audio file. Hence, it proves that a more balanced quality of decoded objects can be maintained compared to the reference approaches. In general, this test validates that the proposed approach is robust to different kinds of audio objects.

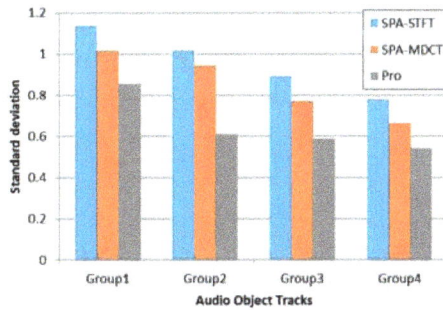

Figure 4. The standard deviation of ODG score of four multi-track audio files.

In the lossy transmission case, the downmix signal which generated by encoder is further compressed using the SQVH at 105.14 kbps, 112.53 kbps and 120.7 kbps, respectively. Each sub-band corresponds to a group of certain $q_{stepsize}$, whose allocation for three types of bitrates can be calculated as shown in Table 2.

Table 2. The $q_{stepsize}$ allocation for three types of bitrates.

The Index of the Bitrate Sub-Band	r			
	1~13	14~26	27~39	40~51
105.14 kbps	$2^{-1.5}$	$2^{-1.0}$	$2^{-0.5}$	$2^{0.0}$
112.53 kbps	$2^{-1.5}$	$2^{-1.0}$	$2^{-1.0}$	$2^{-1.0}$
120.7 kbps	$2^{-1.5}$	$2^{-1.5}$	$2^{-1.5}$	$2^{-1.5}$

The ODG score in three types of bitrates are presented in Figure 5. Condition 'Pro-105', 'Pro-112', 'Pro-120' correspond to compress downmix signal at 105.14 kbps, 112.53 kbps and 120.7 kbps, respectively. It can be observed that the higher quantization precision leads to the better quality of decoded objects but the total bitrates increase as well. Therefore, we cannot pursue a single factor such as high audio quality or low bitrate for transmission [25]. In consequence, we need to make a trade-off between audio quality and total bitrates in practical application scenarios.

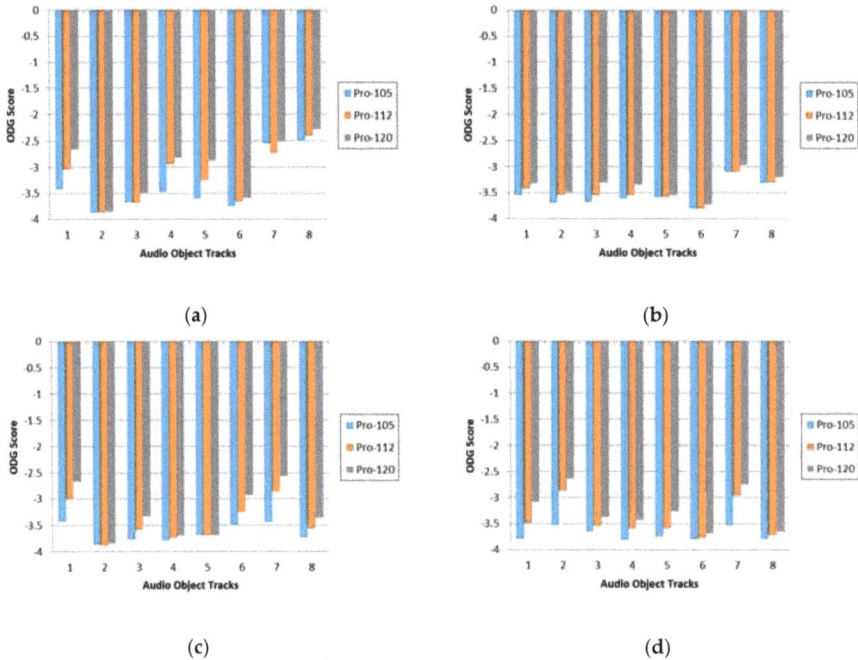

(a)

(b)

(c)

(d)

Figure 5. The ODG score of four multi-track audio files, where each file correspond to three types of bitrates. (**a–d**) represent the results for 4 multi-track audio files.

3.3. Subjective Evaluation

The subjective evaluation is further utilized to measure the perceptual quality of decoded object signals, which consists of four MUltiple Stimuli with Hidden Reference and Anchor (MUSHRA) listening tests. Sennheiser HD600 headphone is used for playback. Note that for the first three tests, each decoded object generated by the corresponding approach is played independently without spatialization.

The first test is the lossless transmission case, aims to make a comparison between our proposed encoding framework and the SPA algorithm. Four group multi-track audio files used in previous experiments are also treated as test data in this section. Condition 'SPA' means the reference approach (the same as condition 'SPA-STFT' in Section 3.2) and condition 'Pro' means the proposed framework. The original object signal is served as the Hidden Reference (condition 'Ref') and condition 'Anchor' is 3.5 kHz low-pass filtered anchor signal. A total of 15 listeners participated in the test.

Results are shown in Figure 6 with 95% confidence intervals. It can be observed that the proposed encoding framework achieves a higher score than the SPA approach with clear statistical significant differences. Moreover, the MUSHRA scores for the proposed framework achieve over 80 indicating 'Excellent' subjective quality compared to the Hidden Reference, which proves that the better perceptual quality can be attained compared to the reference approach.

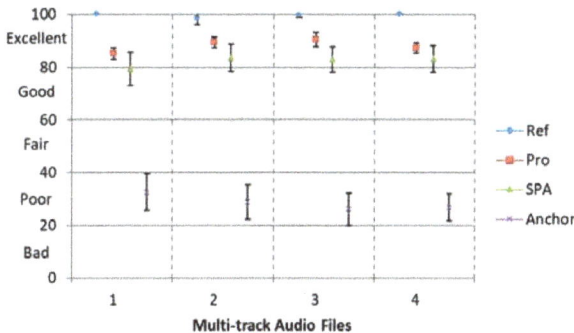

Figure 6. MUltiple Stimuli with Hidden Reference and Anchor (MUSHRA) test results for the SPA framework and the proposed framework with 95% confidence intervals.

For lossy transmission case, the downmix signal encoded at 105 kbps via SQVH corresponds to 'Pro-105'. Condition 'SPA-128' means the reference approach whose downmix signal compressed at the bitrate of 128 kbps using the MPEG-2 AAC codec.

Results are presented in Figure 7 with 95% confidence intervals. Obviously, our proposed encoding scheme has a better perceptual quality and a lower bitrate compared to the SPA approach. That is, when a similar perceptual quality is desired, the proposed method requires less total bitrate than the SPA approach.

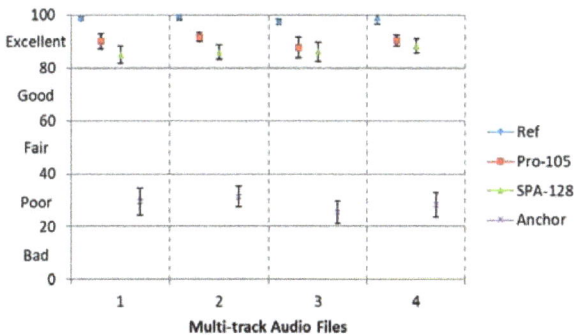

Figure 7. MUSHRA test results for the SPA method encoding at 128 kbps and the proposed approach at 105.14 kbps with 95% confidence intervals.

Furthermore, we evaluate the perceptual quality of the decoded audio objects using our proposed approach, using MPEG-2 AAC to encode each object independently and using Spatial Audio Object Coding (SAOC). The MUSHRA listening test is employed with five conditions, namely, Ref, Pro-105, AAC-30, SAOC and Anchor. The downmix signal in condition 'Pro-105' is further compressed using SQVH at 105.14 kbps. Meanwhile, the side information can be compressed at about 90 kbps [19]. Condition 'AAC-30' is the separate encoding of each original audio object using the MPEG-2 AAC codec at 30 kbps, the total bitrate is almost the same as 'Pro-120' (30 kbps/channel × 8 channels = 240 kbps). Condition 'SAOC' represents the objects are encoded by SAOC. The total SAOC side information rate of input objects is about 40 kbps (5 kbps per object), while the downmix signal generated by SAOC is compressed by the standard audio codec MPEG-2 AAC at the bitrate of 128 kbps.

It is demonstrated in Figure 8 that our proposed approach at 105 kbps possess the similar perceptual quality as separate encoding approach using MPEG-2 AAC. Yet the complexity of separate encoding is much higher than our proposed approach. Furthermore, both our proposed method and separate encoding approach attained a better performance compared with SAOC.

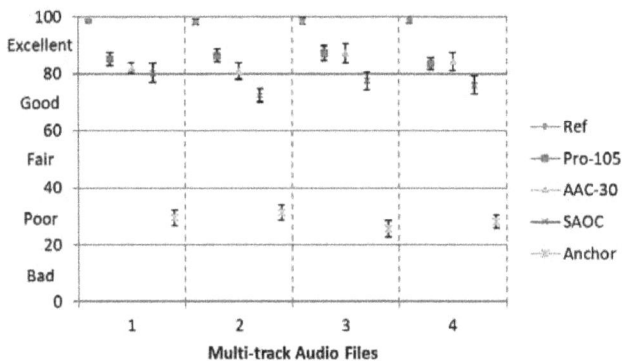

Figure 8. MUSHRA test results for separate AAC (Advanced Audio Coding) encoding at 30 kbps, SAOC (Spatial Audio Object Coding) and our proposed approach at 105 kbps with 95% confidence intervals.

The last test devotes to evaluate the quality of the spatial soundfield generated by positioning the decoded audio objects in different spatial locations, which stands for the real application scenario. Specifically, for each eight-track audio, which are positioned uniformly in a circumference with a center at the listener, i.e., the locations are $0°$, $±45°$, $±90°$, $±135°$, $±180°$, respectively. A binaural signal (test audio data) is created by convoluting each independent decoded audio object signal with the corresponding Head-Related Impulse Responses (HRIR) [31]. The MUSHRA listening test is employed with 6 conditions, namely, Ref, Pro-105, SPA-128, AAC-30, SAOC and Anchor, which are the same as previous tests. Here, Sennheiser HD600 headphone is used for playing the synthesized binaural signal.

It can be observed from Figure 9 that our proposed method can achieve a higher scores compared to all the rest encoding approaches. The results (Figures 8 and 9) also show that the proposed approach achieves a significant improvement over separate encoding method using MPEG-2 AAC for binaural rendering but not in the independently playback scenario. This is due to the spatial hearing theory, which reveals that in each frequency only a few audio objects located at different positions can be perceived by the human ear (i.e., not all audio objects are sensitive at same frequency). In our proposed codec, only the most perceptually important time-frequency instants (not all time-frequency instants) of each audio object are coded with a higher quantization precision, while these frequency components are important for HAS. The coding error produced by our codec can be masked by spatial masking effect to a great extant from the last experiment. However, MPEG-2 AAC encodes all time-frequency instants with a relatively lower quantization precision at 30 kbps. When multiple audio objects were encoded separately by MPEG-2 AAC, there are some coding error that cannot be reduced by spatial masking effect. Hence, the proposed approach shows significant improvements over condition 'AAC-30' for binaural rendering.

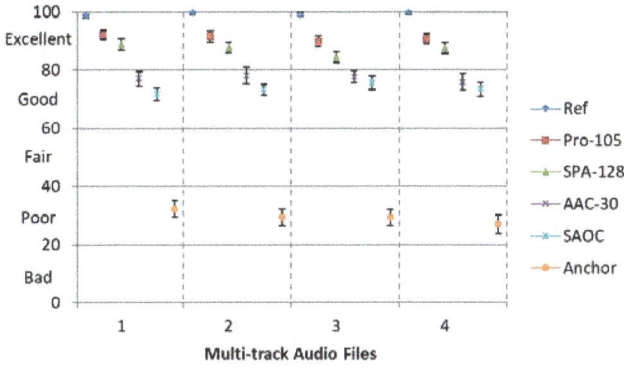

Figure 9. MUSHRA test results with 95% confidence intervals for the soundfield rendering using separate AAC encoding at 30 kbps, SAOC, SPA and our proposed approach at 105 kbps.

From a series of objective and subjective listening test, we prove that the proposed approach can adapt to various bitrates conditions and it is suitable for encoding multiple audio objects in real application scenarios.

4. Conclusions

In this paper, an efficiently encoding approach for multiple audio objects based on intra-object sparsity was presented. Unlike the existing STFT-based compression framework, statistical analysis validated that for the case of tonal solo instruments audio objects possess better energy concentration property in the MDCT domain so that MDCT is selected as basic transform in our encoding scheme. In order to achieve a balanced perceptual quality for all object signals, both psychoacoustic-based and energy balanced *NPTF* allocation strategy algorithm is proposed for obtaining the optimal MDCT coefficients of each object. Moreover, SQVH is utilized to further encode downmix signal at variable bitrates. Objective and subjective evaluations shows that the proposed approach outperforms the existing intra-object based approach and achieves a more balanced perceptual quality when eight simultaneously occurring audio objects were encoded jointly. The results also confirmed that the proposed framework attained higher perceptual quality compared to SAOC. Further research could include the investigation of relative auditory masking threshold, in order to acquire a better perceptual quality amongst all objects.

Acknowledgments: The authors would like to thank the reviewers for their helpful comments. This work has been supported by China Postdoctoral Science Foundation funded project (No. 2017M610731), the Project supported by Beijing Postdoctoral Research Foundation, "Ri xin" Training Programme Foundation for the Talents by Beijing University of Technology.

Author Contributions: Maoshen Jia and Jiaming Zhang contributed equally in conceiving the whole proposed codec architecture, designing and performing the experiments, collecting and analyzing the data, and writing the paper. Changchun Bao corrected some syntax mistake. Xiguang Zheng critically reviewed and implemented final revisions. Maoshen Jia supervised all aspects of this research.

Conflicts of Interest: The authors declare no conflict of interest.

Appendix A. Sparsity Analysis of Audio Signal in the MDCT Domain

Considering that the MDCT is a commonly used time-frequency transform in signal processing, the intra-object sparsity of audio signal in the MDCT domain should be investigated. Thus, a quantitative analysis for sparsity of audio signals both in the MDCT and STFT domain is given in this appendix.

According to the k-sparsity theory interpreted in compressed sensing [32,33], a signal/sequence is regarded as (strict) k-sparse when it contains k nonzero entries with $k \ll K$, where K is the length of the signal or sequence. In addition, a sequence can be considered as an approximate k-sparse if k entries of the sequence occupy the majority of the total amount in magnitude, while the magnitude of other entries are remarkable small. In our previous work [19], we validated that an audio signal is not sparse in time domain, but its STFT coefficients in frequency domain fulfills the approximate k-sparsity. For this reason, STFT is selected as basic transform in our preceding designed object encoding system. The perceptual quality of the decoded signal can achieve a satisfactory level. However, STFT is not an optimum sparseness time-frequency transform. In consideration of the energy compaction property (i.e., a small number of TF instants capture the majority of the energy) of MDCT, therefore, approximate k-sparsity of audio signal in the MDCT domain will be investigated compared to that in the STFT domain by statistical analysis.

Appendix A.1. Measuring the Sparsity of Audio Signal

A time-frequency representations of an audio signal can be obtained by a linear transform. Specifically, for a general dictionary of atoms $D = \{\varphi_l\}$, the linear representation of an audio signal $s_n(m)$ in n^{th} frame can be defined by:

$$S(n, l) = \sum_{m=1}^{M} s_n(m)\varphi_l(m) \tag{A1}$$

where n, m and l represent frame number, time index and frequency index, respectively. M is the length of each frame. Short-time Fourier Transform (STFT) basis functions and Discrete Cosine Transform (DCT) basis functions are ordinarily used as time-frequency atoms in speech and audio signal compression. DCT is widely used in audio coding mainly because of its energy compaction feature. Nevertheless, due to the blocking effect caused by the different quantitative level between frames, the processed signal cannot be perfectly reconstructed by IDCT. Evolved from DCT, MDCT has emerged as an efficaciously tool in high quality audio coding over the last decade because it helps to mitigate the blocking artifacts that deteriorate the reconstruction of transform audio coders with non-overlapped transforms [34]. It should be noted that MDCT can be taken as a filterbank with 50% overlapped window, hence, Time Domain Aliasing Cancellation (TDAC) must be exploited in the practical processing. Meanwhile, the chosen window function must satisfy the TDAC requirement. In this work, a Kaiser-Bessel derived (KBD) window [35] is chosen to meet the computing needs of TDAC and overlap-add algorithms. Particularly, for a finite-length audio signal whose MDCT coefficients are densely concentrated at low indices than the STFT (Short Time Fourier Transform) does, which is called "energy compaction" property [36]. With this prerequisite, a detailed comparative study and analyses of energy compaction feature (a.k.a. sparsity) of different audio objects in the STFT domain and MDCT domain is implemented.

To measure and explore sparsity of audio signal in the time-frequency domain, a measurement addressed as Frame Energy Preservation Ratio (*FEPR*) and the Number of Preserved TF bins (*NPTF*) was proposed in [19]. Specifically, the sparse approximation signal of $S(n, l)$, referred as $S'(n, l)$, contains the maximum K^* TF instants by preserving the portion of TF instants according to their amplitude of $S(n, l)$ while setting the other TF instants to zero, which can be expressed by:

$$S\prime(n, l) = \begin{cases} S(n, l), & \text{if } l \in \mathcal{L} \\ 0, & \text{otherwise} \end{cases} \tag{A2}$$

where $\mathcal{L} \triangleq \{l_1, l_2, \cdots, l_{K^*}\}$, is the set of K^* frequency indices corresponding to the maximum K^* time-frequency instants. Thus, $S'(n, l)$ is a K^*-sparse signal.

Suppose $\theta_n \equiv [S(n, 1), \cdots, S(n, L)]$ is the L-dimensional vector denotes the TF representation of the audio object signal in n^{th} frame, $\theta'_n \equiv [S\prime(n, 1), \cdots, S\prime(n, L)]$ is sparse approximation vector of θ_n. Then, the Frame Energy Preservation Ratio (*FEPR*) can be given by:

$$r_{FEPR}(n) = \frac{\|\theta'_n\|_1}{\|\theta_n\|_1} \tag{A3}$$

where $\|\cdot\|_p$ denotes the l_p-norm.

Afterwards, for arbitrary given r^*_{FEPR}, if there exists a series of subset $\mathcal{L}_i \subset \{1, 2, \cdots, L\}, i = 1, 2, \ldots$, such that the corresponding sparse signal vector $\theta'_{n,i} \equiv [S'_i(n, 1), \cdots, S'_i(n, L)]$. The Number of Preserved TF instants (*NPTF*), written as k, is defined as a function of r^*_{FEPR}:

$$k(r^*_{FEPR}) = \inf \left\{ \|\theta'_{n,i}\|_0 \left| \frac{\|\theta'_{n,i}\|_1}{\|\theta_n\|_1} \geq r^*_{FEPR}, i = 1, 2, \cdots, \right. \right\} \tag{A4}$$

where $\inf\{\cdot\}$ represents the infimum. $k(r^*_{FEPR})$ describes the least achievable preserved TF bins for arbitrary r^*_{FEPR}. Especially, a lower $k(r^*_{FEPR})$ with a certain r^*_{FEPR} means stronger sparsity for an audio signal.

Appendix A.2. Statistical Analysis Results

To reveal the superior properties of MDCT, in each frame, 315 mono audio recordings selected from University of Iowa Music Instrument Samples (Iowa-MIS) audio database [37] sampled at 44.1 kHz and 100 mono speech recordings selected from Nippon Telegraph & Telephone (NTT) database are chosen as the test data. The selected audio recordings contain 7 types of tonal solo instruments. In this statistics work, a 2048-point STFT and MDCT basis with 50% overlapping is applied to form the time-frequency instants. Meanwhile, a KBD window with the size of 2048 points is used as the window function to meet the demand of overlap-add. A statistical analysis of *NPTF* is taken with the *FEPR* ranged from 98% to 80%. Results are shown in Figure A1 with 95% confidence intervals. Note that STFT-domain descriptions corresponding to instruments or speech are respectively denoted by 'Flute-STFT', 'Violin-STFT', 'Sax-STFT', 'Oboe-STFT', 'Trombone-STFT', 'Trumpet-STFT', 'Horn-STFT' and 'Speech-STFT'. In contrast, MDCT-domain representations are respectively regarded as 'Flute-MDCT', 'Violin-MDCT', 'Sax-MDCT', 'Oboe-MDCT', 'Trombone-MDCT', 'Trumpet-MDCT', 'Horn-MDCT' and 'Speech-MDCT'.

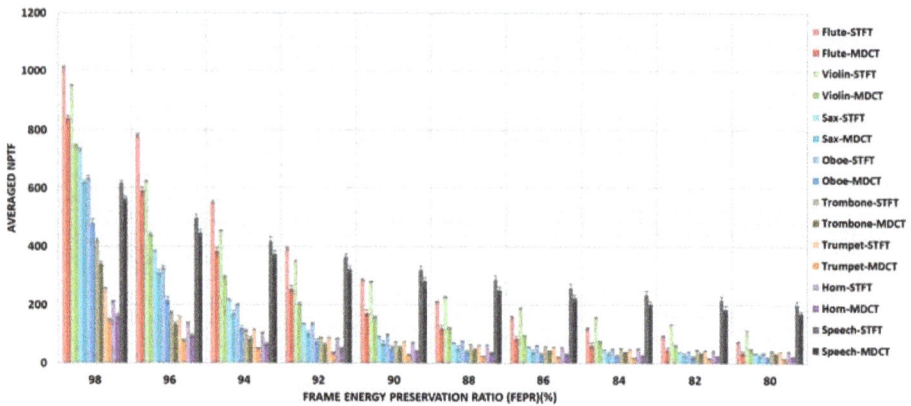

Figure A1. NPTF (Number of Preserved Time-Frequency Bins) results calculated from eight types of audio signals in various FEPR (Frame Energy Preservation Ratio).

Figure A1 indicates that by decreasing *FEPR*, the averaged *NPTF* degrades as well. More precisely, *NPTF* is a convex function as *FEPR* decreases uniformly in terms of all test instruments and speech, that is, audio object or speech signal are sparse both in STFT and MDCT domain. Furthermore, it shows that there exists a noticeable difference between adjacent light color and dark color bars, in other words, the averaged *NPTF* in the MDCT domain is much lower than that in the STFT domain for each instrument and speech with a certain *FEPR*.

While the energy compaction property of MDCT is fairly intuitive, it becomes agnostic as the *FEPR* changes. To measure the disparity between the averaged *NPTF* for MDCT coefficients and STFT coefficients of audio signal with a known *FEPR*, a Normalized Relative Difference Ratio (NRDR) is defined as (k is *NPTF* and r_{FEPR} is *FEPR*):

$$NRDR(r_{FEPR}) = \frac{k(r_{FEPR})_{\text{STFT}} - k(r_{FEPR})_{\text{MDCT}}}{k(r_{FEPR})_{\text{STFT}}} \tag{A5}$$

where $k(r_{FEPR})_{\text{STFT}}$ and $k(r_{FEPR})_{\text{MDCT}}$ are the averaged *NPTF* for an audio signal in the STFT and MDCT domain with a certain *FEPR*, respectively. NRDR is the difference between them. The larger the NRDR is, means that the less *NPTF* needed in the MDCT domain. Then, a statistical bar graph is presented which reflects the relationship between NRDR and *FEPR*.

Results are shown in Figure A2 with different NRDR at r_{FEPR} = 98~80%. It can be observe that the NRDR of all tested audio signals are non-negative, which means that the averaged *NPTF* in the MDCT domain is higher than that in the STFT domain. This result testifies that the performance of MDCT is absolutely dominant for all of the tested 8 items.

Interestingly, we find that NRDR is gradually increasing as r_{FEPR} uniformly decrease from 98% to 88%. When $80\% \leq r_{FEPR} \leq 88\%$, the NRDR maintains at the same level or slightly grow. Videlicet, with the decrement of *FEPR*, the superiority of MDCT is becoming increasingly obvious.

The next phenomenon needs to be noted is that the sparsity of violin and trumpet is particularly evident in the MDCT domain, because their NRDR can reach up to 60% when r_{FEPR} = 80% whilst other instruments can only achieve roughly 45%~55%. Besides, the sparseness of selected speech signals is weaker than all instruments in the MDCT domain but maintain consistency as far as the global regularity.

Hence, the results in Figure A2 confirm that, for all tested signals, MDCT has a better energy compaction capability than STFT to the great extent. It means that audio or speech signal is more sparse in the MDCT domain than in the STFT domain.

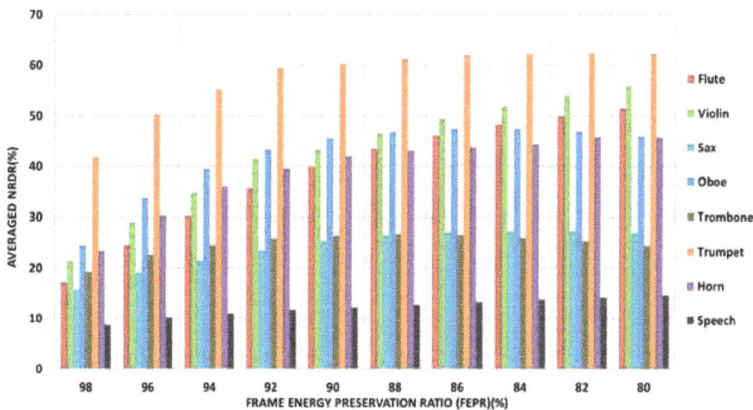

Figure A2. NRDR (Normalized Relative Difference Ratio) of eight types of audio signals under STFT (Short Time Fourier Transform) and MDCT (Modified Discrete Cosine Transform) in various FEPR.

References

1. International Telecommunication Union. *BS.775: Multichannel Stereophonic Sound System with and without Accompanying Picture*; International Telecommunications Union: Geneva, Switzerland, 2006.
2. Bosi, M.; Brandenburg, K.; Quackenbush, S.; Fielder, L.; Akagiri, K.; Fuchs, H.; Dietz, M. ISO/IEC MPEG-2 advanced audio coding. *J. Audio Eng. Soc.* **1997**, *45*, 789–814.
3. Breebaart, J.; Disch, S.; Faller, C.; Herre, J.; Hotho, G.; Kjörling, K.; Myburg, F.; Neusinger, M.; Oomen, W.; Purnhagen, H.; et al. MPEG spatial audio coding/MPEG surround: Overview and current status. In Proceedings of the Audio Engineering Society Convention 119, New York, NY, USA, 7–10 October 2005.
4. Quackenbush, S.; Herre, J. MPEG surround. *IEEE MultiMedia* **2005**, *12*, 18–23. [CrossRef]
5. Cheng, B.; Ritz, C.; Burnett, I. Principles and analysis of the squeezing approach to low bit rate spatial audio coding. In Proceedings of the IEEE International Conference on Acoustics, Speech and Signal Processing (ICASSP), Honolulu, HI, USA, 16–20 April 2007; pp. I-13–I-16.
6. Cheng, B.; Ritz, C.; Burnett, I. A spatial squeezing approach to ambisonic audio compression. In Proceedings of the IEEE International Conference on Acoustics, Speech and Signal Processing (ICASSP), Las Vegas, NV, USA, 31 March–4 April 2008; pp. 369–372.
7. Cheng, B.; Ritz, C.; Burnett, I.; Zheng, X. A general compression approach to multi-channel three-dimensional audio. *IEEE Trans. Audio Speech Lang. Process.* **2013**, *21*, 1676–1688. [CrossRef]
8. Bleidt, R.; Borsum, A.; Fuchs, H.; Weiss, S.M. Object-based audio: Opportunities for improved listening experience and increased listener involvement. In Proceedings of the SMPTE 2014 Annual Technical Conference & Exhibition, Hollywood, CA, USA, 20–23 October 2014.
9. Pulkki, V. Virtual sound source positioning using vector base amplitude panning. *J. Audio Eng. Soc.* **1997**, *45*, 456–466.
10. Dolby Laboratories, "Dolby ATMOS Cinema Specifications". 2014. Available online: http://www.dolby. com/us/en/technologies/dolbyatmos/dolby-atmos-specifications.pdf (accessed on 25 October 2017).
11. Breebaart, J.; Engdegard, J.; Falch, C.; Hellmuth, O.; Hilpert, J.; Holzer, A.; Koppens, J.; Oomen, W.; Resch, B.; Schuijers, E.; et al. Spatial Audio Object Coding (SAOC)—The upcoming MPEG standard on parametric object based audio coding. In Proceedings of the Audio Engineering Society Convention 124, Amsterdam, The Netherlands, 17–20 May 2008.
12. Herre, J.; Purnhagen, H.; Koppens, J.; Hellmuth, O.; Engdegard, J.; Hilper, J.; Villemoes, L.; Terentiv, L.; Falch, C.; Holzer, A.; et al. MPEG Spatial Audio Object Coding—The ISO/MPEG standard for efficient coding of interactive audio scenes. *J. Audio Eng. Soc.* **2012**, *60*, 655–673.

13. Pulkki, V. Directional audio coding in spatial sound reproduction and stereo upmixing. In Proceedings of the Audio Engineering Society Conference: 28th International Conference: The Future of Audio Technology—Surround and Beyond, Piteå, Sweden, 30 June–2 July 2006.

14. Faller, C.; Pulkki, V. Directional audio coding: Filterbank and STFT-based design. In Proceedings of the Audio Engineering Society Convention 120, Paris, France, 20–23 May 2006.

15. Herre, J.; Hilpert, J.; Kuntz, A.; Plogsties, J. MPEG-H 3D audio—The new standard for coding of immersive spatial audio. *IEEE J. Sel. Top. Signal Process.* **2015**, *9*, 770–779. [CrossRef]

16. Zheng, X.; Ritz, C.; Xi, J. Encoding navigable speech sources: An analysis by synthesis approach. In Proceedings of the IEEE International Conference on Acoustics, Speech and Signal Processing (ICASSP), Kyoto, Japan, 25–30 March 2012; pp. 405–408.

17. Zheng, X.; Ritz, C.; Xi, J. Encoding navigable speech sources: A psychoacoustic-based analysis-by-synthesis approach. *IEEE Trans. Audio Speech Lang. Process.* **2013**, *21*, 29–38. [CrossRef]

18. Yilmaz, O.; Rickard, S. Blind separation of speech mixtures via time-frequency masking. *IEEE Trans. Audio Speech Lang. Process.* **2004**, *52*, 1830–1847. [CrossRef]

19. Jia, M.; Yang, Z.; Bao, C.; Zheng, X.; Ritz, C. Encoding multiple audio objects using intra-object sparsity. *IEEE Trans. Audio Speech Lang. Process.* **2015**, *23*, 1082–1095.

20. Yang, Z.; Jia, M.; Bao, C.; Wang, W. An analysis-by-synthesis encoding approach for multiple audio objects. In Proceedings of the IEEE Signal and Information Processing Association Annual Summit and Conference (APSIPA), Hong Kong, China, 16–19 December 2015; pp. 59–62.

21. Yang, Z.; Jia, M.; Wang, W.; Zhang, J. Multi-Stage Encoding Scheme for Multiple Audio Objects Using Compressed Sensing. *Cybern. Inf. Technol.* **2015**, *15*, 135–146.

22. Wang, Y.; Vilermo, M. Modified discrete cosine transform: Its implications for audio coding and error concealment. *J. Audio Eng. Soc.* **2003**, *51*, 52–61.

23. Enqing, D.; Guizhong, L.; Yatong, Z.; Yu, C. Voice activity detection based on short-time energy and noise spectrum adaptation. In Proceedings of the IEEE International Conference on Signal Processing (ICSP), Beijing, China, 26–30 August 2002; pp. 464–467.

24. Painter, T.; Spanias, A. Perceptual coding of digital audio. *Proc. IEEE* **2000**, *88*, 451–515. [CrossRef]

25. Spanias, A.; Painter, T.; Atti, V. *Audio Signal Processing and Coding*; John Wiley & Sons: Hoboken, NJ, USA, 2006; pp. 114 & 274, ISBN 9780470041970.

26. Jia, M.; Bao, C.; Liu, X. An embedded speech and audio coding method based on bit-plane coding and SQVH. In Proceedings of the IEEE International Symposium on Signal Processing and Information Technology (ISSPIT), Ajman, UAE, 11–16 December 2009; pp. 43–48.

27. Xie, M.; Lindbergh, D.; Chu, P. ITU-T G.722.1 Annex C: A new low-complexity 14 kHz audio coding standard. In Proceedings of the IEEE International Conference on Acoustics, Speech and Signal Processing (ICASSP), Toulouse, France, 14–19 May 2006; pp. 173–176.

28. Xie, M.; Lindbergh, D.; Chu, P. From ITU-T G.722.1 to ITU-T G.722.1 Annex C: A New Low-Complexity 14kHz Bandwidth Audio Coding Standard. *J. Multimed.* **2007**, *2*, 65–76. [CrossRef]

29. QUASI Database—A Musical Audio Signal Database for Source Separation. Available online: http://www.tsi.telecomparistech.fr/aao/en/2012/03/12/quasi/ (accessed on 25 October 2017).

30. International Telecommunication Union. *BS.1534: Method for the Subjective Assessment of Intermediate Quality Levels of Coding Systems*; International Telecommunication Union: Geneva, Switzerland, 1997.

31. Gardner, B.; Martin, K. HRTF Measurements of a KEMAR Dummy-Head Microphone. Available online: http://sound.media.mit.edu/resources/KEMAR.html (accessed on 25 October 2017).

32. Candes, E.J.; Wakin, M.B. An introduction to compressive sampling. *IEEE Signal Process. Mag.* **2008**, *25*, 21–30. [CrossRef]

33. Candes, E.J.; Romberg, J.K.; Tao, T. Stable signal recovery from incomplete and inaccurate measurements. *Commun. Pure Appl. Math.* **2006**, *59*, 1207–1223. [CrossRef]

34. Dhas, M.D.K.; Sheeba, P.M. Analysis of audio signal using integer MDCT with Kaiser Bessel Derived window. In Proceedings of the IEEE International Conference on Advanced Computing and Communication Systems (ICACCS), Coimbatore, India, 6–7 January 2017; pp. 1–6.

35. Bosi, M.; Goldberg, R.E. *Introduction to Digital Audio Coding and Standards*; Springer: Berlin, Germany, 2003.

Appl. Sci. **2017**, *7*, 1301

36. Oppenheim, A.V.; Schafer, R.W. *Discrete-Time Signal Processing*, 3rd ed.; Publishing House of Electronics Industry: Beijing, China, 2011; pp. 673–683. ISBN 9787121122026.
37. University of Iowa Music Instrument Samples. Available online: http://theremin.music.uiowa.edu/MIS.html (accessed on 25 October 2017).

![applied sciences logo] applied sciences

MDPI

Article

Wearable Vibration Based Computer Interaction and Communication System for Deaf

Mete Yağanoğlu [1,*] **and Cemal Köse** [2]

[1] Department of Computer Engineering, Faculty of Engineering, Ataturk University, 25240 Erzurum, Turkey
[2] Department of Computer Engineering, Faculty of Engineering, Karadeniz Technical University,
61080 Trabzon, Turkey; ckose@ktu.edu.tr
* Correspondence: yaganoglu@atauni.edu.tr; Tel.: +90-535-445-2400

Academic Editor: Stefania Serafin
Received: 29 September 2017; Accepted: 11 December 2017; Published: 13 December 2017

Abstract: In individuals with impaired hearing, determining the direction of sound is a significant problem. The direction of sound was determined in this study, which allowed hearing impaired individuals to perceive where sounds originated. This study also determined whether something was being spoken loudly near the hearing impaired individual. In this manner, it was intended that they should be able to recognize panic conditions more quickly. The developed wearable system has four microphone inlets, two vibration motor outlets, and four Light Emitting Diode (LED) outlets. The vibration of motors placed on the right and left fingertips permits the indication of the direction of sound through specific vibration frequencies. This study applies the ReliefF feature selection method to evaluate every feature in comparison to other features and determine which features are more effective in the classification phase. This study primarily selects the best feature extraction and classification methods. Then, the prototype device has been tested using these selected methods on themselves. ReliefF feature selection methods are used in the studies; the success of K nearest neighborhood (Knn) classification had a 93% success rate and classification with Support Vector Machine (SVM) had a 94% success rate. At close range, SVM and two of the best feature methods were used and returned a 98% success rate. When testing our wearable devices on users in real time, we used a classification technique to detect the direction and our wearable devices responded in 0.68 s; this saves power in comparison to traditional direction detection methods. Meanwhile, if there was an echo in an indoor environment, the success rate increased; the echo canceller was disabled in environments without an echo to save power. We also compared our system with the localization algorithm based on the microphone array; the wearable device that we developed had a high success rate and it produced faster results at lower cost than other methods. This study provides a new idea for the benefit of deaf individuals that is preferable to a computer environment.

Keywords: wearable computing system; vibrating speaker for deaf; human–computer interaction; feature extraction; speech processing

1. Introduction

In this technological era, information technology is effectively being used in numerous aspects of our lives. The communication problems between humans and information have gradually made machines more important.

One of speech recognition systems' most significant purposes is to provide human–computer communication through speech communication from users in a widespread manner and enable a more extensive use of computer systems that facilitate the work of people in many fields.

Speech is the primary form of communication among people. People have the ability to understand the meaning and to recognize the speaker, gender of speaker, age and emotional situation of the speaker [1]. Voice communication among people starts with a thought and intent activating neural actions generating speech sounds in the brain. The listener receives the speech through the auditory system converting the speech to neural signals that the brain can comprehend [2,3].

Many important computer and internet technology based studies intended to facilitate the lives of hearing impaired individuals are being performed. Through these studies, attempts are being made to improve the living quality of hearing impaired individuals.

The most important problem of hearing impaired individuals is their inability to perceive the point where the sound is coming from. In this study, our primary objective was to enable hearing impaired individuals to perceive the direction of sound and to turn towards that direction. Another objective was to ensure hearing impaired individuals can disambiguate their attention by perceiving whether the speaker is speaking softly or loudly.

Basically, the work performed by a voice recognition application is to receive the speech data and to estimate what is being said. For this purpose, the sound received from the mic, in other words the analogue signal is first converted to digital and the attributes of the acoustic signal is obtained for the determination of required properties.

The sound wave forming the sound includes two significant properties. These properties are amplitude and frequency. While frequency determines the treble and gravity properties of sound, the amplitude determines the severity of sound and its energy. Sound recognition systems benefit from analysis and sorting of acoustic signals.

As shown in Figure 1, our wearable device has also been tested in real time and the results have been compared. In Figure 1, the device is mounted on the clothes of deaf users and it responds instantaneously to vibrations in real time and detects the deaf person.

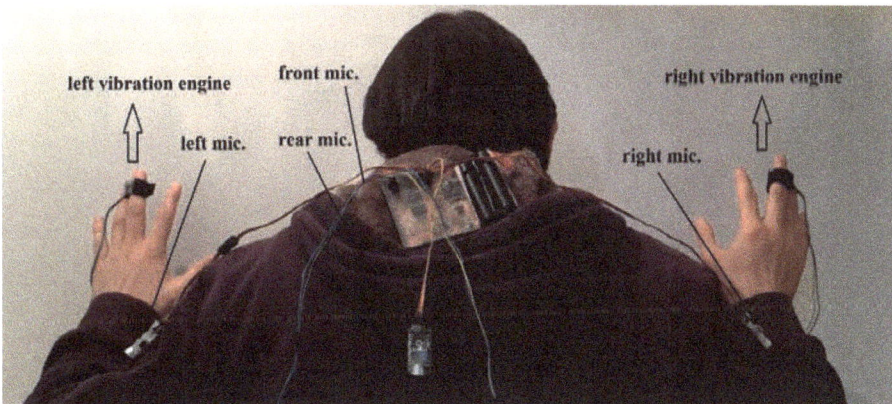

Figure 1. In testing our wearable device on the user in real time.

As can be seen in the system in Figure 2, data obtained from individuals are transferred to the computer via the system we developed. Through this process, the obtained data passes the stages of pre-processing, feature extraction and classification, then the direction of voices is detected; this has also been tested in real time in this study. Subjects were given real time voices and whether they could understand where the voices were coming from was observed.

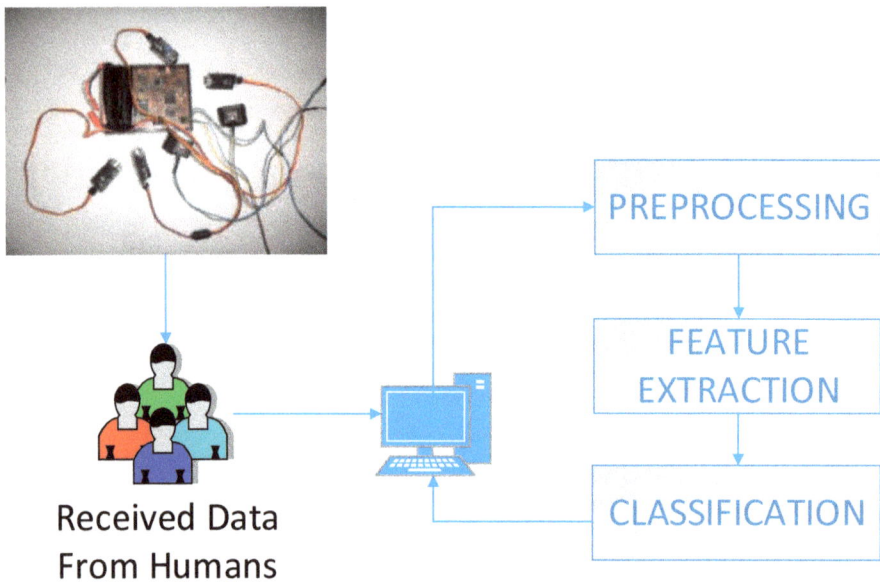

Figure 2. Human–Computer Interface System.

The main purpose of this study was to let people with hearing disabilities hear sounds that were coming from behind such as brake sounds and horn sounds. Sounds coming from behind are a significant source of anxiety for people with hearing disabilities. In addition, hearing the sounds of brakes and horns is important and allows people with hearing disabilities to have safer journeys. It will provide immediate extra perception and decision capabilities in real time to people with hearing disabilities; the aim is to make a product that can be used in daily life by people with hearing disabilities and to make their lives more prosperous.

2. Related Works

Some of the most common problems are the determinations of the age, gender, sensual situation and feasible changing situations of the speaker like being sleepy or drunk. Defining some aspects of the speech signal in a period of more than a few seconds or a few syllables is necessary to create a high number appropriate high-level attribute and to conduct the general machinery learning methods for high-dimensional attributes data. In the study of Pohjalainen et al., researchers have focused on the automatic selection of usable signal attributes in order to understand the assigned paralinguistic analysis duties better and with the aim to improve the classification performance from within the big and non-elective basic attributes cluster [4].

In a period when the interaction between individuals and machines has increased, the definition-detection of feelings might allow the creation of intelligent machinery and make emotions, just like individuals. In voice recognition and speaker definition applications, emotions are at the forefront. Because of this, the definition of emotions and its effect on speech signals might improve the speech performance and speaker recognition systems. Fear type emotion definition can be used in the voice-based control system to control a critical situation [5].

In the study of Vassis et al., a wireless system on the basis of standard wireless techniques was suggested in order to protect the mobile assessment procedure. Furthermore, personalization techniques were implemented in order to adapt the screen display and test results according to the needs of the students [6].

Appl. Sci. **2017**, *7*, 1296

In their study, Shivakumar and Rajasenathipathi connected deaf and blind people to a computer using these equipment hardware control procedures and a screen input program in order to be able to help them benefit from the latest computer technology through vibrating gloves for communication purposes [7].

The window of deaf and blind people opening up to the world is very small. The new technology can be helpful in this, but it is expensive. In their study, Arato et al. developed a very cost-effective method in order to write and read SMS using a smart phone with an internal vibrating motor and tested this. Words and characters were turned into vibrating Braille codes and Morse words. Morse was taught in order to perceive the characters as codes and words as a language [8].

In the study of Nanayakkara et al. the answer to the question whether the tactual and visual knowledge combination can be used in order to increase the music experimentation in terms of the deaf was asked, and if yes, how to use it was explored. The concepts provided in this article can be beneficial in turning other peripheral voice types into visual demonstration and/or tactile input tools and thus, for example, they will allow a deaf person to hear the sound of the doorbell, footsteps approaching from behind, the voice of somebody calling for him, to understand speech and to watch television with less stress. This research shows the important potential of deaf people in using the existing technology to significantly change the way of experiencing music [9].

The study of Gollner et al. introduces a new communication system to support the communication of deaf and blind people, thus consolidating their freedom [10].

In their study, Schmitz and Ertl developed a system that shows maps in a tactile manner using a standard noisy gamepad in order to ensure that blind and deaf people use and discover electronic maps. This system was aimed for both indoor and outdoor use, and thus it contains mechanisms in order to take a broad outline of larger areas in addition to the discovery of small areas. It was thus aimed to make digital maps accessible using vibrations [11].

In their study, Ketabdar and Polzehl developed an application for mobile phones that can analyze the audio content, tactile subject and visual warnings in case a noisy event takes place. This application is especially useful for deaf people or those with hearing disorder in that they are warned by noisy events happening around them. The voice content analysis algorithm catches the data using the microphone of the mobile phone and checks the change in the noisy activities happening around the user. If any change happens and other conditions are encountered, the application gives visual or vibratory-tactile warnings in proportion to the change of the voice content. This informs the user about the incident. The functionality of this algorithm can be further developed with the analysis of user movements [12].

In their study, Caetano and Jousmaki recorded the signals from 11 normal-hearing adults up to 200 Hz vibration and transmitted them to the fingertips of the right hand. All of the subjects reported that they perceived a noise upon touching the vibrating tube and did not sense anything when they did not touch the tube [13].

Cochlear implant (CI) users can also benefit from additional tactile help, such as those performed by normal hearing people. Zhong et al. used two bone-anchored hearing aids (BAHA) as a tactile vibration source. The two bone-anchored hearing aids connected to each other by a special device to maintain a certain distance and angle have both directional microphones, one of which is programmed to the front left and the other to the front right [14].

There are a large number of CI users who will not benefit from permanent hearing but will benefit from the tips available in low frequency information. Wang and colleagues have studied the skill of tactile helpers to convey low frequency cues in the study because the frequency sensitivity of human haptic sense is similar to the frequency sensitivity of human acoustic hearing at low frequencies. A total of 5 CI users and 10 normal hearing participants provide adaptations that are designed for low predictability of words and rate the proportion of correct and incorrect words in word segmentation using empirical expressions balanced against syllable frequency. The results of using the BAHA show that there is a small but significant improvement on the ratio of the tactile helper and correct words, and the word segmentation errors are decreasing. These findings support the use of tactile information in the perceptual task of word segmentation [15].

In the study of Mesaros et al., various metrics recommended for assessment of polyphonic sound event perception systems used in realist cases, where multiple sound sources are simultaneously active, are presented and discussed [16].

In the study of Wang et al., the subjective assessment over six deaf individuals with V-form audiogram suggests that there is approximately 10% recovery in the score of talk separation for monosyllabic Word lists tested in a silent acoustic environment [17].

In the study of Gao et al., a system designed to help deaf people communicate with others was presented. Some useful new ideas in design and practice are proposed. An algorithm based on geometric analysis has been introduced in order to extract the unchanging feature to the signer position. Experiments show that the techniques proposed in the Gao et al. study are effective on recognition rate or recognition performance [18].

In the study of Lin et al., an audio classification and segmentation method based on Gabor wavelet properties is proposed [19].

Tervo et al. recommends the approach of spatial sound analysis and synthesis for automobile sound systems in their study. An objective analysis of sound area in terms of direction and energy provides the synthesis of the emergence of multi-channel speakers. Because of an automobile cabin's excessive acoustics, the authors recommend a few steps to make both objective and perception performance better [20].

3. Materials and Methods

In the first phase, our wearable system was tested and applied in real-time. Our system estimates new incoming data in real time and gives information to the user immediately via vibrations. Our wearable device predicts the direction again as the system responds and redirects the user. Using this method helped find the best of the methods described in the previous section and that method was implemented. Different voices were provided from different directions to subjects and they were asked to guess the direction of each. These results were compared with real results and the level of success was determined for our wearable system.

In the second phase, the system is connected to the computer and the voices and their directions were transferred to a digital environment. Data collected from four different microphones were kept in matrixes each time and thus a data pool was created. The created data passed the stages of preprocessing, feature extraction and classification, and was successful. A comparison was made with the real time application and the results were interpreted.

The developed wearable system (see Figure 3) had four microphone inlets. Four microphones were required to ensure distinguishable differentiation in the four basic directions. The system was first tested using three microphones, but four were deemed necessary due to three obtaining low success rates and due to there being four main directions. They were placed to the right, left, front, and rear of the individual through the developed human–computer interface system. The experimental results showed accuracy improved if four microphones were used instead of three. Two vibration motor outlet units were used in the developed system; placing the vibration motors on the right and left fingertips permitted the indication of the direction of sound by specific vibration frequencies. The most important reason in the preference of fingertips is the high number of nerves present in the fingertips. Moreover, vibration motors placed on the fingertips are easier to use and do not disturb the individual.

The developed system has four Light Emitting Diode (LED) outlets; when sound is perceived, the LED of the outlet in the perceived direction of vibration is lit. The use of both vibration and LED ensures that the user can more clearly perceive the relevant direction is perceived. We use LEDs to give a visual warning. Meanwhile, the use of four different LED lights is considered for the four different directions. If the user cannot interpret the vibrations, they can gain clarity by looking at the LEDs. The role of vibration in this study is to activate the sense of touch for hearing-impaired individuals.

Through touch, hearing-impaired individuals will be able to gain understanding more easily and will have more comfort.

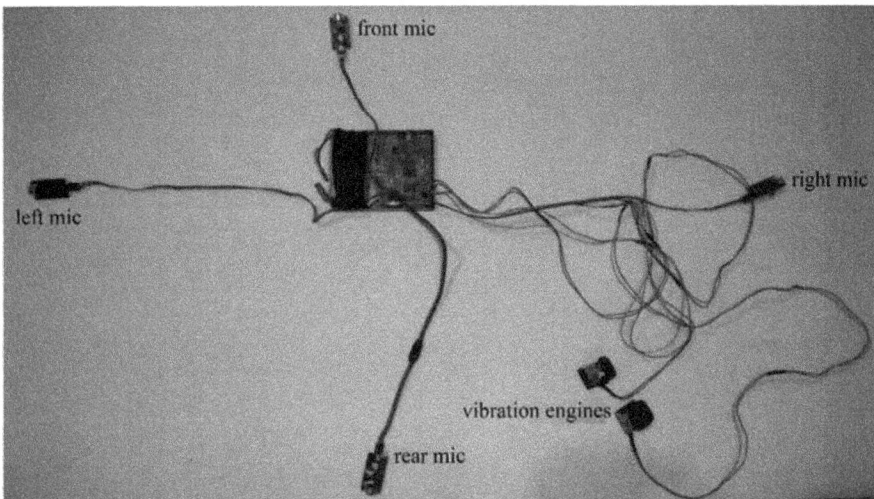

Figure 3. The developed wearable system.

The features of the device we have developed are; ARM-based 32-bit MCU with Flash memory. 3.6 V application supply, 72 MHz maximum frequency, 7 timers, 2 ADCs, 9 com. Interfaces. Rechargeable batteries were used for our wearable device. The batteries can work for about 10 h.

In vibration, individuals are able to perceive the coming sound with a difference of 20 ms, and the direction of coming sound can be determined at 20 ms after giving the vibration. In other words, the individual is able to distinguish the coming sound after 20 ms.

Vibration severities of 3 different levels were applied on the finger:

- 0.5 V–1 V at 1st level for perception of silent sounds
- 1 V–2 V at 2nd level for perception of medium sounds
- 2 V–3 V at 3rd level for the perception of loud sounds

Here, 0.5, 1, 2 and 3 V indicate the intensity of the vibration. This means that if our system detects a loud person, it gives a stronger vibration to the perception of the user. For 50 people with normal hearing, the sound of sea or wind was provided via headphones. The reason for the choice of such sounds is that they are used in deafness tests and were recommended by the attending physician. Those sounds were set to a level (16–60 dB) that would not disturb users; through this, it does not have a distract users.

After the vibration, it was applied in two different stages:

- Low classification level for those below 20 ms,
- High level classification for those 20 ms and above.

Thus, we will be able to perceive whether an individual is speaking loudly or quietly by adjusting the vibration severity. For instance, if there is an individual nearby speaking loudly, the user will be able to perceive it and respond quicker. Through this levelling, a distinction can be made in whether a speaker is yelling. The main purpose of this is to reduce the response time for hearing impaired individuals. It is possible that someone shouting nearby is referring to a problem and the listener should pay more attention.

In the performed study, 50 individuals without hearing impairment, four deaf people and two people with moderate hearing loss were subjected to a wearable system and tested, and significant success was obtained. Normal users could only hear the sea or wind music; the wearable technology we developed was placed on the user's back and tested. The user was aware of where sound was coming from despite the high level of noise in their ear and they were able to head in the appropriate direction. The ears of 50 individuals without hearing impairment were closed to prevent their hearing ability, and the system was started in such a manner that they were unable to perceive where sounds originated. These individuals were tested for five days at different locations and their classification successes were calculated.

Four deaf people and two people with moderate hearing lose were tested for five days in different locations and their results were compared with those of normal subjects based on eight directions during individuals' tests. Sound originated from the left, right, front, rear, and the intersection points between these directions, and the success rate was tested. Four and eight direction results were interpreted in this study and experiments were progressed in both indoor and outdoor environments.

People were used as sound sources in real-time experiments. While walking outside, someone would come from behind and call out, and whether the person using the device could detect them was evaluated. A loudspeaker was used as the sound source in the computer environment.

In this study, there were microphones on the right, left, behind, and front of the user and vibration motors were attached to the left and right fingertips. For example, if a sound came from the left, the vibration motor on the left fingertip would start working. Both the right and left vibration motors would operate for the front and behind directions. For the front direction the right–left vibration motors would briefly vibrate three times. For behind, right, and left directions, the vibration motors vibrate would three times for extended periods. The person who uses the product would determine the direction in approximately 70 ms on average. Through this study, loud or soft low sounding people were recognizable and people with hearing problems could pay attention according to this classification. For example, if someone making a loud sound was nearby, people with hearing problems were able to understand this and react faster according to this understanding.

3.1. Definitions and Preliminaries

There are four microphone inputs, two vibration engine outputs and four LED outputs in the developed system. With the help of vibration engines that we placed on the right and left fingertips, the direction of the voice was shown by certain vibration intervals. When the voice is perceived, if the vibration perceives its direction, the LED that belongs to that output is on. In this study, the system is tested both in real time and after the data are transferred to the computer.

3.1.1. Description of Data Set

There is a problem including four classes:

1. Class: Data received from left mic
2. Class: Data received from right mic
3. Class: Data received from front mic
4. Class: Data received from rear mic

Four microphones were used in this study. Using 4 microphones represents 4 basic directions. The data from each direction is added to the 4 direction tables. A new incoming voice data is estimated by using 4 data tables. In real time, our system predicts a new incoming data and immediately informs the user with the vibration.

3.1.2. Training Data

Training data of the four classes were received and transmitted to matrices. Attributes are derived from training data of each class, and they were estimated for the data allocated to the test.

3.2. Preprocessing

Various preprocessing methods are used in the preprocessing phase. These are: filtration, normalization, noise reduction methods and analysis of basic components. In this study, normalization from among preprocessing methods was used.

Statistical normalization or Z-Score normalization was used in the preprocessing phase. Some values on the same data set having values smaller than 0 and some having higher values indicate that these distances among data and especially the data at the beginning or end points of data will be more effective on the results. By the normalization of data, it is ensured that each parameter in the training entrance set contributes equally to the model's estimation operation. The arithmetic average and standard deviation of columns corresponding to each variable are found. Then the data is normalized by the formula specified in the following equation, and the distances among data are removed and the end points in data are reduced [21].

$$x' = \frac{x_i - \mu_i}{\sigma_i}$$ (1)

It states; x_i = input value; μ_i = average of input data set; σ_i = standard deviation of input data set.

3.3. Method of Feature Extraction

Feature extraction is the most significant method for some problems such as speech recognition. There are various methods of feature extraction. These can be listed as independent components analysis, wavelet transform, Fourier analysis, common spatial pattern, skewness, kurtosis, total, average, variance, standard deviation, polynomial matching [22].

Learning a wider-total attribute indicates utility below: [4,23,24]

- Classification performance stems from the rustication of voice or untrustworthy attributes.
- Basic classifiers that reveal a better generalization skill with less input values in terms of new samplings.
- Understanding the classification problem through application by discovering the relevant and irrelevant attributes.

The main goal of the attributes is collecting as much data as possible without changing the acoustic specialty of speakers sound.

3.3.1. Skewness

Skewness is an asymmetrical measure of distribution. It is also the deterioration degree of symmetry in normal distribution. If the distribution has a long tail towards right, it is called positive skew or skew to right, and if the distribution has a long tail towards left, it is called negative skew or skew to left.

$$S = \frac{\frac{1}{L}\sum_{i=1}^{L}(x_i - \bar{x})^2}{\left(\frac{1}{L}\sum_{i=1}^{L}(x_i - \bar{x})^2\right)^3}$$ (2)

3.3.2. Kurtosis

Kurtosis is the measure of how an adverse inclined distribution is. The distribution of kurtosis can be stated as follows:

$$b = \frac{E(x - \mu)^4}{\sigma^4}$$ (3)

3.3.3. Zero Crossing Rate (ZCR)

Zero Crossing is a term that is used widely in electronic, mathematics and image processing. ZCR gives the ratio of the signal changes from positive to negative or the other way round. ZCR calculates this by counting the sound waves that cut the zero axis [25].

3.3.4. Local Maximum (Lmax) and Local Minimum (Lmin)

Lmax and Lmin points are called as local extremum points. The biggest local maximum point is called absolute maximum point and the smallest of the Lmin point is called absolute minimum point. Lmax starts with a signal changing transformation in time impact area in two dimensional map. Lmax perception correction is made by the comparison of the results of different lower signals. If the Lmax average number is higher, the point of those samplings in much important [26].

3.3.5. Root Mean Square (RMS)

RMS is the square root of the average sum of the signal. *RMS* is a value of 3D photogrammetry and in time the changes in the volume and the shape are considered. The mathematical method for calculating the *RMS* is as follows [27]:

$$X_{RMS} = \sqrt{\frac{x_1^2 + x_2^2 + \ldots + x_N^2}{N}} \tag{4}$$

X is the vertical distance between two points and *N* is the sum of the reference points on the two compared surfaces. *RMS*, is a statistical value which is used for calculating the increasing number of the changes. It is especially useful for the waves that changes positively and negatively.

3.3.6. Variance

Variance is measure of the distribution. It shows the distribution of the data set according to the average. It shows the changing between that moment's value and the average value according to the deviation.

3.4. Classification

3.4.1. K Nearest Neighborhood (Knn)

In Knn, the similarities of the data to be classified with the normal behavior data in the learning cluster are calculated and the assignments are done according to the closest *k* data average and the threshold value determined. An important point is the pre-determination of the characteristics of each class.

Knn's goal is to classify new data by using their characteristics and with the help of previously classified samples. Knn depends on a simple discriminating assumption known as intensity assumption. This classification has been successfully adopted in other non-parametric applications until speech definition [28].

In the Knn algorithm, first the *k* value should be determined. After determination of the *k* value, the calculation of its distance with all the learning samples should be performed and then ordering is performed as per minimum distance. After the ordering operation, which class value it belongs to is found. In this algorithm, when a sample is received from outside the training cluster, we try to find out to which class it belongs. Leave-one-out cross validation (LOOCV) was used in order to select the most suitable *k* value. We tried to find the *k* value by using the LOOCV.

LOOCV consists of dividing the data cluster to *n* pieces randomly. In each *n* repetition, *n* − 1 will be used as the training set and the excluded sampling will be used as the test set. In each of the *n* repetitions, a full data cluster is used for training except a sampling and as test cluster.

LOOCV is normally limited to applications where existing education data is restricted. For instance; any little deviation from tiny education data causes a large scale change in the appropriate model. In such a case, this reduces the deviation of the data in each trial to the lowest level, so adopting a LOOCV strategy makes sense. LOOCV is rarely used for large scale applications, because it is numerically expensive [29].

1000 units of training data were used. 250 data for each of the four classes were derived from among 1000 units of training data. One of 1000 units of training data forms the 1000-1 sub training cluster for validation cluster. Here, the part being specified as the sub training cluster was derived from the training cluster. Training cluster is divided into two being the sub training cluster and the validation cluster. In the validation cluster, the data to be considered for the test are available.

The change of k arising from randomness is not at issue. The values of k are selected as 1, 3, 5, 7, 9, 11 and they are compared with the responses in the training cluster. After these operations, the best k value becomes determined. $k = 5$ value, having the best rate, is selected. Here, the determination of k value states how many nearest values should be considered.

3.4.2. Support Vector Machine (SVM)

SVM was developed by Cortes and Vapnik for the solution of pattern recognition and classification problems [30]. The most important advantage of SVM is that it solves the classification problems by transforming them to quadratic optimization problems. Thus, the number of transactions related to solving the problem in the learning phase decreases and other techniques or algorithm based solutions can be reached more quickly. Due to this technical feature, there is a great advantage on large scale data sets. Additionally, it is based on optimization, classification performance, computational complexity and usability is much more successful [31,32].

SVM is a machine learning algorithm that works by the principle of non-structural risk minimization that is based on convex optimization. This algorithm is an independent learning algorithm that does not need any knowledge of the combined distribution function as data [33].

The aim of SVM is to achieve an optimal separation hyper-plane apart that will separate the classes. In other words, maximizing the distance between the support vectors that belong to different classes. SVM is a machine learning algorithm that was developed to solve multi-class classification problems.

Data sets that can or can't be distinguished as linear can be classified by SVM. The n dimensioned nonlinear data set can be transformed to a new data set as m dimensioned by $m > n$. In high dimensions linear classifications can be made. With an appropriate conversion, data can always be separated into two classes with a hyper plane.

3.5. Feature Selection

ReliefF is the developed version of the Relief statistical model. ReliefF is a widely-used feature selection algorithm [34] that carries out the process of feature selection by handling a sample from a dataset and creating a model based on its nearness to other samples in its own class and distance from other classes [35]. This study applies the ReliefF feature selection method to evaluate every feature in comparison to other features and determine which features are more effective in the classification phase.

3.6. Localization Algorithm Based on the Microphone Array

Position estimation methods in the literature are generally time of arrival (TOA), arrival time difference (TDOA) and received signal strength (RSS) based methods [36]. TDOA-based methods are highly advantageous because they can make highly accurate predictions. TDOA-based methods that use the maximum likelihood approach require a starting value and attempt to achieve the optimal result in an iterative manner [37]. If the initial values are not properly selected, there is a risk of not reaching the optimum result. In order to remove this disadvantage, closed form solutions have been developed.

Closed-loop algorithms utilize the least squares technique widely used for TDOA-based position estimation [38]. In TDOA-based position estimation methods, time delay estimates of the signal between sensor pairs are used. Major difficulties in TDOA estimation are the need for high data sharing and synchronization between sensors. This affects the speed and cost of the system negatively. The traditional TDOA estimation method in the literature uses the cross-correlation technique [39].

3.7. Echo Elimination

The reflection and return of sound wave after striking an obstacle is called echo. The echo causes the decrease of quality and clarity of the audio signal. Finite Impulse Response (FIR) filters are also referred as non-recursive filters. These filters are linear phase filters and are designed easily. In FIR filters, the same input is multiplied by more than one constant. This process is commonly known as Multiple Constant Multiplications. These operations are often used in digital signal processing applications and hardware based architects are the best choice for maximum performance and minimum power consumption.

4. Experimental Results

This study primarily selects the best feature methods and classification methods. Then, the prototype device has been tested using these selected methods on themselves. Meanwhile, tests have also been done on a computer environment and shown comparatively. The ReliefF method aims to find features' values and whether dependencies exist by trying to reveal them. This study selects the two best features using the ReliefF method. The two best feature methods turned out to be the Lmax and ZCR.

The results of the feature extraction method described above are shown in Figure 4 according to the data we took from the data set. As can be seen, the best categorizing method is the Lmax with ZCR using SVM. The results in Figure 4 show the mean values between 1 and 4 m.

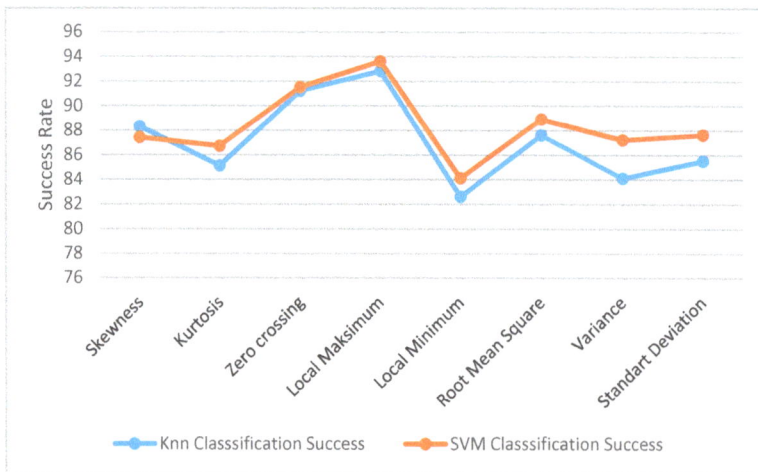

Figure 4. Knn's and SVM's Classification Success of Feature Extraction Methods (SVM: Support Vector Machine; Knn: K Nearest Neighborhood).

As seen in Table 1, the results obtained in real time and data were transferred to the digital environment and compared with the results obtained after the stages of preprocessing, feature extraction and classification. The results in Table 1 show the mean values between 1 and 4 m.

Table 1. Success rate of our system real time and obtained after computer.

	Left	Right	Front	Back
Accurate perception of the system	94.2%	94.3%	92%	92.7%
Accurate perception of the individual	92.8%	93%	91.1%	90.6%

Our wearable device produces results when there are more than 1 person. As can be seen in the Figure 5, people from 1-m and 3-m distances called the hearing-impaired individual. Our wearable device has noticed the individual who is close to him and he has directed that direction. As shown in Figure 5, the person with hearing impairment perceives this when the Person C behind the hearing impaired person calls to himself. 98% success was achieved in the results made in the room environment. It gives visual warning according to the proximity and distance.

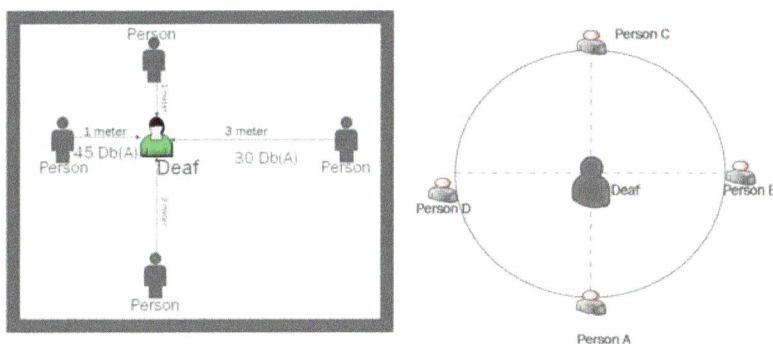

Figure 5. Testing wearable system.

As shown in Table 2, measurements were taken in the room environment, corridor and outside environment. As shown in Table 2, our wearable device was tested in room, the hallway and the exterior with a distance of 1 and 4 m. The success rate is shown by taking the average of the measured values with the sound meter. Each experiment was tested and the results were compared. In Table 2, the average level increase is caused by the increase of noise level in noisy environment and outdoor environment, but the success did not decrease much.

Table 2. Success rate of different environment.

Environment	Measurement	Direction Detection (1 m)	Direction Detection (4 m)	Sound Detection
Room	30 Db (A)	98%	97%	100%
Corridor	34 Db (A)	96%	93%	100%
Outdoor	40 Db (A)	93%	89%	99%

As seen in Table 3, perception of the direction of sound at a distance of ne meter was obtained as 97%. The best success rate was obtained by the sounds received from left and right directions. The success rate decreased with the increase of distance. As the direction increased, two directions were considered in the perception of sound, and the success rate decreased. And the direction where the success rate was the lowest was the front. The main reason for that is the placement of the developed human-computer interface system on the back of the individual. The sound coming from the front is being mixed up with the sounds coming from left or right.

Table 3. Success rate of finding the direction according to the distance.

Distance	Left	Right	Front	Back
1 m	96%	97.2%	92.4%	91.1%
2 m	95.1%	96%	90.2%	90%
3 m	93.3%	93.5%	90%	89.5%
4 m	90%	91%	84.2%	85.8%

As seen in Table 4, when perception of where the sound is coming from is considered, high success was obtained. As can be seen in Table 4, voice perception without checking the direction had great success. Voice was provided at low, normal and high levels and whether our system has perceived correctly or not was tested. Recognition success of the sound without looking at the direction of the source was 100%. It means our system recognized the sounds successfully.

Table 4. Successful perception rate of the voice without considering the distance.

Distance	Success Ratio (Loud Sound)	Success Ratio (Low Sound)
1 m	100%	100%
2 m	100%	100%
3 m	100%	100%
4 m	100%	99%

When individuals without hearing impairment and hearing impaired individuals were compared, Figure 6 compares deaf individuals with normal individuals and moderate hearing lose individuals. As a result of the application in real-time detection of deaf individuals' direction detection has a success rate of 88%. In Figure 6, the recognition of the sound source's direction for normal people, deaf and moderate hearing lose people is shown. As the number of people with hearing problems is low and the ability to teach them is limited, the normal people's number is higher. However, with proper training the number of people with hearing problems can be increased.

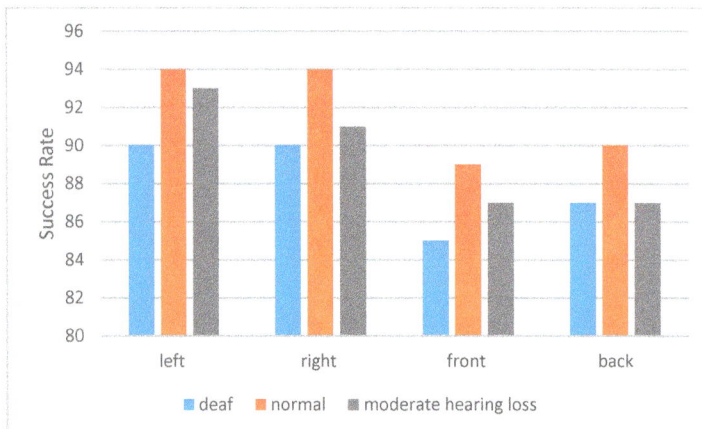

Figure 6. Success of deaf, moderate hearing loss and normal people to perceive the direction of voice.

As shown in Figure 7 in our system the individual speakers talk in real time and it can be determined if it is a shouting voice or a normal voice. Also, the direction of the shouting person or normal talking person can be determined and the classification succeeded. With this application, the normal voice detection is calculated as 95% and shouting voice detection is 91.8%. The results in Figure 7 show the mean values between 1 and 4 m.

It has been conducted by looking at eight directions; therefore a good distinction cannot be made when it is spoken in the middle of two directions. When only four directions are looked, a better success rate is seen. As it can be seen in Table 5, the results based on 4 and 8 directions are compared. The reason of lower rate of success in eight directions is that the direction of the voice could not be determined in intercardinal points. The results in Table 5 show the mean values between 1 and 4 m.

As shown in Table 6, the device we developed is faster than the TDOA algorithm. At the same time, it costs less because it uses less microphones. The device we have developed has been tested in real-time as well as the TDOA algorithm has been tested as a simulation.

The wearable device we developed also provides power management. If there is an echo in the environment, the echo is removed and the success rate is increased by 1.3%. In environments without echo, the echo canceller is disabled.

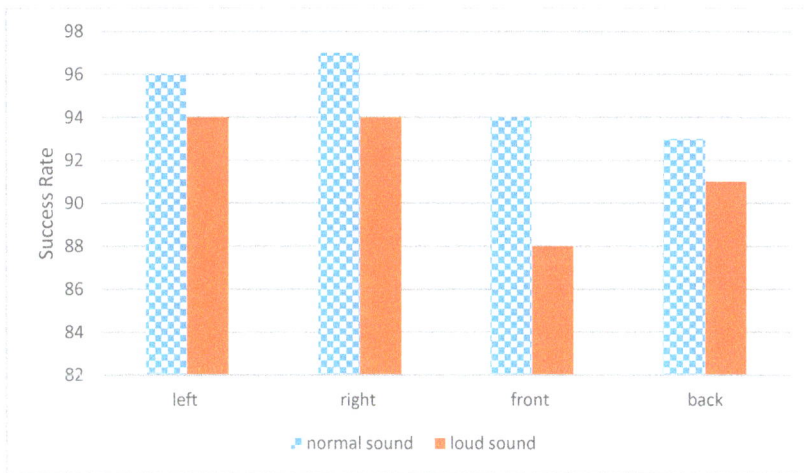

Figure 7. Success rate of loud and normal sound perceptions.

Table 5. Success Rate of 4 and 8 directions.

	Left	Right	Front	Back
4 Direction	94%	94%	92%	92%
8 Direction	91%	90%	88%	85%

Table 6. Compared our system with the localization algorithm based on the microphone array.

	Speed	Cost	Accuracy
TDOA	2.33 s	16 mics	97%
Our study	0.68 s	4 mics	98%

5. Discussion

We want the people with hearing problems to have a proper life at home or in the workplace by understanding the nature of sounds and their source. In this study, a vibration based system was suggested for hearing impaired individuals to perceive the direction of sound.

This study was tested real time. With the help of program we have written, our wearable system on the individual had 94% Success. In close range, SVM and two of the best feature methods are used and 98% successes are accomplished. By decreasing the noise in the area, the success can be increased. One of the most important problems in deaf is that they could not understand where the voice is coming from. This study helped the hearing impaired people to understand where the voice is coming from real time. It will be very useful for the deaf to be able to locate the direction of the stimulating sounds. Sometimes it can be very dangerous for them not to hear the horn of a car which is coming from their backs. No factors affect the test environment. In real-time tests, deaf individuals can determine direction thanks to our wearable devices. In outdoor tests, a decline in the classification success has been observed due to noise.

The vibration-based wearable device we have developed solves the problem of determining the direction from which a voice is coming, which is an important problem for deaf people. A deaf person should be able to sense noise and know its direction in a spontaneous instance. The direction of the voice has been determined in this study and thus it has been ensured that he can senses the direction from which the voice is coming. In particular, voices coming from behind or that a deaf individual cannot see will bother them; however, the device we have developed means that deaf individuals can sense a voice coming from behind them and travel more safely. This study has determined by whether a person is speaking loudly next to a deaf individual. Somebody might increase their tone of voice while speaking to the deaf individual in a panic and therefore this circumstance has been targeted so that the deaf person can notice such panic sooner. The deaf individual will be able to sense whether someone is shouting and if there is an important situation, his perception delay will be reduced thanks to the system we have developed. For instance, the deaf individual will be able to distinguish between a bell, the sound of a dog coming closer, or somebody calling to them; therefore, this can help them live more comfortably and safely with less public stress. In the feedback from deaf individuals using our device, they highlighted that they found the device very beneficial. The fact that they can particularly sense if there is an important voice coming from someone they cannot see makes it feel more reliable. Meanwhile, the fact that they can sense the voices of their parents calling them in real time while they are sleeping at home in particular has ensured that they feel more comfortable.

This study presents a new idea based on vibrating floor to the people with hearing problems who prefer to work with wearable computer related fields. We believe that the information which are present here will be useful for people with hearing problems who are working for system development on wearable processing and human-computer interaction fields. First, the best feature and classification method has been selected in the experimental studies. Using ReliefF from the feature selection algorithms allowed selecting the two best feature methods. Then, both real-time and computer tests were performed. Our tests have been done at a distance of 1–4 m in a (30–40 dB (A)) noisy environment. In the room, corridor, and outside environments, tests were done at a distance of 1–4 m. Whether one speaks with normal voice or screams has also been tested in our studies. The wearable device we have developed as a prototype has provided deaf people with more comfortable lives.

In the study performed, the derivation of training and test data, validation phase and selection of the best attribute derivation method takes time. Especially while determining the attribute, it is required to find the most distinguishing method by using the other methods. While performing operation with multiple data, which data is more significant for us is an important problem in implementations. The attribute method being used may differ among implementations. In such studies, the important point is to derive attributes more than one and to perform their joint use. In this manner, a better classification success can be obtained.

In the study, sound perception will be performed through vibration. A system will be developed for the hearing impaired individuals to perceive both the direction of speaker and what he is speaking of. In this system, first the perception of specific vowels and consonants will be made, and their distinguishing properties will be determined, and perception by the hearing impaired individual through vibration will be ensured.

6. Conclusions

Consequently, the direction of sound is perceived to a large extent. Moreover, it was also determined whether the speaker is shouting or not. In future studies, deep learning, correlation and hidden Markov model will be used and the success of system will tried to be increased. Also, other methods in the classification stage will be used to get the best result. For further studies, the optimum distances for the microphones will be calculated and the voice recognition will be made with the best categorizing agent. Which sounds are most important for deaf people will be determined using our wearable device in future studies; it is important for standard of living which sound is determined, particularly with direction determination. Meanwhile, real-time visualization requires consideration; a wearable device that transmits the direction from which sounds originate will be made into glasses that a deaf individual can wear.

Author Contributions: M.Y. and C.K. conceived and designed the experiments; C.K. performed the experiments; M.Y. analyzed the data; M.Y. and C.K. wrote the paper.

Conflicts of Interest: The authors declare no conflict of interest.

References

1. Milton, A.; Selvi, S.T. Class-specific multiple classifiers scheme to recognize emotions from speech signals. *Comput. Speech Lang.* **2014**, *28*, 727–742. [CrossRef]
2. Huang, X.; Acero, A.; Hon, H.-W.; Reddy, R. *Spoken Language Processing: A Guide to Theory, Algorithm, and System Development*; Prentice Hall PTR: Upper Saddle River, NJ, USA, 2001.
3. Schuller, B.; Steidl, S.; Batliner, A.; Burkhardt, F.; Devillers, L.; MüLler, C.; Narayanan, S. Paralinguistics in speech and language—State-of-the-art and the challenge. *Comput. Speech Lang.* **2013**, *27*, 4–39. [CrossRef]
4. Pohjalainen, J.; Räsänen, O.; Kadioglu, S. Feature selection methods and their combinations in high-dimensional classification of speaker likability, intelligibility and personality traits. *Comput. Speech Lang.* **2015**, *29*, 145–171. [CrossRef]
5. Clavel, C.; Vasilescu, I.; Devillers, L.; Richard, G.; Ehrette, T. Fear-type emotion recognition for future audio-based surveillance systems. *Speech Commun.* **2008**, *50*, 487–503. [CrossRef]
6. Vassis, D.; Belsis, P.; Skourlas, C.; Marinagi, C.; Tsoukalas, V. Secure mobile assessment of deaf and hard-of-hearing and dyslexic students in higher education. In Proceedings of the 17th Panhellenic Conference on Informatics, Thessaloniki, Greece, 19–21 September 2013; ACM: New York, NY, USA, 2013; pp. 311–318.
7. Shivakumar, B.; Rajasenathipathi, M. A New Approach for Hardware Control Procedure Used in Braille Glove Vibration System for Disabled Persons. *Res. J. Appl. Sci. Eng. Technol.* **2014**, *7*, 1863–1871. [CrossRef]
8. Arato, A.; Markus, N.; Juhasz, Z. Teaching morse language to a deaf-blind person for reading and writing SMS on an ordinary vibrating smartphone. In Proceedings of the International Conference on Computers for Handicapped Persons, Paris, France, 9–11 July 2014; Springer: Berlin, Germany, 2014; pp. 393–396.
9. Nanayakkara, S.C.; Wyse, L.; Ong, S.H.; Taylor, E.A. Enhancing musical experience for the hearing-impaired using visual and haptic displays. *Hum.–Comput. Interact.* **2013**, *28*, 115–160.
10. Gollner, U.; Bieling, T.; Joost, G. Mobile Lorm Glove: Introducing a communication device for deaf-blind people. In Proceedings of the 6th International Conference on Tangible, Embedded and Embodied Interaction, Kingston, ON, Canada, 19–22 February 2012; ACM: New York, NY, USA, 2012; pp. 127–130.
11. Schmitz, B.; Ertl, T. Making digital maps accessible using vibrations. In Proceedings of the International Conference on Computers for Handicapped Persons, Vienna, Austria, 14–16 July 2010; Springer: Berlin, Germany, 2010; pp. 100–107.
12. Ketabdar, H.; Polzehl, T. Tactile and visual alerts for deaf people by mobile phones. In Proceedings of the 11th international ACM SIGACCESS Conference on Computers and Accessibility, Pittsburgh, PA, USA, 25–28 October 2009; ACM: New York, NY, USA, 2009; pp. 253–254.
13. Caetano, G.; Jousmäki, V. Evidence of vibrotactile input to human auditory cortex. *Neuroimage* **2006**, *29*, 15–28. [CrossRef] [PubMed]
14. Zhong, X.; Wang, S.; Dorman, M.; Yost, W. Sound source localization from tactile aids for unilateral cochlear implant users. *J. Acoust. Soc. Am.* **2013**, *134*, 4062. [CrossRef]

15. Wang, S.; Zhong, X.; Dorman, M.F.; Yost, W.A.; Liss, J.M. Using tactile aids to provide low frequency information for cochlear implant users. *J. Acoust. Soc. Am.* **2013**, *134*, 4235. [CrossRef]
16. Mesaros, A.; Heittola, T.; Virtanen, T. Metrics for polyphonic sound event detection. *Appl. Sci.* **2016**, *6*, 162. [CrossRef]
17. Wang, Q.; Liang, R.; Rahardja, S.; Zhao, L.; Zou, C.; Zhao, L. Piecewise-Linear Frequency Shifting Algorithm for Frequency Resolution Enhancement in Digital Hearing Aids. *Appl. Sci.* **2017**, *7*, 335. [CrossRef]
18. Gao, W.; Ma, J.; Wu, J.; Wang, C. Sign language recognition based on HMM/ANN/DP. *Int. J. Pattern Recognit. Artif. Intell.* **2000**, *14*, 587–602. [CrossRef]
19. Lin, R.-S.; Chen, L.-H. A new approach for audio classification and segmentation using Gabor wavelets and Fisher linear discriminator. *Int. J. Pattern Recognit. Artif. Intell.* **2005**, *19*, 807–822. [CrossRef]
20. Tervo, S.; Pätynen, J.; Kaplanis, N.; Lydolf, M.; Bech, S.; Lokki, T. Spatial analysis and synthesis of car audio system and car cabin acoustics with a compact microphone array. *J. Audio Eng. Soc.* **2015**, *63*, 914–925. [CrossRef]
21. Suarez-Alvarez, M.M.; Pham, D.-T.; Prostov, M.Y.; Prostov, Y.I. Statistical approach to normalization of feature vectors and clustering of mixed datasets. *Proc. R. Soc. A* **2012**. [CrossRef]
22. Vamvakas, G.; Gatos, B.; Petridis, S.; Stamatopoulos, N. An efficient feature extraction and dimensionality reduction scheme for isolated greek handwritten character recognition. In Proceedings of the 2007 9th International Conference on Document Analysis and Recognition, Parana, Brazil, 23–26 September 2007; IEEE: Piscataway, NJ, USA, 2007; pp. 1073–1077.
23. Blum, A.L.; Langley, P. Selection of relevant features and examples in machine learning. *Artif. Intell.* **1997**, *97*, 245–271. [CrossRef]
24. Reunanen, J. Overfitting in making comparisons between variable selection methods. *J. Mach. Learn. Res.* **2003**, *3*, 1371–1382.
25. Wang, Y.; Liu, Z.; Huang, J.-C. Multimedia content analysis-using both audio and visual clues. *IEEE Signal Process. Mag.* **2000**, *17*, 12–36. [CrossRef]
26. Obuchowski, J.; Wyłomańska, A.; Zimroz, R. The local maxima method for enhancement of time-frequency map and its application to local damage detection in rotating machines. *Mech. Syst. Signal Process.* **2014**, *46*, 389–405. [CrossRef]
27. Moghaddam, M.B.; Brown, T.M.; Clausen, A.; DaSilva, T.; Ho, E.; Forrest, C.R. Outcome analysis after helmet therapy using 3D photogrammetry in patients with deformational plagiocephaly: The role of root mean square. *J. Plast. Reconstr. Aesthet. Surg.* **2014**, *67*, 159–165. [CrossRef] [PubMed]
28. Golipour, L.; O'Shaughnessy, D. A segmental non-parametric-based phoneme recognition approach at the acoustical level. *Comput. Speech Lang.* **2012**, *26*, 244–259. [CrossRef]
29. Cawley, G.C.; Talbot, N.L. Efficient leave-one-out cross-validation of kernel fisher discriminant classifiers. *Pattern Recognit.* **2003**, *36*, 2585–2592. [CrossRef]
30. Cortes, C.; Vapnik, V. Support-vector networks. *Mach. Learn.* **1995**, *20*, 273–297. [CrossRef]
31. Osowski, S.; Siwek, K.; Markiewicz, T. Mlp and svm networks-a comparative study. In Proceedings of the 6th Nordic Signal Processing Symposium, Espoo, Finland, 11 June 2004; IEEE: Piscataway, NJ, USA, 2004; pp. 37–40.
32. Nitze, I.; Schulthess, U.; Asche, H. Comparison of machine learning algorithms random forest, artificial neural network and support vector machine to maximum likelihood for supervised crop type classification. In Proceedings of the 4th GEOBIA, Salzburg, Austria, 7–9 May 2012; pp. 7–9.
33. Soman, K.; Loganathan, R.; Ajay, V. *Machine Learning with SVM and Other Kernel Methods*; PHI Learning Pvt. Ltd.: Delhi, India, 2009.
34. Zhang, J.; Chen, M.; Zhao, S.; Hu, S.; Shi, Z.; Cao, Y. ReliefF-Based EEG Sensor Selection Methods for Emotion Recognition. *Sensors* **2016**, *16*, 1558. [CrossRef] [PubMed]
35. Bolón-Canedo, V.; Sánchez-Marono, N.; AlonsoBetanzos, A.; Benitez, J.M.; Herrera, F. A Review of Microarray Datasets and Applied Feature Selection Methods. *Inf. Sci.* **2014**, *282*, 111–135. [CrossRef]
36. So, H.C. Source localization: Algorithms and analysis. In *Handbook of Position Location: Theory, Practice, and Advances*; John Wiley & Sons: Hoboken, NJ, USA, 2011; pp. 25–66.
37. Dogançay, K. Emitter localization using clustering-based bearing association. *IEEE Trans. Aerosp. Electron. Syst.* **2005**, *41*, 525–536. [CrossRef]

Appl. Sci. **2017**, *7*, 1296

38. Smith, J.; Abel, J. Closed-form least-squares source location estimation from range-difference measurements. *IEEE Trans. Acoust. Speech Signal Process.* **1987**, *35*, 1661–1669. [CrossRef]

39. Knapp, C.; Carter, G. The generalized correlation method for estimation of time delay. *IEEE Trans. Acoust. Speech Signal Process.* **1976**, *24*, 320–327. [CrossRef]

applied
sciences

MDPI

Article

Audio Time Stretching Using Fuzzy Classification of Spectral Bins

Eero-Pekka Damskägg * and Vesa Välimäki

Acoustics Laboratory, Department of Signal Processing and Acoustics, Aalto University,
FI-02150 Espoo, Finland; vesa.valimaki@aalto.fi
* Correspondence: eero-pekka.damskagg@aalto.fi

Academic Editor: Gino Iannace
Received: 3 November 2017; Accepted: 7 December 2017; Published: 12 December 2017

Abstract: A novel method for audio time stretching has been developed. In time stretching, the audio signal's duration is expanded, whereas its frequency content remains unchanged. The proposed time stretching method employs the new concept of fuzzy classification of time-frequency points, or bins, in the spectrogram of the signal. Each time-frequency bin is assigned, using a continuous membership function, to three signal classes: tonalness, noisiness, and transientness. The method does not require the signal to be explicitly decomposed into different components, but instead, the computing of phase propagation, which is required for time stretching, is handled differently in each time-frequency point according to the fuzzy membership values. The new method is compared with three previous time-stretching methods by means of a listening test. The test results show that the proposed method yields slightly better sound quality for large stretching factors as compared to a state-of-the-art algorithm, and practically the same quality as a commercial algorithm. The sound quality of all tested methods is dependent on the audio signal type. According to this study, the proposed method performs well on music signals consisting of mixed tonal, noisy, and transient components, such as singing, techno music, and a jazz recording containing vocals. It performs less well on music containing only noisy and transient sounds, such as a drum solo. The proposed method is applicable to the high-quality time stretching of a wide variety of music signals.

Keywords: audio systems; digital signal processing; music; spectral analysis; spectrogram

1. Introduction

Time-scale modification (TSM) refers to an audio processing technique, which changes the duration of a signal without changing the frequencies contained in that signal [1–3]. For example, it is possible to reduce the speed of a speech signal so that it sounds as if the person is speaking more slowly, since the fundamental frequency and the spectral envelope are preserved. Time stretching corresponds to the extension of the signal, but this term is used as a synonym for TSM. Audio time stretching has numerous applications, such as fast browsing of speech recordings [4], music production [5], foreign language and music learning [6], fitting of a piece of music to a prescribed time slot [7], and slowing down the soundtrack for slow-motion video [8]. Additionally, TSM is often used as a processing step in pitch shifting, which aims at changing the frequencies in the signal without changing its duration [2,3,7,9,10].

Audio signals can be considered to consist of sinusoidal, noise, and transient components [11–14]. The main challenge in TSM is in simultaneously preserving the subjective quality of these distinct components. Standard time-domain TSM methods, such as the synchronized overlap-add (SOLA) [15], the waveform-similarity overlap-add [16], and the pitch-synchronous overlap-add [17] techniques, are considered to provide high-quality TSM for quasi-harmonic signals. When these methods are applied to polyphonic signals, however, only the most dominant periodic pattern of the input waveform is preserved, while other periodic components suffer from phase jump artifacts at the synthesis frame boundaries. Furthermore, overlap-add techniques are prone to transient skipping or duplication when the signal is contracted or extended, respectively. To solve this, transients can be detected and the time-scale factor can be changed during transients [18,19].

Standard phase vocoder TSM techniques [20,21] are based on a sinusoidal model of the input signal. Thus, they are most suitable for processing of signals which can be represented as a sum of slowly varying sinusoids. Even with these kind of signals however, the phase vocoder TSM introduces an artifact typically described as "phasiness" to the processed sound [21,22]. Furthermore, transients processed with the standard phase vocoder suffer from a softening of the perceived attack, often referred to as "transient smearing" [2,3,23]. A standard solution for reducing transient smearing is to apply a phase reset or phase locking at detected transient locations of the input signal [23–25].

As another approach to overcome these problems in the phase vocoder, TSM techniques using classification of spectral components based on their signal type have been proposed recently. In [26], spectral peaks are classified into sinusoids, noise, and transients, using the methods of [23,27]. Using the information from the peak classification, the phase modification applied in the technique is based only on the sinusoidally classified peaks. It uses the method of [23] to detect and preserve transient components. Furthermore, to better preserve the noise characteristics of the input sound, uniformly distributed random numbers are added to the phases of spectral peaks classified as noise. In [28], spectral bins are classified into sinusoidal and transient components, using the median filtering technique of [29]. The time-domain signals synthesized from the classified components are then processed separately, using an appropriate analysis window length for each class. Phase vocoder processing with a relatively long analysis window is applied to the sinusoidal components. A standard overlap-add scheme with a shorter analysis window is used for the transient components.

Both of the above methods are based on a binary classification of the spectral bins. However, it is more reasonable to consider the energy in each spectral bin as a superposition of energy from sinusoidal, noise, and transient components [13]. Therefore, each spectral bin should be allowed to belong to all of the classes simultaneously, with a certain degree of membership for each class. This kind of approach is known as fuzzy classification [30,31]. To this end, in [32], a continuous measure denoted as tonalness was proposed. Tonalness is defined as a continuous value between 0 and 1, which gives the estimated likelihood of each spectral bin belonging to a tonal component. However, the proposed measure alone does not assess the estimation of the noisiness or transientness of the spectral bins. Thus, a way to estimate the degree of membership to all of these classes for each spectral bin is needed.

In this paper, a novel phase vocoder-based TSM technique is proposed in which the applied phase propagation is based on the characteristics of the input audio. The input audio characteristics are quantified by means of fuzzy classification of spectral bins into sinusoids, noise, and transients. The information about the nature of the spectral bins is used for preserving the intra-sinusoidal phase coherence of the tonal components, while simultaneously preserving the noise characteristics of the input audio. Furthermore, a novel method for transient detection and preservation based on the classified bins is proposed. To evaluate the quality of the proposed method, a listening test was conducted. The results of the listening test suggest that the proposed method is competitive against a state-of-the art academic TSM method and commercial TSM software.

The remainder of this paper is structured as follows. In Section 2, the proposed method for fuzzy classification of spectral bins is presented. In Section 3, a novel TSM technique which uses the fuzzy membership values is detailed. In Section 4, the results of the conducted listening test are presented and discussed. Finally, Section 5 concludes the paper.

2. Fuzzy Classification of Bins in the Spectrogram

The proposed method for the classification of spectral bins is based on the observation that, in a time-frequency representation of a signal, stationary tonal components appear as ridges in the time direction, whereas transient components appear as ridges in the frequency direction [29,33].

Thus, if a spectral bin contributes to the forming of a time-direction ridge, most of its energy is likely to come from a tonal component in the input signal. Similarly, if a spectral bin contributes to the forming of a frequency-direction ridge, most of its energy is probably from a transient component. As a time-frequency representation, the short-time Fourier transform (STFT) is used:

$$X[m,k] = \sum_{n=-N/2}^{N/2} x[n+mH_a]w[n]e^{-j\omega_k n}, \tag{1}$$

where m and k are the integer time frame and spectral bin indices, respectively, $x[n]$ is the input signal, H_a is the analysis hop size, $w[n]$ is the analysis window, N is the analysis frame length and the number of frequency bins in each frame, and $\omega_k = 2\pi k/N$ is the normalized center frequency of the kth STFT bin. Figure 1 shows the STFT magnitude of a signal consisting of a melody played on the piano, accompanied by soft percussion and a double bass. The time-direction ridges introduced by the harmonic instruments and the frequency-direction ridges introduced by the percussion are apparent on the spectrogram.

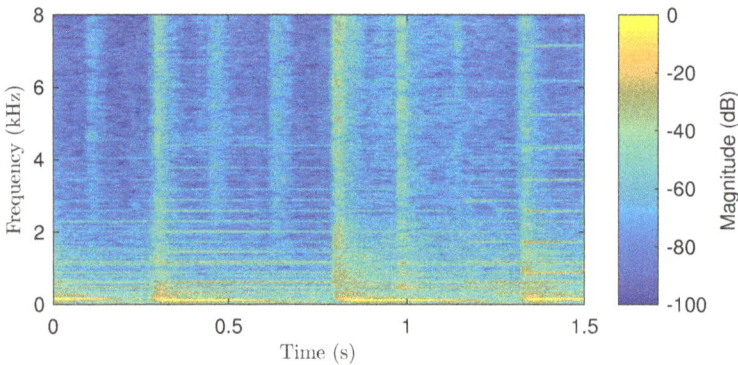

Figure 1. Spectrogram of a signal consisting of piano, percussion, and double bass.

The tonal and transient STFTs $X_s[m,k]$ and $X_t[m,k]$, respectively, are computed using the median filtering technique proposed by Fitzgerald [29]:

$$X_s[m,k] = \text{median}(|X[m-\frac{L_t}{2}+1,k]|, ..., |X[m+\frac{L_t}{2},k]|) \tag{2}$$

and

$$X_t[m,k] = \text{median}(|X[m,k-\frac{L_f}{2}+1]|, ..., |X[m,k+\frac{L_f}{2}]|), \tag{3}$$

where L_t and L_f are the lengths of the median filters in time and frequency directions, respectively. For the tonal STFT, the subscript s (denoting sinusoidal) is used and for the transient STFT the subscript t. Median filtering in the time direction suppresses the effect of transients in the STFT magnitude, while preserving most of the energy of the tonal components. Conversely, median filtering in the frequency direction suppresses the effect of tonal components, while preserving most of the transient energy [29].

The two median-filtered STFTs are used to estimate the tonalness, noisiness, and transientness of each analysis STFT bin. We estimate tonalness by the ratio

$$R_s[m,k] = \frac{X_s[m,k]}{X_s[m,k]+X_t[m,k]}. \tag{4}$$

We define transientness as the complement of tonalness:

$$R_t[m,k] = 1 - R_s[m,k] = \frac{X_t[m,k]}{X_s[m,k] + X_t[m,k]}. \tag{5}$$

Signal components which are neither tonal nor transient can be assumed to be noiselike. Experiments on noise signal analysis using the above median filtering method show that the tonalness value is often approximately $R_s = 0.5$. This is demonstrated in Figure 2b in which a histogram of the tonalness values of STFT bins of a pink noise signal (Figure 2a) is shown. It can be seen that the tonalness values are approximately normally distributed around the value 0.5. Thus, we estimate noisiness by

$$R_n[m,k] = 1 - |R_s[m,k] - R_t[m,k]| = \begin{cases} 2R_s[m,k], & \text{if } R_s[m,k] \le 0.5 \\ 2(1 - R_s[m,k]), & \text{otherwise.} \end{cases} \tag{6}$$

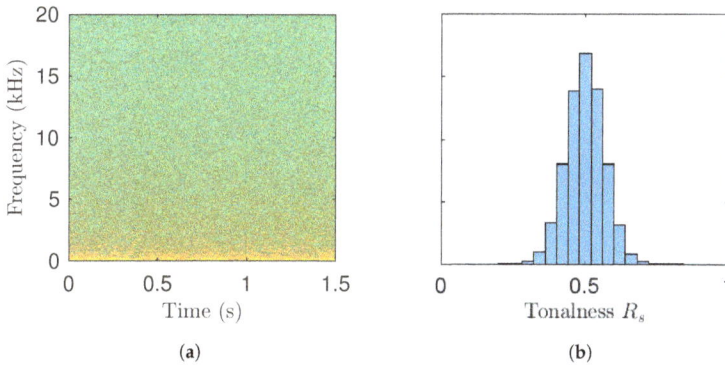

Figure 2. (a) Spectrogram of pink noise and (b) the histogram of tonalness values for its spectrogram bins.

The tonalness, noisiness, and transientness can be used to denote the degree of membership of each STFT bin to the corresponding class in a fuzzy manner. The relations between the classes are visualized in Figure 3.

Figure 4 shows the computed tonalness, noisiness, and transientness values for the STFT bins of the example audio signal used above. The tonalness values in Figure 4a are close to 1 for the bins which represent the harmonics of the piano and double bass tones, whereas the tonalness values are close to 0 for the bins which represent percussive sounds. In Figure 4b, the noisiness values are close to 1 for the bins which do not significantly contribute either to the tonal nor the transient components in the input audio. Finally, it can be seen that the transientness values in Figure 4c are complementary to the tonalness values of Figure 4a.

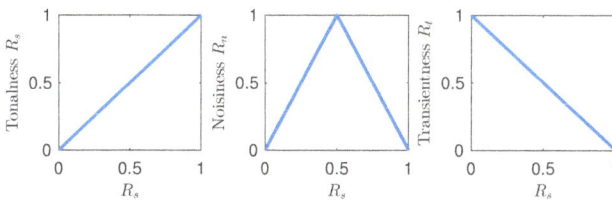

Figure 3. The relations between the three fuzzy classes.

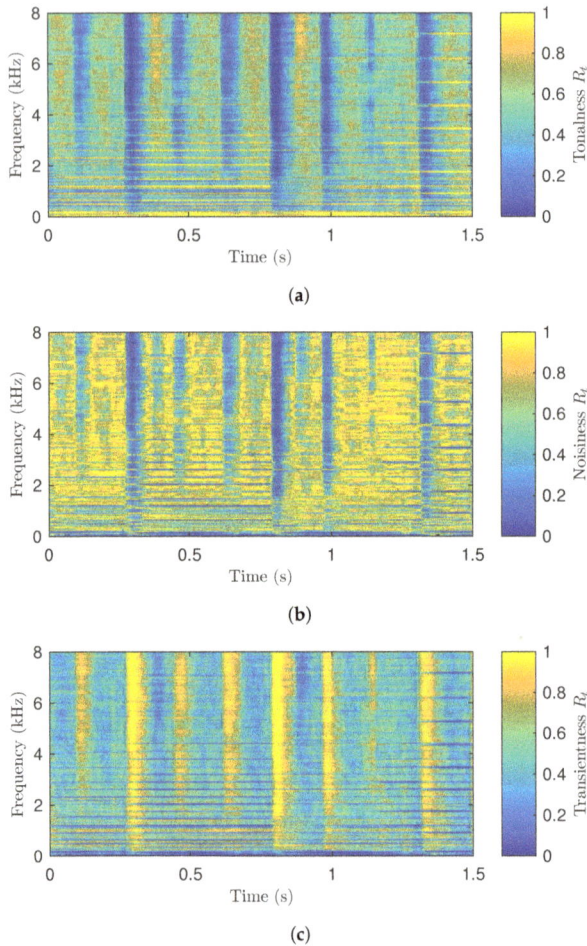

Figure 4. (**a**) Tonalness, (**b**) noisiness, and (**c**) transientness values for the short-time Fourier transform (STFT) bins of the example audio signal. Cf. Figure 1.

3. Novel Time-Scale Modification Technique

This section introduces the new TSM technique that is based on the fuzzy classification of spectral bins defined above.

3.1. Proposed Phase Propagation

The phase vocoder TSM is based on the differentiation and subsequent integration of the analysis STFT phases in time. This process is known as phase propagation. The phase propagation in the new TSM method is based on a modification to the phase-locked vocoder by Laroche and Dolson [21]. The phase propagation in the phase-locked vocoder can be described as follows. For each frame in the analysis STFT (1), peaks are identified. Peaks are defined as spectral bins, whose magnitude is greater than the magnitude of its four closest neighboring bins.

The phases of the peak bins are differentiated to obtain the instantaneous frequency for each peak bin:

$$\omega_{inst}[m,k] = \omega_k + \frac{1}{H_a}\kappa[m,k], \tag{7}$$

where $\kappa[m,k]$ is the estimated "heterodyned phase increment":

$$\kappa[m,k] = \left[\angle X[m,k] - \angle X[m-1,k] - H_a\omega_k\right]_{2\pi}. \tag{8}$$

Here, $[\,\cdot\,]_{2\pi}$ denotes the principal determination of the angle, i.e., the operator wraps the input angle to the interval $[-\pi, \pi]$. The phases of the peak bins in the synthesis STFT $Y[m,k]$ can be computed by integrating the estimated instantaneous frequencies according to the synthesis hop size H_s:

$$\angle Y[m,k] = \angle Y[m-1,k] + H_s\omega_{inst}[m,k], \tag{9}$$

The ratio between the analysis and synthesis hop sizes H_a and H_s determines the TSM factor α. In practice, the synthesis hop size is fixed and the analysis hop size then depends on the desired TSM factor:

$$H_a = \frac{H_s}{\alpha}. \tag{10}$$

In the standard phase vocoder TSM [20], the phase propagation of (7)–(9) is applied to all bins, not only peak bins. In the phase-locked vocoder [21], the way the phases of non-peak bins are modified is known as phase locking. It is based on the idea that the phase relations between all spectral bins, which contribute to the representation of a single sinusoid, should be preserved when the phases are modified. This is achieved by modifying the phases of the STFT bins surrounding each peak such that the phase relations between the peak and the surrounding bins are preserved from the analysis STFT. Given a peak bin k_p, the phases of the bins surrounding the peak are modified by:

$$\angle Y[m,k] = \angle X[m,k] + \left[\angle Y[m,k_p] - \angle X[m,k_p]\right]_{2\pi}, \tag{11}$$

where $\angle Y[m,k_p]$ is computed according to (7)–(9). This approach is known as identity phase locking.

As the motivation behind phase locking states, it should only be applied to bins that are considered sinusoidal. When applied to non-sinusoidal bins, the phase locking introduces a metallic sounding artifact to the processed signal. Since the tonalness, noisiness, and transientness of each bin are determined, this information can be used when the phase locking is applied. We want to be able to apply phase locking to bins which represent a tonal component, while preserving the randomized phase relationships of bins representing noise.

Thus, the phase locking is first applied to all bins. Afterwards, phase randomization is applied to the bins according to the estimated noisiness values. The final synthesis phases are obtained by adding uniformly distributed noise to the synthesis phases computed with the phase-locked vocoder:

$$\angle Y'[m,k] = \angle Y[m,k] + \pi A_n[m,k]\left(u[m,k] - \frac{1}{2}\right), \tag{12}$$

where $u[m,k]$ are the added noise values and $\angle Y[m,k]$ are the synthesis phases computed with the phase-locked vocoder. The pseudo-random numbers $u[m,k]$ are drawn from the uniform distribution $\mathcal{U}(0,1)$. $A_n[m,k]$ is the phase randomization factor, which is based on the estimated noisiness of the bin $R_n[m,k]$ and the TSM factor α:

$$A_n[m,k] = \frac{1}{4}\left[\tanh(b_n(R_n[m,k]-1))+1\right]\left[\tanh(b_\alpha(\alpha - \frac{3}{2}))+1\right], \tag{13}$$

where constants b_n and b_α control the shape of non-linear mappings of the hyperbolic tangents. The values $b_n = b_\alpha = 4$ were used in this implementation.

The phase randomization factor A_n, as a function of the estimated noisiness R_n and the TSM factor α, is shown in Figure 5. The phase randomization factor increases with increasing TSM factor and noisiness. The phase randomization factor saturates as the values increase, so that at most, the uniform noise added to the phases obtains values in the interval $[-0.5\pi, 0.5\pi]$.

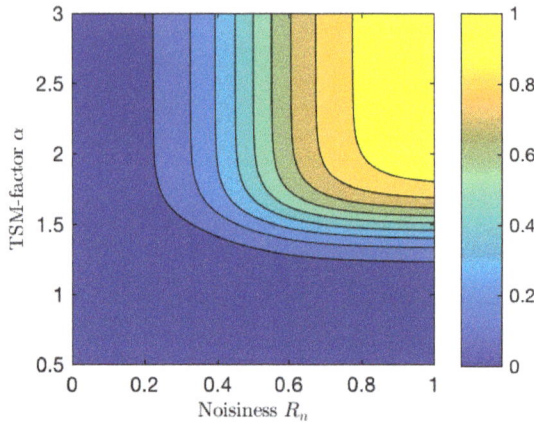

Figure 5. A contour plot of the phase randomization factor A_n, with $b_n = b_\alpha = 4$. TSM: time-scale modification.

3.2. Transient Detection and Preservation

For transient detection and preservation, a similar strategy to [23] was adopted. However, the proposed method is based on the estimated transientness of the STFT bins. Using the measure for transientness, the smearing of both the transient onsets and offsets is prevented. The transients are processed so that the transient energy is mostly contained on a single synthesis frame, effectively suppressing the transient smearing artifact, which is typical for the phase vocoder based TSM.

3.2.1. Detection

To detect transients, the overall transientness of each analysis frame is estimated, and denoted as frame transientness:

$$r_t[m] = \frac{1}{N-1} \sum_{k=1}^{N-1} R_t[m, k]. \tag{14}$$

The analysis frames which are centered on a transient component appear as local maxima in the frame transientness. Transients need to be detected as soon as the analysis window slides over them in order to prevent the smearing of transient onsets. To this end, the time derivative of frame transientness is used:

$$\frac{d}{dm} r_t[m] \approx \frac{1}{H_a} (r_t[m] - r_t[m-1]), \tag{15}$$

where the time derivative is approximated with the backward difference method. As the analysis window slides over a transient, there is an abrupt increase in the frame transientness. These instants appear as local maxima in the time derivative of the frame transientness. Local maxima in the time derivative of the frame transientness that exceed a given threshold are used for transient detection.

Figure 6 illustrates the proposed transient detection method using the same audio excerpt as above, containing piano, percussion, and double bass. The transients appear as local maxima in the frame transientness signal in Figure 6a. Transient onsets are detected from the time derivative of the frame transientness, from the local maxima, which exceed the given threshold (the red dashed line in Figure 6b). The detected transient onsets are marked with orange crosses. After an onset is detected, the analysis frame which is centered on the transient is detected from the subsequent local maxima in the frame transientness. The detected analysis frames centered on a transient are marked with purple circles in Figure 6a.

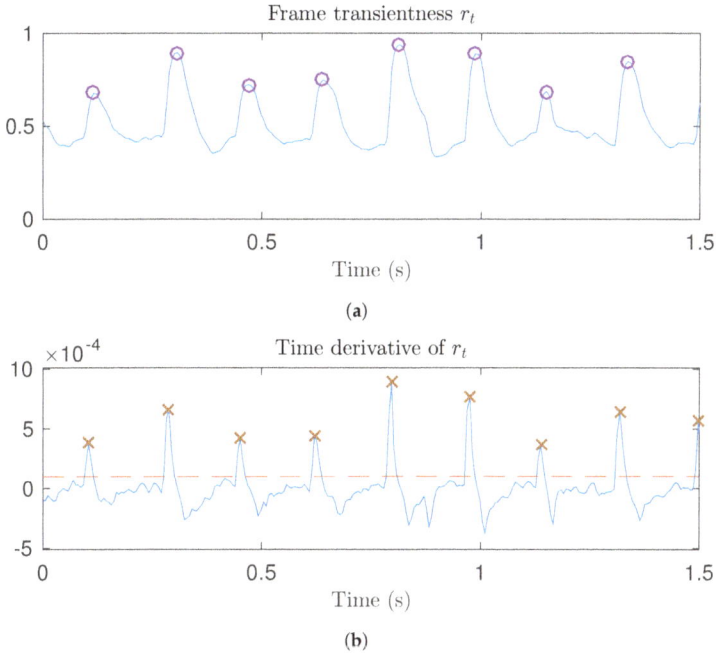

(a)

(b)

Figure 6. Illustration of the proposed transient detection. (**a**) Frame transientness. Locations of the detected transients are marked with purple circles; (**b**) Time derivative of the frame transientness. Detected transient onsets are marked with orange crosses. The red dashed line shows the transient detection threshold.

3.2.2. Transient Preservation

To prevent transient smearing, it is necessary to concentrate the transient energy in time. A single transient contributes energy to multiple analysis frames, because the frames are overlapping. During the synthesis, the phases of the STFT bins are modified, and the synthesis frames are relocated in time, which results in smearing of the transient energy.

To remove this effect, transients are detected as the analysis window slides over them. When a transient onset has been detected using the method described above, the energy in the STFT bins is suppressed according to their estimated transientness:

$$|Y[m,k]| = (1 - R_t[m,k])|X[m,k]|. \tag{16}$$

This gain is only applied to bins whose estimated transientness is larger than 0.5. Similar to [23], the bins to which this gain has been applied are kept in a non-contracting set of transient bins K_t. When it is detected that the analysis window is centered on a transient, as explained above, a phase reset is performed on the transient bins. That is, the original analysis phases are kept during synthesis for the transient bins. Subsequently, as the analysis window slides over the transient, the same gain reduction is applied for the transient bins, as during the onset of the transient (16). The bins are retained in the set of transient bins until their transientness decays to a value smaller than 0.5, or until the analysis frame slides completely away from the detected transient center. Finally, since the synthesis frames before and after the center of the transient do not contribute to the transients' energy, the magnitudes of the transient bins are compensated by

$$|Y[m_t, k_t]| = \frac{\sum_{m \in \mathbb{Z}} w^2[(m_t - m)H_s]}{w^2[0]} \frac{\sum_{k \in K_t} R_t[m_t, k]}{|K_t|} |X[m_t, k_t]|, \tag{17}$$

where m_t is the transient frame index, $|K_t|$ denotes the number of elements in the set K_t, and $k_t \in K_t$, which is the defined set of transient bins.

This method aims to prevent the smearing of both the transient onsets and offsets during TSM. In effect, the transients are separated from the input audio, and relocated in time according to the TSM factor. However, in contrast to methods where transients are explicitly separated from the input audio [13,14,28,34], the proposed method is more likely to keep transients perceptually intact with other components of the sound. Since the transients are kept in the same STFT representation, phase modifications in subsequent frames are dependent on the phases of the transient bins. This suggests that transients related to the onsets of harmonic sounds, such as the pluck of a note while strumming a guitar, should blend smoothly with the following tonal component of the sound. Furthermore, the soft manner in which the amplitudes of the transient bins are attenuated during onsets and offsets should prevent strong artifacts arising from errors in the transient detection.

Figure 7 shows an example of a transient processed with the proposed method. The original audio shown in Figure 7a consists of a solo violin overlaid with a castanet click. Figure 7b shows the time-scale modified sample with TSM factor $\alpha = 1.5$, using the standard phase vocoder. In the modified sample, the energy of the castanet click is spread over time. This demonstrates the well known transient smearing artifact of standard phase vocoder TSM. Figure 7c shows the time-scale modified sample using the proposed method. It can be seen that while the duration of the signal has changed, the castanet click in the modified audio resembles the one in the original, without any visible transient smearing.

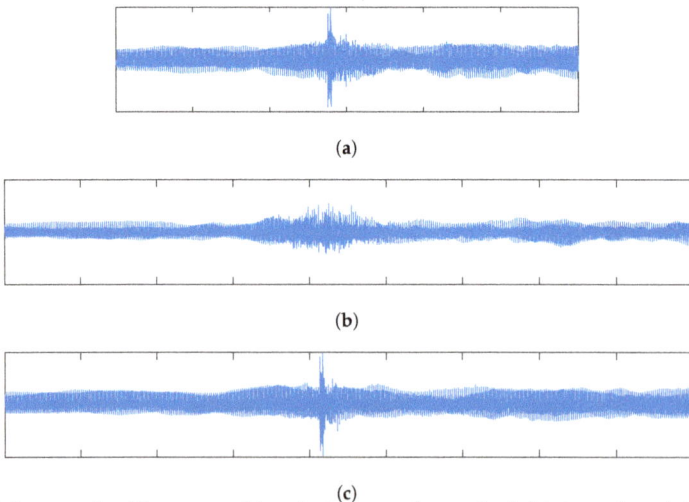

(a)

(b)

(c)

Figure 7. An example of the proposed transient preservation method. (**a**) shows the original audio, consisting of a solo violin overlaid with a castanet click. Also shown are the modified samples with TSM factor $\alpha = 1.5$, using (**b**) the standard phase vocoder, and (**c**) the proposed method.

4. Evaluation

To evaluate the quality of the proposed TSM technique, a listening test was conducted. The listening test was realized online using the Web Audio Evaluation Tool [35]. The test subjects were asked to use headphones. The test setup used was the same as in [28]. In each trial, the subjects were presented with the original audio sample and four modified samples processed with different TSM techniques. The subjects were asked to rate the quality of time-scale modified audio excerpts using a scale from 1 (poor) to 5 (excellent).

All 11 subjects who participated in the test reported having a background in acoustics, and 10 of them had previous experience of participating in listening tests. None of the subjects reported hearing problems. The ages of the subjects ranged from 23 to 37, with a median age of 28. Of the 11 subjects, 10 were male and 1 was female.

In the evaluation of the proposed method, the following settings were used: the sample rate was 44.1 kHz, a Hann window of length $N = 4096$ was chosen for the STFT analysis and synthesis, the synthesis hop size was set to $H_s = 512$, and the number of frequency bins in the STFT was $K = N = 4096$. The length of the median filter in the frequency direction was 500 Hz, which corresponds to 46 bins. In the time direction, the length of the median filter was chosen to be 200 ms, but the number of frames it corresponds to depends on the analysis hop size, which is determined by the TSM factor according to (10). Finally, the transient detection threshold was set to $t_d = 10^{-4} = 0.00010$.

In addition to the proposed method (PROP), the following techniques were included: the standard phase vocoder (PV), using the same STFT analysis and synthesis settings as the proposed method; a recently published technique (harmonic–percussive separation, HP) [28], which uses harmonic and percussive separation for transient preservation; and the élastique algorithm (EL) [36], which is a state-of-the-art commercial tool for time and pitch-scale modification. The samples processed by these methods were obtained using the TSM toolbox [37].

Eight different audio excerpts (sampled at 44.1 kHz) and two different stretching factors $\alpha = 1.5$ and $\alpha = 2.0$ were tested using the four techniques. This resulted in a total of 64 samples rated by each subject. The audio excerpts are described in Table 1. The lengths of the original audio excerpts ranged from 3 to 10 s. The processed audio excerpts and Matlab code for the proposed method are available online at http://research.spa.aalto.fi/publications/papers/applsci-ats/.

Table 1. List of audio excerpts used in the subjective listening test.

Name	Description
CastViolin	Solo violin and castanets, from [37]
Classical	Excerpt from *Bólero*, performed by the *London Symphony Orchestra*
JJCale	Excerpt from *Cocaine*, performed by *J.J. Cale*
DrumSolo	Solo performed on a drum set, from [37]
Eddie	Excerpt from *Early in the Morning*, performed by *Eddie Rabbit*
Jazz	Excerpt from *I Can See Clearly*, performed by the *Holly Cole Trio*
Techno	Excerpt from *Return to Balojax*, performed by *Deviant Species and Scorb*
Vocals	Excerpt from *Tom's Diner*, performed by *Suzanne Vega*

To estimate the sound quality of the techniques, mean opinion scores (MOS) were computed for all samples from the ratings given by the subjects. The resulting MOS values are shown in Table 2. A bar diagram of the same data is also shown in Figure 8.

As expected, the standard PV performed worse than all the other tested methods. For the *CastViolin* sample, the proposed method (PROP) performed better than the other methods, with both TSM factors. This suggests that the proposed method preserves the quality of the transients in the modified signals better than the other methods. The proposed method also scored best with the *Jazz* excerpt. In addition to the well-preserved transients, the results are likely to be explained by the naturalness of the singing voice in the modified signals. This can be attributed to the proposed phase propagation, which allows simultaneous preservation of the tonal and noisy qualities of the singing voice. This is also reflected in the results of the *Vocals* excerpt, where the proposed method also performed well, while scoring slightly lower than HP. For the *Techno* sample, the proposed method scored significantly higher than the other methods with TSM factor $\alpha = 1.5$. For TSM factor $\alpha = 2.0$, however, the proposed method scored lower than EL. The proposed method also scored highest for the *JJCale* sample with TSM factor $\alpha = 2.0$.

Table 2. Mean opinion scores for the audio samples. PV: phase vocoder; HP: harmonic–percussive separation; EL: élastique algorithm; PROP: proposed method.

| | $\alpha = 1.5$ | | | | $\alpha = 2.0$ | | | |
	PV	**HP**	**EL**	**PROP**	**PV**	**HP**	**EL**	**PROP**
CastViolin	1.8	3.8	3.6	**4.1**	1.4	3.6	3.3	**4.1**
Classical	2.3	3.5	**3.7**	3.3	1.6	3.0	**3.7**	2.8
JJCale	2.7	2.5	**3.4**	2.9	1.2	2.5	3.1	**3.2**
DrumSolo	1.5	**3.5**	3.2	2.3	1.7	2.4	**2.5**	1.8
Eddie	1.9	3.1	**4.2**	3.2	1.2	2.2	**3.6**	3.1
Jazz	1.9	**3.6**	3.4	**3.6**	1.5	3.3	2.7	**3.7**
Techno	1.3	2.7	3.3	**4.1**	1.6	2.5	**3.1**	2.7
Vocals	1.7	**3.5**	2.9	3.4	1.5	**3.3**	2.7	3.1
Mean	1.9	3.3	**3.5**	3.4	1.5	2.9	**3.1**	3.1

The proposed method performed more poorly on the excerpts *DrumSolo* and *Classical*. Both of these samples contained fast sequences of transients. It is likely that the poorer performance is due to the individual transients not being resolved during the analysis, because of the relatively long analysis window used. Also, for the excerpt *Eddie*, EL scored higher than the proposed method. Note that the audio excepts were not selected so that the results would be preferable for one of the tested methods. Instead, they represent some interesting and critical cases, such as singing and sharp transients.

The preferences of subjects over the tested TSM methods seem to depend significantly on the signal being processed. Overall, the MOS values computed from all the samples suggest that the proposed method yields slightly better quality than HP and practically the same quality as EL.

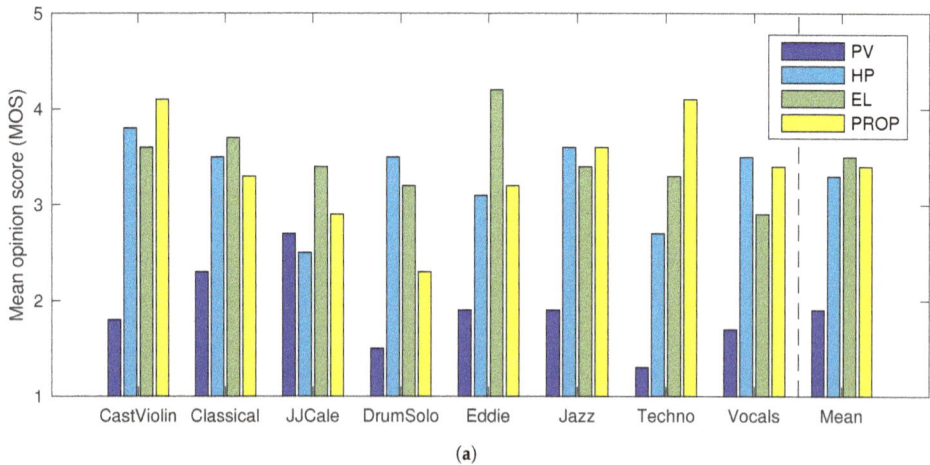

(a)

Figure 8. *Cont.*

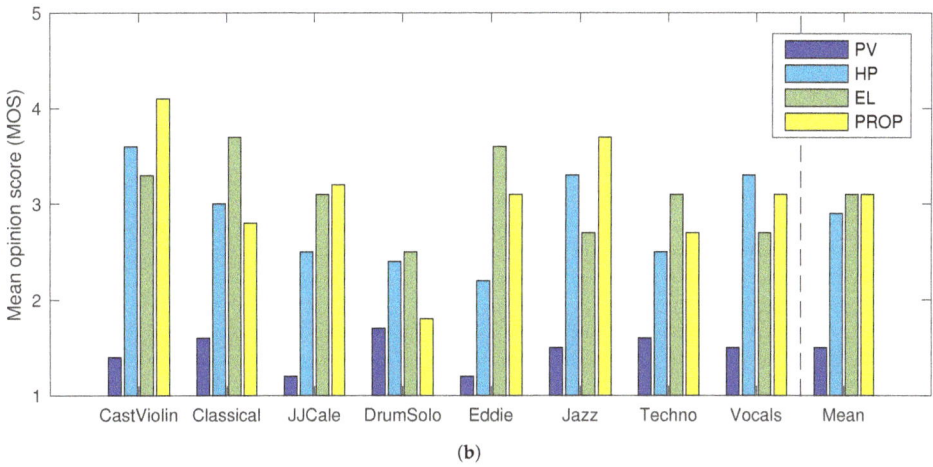

(b)

Figure 8. Mean opinion scores for eight audio samples using four TSM methods for (**a**) medium ($\alpha = 1.5$), and (**b**) large ($\alpha = 2.0$) TSM factors. The rightmost bars show the average score for all eight samples. PV: phase vocoder; HP: harmonic–percussive separation [28]; EL: élastique [36]; PROP: proposed method.

The proposed method introduces some additional computational complexity when compared to the standard phase-locked vocoder. In the analysis stage, the fuzzy classification of the spectral bins requires median filtering of the magnitude of the analysis STFT. The number of samples in each median filtering operation depends on the analysis hop size and the number of frequency bins in each short time spectra. In the modification stage, additional complexity arises from drawing pseudo-random values for the phase randomization. Furthermore, computing the phase randomization factor, as in Equation (13), requires the evaluation of two hyperbolic tangent functions for each point in the STFT. Since the argument for the second hyperbolic tangent depends only on the TSM factor, its value needs to be updated only when the TSM factor is changed. Finally, due to the way the values are used, a lookup table approximation can be used for evaluating the hyperbolic tangents without significantly affecting the quality of the modification.

5. Conclusions

In this paper, a novel TSM method was presented. The method is based on fuzzy classification of spectral bins into sinusoids, noise, and transients. The information from the bin classification is used to preserve the characteristics of these distinct signal components during TSM. The listening test results presented in this paper suggest that the proposed method performs generally better than a state-of-the-art algorithm and is competitive with commercial software.

The proposed method still suffers to some extent from the fixed time and frequency resolution of the STFT. Finding ways to apply the concept of fuzzy classification of spectral bins to a multiresolution time-frequency transformation could further increase the quality of the proposed method. Finally, although this paper only considered TSM, the method for fuzzy classification of spectral bins could be applied to various audio signal analysis tasks, such as multi-pitch estimation and beat tracking.

Acknowledgments: This study has been funded by the Aalto University School of Electrical Engineering. Special thanks go to the experience director of the Finnish Science Center Heureka Mikko Myllykoski, who proposed this study. The authors would also like to thank Mr. Etienne Thuillier for providing expert help in the beginning of this project, and Craig Rollo for proofreading.

Author Contributions: E.P.D. and V.V. planned this study and wrote the paper together. E.P.D. developed and programmed the new algorithm. E.P.D. conducted the listening test and analyzed the results. V.V. supervised this work.

Conflicts of Interest: The authors declare no conflict of interest.

References

1. Moulines, E.; Laroche, J. Non-parametric techniques for pitch-scale and time-scale modification of speech. *Speech Commun.* **1995**, *16*, 175–205.

2. Barry, D.; Dorran, D.; Coyle, E. Time and pitch scale modification: A real-time framework and tutorial. In Proceedings of the International Conference on Digital Audio Effects (DAFx), Espoo, Finland, 1–4 September 2008; pp. 103–110.

3. Driedger, J.; Müller, M. A review of time-scale modification of music signals. *Appl. Sci.* **2016**, *6*, 57.

4. Amir, A.; Ponceleon, D.; Blanchard, B.; Petkovic, D.; Srinivasan, S.; Cohen, G. Using audio time scale modification for video browsing. In Proceedings of the 33rd Annual Hawaii International Conference on System Sciences (HICSS), Maui, HI, USA, 4–7 January 2000.

5. Cliff, D. Hang the DJ: Automatic sequencing and seamless mixing of dance-music tracks. In *Technical Report*; Hewlett-Packard Laboratories: Bristol, UK, 2000; Volume 104.

6. Donnellan, O.; Jung, E.; Coyle, E. Speech-adaptive time-scale modification for computer assisted language-learning. In Proceedings of the Third IEEE International Conference on Advanced Learning Technologies, Athens, Greece, 9–11 July 2003; pp. 165–169.

7. Dutilleux, P.; De Poli, G.; von dem Knesebeck, A.; Zölzer, U. Time-segment processing (chapter 6). In *DAFX: Digital Audio Effects, Second Edition*; Zölzer, U., Ed.; Wiley: Chichester, UK, 2011; pp. 185–217.

8. Moinet, A.; Dutoit, T.; Latour, P. Audio time-scaling for slow motion sports videos. In Proceedings of the International Conference on Digital Audio Effects (DAFx), Maynooth, Ireland, 2–5 September 2013; pp. 314–320.

9. Haghparast, A.; Penttinen, H.; Välimäki, V. Real-time pitch-shifting of musical signals by a time-varying factor using normalized filtered correlation time-scale modification (NFC-TSM). In Proceedings of the International Conference on Digital Audio Effects (DAFx), Bordeaux, France, 10–15 September 2007; pp. 7–13.

10. Santacruz, J.; Tardón, L.; Barbancho, I.; Barbancho, A. Spectral envelope transformation in singing voice for advanced pitch shifting. *Appl. Sci.* **2016**, *6*, 368.

11. Verma, T.S.; Meng, T.H. An analysis/synthesis tool for transient signals that allows a flexible sines+transients+noise model for audio. In Proceedings of the IEEE International Conference on Acoustics, Speech and Signal Processing (ICASSP), Las Vegas, NV, USA, 30 March–4 April 1998; pp. 3573–3576.

12. Levine, S.N.; Smith, J.O., III. A sines+transients+noise audio representation for data compression and time/pitch scale modifications. In Proceedings of the Audio Engineering Society 105th Convention, San Francisco, CA, USA, 26–29 September 1998.

13. Verma, T.S.; Meng, T.H. Time scale modification using a sines+transients+noise signal model. In Proceedings of the Digital Audio Effects Workshop (DAFx), Barcelona, Spain, 19–21 November 1998.

14. Verma, T.S.; Meng, T.H. Extending spectral modeling synthesis with transient modeling synthesis. *Comput. Music J.* **2000**, *24*, 47–59.

15. Roucos, S.; Wilgus, A. High quality time-scale modification for speech. In Proceedings of the IEEE International Conference on Acoustics, Speech, and Signal Processing (ICASSP), Tampa, FL, USA, 26–29 April 1985; Volume 10, pp. 493–496.

16. Verhelst, W.; Roelands, M. An overlap-add technique based on waveform similarity (WSOLA) for high quality time-scale modification of speech. In Proceedings of the IEEE International Conference on Acoustics, Speech, and Signal Processing (ICASSP), Minneapolis, MN, USA, 27–30 April 1993; pp. 554–557.

17. Moulines, E.; Charpentier, F. Pitch-synchronous waveform processing techniques for text-to-speech synthesis using diphones. *Speech Commun.* **1990**, *9*, 453–467.

18. Lee, S.; Kim, H.D.; Kim, H.S. Variable time-scale modification of speech using transient information. In Proceedings of the IEEE International Conference on Acoustics, Speech, and Signal Processing (ICASSP), München, Germany, 21–24 April 1997; Volume 2, pp. 1319–1322.

19. Wong, P.H.; Au, O.C.; Wong, J.W.; Lau, W.H. On improving the intelligibility of synchronized over-lap-and-add (SOLA) at low TSM factor. In Proceedings of the IEEE Region 10 Annual Conference on Speech and Image Technologies for Computing and Telecommunications (TENCON), Brisbane, Australia, 2–4 December 1997; Volume 2, pp. 487–490.
20. Portnoff, M. Time-scale modification of speech based on short-time Fourier analysis. *IEEE Trans. Acoust. Speech Signal Process.* **1981**, *29*, 374–390.
21. Laroche, J.; Dolson, M. Improved phase vocoder time-scale modification of audio. *IEEE Trans. Speech Audio Process.* **1999**, *7*, 323–332.
22. Laroche, J.; Dolson, M. Phase-vocoder: About this phasiness business. In Proceedings of the IEEE ASSP Workshop on Applications of Signal Processing to Audio and Acoustics, New Paltz, NY, USA, 19–22 October 1997.
23. Röbel, A. A new approach to transient processing in the phase vocoder. In Proceedings of the 6th International Conference on Digital Audio Effects (DAFx), London, UK, 8–11 September 2003; pp. 344–349.
24. Bonada, J. Automatic technique in frequency domain for near-lossless time-scale modification of audio. In Proceedings of the International Computer Music Conference (ICMC), Berlin, Germany, 27 August–1 September 2000; pp. 396–399.
25. Duxbury, C.; Davies, M.; Sandler, M.B. Improved time-scaling of musical audio using phase locking at transients. In Proceedings of the Audio Engineering Society 112th Convention, München, Germany, 10–13 May 2002.
26. Röbel, A. A shape-invariant phase vocoder for speech transformation. In Proceedings of the International Conference on Digital Audio Effects (DAFx), Graz, Austria, 6–10 September 2010; pp. 298–305.
27. Zivanovic, M.; Röbel, A.; Rodet, X. Adaptive threshold determination for spectral peak classification. *Comput. Music J.* **2008**, *32*, 57–67.
28. Driedger, J.; Müller, M.; Ewert, S. Improving time-scale modification of music signals using harmonic-percussive separation. *IEEE Signal Process. Lett.* **2014**, *21*, 105–109.
29. Fitzgerald, D. Harmonic/percussive separation using median filtering. In Proceedings of the International Conference on Digital Audio Effects (DAFx), Graz, Austria, 6–10 September 2010; pp. 217–220.
30. Zadeh, L.A. Making computers think like people. *IEEE Spectr.* **1984**, *21*, 26–32.
31. Del Amo, A.; Montero, J.; Cutello, V. On the principles of fuzzy classification. In Proceedings of the 18th International Conference of the North American Fuzzy Information Processing Society, New York, NY, USA, 10–12 June 1999; pp. 675–679.
32. Kraft, S.; Lerch, A.; Zölzer, U. The tonalness spectrum: Feature-based estimation of tonal components. In Proceedings of the International Conference on Digital Audio Effects (DAFx), Maynooth, Ireland, 2–5 September 2013; pp. 17–24.
33. Ono, N.; Miyamoto, K.; Le Roux, J.; Kameoka, H.; Sagayama, S. Separation of a monaural audio signal into harmonic/percussive components by complementary diffusion on spectrogram. In Proceedings of the European Signal Processing Conference (EUSIPCO), Lausanne, Switzerland, 25–29 August 2008; pp. 1–4.
34. Nagel, F.; Walther, A. A novel transient handling scheme for time stretching algorithms. In Proceedings of the Audio Engineering Society 127th Convention, New York, NY, USA, 9–12 October 2009.
35. Jillings, N.; Moffat, D.; De Man, B.; Reiss, J.D. Web Audio Evaluation Tool: A browser-based listening test environment. In Proceedings of the 12th Sound and Music Computing Conference, Maynooth, Ireland, 26 July–1 August 2015; pp. 147–152.
36. Zplane Development. Élastique Time Stretching & Pitch Shifting SDKs. Available online: http://www.zplane.de/index.php?page=description-elastique (accessed on 20 October 2017).
37. Driedger, J.; Müller, M. TSM toolbox: MATLAB implementations of time-scale modification algorithms. In Proceedings of the International Conference on Digital Audio Effects (DAFx), Erlangen, Germany, 1–5 September 2014; pp. 249–256.

applied
sciences

MDPI

Article

Automatic Transcription of Polyphonic Vocal Music [†]

Andrew McLeod [1,*,‡], Rodrigo Schramm [2,‡], Mark Steedman [1] and Emmanouil Benetos [3]

[1]   School of Informatics, University of Edinburgh, Edinburgh EH8 9AB, UK; steedman@inf.ed.ac.uk
[2]   Departamento de Música, Universidade Federal do Rio Grande do Sul, Porto Alegre 90020, Brazil;
     rschramm@ufrgs.br
[3]   School of Electronic Engineering and Computer Science, Queen Mary University of London,
     London E1 4NS, UK; emmanouil.benetos@qmul.ac.uk
*   Correspondence: A.McLeod-5@sms.ed.ac.uk; Tel.: +44-131-650-1000
†   This paper is an extended version of our paper published in R. Schramm, A. McLeod, M. Steedman,
     and E. Benetos. Multi-pitch detection and voice assignment for a cappella recordings of multiple singers.
     In 18th International Society for Music Information Retrieval Conference (ISMIR), pp. 552–559, 2017.
‡   These authors contributed equally to this work.

Academic Editor: Meinard Müller
Received: 31 October 2017; Accepted: 4 December 2017; Published: 11 December 2017

Abstract: This paper presents a method for automatic music transcription applied to audio recordings of a cappella performances with multiple singers. We propose a system for multi-pitch detection and voice assignment that integrates an acoustic and a music language model. The acoustic model performs spectrogram decomposition, extending probabilistic latent component analysis (PLCA) using a six-dimensional dictionary with pre-extracted log-spectral templates. The music language model performs voice separation and assignment using hidden Markov models that apply musicological assumptions. By integrating the two models, the system is able to detect multiple concurrent pitches in polyphonic vocal music and assign each detected pitch to a specific voice type such as soprano, alto, tenor or bass (SATB). We compare our system against multiple baselines, achieving state-of-the-art results for both multi-pitch detection and voice assignment on a dataset of Bach chorales and another of barbershop quartets. We also present an additional evaluation of our system using varied pitch tolerance levels to investigate its performance at 20-cent pitch resolution.

Keywords: automatic music transcription; multi-pitch detection; voice assignment; music signal analysis; music language models; polyphonic vocal music; music information retrieval

1. Introduction

Automatic music transcription (AMT) is one of the fundamental problems of music information retrieval and is defined as the process of converting an acoustic music signal into some form of music notation [1]. A core problem of AMT is multi-pitch detection, the detection of multiple concurrent pitches from an audio recording. While much work has gone into the field of multi-pitch detection in recent years, it has frequently been constrained to instrumental music, most often piano recordings due to a wealth of available data. Vocal music has been less often studied, likely due to the complexity and variety of sounds that can be produced by a singer.

Spectrogram factorization methods have been used extensively in the last decade for multi-pitch detection [1]. These approaches decompose an input time-frequency representation (such as a spectrogram) into a linear combination of non-negative factors, often consisting of spectral atoms and note activations. The most successful of these spectrogram factorization methods have been based on non-negative matrix factorisation (NMF) [2] or probabilistic latent component analysis (PLCA) [3].

While these spectrogram factorisation methods have shown promise for AMT, their parameter estimation can suffer from local optima, a problem that has motivated a variety of approaches that incorporate additional knowledge in an attempt to achieve more meaningful decompositions. Vincent et al. [4] used an adaptive spectral decomposition for multi-pitch detection assuming that the input signal can be decomposed as a sum of narrowband spectra. Kameoka et al. [5] exploited structural regularities in the spectrograms during the NMF process, adding constraints and regularization to reduce the degrees of freedom of their model. These constraints are based on time-varying basis spectra (e.g., using sound states: "attack", "decay", "sustain" and "release") and have since been included in other probabilistic models [6,7]. Fuentes et al. [8] introduced the concept of brakes, slowing the convergence rate of any model parameter known to be properly initialized. Other approaches [7,9,10] avoid undesirable parameter convergence using pre-learning steps, where spectral atoms of specific instruments are extracted in a supervised manner. Using the constant-Q transform (CQT) [11] as the input time-frequency representation, some approaches developed techniques using shift-invariant models over log-frequency [6,10,12], allowing for the creation of a compact set of dictionary templates that can support tuning deviations and frequency modulations. Shift-invariant models are also used in several recent approaches for automatic music transcription [6,13,14]. O'Hanlon et al. [15] propose stepwise and gradient-based methods for non-negative group sparse decompositions, exploring the use of subspace modelling of note spectra. This group sparse NMF approach is used to tune a generic harmonic subspace dictionary, improving automatic music transcription results based on NMF. However, despite promising results of template-based techniques [7,9,10], the considerable variation in the spectral shape of pitches produced by different sources can still affect generalization performance.

Recent research on multi-pitch detection has also focused on deep learning approaches: in [16,17], feedforward, recurrent and convolutional neural networks were evaluated towards the problem of automatic piano transcription. While the aforementioned approaches focus on the task of polyphonic piano transcription due to the presence of sufficiently large piano-specific datasets, the recently released MusicNet dataset [18] provides a large corpus for multi-instrument music suitable for training deep learning methods for the task of polyphonic music transcription. Convolutional neural networks were also used in [19] for learning salience representations for fundamental frequency estimation in polyphonic audio recordings.

Multi-pitch detection of vocal music represents a significant step up in difficulty as the variety of sounds produced by a single singer can be both unique and wide-ranging. The timbre of two singers' voices can differ greatly, and even for a single singer, different vowel sounds produce extremely varied overtone patterns. For vocal music, Bohak and Marolt [20] propose a method for transcribing folk music containing both instruments and vocals, which takes advantage of melodic repetitions present in that type of music using a musicological model for note-based transcription. A less explored type of music is a cappella; in particular, vocal quartets constitute a traditional form of Western music, typically dividing a piece into multiple vocal parts such as soprano, alto, tenor and bass (SATB). In [21], an acoustic model based on spectrogram factorisation was proposed for multi-pitch detection of such vocal quartets.

A small group of methods has attempted to go beyond multi-pitch detection, towards instrument assignment (also called timbre tracking) [9,22,23], where systems detect multiple pitches and assign each pitch to a specific source that produced it. Bay et al. [22] tracked individual instruments in polyphonic instrumental music using a spectrogram factorisation approach with continuity constraints controlled by a hidden Markov model (HMM). To the authors' knowledge, no methods have yet been proposed to perform both multi-pitch detection and instrument/voice assignment on polyphonic vocal music.

An emerging area of automatic music transcription attempts to combine acoustic models (those based on audio information only) with music language models, which model sequences of notes and other music cues based on knowledge from music theory or from constraints automatically derived from symbolic music data. This is in direct analogy to automatic speech recognition systems, which typically combine an acoustic model with a spoken language model. Ryynanen and Klapuri [24], for example, combined acoustic and music language models for polyphonic music transcription, where the musicological model estimates the probability of a detected note sequence. Another example of such an integrated system is the work by Sigtia et al. [16], which combined neural network-based acoustic and music language models for multi-pitch detection in piano music. The system used various types of neural networks for the acoustic component (feedforward, recurrent, convolutional) along with a recurrent neural network acting as a language model for modelling the correlations between pitch combinations over time.

Combining instrument assignment with this idea of using a music language model, it is natural to look towards the field of voice separation [25], which involves the separation of pitches into streams of notes, called voices, and is mainly addressed in the context of symbolic music processing. It is important to note that voice separation, while similar to our task of voice assignment, is indeed a distinct task. Specifically, while both involve an initial step of separating the incoming notes into voices, voice assignment involves a further step of labelling each of those voices as a specific part or instrument, in our case soprano, alto, tenor or bass.

Most symbolic voice separation approaches are based on voice leading rules, which have been investigated and described from a cognitive perspective in a few different works [26–28]. Among these rules, three main principles emerge: (1) large melodic intervals between consecutive notes in a single voice should be avoided; (2) two voices should not, in general, cross in pitch; and (3) the stream of notes within a single voice should be relatively continuous, without long gaps of silence, ensuring temporal continuity.

There are many different definitions of what precisely constitutes a voice, both perceptually and musically, discussed more fully in [25]; however, for our purposes, a voice is quite simply defined as the notes sung by a single vocalist. Therefore, our interest in voice separation models lies with those that separate notes into strictly monophonic voices (i.e., those that do not allow for concurrent notes), rather than polyphonic voices as in [29]. We would also like our chosen model to be designed to be run in a mostly unsupervised fashion, rather than being designed for use with human interaction (as in [30]), and for it not to require background information about the piece, such as time signature or metrical information (as in [31]). While many voice separation models remain that meet our criteria [32–36], the one described in [37] is the most promising for our use because it both (1) achieves state-of-the-art performance and (2) can be applied directly to live performance.

In this work, we present a system able to perform multi-pitch detection of polyphonic a cappella vocal music, as well as assign each detected pitch to a particular voice (soprano, alto, tenor or bass), where the number of voices is known a priori. Our approach uses an acoustic model for multi-pitch detection based on probabilistic latent component analysis (PLCA), which is modified from the model proposed in [21], and an HMM-based music language model for voice assignment based on the model of [37]. Compared to our previous work [38], this model contains a new dynamic dictionary voice type assignment step (described in Section 2.3), which accounts for its increased performance. Although previous work has integrated musicological information for note event modelling [16,20,24], to the authors' knowledge, this is the first attempt to incorporate an acoustic model with a music language model for the task of voice or instrument assignment from audio, as well as the first attempt to propose a system for voice assignment in polyphonic a cappella music. The approach described in this paper focuses on recordings of singing performances by vocal quartets without instrumental accompaniment; to that end, we use two datasets containing a capella recordings of Bach chorales and barbershop quartets. The proposed system is evaluated both in terms of multi-pitch detection and voice assignment, where it reaches an F-measure of over 70% and 50% for the two respective tasks.

The remainder of this paper is organised as follows. In Section 2, we describe the proposed approach, consisting of the acoustic model, the music language model and model integration. In Section 3, we report on experimental results using two datasets comprising recordings of vocal quartets. Section 4 closes with conclusions and perspectives for future work.

2. Proposed Method

In this section, we present a system for multi-pitch detection and voice assignment applied to audio recordings of polyphonic vocal music (where the number of voices is known a priori) that integrates an acoustic model with a music language model. First, we describe the acoustic model, a spectrogram factorization process based on probabilistic latent component analysis (PLCA). Then, we present the music language model, an HMM-based voice assignment model. Finally, a joint model is proposed for the integration of these two components. Figure 1 illustrates the proposed system pipeline.

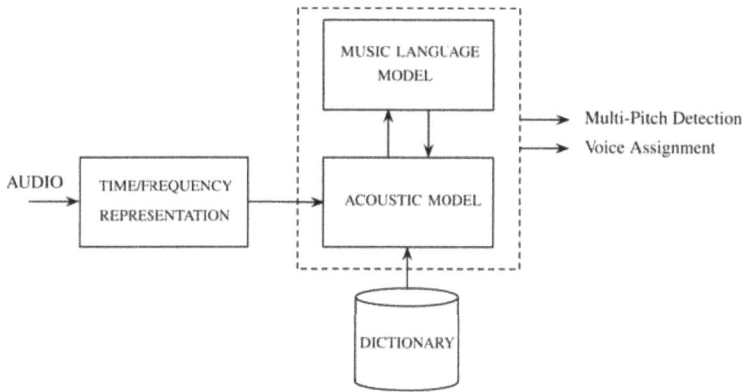

Figure 1. Proposed system diagram.

2.1. Acoustic Model

The acoustic model is a variant of the spectrogram factorisation-based model proposed in [21]. The model's primary goal is to explore the factorization of an input log-frequency spectrogram into components that have a close connection with singing characteristics such as voice type and the vocalization of different vowel sounds. We formulate the model dictionary templates into a six-dimensional tensor, representing log-frequency index, singer source, pitch, tuning deviation with 20 cent resolution, vowel type and voice type. Similarly to [9], the singer source and vowel type parameters constrain the search space into a mixture-of-subspaces, clustering a large variety of singers into a small number of categories. In this model, the voice type parameter corresponds to the vocal part (SATB), where each vocal part is linked to a distinct set of singers (the singer source). For details on the dictionary construction, see Section 2.1.2. As time-frequency representation, we use a normalised variable-Q transform (VQT) spectrogram [39] with a hop size of 20 ms and 20-cent frequency resolution. For convenience, we have chosen a pitch resolution that produces an integer number of bins per semitone (five in this case) and is also close to the range of just noticeable differences in musical intervals [40]. The input VQT spectrogram is denoted as $X_{\omega,t} \in \mathbb{R}^{\Omega \times T}$, where ω denotes log-frequency and t time. In the model, $X_{\omega,t}$ is approximated by a bivariate probability distribution $P(\omega,t)$, which is in turn decomposed as:

$$P(\omega,t) = P(t) \sum_{s,p,f,o,v} P(\omega|s,p,f,o,v)P_t(s|p)P_t(f|p)P_t(o|p)P(v)P_t(p|v) \tag{1}$$

where $P(t)$ is the spectrogram energy (known quantity) and $P(\omega|s,p,f,o,v)$ is the fixed pre-extracted spectral template dictionary. The variable s denotes the singer index (out of the collection of singer subjects used to construct the input dictionary); $p \in \{21,\dots,108\}$ denotes pitch in Musical Instrument Digital Interface (MIDI) scale; f denotes tuning deviation from 12-tone equal temperament in 20-cent resolution ($f \in \{1,\dots,5\}$, with $f = 3$ denoting ideal tuning); o denotes the vowel type; and v denotes the voice type (e.g., soprano, alto, tenor, bass). The contribution of specific singer subjects from the training dictionary is modelled by $P_t(s|p)$, i.e., the singer contribution per pitch over time. $P_t(f|p)$ is the tuning deviation per pitch over time, and finally, $P_t(o|p)$ is the time-varying vowel contribution per pitch . (Although $P_t(o|p)$ is not explicitly used in this proposed approach, it is kept to ensure consistency with the Real World Computing (RWC) audio dataset [41] structure (see Section 2.1.2).) Unlike in [21] (which uses $P_t(v|p)$), this model decomposes the probabilities of pitch and voice type as $P(v)P_t(p|v)$. That is, $P(v)$ can be viewed as a mixture weight that denotes the overall contribution of each voice type to the whole input recording, and $P_t(p|v)$ denotes the pitch activation for a specific voice type (e.g., SATB) over time.

The factorization can be achieved by the expectation-maximization (EM) algorithm [42], where the unknown model parameters $P_t(s|p)$, $P_t(f|p)$, $P_t(o|p)$, $P_t(p|v)$ and $P(v)$ are iteratively estimated. In the expectation step, we compute the posterior as:

$$P_t(s, p, f, o, v|\omega) = \frac{P(\omega|s, p, f, o, v)P_t(s|p)P_t(f|p)P_t(o|p)P(v)P_t(p|v)}{\sum_{s,p,f,o,v} P(\omega|s, p, f, o, v)P_t(s|p)P_t(f|p)P_t(o|p)P(v)P_t(p|v)} \quad (2)$$

In the maximization step, each unknown model parameter is then updated using the posterior from Equation (2):

$$P_t(s|p) \propto \sum_{f,o,v,\omega} P_t(s, p, f, o, v|\omega)X_{\omega,t} \quad (3)$$

$$P_t(f|p) \propto \sum_{s,o,v,\omega} P_t(s, p, f, o, v|\omega)X_{\omega,t} \quad (4)$$

$$P_t(o|p) \propto \sum_{s,f,v,\omega} P_t(s, p, f, o, v|\omega)X_{\omega,t} \quad (5)$$

$$P_t(p|v) \propto \sum_{s,f,o,\omega} P_t(s, p, f, o, v|\omega)X_{\omega,t} \quad (6)$$

$$P(v) \propto \sum_{s,f,o,p,\omega,t} P_t(s, p, f, o, p|\omega)X_{\omega,t} \quad (7)$$

The model parameters are randomly initialised, and the EM algorithm iterates over Equations (2)–(7). In our experiments, we use 30 iterations, as this ensures that the model will converge; in practice, the model converges after about 18 iterations. In order to promote temporal continuity, we apply a median filter to the $P_t(p|v)$ estimate across time, before its normalisation at each EM iteration, using a filter span of 240 ms, a duration of approximately half of one beat in Allegro tempo.

2.1.1. Acoustic Model Output

The output of the acoustic model is a semitone-scale pitch activity tensor for each voice type and a pitch shifting tensor, given by $P(p, v, t) = P(t)P(v)P_t(p|v)$ and $P(f, p, v, t) = P(t)P(v)P_t(p|v)P_t(f|p)$, respectively. By stacking together slices of $P(f, p, v, t)$ for all values of p, we can create a 20-cent resolution time-pitch representation for each voice type v:

$$P(f', v, t) = P\left(f' \ (\text{mod } 5) + 1, \left\lfloor \frac{f'}{5} \right\rfloor + 21, v, t\right) \quad (8)$$

where $f' \in \{0, ..., 439\}$ denotes pitch in 20-cent resolution. The voice-specific 20-cent resolution pitch activation output is given by $P(f', v, t)$, and the overall multi-pitch activations without voice assignment are given by $P(f', t) = \sum_v P(f', v, t)$. The 20-cent resolution multi-pitch activations $P(f', t)$ are converted into multi-pitch detections, represented by a binary matrix $B(f', t)$, through a binarisation process with a fixed threshold L_{th}. Specifically, pitch activations whose values are greater than L_{th} are set to one in matrix $\mathbf{B}$, while all others are set to zero.

This binarised matrix $\mathbf{B}(f', t)$ is then post-processed in order to obtain more accurate pitch activations. In this step, we scan each time frame of the matrix $\mathbf{B}$, replacing the pitch candidates by the position of spectrogram peaks detected from $X_{\omega,t}$ and that are validated by a minimum pitch distance rule:

$$(\Delta_{peaks}(\mathbf{X_t}, \mathbf{B}(f', t)) < T_1) \vee (\Delta_{peaks}(\mathbf{X_t}, \mathbf{B}(f', t - 1)) < T_2), \qquad (9)$$

where $\mathbf{B}(f', t)$ represents each binarised pitch activation at time frame t. The function Δ_{peaks} in (9) indicates the minimum pitch distance between the selected list of peak candidates in X_t and each pitch candidate $\mathbf{B}(f', t)$ and $\mathbf{B}(f', t - 1)$, respectively. In our experiments, we use $T_1 = 1$ and $T_2 = 3$, based on density distributions of $|\Delta_{peaks}|$, which were estimated from measurements in our datasets using the pitch ground truth. The use of the previous frame $(t - 1)$ helps to keep the temporal continuity when a pitch candidate is eventually removed by the L_{th} threshold.

2.1.2. Dictionary Extraction

Dictionary $P(\omega|s, p, f, o, v)$ with spectral templates from multiple singers is built based on English pure vowels (monophthongs), such as those used in the solfège system of learning music: Do, Re, Mi, Fa, Sol, La, Ti and Do. The dictionaries use spectral templates extracted from solo singing recordings in the Musical Instrument Sound subset of the Real World Computing (RWC) database (RWC-MDB-I-2001 Nos. 45–49) [41]. The recordings contain sequences of notes following a chromatic scale, where the range of notes varies according to the tessitura of distinct vocal parts. Each singer sings a scale in five distinct English vowels (/a/, /æ/, /i/, /ɒ/, /u/). In total, we have used 15 distinct singers: 9 male and 6 female, consisting of 3 human subjects for each voice type (bass, baritone, tenor, alto, soprano).

Although the aim of this work is the transcription of vocal quartets, we keep the spectral templates from all five voice types in the dictionary because we do not know in advance the voice types present in each audio recording. This decision allows the dictionary to cover a wider variety of vocal timbres during the spectral decomposition, although not all of the resulting voice assignment probabilities will be used during its integration with the music language model for a single song. Rather, our model dynamically aligns one of the dictionary's voice types to each vocal part in a song. This dynamic dictionary alignment is based on the music language model's voice assignments and is discussed further in Section 2.3.

The fundamental frequency (f_0) sequence from each monophonic recording is estimated using the Probabilistic YIN (PYIN) algorithm [43]. Afterwards, the time-frequency representation is extracted using the VQT, with 60 bins per octave. A spectral template is extracted for each frame, regarding the singer source, vowel type and voice type. In order to incorporate multiple estimates from a common pitch, the set of estimates that fall inside the same pitch bin are replaced by its metrically-trimmed mean, discarding 20% of the samples as possible outliers. The use of the metrically-trimmed mean aims to reduce the influence of possible pitch inaccuracies obtained from the automatic application of the PYIN algorithm. However, there is no guarantee that the final estimate will be free of eventual outliers. The set of spectral templates is then pre-shifted across log-frequency in order to support tuning deviations for ± 20 and ± 40 cent and are stored into a six-dimensional tensor matrix $P(\omega|s, p, f, o, v)$. Due to the available data from the chromatic scales, the resulting dictionary $P(\omega|s, p, f, o, v)$ has some pitch templates missing, as shown in Figure 2a.

To address the aforementioned issue, we have investigated alternative ways to fill out the missing templates in the dictionary, including spectrum estimation by replication [14,44], linear and nonlinear interpolation and a generative process based on Gaussian mixture models (inspired by [45,46]). Following experimentation, we have chosen the replication approach, where existing templates belonging to the same dictionary are used to fill in the missing parts of the pitch scale, as this has been shown to achieve the best performance [47]. In this approach, the spectral shape of a given pitch p_n is repeated (with the appropriate log-frequency shift) over all subsequent pitches

$p \in [p_{n+1}, p_{m-1}]$ until another template is found (the pitch template p_m). Figure 2b illustrates the resulting dictionary templates of one singer example (vowel /a/) from our audio dataset, following the above replication process.

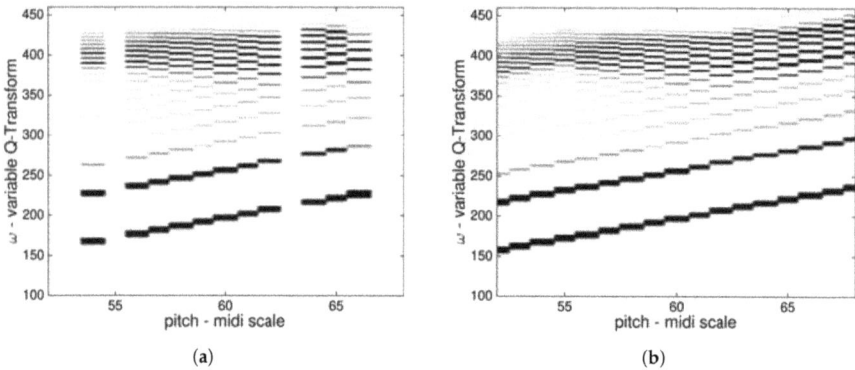

Figure 2. Example from an /a/ vowel utterance (one singer) templates: (**a**) original templates from the variable-Q transform (VQT) spectrogram; (**b**) revised dictionary templates following replication.

2.2. Music Language Model

The music language model attempts to assign each detected pitch to a single voice based on musicological constraints. It is a variant of the HMM-based voice separation approach proposed in [37], where the main change is to the emission function (here it is probabilistic, while in the previous work, it was deterministic). The model separates sequential sets of multi-pitch activations into monophonic voices (of type SATB) based on three principles: (1) consecutive notes within a voice tend to occur on similar pitches; (2) there are minimal temporal gaps between them; and (3) voices are unlikely to cross.

The observed data for the HMM are notes generated from the acoustic model's binarised 20-cent resolution multi-pitch activations $B(f', t)$, where each activation generates a note n with pitch $\text{Pitch}(n) = \lfloor \frac{f'}{5} \rfloor$, onset time $\text{On}(n) = t$ and offset time $\text{Off}(n) = t + 1$. Duplicates are discarded in the case where two 20-cent resolution detections refer to the same semitone pitch. O_t represents this set of observed notes at frame t.

2.2.1. State Space

In the HMM, a state S_t at frame t contains a list of M monophonic voices V_i, $1 \leq i \leq M$. M is set via a parameter, and in this work, we use $M = 4$. In the initial state S_0, all of the voices are empty, and at each frame, each voice may be assigned a single note (or no note). Thus, each voice contains the entire history of the notes, which have been assigned to it from Frame 1 to t. This is necessary because the note history is used in the calculation of the transition probabilities (Section 2.2.2); however, it causes the theoretical state space of our model to blow up exponentially. Therefore, instead of using precomputed transition and emission probabilities, we must use transition and emission probability functions, presented in the following sections.

Conceptually, it is helpful to think of each state as simply a list of M voices. Thus, each state transition is calculated based on the voices in the previous state (though some of the probability calculations require knowledge of individual notes).

2.2.2. Transition Function

A state S_{t-1} has a transition to state S_t if and only if each voice $V_i \in S_{t-1}$ can either be transformed into the corresponding $V_i \in S_t$ by assigning to it a single note with onset time t, or if it is identical to the corresponding $V_i \in S_t$.

This transition from S_{t-1} to S_t can be represented by the variable T_{S_{t-1},N_t,W_t}, where S_{t-1} is the original state, N_t is a list of every note with onset time t assigned to a voice in S_t and W_t is a list of integers, each representing the voice assignment index for the corresponding note N_t. Specifically, N_t and W_t are of equal length, and the i-th integer in W_t represents the index of the voice to which the i-th note in N_t is assigned in S_t. Notice that here, N_t only contains those observed notes that are assigned to a voice in S_t, rather than all observed notes.

The HMM transition probability $P(S_t|S_{t-1})$ is defined as $P(T_{S_{t-1},N_t,W_t})$:

$$P(T_{S_{t-1},N_t,W_t}) = \Psi(W_t) \prod_{i=1}^{|N_t|} C(S_{t-1}, n_i, w_i) P(V_{w_i}, n_i) \tag{10}$$

The first term in the above product is a function representing the voice assignment probability and is defined as follows:

$$\Psi(W) = \prod_{j=1}^{M} \begin{cases} P_v & j \in W \\ 1 - P_v & j \notin W \end{cases} \tag{11}$$

Here, the parameter P_v is the probability that a given voice contains a note in a frame.

$C(S_{t-1}, n, w)$ is a penalty function used to minimize the voice crossings, which are rare, though they do sometimes occur. It returns by default one, but its output is multiplied by a parameter P_{cross}—representing the probability of a voice being out of pitch order with an adjacent voice—for each of the following cases that applies:

1. $w > 1$ and $\text{Pitch}(V_{w-1}) > \text{Pitch}(n)$
2. $w < M$ and $\text{Pitch}(V_{w+1}) < \text{Pitch}(n)$

These cases in fact provide the definition for precisely what constitutes two voices being "out of pitch order". For example, if the soprano voice contains a note at a lower pitch than the alto voice in a given frame, the soprano voice is said to be out of pitch order. Cases 1 and 2 apply when a note is out of pitch order with the preceding or succeeding voice in the state, respectively. $\text{Pitch}(V)$ represents the pitch of a voice and is calculated as a weighted sum of the pitches of its most recent l (a parameter) notes, where each note's weight is twice the weight of the previous note. Here, n_i refers to the i-th note assigned to voice V.

$$\text{Pitch}(V) = \frac{\sum_{i=0}^{\min(l,|V|)} (2^i \text{Pitch}(n_{|V|-i}))}{\sum_{i=0}^{\min(l,|V|)} 2^i} \tag{12}$$

$P(V,n)$ represents the probability of a note n being assigned to a voice V and is the product of a pitch score and a gap score.

$$P(V,n) = \text{pitch}(V,n) \, \text{gap}(V,n) \tag{13}$$

The pitch score, used to minimise melodic jumps within a voice, is computed as shown in Equation (14), where $\mathcal{N}(\mu, \sigma, x)$ represents a normal distribution with mean μ and standard deviation σ evaluated at x, and σ_p is a parameter. The gap score is used to prefer temporal continuity within a voice and is computed using Equation (15), where $\text{Off}(V)$ is the offset time of the most recent note in V, and σ_g and g_{min} are parameters. Both Δ_p and Δ_g return one if V is empty.

$$\text{pitch}(V,n) = \mathcal{N}(\text{Pitch}(V), \sigma_p, \text{Pitch}(n)) \tag{14}$$

$$\text{gap}(V,n) = \max\left(\ln\left(-\frac{\text{On}(n) - \text{Off}(V)}{\sigma_g} + 1\right) + 1, g_{min}\right) \tag{15}$$

2.2.3. Emission Function

A state S_t emits a set of notes with onset at time t, with the constraint that a state containing a voice with a note at onset time t must emit that note. The probability of a state S_t emitting the note set O_t is shown in Equation (16), using the voice posterior $P_t(v|p)$ from the acoustic model.

$$P(O_t|S_t) = \prod_{n \in O_t} \begin{cases} P_t(v = i|p = \rho(n)) & n \in V_i \in S_t \\ 1 & \text{otherwise} \end{cases} \tag{16}$$

Notice that a state is not penalised for emitting notes not assigned to any of its voices. This allows the model to better handle false positives from the multi-pitch detection. For example, if the acoustic model detects more than M pitches, the state is allowed to emit the corresponding notes without penalty. We do, however, penalise a state for not assigning a voice any note during a frame, but this is handled by $\Psi(W)$ from Equation (11).

2.2.4. Inference

To find the most likely final state given our observed note sets, we use the Viterbi algorithm [48] with beam search with beam size b. That is, after each iteration, we save only the $b = 50$ most likely states given the observed data to that point, in order to handle the complexity of the HMM. A simple two-voice example of the HMM being run discriminatively can be found in Figures 3 and 4.

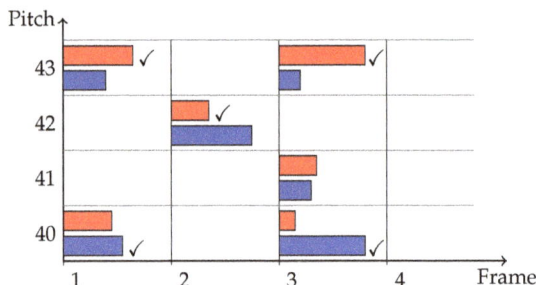

Figure 3. An example of an input to the music language model given a simple song with only two voices. Here, for each detected pitch, there are two bars, representing the relative value of $P_t(p|v)$ for each voice at that frame. (The upper voice is shown in red and the lower voice is shown in blue.) The ground truth voice assignment for each detected note is given by a check mark next to the bar representing the correct voice. Notice that there is a false positive pitch detection at Pitch 41 at Frame 3.

Figure 3 shows example input pitch detections, where empty grid cells represent pitches that have not passed the PLCA's post-processing binarisation step, and the bars in the other cells represent relative values of $P_t(p|v)$ for each colour-coded voice. (The upper voice is shown in red and the lower voice is shown in blue.) There is a check mark next to the bar representing the ground-truth voice assignment for each detected pitch. Notice that there is no check mark in the cell representing Pitch 41 at Frame 3, indicating a false positive pitch detection.

Figure 4 shows the HMM decoding process of the input from Figure 3, using a beam size of two and two voices. Notes are represented as "[pitch, frame]" and are colour-coded based on their ground truth voice assignment. (Notes belonging to the upper voice are shown in red and notes belonging to the lower voice are shown in blue.) Again, notice false positive pitch detection [41, 3]. In this figure, the emission sets O_t are shown on the bottom, and the boxes below each O_t node list the emitted notes in decreasing pitch order. Meanwhile, the voices contained by a state at each time step are listed in the boxes above each S_t node, where voices are listed in decreasing pitch order and are separated by braces. The most likely state hypothesis at each time step is on the bottom row, and each state box (except for

S_0) has an incoming arrow indicating which prior state hypothesis was used to transition into that state. Those state hypotheses with an entirely correct voice assignment are represented by a thick border.

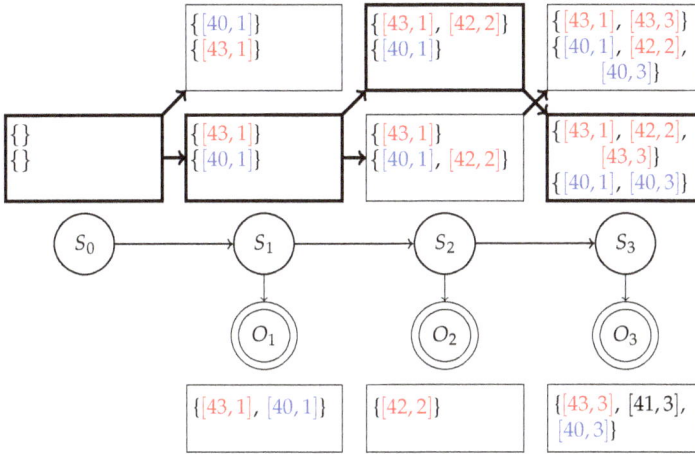

Figure 4. An example of the music language model being run on the detected pitches from Figure 3 with a beam size of two and two voices. Notes are represented as "[pitch,frame]" and are colour-coded based on their ground truth voice assignment. (Notes belonging to the upper voice are shown in red and notes belonging to the lower voice are shown in blue.) The observed note sets are listed beneath each O_t. Notice the false positive pitch detection $[41, 3]$ in O_3. The two most likely state hypotheses at each step are listed in the large rectangles above each state S_t, where the voices are notated with braces. The most likely state hypothesis at each step appears on the bottom row, and each state has an incoming arrow indicating which prior state hypothesis was used to transition into that state. Those state hypotheses with an entirely correct voice assignment are represented by a thick border.

Initially, S_0 contains two empty voices. Next, O_1 is seen, and the most likely voice assignment is also the correct one, assigning the pitches to the voices in decreasing pitch order. The second hypothesis for S_1 is very unlikely: the two voices are out of pitch order with each other, and its values of $P_t(p|v)$ are lower than the correct assignments. Thus, once O_2 is seen at Frame 2, that hypothesis drops out, and both hypothesis S_2 states transition from the most likely S_1 state. However, due to noisy $P_t(p|v)$ estimates from the PLCA, the most likely S_2 contains an incorrect assignment for the note $[42, 2]$, while the second S_2 hypothesis is correct. In S_3, however, these hypotheses flip back, resulting in the correct overall voice assignment for this example input. Notice that the false positive pitch detection $[41, 3]$ is not assigned to any hypothesis state since its values of $P_t(p|v)$ are relatively small. Meanwhile, the $P_t(p|v)$ estimates from the PLCA for the other two pitches are quite good and allow the HMM to correct itself (assuming good parameter settings), judging that the voice $\{[43, 1], [42, 2], [43, 3]\}$ in the higher voice is more likely than the voice $\{[40, 1], [42, 2], [40, 3]\}$ in the lower voice, even given the noisy $P_t(p|v)$ estimates for the note $[42, 2]$.

2.3. Model Integration

In this section, we describe the integration of the acoustic model and the music language model into a single system that jointly performs multi-pitch detection and voice assignment from audio. The pitch activations $P_t(p|v)$ for each voice type from the PLCA dictionary (bass, tenor, baritone, alto and soprano) are quite noisy, resulting in very low accuracy for voice assignment, as can be seen from our results (Table 1, row Schramm and Benetos [21]). However, we have found that a good prior distribution for $P_t(p|v)$ can drive the spectrogram factorisation towards a more meaningful voice assignment. This prior is given by the music language model, and its integration into the system pipeline is performed in two stages.

Table 1. Voice assignment results, where standard deviations are shown in parentheses. The post-processing refinement step described in Section 2.1.1 was also run on the output of all cited methods. For those that do not output any voice assignment information (Klapuri [49], Salamon and Gomez [50], Vincent et al. [4] and Pertusa and Iñesta [51]), the music language model was run once on its output with default settings and $M = 4$. VOCAL4-MP represents our proposed method with the acoustic model only. For VOCAL4-MP and [21], voice assignments are derived from each model's probabilistic voice assignment estimates ($P_t(v|p)$ for [21] and $P_t(p|v)$ for VOCAL4-MP). VOCAL4-VA refers to our fully-integrated model.

Model	Bach Chorales				
	F_{va}	F_s	F_a	F_t	F_b
Klapuri [49]	28.12 (4.38)	24.23 (10.28)	22.98 (11.85)	29.35 (12.43)	35.92 (10.97)
Salamon and Gomez [50]	24.83 (5.31)	30.03 (12.63)	25.24 (10.92)	21.09 (9.91)	22.95 (9.30)
Vincent et al. [4]	18.30 (4.87)	13.43 (7.03)	15.52 (6.50)	17.14 (6.77)	27.10 (8.44)
Pertusa and Iñesta [51]	44.05 (4.60)	40.18 (11.28)	43.34 (7.38)	41.54 (7.02)	50.56 (6.16)
Schramm and Benetos [21]	20.31 (3.40)	20.42 (5.36)	21.27 (4.75)	14.49 (1.37)	25.05 (2.12)
VOCAL4-MP	21.84 (9.37)	12.99 (11.23)	10.27 (10.13)	22.72 (6.72)	41.37 (9.41)
VOCAL4-VA	**56.49 (10.48)**	**52.37 (12.92)**	**49.13 (11.22)**	**53.10 (11.71)**	**71.38 (6.06)**

Model	Barbershop Quartets				
	F_{va}	F_s	F_a	F_t	F_b
Klapuri [49]	20.90 (5.79)	2.53 (4.82)	29.02 (13.25)	7.94 (7.48)	44.09 (14.26)
Salamon and Gomez [50]	20.38 (6.61)	11.14 (10.27)	35.14 (14.04)	8.44 (8.22)	26.81 (13.69)
Vincent et al. [4]	19.13 (8.52)	10.20 (8.25)	17.97 (9.03)	15.93 (8.85)	32.41 (12.41)
Pertusa and Iñesta [51]	37.19 (8.62)	30.68 (13.94)	**36.15 (11.70)**	29.15 (13.90)	52.78 (10.37)
Schramm and Benetos [21]	23.98 (4.34)	24.45 (6.36)	31.61 (6.79)	13.55 (2.18)	26.34 (2.03)
VOCAL4-MP	18.35 (7.56)	2.40 (5.54)	10.56 (13.92)	16.61 (7.31)	43.85 (3.46)
VOCAL4-VA	**49.06 (14.65)**	**41.78 (18.78)**	34.62 (16.29)	**35.59 (16.93)**	**84.25 (6.58)**

Since multi-pitch detections from the acoustic model are the input for the music language model, spurious detections can result in errors during the voice separation process. Therefore, in the first stage, we run the EM algorithm using only the acoustic model from Section 2.1 for 15 iterations to allow for convergence to stable multi-pitch detections. Next, the system runs for 15 more EM iterations, this time also using the music language model from Section 2.2. During each EM iteration in this second stage, the acoustic model is run first, and then, the language model is run on the resulting multi-pitch detections. To integrate the two models, we apply a fusion mechanism inspired by the one used in [52] to improve the acoustic model's pitch activations based on the resulting voice assignments.

The output of the language model is introduced into the acoustic model as a prior to $P_t(p|v)$. During the acoustic model's EM updates, Equation (6) is modified as:

$$P_t^{new}(p|v) = \alpha P_t(p|v) + (1 - \alpha)\phi_t(p|v), \tag{17}$$

where α is a weight parameter controlling the effect of the acoustic and language model and ϕ is a hyperparameter defined as:

$$\phi_t(p|v) \propto P_t^a(p|v)P_t(p|v). \tag{18}$$

$P_t^a(p|v)$ is calculated from the most probable final HMM state $S_{t_{max}}$ using the pitch score $\Delta_p(V, n)$ from the HMM transition function of Equation (14). For V, we use the voice $V_v \in S_{t_{max}}$ as it was at frame $t - 1$, and for n, we use a note at pitch p. The probability values are then normalised over all pitches per voice. The pitch score returns a value of one when the V is an empty voice (thus becoming a uniform distribution over all pitches). The hyperparameter of Equation (18) acts as a soft mask, reweighing the pitch contribution of each voice based on detected pitches from the previous iteration.

Performance depends on a proper alignment between the voice types present in each song and the voice types present in the PLCA dictionary. Therefore, we dynamically assign one of the five voice types present in the dictionary (see Section 2.1.2) to each of the voices extracted by the music language model. During the first integrated EM iteration, the acoustic model's voice probabilities $P_t(p|v)$ are set to a uniform distribution upon input to the music language model. Additionally, we cannot be sure which voice types are present in a given song, so we run the language model with $M = 5$. Here, the acoustic model's detections contain many overtones, and we do not want to simply use $M = 4$, because many of the overtones are actually assigned a slightly greater probability than the correct notes by the acoustic model. Rather, the overtones tend to be higher in pitch than the correct notes and, thus, are almost exclusively assigned to the fifth voice by the HMM. These decisions combined allow the music language model to drive the acoustic model towards the correct decomposition without being influenced by the acoustic model's initially noisy voice type probabilities.

After this initial HMM iteration, we make the dynamic dictionary voice type assignments using the following equation:

$$\text{VoiceType}(V_i) = \arg\max_v \sum_{p,t} P_t(p|v) P_t^a(p|V_i), \tag{19}$$

such that each voice V_i from the HMM is assigned the voice type v from the dictionary that gives the greatest correlation between the (initial, non-uniform) PLCA voice probabilities $P_t(p|v)$ and the HMM voice priors $P_t^a(p|V_i)$. This alignment procedure begins with the HMM's lowest voice and performs a greedy search, such that for each subsequent voice, the arg max only searches over those dictionary voice types not already assigned to a lower HMM voice. This dynamic dictionary voice type assignment allows the model to decide which voice types are present in a given song at runtime. For all subsequent iterations, this voice type assignment is saved and used during integration. Additionally, the HMM is now run with $M = 4$, and the voice type assignment is used to ensure that the PLCA output $P_t(p|v)$ estimates correspond to the correct voice indices in the HMM. This dynamic dictionary type alignment is a novel feature of the proposed model compared to our previous work [38].

We also place certain constraints on the HMM during its first iteration. Specifically, where O_t is the set notes observed at frame t: (1) if $|O_t| \leq M$, each note in O_t must be assigned to a voice in S_t; and (2) if $|O_t| > M$, the voices in S_t must contain exactly the M most likely pitch activations from O_t, according to the $P(p,t)$ from the acoustic model, where ties are broken such that lower pitches are considered more likely (since overtones are the most likely false positives).

The final output of the integrated system is a list of the detected pitches at each time frame that are assigned to a voice in the most probable final HMM state $S_{t_{max}}$, along with the voice assignment for each after the full 30 EM iterations. Figure 6 shows an example output of the integrated system and is discussed more in depth in Section 3.4.

3. Evaluation

3.1. Datasets

We evaluate the proposed model on two datasets of a capella recordings: one of 26 Bach chorales and another of 22 barbershop quartets, in total 104 minutes. (Original recordings are available at http://www.pgmusic.com/bachchorales.htm and http://www.pgmusic.com/barbershopquartet.htm respectively.) These are the same datasets used in [21], allowing for a direct comparison between it and the acoustic model proposed in Section 2.1. Each file is in wave format with a sample rate of 22.05 kHz and 16 bits per sample. Each recording has four distinct vocal parts (SATB), with one part per channel. The recordings from the barbershop dataset each contain four male voices, while the Bach chorale recordings each contain a mixture of two male and two female voices.

A frame-based pitch ground truth for each vocal part was extracted using a monophonic pitch tracking algorithm [43] on each individual monophonic track with default settings. Experiments are conducted using the mix down of each audio file with polyphonic content, not the individual tracks.

3.2. Evaluation Metrics

We evaluate the proposed system on both multi-pitch detection and voice assignment using the frame-based precision, recall and F-measure as defined in the Music Information Retrieval Evaluation eXchange (MIREX) multiple-F0 estimation evaluations [53], with a frame hop size of 20 ms.

The F-measure obtained by the multi-pitch detection is denoted as F_{mp}, and for this, we combine the individual voice ground truths into a single ground truth for each recording. For voice assignment, we simply use the individual voice ground truths and define voice-specific F-measures of F_s, F_a, F_t and F_b for each respective SATB vocal part. We also define an overall voice assignment F-measure F_{va} for a given recording as the arithmetic mean of its four voice-specific F-measures.

3.3. Training

To train the acoustic model, we use recordings from the RWC dataset [41] to generate the six-dimensional dictionary of log-spectral templates specified in Section 2.1, following the procedure described in Section 2.1.2.

For all parameters in the music language model, we use the values reported in [37] that were used for voice separation in the fugues, except that we double the value of σ_p to eight to better handle noise from the acoustic model. We also introduce two new parameters to the system: the voice crossing probability P_{cross} and the voice assignment probability P_v. We use MIDI files of 50 Bach chorales, available at http://kern.ccarh.org/ (none of which appear in the test set), splitting the notes into 20-ms frames, and measure the proportion of frames in which a voice was out of pitch order with another voice and the proportion of frames in which each voice contains a note. This results in values of $P_{cross} = 0.006$ and $P_v = 0.99$, which we use for testing.

To train the model integration weight α, we use a grid search on the range $[0.1, 0.9]$ with a step size of 0.1, maximising F_{va} for each of our datasets. Similarly, the value of the threshold L_{th} that is used for the binarisation of the multi-pitch activations in Section 2.1.1 is based on a grid search on the range $[0.0, 0.1]$ with a step size of 0.01, again maximising F_{va} for each dataset. To avoid overfitting, we employ cross-validation, using the parameter settings that maximise the chorales' F_{va} when evaluating the barbershop quartets, and vice versa; nonetheless, the resulting parameter settings are the same for both datasets: $\alpha = 0.1$ and $L_{th} = 0.01$.

3.4. Results

We use five baseline methods for evaluation: Vincent et al. [4], which uses an adaptive spectral decomposition based on NMF; Pertusa and Iñesta [51], which selects candidates among spectral peaks, validating candidates through additional audio descriptors; Schramm and Benetos [21], a PLCA model for multi-pitch detection from multi-singers, similar to the acoustic model of our proposed system, although it also includes a binary classifier to estimate the final pitch detections from the pitch activations; as well as two multi-pitch detection methods from the Essentia library [54]: Klapuri [49], which sums the amplitudes of harmonic partials to detect pitch presence; and Salamon and Gomez [50], which uses melodic pitch contour information to model pitch detections. For all five of these methods, we also run the post-processing refinement step described in Section 2.1.1 on their output.

We evaluate the above systems against two versions of our proposed model: VOCAL4-MP, using only the acoustic model described in Section 2.1; and VOCAL4-VA, using the fully-integrated model.

From the multi-pitch detection results in Table 2, it can be seen that our integrated model VOCAL4-MP achieves the highest F_{mp} on both datasets. In fact, VOCAL4-VA outperforms VOCAL4-MP substantially, indicating that the music language model is indeed able to drive the acoustic model to a more meaningful factorisation.

Table 2. Multi-pitch detection results, where standard deviations are shown in parentheses. The post-processing refinement step described in Section 2.1.1 was also run on the output of all cited methods. VOCAL4-MP represents our proposed method with the acoustic model only, while VOCAL4-VA refers to our fully-integrated model.

Model	Bach Chorales	Barbershop Quartets
Klapuri [49]	54.62 (3.00)	48.24 (4.50)
Salamon and Gomez [50]	49.52 (5.18)	45.22 (6.94)
Vincent et al. [4]	53.58 (6.27)	51.04 (8.52)
Pertusa and Iñesta [51]	67.19 (3.82)	63.85 (6.69)
Schramm and Benetos [21]	71.03 (3.33)	70.84 (6.17)
VOCAL4-MP	63.05 (3.12)	59.09 (5.07)
VOCAL4-VA	**71.76 (3.51)**	**75.70 (6.18)**

For voice assignment, using each baseline method above that does not output any voice assignment information (Klapuri [49], Salamon and Gomez [50], Vincent et al. [4] and Pertusa and Iñesta [51]), we run our music language model once on its output with default settings and $M = 4$, after the post-processing refinement step. Meanwhile, for Schramm and Benetos [21], as well as VOCAL4-MP, the voice assignments are derived from each model's probabilistic voice assignment estimates ($P_t(v|p)$ for [21] and $P_t(p|v)$ for VOCAL4-MP).

The voice assignment results are shown in Table 1, where it is shown that VOCAL4-VA outperforms the other models, suggesting that a language model is necessary for the task. It is also clear that integrating the language model as we have (rather than simply including one as a post-processing step) leads to greatly improved performance. Specifically, notice that the difference in performance between our model and the baseline methods is much greater for voice separation than for multi-pitch detection, even though we applied our language model to those baseline methods' results as post-processing.

Also interesting to note is that our model performs significantly better on the bass voice than on the other voices. While this is also true of many of the baseline methods, for none of them is the difference as great as with our model. Overtones are a major source of errors in our model, and the bass voice avoids these since it is almost always the lowest voice.

A further investigation into our model's performance can be found in Figure 5, which shows all of the VOCAL4-VA model's F-measures, averaged across all songs in the corresponding dataset after each EM iteration. The first thing to notice is the large jump in performance at Iteration 15, when the language model is first integrated into the process. This jump is most significant for voice assignment, but is also clear for multi-pitch detection. The main source of the improvement in multi-pitch detection is that the music language model helps to eliminate many false positive pitch detections using the integrated pitch prior. In fact, the multi-pitch detection performance improves again after the 16th iteration and then remains relatively stable throughout the remaining iterations.

The voice assignment results follow a similar pattern, though without the additional jump in performance after Iteration 16. In the Bach chorales, the voice separation performance even continues to improve until the end of all 30 iterations. For the barbershop quartets, however, the performance increases until Iteration 20, before decreasing slightly until the end of the process. This slight decrease in performance over the final 10 iterations is due to the alto and soprano voices: F_b and F_t each remain stable over the final 10 iterations, while F_a and F_s each decrease. This difference is likely explained by the acoustic model not being able to properly decompose the alto and soprano voices. The barbershop quartets have no true female voices (i.e., each part is sung by a male vocalist), but the template dictionary's alto and soprano voices are sung by female vocalists; thus, the alto and soprano parts must be estimated through a rough approximation of a spectral basis combination of female voices. Such a rough approximation could be the cause of our model's difficulty in decomposing the alto and soprano voices in the barbershop quartets.

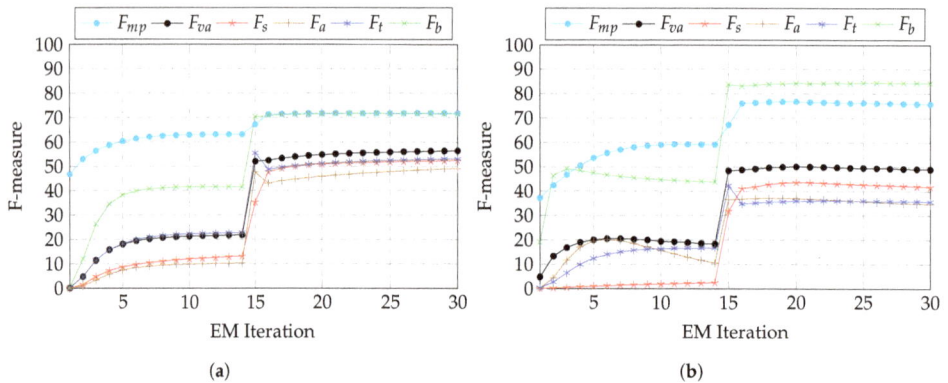

Figure 5. The VOCAL4-VA model's F-measures after each EM iteration, averaged across all songs in each dataset. (**a**) Bach chorales; (**b**) barbershop quartets.

Figure 6 illustrates the output of our proposed system, run on excerpts from both the Bach chorale (a, left) and barbershop quartet (b, right) datasets, for the joint multi-pitch detection and voice assignment tasks. Figure 6(a1),(b1) show the ground truth, using colour to denote vocal part; Figure 6(a2),(b2) show the probabilistic pitch detections from the acoustic model after the 30th EM iteration, summed over all voices ($\sum_{v=1}^{5} P_t(p|v)$), where a darker shade of gray indicates a greater probability; Figure 6(a3),(b3) present the final output of the integrated system, again using colour to denote vocal part.

As mentioned earlier, the bass voice assignment outperforms all other voice assignments in almost all cases, since false positive pitch detections from the acoustic model often correspond with overtones from lower notes that occur in the same pitch range as the correct notes from higher voices. These overtone errors are most commonly found in the soprano voice, for example at around 105 seconds in the Bach chorale excerpt and around 64.5 seconds in the barbershop quartet excerpt, where Figure 6(a2),(b2) clearly show high probabilities for these overtones. It is clear from Figure 6(a3),(b3) that such overtone errors in the soprano voice also lead to voice assignment errors in the lower voices since our system can now assign the correct soprano pitch detections to the alto voice, alto to tenor and tenor to bass.

Another common source of errors (for both multi-pitch detection and voice assignment) is vibrato. The acoustic model can have trouble detecting vibrato, and the music language model prefers voices with constant pitch over voices alternating between two pitches, leading to many off-by-one errors in pitch detection. Such errors are evident throughout the Bach chorale excerpt, particularly in the tenor voice towards the beginning where our system detects mostly constant pitches (both in the acoustic model output and the final output) while the ground truth contains some vibrato. Furthermore, at the end of both excerpts, there is vibrato present, and our system simply detects no pitches rather than the vibrato. This is most evident in the tenor voice of the Bach chorale, but is also evident in the soprano, alto and tenor voices of the barbershop quartet.

Excerpt from Bach Chorale dataset:
"If Thou but Suffer God to Guide Thee"
(Johann Sebastian Bach)

Excerpt from Barbershop Quartet dataset:
"It is a Long, Long Way to Tipperary"
(Jack Judge and Harry H. Williams)

(a1)

(b1)

(a2)

(b2)

(a3)

(b3)

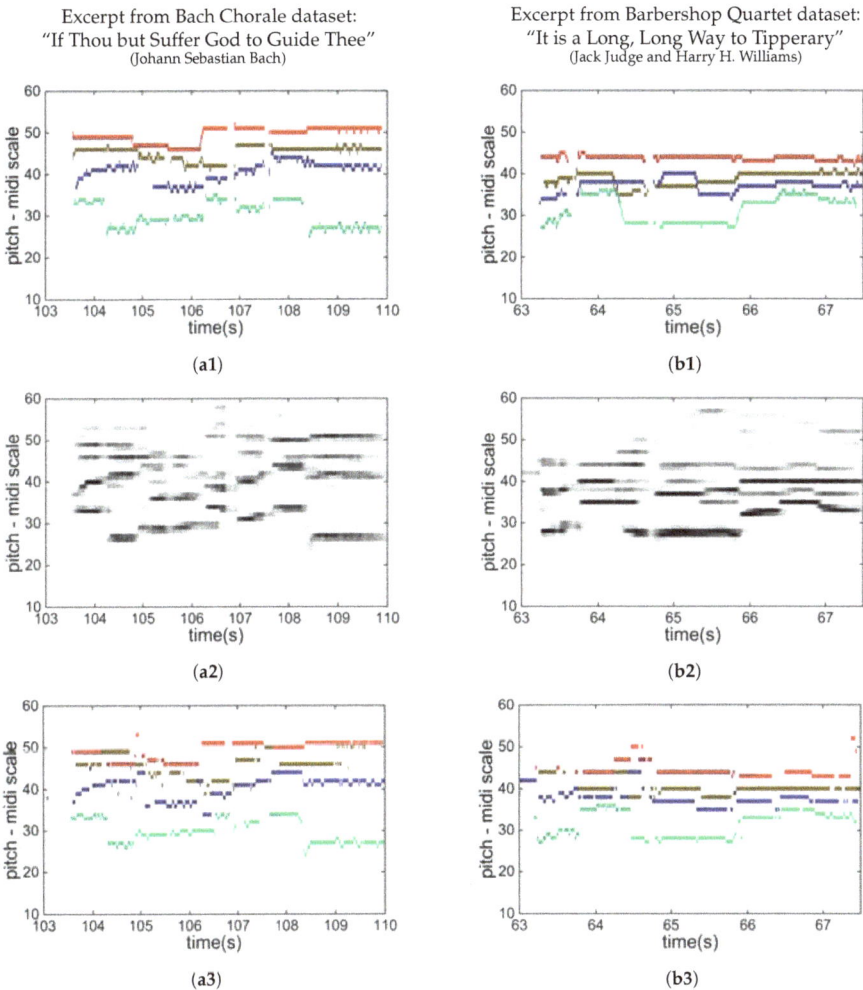

Figure 6. Example system input and output of excerpts from the Bach Chorale (**a**) (left) and barbershop quartet (**b**) (right) datasets. (**a1,b1**) show the ground truth, using colour to denote vocal part (red: soprano; brown: alto; blue: tenor; green: bass). (**a2,b2**) show the probabilistic pitch detections from the acoustic model after the 30th EM iteration, summed over all voices ($\sum_{v=1}^{5} P_t(p|v)$), where a darker shade of gray indicates a greater probability; (**a3,b3**) present the final output of the integrated system, again using colour to denote vocal part.

A closer look at errors from both vibrato and overtones can be found in Figure 7, which shows pitch detections (red) and ground truth (black) for the soprano voice from an excerpt of "O Sacred Head Sore Wounded" from the Bach chorales dataset. Here, errors from overtones can be seen around 108.5 seconds, where the detected pitch 54 is the second partial from the tenor voice (not shown), which is at pitch 42 at that time. Errors from vibrato are evident around 107.75 seconds and 108.6 seconds, where the pitch detections remain at a constant pitch while the ground truth switches between adjacent pitches.

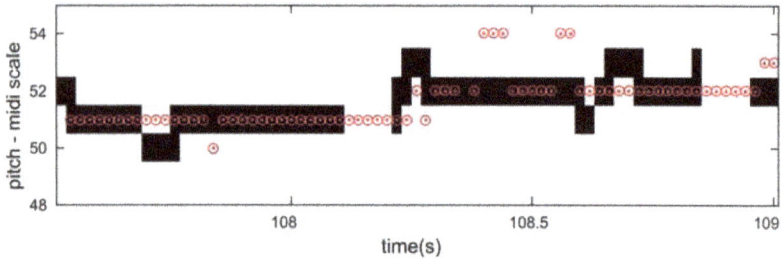

Figure 7. Pitch detections (red) and ground truth (black) for the soprano voice from an excerpt of "O Sacred Head Sore Wounded" from the Bach chorales dataset, showing errors from both vibrato and overtones (from the tenor voice, not shown).

Twenty-Cent Resolution

To further investigate our model's performance, especially on vibrato, we present its performance using 20-cent resolution instead of semitone resolution. Specifically, we divide each semitone into five 20 cent-wide frequency bins. We convert our integrated model's final semitone-based output into these bins using a post-processing step: for each detected pitch, we assign it to the 20-cent bin with the maximum $P_t(f|p)$ value from the acoustic model's final decomposition iteration.

Results are reported in terms of a cent-based pitch tolerance. A tolerance of zero cents means that a pitch detection will only be evaluated as a true positive if it is in the correct 20-cent bin. A tolerance of ± 20 cents means that a pitch detection will be evaluated as a true positive if it is within one bin of the correct bin. In general, a tolerance of $\pm 20k$ cents will count any pitch detection falling within k bins of the correct bin as a true positive.

Figure 8 illustrates our model's performance using different tolerance levels. In general, our model's semitone-based F-measures lie in between its F-measures when evaluated 20-cent resolution at ± 40-cent and ± 60-cent tolerance. This does not sound too surprising as a tolerance of ± 50 cents would approximate a semitone; however, we would have expected our model's performance with 20-cent resolution to be somewhat better than its performance with semitone resolution, as it should reduce errors associated with vibrato that crosses a semitone boundary. This lack of improvement suggests that our model's difficulty in detecting vibrato is not due simply to semitone crossings, but rather, may be a more fundamental issue of vibrato itself.

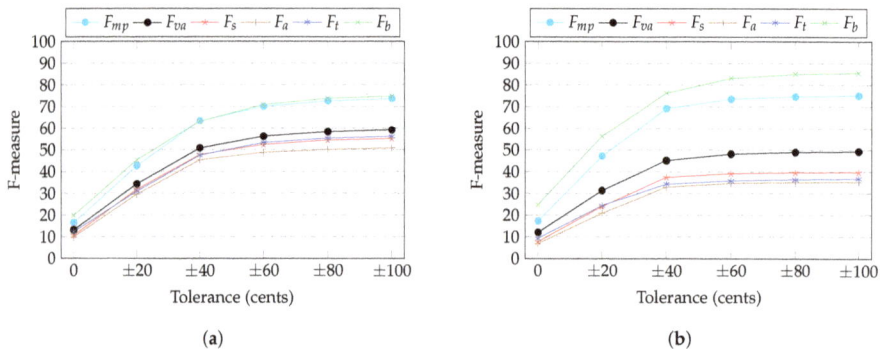

(a)

(b)

Figure 8. Our proposed model's performance on each dataset using pitch tolerance levels from zero cents up to ± 100 cents. (**a**) Bach chorales; (**b**) barbershop quartets.

4. Conclusions

In this paper, we have presented a system for multi-pitch detection and voice assignment for a cappella recordings of multiple singers. It consists of two integrated components: a PLCA-based acoustic model and an HMM-based music language model. To our knowledge, ours is the first system to be designed for the task. (Supporting Material for this work is available at http://inf.ufrgs.br/~rschramm/projects/music/musingers.)

We have evaluated our system on both multi-pitch detection and voice assignment on two datasets: one of Bach chorales and another of barbershop quartets, and we achieve state-of-the-art performance on both datasets for each task. We have also shown that integrating the music language model improves multi-pitch detection performance compared to a simpler version of our system with only the acoustic model. This suggests, as has been shown in previous work, that incorporating such music language models into other acoustic music information retrieval tasks might also be of some benefit, since they can guide acoustic models using musicological principles.

For voice assignment, while our system performs well given the difficulty of the task, there is certainly room for improvement, given that the theoretical upper bound for our model is a perfect transcription if the acoustic model's $P_t(p|v)$ estimates are accurate enough. As overtones and vibrato constitute the main sources of errors in our system, reducing such errors would lead to a great improvement in the performance of our system. Thus, future work will concentrate on methods to eliminate such errors, for example by post-processing steps that examine more closely the spectral properties of detected pitches for overtone classification and the presence of vibrato. Another possible improvement could be found during the dynamic dictionary voice type assignment step. In particular, running a voice type recognition process as a preprocessing step may result in better performance.

We will also investigate the use of incorporating additional information from the acoustic model into the music language model to continue to improve performance. In particular, we currently do not use either the singer subject probabilities $P_t(s|p)$ or the vowel probabilities $P_t(o|p)$ at all, the values of which may contain useful voice separation information. Similarly, incorporating harmonic information such as chord and key information into the music language model could lead to a more informative prior for the acoustic model during integration. Additionally, learning a new dictionary for the acoustic model, for example an instrument dictionary, would allow our system to be applied to different styles of music such as instrumentals or those containing both instruments and vocals, and we intend to investigate the generality of our system in that context.

Another possible avenue for future work is the adaptation of our system to work on the note level rather than the frame level. The music language model was initially designed to do so, but the acoustic model and the integration procedure will have to be adapted as they are currently limited to working on a frame level. Such a note-based system may also eliminate the need for robust vibrato detection, as a pitch with vibrato would then correctly be classified as a single note at a single pitch. An additional benefit to adapting our system to work on the note level would be the ability to incorporate metrical or rhythmic information into the music language model.

Acknowledgments: Andrew McLeod is supported by the University of Edinburgh. Rodrigo Schramm is supported by a UK Newton Research Collaboration Programme Award (Grant No. NRCP1617/5/46). Mark Steedman is supported by EU ERC H2020 Advanced Fellowship GA 742137 SEMANTAX. Emmanouil Benetos is supported by a UK Royal Academy of Engineering Research Fellowship (Grant No. RF/128).

Author Contributions: All authors contributed to this work. Specifically, Rodrigo Schramm and Emmanouil Benetos designed the acoustic model. Andrew McLeod designed the music language model under supervision of Mark Steedman. Andrew McLeod, Rodrigo Schramm and Emmanouil Benetos designed the model integration. Andrew McLeod and Rodrigo Schramm performed the experiments.Andrew McLeod, Rodrigo Schramm and Emmanouil Benetos wrote the paper. All authors proofread the paper.

Conflicts of Interest: The authors declare no conflict of interest.

References

1. Benetos, E.; Dixon, S.; Giannoulis, D.; Kirchhoff, H.; Klapuri, A. Automatic music transcription: Challenges and future directions. *J. Intell. Inf. Syst.* **2013**, *41*, 407–434.
2. Li, D.D.; Seung, H.S. Learning the parts of objects by non-negative matrix factorization. *Nature* **1999**, *401*, 788–791.
3. Shashanka, M.; Raj, B.; Smaragdis, P. Probabilistic latent variable models as nonnegative factorizations. *Comput. Intell. Neurosci.* **2008**, *2008*, 947438.
4. Vincent, E.; Bertin, N.; Badeau, R. Adaptive harmonic spectral decomposition for multiple pitch estimation. *IEEE Trans. Audio Speech Lang. Process.* **2010**, *18*, 528–537.
5. Kameoka, H.; Nakano, M.; Ochiai, K.; Imoto, Y.; Kashino, K.; Sagayama, S. Constrained and regularized variants of non-negative matrix factorization incorporating music-specific constraints. In Proceedings of the 2012 IEEE International Conference on Acoustics, Speech and Signal Processing, Kyoto, Japan, 25–30 March 2012; pp. 5365–5368.
6. Benetos, E.; Dixon, S. Multiple-instrument polyphonic music transcription using a temporally constrained shift-invariant model. *J. Acoust. Soc. Am.* **2013**, *133*, 1727–1741.
7. Benetos, E.; Weyde, T. An efficient temporally-constrained probabilistic model for multiple-instrument music transcription. In Proceedings of the 16th International Society for Music Information Retrieval Conference, Malaga, Spain, 26–30 October 2015; pp. 701–707.
8. Fuentes, B.; Badeau, R.; Richard, G. Controlling the convergence rate to help parameter estimation in a PLCA-based model. In Proceedings of the 22nd European Signal Processing Conference, Lisbon, Portugal, 1–5 September 2014; pp. 626–630.
9. Grindlay, G.; Ellis, D.P.W. Transcribing multi-instrument polyphonic music with hierarchical eigeninstruments. *IEEE J. Sel. Top. Signal Process.* **2011**, *5*, 1159–1169.
10. Mysore, G.J.; Smaragdis, P. Relative pitch estimation of multiple instruments. In Proceedings of the IEEE International Conference on Acoustics, Speech and Signal Processing, Taipei, Taiwan, 19–24 April 2009; pp. 313–316.
11. Brown, J. Calculation of a constant Q spectral transform. *J. Acoust. Soc. Am.* **1991**, *89*, 425–434.
12. Benetos, E.; Dixon, S. A Shift-Invariant Latent Variable Model for Automatic Music Transcription. *Comput. Music J.* **2012**, *36*, 81–94.
13. Fuentes, B.; Badeau, R.; Richard, G. Blind Harmonic Adaptive Decomposition applied to supervised source separation. In Proceedings of the 2012 European Signal Processing Conference, Bucharest, Romania, 27–31 August 2012; pp. 2654–2658.
14. Benetos, E.; Badeau, R.; Weyde, T.; Richard, G. Template adaptation for improving automatic music transcription. In Proceedings of the 15th International Society for Music Information Retrieval Conference, Taipei, Taiwan, 27–31 October 2014; pp. 175–180.
15. O'Hanlon, K.; Nagano, H.; Keriven, N.; Plumbley, M.D. Non-negative group sparsity with subspace note modelling for polyphonic transcription. *IEEE/ACM Trans. Audio Speech Lang. Process.* **2016**, *24*, 530–542.
16. Sigtia, S.; Benetos, E.; Dixon, S. An end-to-end neural network for polyphonic piano music transcription. *IEEE/ACM Trans. Audio Speech Lang. Process.* **2016**, *24*, 927–939.
17. Kelz, R.; Dorfer, M.; Korzeniowski, F.; Böck, S.; Arzt, A.; Widmer, G. On the potential of simple framewise approaches to piano transcription. In Proceedings of the 17th International Society for Music Information Retrieval Conference, New York, NY, USA, 7–11 August 2016; pp. 475–481.
18. Thickstun, J.; Harchaoui, Z.; Kakade, S. Learning features of music from scratch. In Proceedings of the International Conference on Learning Representations (ICLR), Toulon, France, 24–26 April 2017.
19. Bittner, R.M.; McFee, B.; Salamon, J.; Li, P.; Bello, J.P. Deep salience representations for F0 estimation in polyphonic music. In Proceedings of the 18th International Society for Music Information Retrieval Conference, Suzhou, China, 23–27 October 2017; pp. 63–70.
20. Bohak, C.; Marolt, M. Transcription of polyphonic vocal music with a repetitive melodic structure. *J. Audio Eng. Soc.* **2016**, *64*, 664–672.
21. Schramm, R.; Benetos, E. Automatic transcription of a cappella recordings from multiple singers. In Proceedings of the AES International Conference on Semantic Audio, Erlangen, Germany, 22–24 June 2017.

22. Bay, M.; Ehmann, A.F.; Beauchamp, J.W.; Smaragdis, P.; Downie, J.S. Second fiddle is important too: Pitch tracking individual voices in polyphonic music. In Proceedings of the 13th International Society for Music Information Retrieval Conference, Porto, Portugal, 8–12 October 2012; pp. 319–324.

23. Duan, Z.; Han, J.; Pardo, B. Multi-pitch streaming of harmonic sound mixtures. *IEEE/ACM Trans. Audio Speech Lang. Process.* **2014**, *22*, 138–150.

24. Ryynanen, M.P.; Klapuri, A. Polyphonic music transcription using note event modeling. In Proceedings of the IEEE Workshop on Applications of Signal Processing to Audio and Acoustics, New Paltz, NY, USA, 16 October 2005; pp. 319–322.

25. Cambouropoulos, E. Voice and stream: Perceptual and computational modeling of voice separation. *Music Percept. Interdiscip. J.* **2008**, *26*, 75–94.

26. Huron, D. Tone and voice: A derivation of the rules of voice-leading from perceptual principles. *Music Percept.* **2001**, *19*, 1–64.

27. Tymoczko, D. Scale theory, serial theory and voice leading. *Music Anal.* **2008**, *27*, 1–49.

28. Temperley, D. A probabilistic model of melody perception. *Cogn. Sci.* **2008**, *32*, 418–444.

29. Karydis, I.; Nanopoulos, A.; Papadopoulos, A.; Cambouropoulos, E.; Manolopoulos, Y. Horizontal and vertical integration/segregation in auditory streaming: A voice separation algorithm for symbolic musical data. In Proceedings of the Proceedings 4th Sound and Music Computing Conference (SMC'2007), Lefkada, Greece, 11–13 July 2007; pp. 299–306.

30. Kilian, J.; Hoos, H. Voice separation-a local optimization approach. In Proceedings of the 3rd International Conference on Music Information Retrieval, Paris, France, 13–17 October 2002.

31. Kirlin, P.B.; Utgoff, P.E. VOISE: Learning to segregate voices in explicit and implicit polyphony. In Proceedings of the 6th International Conference on Music Information Retrieval, London, UK, 11–15 September 2005; pp. 552–557.

32. Guiomard-Kagan, N.; Giraud, M.; Groult, R.; Levé, F. Improving voice separation by better connecting contigs. In Proceedings of the 17th International Society for Music Information Retrieval Conference, New York, NY, USA, 7–11 August 2016; pp. 164–170.

33. Gray, P.; Bunescu, R. A neural greedy model for voice separation in symbolic music. In Proceedings of the 17th International Society for Music Information Retrieval Conference, New York, NY, USA, 7–11 August 2016; pp. 782–788.

34. Chew, E.; Wu, X. Separating voices in polyphonic music: A contig mapping approach. In Proceedings of the Computer Music Modeling and Retrieval, Esbjerg, Denmark, 26–29 May 2004; pp. 1–20.

35. de Valk, R.; Weyde, T. Bringing 'musicque into the tableture': Machine-learning models for polyphonic transcription of 16th-century lute tablature. *Early Music* **2015**, *43*, 563–576.

36. Duane, B.; Pardo, B. Streaming from MIDI using constraint satisfaction optimization and sequence alignment. In Proceedings of the International Computer Music Conference, Montreal, QC, Canada, 16–21 August 2009; pp. 1–8.

37. McLeod, A.; Steedman, M. HMM-Based Voice Separation of MIDI Performance. *J. New Music Res.* **2016**, *45*, 17–26.

38. Schramm, R.; McLeod, A.; Steedman, M.; Benetos, E. Multi-pitch detection and voice assignment for a cappella recordings of multiple singers. In Proceedings of the 18th International Society for Music Information Retrieval Conference, Suzhou, China, 23–28 October 2017; pp. 552–559.

39. Schörkhuber, C.; Klapuri, A.; Holighaus, N.; Dörfler, M. A matlab toolbox for efficient perfect reconstruction time-frequency transforms with log-frequency resolution. In Proceedings of the AES 53rd Conference on Semantic Audio, London, UK, 26–29 January 2014.

40. Benetos, E.; Holzapfel, A. Automatic transcription of Turkish microtonal music. *J. Acoust. Soc. Am.* **2015**, *138*, 2118–2130.

41. Goto, M.; Hashiguchi, H.; Nishimura, T.; Oka, R. RWC music database: Music genre database and musical instrument sound database. In Proceedings of the 5th International Conference on Music Information Retrieval, Barcelona, Spain, 10–14 October 2004; pp. 229–230.

42. Dempster, A.P.; Laird, N.M.; Rubin, D.B. Maximum likelihood from incomplete data via the EM algorithm. *J. R. Stat. Soc.* **1977**, *39*, 1–38.

43. Mauch, M.; Dixon, S. PYIN: A fundamental frequency estimator using probabilistic threshold distributions. In Proceedings of the 2014 IEEE International Conference on Acoustics, Speech and Signal Processing (ICASSP), Florence, Italy, 4–9 May 2014; pp. 659–663.

44. De Andrade Scatolini, C.; Richard, G.; Fuentes, B. Multipitch estimation using a PLCA-based model: Impact of partial user annotation. In Proceedings of the 2015 IEEE International Conference on Acoustics, Speech and Signal Processing (ICASSP), Queensland, Australia, 19–24 April 2015; pp. 186–190.

45. Goto, M. A real-time music-scene-description system: Predominant-F0 estimation for detecting melody and bass lines in real-world audio signals. *Speech Commun.* **2004**, *43*, 311–329.

46. Kameoka, H.; Nishimoto, T.; Sagayama, S. A multipitch analyzer based on harmonic temporal structured clustering. *IEEE Trans. Audio Speech Lang. Process.* **2007**, *15*, 982–994.

47. Kirchhoff, H.; Dixon, S.; Klapuri, A. Missing template estimation for user-assisted music transcription. In Proceedings of the 2013 IEEE International Conference on Acoustics, Speech and Signal Processing (ICASSP), Vancouver, BC, Canada, 26–31 May 2013; pp. 26–30.

48. Viterbi, A. Error bounds for convolutional codes and an asymptotically optimum decoding algorithm. *IEEE Trans. Inf. Theory* **1967**, *13*, 260–269.

49. Klapuri, A. Multiple fundamental frequency estimation by summing harmonic amplitudes. In Proceedings of the 7th International Conference on Music Information Retrieval, Victoria, BC, Canada, 8–12 October 2006.

50. Salamon, J.; Gomez, E. Melody extraction from polyphonic music signals using pitch contour characteristics. *IEEE Trans. Audio Speech Lang. Process.* **2012**, *20*, 1759–1770.

51. Pertusa, A.; Iñesta, J.M. Efficient methods for joint estimation of multiple fundamental frequencies in music signals. *EURASIP J. Adv. Signal Process.* **2012**, doi:10.1186/1687-6180-2012-27.

52. Giannoulis, D.; Benetos, E.; Klapuri, A.; Plumbley, M.D. Improving instrument recognition in polyphonic music through system integration. In Proceedings of the 39th International Conference on Acoustics, Speech and Signal Processing, Florence, Italy, 4–9 May 2014; pp. 5222–5226.

53. Bay, M.; Ehmann, A.F.; Downie, J.S. Evaluation of multiple-F0 estimation and tracking systems. In Proceedings of the 10th International Society for Music Information Retrieval Conference, Kobe, Japan, 26–30 October 2009, pp. 315–320.

54. Bogdanov, D.; Wack, N.; Gómez, E.; Gulati, S.; Herrera, P.; Mayor, O.; Roma, G.; Salamon, J.; Zapata, J.; Serra, X. Essentia: An Audio Analysis Library for Music Information Retrieval. In Proceedings of the 14th International Society for Music Information Retrieval Conference, Curitiba, Brazil, 4–8 November 2013.

![applied sciences logo] *applied sciences*

MDPI

Article

The Effects of Musical Experience and Hearing Loss on Solving an Audio-Based Gaming Task

Kjetil Falkenberg Hansen [1,*,†] **and Rumi Hiraga** [2,†]

1   Sound and Music Computing, School of Electrical Engineering and Computer Science,
KTH Royal Institute of Technology, Lindstedsvägen 3, 11428 Stockholm, Sweden
2   Industrial Technology Department, Tsukuba University of Technology, 305-8520, Tsukuba, Japan;
rhiraga@a.tsukuba-tech.ac.jp
*   Correspondence: kjetil@kth.se; Tel.: +46-8-7907857
†   These authors contributed equally to this work.

Academic Editor: Tapio Lokki
Received: 23 October 2017; Accepted: 5 December 2017; Published: 10 December 2017

Abstract: We conducted an experiment using a purposefully designed audio-based game called the *Music Puzzle* with Japanese university students with different levels of hearing acuity and experience with music in order to determine the effects of these factors on solving such games. A group of hearing-impaired students ($n = 12$) was compared with two hearing control groups with the additional characteristic of having high ($n = 12$) or low ($n = 12$) engagement in musical activities. The game was played with three sound sets or modes; speech, music, and a mix of the two. The results showed that people with hearing loss had longer processing times for sounds when playing the game. Solving the game task in the speech mode was found particularly difficult for the group with hearing loss, and while they found the game difficult in general, they expressed a fondness for the game and a preference for music. Participants with less musical experience showed difficulties in playing the game with musical material. We were able to explain the impacts of hearing acuity and musical experience; furthermore, we can promote this kind of tool as a viable way to train hearing by focused listening to sound, particularly with music.

Keywords: audio games; educational tools; audio signal processing; computer interfaces; music cognition; perception; training; language

---

1. Introduction

Musical experiences affect persons with hearing loss and hearing persons similarly. Hence, music can provide similar benefits to both groups [1]. However, it is well known that people with hearing impairment listen to music much less. This can be seen, for example, when comparing individuals before and after cochlear implantation [2]. It is also established that even people with only mild or moderate hearing impairment exhibit language disorders [3].

In order to increase the likelihood of people with hearing loss having enjoyable listening experiences, we believe that one solution is exposure to activities involving *focused listening*. Hearing persons focus on the sound itself when they listen to music (musical listening) while they also pay attention to the source or the situation of the sound (everyday listening) [4]. We use "focused listening" for people with hearing loss so that they may listen to sounds, noticing the change along time with its pitch, timbre, and other sound features as hearing persons do.

Thus, playing an audio game where attention to music is required to solve the task—such as one that can be played casually to entertain—would promote actively listening to music. In turn, this voluntary exposure to sound supports language acquisition and development [5], personal development and social grooming [6], and the ability to extract information from the coincidental and surrounding sounds of everyday life [7].

In previous studies, authors and colleagues presented an audio game called the *Music Puzzle* as well as preliminary results from pilot testing (see for instance [8,9]); in the current work, we investigate specifically how hearing acuity and musical experience can impact game-playing achievements. The experiment involved both hearing and hearing-impaired university students.

Hearing impairments, hearing loss, and hearing acuity are closely related terms. Hearing loss is, according to several definitions, "a general term that refers to a reduced auditory acuity" [10]. Auditory acuity is also well defined and "describes how sensitive the auditory system is to sound" [11]. Hearing acuity measured through audiometry will not determine to what extent the person listens to or likes music. Actually, there are many accounts of professional artists with hearing loss who earned great success in music, such as Evelyn Glennie [12], Paul Whittaker [13], and Danny Lane [14]. Another example is the world-touring Gallaudet Dance Company [15], where the members are university students with hearing loss.

Organizations and teachers in different countries manage activities related to teaching music to children with hearing loss (for instance, *Music and the Deaf* [16], and *hear ME now* [17]), and to experiencing music (for instance, initiatives by the Mahler Chamber Orchestra [18]). There are reports on how to accommodate music activities for the hearing-impaired [19], and music education for the hearing-impaired is furthermore an active research area, such as in teaching orchestral music [20].

Even without personal music training or special music activities, many young people with hearing loss enjoy music actively, through dancing, going to karaoke, watching artist promotion videos, playing the drums, or just listening to music. Many of them also like to play music games either on computers or mobile devices, or at video arcades. It has however been shown [21] that interpretation of the communicated emotions in music (arguably music's most important characteristic) is significantly less precise in the hearing-impaired compared to typical listeners, partly due to problems of timbre and pitch perception.

Familiarity with music, gained from exposure, will increase emotional engagement in listening [22]. Also concerning motivations for engaging in musical activities, it was found [23] that although motivations for the hearing-impaired were similar to those of the hearing population, the degree of early exposure to music has an impact on music-making later in life. Musical experiences have also been documented as having positive effects and providing benefits for hearing subjects, for instance, related to language acquisition [24], social interaction [25], and auditory skills in different aspects [26–30]. Without focused listening, the same benefits for language acquisition cannot be achieved [31].

1.1. Hearing Loss, Music Listening, and Music Training

Studies on the relationship between hearing loss and music listening have been performed within several areas. In music therapy, the positive effects of musical interventions on children with hearing loss have been described [32,33]. Much of the recent research on music with hearing loss has focused on the emerging technologies related to cochlear implants, while some studies look particularly at hearing loss with just hearing aids, such as in the description of how people with hearing aids listen to music from an audiological perspective [34,35]. Music perception by cochlear implant users has been observed both by otolaryngology laboratories [36–38] and by psychologists' groups [39,40].

Music perception by people with hearing loss has also been explored from various perspectives: which music elements to use in an experiment, ways to propose music, benefits of cochlear implants and hearing aids, and the age and impairment history of participants. Experiments related to the perception of pitch vary from basic pitch discrimination tasks [41] to memorization [36], singing [1], and recognition [40] of melodies. Experiments related to exploring the role of temporal information for melody recognition have included both tempo and rhythm as well as pitch information [42]. It has been shown that pitch and timbre—when parametrically varied in a synthesized tone signal and with music listening history accounted for—interfere and confuse listeners in discrimination tests [43].

Potentially, musical training can improve timbre perception and identification in cochlear implant patients [44].

Musical training is typically given to people with hearing loss either in the long or short term. Various experiments have investigated possible long-term effects of informal music activities provided to participants at schools [1], and measured the effectiveness of long-term music lessons for improving the perception of environmental sounds [45]. In short-term training for cochlear implant users, little progress was found in terms of music skills [46]. Effects of the training were found in linguistic identification tests after controlled training of combining the acoustic information of a hearing aid with the electric information from a cochlear implant [47]. Some of these results are related to brain development [1,46].

1.2. Games for Training and Special Support, and Audio-Based Games

In recent years, needs for training and skill practice have been studied through so-called serious games or games for learning. Such games have been shown to be both effective and motivational [48,49], and they are generally applied in any type of context. The design of games for persons with physical or cognitive impairments, for instance, does not only have the purpose of giving them opportunities of playing entertaining games, but is also intended to improve logical thinking, cognitive skills, or social skills. Games for children with autism spectrum disorder have, in different studies, been shown to support the development of social skills such as membership, partnership, and friendship [50,51].

For auditory training such as exposing oneself to focused listening, it is reasonable to expect that serious games based on sound would be appropriate. Audio-based games are common both among serious games and among games only for entertainment. However, they differ greatly in design, gameplay, and functionality [52,53]. In particular, there are many examples of such games that have been developed for people with visual impairments and that can be played entirely without a graphical user interface [54,55]. Additionally, there are many general music tutoring games that practice specific skills such as solfège, rhythm, melody, and notation [56–58]. Games for training listening for the hearing-impaired are less common, although some specialize in cochlear implants [59,60].

The above and many other games provide promising interfaces for gameplay involving solving specific musical tasks, or for training in supplementary modalities for the impaired, which is predominantly visual. Instead of adapting these games to sound discrimination training for the hearing-impaired, we suggest methodically focusing on the impaired auditory sensory organ using an alternative game design. The game design is based on focused listening with an elementary graphical interface.

1.3. Aim of the Study

For our studies, we have developed an audio-based game with a simple graphical user interface that provides no visual cues to help solve the game. The game includes musical material, speech in terms of read poems, and mixes of those materials. It is intended that people with hearing loss use focused listening in order to win. We conducted an experiment to explore if the game can be used in auditory training and engaged three participant groups that differed in measured hearing acuity and self-reported music experiences; this way we could investigate the impact these factors have on game playing, but also the impact of speech and language ability since this correlates with hearing acuity.

In addition, we were interested in finding out how the game is played, what makes it enjoyable, and if music is a preferred material in auditory training. In order to resemble an everyday listening situation, the participants played with headphones, and not with e.g., Bluetooth bridging for hearing aids. They adjusted the volume of both hearing aids and the game sounds to their typically preferred level; this way, we explored the impact of their hearing relative to their typical listening conditions.

If the game is appreciated among the experiment participants, it is ready to be used as a formal and informal training tool for a wide group of the hearing-impaired. On the whole, we would be able explain the impact of the above factors, and therefore recommend considering this kind of tool in the future training of hearing.

2. The Music Puzzle

The developed game, called the *Music Puzzle*, has a gameplay which resembles that of a classic jigsaw puzzle. It is developed for Android devices with touchscreen and uses a Pure data real-time audio engine library [61] (see also [8,9]). Our initial idea was to use music, which means that the purpose of the game is to recompose a musically correct piece of music from fragmented parts of a recording. However, the game is not restricted to music, but can use any audio recording.

The complete puzzle to be solved is represented by a ball on the screen, and pressing this ball will play the corresponding sound file (see Figure 1). Then, this ball is divided into smaller pieces or "sound objects" (to be explained shortly) with an identical appearance and they are randomly distributed in the graphical user interface. These smaller puzzle pieces or sound objects can be reassembled into a complete whole. The sound objects can furthermore be manipulated in pitch, and filtered by equalization (from here on referred to as *EQ*); they will, like the bigger ball, play the linked sound upon being pressed. For a video example of the game, please see the supplementary materials.

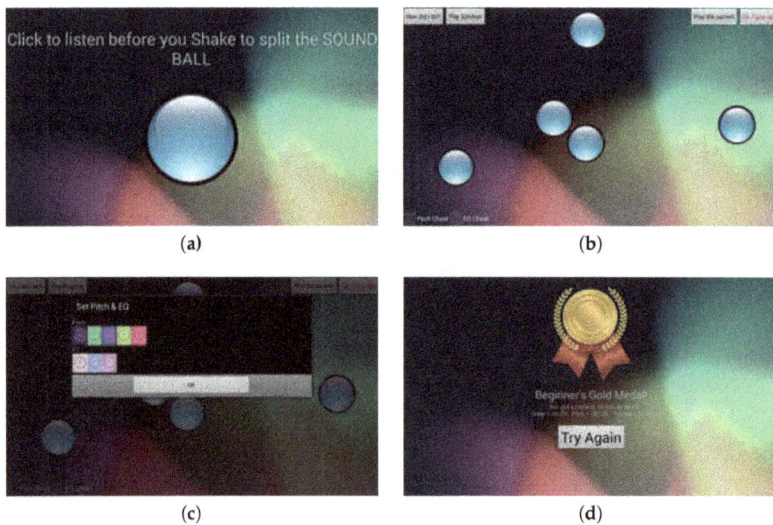

Figure 1. The Music Puzzle gameplay interface as seen on a tablet. (**a**) Initiate a session, listen to the target music piece, and shake the tablet; (**b**) Listen and order sound objects by finger touch. There are four action buttons: *How did I do?* (evaluate current order), *Play Solution* (repeat target piece), *Play the current* (play the order as seen on screen), and *Oh, I give up* (quit the puzzle); (**c**) Adjust pitch and equalization (EQ; filtering) for each object. The radio buttons are randomly colored and ordered so as to not give any visual cues to the solution; (**d**) Completed puzzle with an evaluation.

The player hears the entire puzzle to be solved once, then proceeds to reorder the pieces and change pitch and EQ appropriately. In order to solve the puzzle in its intended music mode, one has to memorize and understand not only melody and rhythm, but also its timbre and possibly other characteristics. With non-musical types of stimuli like speech and environmental sounds, decisions for solving will rely on additional cues for music such as language and meaning.

2.1. Sound Objects

Sound objects are generated as fragments of the original recorded audio following a "shake the tablet" action to mimic the concept of breaking a fragile ball into pieces. The number of objects generated depends on shaking force. Segmentation is done by dividing the whole file into fragments of equal duration. These sound objects are in the game connected with fast crossfades (5 ms) to avoid zero-crossings. A random selection of objects is further modified in pitch or EQ, or both; in the latter cases the task difficulty will naturally increase [43]. Both segmentation and modifications are performed in real time.

The sounds are modified using filtering (EQ) by adjusting the energy of the low and high frequency components in the audio signal; this is done either by a low-pass or a high-pass filter (using the standard Pure data objects lop~ and hip~). The low-pass filter has a cutoff frequency at 2000 Hz, which means everything above this cutoff in the sound is attenuated (i.e., no treble). The high-pass filter has a cutoff frequency at 500 Hz, attenuating everything below (i.e., no bass). In either modification, the important frequency range of 500–2000 Hz is left intact. Cutoff frequencies were determined from experimentation and observation.

The pitch is modified either by −1000 cents, −500 cents, +500 cents, or +1000 cents; a 100-cent change corresponds to a semitone pitch shift. Even pitch changes were determined by experimentation. The modulations are done in the frequency domain which leave durations unaltered (using a modification of the Pure data patch I07-phase.vocoder). All sounds used for the experiment were uncompressed mono audio files sampled at 44.1 kHz to facilitate the frequency-domain manipulations.

2.2. User Interface and Gameplay

Figure 1 shows the different screens of the user interface. First, in Figure 1a, the user listens to the target piece to reconstruct by tapping the large ball. Then, the user shakes the tablet to break apart the target piece into several fragments (sound objects) represented by small identical balls; Figure 1b shows the resulting display after shaking the tablet.

The intended gameplay is to

1. tap and listen to the sounding objects
2. long press and adjust EQ and pitches
3. drag and arrange the objects horizontally from left to right
4. click the menu item to check and evaluate the solution

where the steps can be executed in any order and repeated in a trial-and-error procedure.

The four square buttons at the top of the screen (see Figure 1b) are used for evaluating the order, pitch, and EQ (*How did I do?*), replay the target sound (*Play solution*), play the current arrangement of objects as it appears on the screen (*Play the current*), and there is a final option to end the session (*Oh, I give up*). Two buttons in the lower left are "cheat buttons", described shortly.

EQ and pitch modification dialogs, as seen in Figure 1c, are activated by a long press on an object. Radio buttons are presented in random order and colors so as to not provide visual cues that could help solve the puzzle. Each press on a radio button will play the sound with selected adjustment.

2.3. Game Difficulty

The difficulty level is determined by the number of pieces generated, the pitch shifts, the filtering, and last but not least the characteristics of the sound recording. When the shaking yields many pieces, the game is in most cases more difficult because the durations of the sound objects get shorter and all other factors that increase difficulty are more likely to occur and have a larger impact. In the used version, the game had to be solved perfectly to be counted as accomplished; for other purposes, the threshold for success could be adjusted.

For both modulation types (pitch and EQ), it is possible to set any arbitrary values in a text file, and in this way augment the game with increasing difficulty and levels. However, following a testing phase with intended users, it was not considered necessary at this point.

Sound objects will also constitute a difficulty differentiation. Shake force determines the number of generated objects, and the difficulty naturally increases with more objects. Object duration is determined by their number and also by the target's total duration. Finally, the cut points of the objects may come at any place in the original sound: imagine for instance a drum loop of four bars cut into three objects (easier) compared to four objects (harder), or a piece cut into a large number of objects; some may end up with only silence. In our experiment this was not an issue, but for further use one should apply an automatic analysis of the target sound to avoid unsolvable puzzles.

Alternatively, if the puzzle gets too hard to solve, the pitch and EQ modifications can be automatically corrected using the *Pitch Cheat* and *EQ Cheat* buttons in the left bottom corner of the screen. Furthermore, the user can choose to replay the target sound by clicking *Play solution*. Cheat buttons and the replay function can be deactivated in the settings file.

2.4. Preparation and Data Collection

Before using the game for experimental purposes, the difficulty settings text file should be edited and a collection of pieces should be prepared in folders. Any type and duration of audio recording can be used.

A game session starts when a player listens to the target sound (clicking the large ball) and ends with the final notification in Figure 1d if the participant can build the target sound, or when clicking the "give-up button" . Each game session is recorded in a log file. The log file includes time-stamped information about the session and all the user's actions on the touchscreen.

3. Materials and Methods

An experiment was designed to collect gaming data and user evaluations. The experiment was conducted in accordance with the Declaration of Helsinki and with ethical approval from the involved universities. All included data are anonymous. Each participant was carefully briefed about the experiment and signed a consent form to participate. In the study, we only look at various time measurements, frequencies of interactions, and preference of sound material. In addition, participants were informed about hearing acuity and music listening experiences. We conducted three sets of gaming sessions with a total of 36 participants (14 female). They were university students of ages from 18 to 23, and were recruited into equally-sized groups as follows:

Group	Hearing	Language	Musical experience
HI	Impairment	Japanese	–
NEX	Normal	Japanese	Low
EXP	Normal	Japanese	High

The Japanese hearing-impaired participants (*HI*) group was recruited from a university for hearing-impaired technology students. Eleven participants had profound degrees of hearing loss, while one participant had severe hearing loss [62]. Profound loss is considered to be above 90 dB, and severe in the range 70–90 dB. Hearing loss and acuity are measured with audiometers and expressed in decibels hearing level (dB_{HL}). Because of the human ears' characteristic of perceiving sounds differently depending on frequencies, decibels sound pressure level (dB_{SPL}) cannot show hearing acuity by frequency [63]. Eleven of the participants used hearing aids in the experiment and one was a cochlear implantee. They could use their hearing aids according to their own preferences with two intended benefits: for their comfort, and because this would approximate their typical listening situation. Though the research with hearing-impaired persons tends to focus on either cochlear implanters or users of hearing aids, we do not divide them according to their hearing devices because our research interest is to provide them the opportunities to listen to sounds with joy.

The recruitment of Japanese hearing participants with low musical experience (*NEX*) and Japanese hearing participants with high musical experience (*EXP*) was based on their self-assessment of engagement in musical activities; *NEX* were recruited among students without ongoing music activities, while *EXP* had formal activities. As a simplified measure of music activity, we asked them to rate their musical experience in terms of listening to music in everyday life with respect to five levels ranging from very rare to very often; the question included examples of listening situations. Figure 2 shows the ratio of their music-listening experiences. For *HI*, musical experience was registered, but not used as a qualifier for inclusion in the experiment. There was one hearing-impaired participant who did not listen to music, but otherwise the musical experience of *HI* and *NEX* matched very well (the summed ratios of 'very often' and 'often', and of 'rarely' and 'very rarely' were the same between the two groups). One-way analysis of variance (ANOVA) shows there are significant differences in music-listening experiences ($p = 0.02$), and from multiple comparisons differences are seen between *HI/EXP* and *NEX/EXP*.

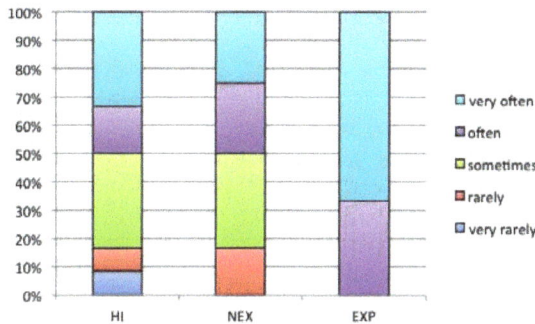

Figure 2. Music-listening experiences of the three participant groups hearing impaired (HI), and normal hearing with low (NEX) and high (EXP) music experience.

3.1. Game Material

We prepared sound sets for four game "levels", each with three game modes consisting of *speech*, *music*, and *mixed* sound material. We will use initial capital letters to denote that the puzzle condition is based on a Music, Speech, or Mixed target piece. The sound sets did not give the game levels increased difficulty; this was handled according to the above. The Speech pieces did not contain any musical sounds, and the Music pieces did not contain any vocals. The Mixed pieces were simply the combination of one Speech and one Music piece sound file mixed to a new mono sound file. All pieces were 15 s long and normalized in Audacity (http://manual.audacityteam.org/man/normalize.html) to have the same peak amplitude.

The speech recordings were from commercially available recordings by Japanese poetry readers, both female and male. Sets 1 and 4 were from old Japanese poems, while Set 2 was from a Japanese translation of an English poem, and the reading of Set 3 was from a Japanese pop song. Most Japanese young people would be familiar with the poems and the pop song.

The music recordings were excerpted from cello performances of well-known compositions. Three were by Japanese composers working in the field of classical music in films, and one was composed by Fauré. They were chosen based on pre-studies of the Music Puzzle, (see e.g., [9]). Before deciding on a recording, we evaluated its suitability by listening to the mix; the main condition was that the speech should be easily legible through the music. Table 1 shows the four sets of sound pieces. The order of sets 2–4 was randomized.

For the experiment, we installed the game and sound material on four Nexus 7 tablets and two Samsung Galaxy tablets. Audio was presented through headphones with large cups so as to fit and accommodate hearing aids; they could choose between Sony MDR-XD200 (closed type), Audio-Technica ATH-AD500X (open type), or their personal headphones if features were comparable.

Table 1. Speech and music material used in the experiments.

Set	Speech Excerpt	Reader	Author	Music Excerpt	Composer
1	Under a cherry tree	Female	*M. Kajii*	Après un réve	*G. Fauré*
2	Do not stand at my grave and weep	Male	*M. Arai*	Nausicaa requiem	*J. Hisaishi*
3	Lemon	Female	*M. Sada*	Always with me	*Y. Kimura*
4	Not losing to the rain	Male	*K. Miyazawa*	Castle in the sky	*J. Hisaishi*

3.2. Procedure

Each experimental session took one hour. First, there was an eight-minute preparation that consisted of reading an explanation of the game purpose, how to play, and a short demonstration. After that, participants gave their consent and other information such as their musical experience and hearing levels. They tried Set 1 as training for 15 min; then they proceeded to play with sets 2–4 for 35 min. We encouraged but did not require them to play with all modes (sound material) in a set, and with no specific order. Finally, participants gave post-descriptions of their preference regarding the sound material and the experienced difficulty.

The experiment took place in a classroom setting with 2–6 students at a time. Each participant had a tablet and headphones. They were instructed to adjust the sound volume to a comfortable level and were free to readjust this setting when necessary. Also, they were allowed to take breaks if needed. They received a token gratitude of about USD10 for their participation.

4. Results

We describe the results of the experiments by comparing the three participant groups in subjective evaluations and the way participants played the game. For comparison of the groups, analysis of variance (ANOVA) was used. Post hoc analyses were performed with the Tukey–Kramer procedure on the independent observations, with the level of statistical significance set at $p < 0.05$.

4.1. Games

Game sessions are divided into the number of sessions played and game achievement. Data were extracted from the log files.

4.1.1. Number of Sessions

The numbers of game sessions during the experiment for each participant group were $HI = 81$, $NEX = 79$, and $EXP = 104$. For *HI*, the ratio of playing Music was larger, the ratio of playing Mixed was smaller, while for Speech there were no differences between the groups. The number of game sessions and ratios of the specific materials are shown in Table 2.

4.1.2. Achievement of Games

Sessions could be ended in three ways: an achieved completed puzzle, the "give up option", and the tablet's back button. A session was considered "achieved" only when the order of sound objects, the EQ, and the pitch were correct—thresholds for correctness can be adjusted in the settings (see Figure 1d). A "give up" exit is recorded when a user presses this action button. When pressing the system's back button, home button or power, the logged action is *exit by back button*; this is an unwanted action but not easily circumvented. It was also observed to happen by mistake.

Table 2(c) shows the ratio of achievement by each participant group. For all three modes (Speech, Music and Mixed), achievement by *HI* was less than for the other groups. In all modes, ANOVA showed significant differences between participant groups. The post hoc test shows that there were differences between *HI* and the other participant groups with all modes.

Table 2. Overview of game sessions with (**a**) the total number of sessions played for each of the three groups hearing impaired (HI), and normal hearing with low (NEX) and high (EXP) music experience, (**b**) the ratio of material played for each mode, (**c**) the ratio of completed sessions for each mode, and (**d**) the ratio of puzzles that were evaluated as easy (ratio = 1) for each participant group and mode.

			HI	NEX	EXP
(a)	*Number of game sessions*	Total	81	79	104
(b)	*Game sessions ratio*	Speech	0.37	0.37	0.34
		Music	0.40	0.32	0.34
		Mixed	0.23	0.32	0.33
(c)	*Achieved sessions ratio*	Speech	0.35	0.93	1.00
		Music	0.25	0.83	0.85
		Mixed	0.23	0.95	1.00
(d)	*Evaluation of difficulty*	Speech	0.53	0.83	0.92
		Music	0.44	0.24	0.66
		Mixed	0.47	0.91	0.91

4.2. Subjective Evaluation

The subjective evaluations were collected from questionnaire data both during and after the sessions. After each game, its difficulty was rated on a five-level scale. A control question queried about which sound material that was heard to confirm that the sounds were played properly. We recorded no errors in determining the mode for the hearing groups, but some for *HI*—this was also an anticipated result and does not imply errors in the playback.

4.2.1. Fondness

The post-activity questionnaire asked participants about the game in terms of enjoyment. The results of rated fondness derived from "how entertaining" the Music Puzzle is and preferences towards material are shown in Figure 3a,b respectively. Questions were answered as follows:

How entertaining was the game? Hearing participants with low reported musical activity (*NEX*) gave the lowest evaluation. Overall, 3 of 12 found it to be "boring" and only half found it to be entertaining. In the other groups, 28 out of 36 gave ratings that the game was entertaining, and only 1 person found it boring.

Which material do you like the best? The three groups showed a difference in their preferred sound mode. The preference for the Music mode was greatest in the *HI* group, while *NEX* clearly preferred the Mixed mode. None of the groups rated the Speech condition highly.

Would you use the game if it was free? With similar distribution across all groups, 69% answered they would use the Music Puzzle if it was free. The game is not currently available in the Android or iOS app stores, and there are no plans to charge for use when it is publicly released.

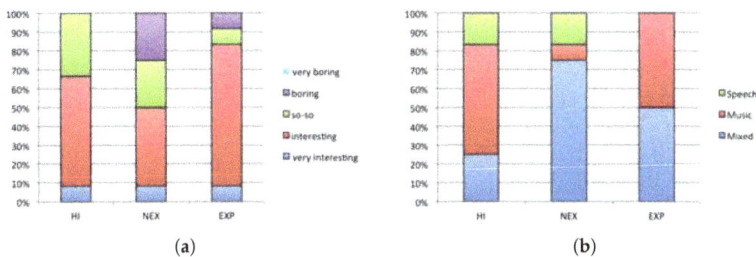

(a)   (b)

Figure 3. Ratings of fondness of playing the game. (**a**) Answers to "how entertaining was the Music Puzzle?" (**b**) Preferred material by each participant group.

4.2.2. Difficulties

We asked participants to rate the difficulty from very difficult to very easy using five levels for each game session. In Table 2(d), the ratios of ratings with low difficulty (the three lowest levels) are elicited for each group and mode. For hearing participants *NEX* and *EXP*, both Mixed and Speech modes were considered easy, while *HI* found the Speech mode to be hard. *NEX* and *EXP* differed in the rating of music stimuli, where *NEX* even rated the Music mode as harder than the *HI* group did. *HI* had similar ratings for the Speech and Music modes, but differed in their rating of the Mixed mode.

The differences between the participant groups were shown by ANOVA in the modes Speech and Mixed, where all *p*-values were small ($p \ll 0.01$). The multiple comparison showed differences between *HI-NEX* and *HI-EXP* in both modes. As described above, two groups were formed for the Speech and Mixed modes: one consisting of *HI* and the other consisting of hearing participants (*NEX* and *EXP*). Table 3 rows (a)–(c) summarize the results of multiple comparison on the number of performed game sessions, the ratio of achieved game sessions, and the subjective evaluation of difficulties, by each mode.

Table 3. Significant differences are shown using asterisks ($p < 0.05$ *, $p < 0.01$ **). The differences between participant groups in (**a**) the ratio of game sessions; (**b**) the ratio of achieved sessions; (**c**) the subjective evaluation of game difficulty; (**d**) clicks on sound objects; (**e**) clicks on "Play Solution"; (**f**) clicks on "Play the current"; (**g**) clicks on "How did I do?"; (**h**) game duration for completing one puzzle; and (**i**) duration per click for a game session measured as inter-onset intervals (IOI).

			HI– NEX	HI– EXP	NEX– EXP
(a)	*Ratio of game sessions*	Speech			
		Music			
		Mixed		*	
(b)	*Ratio of achieved game sessions*	Speech	**	**	
		Music	**	**	
		Mixed	**	**	
(c)	*Difficulty of game*	Speech	**	**	
		Music			
		Mixed	**	**	
(d)	*Clicks on sound objects*	Speech		*	
		Music			
		Mixed			
(e)	*Clicks on Play solution*	Speech	**	**	
		Music	**	**	
		Mixed	**	**	
(f)	*Clicks on Play the current*	Speech			
		Music			
		Mixed			
(g)	*Clicks on "How did I do?" (evaluation)*	Speech		*	
		Music	*	**	
		Mixed	*	*	
(h)	*Game duration*	Speech	**	**	
		Music			
		Mixed	**	**	
(i)	*IOI of clicks*	Speech	**	**	
		Music	**	**	
		Mixed	**	**	

4.3. Interaction

The way participants played Music Puzzle can be described in terms of clicks on sound objects and buttons. We recorded all screen interaction, including the sound objects and interface buttons described earlier. In the following, we analyze game interaction and playing strategies using the number of clicks and time measurements.

4.3.1. Pitch and EQ Cheating

When a user clicks "Pitch Cheat" or "EQ Cheat", pitch alterations and filtering, respectively, are corrected for all sound objects. Since a cheat is persistent and thus available only once in a game session, we looked at the ratio of using cheat buttons in all sessions for each type of material. Figure 4a,b shows these ratios; *HI* used cheat buttons in about the half of the game sessions, and the other groups used these functions more sparingly. There was little difference between the two cheat modes: as it appears, *HI* in particular tended to use both buttons when "cheating".

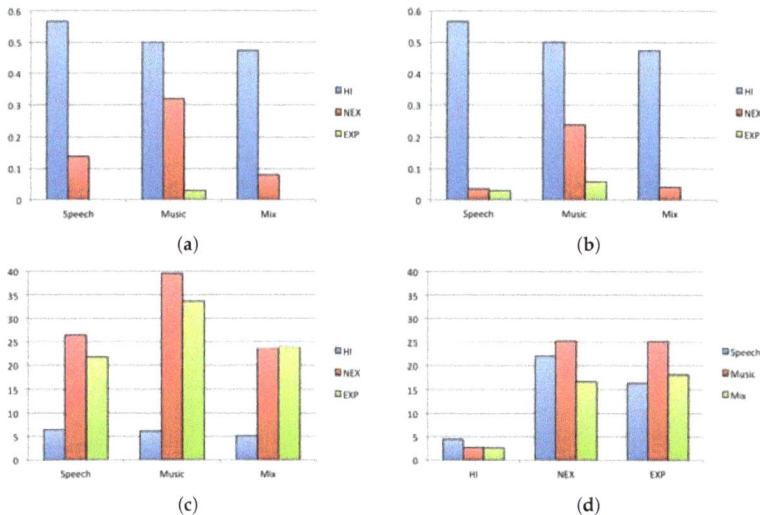

Figure 4. User interaction during play. The ratios of using cheat buttons for correcting all altered pitches (**a**); and filtering (EQ) (**b**); and the number of clicks needed to change pitch (**c**) and EQ (**d**).

4.3.2. Adjusting Pitch and EQ

When a user decides to adjust pitch or EQ (Figure 1c), clicking the radio buttons will play the available variations. The only way to find the correct is by listening, thus the number of clicks can tell how many trials are needed to identify the unaltered one. Figure 4c,d show these numbers, and seemingly, the *HI* perform better than the other participant groups. As we will discuss, this is, however, a consequence of the problems of discriminating timbre and pitch differences among *HI* which leads to activating the cheats. Note that for pitch there are five options, while EQ only has three; this is observable in the figure.

4.3.3. Interaction with Sound Objects and Buttons

The interaction can be divided into compulsory and optional actions. It is necessary to click the sound objects and listen in order to arrange them in the correct order. One can click two or more objects in succession to play a sequence. Furthermore, the evaluation "How did I do?" must be clicked at least once for completing a game. However, players do not need to listen to the target (solution) sound or the current sound during a game session. The function buttons "Play Solution", "Play the current" and "How did I do?" can be clicked any number of times. The plots in Figure 5 show the average number of clicks on sound objects and function buttons.

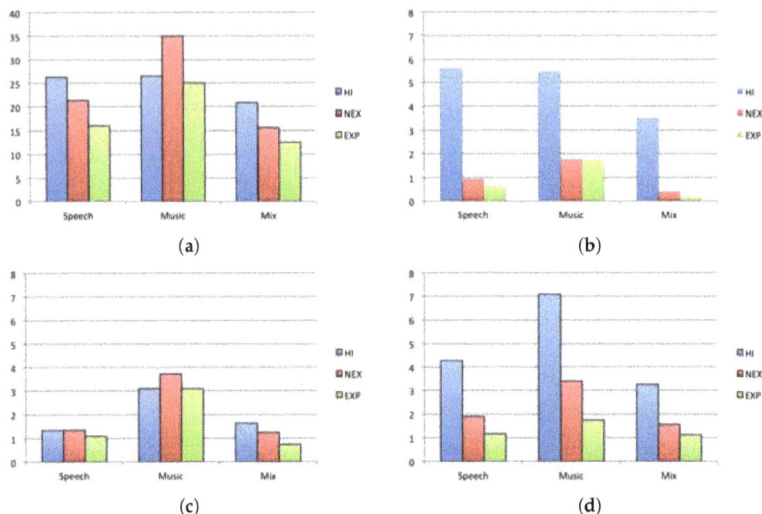

Figure 5. Comparison of the players' interaction during a game session. Number of clicks on (**a**) sound objects; (**b**) "Play Solution" replay target button; (**c**) "Play the Current"; and (**d**) the "How did I do?" evaluation button. Note that the scale is different in (**a**).

Table 3 rows (d)–(g) show the results of multiple comparison to see the differences between participant groups on button clicks. Similar to the difficulties in Table 3(c), the *HI* group is in contrast to the other groups in the tendency to click on "Play Solution". *HI* evaluated the games ("How did I do?") more often than hearing groups. There were no differences for "Play the current", and the gameplay did not require one to click it.

4.3.4. Duration and Speed

Duration is defined here as the time it takes to finish a session, whether the session was successfully achieved or not. We also calculate the time between clicks, or inter-onset intervals (IOI) of clicks. Here we include subsequent clicks of either object clicks, pitch or EQ changes, play solutions, and playing the current. Speed is a reciprocal of the duration per click and represents the swiftness in interaction and gameplay. Figure 6a,b show game duration and IOI by each participant group and for each mode.

Differences between participant groups were found in the duration of game materials Speech and Mixed, and there were also differences in IOI found for all modes. It should be noted that calculating IOI in this kind of game is not trivial because some actions will necessitate longer intervals than others, and the results must be interpreted with some caution. Table 3(h),(i) shows the results of multiple comparison to find the differences between participant groups and material for duration and IOI. The differences between modes on time measurements were $p \ll 0.01$ where the Music mode was different from both the Speech and Mixed modes.

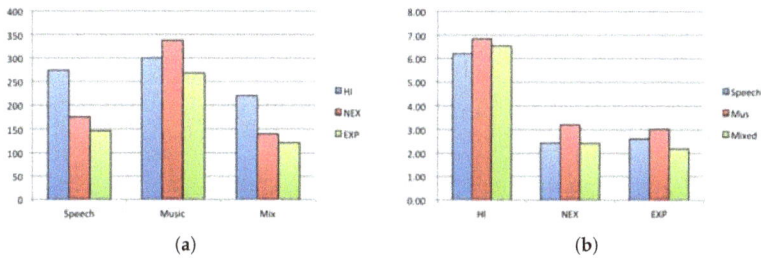

Figure 6. Duration and inter-onset intervals (IOI) in seconds. Durations for (**a**) game sessions; and for (**b**) IOI of clicks (one click to the next).

4.4. Summary

A summary of the effects and found differences that are mentioned in this section is shown in Table 4 (Table 3 shows the results, while Table 4 shows the explanation of the grouping). We consider hearing acuity, level of music experience, and also language proficiency based on the assumption that the hearing-impaired generally show language disorders [3].

Differences in hearing acuity are evident for the *HI* group, and in music experience for the *EXP* group. For language proficiency, *HI* differs from both *NEX* and *EXP*.

Table 4. Summary of significant differences between participant groups. The differences between participant groups concerning the effects of hearing loss, music experiences, and language proficiency.

Effects	HI– NEX	HI– EXP	NEX– EXP
Hearing loss	✓	✓	
Music experiences	✓		✓
Speech and language	✓	✓	

5. Discussion

Through comparing the results of the three participant groups, we discuss the effects hearing acuity, music experience, and language proficiency have on the outcome of playing Music Puzzle. We also consider how people with and without hearing loss enjoy playing the game.

5.1. The Effect of Hearing Acuity

Since hearing loss affects the proficiency of playing Music Puzzle, it follows that the experiment would reveal differences between the *HI* participant group and the others (*NEX* and *EXP*): these are summarized in the first row of Table 4, and more details can be found in Table 3. The following results from multiple comparisons show differences due to hearing loss:

- Lower ratio of the performed Mixed mode (Table 2(b)).
- Lower ratio of achieved sessions for the Speech mode (Table 3(b)).
- Higher ratio of clicks on "Pitch cheat" for Speech and Mixed modes (Figure 4a).
- Higher ratio of clicks on "EQ cheat" for Speech and Mixed modes (Figure 4b).
- Fewer attempts at "change pitch" in all modes (Figure 4c).
- Fewer attempts at "change EQ" in all modes (Figure 4d).
- Higher number of clicks on "How did I do?" for Music and Mixed modes (Table 3(g)).
- Longer inter-onset interval (IOI) of clicks for all modes (Table 3(i)).

These findings lead to considering the following possible interpretations:

Hearing-impairment introduces difficulties in extracting useful cues from both music and speech played simultaneously, and from "listening to" speech. Differences between people with and without hearing loss were found for the Mixed condition in the ratio of achieved sessions. This implies that the overlapping of sounds in speech and music makes the game harder to solve for people with hearing loss, or that those with normal hearing can better utilize the additional cues. In the ratio of the performed modes, differences between people with and without hearing loss were found in Speech. In other words, speech, more than music, is difficult for *HI*. Language proficiency will thus be helpful, but the game's puzzles are still not easily solved by constructing lexical meaning from the fragments; these fragments are short, and the poems relatively intricate.

Hearing-impairment introduces difficulties in distinguishing pitch alterations and filtering. Cheat buttons were used more often by people with hearing loss, and they experienced a greater challenge in correcting pitch and EQ. Cheat buttons were also used by hearing participants when they played with Music material. This implies that remembering nuances of pitch and EQ adjustments in music was also difficult for people without hearing loss.

Persons with hearing-impairment take longer to process a heard sound. The study showed that people with hearing loss wait longer after clicking the sound object button or other buttons before clicking a new one. We know that hearing acuity does not correspond with problems of interacting with computers [64], thus the reasons for timing differences could be: (1) they listen to the whole sound from a sounding object or the effect of other buttons, while people without hearing loss only listen to the start, or listen only a certain extent; (2) they listen to sound then think for a while; or (3) the time to start processing sound could be later for *HI*. Considering that the lengths of the fragments correspond to the intervals recorded by *NEX* and *EXP* (which means these groups would click the next sound object without hesitation), (1) is a less plausible explanation. We conclude that *HI* adopt a focused listening strategy which involves longer time for processing the played sounds.

5.2. The Effect of Music Experiences

We found no distinct differences that could be explained only by music experience in this experiment, as seen in the right column comparing *NEX* and *EXP* in Table 3, but in the next section we will discuss effects that appeared in combination with speech material and language. We should remember that the *less* musically-experienced hearing group in this experiment is comprised of typical university students who still have a comparatively high exposure to music.

As was described in our previous paper [65], one particular individual with hearing loss who had a lot of musical activities was able to achieve all the sessions she tried in that experiment. Thus, introducing a group of people with hearing loss with a lot of musical activities may also yield different results.

5.3. The Effect of Language and Speech

While all the participants were native Japanese speakers, *HI* did not have equivalent language fluency in listening to speech (cf. [3]), and we can thus make two clusters of *HI* and *NEX/EXP*. Effects of language can be found in the differences shown in the row titled "Speech and language" in Table 4. One effect is found in the number of clicks on sound objects in playing with Speech material as shown in Table 3(d). This shows the difficulties of using language cues in solving the puzzle, for instance from remembering fragments of a spoken sentence. Other effects are found in the following cases when playing Speech and Mixed materials:

- Subjective evaluation of difficulties (Table 3(c)).
- Time to complete a game session (Table 3(h)).

These show no differences for music listening in the two clusters; while speech will be more problematic for hearing-impaired who can use fewer cues in constructing a meaningful whole, solving for music is comparable for *HI* and *NEX/EXP*. Even the third row in Table 4 shows no differences between *NEX* and *EXP*. Differences were found with *HI* for the duration of game completion when playing in the Mixed and Speech modes (Table 3(h)). However, *NEX* took longer to play the Music mode than *EXP*, rating it more difficult (Table 2(d)). They also liked it less (Figure 3). This implies that musical experiences affect play after all.

5.4. General Discussion

The game was generally well received, as seen in Figure 3a; in fact, regardless of hearing loss, 70% claim they would use it—although this is a speculative measurement, the proportionality between groups is illustrious. As could be expected, the results of this experiment show that people with hearing loss could not complete puzzles to the same extent as the hearing control groups. However, they enjoyed the puzzles and liked the Music mode best among the three sound materials. This implies that people with hearing loss have good motivations for music listening through the game. It is worth noting that the experiment allowed the participants to choose modes quite freely, and that an alternative test design would highlight other issues in addition to preference.

A related and not as expected finding was that *NEX* and *EXP* rated the Music mode as rather difficult. *NEX* did not prefer the Music mode, while *EXP* did. Could it be that this attitude in *NEX* was caused by believing that there were external expectations about understanding music that they needed to fulfill? Possibly they show anxiety in terms of making mistakes which are not seen as clearly for *EXP* who would likely have a more analytical approach towards music listening.

The number of clicks for adjusting pitch was much higher for the Music condition than the others. This should mean that pitch manipulations in the Speech condition were more easily detected (our material had both male and female readers). Because Japanese is a pitch-accent language [66], prosodic cues are probably used in solving the puzzle; this would need to be investigated further through speech material with fewer pitch variations.

As seen, the *HI* spent much more time on pitch and EQ adjustments for all conditions. The most probable reason that we can see is that the task was just too hard, and they simply gave up and used cheat buttons. During the experiment design and set-up phase, it became very clear that our initial values for manipulations were far too subtle. The implemented pitch shifts of 5 and 10 semitones, and the filter thresholds of 2000 Hz for low-pass and 500 Hz for high-pass are to a normal hearing person easily identifiable, except possibly for pitch in music pieces with solo instruments. Even the maximum possible number of generated objects was reduced from around 18–20 to around 6–7 pieces.

The description of the waiting time between two clicks relates to the way people with hearing loss are playing the game. Currently, though it is not clear what the reason is for them to wait longer after clicks, this could depend on whether they remember any elements of music. If this is so, then it is helpful to understand *what* they remember in helping to enjoy music more.

5.5. Further Development

In its current gameplay design and aesthetics, the game leaves much to be desired in order to compete with the attractiveness of trending games on the market. However, the functionality worked according to planned use, and the material was sufficient for the scope of testing. From here, as we now consider the concept to be verified to be beneficial as a training tool, the Music Puzzle will be subject to changes in: (1) graphical design and interaction; (2) game types and difficulty progression; (3) sound material and transformations; (4) logging and social connectivity; (5) targeted training recommendations; and likely (6) a platform change to Web Audio (see http://www.w3.org/TR/webaudio/).

6. Conclusions

The *Music Puzzle*—an audio-based puzzle game—had the purpose of giving persons with hearing loss an effective and entertaining alternative exposure to focused listening. Research has shown that focused listening is beneficial for training listening ability, and also language development. The game was tailored for hearing impairments, but was designed to be engaging for everyone. The game does not require much from the user in terms of previous training or gaming experiences, and was designed to be inclusive for the hearing-impaired (the game has since then been developed with alternative interfaces and for other purposes not reported here). Despite the fact that people with hearing loss could not complete nearly as many started games as their hearing peers, the task was still not found to be too difficult and it can give new prospects for voluntary, focused listening to music or other sounds.

Care should be taken in selecting sound materials and in designing the gameplay to accommodate differences in processing time for sounds. Although speech and language are important objectives for training, music was both found to have appreciated qualities and was preferred by the target user group. Music was also found to be more challenging as a game task in general.

Many music training programs require instruction from a professional in an equipped, dedicated space, but the Music Puzzle is a game designed to be used at leisure. Any persons with access to commonplace technology such as smartphones or tablets can play at any place alone at any time. This way, opportunities for listening to music attentively increase with no special resources: the game can be acquired and used for free, expanded upon, and used for different purposes. With development of its design, the Music Puzzle is a conceptually different and attractive audio game.

Supplementary Materials: Supplementary materials can be accessed at: http://www.mdpi.com/2076-3417/7/12/1278/s1.

Acknowledgments: This work was supported by JSPS KAKENHI Grant Number 26282001.

Author Contributions: K.F.H. and R.H. contributed equally in conceiving, designing and performing the experiments, analyzing the data, and writing the paper.

Conflicts of Interest: The authors declare no conflict of interest.

Abbreviations

The following abbreviations are used in this manuscript:

HI	Japanese hearing-impaired participants
NEX	Japanese hearing participants with low musical experience
EXP	Japanese hearing participants with high musical experience
IOI	Inter-onset interval
EQ	Equalisation

References

1. Torppa, R.; Huotilainen, M.; Leminen, M.; Lipsanen, J.; Tervaniemi, M. Interplay between singing and cortical processing of music: A longitudinal study in children with cochlear implants. *Front. Psychol.* **2014**, *5*, 1389.
2. Looi, V.; Gfeller, K.; Driscoll, V.D. Music appreciation and training for cochlear implant recipients: A review. *Semin. Hear.* **2012**, *33*, 307–334.
3. Delage, H.; Tuller, L. Language Development and Mild-to-Moderate Hearing Loss: Does Language Normalize With Age? *J. Speech Lang. Hear. Res.* **2007**, *50*, 1300.
4. Gaver, W.W. What in the World Do We Hear?: An Ecological Approach to Auditory Event Perception. *Ecol. Psychol.* **1993**, *5*, 1–29.
5. Tremblay, K.; Kraus, N.; McGee, T.; Ponton, C.; Otis, B. Central auditory plasticity: changes in the N1-P2 complex after speech-sound training. *Ear Hear.* **2001**, *22*, 79–90.
6. Schäfer, T.; Sedlmeier, P.; Städtler, C.; Huron, D. The psychological functions of music listening. *Front. Psychol.* **2013**, *4*, doi:10.3389/fpsyg.2013.00511.
7. Song, J.H.; Skoe, E.; Banai, K.; Kraus, N. Training to Improve Hearing Speech in Noise: Biological Mechanisms. *Cereb. Cortex* **2011**, *22*, 1180–1190.
8. Hansen, K.F.; Hiraga, R.; Li, Z.; Wang, H. Music Puzzle: An audio-based computer game that inspires to train listening abilities. In *Advances in Computer Entertainment*; *Lecture Notes in Computer Science*;

Reidsma, D., Katayose, H., Nijholt, A., Eds.; Springer International Publishing: New York, NY, USA, 2013; Volume 8253, pp. 540–543.

9. Hiraga, R.; Hansen, K.F. Sound preferences of persons with hearing loss playing an audio-based computer game. In Proceedings of the 3rd ACM International Workshop on Interactive Multimedia on Mobile & Portable Devices, Barcelona, Spain, 22 October 2013; ACM: New York, NY, USA, 2013; pp. 25–30.

10. Venail, F.; Camilleri, M.; Lorenzi, A. What's a Hearing Impairment? A tinnitus? Available online: http://www.cochlea.org/en/impairment (accessed on 30 November 2017).

11. McCullagh, J. Auditory Acuity. In *Encyclopedia of Autism Spectrum Disorders*; Springer: New York, NY, USA, 2013; p. 312.

12. Glennie, E. Teach the World to Listen. Available online: http://www.evelyn.co.uk (accessed on 30 November 2017).

13. Whittaker, P. Dr. Paul Whittaker OBE. Available online: http://www.paulwhittaker.org.uk/ (accessed on 30 November 2017).

14. Lane, D. Pianist & Artistic Director. Available online: https://britishmusiccollection.org.uk/article/artistic-director-music-and-deaf-danny-lane (accessed on 30 November 2017).

15. Gallaudet Dance Company. Available online: http://www.gallaudet.edu/department-of-art-communication-and-theatre/gallaudet-dance-company (accessed on 30 November 2017).

16. Enriching Lives through Music. Available online: http://www.matd.org.uk (accessed on 30 November 2017).

17. Music for Little Ears. Available online: http://hear-me-now.org/preschool-music-class/ (accessed on 30 November 2017).

18. Feel the Music. Available online: http://mahlerchamber.com/learning/education-and-outreach/feel-the-music-programme (accessed on 30 November 2017).

19. NDCS Resource: How to Make Music Activities Accessible for Deaf Children and Young People. Available online: http://www.ndcs.org.uk/document.rm?id=8830 (accessed on 30 November 2017).

20. Hash, P.M. Teaching instrumental music to deaf and hard of hearing students. *Res. Issues Music Educ.* **2003**, *1*, 1–8.

21. Darrow, A.A. The role of music in deaf culture: Deaf students' perception of emotion in music. *J. Music Ther.* **2006**, *43*, 2–15.

22. Pereira, C.S.; Teixeira, J.; Figueiredo, P.; Xavier, J.; Castro, S.L.; Brattico, E. Music and emotions in the brain: Familiarity matters. *PLoS ONE* **2011**, *6*, e27241.

23. Fulford, R.; Ginsborg, J.; Goldbart, J. Learning not to listen: The experiences of musicians with hearing impairments. *Music Educ. Res.* **2011**, *13*, 447–464.

24. Brandt, A.K.; Slevc, R.; Gebrian, M. Music and early language acquisition. *Front. Psychol.* **2012**, *3*, doi:10.3389/fpsyg.2012.00327.

25. Kirschner, S.; Tomasello, M. Joint music making promotes prosocial behavior in 4-year-old children. *Evolut. Hum. Behav.* **2016**, *31*, 354–364.

26. Fujioka, T.; Ross, B.; Kakigi, R.; Pantev, C.; Trainor, L.J. One year of musical training affects development of auditory cortical-evoked fields in young children. *Brain* **2006**, *129*, 2593–2608.

27. Kraus, N.; Chandrasekaran, B. Music training for the development of auditory skills. *Nat. Rev. Neurosci.* **2010**, *11*, 599–605.

28. Shahin, A.J.; Roberts, L.E.; Chau, W.; Trainor, L.J.; Miller, L.M. Music training leads to the development of timbre-specific gamma band activity. *Neuroimage* **2008**, *41*, 113–122.

29. Strait, D.L.; Slater, J.; O'Connell, S.; Kraus, N. Music training relates to the development of neural mechanisms of selective auditory attention. *Dev. Cognit. Neurosci.* **2015**, *12*, 94–104.

30. Tierney, A.T.; Krizman, J.; Kraus, N. Music training alters the course of adolescent auditory development. *Proc. Natl. Acad. Sci. USA* **2015**, *112*, 10062–10067.

31. Jäncke, L.; Sandmann, P. Music listening while you learn: No influence of background music on verbal learning. *Behav. Brain Funct.* **2010**, *6*, 3.

32. Barton, C. Music and literacy development in young children with hearing loss: A duet. *Imagine* **2011**, *2*, 53–55.

33. Gfeller, K.; Driscoll, V.; Kenworthy, M.; Van Voorst, T. Music therapy for preschool cochlear implant recipients. *Music Ther. Perspect.* **2011**, *29*, 39–49.

34. Chasin, M.; Hockley, N.S. Some characteristics of amplified music through hearing aids. *Hear. Res.* **2014**, *308*, 2–12.

35. Chasin, M.; Russo, F.A. Hearing aids and music. *Trends Amplif.* **2004**, *8*, 35–47.

36. Hopyan, T.; Peretz, I.; Chan, L.P.; Papsin, B.C.; Gordon, K.A. Children using cochlear implants capitalize on acoustical hearing for music perception. *Front. Psychol.* **2012**, *3*, 425.

37. Limb, C.J.; Roy, A.T. Technological, biological, and acoustical constraints to music perception in cochlear implant users. *Hear. Res.* **2014**, *308*, 13–26.

38. Roy, A.T.; Jiradejvong, P.; Carver, C.; Limb, C.J. Assessment of sound quality perception in cochlear implant users during music listening. *Otol. Neurotol.* **2012**, *33*, 319–327.

39. Nakata, T.; Trehub, S.E.; Kanda, Y. Effect of cochlear implants on children's perception and production of speech prosody. *J. Acoust. Soc. Am.* **2012**, *131*, 1307.

40. Vongpaisal, T.; Trehub, S.E.; Schellenberg, E.G. Song recognition by children and adolescents with cochlear implants. *J. Speech Lang. Hear. Res.* **2006**, *49*, 1091–1103.

41. Sucher, C.M.; McDermott, H.J. Pitch ranking of complex tones by normally hearing subjects and cochlear implant users. *Hear. Res.* **2007**, *230*, 80–87.

42. Kong, Y.Y.; Cruz, R.; Jones, J.A.; Zeng, F.G. Music perception with temporal cues in acoustic and electric hearing. *Ear Hear.* **2004**, *25*, 173–185.

43. Caruso, V.C.; Balaban, E. Pitch and timbre interfere when both are parametrically varied. *PLoS ONE* **2014**, *9*, e87065.

44. Macherey, O.; Delpierre, A. Perception of musical timbre by cochlear implant listeners: a multidimensional scaling study. *Ear Hear.* **2013**, *34*, 426–436.

45. Rochette, F.; Moussard, A.; Bigand, E. Music lessons improve auditory perceptual and cognitive performance in deaf children. *Front. Hum. Neurosci.* **2014**, *8*, 488.

46. Petersen, B.; Weed, E.; Sandmann, P.; Brattico, E.; Hansen, M.; Sørensen, S.D.; Vuust, P. Brain responses to musical feature changes in adolescent cochlear implant users. *Front. Hum. Neurosci.* **2015**, *9*, 7.

47. Zhang, T.; Dorman, M.F.; Fu, Q.J.; Spahr, A.J. Auditory training in patients with unilateral cochlear implant and contralateral acoustic stimulation. *Ear Hear.* **2013**, *33*, e70–e79.

48. Wouters, P.; van Nimwegen, C.; van Oostendorp, H.; van der Spek, E.D. A meta-analysis of the cognitive and motivational effects of serious games. *J. Educ. Psychol.* **2013**, *105*, 249–265.

49. Boyle, E.A.; Hainey, T.; Connolly, T.M.; Gray, G.; Earp, J.; Ott, M.; Lim, T.; Ninaus, M.; Ribeiro, C.; Pereira, J. An update to the systematic literature review of empirical evidence of the impacts and outcomes of computer games and serious games. *Comput. Educ.* **2016**, *94*, 178–192.

50. Andersson, U.; Josefsson, P.; Pareto, L. Challenges in designing virtual environments training social skills for children with autism. *Int. J. Disabil. Hum. Dev.* **2006**, *5*, 105–111.

51. Boyd, L.E.; Ringland, K.E.; Haimson, O.L.; Fernandez, H.; Bistarkey, M.; Hayes, G.R. Evaluating a Collaborative iPad Game's Impact on Social Relationships for Children with Autism Spectrum Disorder. *ACM Trans. Access. Comput.* **2015**, *7*, 1–18.

52. Friberg, J.; Gärdenfors, D. Audio games. In Proceedings of the 2004 ACM SIGCHI International Conference on Advances in Computer Entertainment Technology—ACE'04, Singapore, 3–4 June 2004; ACM Press: New York, NY, USA, 2004.

53. Rovithis, E. A classification of audio-based games in terms of sonic gameplay and the introduction of the audio-role-playing-game: Kronos. In Proceedings of the 7th Audio Mostly Conference on A Conference on Interaction with Sound—AM'12, Corfu, Greece, 26–28 September 2012; ACM Press: New York, NY, USA, 2012.

54. Carvalho, J.; Guerreiro, T.; Duarte, L.; Carriço, L. Audio-based puzzle gaming for blind people. In Proceedings of the Mobile Accessibility Workshop at MobileHCI (MOBACC), Linz, Austria, 11–13 July 2012.

55. Brieger, S. Sound Hunter: Developing a Navigational HRTF-Based Audio Game for People with Visual Impairments. In *Proceedings of the Sound and Music Computing Conference*; Bresin, R., Ed.; Logos Verlag: Berlin, Germany, 2013; pp. 245–252.

56. Jaime, J.; Barbancho, I.; Urdiales, C.; Tardón, L.J.; Barbancho, A.M. A new multiformat rhythm game for music tutoring. *Multimed. Tools Appl.* **2015**, *75*, 4349–4362.

57. Baratè, A.; Ludovico, L.A. Serious games for music education. A mobile application to learn clef placement on the stave. In Proceedings of the International Conference on Computer Supported Education (CSEDU), Aachen, Germany, 6–8 May 2013; SciTe Press: Setubal, Portugal. 2013; pp. 234–237.

58. Respino, J.; Juana, S.J.; Solamo, M.; Feria, R. Pitch paradise: A mobile game as an educational tool for music. In Proceedings of the 2011 9th International Conference on Education and Information Systems, Technologies and Applications (EISTA), Orlando, FL, USA, 19–22 July 2011.

59. Zhou, Y.; Sim, K.C.; Tan, P.; Wang, Y. MOGAT. In Proceedings of the 20th ACM International Conference on Multimedia —MM'12, Nara, Japan, 29 October–2 November 2012; ACM Press: New York, NY, USA, 2012.

60. Duan, Z.; Gupta, C.; Percival, G.; Grunberg, D.; Wang, Y. SECCIMA: Singing and ear training for children with cochlear implants via a mobile application. In Proceedings of the 14th Sound and Music Computing Conference, Espoo, Finland, 5–8 July 2017, pp. 200–207.

61. Brinkmann, P.; Kirn, P.; Lawler, R.; McCormick, C.; Roth, M.; Steiner, H.C. Embedding Pure Data with libpd. In Proceedings of the Pure Data Convention, Weimar, Norway, 30 May–1 June 2011; Volume 291.

62. How to Read an Audiogram and Determine Degrees of Hearing Loss. Available onlne: http://www.nationalhearingtest.org/wordpress/?p=786 (accessed on 30 November 2017).

63. Bauman, N. Understanding the Difference between Sound Pressure Level (SPL) and Hearing Level (HL) in Measuring Hearing Loss. Available onlne: http://hearinglosshelp.com/blog/understanding-the-difference-between-sound-pressure-level-spl-and-hearing-level-hl-in-measuring-hearing-loss/ (accessed on 30 November 2017).

64. Maiorana-Basas, M.; Pagliaro, C.M. Technology Use Among Adults Who Are Deaf and Hard of Hearing: A National Survey. *J. Deaf Stud. Deaf Educ.* **2014**, *19*, 400–410.

65. Hiraga, R.; Hansen, K.F.; Kano, N.; Matsubara, M.; Terasawa, H.; Tabuchi, K. Music perception of hearing-impaired persons with focus on one test subject. In Proceedings of the 2015 IEEE International Conference on Systems, Man, and Cybernetics, Kowloon Tong, Hong Kong, China, 9–12 October 2015; pp. 2407–2412.

66. Tsujimura, N. *An Introduction to Japanese Linguistics*; John Wiley & Sons: New York, NY, USA, 2013.

applied
sciences

MDPI

Article

Optimization of Virtual Loudspeakers for Spatial Room Acoustics Reproduction with Headphones

Otto Puomio *, Jukka Pätynen and Tapio Lokki

Department of Computer Science, Aalto University School of Science, P.O. Box 13300, 00076 Aalto, Finland;
jukka.patynen@aalto.fi (J.P.); tapio.lokki@aalto.fi (T.L.)
* Correspondence: otto.puomio@aalto.fi; Tel.: +358-45-678-5213

Received: 31 October 2017; Accepted: 5 December 2017; Published: 9 December 2017

Abstract: The use of headphones in reproducing spatial sound is becoming more and more popular. For instance, virtual reality applications often use head-tracking to keep the binaurally reproduced auditory environment stable and to improve externalization. Here, we study one spatial sound reproduction method over headphones, in particular the positioning of the virtual loudspeakers. The paper presents an algorithm that optimizes the positioning of virtual reproduction loudspeakers to reduce the computational cost in head-tracked real-time rendering. The listening test results suggest that listeners could discriminate the optimized loudspeaker arrays for renderings that reproduced a relatively simple acoustic conditions, but optimized array was not significantly different from equally spaced array for a reproduction of a more complex case. Moreover, the optimization seems to change the perceived openness and timbre, according to the verbal feedback of the test subjects.

Keywords: Spatial audio; Spatial sound reproduction; SDM; Headphone reproduction; Optimization

1. Introduction

Spatial audio aims to reproduce a believable illusion for a listener being in a real acoustic space by electronic means [1]. Dozens of different ways exist to record or artificially create the spatial sound signals, which are further reproduced with an array of loudspeakers or with headphones [2]. For good spatial resolution, a high number of reproduction loudspeakers are often used in research facilities or in special venues, but such arrays are impractical in domestic or other daily listening environments. Therefore, the headphone reproduction of spatial sound is gaining interest. Commonly, the headphone-based spatial sound is implemented by virtualizing the reproduction loudspeaker array with the Head-Related Transfer Functions (HRTFs) [3]. In essence, HRTFs are applied to virtually position the sound sources around the listener. The resulting binaural rendering can sound very convincing at the best, but, for some users, the sound is localized inside the head. To achieve better externalization, head-tracking devices are often used to compensate the movement of the listener's head and to keep the reproduced auditory environment stable [4,5].

This work studies the virtual loudspeaker positioning in headphone-based spatial sound reproduction. The sound rendering is based on the Spatial Decomposition Method (SDM) [6], which in essence analyzes directional information in spatial room impulse responses (RIRs). In brief, the SDM uses a compact array of microphones in a RIR measurement. Based on the time difference of arrivals between microphone pairs in short time windows, it estimates the direction of arrival (DOA) for each audio sample in the captured RIR. Therefore, the SDM allows a wide range of perceptual room acoustics studies, and it has been recently applied to study concert halls [7,8], studio control rooms [9], car cabins [10,11], as well as stage acoustics for musicians [12].

To reproduce the spatial sound analyzed with the SDM, audio samples in RIR are assigned to the loudspeakers of a given reproduction array. The assignment yields a number of sparse impulse responses equal to the number of loudspeakers used. The spatial sound is finally rendered by convolving sound signals with each of these sparse impulse responses. That is, the measured RIR is distributed spatially as convolution reverberators for a defined reproduction loudspeaker array.

Although the direction of arriving sound is accurately estimated in the analysis, the practical limits in the real or virtual loudspeaker array introduce varying amounts of angular error to the spatial sound reproduction. In the worst case, the mismatch between original and synthesized DOA may change the spatial image of the acoustic space drastically. In theory, all angular errors could be avoided and the analyzed sound field could be reproduced perfectly by assigning each sample to its own loudspeaker. However, this kind of approach is physically infeasible to implement, especially for real room-acoustic conditions.

One popular method to reduce the angular error in spatial sound synthesis is the Vector Base Amplitude Panning (VBAP) [13]. VBAP weights each sound sample between the three closest reproduction loudspeakers so that the resulting sound appears to arrive from the original direction. When applied to all samples in the RIR, the spatial image of the space is reproduced correctly. However, it is known that amplitude panning of samples corresponds to time-averaging of multiple HRTFs, leading to low-pass filter effect on the perceived sound [14].

Earlier studies employing SDM [7–12] have utilized a more straightforward synthesis method, namely Nearest Loudspeaker Synthesis (NLS), partially to circumvent the potential spectral issues with VBAP. NLS distributes each audio sample of a single monaural RIR to the nearest reproduction loudspeaker based on the estimated DOA information. Since only one reproduction loudspeaker at a time is involved in reproducing the RIR sample, this approach is free from the effects of HRTF averaging, but at the cost of increased angular error.

The aforementioned studies have used predetermined physical loudspeaker arrays for reproduction. Similar studies could still benefit from increased fidelity of spatial sound reproduction either by increasing the number of loudspeakers or by using the existing loudspeakers more efficiently. However, in many cases, adding loudspeakers is less feasible than optimizing the existing physical setup. When the reproduction of spatial audio is substituted with headphone listening, as in this study, the optimization becomes even more sensible. In theory, headphones enable the use of practically an unlimited number of virtual loudspeakers in any given direction. However, since the computational cost is directly proportional to the number of spatial channels rendered for headphones, using a smaller number of virtual loudspeakers is preferred. This requirement justifies smarter allocation of resources, which, in this case, means optimized virtual loudspeaker positions.

This paper presents a method of determining a room-specific virtual loudspeaker array that reproduces spatial sound perceptually more efficiently than a predetermined conventional array. The proposed method aims to minimize the directional error of NLS as well as to enhance labor-to-quality ratio of the rendering. In other words, spatial sound can be reproduced either more accurately with the same number of virtual loudspeakers, or at the same quality with a reduced channel number. These so-called optimized loudspeaker setups are compared with uniformly distributed setups to measure how recognizable are the differences in reproduction.

The paper is structured as follows. First, Section 2 outlines the structure of the position optimization system. Next, Section 3 describes the listening test, followed by the results in Section 4. Finally, more detailed analysis is presented with discussion in Section 5 and wrapped up in Section 6.

2. Virtual Loudspeaker Position Optimization

The sound reproduction system used in this paper is similar to one presented by Tervo et al. [10]. The system performs SDM analysis and NLS reproduction as described in Section 1, resulting in a sparse impulse response (IR) for each reproduction loudspeaker. Finally, those sparse IRs are convolved with the audio signals to create the spatial sound. The used version of the SDM also post-equalizes the loudspeaker IRs as the rapid channel changes of the IR cause whitening of the signal. It should be noted that the optimization of the reproduction loudspeaker positions is done for SDM data before the convolution step. In other words, the optimization requires a spatial sound reproduction technique that has information on the spatial IR for each sound source, and thus cannot be applied to an arbitrary spatial sound reproduction technique.

The NLS requires information on the reproduction loudspeaker positions in order to be able to distribute the samples properly. As mentioned in Section 1, the positions are usually static and not changed according to the acoustics of the space being reproduced. The method presented in this section replaces these static loudspeaker positions with optimal ones that are calculated from the RIR of the room being reproduced. The system used in reproduction is therefore the same as described above except for the way the loudspeaker setup is determined.

Figure 1 outlines the position optimization process which is done in two steps in its most basic form. First, SDM samples including RIR pressure and directional metadata (azimuth and elevation) are weighted according to their energy as well as their spatiotemporal properties. Then, loudspeaker positions are initialized based on the calculated weights and the final positions are obtained by clustering weighted DOAs iteratively until convergence. The clustering is computationally heavy by default due to a large number of RIR samples. To accelerate this process, the weighting data are reduced to a discrete number of equidistantly spaced points on the surface of a unit sphere, creating a downsampled spatial map of weights. Here, this operation is called spatial downsampling.

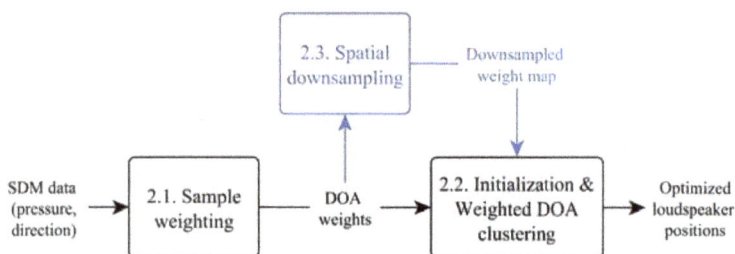

Figure 1. Position optimization system outline. Spatial Decomposition Method (SDM) generated data containing directions of arrival (DOA) the omnidirectional pressure values for each sample are provided to the algorithm, from which the optimized loudspeaker positions are approximated. The numbers inside the boxes refer to the corresponding sections in this paper, spatial downsampling part (in blue) working as an extra acceleration component for the main algorithm (in black).

These steps are described in detail in the following subsections. First, the weighting and clustering operations are described, followed by the downsampling step.

2.1. Sample Weighting

Even though loudspeaker position optimization is calculated from the RIR, optimal results are not achieved with the pressure values of the IR alone. The early part of the RIR, which includes the direct sound and early reflections, can be assumed to contain the most important perceptual information about the acoustic space. As opposed to the late part, the central role of the early part is supported by its significance in the identification of the acoustic space [15]. This is why it is reasonable to emphasize the early part of the response through directionally accurate reproduction. In addition, some of the DOAs calculated by the SDM describe the actual incoming direction of sound more accurately than others, which is discussed below in more detail. To take these presented properties into account in optimization, the first step in the process is to weight each pressure sample in the RIR accordingly.

The weighting is based on SDM data points—in other words, pressure and DOA of each audio sample in the RIR. As a whole, this information is called SDM data. The weighting $\mathbf{w} = \{w_1 \ldots w_N\}$ can be seen as a mapping of this data:

$$\left\{ f : (\mathbf{p}, \mathbf{u}_{doa}) \mapsto \mathbf{w} \quad \middle| \quad \mathbf{p}, \mathbf{w} \in \mathbb{R}^N, \, \mathbf{u}_{doa} \in \mathbb{R}^{N \times 3}, \, ||\mathbf{u}_{doa,n}|| = 1, \, w_n \in [0,1] \, \forall n = 1 \ldots N \right\} \quad (1)$$

where $\mathbf{p}$ and $\mathbf{u}_{doa}$ are the pressure and DOA values of the SDM data, respetively; and N is the length of the IR in samples. In short, each SDM data point generates one scalar weight to be associated with itself.

Each weight consists of four distinct subweights named energy, delay, gradient and direction that are described in more detail below.

1. Energy weighting corresponds to the most traditional form of weighting. Each SDM data point is weighted according to its energy:

$$w_{E,n} = \frac{p_n^2}{\max_n(p_n^2)} \quad (2)$$

where p_n is the omnidirectional pressure value of the nth data point. This gives more weight to data points with more incident energy.

2. Delay weighting emphasizes the SDM data points that locate in the earlier part of the RIR. Weighting is computed as a normalized backward Schroeder integral [16] over the RIR:

$$w_{T,n} = \frac{\sum_{k=n}^{N} p_k^2}{\sum_{k=1}^{N} p_k^2} \quad (3)$$

This causes the direct sound and early reflections with distinctively more energy to be weighted more than samples in the later part of the response. This aspect is linked to psychocaoustics, as the early part of the RIR is perceptually more important than the late reverberation [15].

3. Gradient weighting is used as a reliability measure to the SDM data points. The reliability of one SDM data point is dependent on the data points directly before and after it in time. If DOAs of the neighboring data points are close to the DOA of the current point, the point is given a large weight, and the greater the distance to its neighbors, the smaller the weight. The weight of one SDM data point is resolved as a mean of the distances between the DOA of the point and the DOAs of the previous and next data points:

$$w_{G,n} = 10^{(\min_n(d_{G,n}) - d_{G,n})/10} \quad (4)$$

$$d_{G,n} = \frac{||\mathbf{u}_{doa,n} - \mathbf{u}_{doa,n-1}|| + ||\mathbf{u}_{doa,n+1} - \mathbf{u}_{doa,n}||}{2h} \quad (5)$$

where h is the size of the time step between two SDM samples. The reasoning for this procedure is based on the fact that the SDM provides weighted average of the true DOAs in case of overlapping plane waves [6,10]. The set of samples over which the DOA changes can then be considered as spatially less accurate than samples with more steady DOA estimate.

4. Direction weighting emphasizes the data points that have a lot of energy arriving from their general direction regardless of the temporal information. The operation can be thought as a low-pass filter for directions; a single high-energy sample does not get a large weight unless there are more high energy samples in the same spherical sector. Conversely, the sectors with mainly low-energy samples are given a small weight. Perceptually, this operation can be thought as simulating the limits of perception. A single high energy sample cannot be heard separately but a longer period of time is needed to generate a perceivable acoustic event. Therefore, directions with more high-energy samples should be prioritized when searching for optimal loudspeaker positions. There is also a benefit in algorithmic means as the influence of high-energy sectors are

spread, making it easier for the optimization algorithm to iterate to the directions with higher energy density. Similar to the spatial downsampling presented later in Section 2.3, the general energy directions are approximated by calculating an energy map. First, an equidistant grid of points is created on the surface of the unit sphere, representing DOAs in the listener space. Then, each SDM data point is assigned to the closest point in this grid, measuring the distance from the DOA of the data point to the DOA of the grid point. When this nearest-neighbor search is ready, the energies of the assigned data points are accumulated grid point wise. The operation results in an energy map where the energy of different incoming directions has been approximated. Finally, the weight of the SDM data point is calculated from this map by interpolation:

$$w_{D,n} = \frac{\sum_{i=1}^{3} E_{\min(n,i)} * d_{\min(n,i)}}{\sum_{i=1}^{3} d_{\min(n,i)}} \tag{6}$$

$$d_{\min(n,i)} = i\text{th smallest value of a set} \quad \{||\mathbf{u}_{\text{doa},n} - \mathbf{u}_{\text{grid},m}||, \; m = 1 \ldots M\} \tag{7}$$

where $\mathbf{u}_{\text{grid},m}$ is the DOA of mth grid point, M is the number of grid points and $E_{\min(n,i)}$ is the energy value associated with the ith closest grid point.

These four subweights form the final weight vector $\mathbf{w}_{final}$ through the following equation:

$$\mathbf{w}_{final} = \prod_{i=1}^{4} \mathbf{w}_i^{c_i} \tag{8}$$

where $\mathbf{w}_i$ is a vector containing the values of one of the subweights described above and c_i is the corresponding mixing coefficient. As all $\mathbf{w}_i$ are equalized in range $[0, 1]$, mixing coefficients practically adjust how much each partial weight vector reduces the resulting weights. Through the rest of the position optimization process, the weights are combined with their corresponding DOAs and used as a replacement for the pressure of the IR.

2.2. Initialization and Weighted DOA Clustering

As described before, NLS assigns each SDM data point to the closest loudspeaker in the reproduction array, resulting in a set of sparse IRs. However, the end result is not optimal as more important data points, for example early reflections, are allowed as much directional error as less direction-critical samples in the late reverberation. The weighting presented in the previous section solves the importance problem of different samples, but does not determine the actual virtual loudspeaker positions by itself. Therefore, weighted K-means clustering algorithm is used to find the most optimal loudspeaker positions based on the weighting data.

The aim of clustering is to find structure in the data iteratively, and the K-means algorithm is one of the most widely used clustering methods. The idea of the method is to minimize the Euclidean distance of the cluster data points by first assigning data vectors to cluster nodes and then update the node location according to the assigned vectors, most commonly to data mean. When these two operations are repeated, the result starts to converge towards the concentrations of data. However, the conventional K-means algorithm does not work in this case, as our data are not equally weighted. Instead, we use the weighted mean of the allocated samples to determine the cluster node position—or, as in this case, a new loudspeaker position. This causes the algorithm to position the node closer to the more weighted areas, effectively reducing the spatial error of directionally more important parts. The weights used by this algorithm are calculated as described in Section 2.1.

The downside of the K-means algorithm is that the method is prone to find only a local optimum. To circumvent this problem, a good initialization of loudspeaker positions is required. Smart initialization not only boosts the probability of finding more optimal solution than a random one, but also helps the K-means to converge faster and find that solution in less time.

The initialization step is implemented in the presented system as follows. First, a weight map similar to the one used in direction weighting in Section 2.1 is generated. There is one major difference:

instead of using an equidistantly spaced point grid, the applied grid has equidistantly spaced points in the azimuth direction, but the elevation spacing is cosine-weighted. This procedure is applied to have every grid point to accumulate energy from approximately equally sized area. This kind of grid is also easier to sample from, as there is an equal number of azimuth points at each elevation angle. After equalizing the values of this map, it becomes a probability distribution function (pdf) which is later sampled from for new loudspeaker positions.

The initial loudspeaker positions are defined with Monte Carlo sampling. New positions are drawn from the weight distribution one-by-one and each new position is compared with all other positions already selected. If the new position is too close to any of the older locations, the sample is discarded and a new draw is made. If the new candidate is valid, it is stored and the sampling pdf is altered with Von Mises-Fisher distribution [17] so that the proximity of the new position is picked less probably in the next iteration round. After drawing all the initial positions, the final positions are determined with weighted K-means clustering described above.

Occasionally, virtual loudspeaker positions tend to cluster during weighted K-means step iteration. In extreme cases, the final setup has two or more virtual loudspeakers positioned within a few degrees of each other. To eliminate such cases, each position has a repulsion area around itself. When two or more loudspeakers are moved too close to each other, the one with the most weight keeps its position while the others are relocated. The new positions are determined so that the relocated loudspeakers are as far from the other loudspeakers as possible. The repulsion area is gradually reduced to ensure that the algorithm converges even when there are lots of virtual loudspeakers to relocate.

Finally, the optimization algorithm described above is summarized in Algorithm 1. First, the initial loudspeaker positions are sampled from the weight map that is generated from the weights (Section 2.3). The initial distances are controlled by rejecting the samples that are too close to previously sampled positions. When all the initial locations have been sampled successfully, the final positions are iterated by using weighted K-means clustering. Again, the distances of the iterated loudspeaker positions are monitored and loudspeakers with less weight are relocated in the case they come too close to more weighted ones. This relocation area is gradually reduced by the algorithm, which leads to convergence. The final result is then a set of virtual loudspeaker positions that are located close to the most weighted samples in their reproduction region.

Algorithm 1 Virtual loudspeaker optimization by using weighted DOA clustering.

1: **function** OPTIMIZELOUDSPEAKERPOSITIONS($\mathbf{w}, \mathbf{u}_{\text{doa}}, N_{\text{ls}}, d_{\text{repulsion}}$)
2:　　$\mathbf{W}_{\text{map}} \leftarrow$ CalculateWeightMap($\mathbf{w}, \mathbf{u}_{\text{doa}}$)　　　　　　　　▷ **Initialize loudspeaker positions**
3:　　$\mathbf{u}_{\text{ls}} \leftarrow$ SampleWeightMap($\mathbf{W}_{\text{map}}, N_{\text{ls}}$)
4:　　**repeat**　　　　　　　　　　　　　　　　　　　　　　▷ **Calculate weighted K-means**
5:　　　　$\mathbf{u}_{\text{ls,old}} \leftarrow \mathbf{u}_{\text{ls}}$
6:　　　　$\mathbf{c}_{\text{ls}} \leftarrow \mathbf{0}^{N \times 1}$
7:　　　　**for** $n \leftarrow 1$ **to** N **do**
8:　　　　　　$\mathbf{c}_{\text{ls},n} \leftarrow \arg\min_i(||\mathbf{u}_{\text{ls},i} - \mathbf{u}_{\text{doa},n}||)$　▷ find the closest loudspeaker to the SDM data point
9:　　　　**end for**
10:　　　　**for** $i \leftarrow 1$ **to** N_{ls} **do**
11:　　　　　　$(\mathbf{w}_{\text{cls}}, \mathbf{u}_{\text{cls}}) \leftarrow (\mathbf{w}, \mathbf{u}_{\text{doa}})|_{\mathbf{c}_{\text{ls},n}=i}$　　　▷ assign the data point to the ith loudspeaker
12:　　　　　　$\mathbf{u}_{\text{ls},i} \leftarrow \sum_k (w_{\text{cls},k}\mathbf{u}_{\text{cls},k}) / \sum_k (w_{\text{cls},k})$　　　▷ weighted mean of the assigned DOAs
13:　　　　**end for**
14:　　　　$\mathbf{d}_{\text{closest}} \leftarrow \mathbf{0}^{N_{\text{ls}} \times 1}$　　　　　　　　　　　　　　　▷ **Apply repulsion area**
15:　　　　**for** $i \leftarrow 1$ **to** N_{ls} **do**
16:　　　　　　$d_{\text{closest},i} \leftarrow \min_{j \neq i}(||\mathbf{u}_{\text{ls},i} - \mathbf{u}_{\text{ls},j}||)$　　　　　　▷ distance to the closest neighbor
17:　　　　**end for**
18:　　　　**for all** $d_{\text{closest},i} < d_{\text{repulsion}}$, from smallest to largest **do**
19:　　　　　　$\mathbf{u}_{\text{v}} \leftarrow$ vertices of a Voronoi diagram of $\mathbf{u}_{\text{ls}}$　　　　▷ potential furthest points
20:　　　　　　$\mathbf{u}_{\text{ls},i} \leftarrow \arg\max_{\mathbf{u}_{\text{v},j}}(\min_i(||\mathbf{u}_{\text{v},j} - \mathbf{u}_{\text{ls},i}||))$　　▷ Select the $\mathbf{u}_{\text{v},j}$ furthest from all $\mathbf{u}_{\text{ls}}$
21:　　　　**end for**
22:　　　　reduce $d_{\text{repulsion}}$
23:　　**until** all $||\mathbf{u}_{\text{ls}} - \mathbf{u}_{\text{ls,old}}|| <$ threshold
24:　　**return** $\mathbf{u}_{\text{ls}}$
25: **end function**

2.3. Spatial Downsampling

The optimization process described in Sections 2.1 and 2.2 is already a working solution, which finds the optimized virtual loudspeaker positions from the SDM data. However, the process is slow due to the amount of data. The complexity of K-means algorithm is directly proportional to the number of data points clustered. As an IR of a regular concert hall is approximately two seconds long, there are 96,000 data points after SDM analysis when using 48 kHz sampling rate. Finding the optimal positions over the whole data is possible, but requires long computation time to complete. However, a considerable speedup is possible by reducing the number of data points. If the reduced data are used to find a coarse approximation of the positions, and the whole data are only used to fine-tune the result, a considerable speedup may be achieved. Here, this data reduction step is called spatial downsampling.

The implementation of spatial downsampling is similar to direction weighting and initialization of loudspeaker positions described before. The SDM data points are condensed to equidistant point grid on the unit sphere. The difference is that the energy of each data point is distributed between the three closest grid points in order to get more accurate approximation of the surrounding energy field. The distribution is calculated by determining barycentric coordinates of the data point with respect to those three closest grid points. The reduced point grid is then used instead of SDM data to initialize and optimize the virtual loudspeaker positions. After the optimization algorithm has converged, the reduced data are replaced by the original SDM data and the final optimized positions are fine-tuned from the reduced data optimum. The processing time is shortened due to reduced data, but the result is the same as without the reduction due to the fine-tuning step.

3. Perceptual Evaluation with a Listening Test

The performance of the virtual loudspeaker location optimization and its effect on the perceived spatial sound reproduction was evaluated with a subjective listening test. The aim of the listening test was to examine how perceivable the differences are in two typical use cases of the SDM-based spatial sound. The first case was a stereo music played in a dry studio control room and the second case was an orchestra music performance set to the acoustics of a concert hall. It should be noted that the first case has two sound sources in a small space and the second one has 24 sound sources in a large space.

3.1. Listening Test Setup and Sound Signals

The listening test setup is illustrated in Figure 2. The setup consisted of a desk, a computer and noise canceling headphones (Bose QuietComfort 25) and was located in the corner of a quiet open plan office. The listener's head was tracked with a commercial tracking system (Optitrack V100 and TrackingTools software, version 2.5.3; 2012 by NaturalPoint Inc., Corvallis, OR, USA) that utilizes six infrared cameras surrounding the listening space. To reduce visual distractions, the front field of view of the subject was obscured with a curtain. The listening test program was built on the Unity engine and it utilized head tracking in six degrees of freedom. HRTFs were generated from a scanned human head with a fast boundary element method [18] in far field. All the participants used the same HRTF set containing 836 directions. No interpolation was used between the HRTFs and the filter was swapped without cross-fading.

The experiment consisted of two test sets. Both sets introduced different listening conditions in order to compare loudspeaker optimization performance. The acoustic spaces were a studio control room in Helsinki, Finland and Musikverein concert hall, Vienna, Austria, later referred to as a "small room" and a "concert hall", respectively. The small room RIRs had been measured with stereo pair of loudspeakers, whereas the acoustic response of the concert hall had been captured by using loudspeaker orchestra [7]. Based on these prior measurements, the SDM analysis had been done for both rooms in advance.

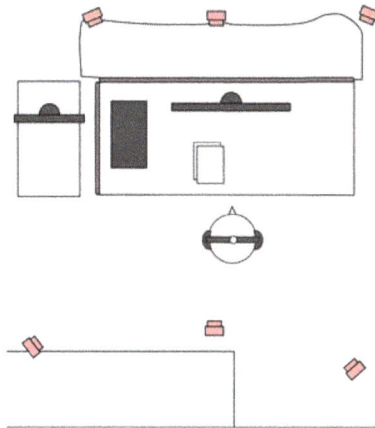

Figure 2. A sketch of the listening test setup. The listener (in the middle) is surrounded by six infrared cameras (red) that track the movements of the noise-canceling headphones. The field of view has been obscured with a curtain.

Both sets contained four different virtual loudspeaker setups; two position optimized setups and two uniformly distributed setups with fixed virtual loudspeaker positions for direct sounds. In Figures 3 and 4, all these conditions have been illustrated for both spaces. The optimized loudspeaker setups were visually inspected for potential loudspeaker clusters, and uniform setups were based on Platonic solids.

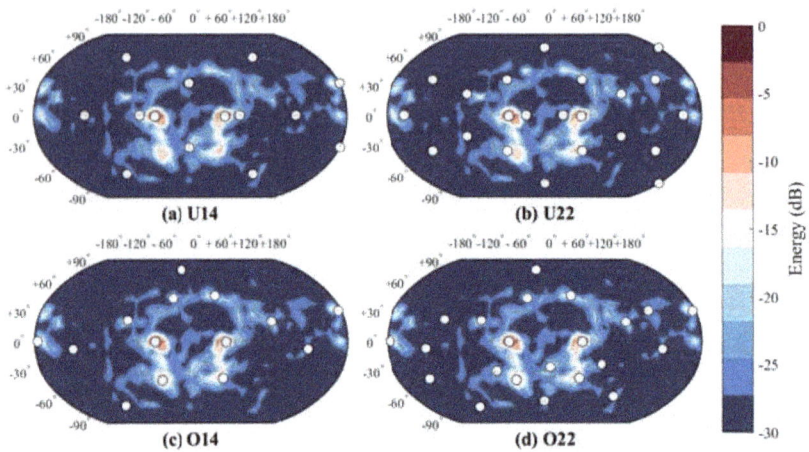

Figure 3. Loudspeaker setups (white circles) overlaid with the spatial map of overall sound energy in the small room case: (**a**) uniform setup with 14 loudspeakers; (**b**) uniform setup with 22 loudspeakers; (**c**) optimized setup with 14 loudspeakers; and (**d**) optimized setup with 22 loudspeakers.

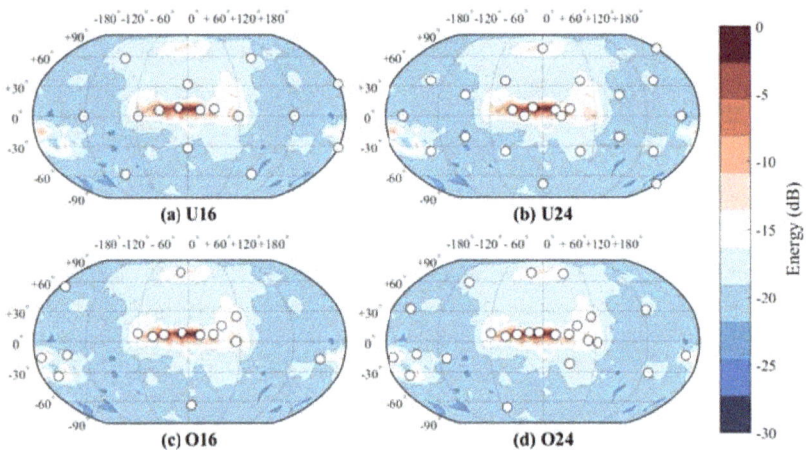

Figure 4. Loudspeaker setups used in the concert hall samples: (**a**) uniform setup with 16 loudspeakers; (**b**) uniform setup with 24 loudspeakers; (**c**) optimized setup with 16 loudspeakers; and (**d**) optimized setup with 24 loudspeakers.

The sound signals used were typical for both rooms. Since the small room was measured with traditional stereo setup, a stereo signal was used. In our case, a 34-s excerpt from the beginning of Céline Dion's song "Because You Loved Me" was selected. For the concert hall, the music signal selected for the loudspeaker orchestra was Jean Sibelius's Lemminkäinen suite, 1st part, 851–885 s. The orchestra signals were captured in professional recording with 21 close microphones for different instrument groups as quasi-anechoic material, which were then mapped to 24 measurement source channels in the loudspeaker orchestra.

The reproduction levels of all presented stimuli were equalized. This was accomplished by computing the equivalent level L_{eq} of combined virtual loudspeaker channels as

$$L_{eq} = 20 \log \left(\sqrt{\frac{\sum_{i=1}^{M} \sum_{n=1}^{N} x_{i,n}^2}{N}} \right) \quad (9)$$

where M is the number of channels, N is the length of one signal and $x_i = \{x_{i,1} \dots x_{i,N}\}$ is the input signal of channel i. This level was then used to calculate level alignment coefficient C_{eq}:

$$C_{eq} = 10^{((L_{eq,0} - L_{eq})/20)} \quad (10)$$

where $L_{eq,0} = -25$ dB was the target level. Finally, all output channels were multiplied by C_{eq} to get the level aligned multichannel sound.

3.2. Listening Test Method

The listening test was executed as an ABX discrimination test. Subjects were asked to answer a question: "Which one of the samples A or B is the reference X?". Moreover, they were asked to write down the criterion that they used to discriminate the reference from the odd sample. If they could not discriminate the samples from each other, they were asked to choose their answer at random. Full comparison between four conditions forms six pairs, which were repeated four times. Thus, a total of 24 stimulus triplets were presented for both rooms. A preference test was also considered as an option for the listening test. However, the differences between some of the samples were found to be small during preliminary tests, which made the asking for a preference unfeasible. Preference is also a matter of taste, which would have required more participants in the test, in case there would have been more than one preference group.

The listening test started with the participant reading and signing a paper of informed consent describing the test procedure, possible harm, and data policy. Then, the subject did both test sets in randomized order. Before each set, there were a training set of four ABX triplets, during which the listener was instructed to adjust the listening volume to a reasonable level. After the training set, the subject was asked to keep the volume at the same level during the test set. The subject was given an option to take a break between the test sets, and after completing the experiment, a small non-monetary compensation was offered to the subjects.

3.3. Statistical Analysis

The ABX test is designed to detect the small differences of the compared signals. Therefore the difference should not be detected by all the participants or otherwise the test loses its purpose. The contrary also holds: if the difference is too small, all participants have to guess and the result of the experiment is statistical noise. That being said, an approximation to the ability to distinguish the sound samples is needed to determine the expected detection rate over the listener population. In other words, the experimenter should decide how big portion of the subjects has truly heard the difference in the samples. When determining the threshold, one should also keep in mind the time and resources that can be used—smaller threshold needs more subjects and vice versa. The limit selected for this experiment is a compromise between distinction and the number of subjects; the expected detection threshold was set at one third of the population.

The determined threshold cannot be used directly to determine how large proportion of subjects has successfully discriminated the difference. Instead, the determined threshold needs to be adjusted to calculate the proportion P_{obs} with the rearrangement of Abbot's formula [19]:

$$P_{obs} = P_d + P_0(1 - P_d) \quad (11)$$

where P_d is the proportion of subjects that truly noticed the difference and P_0 is the proportion of the population that got the test right by chance. P_{obs} basically tells us the proportion of subjects that should score the sample pair right to prove the alternate hypothesis right.

After calculating the required observed proportion, the required number of subjects can be approximated with the following equation [19]:

$$N = \left[\frac{Z_\alpha \sqrt{p_0 q_0} + Z_\beta \sqrt{p_a q_a}}{p_0 - p_a} \right]^2 \tag{12}$$

where Z_α and Z_β are the corresponding Z-scores of selected false positive and negative rates; p_0 is the chance probability; p_a is the chosen probability for an alternate hypothesis; and $q_0 = 1 - p_0$, $q_a = 1 - p_a$. However, required subject count becomes prohibitively large if our selected detection threshold is used for a simple ABX test setup. The reason for this is the chance of a subject guessing the correct sample—for the used test, the chance rate is 50 percent per evaluated pair. According to the Equation (12), 94 participants would be needed to ensure the significance of the effect. To reduce the number of required subjects, replications of the same samples were used. The strategy was to make the participant evaluate the same sample pair multiple times. The subject was required to get all the replications right for the test case in order to count the sample as properly discriminated. The number of replications in this test was set at four times per sample pair, effectively reducing the chance rate to 6.25 percent and the number of required participants to 15 people.

To calculate Z-score for the results, the proportion of true discriminators should be calculated from the results. The equation for calculating the adjusted proportion P_{adj} can be derived from Equation (11):

$$P_{adj} = \frac{P_{obs} - P_0}{1 - P_0}. \tag{13}$$

From P_{adj}, Z-score can be calculated with a binomial test for proportions [19]:

$$z = \frac{P_{adj} - p_0 - 1/(2N)}{\sqrt{p_0 q_0 / N}} \tag{14}$$

where N is the number of subjects. Finally, 95 percent confidence intervals were calculated [19]:

$$CI_{95\%} = P_{adj} \pm Z_{95\%} SE_{P_{adj}} \tag{15}$$

where $Z_{95\%} = 1.645$ is the Z-score for 95 percent confidence interval and $SE_{P_{adj}}$ is the standard error for the adjusted proportion of discriminators [19]:

$$SE_{P_{adj}} = \sqrt{\frac{P_{adj}(1 - P_{adj})}{N}}. \tag{16}$$

4. Results

A total of 17 subjects participated in the listening test out of which 15 completed the whole experiment. The population consisted of acoustics experts and an audio engineer. Three people reported some kind of hearing defects: small dips, slight oversensitivity and 10 dB hearing threshold difference between right and left ears. However, these defects did not affect the test performance of these particular subjects, which is why their data was not excluded from the results. Both subjects that did not finish reported hearing clicking sounds during cross-fade, thus preventing them from focusedly discriminating the stimuli. However, only a few participants who finished the test reported observing this phenomenon or being distracted by it.

4.1. Discrimination

Discrimination results for all test cases are presented in Figure 5, each case presenting the chance-corrected discrimination rate (a cross) and its one-tailed 95 percent Confidence Interval (CI). In addition, the selected detection threshold $P_d = 33.33\%$ and chance rate $P_0 = 6.25\%$ have been visualized. The numerical values of the results are tabulated in Table 1.

In the small room case, five out of six comparisons were significantly recognizable. Especially U14 has been clearly separated from the others. Both cases involving U22 were also recognized by a smaller margin; the cases are clearly over the chance rate, but may be heard less than third of the population within the limits of the confidence intervals. Comparison between optimized setups did not cross P_d, thus reaching significant similarity. In other words, less than one third of the population can ever discriminate the setups from each other.

On the contrary, none of the concert hall comparisons were significantly recognizable. Comparison of uniform setups appeared to be the most recognizable out of these setups, but further experiments are needed to conclude how recognizable the case is in the end. The rest of the comparisons and their CIs did not cross P_d, again reaching significant similarity. In particular, comparison between O16 and uniform setups could not be recognized, probably because the direct sound was reproduced similarly in all of them.

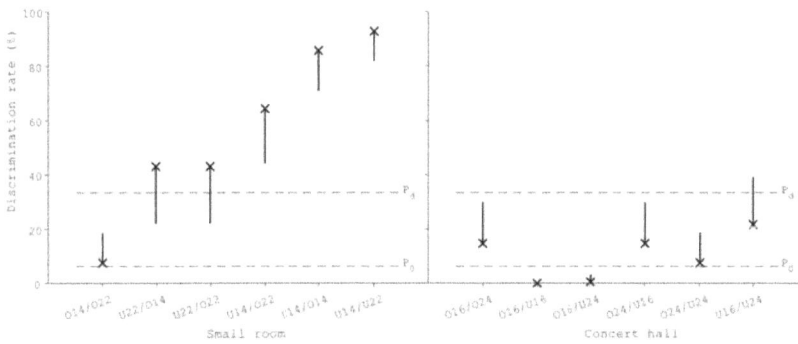

Figure 5. The discrimination rates of different listening conditions and their one-sided 95% confidence intervals (P_d set discrimination level, P_0 chance rate).

Table 1. Results of the listening test and the frequencies of the attributes elicited by 15 subjects for correctly rated pairs.

Space	Small Room						Concert Hall					
Sample A	O14	O14	O14	O22	O22	U14	O16	O16	O16	O24	O24	U16
Sample B	O22	U14	U22	U14	U22	U22	O24	U16	U24	U16	U24	U24
Subjects 4/4 correct	2	13	7	10	7	14	3	0	1	3	2	4
4/4 proportion (%)	13.3	86.7	46.7	66.7	46.7	93.3	20.0	0.0	6.7	20.0	13.3	26.7
P_{adj} (%)	7.6	85.8	43.1	64.4	43.1	92.9	14.7	0.0	0.4	14.7	7.6	21.8
CI. lower (%)	-	70.9	22.1	44.1	22.1	82.0	-	-	-	-	-	-
CI. upper (%)	18.8	-	-	-	-	-	29.7	0.0	3.3	29.7	18.8	39.3
Attribute	**Frequency of attributes elicited on the difference between samples A and B**											
image shift	4	15	9	13	9	12	4	2	6	2	3	2
reverberance	-	7	1	5	7	15	4	3	5	4	3	2
width	-	6	2	4	6	4	2	7	3	4	3	3
spectral balance	2	9	5	3	1	2	4	1	2	-	1	7
timbre	1	3	3	4	4	3	-	1	2	4	4	3
spatial impression	1	3	6	2	2	5	3	1	-	-	2	4
envelopment	-	4	3	5	1	3	2	2	3	1	2	3
bass	1	3	2	2	4	4	1	1	2	-	-	1
loudness	1	1	-	-	1	-	6	3	1	3	1	2
brightness	1	1	-	2	-	3	2	1	-	1	1	-
distance	-	1	-	2	1	1	1	-	1	1	2	-
size of space	-	-	-	-	-	-	4	1	-	1	1	-
focus	1	-	-	-	2	2	-	-	-	-	-	-
warmth	-	-	-	1	-	2	-	-	-	-	-	-
openness	-	-	-	1	-	-	-	-	-	-	2	-
dynamic range	-	-	-	-	-	-	-	-	-	1	1	-
clarity	-	-	-	-	-	1	-	-	-	-	-	-
source presence	-	-	1	-	-	-	-	-	-	-	-	-
spatial balance	-	-	-	-	1	-	-	-	-	-	-	-
Total	12	53	32	44	39	57	33	23	25	22	26	27

4.2. Discrimination Criteria

The discrimination criteria reported by subjects on a paper form were first encoded into electrical form. These answers were then interpreted and translated into English by using the wheel of concert hall acoustics [20], each answer translating into one or more terms. Then, the resulting attributes were sorted and accumulated so that the frequencies of all individual attributes could be reported for every sample pair separately. Only those attributes were included in the analysis whose corresponding discrimination was correct.

The results of the analysis have been reported in order of frequency in Table 1. The most frequent perceptual discriminating factors between conditions were image shift, reverberance, and width. The small room had also reported differences in spectral balance, spatial impression and timbre. In the concert hall renderings, loudness differences were reported in addition to the attributes used also for the small room.

Image shift was most frequently reported as the discriminating attribute in the small room. Uniform setup with 14 loudspeakers (U14) systematically collected the most reports, which reflects the high discrimination rate. In addition, the difference in reverberance made the discrimination of uniform setups (U14 vs. U22) even easier. Optimized setup with 14 loudspeakers (O14) had notable differences in spectral balance when compared with uniform setups. Most frequent reports on spatial impression concentrated in comparison of U22 to both O14 and U14.

The elicited attributes in concert hall pairs were not as distinct as in the case of the small room. The most evident differences were width in O16 vs. U16 and spectral balance in U16 vs. U24. Otherwise the distribution of answers was more uniform. In O16 vs. O24 loudness appeared the most frequent factor, as image shift and reverberance did in the case of O16 vs. U24. In the rest of the cases, there was no consensus on discriminative attributes.

5. Discussion

As the results show, the listening conditions drastically affect how well the differences in virtual loudspeaker positioning are noticed in binaural reproduction. The small room case appeared very susceptible to perceived image shifts and altered reverberance when the reproduction loudspeaker configuration was changed, whereas the changes in the setups were barely noticeable with the concert hall case.

As visualized in Figure 3, perceived image shifts in the small room are explainable by the changes in the directions of early reflections. There are four strong reflections visible above the direct sounds, as well as reflections from the desk. The virtual loudspeakers of U14 that reproduce those directions are far from the reflections, whereas optimized setups have co-located loudspeakers for each of them. In U22, virtual loudspeakers are also off from those directions, but they are closer to them than in U14. This partly explains why U22 is as easy to discriminate from the optimized setups as U14 is. Visual inspection also suggests that O14 is a subset of O22, explaining why they are difficult to discriminate from each other. In short, strong early reflections affect the spatial image drastically, therefore requiring precise reproduction of their directions.

A notable aspect in the concert hall case is that the directions of early reflections did not play as large role as they did in the small room. The direct sound reproduction loudspeakers were positioned identically between O16 and uniform setups, making discrimination between the setups hard. O24 had more loudspeakers optimized to reproduce direct sounds, thus being more distinctive than O16 when compared with uniform setups. Finally, U16 and U24 could probably have been discriminated from each other because of the combined effect of differences in the reproduction of side, ceiling and back wall reflections. To summarize, direct sounds dominate the spatial impression, allowing more drastic changes in reflection directions before the difference is heard.

The small room case had only two sound sources with 60 degree separation. The reflections are then relatively scarce, therefore likely more audible and susceptible to changes in reproduction. The concert hall case in turn had 24 densely located sound sources, each having their own set of reflections and source signals. As the source positions are close to each other at the stage, most of the reflections are coming from adjacent directions. However, as these directions are not precisely the same, the reflections are spread over a larger area than in the two-channel case, therefore making the precise direction of the reflection fuzzier. However, more experiments are needed in different spaces to prove this theory.

Another difference between the cases is the strength of the reverberation. Small room has only little if any diffuse reverberation, whereas the concert hall has prominently longer and more diffuse reverberation. Strong reverberant field may reduce the importance of correct spatial reproduction of earliest reflections. This psychoacoustic factor remains hypothetical and also needs more research.

6. Conclusions

The aim of this study was to present a method of optimizing virtual loudspeaker positions for spatial sound reproduction with head-tracked headphones. The rendering of spatial sound was implemented with the Spatial Decomposition Method in combination with nearest loudspeaker synthesis to analyze and reproduce measured room impulse responses in perceptual room acoustics studies. These studies have earlier used a static loudspeaker array to synthesize spatial impulse responses convolved with sound signals. However, a static reproduction array is not an optimal way of reproduction when concerning the arbitrary directions of early reflections. This is especially the case with headphones where the loudspeakers are virtualized with the help of HRTFs.

The implementation was evaluated with a discriminative listening test, which consisted of one simple and one complex auditory scene. They were a small room with stereo loudspeaker audio and a concert hall with 24 channel orchestra music. The results implied that optimization of virtual loudspeaker positions is important in a small room. This is because misplaced early reflections cause image shifting and change perception of reverberance. However, this effect was not so large in the concert hall case. In any case, the number of virtual loudspeakers can be reduced to minimize the real-time computation required for HRTF processing, if the directions of direct sounds and early reflections are accurately reproduced. The study for the minimum number of virtual loudspeakers needed without deteriorating the sound quality is left for future work.

Acknowledgments: The research has been funded by Academy of Finland (Project Nos. 296393 and 289300).

Author Contributions: Otto Puomio designed and implemented the algorithms, implemented and performed the listening test and wrote the paper. Jukka Pätynen was involved in the design of algorithms and listening test. Tapio Lokki contributed to listening test design, data analysis, and writing the paper.

Conflicts of Interest: The founding sponsors had no role in the design of the study; in the collection, analyses, or interpretation of data; in the writing of the manuscript, and in the decision to publish the results.

Abbreviations

The following abbreviations are used in this manuscript:

SDM	Spatial Decomposition Method
IR	Impulse Response
RIR	Room Impulse Response
NLS	Nearest Loudspeaker Synthesis
HRTF	Head Related Transfer Function
VBAP	Vector Base Amplitude Panning
O14/O16/O22/O24	Optimized loudspeaker setup with 14/16/22/24 loudspeakers
U14/U16/U22/U24	Uniform loudspeaker setup with 14/16/22/24 loudspeakers
CI	Confidence interval
pdf	Probability distribution function

References

1. Brandenburg, K.; Werner, S.; Klein, F.; Sladeczek, C. The Technology of Binaural Listening & Understanding: Auditory illusion through headphones: History , challenges and new solutions Auditory illusion through headphones: History , challenges and new solutions. In Proceedings of the 22nd International Congress on Acousitcs, Buenos Aires, Argentina, 5–9 September 2016.
2. Hacihabiboglu, H.; De Sena, E.; Cvetkovic, Z.; Johnston, J.; Smith, J.O., III. Perceptual Spatial Audio Recording, Simulation, and Rendering: An overview of spatial-audio techniques based on psychoacoustics. *IEEE Signal Process. Mag.* **2017**, *34*, 36–54.
3. Möller, H. Fundamentals of binaural technology. *Appl. Acoust.* **1992**, *36*, 171–218.
4. Brimijoin, W.O.; Boyd, A.W.; Akeroyd, M.A. The contribution of head movement to the externalization and internalization of sounds. *PLoS ONE* **2013**, *8*, e83068.
5. Hendrickx, E.; Stitt, P.; Messonnier, J.C.; Lyzwa, J.M.; Katz, B.F.; de Boishéraud, C. Influence of head tracking on the externalization of speech stimuli for non-individualized binaural synthesis. *J. Acoust. Soc. Am.* **2017**, *141*, 2011–2023.
6. Tervo, S.; Pätynen, J.; Kuusinen, A.; Lokki, T. Spatial decomposition method for room impulse responses. *J. Audio Eng. Soc.* **2013**, *61*, 17–28.
7. Lokki, T.; Pätynen, J.; Kuusinen, A.; Tervo, S. Concert hall acoustics: Repertoire, listening position and individual taste of the listeners influence the qualitative attributes and preferences. *J. Acoust. Soc. Am.* **2016**, *140*, 551–562.
8. Pätynen, J.; Lokki, T. Concert halls with strong and lateral sound increase the emotional impact of orchestra music. *J. Acoust. Soc. Am.* **2016**, *139*, 1214–1224.

9. Tervo, S.; Laukkanen, P.; Pätynen, J.; Lokki, T. Preference of critical listening environment among sound engineers. *J. Audio Eng. Soc.* **2014**, *62*, 300–314.

10. Tervo, S.; Pätynen, J.; Kaplanis, N.; Lydolf, M.; Bech, S.; Lokki, T. Spatial Analysis and Synthesis of Car Audio System and Car-Cabin Acoustics with a Compact Microphone Array. *J. AES* **2015**, *63*, 914–925.

11. Kaplanis, N.; Bech, S.; Tervo, S.; Pätynen, J.; Lokki, T.; Van Waterschoot, T.; Jensen, S.H. A rapid sensory analysis method for perceptual assessment of automotive audio. *AES J. Audio Eng. Soc.* **2017**, *65*, 130–146.

12. Amengual Gari, S.; Kob, M.; Lokki, T.; Pätynen, J.; Välimäki, V. Investigations on Stage Acoustic Preferences of Solo Trumpet Players using Virtual Acoustics. In Proceedings of the 14th Sound and Music Computing Conference, Espoo, Finland, 5–8 July 2017.

13. Pulkki, V. Virtual Sound Source Positioning Using Vector Base Amplitude Panning. *J. Audio Eng. Soc.* **1997**, *45*, 456–466.

14. Pätynen, J.; Tervo, S.; Lokki, T. Amplitude panning decreases spectral brightness with concert hall auralizations. In Proceedings of the 55th International Conference of the Audio Engineering Society on Spatial Audio, Helsinki, Finland, 27–29 August 2014; pp. 1–8.

15. Haapaniemi, A.; Lokki, T. Identifying concert halls from source presence vs room presence. *J. Acoust. Soc. Am.* **2014**, *135*, EL311–EL317.

16. Schroeder, M.R. New Method of Measuring Reverberation Time. *J. Acoust. Soc. Am.* **1965**, *37*, 409–412.

17. Fisher, N.I.; Lewis, T.; Embleton, B.J. *Statistical Analysis of Spherical Data*; Cambridge University Press: Cambridge, UK, 1987.

18. Huttunen, T.; Vanne, A. *End-to-End Process for HRTF Personalization*; Audio Engineering Society Convention 142; Audio Engineering Society: Berlin, Germany, 2017.

19. Lawless, H.T.; Heymann, H. *Sensory Evaluation of Food: Principles and Practices*, 2nd ed.; Springer Science & Business Media: Berlin, Germany, 2010.

20. Kuusinen, A.; Lokki, T. Wheel of Concert Hall Acoustics. *Acta Acust. United Acust.* **2017**, *103*, 185–188.

![applied sciences logo] *applied* *sciences*

MDPI

Article

Melodic Similarity and Applications Using Biologically-Inspired Techniques

Dimitrios Bountouridis [1],*, Daniel G. Brown [2], Frans Wiering [1] and Remco C. Veltkamp [1]

[1] Department of Information and Computing Sciences, Utrecht University, 3584 CC Utrecht, The Netherlands; f.wiering@uu.nl (F.W.); r.c.veltkamp@uu.nl (R.C.V.)

[2] David R. Cheriton School of Computer Science, University of Waterloo, Waterloo, ON N2L 3G1, Canada; dan.brown@uwaterloo.ca

* Correspondence: d.bountouridis@uu.nl; Tel.: +31-30-253-1172

Academic Editor: Meinard Müller
Received: 30 September 2017; Accepted: 27 November 2017; Published: 1 December 2017

Abstract: Music similarity is a complex concept that manifests itself in areas such as Music Information Retrieval (MIR), musicological analysis and music cognition. Modelling the similarity of two music items is key for a number of music-related applications, such as cover song detection and query-by-humming. Typically, similarity models are based on intuition, heuristics or small-scale cognitive experiments; thus, applicability to broader contexts cannot be guaranteed. We argue that data-driven tools and analysis methods, applied to songs known to be related, can potentially provide us with information regarding the fine-grained nature of music similarity. Interestingly, music and biological sequences share a number of parallel concepts; from the natural sequence-representation, to their mechanisms of generating variations, i.e., oral transmission and evolution respectively. As such, there is a great potential for applying scientific methods and tools from bioinformatics to music. Stripped-down from biological heuristics, certain bioinformatics approaches can be generalized to any type of sequence. Consequently, reliable and unbiased data-driven solutions to problems such as biological sequence similarity and conservation analysis can be applied to music similarity and stability analysis. Our paper relies on such an approach to tackle a number of tasks and more notably to model global melodic similarity.

Keywords: melodic similarity; alignment; stability; variation; bioinformatics

---

1. Introduction

In 2016, digital music revenues overtook physical revenues for the first time (www.ifpi.org/downloads/GMR2016.png), a testament to the music industry's adaptability to the digital age. Listeners are currently able to stream and explore massive collections of music such as Spotify's (www.spotify.com) library of around 30 million tracks. Such a development has changed not only the way people listen to music, but also the way they interact with it. According to a 2015 survey (www.midiaresearch.com/blog/midia-chart-of-the-week-music-discovery), 35% of users of streaming services use them to discover new songs and artists, new and exciting music for their unique personal taste or listening habits. At the same time, the proliferation of digital music services has raised the listeners' interest in the accompaniment chords (www.chordify.com), the lyrics (www.musixmatch.com), the original versions of a cover, the sample (loop) (www.whosampled.com) that a song uses and many more scenarios that service providers cannot deal with manually.

This development brings Music Information Retrieval (MIR) to the centre of attention. The field includes research about accurate and efficient computational methods, applied to various music retrieval and classification tasks such as melody retrieval, cover song detection, automatic chord extraction and of course music recommendation. Such applications require us to build representations of previously seen classes (e.g., sets of covers of the same song), which can be only compared to a query (e.g., a cover song whose original is unknown) by means of a meaningful music similarity function. A robust MIR system should model the fuzziness and uncertainty of the differences between two musical items perceived as similar. As Van Kranenburg argues specifically about folk song melodies: "knowledge about the relation between a desired melody and the way this melody is sung from memory" can increase the robustness of melody retrieval tasks [1].

However, this "knowledge", the exact mechanics of perceived similarity, is still unknown or incomplete [2]. This is not surprising considering music's inherently complex nature [3,4]. The perceived similarity between two musical pieces is known to be subjective: judgements of different individuals can vary significantly. Marsden [5] argues that similarity involves interpretation, which by itself is a personal creative act. Ellis et al. [6] argue that the individual perception of similarity can show variation depending on the listener's mood or familiarity with the musical culture and can even change through time. The individual interpretation can be affected also by the multidimensionality of music, since similarity between two songs can be a function of timbre, melody, rhythm, structure or indeed any combinations of those (or other) dimensions. To make matters worse, music similarity is known to be contextual, thus depending on the circumstances of comparison. Deliège [7] argues that similarity can appear as stable only when the context, "the structure of the natural world or a specific cultural system" is quite stable itself.

To overcome, or avoid addressing the aforementioned issues, many MIR approaches to similarity rely on cognition studies, expert heuristics, music theory or formalized models in general. Cognition studies are scientifically well-founded, but often cannot capture the general consensus due to practical limitations, such as access to a sufficient number of participants that fit a certain profile for the study. Expert knowledge, on the other hand, can be a valuable source of information, but with regard to music, expert knowledge cannot fully explain its highly complex nature and the sophisticated human perception. In addition, heuristic approaches have the risk of being descriptive rather than predictive. Formalized models founded on music theory typically neglect that it is not a theory of music perception of similarity. In addition, such models have the highest risk of being solely descriptive, thus not providing us with new knowledge. To their defence, all such approaches can have a certain practical validity, but limited explanatory power, as long as they are evaluated only on a reliable ground-truth and are applied to narrow contexts. Human ratings of similarity are highly problematic with studies showing that subjects are inconsistent with each other and even with themselves [8,9]. Regarding the assessment of similarity between song-triads particularly, Tversky [10] argues that subjects are affected by the song order of appearance and even the song popularity. Regarding the context, a one-fits-all model of similarity is impossible, and as Marsden argues: "the best one can hope for is a measure which will usefully approximate human judgements of similarity in a particular situation" [5].

As long as music cognition fails to provide us a blueprint of how to develop a computational, generalizable model of music similarity, we are required to explore alternative, data-driven approaches that aim to model the knowledge extracted from the data and the data relations. Data-driven music similarity is not a new concept in MIR, but such studies [11,12] have focused on high-level similarity (genre, artist) where listeners' opinions are fuzzy. Approaches on more fine-grained music similarity at the note or chord level, such as the work of Hu et al. [13], are scarce for a legitimate reason: in order for the data relations to be bias-free and visible, the data need to be organized in a proper-for-knowledge-extraction form. Properly annotated and disambiguated corpora of note-to-note or chord-to-chord relationships are extremely hard to find.

Fortunately, algorithms that properly organize sequential data have been widely used and are fundamental in the field of bioinformatics. One of the most notable algorithms from the vast bioinformatics toolbox, pairwise sequence alignment via dynamic programming, has been successfully adapted by MIR to compare musical items such as melodies [14] or chord sequences [15]. On closer look, musical and biological sequences are not as unrelated as one might think: even as early as the 1950s, it had been observed that they share a number of resembling concepts [16]. Krogh states that "the variation in a class of sequences can be described statistically, and this is the basis for most methods used in biological sequence analysis" [17]. By acknowledging that the variation of certain quantifiable musical features in a group of related music sequences can be described statistically, as well [18], we gain access to a number of sophisticated, data-driven approaches and bias-free tools that can be adopted from bioinformatics, allowing the modelling of music similarity.

1.1. From Bioinformatics to MIR

Bioinformatics use statistical and computational techniques to connect molecular biology and genetics. Bioinformatics deal with different types of data. DNA sequences carry most of the inherited genetic information of living organisms. These sequences can be represented as a string over a four-letter alphabet {A,C,G,T}, where each symbol represents a nucleotide base. DNA sequences can be as long as several billion symbols, depending on the organism. The instructions to form proteins, which are essential to living organisms, are encoded in the DNA in the form of subsequences or sections called genes. Through a translation process, certain genes are mapped into long chains of amino acids, which fold into three-dimensional protein structures. For computational purposes, proteins can likewise be considered as strings of characters (typically several hundred symbols) from a 20-letter alphabet (since there are 20 different common amino acids).

Music, unlike static forms of art, has a temporal nature. As such, music perception relies on temporal processing [19]. As Gurney argues regarding melodies specifically: "The elements are units succeeding one another in time; and though each in turn, by being definitely related to its neighbours, is felt as belonging to a larger whole" [20]. The same idea actually holds for other music elements, such as chords (notes sounding almost simultaneously) or rhythm. It is therefore not surprising that certain music items, such as symbolic scores, chord transcriptions and others, similarly to DNA or proteins, can be naturally represented as sequences of characters from a finite alphabet. When it comes to music applications, the importance of sequence representation has been demonstrated most notably by Casey and Slaney [21] and by numerous other works that adopted it over the years.

A core assumption of molecular biology is that of homology: related sequences diverge from a common ancestor through random processes, such as mutation, insertion, deletion, and more complex events, aided by natural selection. This process of genetic variation provides the basis for the biodiversity of organisms. Homologues might share preferentially "conserved" regions, subjected to fewer mutations compared to the rest of the sequence [22], which are considered crucial for the functionality of a protein [23]. Similarly, a fundamental observation in music is that music information passing orally, or in other form, can be subjected to noise. Due to our limited cognitive capacity, or for artistic purposes, a musical piece can change throughout a network of musical actors. A folk song that has been transmitted from mouth to mouth and from generation to generation, might differ dramatically from its original version. Even recorded songs can differ when covered by other artists or performed live. There is a strong resemblance to biological evolution since music homologues can occur by altering, inserting, deleting or duplicating music elements to a certain extent [16]. Intuitively also, certain salient parts of a melody or a chord progression are less likely to mutate, thus remaining "conserved", in an alternative version.

Identifying similarity is crucial not only for MIR, but for bioinformatics applications, as well. Finding homologues through sequence-similarity search is key. Besides the systematic organization, homologue search can help relate certain characteristic behaviours of a poorly-known protein sequence [24]. In addition, experimental results on model species can be applied to humans. Pairwise sequence alignment is the most popular method for assessing the similarity of two sequences. The idea is to introduce gaps '-' to sequences so that they share the same length, while placing "related" sequence elements in the same positions. As such, pairwise alignment aims to find the optimal alignment with respect to a scoring function that optimally captures the evolutionary relatedness between amino acids (how probable it is for one amino acid to be mutated to another). Another important bioinformatics application is finding conserved regions or patterns among multiple homologue sequences which allows for the estimation of their evolutionary distance, for phylogenetic analysis and more. This is achieved by aligning three or more sequences simultaneously, a process typically called Multiple Sequence Alignment (MSA).

1.2. Contribution

In this paper, we argue that MIR can benefit immensely by exploring the full potential of tools, methods and knowledge from the field of bioinformatics and biological sequence analysis, particularly considering melodic-similarity related applications. Despite the high resemblance of concepts (see Table 1), MIR has yet to fully adopt sophisticated solutions such as multiple sequence alignment. As Van Kranenburg suggested, there is a potential for MIR to harvest the bioinformatics' long history of algorithm development, improvement and optimization for biological sequence analysis [1].

Table 1. Shared concepts and terms between music and bioinformatics.

Music	Bioinformatics
Melodies, chord progressions	DNA, proteins
Oral transmission, cover songs	Evolution
Variations, covers	Homologues
Tune family, clique	Homology, family
Cover song identification, melody retrieval	Homologue detection
Stability	Conservation

Our previous works on aligning polyphonic voices [25] and melody retrieval [26] more notably, briefly touched on the relationship between MIR and bioinformatics. However, their ideas and bioinformatics-inspired solutions facilitated the work presented in this paper. As such, this paper's contribution relies first on establishing a strong connection between musical and biological sequences. This allows us to adopt analysis pipelines and algorithms from bioinformatics to: (a) gain new insights regarding music similarity by performing a stability analysis, and (b) present novel solutions for tackling melody retrieval by modelling global similarity. Most importantly, our pipelines are purely data-driven and free of heuristics, as opposed to other MIR methods. To validate the generalization-ability of our approach, we apply it to two melodic datasets of different music. As such, we diverge from previous MIR studies that focused on a specific subset of all possible music. In addition, previous work on datasets of chord sequences [27] also supports the usability of this approach to more than melodic data.

The remainder of this paper is organized as follows: Section 2 acts as an introduction the fundamental sequence comparison and analysis tools derived from bioinformatics. Section 3 describes the musical datasets used in our work. From there on, we apply the bioinformatics methods and tools to the datasets. Section 4 investigates the concept of "meaningful" alignments, while Section 5 uses the findings of 4 to present an analysis of music stability. Section 6 tackles the problems of modelling global similarity. Finally, Section 7 discusses the conclusions of this paper.

2. Methods and Tools

This section aims to describe the fundamental methods used in biological sequence analysis: pairwise alignment and multiple sequence alignment. Understanding their mechanics and limitations is crucial for successfully applying them to MIR tasks. However, the reader familiar with these methods can skip to Section 3 directly.

2.1. Pairwise Alignment

An intuitive method for DNA or protein sequence comparison is the Levenshtein (or Edit) distance, which computes the minimal number of one-symbol substitutions, insertions and deletions to transform one sequence into the other. Such operations can be naturally mapped to the biological process of mutation. Given a cost for each operation, the weighted Levenshtein distance can be computed using dynamic programming. The major drawback of the Levenshtein distance is that it captures the divergence of the two sequences rather than their relatedness or, the important to this paper, similarity. In addition, it does not allow for identifying conserved regions between the sequences, since it is a purely mathematical distance function. As such, computing the similarity of two DNA or protein sequences is typically performed using alignment, the converse to Edit distance. During alignment, gaps '-' that represent symbols that were deleted from the sequences via the process of evolution [28], are introduced in the sequences, until they have the same length and the amount of "relatedness" between symbols at corresponding positions is maximized.

More formally, consider two sequences over an alphabet of symbols $\mathcal{A}$, $X := x_1, x_2, .., x_n$ and $Y := y_1, y_2, .., y_m$ with all $x_i, y_i \in \mathcal{A}$. An alignment A of X and Y, consists of two sequences X' and Y' over $\{-\} \cup \mathcal{A}$, such that $|X'| = |Y'| = L$, where if we remove all '-' from X', Y' we are left with X and Y respectively. The number of possible alignments A for a pair of sequences is exponential in n and m, so an optimal alignment should be selected given a scoring function that typically derives from a model of "relatedness" between the symbols of $\mathcal{A}$, where the goal is to put similar symbols at the same position. The most typical such scoring function is the alignment score:

$$c(A) = \sum_{p=1}^{L} v(x'_p, y'_p) \tag{1}$$

where $v : \mathcal{A} \times \mathcal{A} \to \mathbb{R}$. The scoring function v is typically encoded as an $|\mathcal{A}| \times |\mathcal{A}|$ matrix called the substitution matrix. Most pairwise alignment methods use a Dynamic Programming (DP) method, credited to Needleman and Wunsch [29], which computes the optimal (highest scoring) alignment by filling a cost matrix D recursively:

$$D(i,j) = max \begin{cases} D(i-1, j-1) + v(x_i, y_j) \\ D(i-1, j) - \gamma \\ D(i, j-1) - \gamma \end{cases} \tag{2}$$

where γ is the gap penalty for aligning a symbol to a gap. An extension uses an affine gap penalty based on the assumption that the occurrence of consecutive deletions/insertions is more probable than the occurrence of the same amount of isolated mutations [28]: for a gap of length z, the gap penalty would be:

$$\gamma(z) = -d - (z-1)e \tag{3}$$

where d and e are the gap open and gap extension penalties respectively. To optimize an alignment that uses an affine gap penalty requires a slightly more complex DP algorithm [30]. In the simple non-affine gap case, the score of the optimal alignment is stored in $D(n, m)$, while the alignment itself can be obtained by backtracking from $D(n, m)$ to $D(0, 0)$. The Needleman and Wunsch approach is a global alignment method, since it aims to find the best score among alignments of full-length sequences. On the other hand, the local alignment framework, first optimized by Smith and Waterman [31], aims to find the highest scoring alignments of partial sequences by tracking back from $max(D(i, j))$ instead of $D(n, m)$, and by forcing all $D(i, j)$ to be non-negative. Local alignment allows for the identification of substrings (patterns) of high similarity.

When affine gaps are not considered, meaningful, high-quality alignments are solely dependent on the knowledge captured by the substitution matrix used [30]: optimal alignments with good scoring matrices will assign high scores to pairs of related sequences, while giving a low alignment score to unrelated sequences. More formally, given the two sequences X, Y their alignment score $c(A)$ should represent the relative likelihood that the two sequences are related as opposed to being unrelated (aligned by chance). This is typically modelled by a ratio, denoted as odds ratio:

$$\frac{P(X,Y|M)}{P(X,Y|R)} \tag{4}$$

where M is a probabilistic model of related sequences and R is a model generating unrelated sequences. If q_a is the frequency of a symbol a, and both X and Y have the same distribution under R, then for the random alignment case aligned pairs happen independently, which translates to:

$$P(X,Y|R) = P(X|R)P(Y|R) = \prod_i q_{x_i} \prod_j q_{y_j} \tag{5}$$

For the matching case, where aligned pairs happen with a joint probability p, the probability for the alignment is:

$$P(X,Y|M) = \prod_i p_{x_i y_i} \tag{6}$$

In order to get an additive scoring system, it is standard practice to get the logarithm of Equation (3), which after substitution becomes:

$$log\frac{P(X,Y|M)}{P(X,Y|R)} = \sum_i log\left(\frac{p_{x_i y_i}}{q_{x_i} q_{y_i}}\right) \tag{7}$$

A substitution matrix can be considered nothing more than a matrix arrangement of the $log(p_{x_i y_i}/q_{x_i} q_{y_i})$ values (scores) of all possible pairwise symbol combinations.

Sequence alignment via dynamic programming and its time-series counterpart, Dynamic Time Warping (DTW), have been fundamental tools for many MIR tasks since first being applied in a melody retrieval task by Mongeau and Sankoff [32]. Alignment, despite being considered an ill-posed problem for strongly deviating versions of a musical piece [33], has proven to be very useful for identification or classification tasks where strong similarities are present [1,34] and high scoring alignment has been shown to correlate well with human judgements [35,36]. It has been used for cover song detection [37], pattern mining [38], extensively for query-by-humming [14,39] and in other MIR tasks. Interestingly, DTW has been extended to align items that cannot be naturally represented as single sequences, such as polyphonic music [40] or audio [41,42]. Consequently, alignment has been also key to finding correspondences among related music items of not the same format (typically called music synchronization): it has been used for score following, the task of aligning different music representations such as audio and score or MIDI (Musical Instrument Digital Interface) [41,43]. Describing alignment's numerous MIR applications exceeds the scope of this study. However, a complete overview of DTW in music up until 2007 can be found in the work of Müller [44].

2.2. Multiple Sequence Alignment

A multiple sequence alignment inserts gaps into more than two sequences over an alphabet so that they have the same length and the relatedness between symbols in the same columns is maximized. Formally, given k sequences $s_1, s_2, ..., s_k$ over an alphabet $\mathcal{A}$ and a gap symbol '-' $\notin \mathcal{A}$, and let $g : (\{-\} \cup \mathcal{A})^* \to \mathcal{A}^*$ be a mapping that removes all gaps from a sequence containing gaps. A multiple sequence alignment A consists of k sequences $s_1', s_2', ..., s_k'$ over $\{-\} \cup \mathcal{A}$ such that $g(s_i') = s_i$ for all i, $(s_{1,p}', s_{2,p}', ..., s_{k,p}') \neq (-, ..., -)$ for all p, and $|s_i'| = L$ for all i.

Similar to pairwise alignment, there is a great number of possible MSAs for a single input of sequences [30]. We typically want to pick the most "meaningful" considering our task at hand. More formally: given an objective scoring function $c : A \to \mathbb{R}$ that maps each alignment to a real number, we are interested in $A' = \arg\max_A (c(A))$. There are many such functions [28], but the most widely used is the Weighted Sum-Of-Pairs (WSOP or SOP) [45], a summing of scores of all symbol-pairs per column. Let m_i^j be the i-th column j-th row of A, the SOP is defined as such:

$$c(A) = \sum_i \sum_{k<l}^{L} w_{k,l} v(m_i^k, m_i^l) \tag{8}$$

where $w_{k,l}$ is a weight assigned to the pair of sequences k, l and $k < l$ corresponds to an iteration over all pairs of rows in the column. Naturally, the objective function can be adapted to accommodate affine gaps. Computing the optimal MSA is unfortunately NP-complete [46] and cannot be used in realistic scenarios that include numerous and long sequences. Therefore in the field of bioinformatics, heuristic approaches that give good alignments, though not guaranteed to be optimal, have been developed. According to Kemena and Notredame [47], more than 100 different MSA algorithms have been proposed over the last 30 years but discussing them in detail exceeds the scope of this paper.

MSA algorithms have found a rather small application in MIR. Liu [48] uses the progressive alignment algorithm to compare different music performances represented as strings derived from chroma features (distribution of the signal's energy across a set of pitch classes). In a similar manner Wang et al. [49] showed that progressive alignment of multiple versions can stabilize the comparison for hard-to-align recordings that can lead to an increase in alignment accuracy and robustness. Finally in a tangential task, Knees et al. [50] use a progressive alignment approach to align multiple lyrics gathered from various online sources.

3. Melodic Sequence Data

Music comprises sound events that can be pitched or unpitched (percussive) with either stable or unstable pitch. In the context of this paper we consider the tone, a fixed frequency sound (pitch), to be the most important musical element. In music notation (scores), tones are represented as notes with accompanying duration values. A series of notes arranged in time and perceived as a distinct group or idea, is what we roughly define as a melody, although years of musicological studies have failed to agree on a consensus definition. Poliner et al. [51] define it as "the single (monophonic) pitch sequence that a listener might reproduce if asked to whistle or hum a piece of polyphonic music, and that a listener would recognize as being the essence of that music when heard in comparison." As Kim et al. [52] also mention, one can recognize a song (out of all known songs) just by its melody even though it might have been corrupted with noise or cut short. This observation is a testament to melody's importance to music perception. As such, melodies have been at the centre of musicological research [53] and music cognition [54]. In MIR, melody extraction from audio has been an active research topic, since melodies can act as robust and efficient features for song retrieval [55]. Query-by-humming, i.e., retrieving similar items using a sung melody as a query, has been also an important, on-going MIR task [56,57].

When it comes to comparing melodies in terms of their similarity, sequence representation is key; we need to carefully select the music features that we will represent as sequences [47]. As Volk et al. [2] argue based on relevant studies, music similarity works on many dimensions, such as melodic, rhythmic or harmonic, but the musicological insights regarding the relative importance of each dimension are insufficient. The works of Van Kranenburg [1] and Hillewaere et al. [58] revealed the importance of the pitch dimension, so our work considers melodies as pitch-contours, meaning series of relative pitch transitions constrained to the region between $+11$ and -11 semitones (folded to one octave so that a jump of an octave is treated as unison and therefore ignored). Besides their simplicity and key-invariance, pitch contours have been found to be more significant to listeners for assessing melodic similarity than alternative representations [59]. In our work, all sequences of pitch transitions are mapped to an extension of the 20-letter alphabet that is used to represent the naturally occurring amino acid for ease of adaptation.

3.1. Datasets

Reliable analysis and modelling of similarity requires first and foremost datasets of unambiguous relationships between music items. Marsden [5] among others, makes a strong case regarding the validity of similarity ranking annotations, considering the paradigm differences of the listening experiments that generated them. However, he is more supportive to binary or definite annotations of similarity, such as songs known to be covers, or songs known to be related from musicological studies. Such data can be used to verify a computational model with regard to its retrieval or classification performance, since the distance for music items within a category should be less than the distance of items belonging to different categories. As such, this paper uses two datasets of symbolically represented melodies of varying size and nature, containing melodies that are considered related (e.g., covers of the same song) grouped into definite groups called either families, classes or cliques. Summary statistics for both sets are presented in Table 2.

The Annotated Corpus of the Meertens Tune Collections [60], or TUNEFAM-26, is a set of 360 Dutch folk songs grouped into 26 "tune families" by Meertens Institute experts. Each contains a group of melody renditions related through an oral transmission process. For this dataset, expert annotators assessed the perceived similarity of every melody over a set of dimensions (contour, rhythm, lyrics, etc.) to a set of 26 prototype "reference melodies". In addition, the dataset contains 1426 annotated motif occurrences grouped into 104 classes, where "motifs" correspond to recurring patterns inside the melodies of a tune family. The Cover Song Variation dataset [61], or CSV-60, is a set of expert-annotated, symbolically-represented vocal melodies derived from matching structural segments (such as verses and choruses) of different renditions of sixty pop and rock songs. CSV-60 is inherently different from TUNEFAM-26 in two ways. First, the grouping of melodies into classes is certain: the songs were pre-chosen as known covers of songs of interest. Secondly, cover songs are typically not a by-product of an oral transmission process since cover artists have access to the original version.

Table 2. Summary statistics for the datasets considered in our work. We also present the Area Under the Curve (AUC) value for the Receiver Operating Characteristic curve (ROC) on the Percentage Sequence Identity (PID). Given two aligned sequences, the PID score is simply the number of identical positions in the alignment divided by the number of aligned positions [62]. The higher the AUC PID the more similar the sequences are in a clique compared to the whole dataset. It should be noted that the alphabet size presented corresponds to the number of unique symbols appearing in the dataset.

Summary statistics	TUNEFAM-26	CSV-60
Number of cliques	26	60
Clique Size median (var)	13.0 (4.016)	4.0 (1.146)
Sequence Length median (var)	43.0 (15.003)	26.0 (10.736)
AUC PID	0.84	0.94
Alphabet Size	22	22

4. Multiple Sequence Alignment Quality for Melodic Sequences

This paper's main approach on modelling melodic similarity relies on capturing the variation among two or more perceived-as-similar melodies. For that we need trustworthy, "meaningful" alignments of related music sequences, such that the statistical properties of the alignment can inform us about the note-to-note relationships. Since such data can be hard to find, we are required to align related sequences using alignment algorithms. Alignment, pairwise or otherwise, with notable exceptions [63,64] has been typically used as an out-of-the-box tool to align instances of music sequences, with the sole purpose of using its score output further in a retrieval pipeline. The quality or musical appropriateness of the alignment of symbols themselves has always been evaluated via a proxy, i.e., some kind of music retrieval scenario. As long as the alignment-pipeline outperformed other approaches, its utility was considered significant. The major problem however, is that outside the proxy strategy, there are no studies or musical intuition to prefer one alignment over the other.

Identifying the features that make a "good", meaningful alignment is an intricate task, not only for musical but biological sequences as well. Interestingly, proteins are folded into diverse and complex three-dimensional structures. Structure motifs (not to be confused with the homonym musical concept) diverge slower in the evolutionary time scale than sequences, and consequently homology detection among highly divergent sequences is easier in the structural than the sequence domain, though the actual algorithms for three-dimensional shape alignment are complex. As such, structure motifs have been used to aid the alignment of highly-divergent sequences [65]. In addition, reference alignments produced from biological information, such as a conserved structure, have been frequently used to assess the quality of an MSA [66].

We argue that similar to biological sequences, a "good" meaningful alignment of musical significance, can be only evaluated via a trustworthy reference alignment. Previous related work [67] generated "trustworthy" alignments of the CSV-60 set by using a progressive alignment algorithm extended on three musical dimensions (pitch, onset, duration). Bountouridis and Van Balen's choice was based largely on intuition, since there is no literature supporting those three dimensions. Prätzlich and Müller [64] investigated the evaluation of music alignment by using solely triplets of recordings of the same piece and made clear that there are theoretical considerations of alignment quality-assessment without a reference alignment. Therefore, the question becomes whether there exists a musical analogy to the protein structure motifs.

In musicology shared, transformed but yet recognizable musical patterns are called "variations" and according to musicological and cognitive studies, variations are essential to the human perception of music similarity [2]. Specifically when it comes to classifying folk songs into tune families, i.e., groups of songs with a common ancestor, Cowdery [68] considers the shared patterns to be a key criterion. An annotation study on Dutch folk songs by Volk and Van Kranenuburg [4] also supported this claim by proving that shared, stable musical patterns, called motifs were important for the expert assessment of music similarity. Consequently, we can theoretically use the motif alignment as reference for evaluating the quality of musical sequence alignment. For example, consider the following sequences with expert annotated motifs "AB" (red) and "AFF" (cyan): AFFGABBBBC, ABDDBBC and AFFABB. Two possible alignments with equal SOP scores are:

```
AFFGABB-BBC    AFFGABB-BBC
----ABDDBBC    A----BDDBBC
AFF-ABB----    AFF-ABB----
```

From a musicological perspective though, the first alignment is considered of higher quality, since it aligns perfectly those subsequences that are annotated as same-label motifs. It is of high importance to investigate which MSA algorithms and settings are optimal with regard to motif alignment (for example, which algorithm would be more likely to generate the first alignment rather than the second). The following paragraphs describe the appropriate experiments to answer such question.

Our experiment pipeline comprises aligning a group of related sequences (that include motifs) using different motif-agnostic MSA strategies, and then comparing the resulting alignment of motifs to a reference optimal motif alignment. The comparison is not based on a distance function between the alignments, but rather on assigning a score to both of them. Besides the different MSA strategies (to be discussed in Section 4.3), the pipeline requires the following: first, a motif alignment scoring function that is well-founded (see Section 4.1). Secondly, it requires a dataset of musical sequences that contain annotated motifs for each clique, combined with trustworthy alignments of these motifs that would act as a reference (see Section 4.2).

4.1. Motif Alignment Scoring

The only information available to compute a meaningful motif-based MSA score is the motifs' position in the sequence, length and notes they contain. Due to the lack of knowledge regarding which pairs of pitches should be aligned together, the motif alignment scoring method cannot be founded on the pitch dimension. We are confident for only one thing: the notes belonging to same-labelled motifs should be somehow aligned. As a consequence, we focus on an intuitive scoring function that is maximized when same-labelled motifs are maximally overlapped. Given a function $label(x_i)$ that returns the motif label of the i-th note of a sequence X, the WSOP score (denoted motif-SOP) of an MSA is based on the following scoring function:

$$v(x_m, y_m) = \begin{cases} +1 & \text{if } label(x_m) = label(y_m) \\ -1 & \text{if } label(x_m) \neq label(y_m) \\ 0 & \text{if } label(x_m) = \varnothing \text{ or } label(y_m) = \varnothing \end{cases} \quad (9)$$

In other words, we only penalize those alignments that align notes belonging to different motifs. Alignment between notes not belonging to any motif ($label(x_i) = \varnothing$), and labelled notes are considered neutral since no studies or intuition suggests otherwise. The particular scoring function would assign the same motif-SOP score for both the following alignments, since only the alignment of motif labels (represented as colours) is taken into consideration:

```
AFFGABB-BBC    -AFFGABB-BBC
----ABDDBBC    -----ABDDBBC
-AFFABB----    A-FF-ABB----
```

4.2. Dataset and Reference Motif Alignments

The TUNEFAM-26 dataset is the best benchmark for our experiment, since it contains related melodies (grouped into tune families) with a number of subsequences annotated by experts and uniquely labelled as motifs (see Figure 1). It is however, not the optimal benchmark since the expert annotated motifs of the same label, which can be of different lengths, do not come pre-aligned; we know which sub-sequences in the family's melodies are motifs, but we do not know their note-to-note alignment. Since there are no trustworthy motif alignments, the optimal alignment should be a by-product of the motif-SOP function and the intuition behind it, i.e., the reference alignment should be the one that maximizes the motif-SOP score. In order to acquire that for each family, through visual inspection, we manually align the motif variations. At the same time, we consider the motif-SOP score of the original unaligned sequences as the lower bound min_{mSOP}, i.e., the worst possible scenario. The min_{mSOP} and max_{mSOP} scores allow us to normalize any motif-SOP score to a meaningful $[0, 1]$ range.

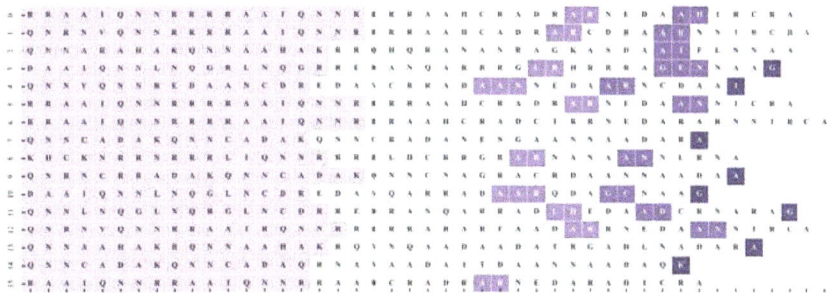

Figure 1. The 15 unaligned sequences of the tune family "Daar ging een heer". Colours correspond to motif labels. White colour indicates no motif label.

4.3. Multiple Sequence Alignment Algorithms and Settings

From the numerous MSA algorithms, we selected three based on many factors including simplicity, popularity or quality of results on several bioinformatic benchmarks. One of the simplest approaches to MSA, named "star" alignment, aims at employing only pairwise alignments for building the final

MSA. The idea is to first find the most "central" among the sequences, pairwise align it to each one of the rest and then combine the pairwise alignments into a single MSA. This method does not necessarily attempt to optimize the objective function (see Section 2.2) and as such is rarely used. In our case, star alignment can act as a naive baseline for the more sophisticated algorithms to be compared against.

Progressive Alignment (PA) [69] is one of the most popular and intuitive approaches, and it comprises three fundamental steps. At first, all pairwise alignments between sequences are computed to determine the similarity between each pair. At the second step, a similarity tree (guide tree) is constructed using a hierarchical clustering method, which in biological sequences is sometimes used to attempt to identify evolutionary relationships between taxa. Finally, working from the leaves of the tree to the root, one aligns alignments, until reaching the root of the tree, where a single MSA is built. The drawback of PA, is that incorrect gaps (especially those at early stages) are retained throughout the process since the moment they are first inserted (the "once a gap, always a gap" rule). Iterative refinement methods [70,71] aim to tackle this problem by iteratively removing each sequence and realigning it with a profile created from the MSA of the rest, until an objective function has been maximized. Our experiments use the PA-based T-COFFEE software (Tree-based consistency objective function for alignment evaluation) [72]. T-COFFEE aims to tackle the problem by making better use of information in the early stages . It uses an objective function (called COFFEE [73]) that first builds a library of all optimal pairwise alignments and secondly, scores a multiple sequence alignment by measuring its consistency with the library: how many of the aligned pairs in the MSA appear in the library.

Locating very similar short and shared sub-regions between large sequences has been in important task in bioinformatics. Such segments can efficiently reduce MSA runtimes and as a consequence, MSA solutions that incorporate some of form of segmentation, such as DIALIGN[74] and MAFFT [75], have found successful application. MAFFT in particular, is a progressive alignment method at its core, but incorporates the Fast Fourier Transform (FFT) for biological sequences. In addition, MAFFT allows the usage of the iterative refinement method. For non-biological sequences, MAFFT offers a "text" alignment option that excludes biological and chemically-inspired heuristics from its pipeline. In such a case, segmenting the sequences becomes a by-product of MAFFT's objective function that incorporates both a WSOP and a COFFEE-like scoring. According to MAFFT's website (mafft.cbrc.jp/alignment/software), "the use of the WSOP score has the merit that a pattern of gaps can be incorporated into the objective function".

MAFFT offers three different strategies for the initial pairwise alignment, that behave differently with regard to the structure of the sequences. Local alignment with affine gap costs `localpair` is appropriate for unaligned sequences centred around a conserved region. The `genafpair` strategy uses local alignment with generalized affine gap costs [76] and is appropriate for sequences with several conserved sub-sequences in long unalignable regions. Global alignment with affine gap costs `globalpair` is appropriate for throughout alignable sequences. A lesser known option, which can be applied on top of `localpair` and `globalpair` strategies, is `allowshift` which is appropriate for sequences that are largely similar but contaminated by small dissimilar regions.

Each MSA algorithm aims to find the alignment that maximizes the SOP score on the Identity (ID) scoring scheme, i.e., $v(x,y) = +1$ if $x = y$ and $v(x,y) = -1$ if $x \neq y$. As a matrix, the ID scheme has +1 in the diagonal and -1 otherwise. The effect and importance of gap penalties, or gap settings (see Equation (3)), is well known for biological sequences [77] and for musical sequences as well [78]. Understanding their behaviour with regard to the MSA is crucial, especially when different matrices are used. Since literature suggests setting them empirically [77] and the ID matrix is used on each MSA algorithm in our case, we experiment with a only a small variety of gap settings. At the same time, we keep in mind that there is no guarantee that these settings optimize the performance of all MSA algorithms. Regarding T-COFFEE, such penalties are not essential when building the MSA, since in theory the penalties are estimated from the library of pairwise alignments. In practice, it is suggested to experiment with different settings while keeping in mind that the penalties are not related to the substitution matrix. Gap open can be in the range of $[0, -5000]$ and gap extension in the range of $[-1, -10]$.

4.4. Results

For each clique of sequences we generated a reference motif alignment manually, and computed its motif-SOP score (called S_{ref}). At the same time, for each motif-agnostic configuration (MSA algorithm, gap settings), we aligned the melodic sequences. Each resulting alignment was also assigned a motif-SOP score (called S_{auto}). In order to identify the best MSA configuration with respect to motif alignment, we compute its normalized motif-SOP score S_{ref}/S_{auto}.

Before proceeding into the quantitative results, it is worth visually examining the alignments created by the MSA algorithms. Figure 2 presents different alignments of the tune family "Daar_ging_een_heer_1" for a number of configurations. Regarding quantitative results, Figure 3 and Table 3 present the normalized motif-SOP score for different configurations. There are a number of observations that become immediately apparent: first, the normalized motif-SOP score can be less than zero, since the original unaligned sequences, that act as the lower bound, may include correctly aligned motifs by random chance (see Figure 1). Secondly as expected, star alignment is the worst performing algorithm across all gap settings. Regarding the relative performance of the configurations themselves (excluding star alignment), a Friedman significance test showed that there was a statistically significant difference in motif-SOP depending on the configuration with $p = 0.041$. However, post hoc analysis with Wilcoxon signed-rank tests and Bonferroni correction, revealed that there were no significant differences among any pair of configurations.

Regarding the overall performance of the algorithms themselves, a Friedman significance test showed that there was a statistically significant difference in normalized motif-SOP depending on the algorithm with $p < 10^{-6}$. Post hoc analysis with Wilcoxon signed-rank tests and a Bonferroni correction resulted in a significance level set at $p < 0.003$. p values for all possible pairs are presented in Table 4. It is clear that MAFFT, run with the `globalpair` strategy, outperforms T-COFFEE and that of the three MAFFT strategies, `globalpair` performs the best. A Wilcoxon signed-rank test between all the MAFFT algorithms using and not using the `allowshift` option, revealed that the `allowshift` option does not have a significant impact on the results, $p = 0.11$.

Finally, regarding the gap settings, significance tests showed that for MAFFT and T-COFFEE in general, there is a significant difference depending on the gap penalties used. For T-COFFEE in particular, large gap settings such as $(-60, -3)$ or $(-40, -2)$ are not recommended. For MAFFT on the other hand, small gap penalties, such as $(-0.8, -0.5)$ should be avoided. However it should be noted that for each particular MAFFT strategy, gap settings have no significant effect.

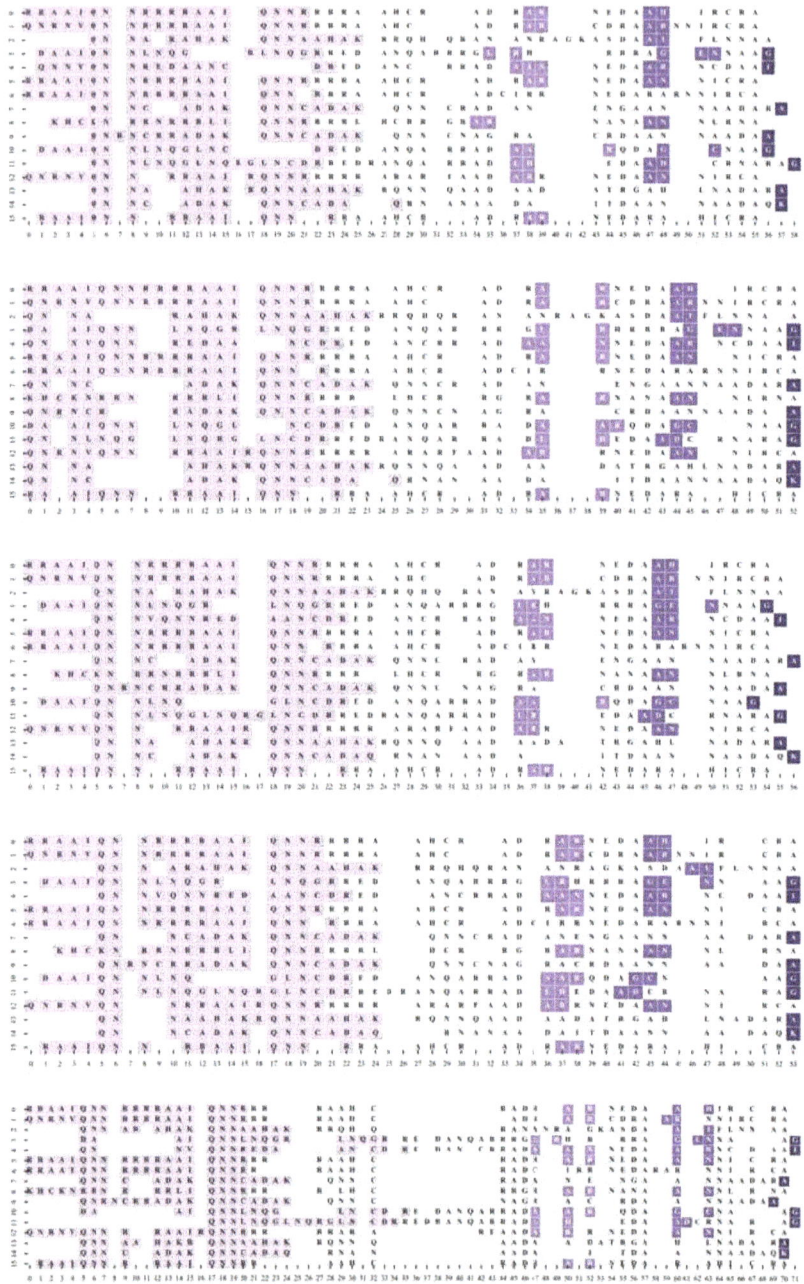

Figure 2. Automatically aligned melodic sequences of the tune family "Daar ging een heer" using the following configurations (**top–bottom**): MAFFT-genafpair-4.-2., MAFFT-globalpair-allowshift-4.-2., MAFFT-loacalpair-2.-1.5, MAFFT-localpair-allowshift-3.-1.5 and T-COFFEE-8-0.5. Colours correspond to motif labels. White colour indicates no motif label.

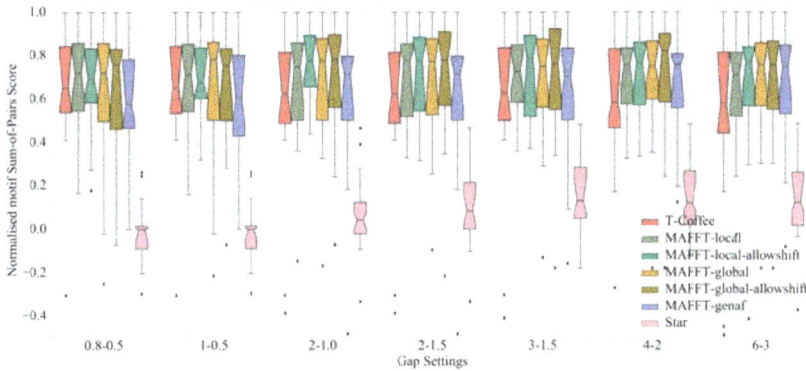

Figure 3. Normalized motif-Sum-Of-Pairs (SOP) score (*y*-axis) for different gap settings (*x*-axis) and Multiple Sequence Alignment (MSA) algorithms.

Table 3. Median (standard deviation) normalized motif-SOP scores for different MSA algorithms and gap settings. For T-COFFEE, the gap open values are multiplied by 10.

Algorithm	0.8–0.5	1–0.5	2–1.0	2–1.5	3–1.5	4–2	6–3
MAFFT-genaf	0.57 (0.95)	0.61 (0.83)	0.71 (0.68)	0.71 (0.68)	0.70 (0.61)	0.76 (0.53)	0.75 (0.80)
MAFFT-global	0.72 (0.60)	0.78 (0.60)	0.77 (0.49)	0.77 (0.48)	0.75 (0.50)	0.75 (0.41)	0.76 (0.25)
MAFFT-global-allowshift	0.76 (0.83)	0.75 (0.70)	0.78 (0.49)	0.78 (0.47)	0.76 (0.46)	0.82 (0.40)	0.76 (0.26)
MAFFT-local	0.72 (0.58)	0.71 (0.57)	0.75 (0.50)	0.78 (0.45)	0.73 (0.46)	0.76 (0.38)	0.71 (0.24)
MAFFT-local-allowshift	0.69 (0.68)	0.67 (0.72)	0.78 (0.60)	0.77 (0.45)	0.79 (0.45)	0.77 (0.35)	0.77 (0.29)
T-COFFEE	0.65 (0.72)	0.65 (0.72)	0.62 (0.78)	0.62 (0.78)	0.63 (0.80)	0.58 (0.95)	0.58 (1.04)
Star	0.00 (0.49)	0.00 (0.48)	0.04 (0.33)	0.08 (0.37)	0.13 (0.37)	0.12 (0.29)	0.12 (0.24)

Table 4. *p* values of the Wilcoxon signed-rank tests for pairs of algorithms with regard to the normalized motif-SOP score. "-a" indicates the `allowshift` option. *p*-values larger than 0.05 are not presented.

Algorithm	MAFFT-genaf	MAFFT-global	MAFFT-global-a	MAFFT-local	MAFFT-local-a
MAFFT-global	$< 10^{-6}$				
MAFFT-global-a	$< 10^{-5}$				
MAFFT-local					
MAFFT-local-a	$< 10^{-3}$				
T-COFFEE		$< 10^{-4}$	$< 10^{-4}$		

4.5. Discussion

In this section, we first established a measure of MSA quality based on motifs. Secondly, we evaluated different MSA algorithms and gap settings on a dataset of folk song melodies. Despite the small dataset of 26 tune families, the results offer strong proof about the benefits of the MSA algorithms, and MAFFT in particular. Regarding MAFFT's success, we hypothesize that it can be attributed to its objective function that results to gap-free segments. According to Margulis [79], the phrase structure of a melody is of major importance for the human perception of variation patterns. By treating the located sub-regions as gap-free segments, MAFFT can be the closest to partitioning melodies into perceptually meaningful units without using heuristics or expert knowledge.

In general, by establishing a reliable strategy to align multiple instances of melodies, we eliminate the prerequisite to invent a retrieval/classification proxy to assess the quality of an alignment. We can also now benefit from both the alignment score and the alignment's structure itself. Particularly regarding the latter, since the alignment of notes is musically significant, we can now extract knowledge about their relationships. For example, we can perform reliable analysis on notions such as stability (as we do in the following Section 5) or generate models of similarity (as we do in Section 6).

5. Analysis of Melodic Stability

It has been theorized that our perception and memorization of melodies is dynamic, meaning that certain musical events throughout a melody's length, can be perceived as more stable (resistant to change) than others depending on the context [80]. Klusen et al. [81] showed that every note in a melody can be altered in an oral transmission scenario, but some notes are more stable than others. Numerous studies from cognition [80,82,83] to corpus-based analysis points-of-view [84], have also evaluated the importance of certain musical factors with regard to their influence on the perceived music stability. Since the alteration of stable elements can affect the process of recognition, stability is also a key component for understanding music similarity. Unsurprisingly, stability and music variation (stability's counterpart) have been at the core of both musicology and MIR. From a musicological perspective, knowledge of the mechanics of those concepts would allow researchers to trace, classify or possibly even pinpoint in time variations of songs. Similarly, scientists from computational disciplines may use knowledge of stable musical elements to improve the automatic classification and retrieval of musical objects, such as the work of Van Balen et al. [85]. Therefore, before proceeding into modelling music similarity (see Section 6), it is worth investigating the complementary concept of music stability.

Interestingly, conservation is at the centre of biological sequence analysis, in the same way that stability is at the core of musicology or MIR. As Valdar [23] nicely describes, a multiple sequence alignment of protein homologue sequences (together with the phylogeny) is a historical record that tells a story about the evolutionary processes applied and how they shaped a protein through time. Useful and important regions of a protein sequence often appear as "conserved" columns in the MSA, and major sequence events that appear on a phylogenetic tree often correspond to epochal moments in natural history.

In this section we argue that, much as an MSA of protein homologues can inform us about the statistical properties of the evolutionary processes, an MSA of related melodies can provide us with valuable information regarding the processes of musical variation. We aim to determine and analyse regions of less variation inside a selection of related melodies, or in other words, regions of melodic stability. Analysing stability requires trustworthy MSAs such that the assignment of corresponding notes across different versions can be directly observed by looking at the MSA's columns. The findings of Section 4 allows us to be confident regarding the results of a stability analysis since it can be conducted on high-quality, musically meaningful alignments.

5.1. Setup

We are interested in applying the best alignment configuration (as established on Section 4) to the TUNEFAM-26 and CSV-60 melodic datasets. We can later perform an analysis on the aligned cliques (tune families or cover song melodies) by using an appropriate measure of stability applied on each column of the MSA. The results from Section 4 have indicated that the best MSA algorithm for melodic sequences is MAFFT, while its `globalpair` and `localpair` strategies are indistinguishable in terms of alignment quality. Gap settings have little or no effect per strategy, MAFFT options and gap penalties had minimal effect on alignment quality, so we explored several parameterizations: MAFFT-globalpair with $(-4,-2)$ gap penalties, MAFFT-globalpair-allowshift with $(-4,-2)$ gap penalties and MAFFT-localpair-allowshift with $(-2,-1)$ gap penalties.

A quantitative measure of stability, suitable for music sequences, does not exist as a result of the lack of supporting literature and research. Nevertheless, Bountouridis and Van Balen [67] use a probabilistic interpretation of the WSOP measure that aims to answer the following question: given that we observe a single, randomly chosen melodic element, what is the probability for this element to appear unchanged when we observe a new, unseen a variation of it. In practice, given a set of k aligned sequences of length m such as $S_i : s_{i,1}, s_{i,2}, ..., s_{i,m}$, the stability of the non-gap symbol e in position j is defined as:

$$stab(e,j) = \frac{\sum_{i=1}^{k} |s_{i,j} = e| - 1}{k - 1} \tag{10}$$

while the stability of the j-th MSA column is simply $PS_j = \sum stab(e,j)$ over all unique e.

It is worth examining the related bioinformatics literature regarding the equivalent concept of conservation scores. Valdar [23] mentions that "there is no rigorous mathematical test for judging a conservation measure". A scoring method can be only judged with respect to biochemical intuition, and therefore a number of conservation scores have been proposed through the years [22]. The same authors list a number of intuitive prerequisites that a conservation score should fulfil, including sequence weighting (to avoid bias due to near-duplicate sequences) or the consideration of prior amino acid frequencies. However, applying the same prerequisites to music sequences is not supported by any musical literature. Consequently, our analysis adopts two widely used and interpretable conservation scores from bioinformatics: the WSOP score (already discussed thoroughly) and Information Content (IC). Based on Shannon's entropy, the IC score of the j-th column is defined as such:

$$IC_j = \sum_{i=1}^{N_a} P_{e,j} log(\frac{P_{e,j}}{Q_e}) \tag{11}$$

where N_a is alphabet size, $P_{e,j}$ is the frequency of a particular symbol e in the j-th column, while Q_e is the expected frequency of symbol e in the dataset (prior). It should be noted that symbols in a column with zero frequency are not taken into account.

5.2. Analysis

The next paragraphs present a brief analysis on stability and variation with regard to two music dimensions, position and pitch intervals. However, it is possible to extend the analysis to dimensions such as note durations [67] or interval n-grams. Janssen et al. [84] on their corpus-based analysis on the TUNEFAM-26 dataset, investigated stability with regard to global features related to memorability, i.e., a phrase length, its position in the melody, its repetitiveness and others.

We hypothesize that certain parts of a melody, such as the beginning or end, are more robust to variations. We are therefore interested in the stability with regard to a note's relative position in the melody. Each column j of an MSA has a computed stability score. Each i-th index of a sequence in the MSA is assigned the stability score of its corresponding column. It should be noted that due to gaps, the i-th index of two different sequences may not correspond to the same j column. For each dataset (TUNEFAM-26 and CSV-60) we accumulate all the position versus stability data, where position corresponds to the i-th index normalized to the $[0,1]$ range. Figures 4 and 5 present the stability scores using different scoring methods (computed over three different alignment configurations) versus the relative position of a note (interval in our case) for the TUNEFAM-26 and CSV-60 datasets respectively. The corresponding gap ratio of the MSA versus the note position is also presented as a reference, since all conservation scores are affected by the amount of gaps per column.

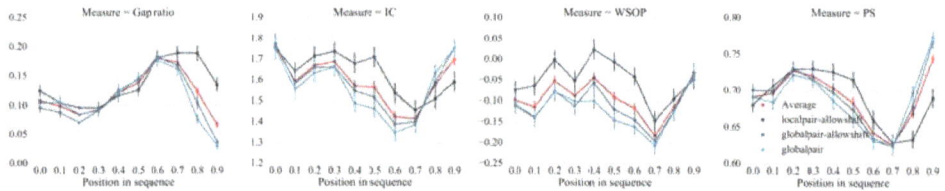

Figure 4. Position versus various stability scores (Information Content (IC), Weighted Sum-Of-Pairs (WSOP) and PS) for the TUNEFAM-26 dataset using three different alignment configurations. Position versus gap ratio is also presented (first to the left). Points are quantised to 10 bins.

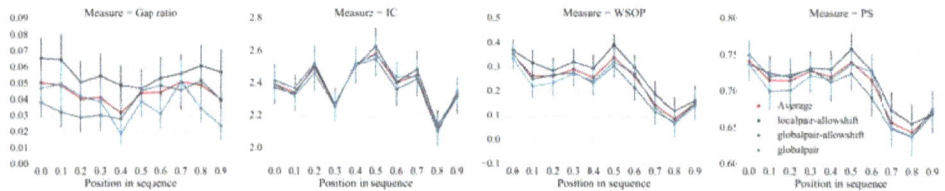

Figure 5. Position versus various stability scores (IC, WSOP and PS) for the CSV-60 dataset using three different alignment configurations. Position versus gap ratio is also presented (first to the left). Points are quantised to 10 bins.

For both datasets there are a number of observations (trends) that become immediately apparent: first, there is a strong indication that roughly the first half of a melody (up until 60% of its length) is more stable than the remaining. The downward slope after position 0.6 is prominent in both datasets and on all different stability scoring methods. This observation seems to agree with findings of Janssen et al. [84]; stable phrases occur relatively early in the melody. Secondly, the stability towards the final notes of a melody seems to be increasing. For the TUNEFAM-26 dataset in particular, the final 20% of the melody is very stable. The trend is less obvious on the CSV-60 dataset. However, it should be reminded that TUNEFAM-26 contains whole folk tune melodies, while CSV-60 contains melodies corresponding to structural segments of pop/rock songs; we cannot expect certain trends to be completely shared by both sets.

A potential explanation for this trend would be that artists interpreting a song creatively start out with an emphasis on the familiar material and take more liberty as the melody or segment progresses, introducing more variation along the way. But in contrast to the findings of Bountouridis and Van Balen, our results indicate that artists end with a familiar set of notes (for folk tunes more notably). This can be potentially attributed to the capacity of our short-memory; after a considerable part of varied material, our brain requires familiarity as to identify the whole piece as a variation of the original. For the CSV-60 dataset, since the melodies are shorter, the effect of short-term memory's capacity is weaker thus explaining the less obvious trend.

We now turn our focus to pitch intervals. We hypothesize that certain pitch intervals are more stable than others, i.e., certain transitions are less likely to be varied in a variation of the melody. To test our hypothesis, we need to measure the overall stability for each interval, while avoiding biases related to their likelihood of appearing in a sequence or column. We use the Information Content measure, computed for each symbol (note interval) e in the j-th index of the MSA as such:

$$IC_j(e) = P_{e,j} log(\frac{P_{e,j}}{Q_e}) \tag{12}$$

where $P_{e,j}$ is the frequency of a particular symbol e in the jth column, while Q_e is the expected frequency of symbol e (prior).

Figure 6 presents the overall stability scores per interval for the whole TUNEFAM-26 and CSV-60 datasets, in addition to their interval distribution. We show the results for the 13 most frequent intervals, since the remaining are too scarce for reliable analysis. Starting our analysis from the interval distribution profiles, we observe that they agree with Schoenberg's "singableness" hypothesis, that posits (among others) that a melody consists of more stepwise than leap intervals as a result of the human voice's nature [86]. The scarcity of chromatic jumps can be explained if we consider them as short excursions from the scale, which offer difficulties as well according to Schoenberg.

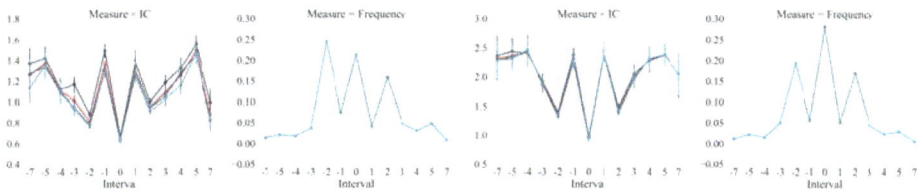

Figure 6. Pitch interval versus IC stability for the TUNEFAM-26 (left) and CSV-60 (right) datasets using three different alignment configurations. Interval frequencies per dataset are also presented. Results for only the 13 most frequent intervals are presented.

Regarding the stability-per-interval profiles, on first look, they are quite similar for the two datasets. Interestingly, the variance seems proportional to the interval's frequency despite the fact that our stability measure IC is normalized for the expected frequency per interval. On closer look and regarding the TUNEFAM-26 dataset, the ±1 and ±5 intervals are significantly more stable than the ±3, ±4 intervals of similar frequency of appearance. In addition, the +7 interval is as stable as the very frequent ±2 intervals. Therefore, we conclude that there is something inherently salient about the ±1 and ±5 intervals (at least in the TUNEFAM-26 dataset), but it is unsafe to make hypothesis regarding why this is the case. It should be noted that the findings of Janssen et al. [84] indicated that stable phrases are likely to comprise (among others) small pitch intervals and little surprising melodic material. However, their analysis approach is focused on stable phrases' global features, while ours on note-level features. Therefore, a direct comparison of findings, at least for pitch intervals, cannot be performed.

6. Data-Driven Modelling of Global Similarity

The findings of our stability analysis validated the intuitive hypothesis that some notes are more likely to be altered in a melodic variation than others. As such, any fine-grained melodic similarity function needs to accommodate for that fact by integrating meaningful scores for any pair of notes. In pairwise alignment via dynamic programming, integrating domain knowledge is only possible through the substitution matrix, which constitutes a model of global similarity, since it identifies notes commonly changed into other notes. Van Kranenburg [1] extended the DTW scoring function to include multiple musical dimensions, such as inner-metric analysis or phrase boundaries. On a melody classification task, he showed that expert-based heuristics could achieve almost perfect results. De Haas [87] showed that with regard to chord sequence similarity, local alignment with a substitution matrix based on simple heuristics [15], significantly outperforms his more sophisticated geometric model that takes into consideration the temporal relations between chords. Despite their success, the major concern with such approaches is their reliance on heuristics with known issues, such as limited generalization (see Section 1).

Interestingly in bioinformatics, the problem of meaningful substitution matrices, has been addressed following a data-driven approach. The major difficulty of the scoring matrix calculation is the computation of the joint probability $p_{x_iy_i}$ (see Equation 7) that expresses the likelihood of the two symbols at homologous sites. In bioinformatics, the key idea for solving this problem is that trusted alignments of related sequences can provide information regarding the mutability of symbols. One of the most widely-used matrices for protein comparison, BLOSUM [88], is actually derived from a large number of manually constructed, expert-aligned amino-acid sequences by counting how often certain amino-acids are substituted (mutated).

It follows naturally to investigate the potential of data-driven approaches in the MIR domain as well. Hirjee and Brown [89,90] generated a data-driven phoneme substitution matrix from misheard lyrics, gathered from online sources, and successfully applied it on a lyrics retrieval task. Similarly, Bountouridis et al. [27] used online sources to generate a chord similarity matrix for the task of cover song detection. Hu et al. [13] on the other hand, based their approach on pairs of aligned sung and reference melodies for the task of query-by-humming, but failed to significantly outperform a simple heuristic matrix. This might be attributed to the lack of experimentation with gap penalties or the noisy frame-based instead of note-based representation. Another major drawback for them was the amount of data, which consisted of only 40 sung melodies. We argue that expert-based alignments are generally problematic due to their limited quantity. Online sources have been shown to be potential solutions for lyrics or chords, but their existence cannot be guaranteed for all possible musical items such as melodies.

To eliminate the need for trustworthy pre-aligned melodic variations, in this section we propose the usage of trusted alignment algorithms as discussed in Section 4. Alignments generated by such algorithms can provide us with the appropriate information to generate a substitution matrix by computing log odds ratios for any pairs of symbols. While trusted alignment algorithms reduce the need for expert or crowd-sourced alignments, they still require melodies grouped (by experts preferably) into related cliques or tune families. These are still hard to find and as such, the applicability of our approach in real-life scenarios can be limited. Interestingly, in the same way that melody cliques contain melodic variants, melodies themselves may contain short recurring fragments, intra-song motifs. Such motifs may appear in variations throughout the melody. It is therefore also possible to generate a model of similarity among intra-song motifs if properly aligned. We hypothesize that intra-song motivic similarity can approximate the melodic similarity, or in other words, independent melodies contain enough information to explain variations in melodic cliques.

In the following paragraphs we present two data-driven approaches for capturing global similarity realized as substitution matrices for the TUNEFAM-26 and CSV-60 datasets. First, a matrix generated by alignments of melodic variations belonging to a clique (denoted simply melodic similarity). Secondly, matrices generated from different alignments of individual melodies with themselves (denoted intra-song motivic similarity). In order to assess their quality, we later perform an experiment to evaluate their retrieval performance.

6.1. Generating Substitution Matrices

Before discussing the alignments used, we explain the general process of converting them into a scoring system (a substitution matrix). The SubsMat package from the bioinformatics library Biopython provides routines for creating the typical log-odds substitution matrices. For our data, we firstly create the Accepted Replacement Matrix (ARM), meaning the counted number of replacements (confusions) according to the alignments. In order to avoid matrix entries of value zero, we apply pseudo-counts, meaning we add one to each entry. We generate the log-odds matrix M by applying a function that builds the observed frequency matrix from the ARM. We use the default settings: log base $b = 10$ and a multiplication factor f of 10. For two symbols x and y, their corresponding log-odds score is:

$$M(x,y) = log_b\left(\frac{p_{xy}}{q_xq_y}\right) \times f \tag{13}$$

with $M(x,y)$ rounded to the nearest integer. We normalize the matrix by dividing each of its elements with $max(M(x,y))$, so that the maximum score assigned to a pair of symbols is one.

6.2. Computing the Alignments for Melodic and Intra-Song Motivic Similarity

For the modelling of melodic similarity, the results from Section 4 have indicated that, although MAFFT is the best alignment strategy, the differences between various configurations are rather insignificant. Therefore, instead of generating a substitution matrix from clique alignments of one configuration only, we decided to use the following: MAFFT-globalpair with $(-4,-2)$ gap penalties, MAFFT-globalpair-allowshift with $(-4,-2)$ gap penalties and MAFFT-localpair-allowshift with $(-2,-1)$ gap penalties. The melodic similarity matrices generated for the TUNEFAM-26 and CSV-60 datasets are denoted TFAM-matrix and CSV-matrix respectively.

For the modelling of intra-song motivic similarity, the idea is to align each sequence with artificial versions of itself, such that all possible instances of intra-song motifs are aligned. In such a context, a useful and informative version of a sequence is one that when aligned to the original, maximizes the overlap between different instances of perceived-as-similar motifs. This informativeness criterion partially agrees with Hertz's and Stormo's definition of interesting alignments: those whose symbol frequencies most differ from the a priori probabilities of the symbols [91]. However, since informativeness can be erroneously biased, we are interested in alignments that at the same time minimize the overlap between perceptually different motifs.

Let us consider an example sequence S_o with two known motif instances "ABF" (cyan), "AGG" (green) of label L_1 and one motif instance "KLM" (red) of label L_2: XXABFXXXAGGXXXKLM. Furthermore, consider three versions of the S_o sequence based on arbitrary splitting in segments and further duplication or shuffling: KLXXXXABFXXXAGGXF, XAGGXXABFXAGGX and AGGXXKLMXXXABFXXX. Three possible pairwise alignments of the versions with the original are:

```
----XXABFXXXAGGXXXKLM    XXABFXXXAGGXXXKLM-    XXABFXXX-----AGGXXXKLM
LMXXXXABFXXXAGG--XK--    X-AGGXX-ABFX--AGGX    --AGGXXXKLMXXABFXXX---
```

The first example contains alignment of same-label motif instances with themselves (e.g., ABF to ABF), which provide no new information regarding their variation and therefore is of no value. The second alignment matches different instances of same-label motifs (e.g., ABF to AGG) but incorrectly aligns different-label motifs (e.g., AGG to KLM). It is only the third case that satisfies our criteria of a useful version of a sequence.

In order to identify the method that can be better used in practice to align any intra-song motifs (where the actual motifs are unknown), we design a simple experiment: we select all single sequences from the TUNEFAM-26 dataset that contain annotated motifs with two instances and devise three version-creation methods based on intuition. We then pairwise-align each original sequence to its different versions using different configurations of motif-agnostic alignment algorithms. In our experiment, the usefulness criteria are formulated as such: we are given the set L of all motif labels in a sequence S and $M_k = \{m_1^k, m_2^k, ..., m_j^k\}$, the set of all instances of intra-song motifs of label $k \in L$. We are interested in generating and pairwise-aligning different sequence versions with S, such that average relative likelihood R_M that the different instances $\in M_k$, $\forall\, k$ are aligned as opposed to be aligned by chance, is greater than one and maximal:

$$r_M^k = \sum_{i,j\; j \neq i} \frac{p_{m_i^k m_j^k}}{q_{m_i^k} q_{m_j^k}} \qquad R_M = \frac{1}{L} \sum_{k \in L} r_M^k \tag{14}$$

At the same time the average relative likelihood R_{NM} that any instances of different-labels motifs are aligned as opposed to be aligned by chance should be less than one and minimal:

$$R_{NM} = \frac{1}{L} \sum_{k,l \; k \neq l} \sum_{i,j} \frac{p_{m_i^k m_j^l}}{q_{m_i^k} q_{m_j^l}} \tag{15}$$

In practice, we are interested in the setup (version method plus alignment configuration) that maximizes $R_M - R_{NM}$. We experiment with three different automatic methods for version creation. Each method generates θ versions of the original sequence which is then pairwise-aligned to the original. We experiment with $\theta = \{4, 8, 12, 16\}$. The automatic methods for version creation are as follows:

1. Permutations: The original sequence is first split into n same-size segments. Each version is one of the $n!$ rearrangements of the segments. In our case n is arbitrarily set to four. Although automatic melody segmentation algorithms could have been used, we decided to used a fixed number of segments for the sake of simplicity.
2. Halves: The original sequence is iteratively split in subsequences of half size until their length is equal to four or their number is equal to θ. Each version is a sequence of length equal to the original, created by the concatenation of one of the subsequences.
3. Halves and shifts: A set of versions created by shifting the sequence by $1/k$ of its length to the right k times, resulting to k versions. The idea is to fuse the current set with the halves. We do that by randomly selecting $\theta/2$ versions from the halves method and $\theta/2$ versions from the current set.

The different versions are pairwise-aligned to the original using the following alignment configurations: MAFFT-globalpair with $(-4, -2)$ gap penalties, MAFFT-globalpair-allowshift with $(-4, -2)$ gap penalties and MAFFT-localpair-allowshift with $(-2, -1)$ gap penalties.

The R_M, R_{NM} and $R_M - R_{NM}$ figures for each version-creation method over all θ and for each alignment configuration, are presented in Figure 7. We notice that R_M is greater than one and R_{NM} is less than 1 for most setups, meaning that useful alignments are indeed generated. However, the versions created with the halves method ($\theta = \{4, 6\}$) and aligned to the original with localpair-allowshift with $(-2, -1)$ gap penalties, achieve the highest $R_M - R_{NM}$ (see the third column, second row in Figure 7). As such, we generate matrices (denoted halves-θ:4 and halves-θ:6 for both datasets) based on this configurations.

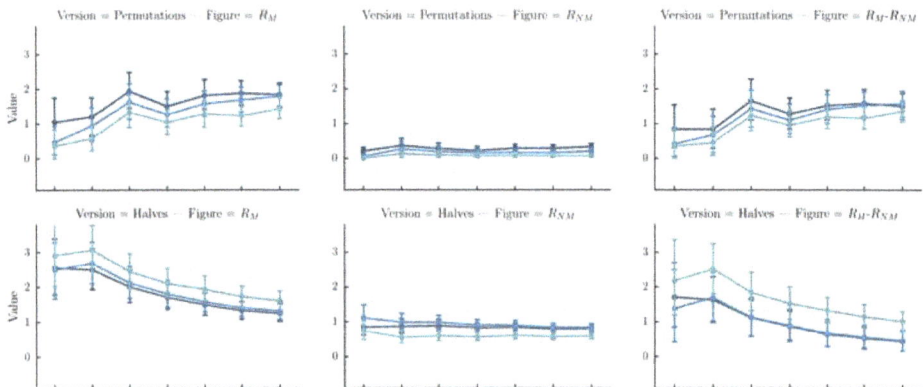

Figure 7. Cont.

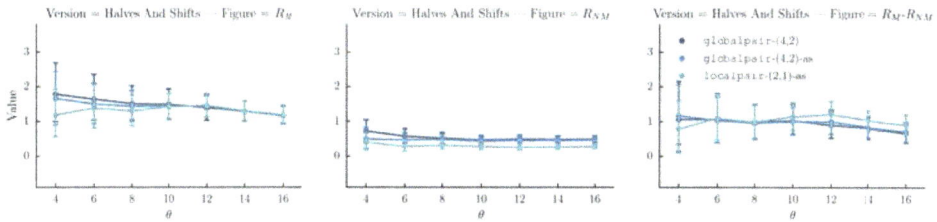

Figure 7. The average relative likelihood R_M (first column) that the different motif instances are aligned as opposed to be aligned by chance, the average relative likelihood R_{NM} (middle column) that any instances of different-labels motifs are aligned as opposed to be aligned by chance, and $R_M - R_{NM}$ (last column) figures for each version-creation method (per row) over all θ and for each alignment configuration.

6.3. Experimental Setup

We are interested in evaluating whether the scoring matrices generated from alignments using the methods of the previous Section 6.2, outperform the the standard ± 1 scoring matrix on the TUNEFAM-26 and CSV-60 datasets. In the retrieval case, we want to rank higher those melodies belonging to the same tune family or clique as the query. In the classification task, we want the tune family or clique of the highest ranked melody to correspond to the query's (that is, we are doing a k-Nearest Neighbour (kNN) classification experiment with $k = 1$).

Regarding the gap settings for this experiment, we should be extremely careful: the significant variation among the distribution of scores in between the matrices, renders the effect of the gap settings unpredictable, which can be problematic when aiming for a fair matrix comparison. Intuitively, there are two possible solutions: either compute the optimal gap settings per matrix, e.g., via a training process that optimizes the sensitivity (true positive rate) and selectivity (true negative rate) [92], or present their performance across a set of different penalties. The first approach is suitable for large datasets but is prone to over-fitting, and lacks a proper theoretical framework [93]. The second approach resembles the task of systematically comparing classifiers , which allows for a more complete view of each matrix by exploring the effect of the gap settings. Such an approach follows an intuitive classifier quality principle that agrees with our goal to develop generalizable solutions: "if a good classification is achieved only for a very small range in the parameter space, then for many applications it will be very difficult to achieve the best accuracy rate provided by the classifier" [94].

Picking a range of gap settings for each matrix that fairly represent its quality is not trivial. To solve the problem of fair matrix comparison, we need a meaningful intermediate mapping between two gap spaces $G_A \in \mathbb{R}^2$ and $G_B \in \mathbb{R}^2$ that work on matrices A and B respectively; or a single function $f : \mathbb{R}^n \to \mathbb{R}$ under which (G_A, A) and (G_B, B) have the same image (are equivalent). Given two sequences to be aligned, we argue that two settings $(g_a \in G_A, A)$ and $(g_b \in G_B, B)$ are equivalent and comparable only when they are of same flexibility, meaning they result to alignments of equal length relative to the original sequences (which translates to equal ratio of gaps to non-gap symbols for both settings). This idea is based on the observation that for two settings that result to the same amount of gaps, the alignment quality is solely dependent on the matrices used; as such, the matrices can be compared fairly. To compute the flexibility values for each of the TUNEFAM-26 and CSV-60 datasets, we randomly selected a subset of 50 sequences and pairwise aligned them using a range of different gap settings per matrix ($d, e \in [0.1, 2.0]$ with 0.1 intervals and $e \le 0.5d$). We used subsets instead of whole datasets for efficiency reasons, while the gap boundaries 0.1 and 2.0 are considered typical. For each alignment of sequences s_1 and s_2 of length l_1 and l_2 respectively, we computed the gap to non-gap ratio $r = (n_g - |l_1 - l_2|)/(l_1 + l_2)$, where n_g corresponds to the amount of gaps in the alignment. The average r over all pairwise alignments using a gap setting on the matrix is what we consider the setting's flexibility for that particular dataset. Given the mapping of each gap setting to the flexibility space, we can now fairly compare matrices by investigating their retrieval performance across different flexibility values.

6.4. Results

Figure 8a,b present the average precision and classification accuracy per substitution matrix over a range of flexibility values for the TUNEFAM-26 and Csv-60 datasets respectively. For the TUNEFAM-26 dataset and starting from the performance of the TFAM-matrix, we observe that it significantly increases the retrieval performance across all gap settings. In average, the TFAM-matrix increases the mean average precision from ID's 0.65 to 0.69, indicating that some meaningful similarity/variation knowledge has been indeed captured. The Csv-matrix presents a higher retrieval performance than the ID matrix, but the significance is not constant across all flexibilities. The same holds for the intra-song motivic matrices halves-θ:4 and halves-θ:6. If we concatenate the average precision scores over all flexibilities per matrix, besides the TFAM-matrix (see Figure 8a (top-right)) and perform a Friedman test, we discover that there is a significant difference between the four matrices. Post hoc analysis shows that the difference is due to the difference in between all pairs of matrices except halves-θ:4 and halves-θ:6. With regard to the classification accuracy, we do not observe a significant difference among the matrices.

For the Csv-60 dataset, the differences between matrices are more accentuated even through visual inspection. The Csv-matrix and learned matrix from the folk tunes collection TFAM-matrix, significantly outperform ID across almost all flexibilities. The implication of their similar performance in average will be discussed in the next section. Regarding the intra-song motivic matrices, both present significantly better performance than ID. Excluding Csv-matrix, a Friedman test with post hoc analysis on the concatenated average precision, reveals significant difference between all pairs of matrices except for the halves-θ:4 (0.74) and halves-θ:6 (0.75).

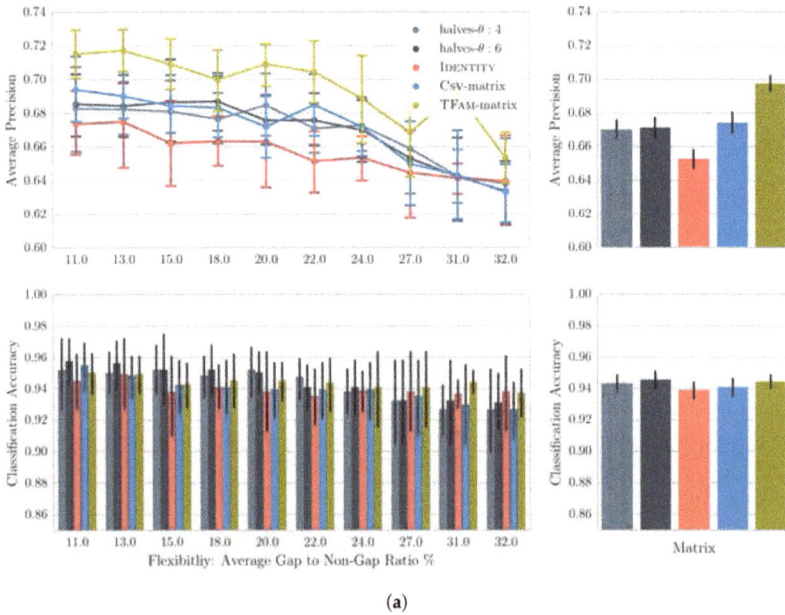

(a)

Figure 8. *Cont.*

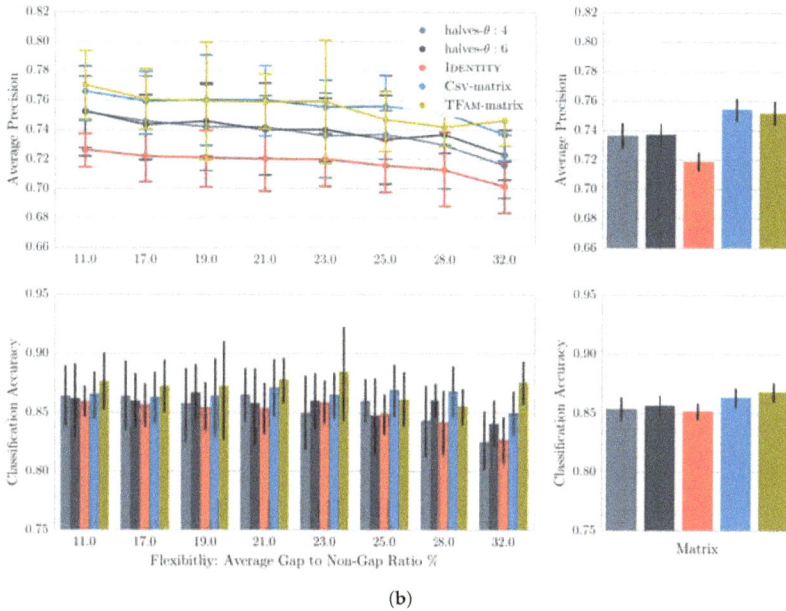

(b)

Figure 8. Average precision and classification accuracy for each matrix over a range of flexibility values (left) for the TUNEFAM-26 (**a**) and CSV-60 (**b**) datasets. The average precision and accuracy over all flexibilities per matrix are also presented on the right.

6.5. Discussion

The results offer a number of interesting findings that are secondary to our main question, e.g., the insignificant difference among matrices for the classification task implies the existence of almost-duplicates for each query. Or the inverse relation between the retrieval performance of each matrix to the flexibility value, indicates that real-life retrieval systems should aim for gap settings of low flexibility. However, most importantly, our results strongly suggest that data-driven matrices, learned from either melody variations or intra-song motif variations, capture some meaningful relationships between notes that can find application in melody retrieval. In the case of TUNEFAM-26, the results are obviously not impressive despite their statistical significance. Van Kranenburg's heuristics on the same dataset and task, pushed the MAP and classification accuracy to 0.85 and 0.98 respectively [1]. However, Van Kranenburg used only one arbitrarily selected gap setting (-0.8, -0.5), thus leaving the effect of gap settings uninvestigated. In our case however, we established a fairer framework for comparing matrices. In addition compared to our data-driven approach, Van Kranenburg had to experiment with a large number of heuristics to find the optimal. For the CSV-60 dataset, and in contrast to TUNEFAM-26, learning note relationships from folk tune variations or intra-song motifs seems to have a much more very positive effect in the overall retrieval performance. The reason behind this difference is unclear, but we can speculate based on intuition. In general, we observe that the vertical variation, i.e., among melodies belonging to the same family/clique, in the TUNEFAM-26 is more informative than the vertical variation in CSV-60. This explains why the TFAM-matrix is successful on both datasets, while CSV-matrix is only successful on CSV-60. Probably, tune families contain an adequate amount of melodic variations that allows for the generation of an informative matrix. At the same time the horizontal variation, i.e., among intra-song motifs, is similarly informative in both datasets. This explains why the performance of halves-θ:4 and halves-θ:6 matrices lies in between that of the ID and the best performing matrix for each dataset.

In summary, the results indicate that vertical variation models are more beneficial in a retrieval scenario. At the same time, the captured relationships of the horizontal models seem inadequate to approximate their performance. This implies that the way a song varies across its length does not follow the same principles as its variation through time, but further confirmation with note-to-note alignments of intra-song motifs and melodic variations is required. Nevertheless, the modelling of horizontal variation can be considered highly appropriate for practical scenarios of melody retrieval and classification where clique information is unavailable.

7. Conclusions

Modelling music similarity is a fundamental, but intricate task in MIR. Most previous works on music similarity, practical or theoretical, relied heavily on heuristics. In contrast, our work focused on acquiring knowledge on music and melodic similarity in particular from the data itself. Since data-driven methods and tools have been under development for years in bioinformatics, and since biological and music sequence share resembling concepts, we investigated their applicability inside a musical context.

First, we tackled the concept of meaningful and musically significant alignments of related melodies, by applying the bioinformatics structural alignment metaphor to music motifs. Our results revealed that the MAFFT multiple alignment algorithm, which uses gap-free sections as anchor points, is a natural fit for multiple melodic sequences; a strong indication of the importance of musical patterns for melodic similarity. Trusted MSA techniques made it possible to organize melodic variations such that melodic stability/variation can be analysed. We argue that our stability analysis findings are free of heuristics or biases that might have been introduced following other approaches.

Secondly, we investigated the modelling of global melodic similarity. We captured the probability of one note to be changed to another in a variation and created musically appropriate note-substitution scoring matrices for melodic alignment. We then put these matrices successfully to the test by designing retrieval and classification tasks. Our data-driven modelling of music similarity outperforms the naive ± 1 matrix, indicating that indeed some novel knowledge was captured. Additionally, we showed that variations inside a melody can be an alternative source for modelling the similarity of variations among tune families or cliques of covers.

In general, we showed that bioinformatics tools and methods can find successful application in music, to answer in a reliable, data-driven way a number of important, on-going questions in MIR. We argue data-driven approaches, such as ours, constitute an ideal balance between the two occasionally contradicting goals of MIR, problem solving and knowledge acquisition. Unfortunately, in the current age of big data, the potential in exploring musical relationships that can aid both the digital music services and our understanding of music itself remains largely idle. We hope that our work will stimulate future research to focus on a more constructive direction.

Author Contributions: Dimitrios Bountouridis and Daniel G. Brown both developed the relationship to bioinformatics applications and designed the experiments. Dimitrios Bountouridis performed the experiments, analysed the data and wrote the paper. Daniel G. Brown, Frans Wiering and Remco C. Veltkamp contributed to the writing of the paper.

Conflicts of Interest: The authors declare no conflict of interest. The founding sponsors had no role in the design of the study; in the collection, analyses or interpretation of data; in the writing of the manuscript; nor in the decision to publish the results.

References

1. Van Kranenburg, P. A Computational Approach to Content-Based Retrieval of Folk Song Melodies. Ph.D. Thesis, Utrecht University, Utrecht, The Netherlands, 2010.
2. Volk, A.; Haas, W.; Kranenburg, P. Towards modelling variation in music as foundation for similarity. In Proceedings of the International Conference on Music Perception and Cognition, Thessaloniki, Greece, 23–28 July 2012; pp. 1085–1094.

3. Pampalk, E. Computational Models of Music Similarity and Their Application to Music Information Retrieval. Ph.D. Thesis, Vienna University of Technology, Vienna, Austria, 2006.

4. Volk, A.; Van Kranenburg, P. Melodic similarity among folk songs: An annotation study on similarity-based categorization in music. *Music. Sci.* **2012**, *16*, 317–339.

5. Marsden, A. Interrogating melodic similarity: A definitive phenomenon or the product of interpretation? *J. New Music Res.* **2012**, *41*, 323–335.

6. Ellis, D.P.; Whitman, B.; Berenzweig, A.; Lawrence, S. The quest for ground truth in musical artist similarity. In Proceedings of the International Society of Music Information Retrieval Conference, Paris, France, 13–17 October 2002; pp. 170–177.

7. Deliège, I. Similarity perception categorization cue abstraction. *Music Percept.* **2001**, *18*, 233–244.

8. Novello, A.; McKinney, M.F.; Kohlrausch, A. Perceptual evaluation of music similarity. In Proceedings of the International Society of Music Information Retrieval, Victoria, BC, Canada, 8–12 October 2006; pp. 246–249.

9. Jones, M.C.; Downie, J.S.; Ehmann, A.F. Human similarity judgements: Implications for the design of formal evaluations. In Proceedings of the International Society of Music Information Retrieval, Vienna, Austria, 23–30 September 2007; pp. 539–542.

10. Tversky, A. Features of similarity. *Psychol. Rev.* **1977**, *84*, 327–352.

11. Lamere, P. Social tagging and music information retrieval. *J. New Music Res.* **2008**, *37*, 101–114.

12. Berenzweig, A.; Logan, B.; Ellis, D.P.; Whitman, B. A large-scale evaluation of acoustic and subjective music-similarity measures. *Comput. Music J.* **2004**, *28*, 63–76.

13. Hu, N.; Dannenberg, R.B.; Lewis, A.L. A probabilistic model of melodic similarity. In Proceedings of the International Computer Music Conference, Göteborg, Sweden, 16–21 September 2002.

14. Hu, N.; Dannenberg, R.B. A comparison of melodic database retrieval techniques using sung queries. In Proceedings of the 2nd ACM/IEEE-Cs Joint Conference on Digital Libraries, Portland, OR, USA, 13–17 July 2002; pp. 301–307.

15. Hanna, P.; Robine, M.; Rocher, T. An alignment based system for chord sequence retrieval. In Proceedings of the 9th ACM/IEEE-Cs Joint Conference on Digital Libraries, Austin, TX, USA, 14–19 June 2009; pp. 101–104.

16. Bronson, B.H. Melodic stability in oral transmission. *J. Int. Folk Music Counc.* **1951**, *3*, 50–55.

17. Krogh, A. An introduction to hidden markov models for biological sequences. *New Compr. Biochem.* **1998**, *32*, 45–63.

18. Bascom, W. The main problems of stability and change in tradition. *J. Int. Folk Music Counc.* **1959**, *11*, 7–12.

19. Drake, C.; Bertrand, D. The quest for universals in temporal processing in music. *Ann. N. Y. Acad. Sci.* **2001**, *930*, 17–27.

20. Gurney, E. *The Power of Sound*; Cambridge University Press: Cambridge, UK, 2011.

21. Casey, M.; Slaney, M. The importance of sequences in musical similarity. In Proceedings of the International Conference On Acoustics, Speech and Signal Processing, Toulouse, France, 14–19 May 2006; pp. 5–8.

22. Capra, J.A.; Singh, M. Predicting functionally important residues from sequence conservation. *Bioinformatics* **2007**, *23*, 1875–1882.

23. Valdar, W.S. Scoring residue conservation. *Proteins Struct. Funct. Bioinform.* **2002**, *48*, 227–241.

24. Luscombe, N.M.; Greenbaum, D.; Gerstein, M. What is bioinformatics? A proposed definition and overview of the field. *Methods Inf. Med.* **2001**, *40*, 346–358.

25. Bountouridis, D.; Wiering, F.; Brown, D.; Veltkamp, R.C. Towards polyphony reconstruction using multidimensional multiple sequence alignment. In Proceedings of the International Conference on Evolutionary and Biologically Inspired Music and Art, Amsterdam, The Netherlands, 19–21 April 2017; pp. 33–48.

26. Bountouridis, D.; Brown, D.; Koops, H.V.; Wiering, F.; Veltkamp, R. Melody retrieval and classification using biologically-inspired techniques. In Proceedings of the International Conference on Evolutionary and Biologically Inspired Music and Art, Amsterdam, The Netherlands, 19–21 April 2017; pp. 49–64.

27. Bountouridis, D.; Koops, H.V.; Wiering, F.; Veltkamp, R. A data-driven approach to chord similarity and chord mutability. In Proceedings of the International Conference on Multimedia Big Data, Taipei, Taiwan, 20–22 April 2016; pp. 275–278.

28. Nguyen, K.; Guo, X.; Pan, Y. *Multiple Biological Sequence Alignment: Scoring Functions, Algorithms and Evaluation*; John Wiley & Sons: Hoboken, NJ, USA, 2016.

29. Needleman, S.B.; Wunsch, C.D. A general method applicable to the search for similarities in the amino acid sequence of two proteins. *J. Mol. Biol.* **1970**, *48*, 443–453.
30. Durbin, R.; Eddy, S.R.; Krogh, A.; Mitchison, G. *Biological Sequence Analysis: Probabilistic Models of Proteins and Nucleic Acids*; Cambridge University Press: Cambridge, UK, 1998.
31. Smith, T.F.; Waterman, M.S. Identification of common molecular subsequences. *J. Mol. Biol.* **1981**, *147*, 195–197.
32. Mongeau, M.; Sankoff, D. Comparison of musical sequences. *Comput. Humanit.* **1990**, *24*, 161–175.
33. Ewert, S.; Müller, M.; Dannenberg, R.B. Towards reliable partial music alignments using multiple synchronization strategies. In Proceedings of the International Workshop on Adaptive Multimedia Retrieval, Madrid, Spain, 24–25 September 2009; pp. 35–48.
34. Serra, J.; Gómez, E.; Herrera, P.; Serra, X. Chroma binary similarity and local alignment applied to cover song identification. *Audio Speech Lang. Process.* **2008**, *16*, 1138–1151.
35. Müllensiefen, D.; Frieler, K. Optimizing measures of melodic similarity for the exploration of a large folk song database. In Proceedings of the International Society of Music Information Retrieval, Barcelona, Spain, 10–15 October 2004; pp. 1–7.
36. Müllensiefen, D.; Frieler, K. Cognitive adequacy in the measurement of melodic similarity: Algorithmic vs. human judgements. *Comput. Musicol.* **2004**, *13*, 147–176.
37. Sailer, C.; Dressler, K. Finding cover songs by melodic similarity. In Proceedings of the Annual Music Information Retrieval Evaluation Exchange, Victoria, BC, Canada, 8–12 September 2006. Available online: www.music-ir.org/mirex/abstracts/2006/CS_sailer.png (accessed on 28 November 2017).
38. Ross, J.C.; Vinutha, T.; Rao, P. Detecting melodic motifs from audio for hindustani classical music. In Proceedings of the International Society of Music Information Retrieval, Porto, Portugal, 8–12 October 2012; pp. 193–198.
39. Salamon, J.; Rohrmeier, M. A quantitative evaluation of a two stage retrieval approach for a melodic query by example system. In Proceedings of the International Society of Music Information Retrieval, Kobe, Japan, 26–30 October 2009; pp. 255–260.
40. Hu, N.; Dannenberg, R.B.; Tzanetakis, G. Polyphonic audio matching and alignment for music retrieval. In Proceedings of the Workshop in Applications of Signal Processing to Audio and Acoustics, New Paltz, NY, USA, 19–22 October 2003; pp. 185–188.
41. Ewert, S.; Müller, M.; Grosche, P. High resolution audio synchronization using chroma onset features. In Proceedings of the International Conference On Acoustics, Speech and Signal Processing, Taipei, Taiwan, 19–24 April 2009; pp. 1869–1872.
42. Balke, S.; Arifi-Müller, V.; Lamprecht, L.; Müller, M. Retrieving audio recordings using musical themes. In Proceedings of the International Conference on Acoustics, Speech and Signal Processing, Shanghai, China, 20–25 March 2016; pp. 281–285.
43. Raffel, C.; Ellis, D.P. Large-scale content-based matching of midi and audio files. In Proceedings of the International Society of Music Information Retrieval, Malaga, Spain, 26–30 October 2015; pp. 234–240.
44. Müller, M. *Information Retrieval for Music and Motion*; Springer: Berlin, Germany, 2007.
45. Thompson, J.D.; Higgins, D.G.; Gibson, T.J. Clustal W: Improving the sensitivity of progressive multiple sequence alignment through sequence weighting, position-specific gap penalties and weight matrix choice. *Nucleic Acids Res.* **1994**, *22*, 4673–4680.
46. Wang, L.; Jiang, T. On the complexity of multiple sequence alignment. *J. Comput. Biol.* **1994**, *1*, 337–348.
47. Kemena, C.; Notredame, C. Upcoming challenges for multiple sequence alignment methods in the high-throughput era. *Bioinformatics* **2009**, *25*, 2455–2465.
48. Liu, C.C. Towards automatic music performance comparison with the multiple sequence alignment technique. In Proceedings of the International Conference on Multimedia Modelling, Huangshan, China, 7–9 January 2013; pp. 391–402.
49. Wang, S.; Ewert, S.; Dixon, S. Robust joint alignment of multiple versions of a piece of music. In Proceedings of the International Society of Music Information Retrieval, Taipei, Taiwan, 27–31 October 2014; pp. 83–88.
50. Knees, P.; Schedl, M.; Widmer, G. Multiple lyrics alignment: Automatic retrieval of song lyrics. In Proceedings of the International Society of Music Information Retrieval, London, UK, 11–15 October 2005; pp. 564–569.
51. Poliner, G.E.; Ellis, D.P.; Ehmann, A.F.; Gómez, E.; Streich, S.; Ong, B. Melody transcription from music audio: Approaches and evaluation. *IEEE Trans. Audio Speech Lang. Process.* **2007**, *15*, 1247–1256.

52. Kim, Y.E.; Chai, W.; Garcia, R.; Vercoe, B. Analysis of a contour-based representation for melody. In Proceedings of the International Society of Music Information Retrieval, Plymouth, MA, USA, 23–25 October 2000.

53. Huron, D. The melodic arch in western folksongs. *Comput. Musicol.* **1996**, *10*, 3–23.

54. Margulis, E.H. A model of melodic expectation. *Music Percept. Interdiscip. J.* **2005**, *22*, 663–714.

55. Salamon, J.; Gómez, E.; Ellis, D.P.; Richard, G. Melody extraction from polyphonic music signals: Approaches, applications, and challenges. *IEEE Signal Process. Mag.* **2014**, *31*, 118–134.

56. Suyoto, I.S.; Uitdenbogerd, A.L. Simple efficient n-gram indexing for effective melody retrieval. In Proceedings of the Annual Music Information Retrieval Evaluation Exchange, London, UK, 14 September 2005. Available online: pdfs.semanticscholar.org/4103/07d4f5398b1588b04d2916f0f592813a3d0a.png (accessed on 28 November 2017).

57. Ryynanen, M.; Klapuri, A. Query by humming of midi and audio using locality sensitive hashing. In Proceedings of the International Conderence on Acoustics, Speech and Signal Processing, Las Vegas, NV, USA, 31 March–4 April 2008; pp. 2249–2252.

58. Hillewaere, R.; Manderick, B.; Conklin, D. Alignment methods for folk tune classification. In Proceedings of the Annual Conference of the German Classification Society on Data Analysis, Machine Learning and Knowledge Discovery, Hildesheim, Germany, 1–3 August 2014; pp. 369–377.

59. Gómez, E.; Klapuri, A.; Meudic, B. Melody description and extraction in the context of music content processing. *J. New Music Res.* **2003**, *32*, 23–40.

60. Van Kranenburg, P.; de Bruin, M.; Grijp, L.; Wiering, F. *The Meertens Tune Collections*; Meertens Online Reports; Meertens Institute: Amsterdam, The Netherlands, 2014.

61. Bountouridis, D.; Van Balen, J. The cover song variation dataset. In Proceedings of the International Workshop on Folk Music Analysis, Istanbul, Turkey, 12–13 June 2014.

62. Raghava, G.; Barton, G. Quantification of the variation in percentage identity for protein sequence alignments. *BMC Bioinform.* **2006**, *7*, 415–419.

63. Ewert, S.; Müller, M.; Müllensiefen, D.; Clausen, M.; Wiggins, G.A. Case study "Beatles songs" what can be learned from unreliable music alignments? In Proceedings of the Dagstuhl Seminar, Dagstuhl, Germany, 15–20 March 2009.

64. Prätzlich, T.; Müller, M. Triple-based analysis of music alignments without the need of ground-truth annotations. In Proceedings of the International Conference on Acoustics, Speech and Signal Processing, Shanghai, China, 20–25 March 2016; pp. 266–270.

65. Pei, J.; Grishin, N.V. Promals: Towards accurate multiple sequence alignments of distantly related proteins. *Bioinformatics* **2007**, *23*, 802–808.

66. Blackburne, B.P.; Whelan, S. Measuring the distance between multiple sequence alignments. *Bioinformatics* **2012**, *28*, 495–502.

67. Bountouridis, D.; Van Balen, J. Towards capturing melodic stability. In Proceedings of the Interdisciplinary Musicology Conference, Berlin, Germany, 4–6 December 2014.

68. Cowdery, J.R. A fresh look at the concept of tune family. *Ethnomusicology* **1984**, *28*, 495–504.

69. Hogeweg, P.; Hesper, B. The alignment of sets of sequences and the construction of phyletic trees: An integrated method. *J. Mol. Evol.* **1984**, *20*, 175–186.

70. Berger, M.; Munson, P.J. A novel randomized iterative strategy for aligning multiple protein sequences. *Comput. Appl. Biosci. Cabios* **1991**, *7*, 479–484.

71. Gotoh, O. Optimal alignment between groups of sequences and its application to multiple sequence alignment. *Comput. Appl. Biosci. Cabios* **1993**, *9*, 361–370.

72. Notredame, C.; Higgins, D.G.; Heringa, J. T-coffee: A novel method for fast and accurate multiple sequence alignment. *J. Mol. Biol.* **2000**, *302*, 205–217.

73. Notredame, C.; Holm, L.; Higgins, D.G. Coffee: An objective function for multiple sequence alignments. *Bioinformatics* **1998**, *14*, 407–422.

74. Morgenstern, B.; Dress, A.; Werner, T. Multiple dna and protein sequence alignment based on segment-to-segment comparison. *Proc. Natl. Acad. Sci. USA* **1996**, *93*, 12098–12103.

75. Katoh, K.; Misawa, K.; Kuma, K.i.; Miyata, T. Mafft: A novel method for rapid multiple sequence alignment based on fast fourier transform. *Nucleic Acids Res.* **2002**, *30*, 3059–3066.

76. Altschul, S.F. Generalized affine gap costs for protein sequence alignment. *Proteins Struct. Funct. Genet.* **1998**, *32*, 88–96.
77. Carroll, H.; Clement, M.J.; Ridge, P.; Snell, Q.O. Effects of gap open and gap extension penalties. In Proceedings of the Biotechnology and Bioinformatics Symposium, Provo, Utah, 20–21 October 2006; pp. 19–23.
78. Dannenberg, R.B.; Hu, N. Understanding search performance in query-by-humming systems. In Proceedings of the Conference of the International Society of Music Information Retrieval, Barcelona, Spain, 10–15 October 2004.
79. Margulis, E.H. Musical repetition detection across multiple exposures. *Music Percept. Interdiscip. J.* **2012**, *29*, 377–385.
80. Bigand, E.; Pineau, M. Context effects on melody recognition: A dynamic interpretation. *Curr. Psychol. Cogn.* **1996**, *15*, 121–134.
81. Klusen, E.; Moog, H.; Piel, W. Experimente zur mündlichen Tradition von Melodien. *Jahrbuch Fur Volksliedforschung* **1978**, *23*, 11–32.
82. Bigand, E. Perceiving musical stability: The effect of tonal structure, rhythm, and musical expertise. *J. Exp. Psychol. Hum. Percept. Perform.* **1997**, *23*, 808–822.
83. Schmuckler, M.A.; Boltz, M.G. Harmonic and rhythmic influences on musical expectancy. *Atten. Percept. Psychophys.* **1994**, *56*, 313–325.
84. Janssen, B.; Burgoyne, J.A.; Honing, H. Predicting variation of folk songs: A corpus analysis study on the memorability of melodies. *Front. Psychol.* **2017**, *8*, 621.
85. Van Balen, J.; Bountouridis, D.; Wiering, F.; Veltkamp, R. Cognition-inspired descriptors for scalable cover song retrieval. In Proceedings of the International Society of Music Information Retrieval, Taipei, Taiwan, 27–31 October 2014.
86. Schoenberg, A.; Stein, L. *Fundamentals of Musical Composition*; Faber: London, UK, 1967.
87. De Haas, W.B.; Wiering, F.; Veltkamp, R. A geometrical distance measure for determining the similarity of musical harmony. *Int. J. Multimed. Inf. Retr.* **2013**, *2*, 189–202.
88. Henikoff, S.; Henikoff, J.G. Amino acid substitution matrices from protein blocks. *Proc. Natl. Acad. Sci. USA* **1992**, *89*, 10915–10919.
89. Hirjee, H.; Brown, D.G. Rhyme analyser: An analysis tool for rap lyrics. In Proceedings of the International Society of Music Information Retrieval, Utrecht, The Netherlands, 9–13 August 2010.
90. Hirjee, H.; Brown, D.G. Solving misheard lyric search queries using a probabilistic model of speech sounds. In Proceedings of the International Society of Music Information Retrieval, Utrecht, The Netherlands, 9–13 August 2010; pp. 147–152.
91. Hertz, G.Z.; Stormo, G.D. Identifying dna and protein patterns with statistically significant alignments of multiple sequences. *Bioinformatics* **1999**, *15*, 563–577.
92. Yamada, K.; Tomii, K. Revisiting amino acid substitution matrices for identifying distantly related proteins. *Bioinformatics* **2013**, *30*, 317–325.
93. Long, H.; Li, M.; Fu, H. Determination of optimal parameters of MAFFT program based on BAliBASE3.0 database. *SpringerPlus* **2016**, *5*, 736–745.
94. Amancio, D.R.; Comin, C.H.; Casanova, D.; Travieso, G.; Bruno, O.M.; Rodrigues, F.A.; da Fontoura Costa, L. A systematic comparison of supervised classifiers. *PLoS ONE* **2014**, *9*, e94137.

![applied sciences logo] *applied sciences*

MDPI

Article

Exploring the Effects of Pitch Layout on Learning a New Musical Instrument †

Jennifer MacRitchie [1,2,*,‡] and **Andrew J. Milne** [1,‡]

1   The MARCS Institute for Brain, Behaviour and Development, Western Sydney University, Penrith 2751, Australia; a.milne@westernsydney.edu.au

2   School of Humanities and Communication Arts, Western Sydney University, Penrith 2751, Australia

*   Correspondence: j.macritchie@westernsydney.edu.au; Tel.: +61-2-9772-6166

†   This article is a re-written and expanded version of "Evaluation of the Learnability and Playability of Pitch Layouts in New Musical Instruments." In Proceedings of the 14th Sound and Music Computing Conference, Espoo, Finland, 5–8 July 2017; pp. 450–457.

‡   These authors contributed equally to this work.

Academic Editor: Vesa Valimaki

Received: 27 October 2017; Accepted: 21 November 2017; Published: 24 November 2017

Featured Application: The results obtained in this paper are applicable to the design of new musical instruments intended to facilitate the learning and playing of music.

Abstract: Although isomorphic pitch layouts are proposed to afford various advantages for musicians playing new musical instruments, this paper details the first substantive set of empirical tests on how two fundamental aspects of isomorphic pitch layouts affect motor learning: *shear*, which makes the pitch axis vertical, and the *adjacency* (or *nonadjacency*) of pitches a major second apart. After receiving audio-visual *training tasks* for a scale and arpeggios, performance accuracies of 24 experienced musicians were assessed in *immediate retention tasks* (same as the training tasks, but without the audio-visual guidance) and in a *transfer task* (performance of a previously untrained nursery rhyme). Each participant performed the same tasks with three different pitch layouts and, in total, four different layouts were tested. Results show that, so long as the performance ceiling has not already been reached (due to ease of the task or repeated practice), adjacency strongly improves performance accuracy in the training and retention tasks. They also show that shearing the layout, to make the pitch axis vertical, worsens performance accuracy for the training tasks but, crucially, it strongly improves performance accuracy in the transfer task when the participant needs to perform a new, but related, task. These results can inform the design of pitch layouts in new musical instruments.

Keywords: sound and music computing; new musical instruments; pitch layouts; perception and action; motor learning

---

1. Introduction

Designers of new musical instruments can often be concerned with ensuring accessibility for users either with no previous musical experience, or for those who already have training in another instrument, so that they can easily alter or learn new techniques. Several claims regarding the optimal pitch layout of new musical instruments or interfaces have been made, but as yet there is little empirical investigation of the factors that may enhance or disturb learning and performance on these devices. Our previous conference paper detailed the impact of adjacency and shear on pitch accuracy for the transfer task [1]; in this paper, we take a more comprehensive approach by also considering timing accuracy, and the training and retention tasks.

1.1. Isomorphic Layout Properties

Since the nineteenth century, numerous music theorists and instrument builders have conjectured that *isomorphic pitch layouts* provide important advantages over the conventional pitch layouts of traditional musical instruments [2–5]. Indeed, a number of new musical interfaces have used isomorphic layouts (e.g., Array Mbira [6], Thummer [7], AXiS-49 [8], Musix Pro [9], LinnStrument [10], Lightpad Block [11], Terpstra [12]).

An isomorphic layout is one where the spatial arrangement of any set of pitches (a chord, a scale, a melody, or a complete piece) is invariant with respect to musical transposition. This contrasts with conventional pitch layouts on traditional musical instruments; for example, on the piano keyboard, playing a given chord or melody in a different transposition (e.g., in a different key) typically requires changing fingering to negotiate the differing combinations of vertically offset black and white keys.

Isomorphic layouts also have elegant properties for microtonal scales, which contain pitches and intervals "between the cracks" of the piano keyboard [13]. Although strict twelve-tone equal temperament (12-TET) is almost ubiquitous in contemporary Western music, different tunings are found in historical Western and in non-Western traditions. Isomorphic layouts may, therefore, facilitate the performance of music both within and beyond conventional contemporary Western traditions.

One elegant property relevant to non-standard tunings is that, unlike the piano keyboard, isomorphic layouts do not have an immutable periodicity in their spatial structure. On the piano keyboard, only scale systems that repeat every twelve pitches can be intuitively mapped to its keys. Conversely, isomorphic layouts provide consistent spatial representations of scales regardless of their periodicity. This matters because there are many useful tuning systems that do not repeat every 12 chromatic pitches, such as meantone tunings in 19-TET or 31-TET, which are suitable for conventional Western music but provide better approximations to just intonation than the standard 12-TET; Bohlen-Pierce scales, which repeat every 13 equal divisions of the 3/1 tritave (instead of the standard 2/1 octave); the Javanese Pelog system, which is often approximated by a 7-pitch scale in 9-TET; and numerous other scale systems [14].

In this paper, we do not compare isomorphic and non-isomorphic layouts. Instead, we focus on how different isomorphic layouts impact on learning. This is because there are an infinite number of unique isomorphic layouts (and a large number that are practicable for conventionally tuned diatonic-chromatic music): they all share the property of transpositional invariance (by definition) but they differ in a number of other ways that may plausibly impact their usability. For example, successive scale pitches, such as C, D, and E, are spatially adjacent in some isomorphic layouts while in others they are not; additionally, in some isomorphic layouts, pitches are perfectly correlated to a horizontal or vertical axis while in others they are not [15]. With respect to the instrumentalist, the "horizontal" axis runs from left to right, the "vertical" axis from bottom to top or from near to far. In some layouts, octaves may be vertically or horizontally aligned; in others, they are slanted. Properties such as a vertical pitch axis or a vertical octave axis, or adjacent major seconds, and so forth, may be conjectured as desirable (or undesirable): either way, they are typically non-independent because changing one (e.g., pitch axis orientation) may change another (e.g., octave axis orientation). Choosing an optimal layout thus becomes a non-trivial task that requires knowledge of the relative importance of the different properties. However, due to their non-independence, it is challenging to investigate the relative importance of these features experimentally.

To address this, the experiment presented in this paper explores how two independent spatial transformations of isomorphic layouts—shear and adjacency—impact on learning in a set of melody retention and transfer tasks. The shear is used to manipulate the angles of the pitch axis and major second axis, while keeping the octave axis constant; the adjacency manipulation determines whether or not major seconds are spatially adjacent. These two transformations enable us to test our hypotheses that adjacent major seconds and a vertical pitch axis facilitate the learning and playing of melodies.

The four layouts that result from these transformations are illustrated in Figure 1a–d. Each figure shows how pitches are positioned, and the orientation of three axes that we hypothesize will impact on the layout's usability. Each label indicates whether the layout has adjacent major seconds or not (*A* and *A'*, respectively) and whether it is sheared or not (*S* and *S'*, respectively). The three axes are the *pitch axis*, the *octave axis*, and the *major second axis*, as now defined (the implications of these three axes, and why they may be important, are detailed in Section 1.1.2).

- The *pitch axis* is any axis onto which the orthogonal (perpendicular) projections of all button centres are proportional to their pitch; for any given isomorphic layout, all such axes are parallel [16] (see the caption for Figure 1 for a practical demonstration of how this works).
- The *octave axis* is here defined as any axis that passes through the closest button centres that are an octave apart.
- The *major second axis* (*M2 axis*, for short) is here defined as any axis that passes through the closest button centres that are a major second apart.

When considering tunings different to 12-TET (e.g., meantone or Pythagorean), alternative—but more complex—definitions for the octave and M2 axes become useful.

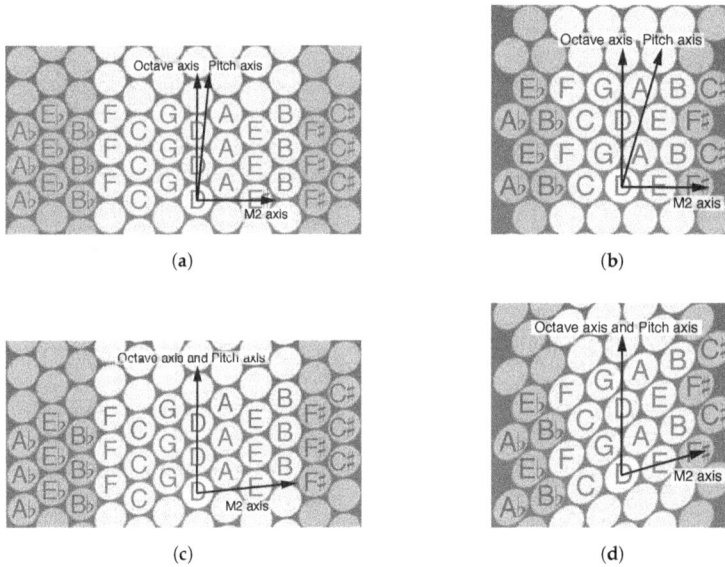

Figure 1. The four isomorphic layouts tested in the experiment. They have differently angled pitch axes and major seconds axes. An easy way to understand the meaning of the pitch axis is to place a ruler on any of the above subfigures so that it is at right angles to the pitch axis. If the ruler is then slid in the direction of the pitch axis, with its angle kept constant, the button-centres passing under the ruler's edge will always be encountered in ascending pitch order. This occurs only when the ruler is oriented and moved, in this way, with the pitch axis. This means that, when the pitch axis is vertical, as in (**c**,**d**), the pitch of each button is proportional to its vertical position. In all four layouts, the octave axis is vertical; that is, buttons that are an octave apart are vertically aligned. (**a**) *A'S'*: nonadjacent M2s, unsheared; (**b**) *AS'* (the Wicki layout [4]): adjacent M2s, unsheared; (**c**) *A'S*: nonadjacent M2s, sheared; (**d**) *AS*: adjacent M2s, sheared.

1.1.1. Adjacent (*A*) or NonAdjacent (*A'*) Seconds

Scale steps (i.e., major and minor seconds) are, across cultures, the commonest intervals in melodies [17]. It makes sense for such musically privileged intervals also to be spatially privileged. An obvious way of spatially privileging intervals is to make their pitches adjacent: this makes transitioning between them physically easy, and makes them visually salient. However, when considering bass or harmony parts, scale steps may play a less important role. This suggests that differing layouts might be optimal for differing musical uses.

The focus of this experiment is on melody so, for any given layout, we tested one version where all major seconds are adjacent and an adapted version where they are nonadjacent (minor seconds were nonadjacent in both versions). Both types of layouts have been used in new musical interfaces; for example, the Thummer (which used the Wicki layout (Figure 1b) had adjacent major seconds, while the AXiS-49 (which uses a *Tonnetz*-like layout [18]) has nonadjacent seconds but adjacent thirds and fifths.

1.1.2. Sheared (S) or Unsheared (S')

We conjecture that having any of the above-mentioned axes (pitch, octave, and M2) perfectly horizontal or perfectly vertical makes the layout more comprehensible: if the pitch axis is vertical or horizontal (rather than slanted), it allows for the pitch of buttons to be more easily estimated by sight, thereby enhancing processing fluency. Similar advantages hold for the octave and M2 axes: scales typically repeat at the octave, while the major second is the commonest scale-step in both the diatonic and pentatonic scales that form the backbone of most Western music.

However, changing the angle of one of these axes requires changing the angle of one or both of the others, so their independent effects can be hard to disambiguate. A way to gain partial independence of axis angles is to shear the layout parallel with one of the axes—the angle of the parallel-to-shear axis will not change while the angles of the other two will. A *shear* is a spatial transformation in which points are shifted parallel to an axis by a distance proportional to their distance from that axis. (For example, shearing a rectangle parallel to an axis running straight down its middle produces a parallelogram; the sides that are parallel to the shear axis remain parallel to it, while the other two sides rotate). As shown by comparing Figure 1a with Figure 1c, or by comparing Figure 1b with Figure 1d, we used a shear parallel with the octave axis to create two versions of the nonadjacent layout and two versions of the adjacent layout: each unsheared version ($A'S'$ or AS') has a perfectly horizontal M2 axis but a slanted (non-vertical) pitch axis; each sheared version version ($A'S$ or AS) has a slanted (non-horizontal) M2 axis but a vertical pitch axis. In both cases the octave axis was vertical.

In this investigation, therefore, we remove any possible impact of the octave axis orientation; we cannot, however, quantitatively disambiguate between the effects of the pitch axis and the M2 axis.

Unsheared layouts are common in new musical interfaces because these typically use buttons arranged in a perfectly square or hexagonal array; we are not aware of a hardware interface that makes use of shear to make the pitch axis vertical or horizontal (although this is a design feature of the software MIDI sequencer Hex [15]).

1.2. Motor Skill Learning in Music Performance

Learning a new musical instrument requires a number gross and fine motor skills in order to physically play a note. This is often carried out in tandem with sensory processing of feedback from the body and of auditory features (e.g., melody, rhythm, timbre) in order to learn how to play specific sequences [19]. For the purposes of our experiment, by using musically-trained participants and sequences familiar to those musicians such as scales and arpeggios, we reduce this to a motor learning problem. How best can musicians learn to play on a new pitch layout?

In learning a motor skill there are three general stages [20]:

- A *cognitive stage*, encompassing the processing of information and detecting patterns. Here, various motor solutions are tried out, and the performer finds which solutions are most effective.
- A *fixation stage*, when the general motor solution has been selected, and a period commences where the patterns of movement are perfected. This stage can last months, or even years.

- An *autonomous stage*, where the movement patterns do not require as much conscious attention on the part of the performer.

Essentially, learning the motor-pitch associations of a new instrument requires the performer to perceive and remember pitch patterns. Once these pitch patterns are learned, the performer becomes more focused on eliminating various sources of motor error. Because achieving motor autonomy is a lengthy process—one that can seldom be captured by short-term experiments—our current study focuses on only the first two elements of motor learning.

Learning a pattern of actions and their associated responses can be affected by pre-existing action-response representations: essentially, the anticipated effects of an action have an influence on the performance of that action; for example, reaction time is faster when participants are instructed to press a button forcefully and this elicits a loud tone, rather than when the effect is not compatible with the action (e.g., a soft tone) [21]. Therefore, it may also hold that pre-existing expectations of the pitch effects of a sequence of actions may have an influence on the performance of that sequence.

Research into the Spatial-Musical Association of Response Codes (or SMARC effect) demonstrates not only a vertical alignment (increasing pitch height is mapped vertically from low to high), but also a horizontal alignment (increasing pitch height is mapped horizontally from left to right) in musically trained participants [22]. This horizontal effect is far more subtle in non-musicians [23] and in some cases non-existent [24,25], suggesting that musical training enhances this particular spatial dimension. It is posited that this may be a learned-association effect [26]. These pitch representations have been shown to influence motor planning and action. Keller and colleagues found that, for a sequence of three consecutive keypresses, timing was more accurate when the produced tones were compatible with the pre-existing associations that increasing vertical movement results in an increase in pitch height [27,28]. This appears to be evident across different levels of expertise (non-musicians and trained musicians), although, as expected, training enhances the strength of this existing representation [29]. We investigate only 2-dimensional pitch layouts, so do not consider the implications of Shepard's helical model of pitch perception [30], which requires a cylindrical—hence 3-dimensional—form [31].

The tendency in the pitch-motor representation literature has been to reverse or scramble pitches from the traditional down-to-up or left-to-right assignment. Although many new pitch layouts may not violate this basic learned pitch-motor association, adjustments to the learned general motor pattern may still be required depending on the spacing of intervals, and the precise orientation of the pitch axis. Stewart and colleagues [26] demonstrated an effect on reaction time in a task using "normal" versus "stretched" representations of pitch along a horizontal axis (sequences which did or did correspond to a learned pattern of movement that could be played with the fingers of a single hand). This suggests that, despite their similarity to other layouts (both "normal" and "stretched" satisfied the left-right horizontal sequence), the patterns of notes may have fundamentally changed for the performer, and so require a certain amount of motor learning in this new (but clearly related) task.

It seems plausible then that certain aspects of a new layout, within the realm of satisfying the vertical and horizontal SMARC effects, will facilitate such learning, while others may hinder it. These aspects may be related to (a) previously learned pitch-motor mappings; (b) ergonomic issues, such as the physical ease of making the motions required to play the target pitches, and also from (c) processing fluency, such as how easy it is to see or sense, by proprioception, musical features that are relevant to the task. As detailed in Sections 1.1.1 and 1.1.2, in this experiment, we focus on the last of these and, in particular, on two musical attributes that are important for melodies and two spatial attributes that have a plausible impact on processing fluency. The musical attributes are major seconds (important because of their prevalence in melodies and musical scales) and pitch height. The spatial attributes are verticality (we hypothesize that perfectly vertical, or horizontal, lines are easier to imagine than are slanted lines) and adjacency (we hypothesize that it is generally easier to find a spatially adjacent pitch than one that is separated). The experimental manipulation, therefore, involves participants learning and playing pitch layouts with vertical versus slanted pitch axes, and adjacent versus nonadjacent major seconds.

To test how well the participants have learned the new layouts and perfected their motor pattern, we are particularly interested in the transfer of learning from one task to another. For instance, a piano player will practice scales not only to achieve good performance of scales, but also to fluently play scale-like passages in other musical pieces. In our study, we designed a training and testing paradigm for the different pitch layouts such that the transfer task involved a previously unpracticed, but familiar (in pitch) melody.

1.3. Study Design

For this experiment, we were interested in examining how features of a pitch layout affected performance accuracy in the learning of a new motor pattern, how this skill was retained at test immediately after training, and performance accuracy in transfer of this skill to a new, untrained task. Musically experienced participants played three out of the four layouts under consideration (see Figure 1): all 24 participants played both AS' and AS, with 12 participants each playing either $A'S$ or $A'S'$.

The independent variables were

- *Adjacency* $\in \{0, 1\}$, where 0 is the code for a layout with non-adjacent major seconds ($A'S'$ or $A'S$), and 1 is the code for a layout with adjacent major seconds (AS' or AS).
- *Shear* $\in \{0, 1\}$, where 0 is the code for an unsheared layout ($A'S'$ or AS'), and 1 is the code for a sheared layout ($A'S$ or AS).
- *LayoutNo* $\in \{0, 1, 2\}$, where 0 is the code for the first layout played by a participant, 1 is the code for the second layout they played, and 2 is the code for the third and final layout they played.
- *PerfNo* $\in \{0, 1, 2, 3\}$, where 0 is the code for their first performance of a given layout, 1 is the code for their second performance of a given layout, 2 is the code for their third performance of a given layout, 3 is the code for their fourth performance of a given layout. Note that participants gave three performances for the training, two performances for the immediate retention tasks, and four performances for the transfer task.

Each participant played the layouts in one of four different sequences, and each such sequence was played by 6 participants:

- AS' then $A'S'$ then AS
- AS' then $A'S$ then AS
- AS then $A'S'$ then AS'
- AS then $A'S$ then AS'.

This means that the nonadjacent seconds layouts ($A'S'$ and $A'S$) were always presented second, and that participants who started with the unsheared adjacent layout (AS') finished with the sheared adjacent layout (AS), and vice versa.

In each such layout, participants received an equivalent training and testing program: first for the C major scale, then for arpeggios of all triads in C major. The scale task was used to support the learning of the spatial patterns of seconds in the diatonic scale; the arpeggios to support the learning of the spatial patterns of larger intervals such as thirds and fourths in the diatonic scale. Immediate retention (performance without any audiovisual training) was tested after each task. The transfer task required participants to perform a well-known melody (*Frère Jacques*) for which they had received no prior training. This melody contains numerous major and minor seconds but also larger intervals. Participants were given 20 s to practice before their performances were recorded. These procedures are further detailed in Sections 2.2–2.4.

Participants' preferences were elicited in a semi-structured interview, a detailed analysis of which is available in [1]. The current paper will fully describe the results of the performances of training and testing materials (both retention and transfer tasks), assessed for their inaccuracy in terms of number of incorrect notes as well as the timing of the performed notes in comparison to either the audiovisual sequence (training) or the metronome beat (retention and transfer).

2. Results

2.1. Modelling Approach

In order to determine effect sizes and significances of the independent variables (*Adjacency, Shear, LayoutNo, PerfNo*), these variables were regressed on the dependent variable *Inaccuracy*. The method used to calculate *Inaccuracy* (detailed in Section 4.4) takes account of both pitch errors and timing errors in participants' performances. Due to fundamental differences in the training tasks compared with the test tasks (retention and transfer), the method for calculating their respective *Inaccuracy* values differs (see Section 4.4). In both cases, however, *Inaccuracy* takes only positive values, because a value of 0 implies a perfect performance.

Initially, linear mixed effects models were fitted to the *Inaccuracy* data. (*Mixed effects models* are regressions where participants' coefficients are treated as samples from a single, and best-fitting, multivariate normal distribution. Given this distribution, the means of the coefficients are termed *fixed effects*, their (co)variances are termed *random effects*. For within subject designs, mixed effects models are widely recommended to avoid Type I errors [32]).

However, the residuals were skewed and heteroscedastic. For this reason, a generalized linear mixed effects model with a gamma distribution and a log link was used—this combination of distribution and link function providing the best fit to the data. (*Generalized linear models* are designed to cope with data that do not meet the assumptions required by linear regression models; notably, that the residuals are homoscedastic and symmetric. These violations typically occur when the data can take only a subset of real values, such as *Inaccuracy*, which can take only positive values. For positive-valued data, the gamma distribution with a log link is commonly used. The link function transforms the mean of the chosen distribution so that it is linearly related to the independent variables [33].)

The log link means that the exponential of each predictor's fixed effect value represents the multiplicative factor by which *Inaccuracy* changes for each unit increase of that predictor (all else being equal). For example, a fixed effect of 0.5 means that a unit increase in that predictor multiplies *Inaccuracy* by $\exp(0.5) = 1.649$; put differently, *Inaccuracy* increases by 64.9%. In the subsequent tables, this exponential value is shown in the "Factor" column.

In each model, the intercept was included as a random effect grouped by participant to allow it to take account of participants' differing abilities. Generally, maximal random effects structures [32] (with all fixed effects and their interactions included as random effects) were attempted, but such models failed to converge, even after removing interactions. The resulting models, therefore, assume that the independent variables are invariant across participants in the population, but that participants' overall ability does vary.

The experimental design was unbalanced because nonadjacent layouts occurred only during the second layout (i.e., *Adjacency* $= 0 \iff$ *LayoutNo* $= 1$). For this reason, the predictor *Adjacency* was not included in any interactions. Including *Adjacency* in interactions results in rank-deficient design matrices or in conditional effects for lower-order terms that are hard to interpret; for example, if the interactions *Shear:Adjacency* and *Shear:LayoutNo* are included, then the conditional effect for *Shear* refers to the effect of Shear when *LayoutNo* $= 0$ and *Adjacency* $= 0$, which never actually occurs in the experiment. All other possible interactions were included in the initial model. To simplify interpretation of lower-order terms, any nonsignificant 3-way interactions were removed from the model, and the model was refitted. For the same reason, any nonsignificant 2-way interactions not part of a 3-way interaction were then removed, and the model refitted.

Classical *t*-tests of significance in mixed effects models are considered to be anti-conservative (prone to incorrectly low *p*-values); for this reason, each predictor's *p*-value (as shown in the subsequent summary tables) was estimated through theoretical ($\chi^2(1)$) tests of the likelihood ratio between the model with that predictor and the model without that predictor. All models were fitted with the `lme-4` package in R, and *p*-values estimated by the `drop1` function using the `Chisq` option, or the `anova` function. The r^2 value shown in each table is the squared correlation between the model's predictions and the observed data, hence loosely analogous to R^2 in linear regression.

2.2. Training Performances

Participants performed two training tasks: (1) Scales, and (2) Arpeggios. For each layout, participants completed three training performances of the scale, and three of the arpeggios. The accuracy of performances in the two training tasks, averaged across participants, are summarized in Figures 2 and 3 as a function of the independent variables detailed in Section 1.3. The 95% confidence intervals were obtained with 100,000 bootstrap samples, calculated and plotted in MATLAB. The generalized linear mixed effects models used to estimate the effects' sizes and significances of *Adjacency*, *Shear*, *LayoutNo*, and *PerfNo* are summarized in Tables 1 and 2.

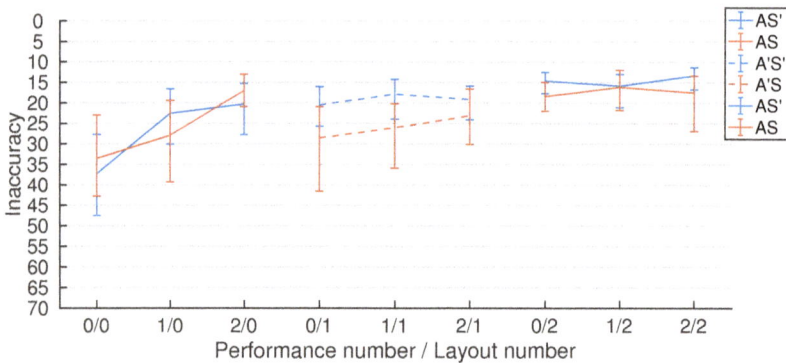

Figure 2. Inaccuracies, averaged across participants, for the scale training task. The higher the line, the more accurate the average performance. The bootstrapped confidence intervals cover a 95% range. Adjacent layouts have solid lines, nonadjacent have dashed lines. Sheared layouts have orange lines, unsheared have blue lines. The first performance is coded 0, the second is coded 1, ...; the first layout is coded 0, the second is coded 1.

Table 1. Generalized linear mixed effects model for the scale training task.

Fixed Effect	Estimate	Factor	*p*-Value
(Intercept)	3.48	32.61	<0.001 ***
Adjacency	−0.06	0.94	0.263
Shear	0.13	1.14	0.026 *
LayoutNo	−0.37	0.69	0.001 ***
PerfNo	−0.27	0.77	<0.001 ***
LayoutNo:PerfNo	0.12	1.13	0.003 **
Log Likelihood	−727.67		
Num. obs.	208		
Num. groups: ID	24		
Var: ID (Intercept)	0.04		
Var: Residual	0.17		
r^2	0.44		

*** $p < 0.001$, ** $p < 0.01$, * $p < 0.05$.

The significant coefficient for *Shear* (1.14) indicates that sheared layouts worsen performance accuracy in the scale training task: they multiply inaccuracy by 1.14 (increase it by 14%). Looking at

Figure 2, it is apparent that this effect is arising mainly in the non-adjacent layouts—the accuracies for $A'S$ (dashed orange) are lower than the accuracies for $A'S'$ (dashed blue). Note that the interaction between *Adjacency* and *Shear* was not tested for the reasons explained in Section 2.1.

The significant coefficients for *LayoutNo* (0.69), *PerfNo* (0.77), and their interaction *LayoutNo:PerfNo* (1.13), indicate that accuracy in the scale training task improves over successive layouts and performances, but less so when either is high. In Figure 2, the positive gradient of the lines in the first layout block (0/0, 1/0, 2/0) appear greater than those of the second (0/1, 1/1, 2/1) and third (0/2, 1/2, 2/2) blocks; the inaccuracy measured at the beginning of each layout block (performance 0) also appears to increase across layout blocks 0–2, but less so with performances 1 and 2 (the middles and ends of each layout block).

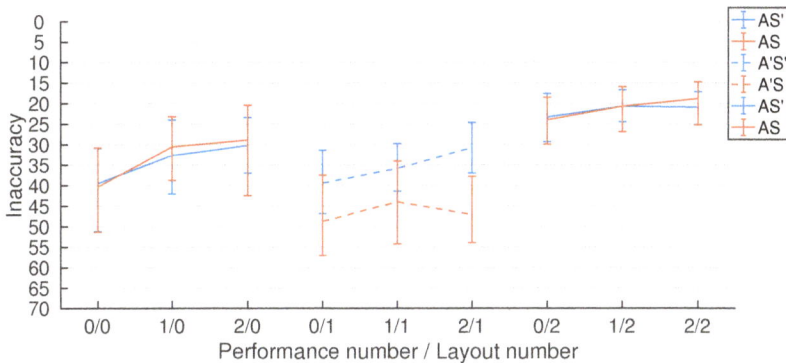

Figure 3. Inaccuracies, averaged across participants, for the arpeggio training task. The higher the line, the more accurate the average performance. The bootstrapped confidence intervals cover a 95% range. Adjacent layouts have solid lines, nonadjacent have dashed lines. Sheared layouts have orange lines, unsheared have blue lines. The first performance is coded 0, the second is coded 1, ...; the first layout is coded 0, the second is coded 1.

Table 2. Generalized linear mixed effects model for the arpeggio training task.

Fixed Effect	Estimate	Factor	*p*-Value
(*Intercept*)	3.92	50.36	<0.001 ***
Adjacency	−0.43	0.65	<0.001 ***
Shear	0.08	1.08	0.037 *
LayoutNo	−0.22	0.80	<0.001 ***
PerfNo	−0.09	0.91	<0.001 ***
Log Likelihood	−738.65		
Num. obs.	209		
Num. groups: ID	24		
Var: ID (Intercept)	0.05		
Var: Residual	0.08		
r^2	0.78		

*** $p < 0.001$, ** $p < 0.01$, * $p < 0.05$.

The significant coefficient for *Adjacency* (0.65) indicates that adjacent layouts strongly improve performance accuracy in the arpeggio training task: they multiply inaccuracy by 0.65 (decrease it by 35%). This is visible in Figure 3 where the dashed lines in the second layout block are lower than an imaginary line connecting the first and third blocks.

The significant coefficient for *Shear* (1.08) indicates that sheared layouts slightly worsen accuracy in the arpeggio training task: they multiply inaccuracy by 1.08 (increase it by 8%). Looking at Figure 3, it is apparent that this effect is arising mainly in the non-adjacent layouts—the accuracies for $A'S$ (dashed orange) are lower than the accuracies for $A'S'$ (dashed blue). Note that the interaction between *Adjacency* and *Shear* was not tested for the reasons explained in Section 2.1.

The significant coefficients for *LayoutNo* (0.80) and *PerfNo* (0.91) indicate that accuracy in the scale training task improves over successive layouts and performances (inaccuracy is reduced by 20% and 9%, respectively). This is shown, in Figure 3, by the positive gradient across the three performances in the majority of the layouts presented, and the general increase in accuracy across the three layout blocks (bearing in mind the drop in the second block resulting from the effects of non-adjacency).

2.3. Immediate Retention

Performances to evaluate immediate retention were recorded for the two trained tasks (1) Scales; (2) Arpeggios. Test performance results are reported here only for item 1 because most of the test performances of item 2 had too many errors in pitch and timing to be reliably tracked with respect to the target pitches. The distinctly lower performance accuracy here may reflect the greater musical complexity of the arpeggios compared to the scales and therefore increased difficulty in memorising the sequence (compare Figures 7 and 8 in Section 4.2.2). It may also be because the tested pitch layouts are not as well suited to the arpeggios as they are to the scales. The results for the arpeggio training task—detailed in Section 2.2—are, however, still useful in the comparison with the scale training task.

For each layout, participants completed two retention performances of the scale without audiovisual support. The accuracy of these performances, averaged across participants, are summarized in Figure 4. The generalized linear mixed effects model used to estimate the effects' sizes and significances of *Adjacency*, *Shear*, *LayoutNo*, and *PerfNo* is summarized in Table 3.

The significant coefficient for *Adjacency* (0.58) indicates that adjacent layouts strongly improve performance accuracy in the scale retention task: they multiply inaccuracy by 0.58 (decrease it by 42%). In Figure 4, note how the accuracies for the non-adjacent layouts $A'S'$ and $A'S$ (dashed lines) are lower than the accuracies in the adjacent layouts.

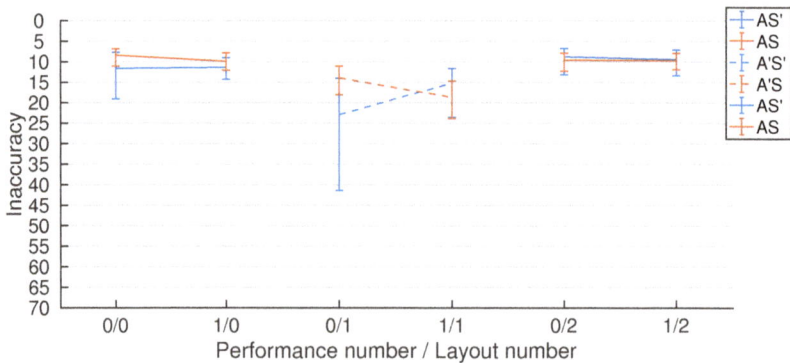

Figure 4. Inaccuracies, averaged across participants, for the scale retention task. The higher the line, the more accurate the average performance. The bootstrapped confidence intervals cover a 95% range. Adjacent layouts have solid lines, nonadjacent have dashed lines. Sheared layouts have orange lines, unsheared have blue lines. The first performance is coded 0, the second is coded 1; the first layout is coded 0, the second is coded 1.

Table 3. Generalized linear mixed effects model for immediate retention of the scale task.

Fixed Effect	Estimate	Factor	*p*-Value
(Intercept)	2.85	17.32	<0.001 ***
Adjacency	−0.54	0.58	<0.001 ***
Shear	−0.08	0.92	0.305
LayoutNo	−0.04	0.96	0.412
PerfNo	0.04	1.04	0.581
Log Likelihood	−437.37		
Num. obs.	142		
Num. groups: ID	24		
Var: ID (Intercept)	0.06		
Var: Residual	0.26		
r^2	0.38		

*** $p < 0.001$, ** $p < 0.01$, * $p < 0.05$.

2.4. Transfer

Transfer of learning was evaluated by performances of a separate test melody that participants had not received training for: *Frère Jacques*. For each layout, participants performed the transfer task for four consecutive performances.

The accuracy of performances for the transfer task, averaged across participants, are summarized in Figure 5. The generalized linear mixed effects model used to estimate the effects' sizes and significances of *Adjacency*, *Shear*, *LayoutNo*, and *PerfNo* is summarized in Table 4.

The significant coefficient for *Shear* (0.61) indicates that, for the first layout, which is always adjacent, sheared layouts strongly improve accuracy in the transfer task: they multiply inaccuracy by 0.61 (decrease it by 39%). The significant coefficient for the interaction *LayoutNo:Shear* (1.39) indicates that the positive effect of shear vanishes in the second and third layouts. In Figure 5, note how, in the first layout block (0/0, 1/0, 2/0, 3/0), the accuracies for the sheared layout *AS* (orange) are higher than the unsheared layout *AS'* (blue) but, in the second or third layout blocks, the accuracies for the sheared layouts (*AS*, *A'S*—orange) are, if anything, slightly lower than those for the unsheared layouts (*AS'*, *A'S'*—blue).

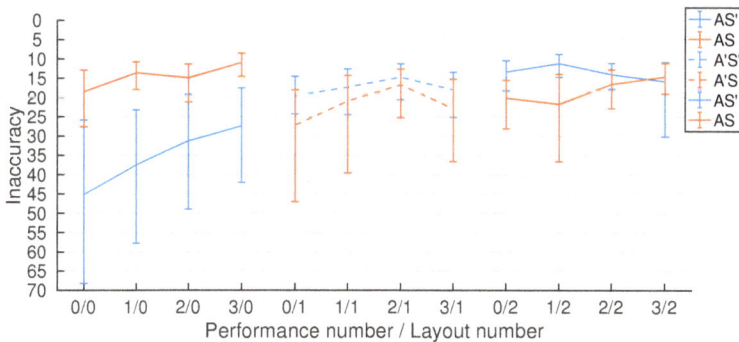

Figure 5. Inaccuracies, averaged across participants, for the transfer task (*Frère Jacques*). The higher the line, the more accurate the average performance. The bootstrapped confidence intervals cover a 95% range. Adjacent layouts have solid lines, nonadjacent have dashed lines. Sheared layouts have orange lines, unsheared have blue lines. The first performance is coded 0, the second is coded 1, …; the first layout is coded 0, the second is coded 1.

Table 4. Generalized linear mixed effects model for the transfer task (*Frère Jacques*).

Fixed Effect	Estimate	Factor	*p*-Value
(*Intercept*)	3.43	30.85	<0.001 ***
Adjacency	−0.05	0.96	0.408
Shear	−0.50	0.61	<0.001 ***
LayoutNo	−0.38	0.68	<0.001 ***
PerfNo	−0.13	0.88	<0.001 ***
LayoutNo:Shear	0.33	1.39	0.002 **
LayoutNo:PerfNo	0.07	1.07	0.021 *
Log Likelihood	−996.89		
Num. obs.	288		
Num. groups: ID	24		
Var: ID (Intercept)	0.11		
Var: Residual	0.26		
r^2	0.66		

*** $p < 0.001$, ** $p < 0.01$, * $p < 0.05$.

The significant coefficients for *LayoutNo* (0.68), *PerfNo* (0.88), and their interaction *LayoutNo:PerfNo* (1.07) indicate that accuracy in the transfer task improves over successive layouts and performances, but less so when either is high. In Figure 5, the positive gradient of the slopes in the first layout block (0/0, 1/0, 2/0, 3/0) appear greater than those of the second (0/1, 1/1, 2/1, 3/1) and third (0/2, 1/2, 2/2, 3/2) blocks; the inaccuracy measured at the beginning of each layout block (performance 0) also appears to increase across layout blocks 0–2, but less so with later performances.

3. Discussion

3.1. PerfNo and LayoutNo

Three out of the four tasks (the two training tasks and the transfer task) show a strong positive effect of *PerfNo* and *LayoutNo*. Both effects are indicative of learning. The former indicates that the more a participant plays on a given layout the better they get (until the ceiling is reached). The latter indicates that learning in one layout extends to different layouts. This is unsurprising given the similarity of the four different layouts: they all exhibit a "3 + 4" scale pattern where, to play a major scale, a row of three buttons is played from left to right, then there is a "carriage return" to the next row above where a row of four buttons is played followed by a "carriage return" to the next row above, and the pattern starts again (in the next higher octave). The immediate retention of the scale task does not show any impact of *PerfNo* or *LayoutNo*. This is clearly due to the simplicity of the task—participants are close to their ceiling from the very start.

3.2. Adjacency

In two out of the four tasks (the arpeggio training task and the transfer task), *Adjacency* has a strong positive effect (35% and 42% decreases in inaccuracy). The statistical results from the other two tasks (see also Figures 2 and 5) also hint at this positive effect but the values are not significant. It seems that any impact of adjacency in the transfer task is swamped by the learning that has already occurred by the time the adjacent layout is played (remember that an adjacent layout always came second for all participants and never first or third). A future experiment that balances the design, by also having nonadjacent layouts first and third, will have greater statistical power to detect the impact of adjacency in this context. Judging the results in total, we have strong evidence that adjacent major seconds improve playing accuracy in a variety of one-handed melodic tasks. Although this could be a result of the high number of major seconds present in both the scales and the nursery rhyme chosen for the transfer task, the positive effect of adjacency effect is also seen in the arpeggio task, which has a large emphasis on other intervals—notably major and minor thirds, and perfect fourths.

Here, we might have expected the adjacent layout to be no more useful than the nonadjacent layout. However, the findings suggest that successfully learning major seconds may also provide a useful foundation for learning and playing other intervals. Hence, adjacent major seconds may be a broadly useful property.

3.3. Shear

The impact of *Shear* is a little more complicated. In both of the training tasks, it has a significant negative effect on playing accuracy (14% and 8% increases in inaccuracy). It has no significant effect in the immediate retention of the scale task. However, for the principal data collected in this experiment—the *Frère Jacques* transfer task—it has a strong positive effect for the first (and adjacent) layout presented. The reason this effect does not carry through to the second and third layouts presented may be due to performances hitting ceiling (as appears to be the case by looking at Figure 5); they may also be due to shear having a positive effect with adjacent layouts but not with nonadjacent layouts (as explained in Section 2.1, due to the experimental design, interactions with *Adjacency* were not included and, hence, not directly tested).

It remains to account for why shear may worsen performances in simple training tasks, but improve them when playing a new melody. We hypothesize that the slanted runs of major seconds in the sheared layouts provide an impediment to accuracy due to the slant making movement physically more difficult or confusing due to its unfamiliarity compared with the piano keyboard. However, in a task like playing a new melody, which requires finding correct pitches without having previously learned a sequential pattern, and has a more complex and variable sequence of movements, the additional clarity of the vertical pitch axis in the sheared layouts becomes more important and trounces the previously mentioned disadvantages. This hypothesis matches the outcomes we see, but would require further testing using differing playing tasks with differing requirements for quickly accessing and recognizing relative pitch heights.

3.4. Limitations

There are some limitations of the current experiment. First, the nonadjacent layouts ($A'S$ and $A'S$) occurred only second, which means that in the transfer task the impact of adjacency was overwhelmed by learning effects; it also means that the full range of interactions between the variables could not be tested. Secondly (as detailed in Section 1.1.2), by shearing parallel with the octave axis, the angles of the pitch axis and the angle of the major second axis covaried, so these two effects cannot be disambiguated; furthermore, the angle of the octave axis was invariant (it was always vertical) so not tested. Thirdly, only melodies were tested—we would expect different results for bass lines (where perfect fifths and fourths are prevalent) or for harmony, where thirds are also prevalent. Fourthly, all of the pitch layouts tested were rather similar in their overall form—this similarity is a natural consequence of good experimental design, which requires changing only the variables of interest across conditions, but it may limit generalizability. Finally, other covarying aspects of the layouts were not independently tested (e.g., adjacencies or angles of other intervals).

Most of these limitations can be overcome by conducting further experiments, although it would not be feasible to address all of them at once. For instance, the nonadjacent layouts could be presented first and third; the shears could be made parallel with the pitch axis or with the major second axis—doing both, in conjunction with this experiment, would enable the effects of the pitch, octave, and M2 axes' angles to be fully disambiguated; differing musical contexts, such as chord progressions or bass lines, could be trained and tested; a wider range of layouts, with configurations distinctly different from those considered here, could be used.

With respect to the results, limitations can be seen in the ceiling effect achieved quickly for some of the tasks, particularly with the second and third layouts played. The experiment allowed for training with audiovisual performances for 4 separate instances (one watching the audiovisual sequence, and the remaining three playing along with the sequence). For the scale task, this was sufficient for participants to excel in the immediate retention test; conversely, for the arpeggio task, this may not have been enough training because performances in the immediate retention were too poor to analyse. For the transfer task, 20 s of exploration was allowed prior to the first performance on each layout. This is not a standard figure, but was used to avoid a floor performance for this untrained task. However, for the second and third layouts, this may have helped participants too much because ceiling performance was reached quickly. Selecting training period lengths and tasks with the appropriate difficulty are important in the design of learning experiments. A task of appropriate difficulty can ensure that a learning effect can be measured.

3.5. Summary

In broad terms, the results demonstrate that the precise form of a pitch layout has a crucial impact on its effectiveness, hence the importance of uncovering the underlying properties—such as interval adjacency and axis angles—that account for these effects on motor learning. With regard to these two properties, the data support our hypotheses that spatially privileging major seconds by making them adjacent, and making the pitch axis salient by giving it a vertical orientation, facilitates motor learning for melodies played on novel pitch layouts.

The two historically established pitch layouts that have adjacent major seconds and, for 12-TET, also have a pitch axis that is close to vertical or horizontal (and so require only a small shear to perfectly align them) are the Wicki layout [4] and the Bosanquet layout [2,3]. The results obtained here, therefore, suggest that appropriately sheared versions of these two layouts (e.g., the *AS* layout, illustrated in Figure 1d, which is a sheared Wicki layout) are optimal for playing melodies.

4. Methods

The methods and materials are, for the most part, the same as in [1]. Previously, only the number of correct notes was reported for the transfer task. In this paper, we have developed a new method (detailed in Section 4.4) for measuring the inacuracies of participants' performances during the training, immediate retention, and transfer tasks.

4.1. Participants

Twenty-four participants were recruited (mean age = 26, age range: 18–44) with at least 5 years of musical experience on at least one instrument (excluding the voice). Ethical approval for this experiment was obtained via the Western Sydney University Human Research Ethics Committee (approval number: H10487).

4.2. Materials

4.2.1. Hardware and Software

The software sequencer Hex [15] was modified to function as a multitouch MIDI controller, and presented on an Acer touchscreen notebook, as shown in Figure 6. Note names were not shown on the interface, but middle D was indicated with a subtly brighter button to serve as a global reference. The position of middle C was indicated to participants, this being the starting pitch of every scale, arpeggio, or melody they played.

In order to present training sequences effectively, both aurally and visually, Hex's virtual buttons were highlighted in time with a MIDI sequence. All training sequences were at 90 bpm and introduced by a two-bar metronome count.

Figure 6. The multitouch interface used in the experiment.

4.2.2. Musical Tasks

Melodies for musical tasks were chosen to be single-line sequences to be performed solely with the right hand. The training melodies consisted of a set of C major scales (Figure 7) and a set of arpeggios (again only using the notes of the C major scale—Figure 8) spanning two octaves, and all starting and ending on middle C. The well-known nursery rhyme *Frère Jacques* was used as the transfer task melody (Figure 9); it too was played in C major and began and ended on middle C.

Figure 7. The scale training and retention task.

Figure 8. The arpeggio training and retention task.

Figure 9. The transfer task: *Frère Jacques.*

4.3. Procedure

The layouts were presented in four different sequences, with each sequence played by 6 participants: AS' then $A'S'$ then AS; or AS' then $A'S$ then AS; or AS then $A'S'$ then AS'; or AS then $A'S$ then AS'.

4.3.1. Training Paradigm and Testing of Immediate Retention

For each of their three layouts, participants were directed through a 15 min training and testing paradigm involving (1) scales and (2) arpeggios. For each stage, this involved:

1. watching the sequence once as demonstrated by audiovisual highlighting
2. playing along with the audiovisual highlighted sequence three times (training)
3. reproducing the sequence in the absence of audiovisual highlighting, for two consecutive performances (immediate retention task)

All demonstration sequences and participant performances were played in time with a 90 bpm metronome, and recorded as MIDI files.

4.3.2. Transfer Task

A final production task asked participants to play a well-known nursery rhyme—*Frère Jacques*. Participants first heard an audio recording of the nursery rhyme to confirm their knowledge of the melody. They were then given 20 s initially to explore the layout and find the correct notes before giving four consecutive performances. Again, these performances were instructed to be played in time with a 90 bpm metronome. Although this represents a fairly simple task, the nursery rhyme was chosen as it facilitated measurement of participants' skill with each particular layout. We assume that as the participants' memory for the melody was intact, their performance would only be affected by their memory of the layout itself.

4.4. Measuring Performance Inaccuracy

Separate measures were developed to assess performance inaccuracy in the training tasks (playing along with the audiovisual highlighted sequence), and performance inaccuracy in the retention and transfer tasks (playing along with a metronome beat). Both measures first calculate an *Accuracy* score by taking account of both the number of "correct" notes played, and the timing of these notes in comparison to either the audiovisual sequence (in the training tasks), or the audio metronome beat (in the retention and transfer tasks). This is then converted into an *Inaccuracy* score by subtracting *Accuracy* from the maximum possible *Accuracy* value (as would be achieved by a flawless performance).

4.4.1. Training Tasks

In the training tasks, the participant played along with the audiovisual highlighted sequence. Here, a "correct" performance would constitute playing the correct pitches at the correct times in the sequence; that is, matching the sequence being produced aurally and visually onscreen. For each *target note* in the audiovisual demo, a window of 666 ms (the interonset interval at 90 bpm), centred around the target note, was created. The first performed note within this window to match the target's MIDI note number was used. In this way, inserted notes which did not match the expected MIDI note were not penalized if the "correct" note was played at some point in the time window. If no matching MIDI note number was identified within this window, a zero score was allocated for this target note in the sequence. The first matching MIDI note number was then assigned a score reflecting its timing accuracy—a value of 1, if played at the same time as the target; linearly reducing to 0 as the timing error increases to the boundary of the window ($\pm$333 ms). The resulting scores are summed. As a final step, the total is normalized by dividing it by the number of notes in the target sequence and multiplied by 100. This puts accuracy onto the same scale for target sequences of different lengths, such that 100 always means the target sequence has been performed perfectly. The formula to calculate a participant's inaccuracy in the training sequences of the scales and arpeggios is

$$Inaccuracy = 100 - Accuracy, \text{ where}$$

$$Accuracy = \frac{100}{N} \sum_{n=1}^{N} 1 - \frac{|t_p - t_n|}{333}, \tag{1}$$

and

- N is the total number of notes in the target sequence
- t_n is the time in milliseconds of nth target note
- t_p is the time in milliseconds of first performed note within the $t_n \pm 333$ time window that matches the pitch of the target note.

4.4.2. Retention and Transfer Tasks

In the immediate retention and transfer tasks, the participant played the test sequence along with a 90 bpm metronome. Here, a "correct" performance would constitute playing the correct notes in the expected sequence, as before, but with one caveat related to penalizing errors. Because these tasks were accompanied only aurally by a metronome beat, the performer might add an extra note or leave a gap so that all subsequent notes are then played one metronome beat late. In a situation such as this, it is reasonable to penalize the first late note, but not the subsequent notes, which are then correct subject to a delay. Similarly, a performer might skip a note, hence all subsequent notes will be a metronome beat early. As before, it is reasonable only to penalize the first pitch error.

To establish which notes of the performed sequence were "correctly" pitched, we used the Note Time Playing Path software [34], which uses a windowing process to identify where extra, skipped, or substituted (wrongly pitched) notes occur. For performances where there were large numbers of pitch errors (>5), this matching process was visually confirmed.

In order to count the total number of errors, we took a novel approach where only the first pitch error in a consecutive sequence of pitch errors was included. For example, consider a performer who plays the first four notes of a scale correctly (C, D, E, F). After this he/she plays three wrong notes in a row (D, D, E), but then returns to the original scale sequence one note after the last correct note (A, B). In this instance of four consecutive errors—three wrong notes (D, D, E) and one deletion (G)—only the first error is counted. The final penalty is then calculated by the total number of such "first" errors, designated by the letter E in Equation (2). This final error value is subtracted from an accuracy score calculated in the same manner as Equation (1) with the exception that it refers only to the correct notes in the performance. As before, the accuracy value is normalized by $100/N$ to ensure a perfect performance of a melody of any length has an accuracy of 100, which means the formula to calculate a participant's inaccuracy in the retention and transfer tasks is

$$Inaccuracy = 100 - Accuracy, \text{ where}$$

$$Accuracy = \frac{100}{N} \left(\left(\sum_{c=1}^{C} 1 - \frac{|t_c - t_m|}{333} \right) - E \right), \tag{2}$$

and

- N is the total number of notes in the target sequence.
- C is the total number of notes in the corrected performance.
- t_c is the time in milliseconds of the cth note in the corrected performance.
- t_m is the time of the metronome beat closest to the performed note t_c.
- E is the number of errors calculated as explained above.

We have defined two versions of *Inaccuracy*—one for the training tasks, one for the retention and transfer tasks. The context (e.g., in Section 2) will always make clear which of these is being used.

Acknowledgments: Andrew Milne is the recipient of an Australian Research Council Discovery Early Career Researcher Award (project number DE170100353) funded by the Australian Government. The authors would like to thank the Anthony Prechtl for specially recoding Hex to make it respond to multitouch input.

Author Contributions: Drs MacRitchie and Milne collaboratively conceived, designed and conducted the experiments reported in this article. Data was analysed and interpreted together. Both authors collaboratively wrote the paper.

Conflicts of Interest: The authors declare no conflict of interest.

References

1. MacRitchie, J.; Milne, A.J. Evaluation of the Learnability and Playability of Pitch Layouts in New Musical Instruments. In Proceedings of the 14th Sound and Music Computing Conference, Espoo, Finland, 5–8 July 2017; pp. 450–457.
2. Bosanquet, R.H.M. *Elementary Treatise on Musical Intervals and Temperament*; Macmillan: London, UK, 1877.
3. Helmholtz, H.L.F. *On the Sensations of Tone as a Physiological Basis for the Theory of Music*; Dover: New York, NY, USA, 1877.
4. Wicki, K. Tastatur für Musikinstrumente. Swiss Patent 13329, 30 October 1896.
5. Keislar, D. History and principles of microtonal keyboards. *Comput. Music J.* **1987**, *11*, 18–28.
6. Array Instruments. The Array Mbira. Available online: http://www.arraymbira.com (accessed on 24 October 2017).
7. Paine, G.; Stevenson, I.; Pearce, A. Thummer Mapping Project (ThuMP) Report. In Proceedings of the 2007 International Conference on New Interfaces for Musical Expression (NIME07), New York, NY, USA, 6–10 June 2007; pp. 70–77.
8. C-Thru Music. The AXiS-49 USB Music Interface. Available online: http://www.c-thru-music.com/cgi/?page=prod_axis-49 (accessed on 24 October 2017).
9. Park, B.; Gerhard, D. Rainboard and Musix: Building Dynamic Isomorphic Interfaces. In Proceedings of the International Conference on New Interfaces for Musical Expression, Daejeon, Korea, 25–31 May 2013; pp. 319–324.
10. Roger Linn Design. LinnStrument: A Revolutionary Expressive Musical Performance Controller. Available online: http://www.rogerlinndesign.com/linnstrument.html (accessed on 24 October 2017).
11. ROLI. ROLI | BLOCKS. Available online: https://roli.com/products/blocks (accessed on 24 October 2017).
12. Terpstra, S.; Horvath, D. Terpstra Keyboard. Available online: http://terpstrakeyboard.com (accessed on 24 October 2017).
13. Milne, A.J.; Sethares, W.A.; Plamondon, J. Isomorphic controllers and Dynamic Tuning: Invariant fingering over a tuning continuum. *Comput. Music J.* **2007**, *31*, 15–32.
14. Huygens Fokker Foundation. Microtonality - Scales. Available online: http://www.huygens-fokker.org/microtonality/scales.html (accessed on 24 October 2017).
15. Prechtl, A.; Milne, A.J.; Holland, S.; Laney, R.; Sharp, D.B. A MIDI sequencer that widens access to the compositional possibilities of novel tunings. *Comput. Music J.* **2012**, *36*, 42–54.
16. Milne, A.J.; Sethares, W.A.; Plamondon, J. Tuning continua and keyboard layouts. *J. Math. Music* **2008**, *2*, 1–19.
17. Vos, P.G.; Troost, J.M. Ascending and descending melodic intervals: Statistical findings and their perceptual relevance. *Music Percept.* **1989**, *6*, 383–396.
18. Euler, L. *Tentamen Novae Theoriae Musicae ex Certissismis Harmoniae Principiis Dilucide Expositae*; Saint Petersbury Academy: Saint Petersburg, Russia, 1739.
19. Zatorre, R.J.; Chen, J.L.; Penhune, V.B. When the brain plays music: Auditory-motor interactions in music perception and production. *Nat. Rev. Neurosci.* **2007**. doi:10.1038/nrn2152.
20. Schmidt, R.A.; Lee, T.D. *Motor Control and Learning*, 5th ed; Human Kinetics: Champaign, IL, USA, 2011.
21. Kunde, W.; Koch, I.; Hoffmann, J. Anticipated action effects affect the selection, initiation, and execution of actions. *Q. J. Exp. Psychol. Sect. A* **2004**, *57*, 87–106.
22. Rusconi, E.; Kwan, B.; Giordano, B.L.; Umiltà, C.; Butterworth, B. Spatial representation of pitch height: The SMARC effect. *Cognition* **2006**, *99*, 113–129.

23. Hartmann, M. Non-musicians also have a piano in the head: Evidence for spatial–musical associations from line bisection tracking. *Cogn. Process.* **2017**, *18*, 75–80.

24. Pitteri, M.; Marchetti, M.; Priftis, K.; Grassi, M. Naturally together: Pitch-height and brightness as coupled factors for eliciting the SMARC effect in non-musicians. *Psychol. Res.* **2017**, *81*, 243–254.

25. Lega, C.; Cattaneo, Z.; Merabet, L.B.; Vecchi, T.; Cucchi, S. The effect of musical expertise on the representation of space. *Front. Hum. Neurosci.* **2014**, *8*, 250.

26. Stewart, L.; Verdonschot, R.G.; Nasralla, P.; Lanipekun, J. Action-perception coupling in pianists: Learned mappings or spatial musical association of response codes (SMARC) effect? *Q. J. Exp. Psychol.* **2013**, *66*, 37–50.

27. Keller, P.E.; Dalla Bella, S.; Koch, I. Auditory imagery shapes movement timing and kinematics: Evidence from a musical task. *J. Exp. Psychol. Hum. Percept. Perform.* **2010**, *36*, 508–513.

28. Keller, P.E.; Koch, I. The planning and execution of short auditory sequences. *Psychon. Bull. Rev.* **2006**, *13*, 711–716.

29. Pfordresher, P.Q. Musical training and the role of auditory feedback during performance. *Ann. N. Y. Acad. Sci.* **2012**, *1252*, 171–178.

30. Shepard, R.N. Geometrical approximations to the structure of musical pitch. *Psychol. Rev.* **1982**, *89*, 305–333.

31. Hu, H.; Gerhard, D. Appropriate Isomorphic Layout Determination Using 3-D Helix Lattices. In Proceedings of the 43rd International Computer Music Conference/The 6th Electronic Music Week, Shanghai, China, 16–20 October 2017.

32. Barr, D.J.; Levy, R.; Scheepers, C.; Tilyc, H.J. Random effects structure for confirmatory hypothesis testing: Keep it maximal. *J. Mem. Lang.* **2013**, *68*, 255–278.

33. McCullagh, P.; Nelder, J.A. *Generalized Linear Models*, 2nd ed.; Chapman and Hall/CRC: Boca Raton, FL, USA, 1989.

34. Jayanthakumar, J.; Schubert, E.; de Graaf, D. *NTPP (Note Time Playing Path) MIDI Performance Analyser*; Version 0.1; University of New South Wales: Sydney, Australia, 2011.

applied sciences

MDPI

Article

EigenScape: A Database of Spatial Acoustic Scene Recordings

Marc Ciufo Green *and Damian Murphy

Audio Lab, Department of Electronic Engineering, University of York, Heslington, York YO10 5DQ, UK;
damian.murphy@york.ac.uk
* Correspondence: marc.c.green@york.ac.uk; Tel.: +44-190-4324-231

Academic Editor: Tapio Lokki
Received: 23 October 2017; Accepted: 8 November 2017; Published: 22 November 2017

Abstract: The classification of acoustic scenes and events is an emerging area of research in the field of machine listening. Most of the research conducted so far uses spectral features extracted from monaural or stereophonic audio rather than spatial features extracted from multichannel recordings. This is partly due to the lack thus far of a substantial body of spatial recordings of acoustic scenes. This paper formally introduces EigenScape, a new database of fourth-order Ambisonic recordings of eight different acoustic scene classes. The potential applications of a spatial machine listening system are discussed before detailed information on the recording process and dataset are provided. A baseline spatial classification system using directional audio coding (DirAC) techniques is detailed and results from this classifier are presented. The classifier is shown to give good overall scene classification accuracy across the dataset, with 7 of 8 scenes being classified with an accuracy of greater than 60% with an 11% improvement in overall accuracy compared to use of Mel-frequency cepstral coefficient (MFCC) features. Further analysis of the results shows potential improvements to the classifier. It is concluded that the results validate the new database and show that spatial features can characterise acoustic scenes and as such are worthy of further investigation.

Keywords: soundscape; acoustic environment; acoustic scene; ambisonics; spatial audio; Eigenmike; machine learning; dataset; recordings

1. Introduction

Since machine listening became an eminent field in the early 1990s, the vast majority of research has focused on automatic speech recognition (ASR) [1] and computational solutions to the well-known 'cocktail party problem'—the "ability to listen to and follow one speaker in the presence of others" [2]. This is now a mature field of study, with robust speech recognition systems featured in most modern smartphones. There has also been a great deal of research on music information retrieval (MIR) [3], a technology with applications in intelligent playlist algorithms used by online music streaming services [4]. There has been comparatively little research investigating the automatic recognition of general acoustic scenes or acoustic events, though there has been an increase in interest in this area in recent years, largely due to the annual Detection and Classification of Acoustic Scenes and Events (DCASE) challenges established in 2013 [5].

The DCASE challenges have attracted a large number of submissions designed to solve the problem of acoustic scene classification (ASC) or acoustic event detection (AED). A typical ASC or AED system requires a feature extraction stage in order to reduce the complexity of the data to be classified. The key is the coarsening of the available data such that similar sounds will yield similar features (generalisation), yet the features should be distinguishable from those yielded by different types of sounds (discrimination). Generally, the audio is split into frames and some kind of mathematical transform is applied in order to extract a feature vector from each frame. Features extracted from labelled recordings (training data) are used to train some form of classification algorithm, which can then be used to return labels for new unlabelled recordings (testing data). See [6] for a thorough overview of this process.

The systems submitted to DCASE all identify acoustic scenes and events based upon features extracted from monaural or stereophonic recordings. A small number of systems have used spatial features extracted from binaural recordings [7–9], but the potential for extracting features using more sophisticated spatial recordings remains almost completely unexplored. This is due to a number of factors, including inheritance of techniques from ASR and MIR and the envisioned applications of ASC and AED.

A majority of the early research into ASR approached the problem with the aim of emulating elements of human sound perception. This "biologically relevant" [1] approach can be seen in the popular Mel-frequency cepstral coefficient (MFCC) features, which use a mel-scaled filter bank in order to crudely emulate the human cochlear response [10]. A more fundamental self-imposed limitation of this approach is the use of one- or two-microphone recordings. Although, on introducing the DCASE challenge, Stowell et al. stated that "human-centric aims do not directly reflect our goal... which is to develop systems that can extract semantic information about the environment around them from audio data" [5], it is natural to inherit techniques from more mature related fields.

The most commonly stated applications of ASC and AED technologies include adding context-awareness to smart devices, wearable technology, or robotics [6] where mounting of spatial microphone arrays would perhaps be more impractical. Another application is automatic labelling of archive audio, where the majority of recordings will be in mono or stereo format [5,6].

Some lesser-considered applications of ASC and AED technology involve the holistic analysis of acoustic scenes in and of themselves. The focus here is gaining a greater understanding of the constituent parts of acoustic scenes and how they change over time. This has potential applications in acoustic ecology research for natural environments, re-synthesis of acoustic scenes for virtual reality, and in obtaining more detailed measures for urban environmental sound than the prevailing L_{Aeq} sound level metric. The L_{Aeq} measurement aggregates all sound present in a scene into one single sound level figure. This disregards the variety of sources of the sounds, influencing much environmental sound legislation to focus on its suppression—an "environmental noise approach" [11]. A machine listening system could consider the content of an acoustic scene as well as absolute sound levels. This information could be used to create more subtle metrics regarding urban sound, taking into account human perception and preference—a "soundscape approach" [11]. This kind of system was proposed by Bunting et al. [12], but despite some promising work involving source separation in Ambisonic audio [13], published results from that project have been limited. The term *soundscape* is used here according to the ISO definition, meaning "the acoustic environment of a place, as perceived by people, whose character is the result of the action and interaction of natural and/or human factors" [14]. This emphasis on perception is apt in this case, but a subjective perceptual construct is clearly not what a machine listening system will receive as input for analysis. We therefore use the term 'acoustic scene' when discussing recordings.

Another potential application of such a system is assisting in the synthesis of acoustic scenes for experimental purposes. If a researcher or organisation wishes to obtain detailed data on human perception of environmental sound, one technique that can be used is a sound walk, in which listening tests can be conducted in situ at a location of interest. This gives the most realistic stimulus possible, direct from the environment itself. Results gained using this technique are therefore as representative as possible of subjects' reactions with respect to the real-world acoustic environment, a factor known as 'ecological validity' [15,16]. The key disadvantages are that this method is not repeatable [15] and can be very time-consuming [17]. An alternative is laboratory reproduction of acoustic scenes, presented either binaurally [18] or using Ambisonics [19,20]. These are less time-consuming and more repeatable [15], but the clear disadvantage is the potential for reduced ecological validity of the results, which leads to the criticism that lab results "ought to be validated in situ" [21]. A key issue is how to condense an urban sound recording into a shorter format whilst retaining ecological validity. Methods for this have included selection of small clips at random [15] or manually arranging a acoustic scene "composition" in order to "create a balanced impression" [19]—essentially condensing the acoustic scene by ear. Whilst manual composition of a stimulus is undoubtably more robust than presentation of a random short clip that may or may not be representative of the acoustic scene as a whole, it is not an optimal process. The subjective recomposition of an acoustic scene by a researcher introduces a source of bias that could be reflected in the results. A machine listening system could effectively bypass this issue by providing detailed analysis that could assist with synthesis of shorter clips that remained statistically representative of real acoustic scenes.

The limitation to low channel counts is less applicable given these applications of machine listening technology. Spatial recordings offer the potential for a rich new source of information that could be utilised by machine listening systems and higher channel counts offer the opportunity for sophisticated source separation [13,22] which could assist with event detection.

The lack of research into classification using spatial audio features could also be due to the fact that there has been, as yet, no comprehensive database of spatially-recorded acoustic scenes. Any modern database of recordings intended for use in ASC research must contain many examples of each location class. This is to avoid the situation whereby classification results are artificially exaggerated due to test clips being extracted from the same longer recordings as clips used to train classifiers, as exemplified in [23].

A similar phenomenon has been seen in MIR research where classifiers were tested on tracks from the same albums as their training material [24]. The TUT Database [25], used in DCASE challenges since 2016, fulfils this criterion. It features recordings of 15 different acoustic scene classes made across a wide variety of locations, with details provided in order to avoid any crossover in locations between the training and testing sets. This database was recorded using binaural in-ear microphones. The DCASE 2013 AED task [5] used a small set of office recordings made in Ambisonic B-format (though only stereo versions were released as part of the challenge). Since it was intended for AED, this dataset features recordings of office environments only, not the wide range of locations needed for ASC work. The DEMAND database [26] features spatial recordings of six different acoustic scene classes, each recorded over three different locations. This is a substantial amount of data, but potentially still too small a collection for effective classifier training and validation. The recordings were made using a custom-made 16-channel microphone grid, which offers potential for spatial information extraction, though techniques developed using this data might not be generalisable to other microphone setups. This paper introduces EigenScape, a database of fourth-order Ambisonic recordings of a variety of urban and natural acoustic scenes for research into acoustic scene and event detection. The database and associated materials are freely available—see Supplementary Materials for the relevant URLs.

The paper is organised as follows: Section 2 covers the technical details of the recording process, provides information on the recorded data itself, and describes the baseline classification used for initial analysis of the database. Section 3 gives detailed results from the baseline classifier and offers some analysis of its behaviour and the implications this has for the dataset. Section 4 offers some additional discussion of the results, details potential further work, and concludes the paper.

2. Materials and Methods

2.1. Recording

EigenScape was recorded using the mh Acoustics EigenMike [27], a 32-channel spherical microphone array capable of recording up to fourth-order Ambisonic format. In Bates' Ambisonic microphone comparisons [28,29] the EigenMike is among the lowest rated in terms of perceptual audio quality, rated as sounding "dull" compared to other microphones. Conversely, directional analysis shows the EigenMike gives the highest directional accuracy of any of the microphones tested, including the popular first-order Ambisonic Soundfield MKV and Core Sound TetraMic. It should be noted that the analysis in [28,29] used only the first-order output from the EigenMike (for parity with the other microphones), disregarding the higher-order channels. Since the dataset presented in this paper is primarily aimed at machine (rather than human) listening, and the EigenMike can record far more detailed spatial information than first-order microphones whilst retaining a relatively portable form factor, the EigenMike was chosen for this task.

Recordings were made using the proprietary EigenMike Microphone Interface Box and EigenStudio recording application [27]. Recordings were made at 24-bit/48 kHz resolution and the files use Ambisonic Channel Number (ACN) ordering [30]. All the recordings used a gain level of +25 dB set within the EigenStudio software as the ambient sound at many recording locations did not yield an adequate recording level at lower gain levels. The only exception to this is the recording labelled 'TrainStation-08', which used only +5 dB gain as very high level locomotive engine noise present at that location caused severe clipping at +25 dB.

For the majority of the recordings, the EigenMike was mounted in a Rycote windshield designed for use with the SoundField ST350 microphone [31]. Although the windshield was not designed for the EigenMike, care was taken to rigidly mount the microphone and the shield was effective in cancelling wind noise. The first few recordings used a custom-made windshield, but this was switched for the Rycote as the set-up time proved far too long. One indoor recording did not use any windshield. The discrepancies in windshield use and gain level should be negligible by comparison to the wide variety of sounds present in the scenes, especially when coarse features are extracted for use in a machine listening system. Such a system should be robust to the small spectral changes incurred by use of different windshields and to differences in ambient sound level between scenes. Indeed, the DARES project [32] used entirely different recording setups for indoor and outdoor recordings and this was judged to have "minimal influence on the quality of the database". Nevertheless, these discrepancies are noted in metadata provided for EigenScape.

To make these recordings, the microphone was mounted on a standard microphone stand set to around head height. A Samsung Gear 360 camera [33] was also mounted to the tripod, recording video in order to assist with future annotation of events within scenes where the sound might be ambiguous. Figure 1 shows the full recording apparatus.

Figure 1. The setup used to record the EigenScape database: mh-Acoustics Eigenmike within a Rycote windshield, a Samsung Gear 360 camera, an Eigenmike Microphone Interface Box, and an Apple MacBook Pro. The equipment is shown here at Redcar Beach, UK: 54°37′16″ N, 1°04′50″ W.

2.2. Details

Eight different examples each of eight different classes of acoustic scene were recorded for a total of 64 recordings. All recordings are exactly 10 minutes in length. The uniform recording duration facilitates easy segmentation into clips of equal length (e.g., 20 segments, each 30 s long). Basic segmentation tools are available with the dataset in order to assist with this. The recordings were planned out specifically to create a completely evenly-weighted dataset between the various scenes and to facilitate easy partitioning into folds (e.g., six recordings used for training, the other two used for testing).

The location classes were inspired by the classes featured in the TUT database: lakeside beach, bus, cafe/restaurant, car, city center, forest path, grocery store, home, library, metro station, office, urban park, residential area, train, and tram [25], but restricted to open public spaces, reflecting the shifted focus of this work towards acoustic scene analysis. The eight classes in EigenScape are as follows: Beach, Busy Street, Park, Pedestrian Zone, Quiet Street, Shopping Centre, Train Station, and Woodland. These location classes were chosen to give a good variety of acoustic environments found in urban areas and to be relatively accessible for the recording process. The recordings were made at locations across the North of England in May 2017. An online map has been created showing all the recording sites and is listed in Supplementary Materials. Basic location details are included in the dataset metadata, along with recording dates and times. Although individual consent is not required for recording in public spaces, permissions of the relevant local authorities or premises management was sought where possible. Some locations would not allow tripod-based recordings, so the microphone stand was held as a monopod. These are noted in the metadata.

A little over 10 minutes was recorded at each location, with a short amount of time removed from the beginning and end of each file post-recording. This removed the experimenter noise incurred by activating and deactivating the equipment and achieved the exactly uniform length of the audio clips. During recording, every effort was made to minimise sound introduced to the scene by the experimenter or equipment. It should be noted that occasionally a curious passerby would ask about what was happening. This was fairly unavoidable in busier public places, but since conversation is part of the acoustic scenes of such locations, these incidents should not affect feature extraction too much. Discretion is advised if these recordings are used for listening tests or as background ambiences in sound design work.

The complete dataset has been made available online for download. The full database is presented in uncompressed WAV format within a series of ZIP files organised by class. Since each recording is 10 minutes of 25 tracks at 24-bit/48 kHz, the whole set is just under 140 GB in size. As this could potentially be very taxing on disk space and problematic to download on slower internet connections, a second version of the dataset was created for easier access. This second version consists of all the recordings, but limited to the first-order Ambisonic channels (four tracks) and losslessly compressed to FLAC format within a single ZIP file. This results in a much more manageable size of 12.6 GB, whilst still enabling spatial audio analysis and reproduction. This is also in accordance with the UK Data Service's recommended format for audio data [34].

2.3. Baseline Classification

To create a baseline for this database that utilises spatial information whilst maintaining a level of parity with the MFCC-Gaussian mixture model (GMM) baseline typically used in DCASE challenges [5,6,25], the audio was filtered into 20 mel-spaced frequency bands (covering the frequency range up to 12 kHz) using a bank of bandpass finite impulse response (FIR) filters. The filters each used 2048 taps and were designed using hamming windows. Estimate direction of arrival (DOA) estimates to be used as features were extracted from each band using directional audio coding (DirAC) analysis [35–37] as follows:

$$D = -PU \tag{1}$$

where P contains the 20 mel-filtered versions of the zeroth-order Ambisonic channel (W) of the recording and U contains the filtered versions of the first-order Ambisonic bi-directional X, Y and Z-channels in a three-dimensional matrix. Resultant matrix D contains instantaneous DOA estimates for each frequency band. Mean values of D were calculated over a frame length of 2048 samples, with 25% overlap between frames. Angular values for azimuth θ and elevation ϕ were derived from this as follows [38]:

$$\theta = \arctan\left(\frac{X}{Y}\right) \tag{2}$$

$$\phi = \arccos\left(\frac{Z}{||D||}\right) \tag{3}$$

where X, Y, and Z are the X, Y and Z channel matrices extracted from D. These angular values were used as features. Diffuseness values were also used as features, and were calculated as follows [36]:

$$\psi = 1 - \frac{|| - D||}{c\{E\}} \tag{4}$$

where $\{.\}$ represents the mean-per-frame values described previously, c is the speed of sound, and:

$$E = \frac{1}{2}\rho_0\left(\frac{P^2}{Z_0^2} + ||U||^2\right) \tag{5}$$

where Z_0 is the characteristic acoustic impedance and ρ_0 is the mean density of air.

The database was split into four folds for cross-validation. In each fold, six location class recordings were used for training, with the remaining two used for testing. The extracted DirAC features from each frame of the training audio were used to train a bank of 10-component GMMs (one per scene class). The test audio was cut into 30-s segments (40 segments in total for testing). Features were extracted from these segments, and each GMM gave a probability score for the frames. These scores were summed across frames from the entire 30-s segment, with the segment classified according to the model which gave the highest total probability score across all frames.

3. Results

Initial analysis of this dataset previously published as part of the DCASE 2017 workshop [39] compared classification accuracies achieved using the DirAC features to those achieved when using MFCCs. In addition, classifiers were trained using individual DirAC features—azimuth, elevation and diffuseness—and a classifier was trained using a concatenation of all MFCC and DirAC features. Figure 2 shows the mean and standard deviation classification accuracies achieved across all scenes using these various feature sets. It can be seen that using all DirAC features to train a GMM classifier gives a mean accuracy of 64% across all scene classes, whereas MFCC features give a 58% mean accuracy (averaged across all folds). Azimuth data alone is much less discriminative between scenes, giving an accuracy of 43% on average, which is markedly worse than MFCCs. Elevation data, on the other hand, gives similar accuracies, and diffuseness data gives slightly better accuracies than MFCCs. The low accuracy when using azimuth data is probably attributable to the fact that azimuth estimates will be affected by the orientation of the microphone array relative to the recorded scene, whereas elevation and diffuseness should be rotation-invariant. A new classifier using elevation and diffuseness values only was therefore trained and gave an average classification accuracy of 69%, which is the best performance that was achieved. The elevation/diffuseness (E/D)—GMM classifier was therefore adopted as the baseline classifier and all further results reported here are derived from it.

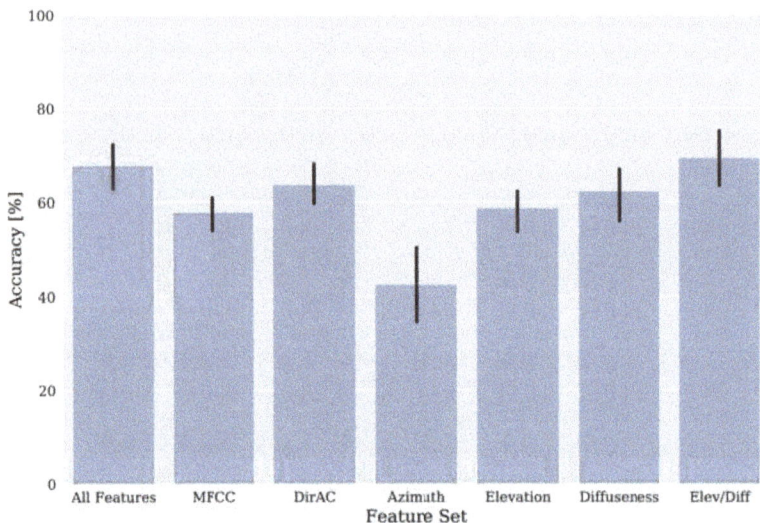

Figure 2. Mean and standard deviation classification accuracies across all folds for the entire dataset using various different feature sets (from [39]). MFCC: Mel-frequency cepstral coefficient; DirAC: directional audio coding.

Figure 3 shows the mean and standard deviation classification accuracies from the baseline for each acoustic scene class. As previously mentioned, the mean accuracy across all scene classes is 69%. The low standard deviation (7%) indicates the dataset as a whole gives features that are fairly consistent across all folds. All of the scene classes except Beach are classified with mean accuracies above 60%. In fact, if the Beach class is discounted, the overall mean accuracy rises by 9%. Busy Street, Pedestrian Zone and Woodland are classified particularly well, at 86%, 97% and 85% accuracy, respectively. Looking at the standard deviation values for accuracy across folds could give some indication of the within-class variability between the different scene recordings. The very low standard deviation in Pedestrian Zone accuracies of 4% implies that the Pedestrian Zone recordings have very similar sonic characteristics, that is, they give very consistent features. Busy Street, Park and Train Station could be said to be moderately consistent, whereas Quiet Street, Shopping Centre and Woodland show more variability between the various recordings. The drastically lower accuracy of the Beach scene classification is very anomalous. It could be that as the primary sound source at a beach will likely be widespread and diffuse broadband noise from the ocean waves, this could yield indistinct features that could be difficult for the classifier to separate from other scenes.

Figure 4 shows confusion matrices (previously published in [39]), which indicate classifications made by the MFCC and E/D classifiers averaged across all folds. Rows indicate the true classes and columns indicate the labels returned by the classifiers. The E/D matrix features a much more prominent leading diagonal and confusion is much less widespread than in the MFCC matrix, clearly indicating that the E/D classifier outperforms the MFCC classifier in the vast majority of cases. Beach is the only class in which the MFCC classifier significantly outperforms the E/D classifier. The most commonly-returned labels for the Beach scene by the E/D classifier are Quiet Street and Busy Street, perhaps due to the aforementioned broadband noise from ocean waves yielding spatial features similar to that of passing cars. This interpretation is corroborated by Figure 5, which shows elevation estimates extracted using Equation (3) from 30-s segments of Beach, Quiet Street and Train Station recordings as heat maps for comparison. The Beach and Quiet Street plots both show large areas across time and frequency where elevation estimates remain broadly consistent at around 90°, indicating the presence of broadband noise sources dominating around that angle. The Train Station plot, on the other hand, shows much more erratic changes in elevation estimates across time, and indeed there is no confusion between Beach and Train Station using the E/D classifier.

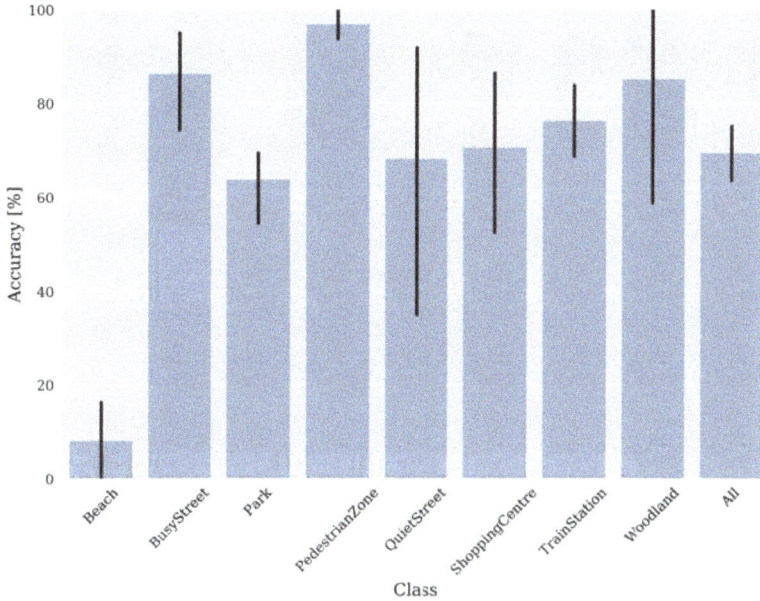

Figure 3. Mean and standard deviation classification accuracies across all folds for each scene class using the elevation/diffuseness–Gaussian mixture model (E/D-GMM) classifier.

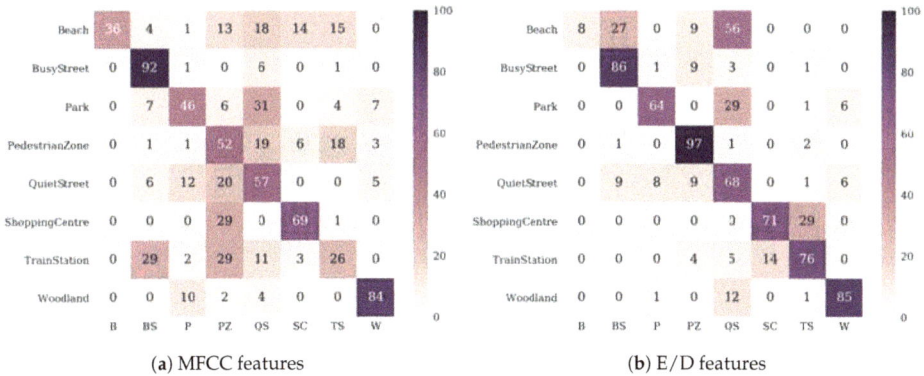

(a) MFCC features

(b) E/D features

Figure 4. Confusion matrices of classifiers trained using MFCC features and elevation/diffuseness features extracted using DirAC. Figures indicate classification percentages across all folds (from [39]).

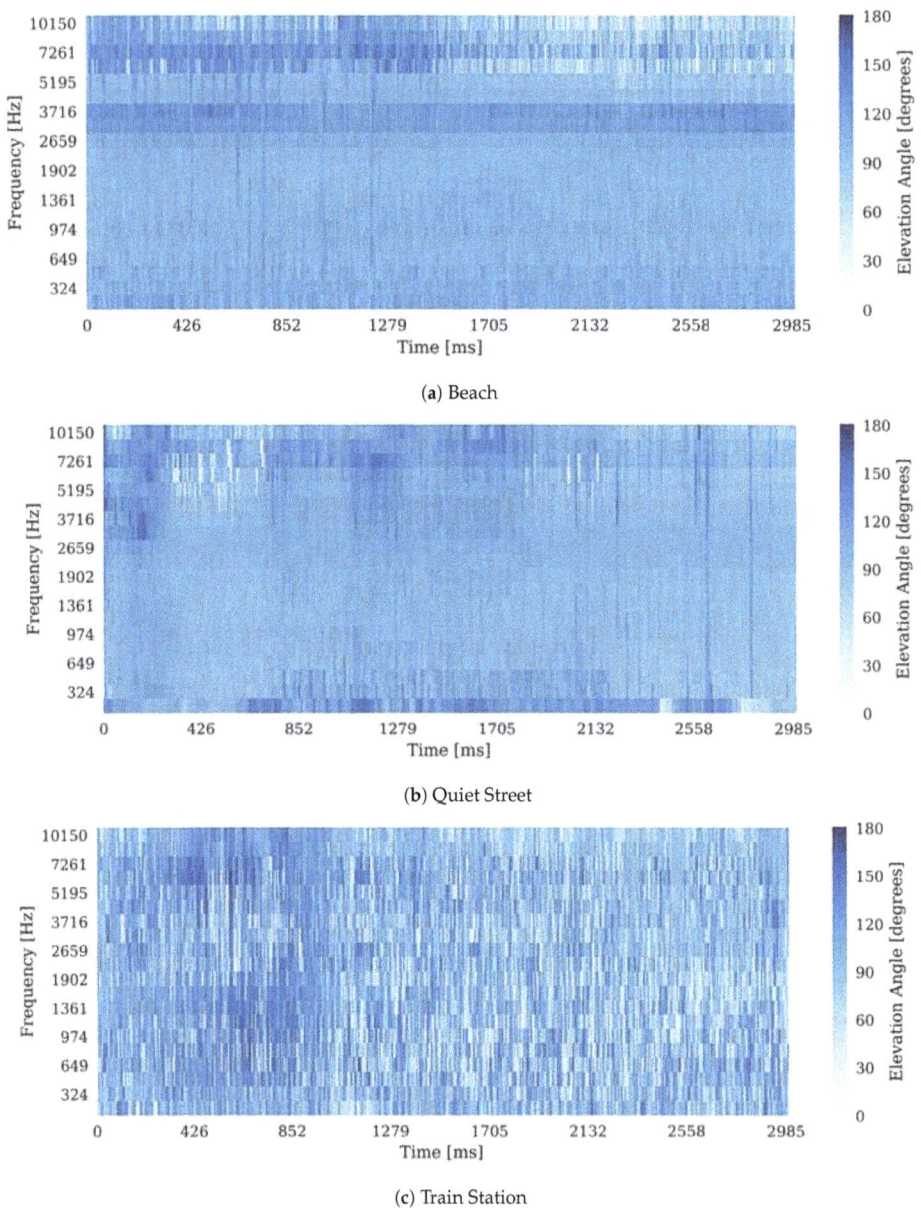

(a) Beach

(b) Quiet Street

(c) Train Station

Figure 5. Heat maps depicting elevation estimates extracted from 30-s segments of Beach, Quiet Street and Train Station recordings.

It is interesting to consider instances where the E/D classifier considerably outperforms the MFCC classifier, such as with Pedestrian Zone, which is classified with 97% accuracy by the E/D classifier, whereas the MFCC classifier only manages 52%. This indicates that the spatial information present in pedestrian zones is much more discriminative than the spectral information, which seems to share common features with both quiet streets and train stations. Further to this, it is interesting to investigate the instances where there is significant confusion present in both classifiers. Park, for instance, is most commonly misclassified as Quiet Street by both classifiers. This is probably due to the fact that both Park and Quiet Street scenes are both characterised as being relatively quiet locations, yet are still in the midst of urban areas. These recordings tend to contain occasional human sound and low-level background urban 'hum' (as opposed to Woodland, which tends to lack this). In other cases, however, the specific misclassifications do not always correspond. The most common misclassification of the Shopping Centre by the MFCC classifier is the Pedestrian Zone, a result perhaps caused by prominent human sound found in both locations. In contrast to this, for the E/D classifier the most common misclassification of the Shopping Centre is Train Station, and in fact there is no confusion with the Pedestrian Zone at all. This could be due to the similarity in acoustics between the large reverberant indoor spaces typical of train stations and shopping centres, which could have an impact on the values calculated for elevation and diffuseness.

Figure 6 shows receiver operating characteristic (ROC) curves for the individual models trained to identify each location class. These curves evaluate each GMM's performance as a one-versus-rest classifier. The curves were generated by comparing the scores generated by each model with the ground-truth labels for each scene and calculating the probabilities that a certain score will be given to a correct clip (true positive) or will be given to a clip from another scene (false positive). These pairs of probabilities are calculated for every score output from the classifier and when plotted, form the ROC curve. The larger the area under the curve (AUC), the better the classifier. The curves shown in Figure 6 show the mean ROC across the four folds. It can clearly be seen that the AUC values do not follow the pattern of the classification accuracies shown in Figure 3. This discrepancy is most stark in Figure 6a, which shows the Beach model to be the best individual classifier, with an AUC of 0.95. This indicates that the Beach model is individually very good at telling apart Beach clips from all other scenes. The very low Beach classification accuracy from the system as a whole could be explained by the fact that all the other scene models have lower AUC values than the Beach model, which suggests greater tendencies in the other models to give incorrect scenes higher probability scores.

It should be noted here that points on the ROC curves do not indicate absolute score levels. For instance, a false positive point on any given curve will not necessarily be reached at the same absolute probability score as that point on any other curve. It is therefore possible that the Beach model tends to give lower probability scores in general than the other models, and is therefore most of the the time 'outvoted' by other models.

These results suggest that classification accuracies could be improved by using the AUC values from each model to create confidence weightings to inform the decision making process beyond the basic summing of probability scores. A lower score from the Beach model could, for instance, carry more weight than from the Train Station model, which has an AUC of 0.58, indicating performance at only slightly higher than chance levels.

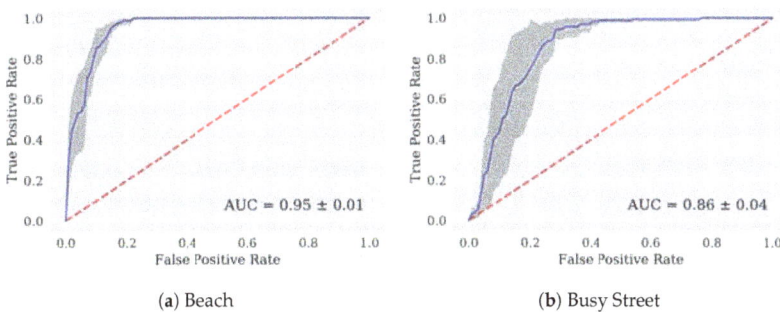

(a) Beach

(b) Busy Street

Figure 6. *Cont.*

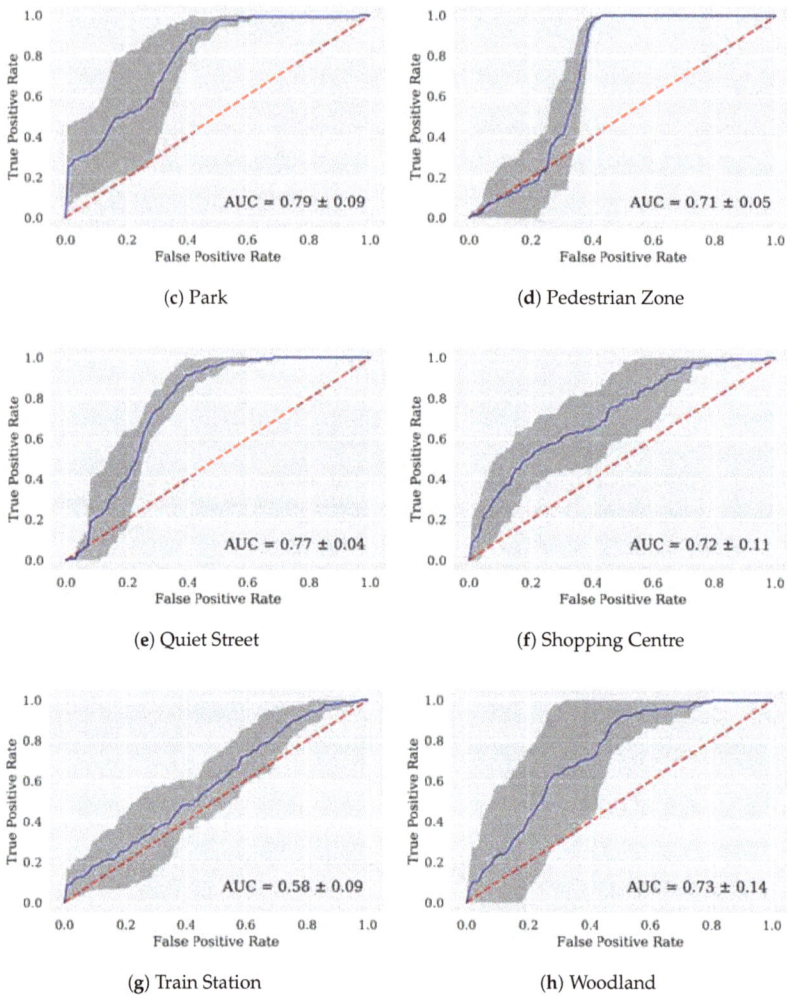

Figure 6. Receiver operating characteristics (ROC) curves for each scene classifier, showing mean (solid line) and standard deviation (grey area) of the curves calculated using results across all folds. Dotted line represents chance performance. AUC: area under the curve.

4. Discussion

The results presented in Section 3 indicate that the collation of EigenScape has been successful in that this classification exercise shows the suitability of this dataset for segmentation and cross-validation. The good, but not perfect, degree of accuracy shown by the baseline E/D-GMM classifier is very significant in that it goes some way towards showing the validity of this dataset in terms of providing a good variety of recordings. Recordings within a class label are similar enough to be grouped together by a classifier, whilst retaining an appropriate degree of variation.

These results suggest that DirAC spatial features extracted from Ambisonic audio could be viable and useful features to use for acoustic scene identification. The simplicity of the classifier used here indicates that higher accuracies could be gleaned from these features, perhaps by using a more sophisticated decision-making process, or simply more sophisticated models. Utilising temporal features could be a compelling next step in this work. It would be especially interesting to investigate whether Δ-azimuth values could be more discriminative than the azimuth values themselves, being perhaps less dependent on microphone orientation. It is also worth noting that all spatial analysis of this dataset so far has used only the first-order Ambisonic channels for feature extraction. The fourth-order channels present in this database provide much higher spatial precision that could enable more sophisticated feature extraction. The high channel count should also facilitate detailed source separation that could be used for polyphonic event detection work. Event detection within scenes should be a key area of research with this dataset moving forwards.

The size and scope of this database are such that there is a lot more knowledge to be gained than has been presented here. The findings of this paper are important initial results that indicate the investigation of spatial audio features could be a fertile new area in machine listening, especially with a view to applications in environmental sound monitoring and analysis.

Supplementary Materials: The EigenScape database is provided freely to inspire and promote research work and creativity. Please cite this paper in any published research or other work utilising this dataset. EigenScape Dataset: http://doi.org/10.5281/zenodo.1012809; Baseline code and segmentation tools: https://github.com/marc1701/EigenScape; Recording Map: http://bit.ly/EigenSMap.

Acknowledgments: Funding was provided by a UK Engineering and Physical Sciences Research Council (EPSRC) Doctoral Training Award, the Department of Electronic Engineering at the University of York, and in part by Digital Creativity Labs, EPSRC Grant Number: EP/M023265/1.

Author Contributions: M.G. conducted the recording work, created the baseline classifier, conducted analysis and wrote the paper. D.M. supervised the project, provided initial ideas and guidance and has acted as editor and secondary contributor to this paper.

Conflicts of Interest: The authors declare no conflict of interest.

Abbreviations

The following abbreviations are used in this manuscript:

ASR	Automatic Speech Recognition
MIR	Music Information Retrieval
DCASE	Detection and Classification of Acoustic Scenes and Events
ASC	Acoustic Scene Classification
AED	Acoustic Event Detection
MFCC	Mel-Frequency Cepstral Coefficients
DOA	Direction of Arrival
DirAC	Directional Audio Coding
GMM	Gaussian Mixture Model
E/D	Elevation/Diffuseness
ROC	Receiver Operating Characteristic
AUC	Area Under the Curve

References

1. Wang, D. *Computation Auditory Scene Analysis: Principles, Algorithms and Applications*; Wiley: Hoboken, NJ, USA, 2006.
2. Cherry, C. *On Human Communication: A Review, a Survey, and a Criticism*; MIT Press: Cambridge, MA, USA, 1978.
3. Raś, Z. *Advances in Music Information Retrieval*; Springer-Verlag: Berlin/Heidelberg, Germany, 2010.
4. The Magic that Makes Spotify'S Discover Weekly Playlists So Damn Good. Available online: https://qz.com/571007/the-magic-that-makes-spotifys-discover-weekly-playlists-so-damn-good/ (accessed on 18 September 2017).
5. Stowell, D.; Giannoulis, D.; Benetos, E.; Lagrange, M.; Plumbley, M.D. Detection and Classification of Acoustic Scenes and Events. *IEEE Trans. Multimed.* **2015**, *17*, 1733–1746.
6. Barchiesi, D.; Giannoulis, D.; Stowell, D.; Plumbley, M.D. Acoustic Scene Classification: Classifying environments from the sounds they produce. *IEEE Signal Process. Mag.* **2015**, *32*, 16–34.

7. Adavanne, S.; Parascandolo, G.; Pertilä, P.; Heittola, T.; Virtanen, T. Sound Event Detection in Multisource Environments Using Spatial and Harmonic Features. In Proceedings of the Detection and Classification of Acoustic Scenes and Events, Budapest, Hungary, 3 September 2016.

8. Eghbal-Zadeh, H.; Lehner, B.; Dorfer, M.; Widmer, G. CP-JKU Submissions for DCASE-2016: A Hybrid Approach Using Binaural I-Vectors and Deep Convolutional Neural Networks. In Proceedings of the Detection and Classification of Acoustic Scenes and Events, Budapest, Hungary, 3 September 2016.

9. Nogueira, W.; Roma, G.; Herrera, P. Sound Scene Identification Based on MFCC, Binaural Features and a Support Vector Machine Classifier; Technical Report; IEEE AASP Challenge on Detection and Classification of Acoustic Scenes and Events; IEEE: Piscataway, NJ, USA, 2013. Available online: http://c4dm.eecs.qmul. ac.uk/sceneseventschallenge/abstracts/SC/NR1.png (Accessed on 6 January 2017).

10. Mel Frequency Cepstral Coefficient (MFCC) Tutorial. Available online: http://practicalcryptography. com/miscellaneous/machine-learning/guide-mel-frequency-cepstral-coefficients-mfccs/ (accessed on 18 September 2017).

11. Brown, A.L. Soundscapes and environmental noise management. *Noise Control Eng. J.* **2010**, *58*, 493–500.

12. Bunting, O.; Stammers, J.; Chesmore, D.; Bouzid, O.; Tian, G.Y.; Karatsovis, C.; Dyne, S. Instrument for Soundscape Recognition, Identification and Evaluation (ISRIE): Technology and Practical Uses. In Proceedings of the EuroNoise, Edinburgh, UK, 26–28 October 2009.

13. Bunting, O.; Chesmore, D. Time frequency source separation and direction of arrival estimation in a 3D soundscape environment. *Appl. Acoust.* **2013**, *74*, 264–268.

14. International Standards Organisation. *ISO 12913-1:2014—Acoustics—Soundscape—Part 1: Definition and Conceptual Framework*; International Standards Organisation: Geneva, Switzerland, 2014.

15. Davies, W.J.; Bruce, N.S.; Murphy, J.E. Soundscape Reproduction and Synthesis. *Acta Acust. United Acust.* **2014**, *100*, 285–292.

16. Guastavino, C.; Katz, B.F.; Polack, J.D.; Levitin, D.J.; Dubois, D. Ecological Validity of Soundscape Reproduction. *Acta Acust. United Acust.* **2005**, *91*, 333–341.

17. Liu, J.; Kang, J.; Behm, H.; Luo, T. Effects of landscape on soundscape perception: Soundwalks in city parks. *Landsc. Urban Plan.* **2014**, *123*, 30–40.

18. Axelsson, Ö.; Nilsson, M.E.; Berglund, B. A principal components model of soundscape perception. *J. Acoust. Soc. Am.* **2010**, *128*, 2836–2846.

19. Harriet, S.; Murphy, D.T. Auralisation of an Urban Soundscape. *Acta Acust. United Acust.* **2015**, *101*, 798–810.

20. Lundén, P.; Axelsson, Ö.; Hurtig, M. On urban soundscape mapping: A computer can predict the outcome of soundscape assessments. In Proceedings of the Internoise, Hamburg, Germany, 21–24 August 2016; pp. 4725–4732.

21. Aletta, F.; Kang, J.; Axelsson, Ö. Soundscape descriptors and a conceptual framework for developing predictive soundscape models. *Landsc. Urban Plan.* **2016**, *149*, 65–74.

22. Bunting, O. Sparse Seperation of Sources in 3D Soundscapes. Ph.D. Thesis, University of York, York, UK, 2010.

23. Aucouturier, J.J.; Defreville, B.; Pachet, F. The Bag-of-frames Approach to Audio Pattern Recognition: A Sufficient Model for Urban Soundscapes But Not For Polyphonic Music. *J. Acous. Soc. Am.* **2007**, *122*, 881–891.

24. Lagrange, M.; Lafay, G. The bag-of-frames approach: A not so sufficient model for urban soundscapes. *J. Acoust. Soc. Am.* **2015**, *128*, doi:10.1121/1.4935350.

25. Mesaros, A.; Heittola, T.; Virtanen, T. TUT Database for Acoustic Scene Classification and Sound Event Detection. In Proceedings of the 24th European Signal Processing Conference (EUSIPCO), Budapest, Hungary, 28 August–2 September 2016.

26. Joachim Thiemann, N.I.; Vincent, E. The Diverse Environments Multi-channel Acoustic Noise Database (DEMAND): A database of multichannel environmental noise recordings. In Proceedings of the Meetings on Acoustics, Montreal, QC, Canada, 2–7 June 2013; Volume 19.

27. MH Acoustics. *em32 Eigenmike® Microphone Array Release Notes*; MH Acoustics: Summit, NJ, USA, 2013.

28. Bates, E.; Gorzel, M.; Ferguson, L.; O'Dwyer, H.; Boland, F.M. Comparing Ambisonic Microphones—Part 1. In Proceedings of the Audio Engineering Society Conference: 2016 AES International Conference on Sound Field Control, Guildford, UK, 18–20 July 2016.

29. Bates, E.; Dooney, S.; Gorzel, M.; O'Dwyer, H.; Ferguson, L.; Boland, F.M. Comparing Ambisonic Microphones—Part 2. In Proceedings of the 142nd Convention of the Audio Engineering Society, Berlin, Germany, 20–23 May 2017.

30. MH Acoustics. *Eigenbeam Data Specification for Eigenbeams Eigenbeam Data Specification for Eigenbeams Eigenbeam Data Specification for Eigenbeams Eigenbeam Data: Specification for Eigenbeams*; MH Acoustics: Summit, NJ, USA, 2016.

31. Soundfield. *ST350 Portable Microphone System User Guide*; Soundfield: London, UK, 2008.

32. Van Grootel, M.W.W.; Andringa, T.C.; Krijnders, J.D. DARES-G1: Database of Annotated Real-world Everyday Sounds. In Proceedings of the NAG/DAGA Meeting, Rotterdam, The Netherlands, 23–26 March 2009.

33. Samsung Gear 360 Camera. Available online: http://www.samsung.com/us/support/owners/product/gear-360-2016 (accessed on 8 September 2017).

34. UK Data Service—Recommended Formats. Available online: https://www.ukdataservice.ac.uk/manage-data/format/recommended-formats (accessed on 11 September 2017).

35. Pulkki, V. Directional audio coding in spatial sound reproduction and stereo upmixing. In Proceedings of the AES 28th International Conference, Pitea, Sweden, 30 June–2 July 2006.

36. Pulkki, V. Spatial Sound Reproduction with Directional Audio Coding. *J. Audio Eng. Soc.* **2007**, *55*, 503–516.

37. Pulkki, V.; Laitinen, M.V.; Vilkamo, J.; Ahonen, J.; Lokki, T.; Pihlajamäki, T. Directional audio coding—Perception-based reproduction of spatial sound. In Proceedings of the International Workshop on the Principle and Applications of Spatial Hearing, Miyagy, Japan, 11–13 November 2009.

38. Kallinger, M.; Kuech, F.; Shultz-Amling, R.; Galdo, G.D.; Ahonen, J.; Pulkki, V. Analysis and adjustment of planar microphone arrays for application in Directional Audio Coding. In Proceedings of the 124th Convention of the Audio Engineering Society, Amsterdam, The Netherlands, 17–20 May 2008.

39. Green, M.C.; Murphy, D. Acoustic Scene Classification Using Spatial Features. In Proceedings of the Detection and Classification of Acoustic Scenes and Events, Munich, Germany, 16–17 November 2017.

applied
sciences

MDPI

Article

Identifying Single Trial Event-Related Potentials in an Earphone-Based Auditory Brain-Computer Interface

Eduardo Carabez *,†, Miho Sugi †, Isao Nambu †and Yasuhiro Wada †

Department of Electrical Engineering, Nagaoka University of Technology, 1603-1, Kamitomioka Nagaoka, Niigata 940-2188, Japan; m.sugi1229@gmail.com (M.S.); inambu@vos.nagaokaut.ac.jp (I.N.); ywada@nagaokaut.ac.jp (Y.W.)
* Correspondence: eduardo@stn.nagaokaut.ac.jp; Tel.: +81-258-47-5349
† These authors contributed equally to this work.

Academic Editor: Vesa Valimaki
Received: 20 October 2017; Accepted: 17 November 2017; Published: 21 November 2017

Abstract: As brain-computer interfaces (BCI) must provide reliable ways for end users to accomplish a specific task, methods to secure the best possible translation of the intention of the users are constantly being explored. In this paper, we propose and test a number of convolutional neural network (CNN) structures to identify and classify single-trial P300 in electroencephalogram (EEG) readings of an auditory BCI. The recorded data correspond to nine subjects in a series of experiment sessions in which auditory stimuli following the oddball paradigm were presented via earphones from six different virtual directions at time intervals of 200, 300, 400 and 500 ms. Using three different approaches for the pooling process, we report the average accuracy for 18 CNN structures. The results obtained for most of the CNN models show clear improvement over past studies in similar contexts, as well as over other commonly-used classifiers. We found that the models that consider data from the time and space domains and those that overlap in the pooling process usually offer better results regardless of the number of layers. Additionally, patterns of improvement with single-layered CNN models can be observed.

Keywords: convolutional neural networks (CNN); auditory brain-computer interface (BCI); P300; virtual sound; electroencephalogram (EEG); pool strategies; classification

1. Introduction

Brain-computer interfaces (BCI) provide a way for their users to control devices by basically interpreting their brain activity [1]. BCI have enormous potential for improving quality of life, particularly for those who have been affected by neurological disorders that partially or fully impede their motor capacities. In severe conditions such as complete locked-in syndrome (CLIS), patients are unable to willfully control movements of the eye or any other body part. In such cases, BCI based on only auditory cues are a viable option for establishing a communication channel [2]. BCI can be seen as module-based devices, where at least two essential parts can be recognized: the brain activity recording module and the brain activity classification one.

To record brain activity, electroencephalography (EEG)-based technologies are often used because they are noninvasive, portable, produce accurate readings and are affordable compared with other methods [3–5]. Within the EEG readings, we can find some recognizable patterns, the P300 event-related potential (ERP) being of particular interest. The P300 is a positive deflection that can be observed in the brain activity of a subject, and it can be elicited via cue presentation following the oddball paradigm, an experimental setting in which sequences of regular cues are interrupted by irregular ones in order to evoke recognizable patterns within the brain activity of the subject. The P300 occurs between 250 and 700 ms after the presentation of an irregular cue in an experimental setting in which the participant is asked to attend to a particular cue (an irregular one in the oddball paradigm). The P300 has been exploited in many ways to produce a number of functional applications [6,7]. Although the specific technology used for recording the brain activity is closely tied to the final performance of the classifier used, [8,9] demonstrated that training and motivation have a positive and visible impact on the shape and appearance of the P300. Experimental setups using EEG and the P300 have been widely used in the development of BCI [10–13].

For data classification, machine learning models such as artificial neural networks (ANN) and support vector machines (SVM) have not only been used widely but have also produced satisfactory results in many BCI applications [14–18]. In recent years, the implementation of convolutional neural networks (CNN) for classification purposes in tasks such as image and speech recognition has been successful [19,20]. As a result, CNN have become an increasing topic of focus in various research fields, especially those involving multidimensional data. The CNN topology enables dimensional reduction of the input while also extracting relevant features for classification. For BCI, CNN have successfully been used for rapid serial visual presentation (RSVP) tasks [21], as well as for navigation in a virtual environment [22]. CNN consist of an arrangement of layers where the input goes through a convolution and a sub-sampling process called pooling, generating in this way features and reducing the size of the needed connections.

In the work of [15], which serves as a major inspiration and reference for the present study, the authors advise against the use of CNN models that mix data from multiple dimensions during the processes of the convolution layer for classification purposes in BCI. However, for our research, we found that considering data from both the time and space domains for the pooling process of the convolution layer results in better CNN classification accuracy. Additionally, we tested pool processes with and without overlapping to assess whether this difference in processing impacts CNN performance. These overlapping approaches were explored for image classification in the work of [23], who reported better performance in the overlapping case, and with respect to speech-related tasks in [24], who found no difference between the approaches and stated that it might depend strictly on the data being used.

In this study, we present and test 18 different CNN models that use the above-mentioned approaches for the pooling process, but also different numbers of convolution layers to classify whether the P300 is present or absent in single-trial EEG readings from an auditory BCI experimental setup. For the experiment, nine subjects were presented with auditory stimuli (100 ms of white noise) for six virtual directions following the oddball paradigm and were asked to attend to the stimuli coming from a specific direction at a time and count in silence every time this happened to potentially increase the correct production of the P300. The BCI approach followed in this work is a reproduction of the one presented in [25] as it has relevant characteristics for auditory BCI (especially portability) such as the use of earphones to present the auditory stimuli and the capacity to simulate sound direction through them. Unlike the work of [25], which considers only one trial interval of 1100 ms between stimuli presentation, we considered variant time intervals (200, 300, 400 and 500 ms) between presentations of the auditory stimuli for all 18 CNN models to evaluate the extent to which this variation could affect the performance of the classifier.

This paper is organized as follows: Section 2 contains the information regarding the conformation of the dataset used, such as the experimental setup and data processing. The structure of proposed CNN models, specific parameters considered for this study and the details of the selected models are described in Section 3. A summary of the obtained results is presented in Section 4, with a strong focus on the similarities between the observed patterns in the performance of the structures. Finally, in Sections 5 and 6, we discuss our results and ideas for future work.

2. Experimental Setup and Production of the Datasets

2.1. Experiment

The dataset used for this study corresponds to the evoked P300 waves of nine healthy subjects (8 men, 1 woman) on an auditory BCI paradigm. A digital electroencephalogram system (Active Two, BioSemi, Amsterdam, Netherlands) was used to record the brain activity at 256 Hz. The device consists of 64 electrodes distributed over the head of the subject by means of a cap with the distribution shown in Figure 1a. This study was approved by the ethics board of the Nagaoka University of Technology. All subjects signed consent forms that contained detailed information about the experiment, and all methods complied with the Declaration of Helsinki.

By using the out of the head sound localization method [26], the subjects were presented with stimuli (100 ms of white noise) from six different virtual directions via earphones followed by an interval in which no sound was produced (silent interval). Figure 1b shows the six virtual direction positions relative to the subject.

We refer to one stimulus and one corresponding silent interval as a trial. Four different trial lengths (200, 300, 400, and 500 ms) were considered in order to analyze the impact that the speed of the stimuli presentation could have on the identification of the P300 wave.

For the creation of this dataset, each subject completed a task, which was comprised of a collection of 12 sessions, for each of the proposed trial lengths. Figure 1c illustrates the conformation of a task. Each session had as the attention target a fixed sound direction that changed clockwise from one session to another starting from Direction 1 (see Figure 1b). Subjects were asked to attend only to the stimuli perceived to be coming from the target direction and to count in silence the number of times it was produced. The subjects performed this experiment with their eyes closed.

In each session, around 180 pseudo-randomized trials were produced, meaning that for every six trials, sound from each direction was produced at least once and that stimuli coming from the target direction were never produced subsequently to avoid overlapping of the P300 wave. Thus, of the approximately 180 trials contained in each session, only a sixth of these would contain the P300 wave corresponding to the target stimuli.

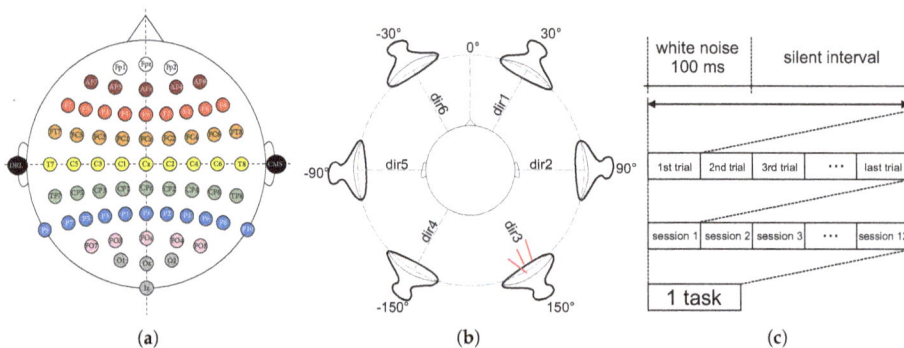

Figure 1. (a) 64 electroencephalogram (EEG) channel layout used in the experiments. Reference electrodes attached to the ears; (b) Virtual disposition of the six sound directions with respect to the user. A stimulus is being produced from Direction 3; (c) Task constitution.

2.2. EEG Data Preprocessing and Accommodation

EEG data preprocessing is conducted as follows: The recorded EEG data are baseline corrected and filtered. Baseline correction is carried out using a first order Savitzky–Golay filter to produce a signal approximation that is then subtracted from the original signal. In that case, the baseline correction is conducted for the period from −100 ms before the stimulus onset until the end of the trial (i.e., end of the silent period after the stimulus offset). This then becomes an example in the training or testing datasets.

For the filtering process, we use Butterworth coefficients to make a bandpass filter with low and high cutoff frequencies of 0.1 Hz and 8 Hz, respectively. Once the correction and filtering are completed, the data are then down-sampled to 25 Hz to reduce the size of the generated examples.

To generate the training and test sets that will be input into the CNN, trials are divided into two groups, randomly: those with and without the target stimuli. Each trial constitutes an example in the training or test set, so there are around 180 examples for each session. As there are 12 sessions for each task and a sixth of the trials correspond to when the stimuli were heard, a total of approximately 360 target and around 1800 non-target examples can be obtained for a single subject in one task. The target and non-target examples are distributed as closely as possible into a 50/50 relation among the training and test sets. Regardless of the trial length, the examples have a matrix shape of 28 × 64, which corresponds to 1100 ms of recordings along the 64 EEG channels after the stimuli were presented. This is done to assure each example contains the same amount of information.

3. Convolutional Neural Networks

This particular neural network architecture is a type of multilayer perceptron with a feature-generation and a dimension-reduction -oriented layer, which, together, compose what is called a convolutional layer. Unlike other layer-based neural networks, the CNN can receive a multidimensional input in its original form, process it and successfully classify it without a previous feature extraction step. This is possible because the features are generated within the CNN layers, preventing possible information loss caused by user-created features or data rearrangement. Figure 2 shows the process an input experiences before classification by one of our proposed CNN models.

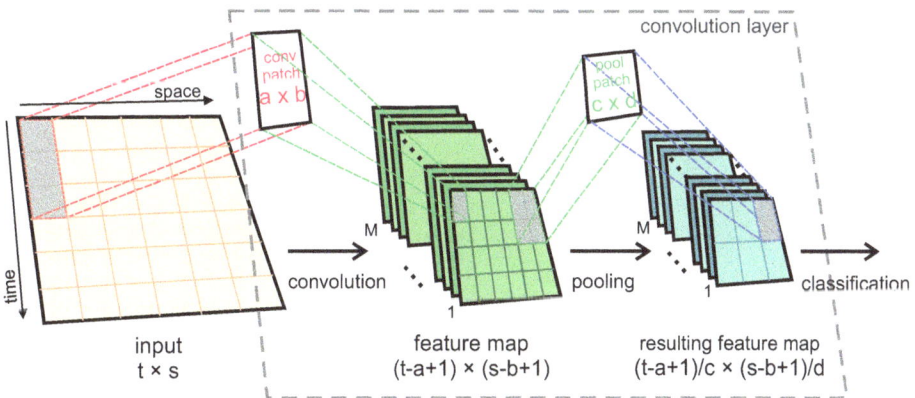

Figure 2. Structure of a convolutional neural network (CNN) depicting the results of applying the convolution and pooling processes in the convolution layer for the input. Default pooling (non-overlapping) is shown in this figure.

The convolution and pooling processes consist of applying patches (also known as kernels) to the input or the result from the previous patch application to extract features and reduce their sizes. For our study, a number $M = 64$ of feature maps is produced as a result of the applications of such patches, each producing a feature map different from the other ones as the weights of the patches change. If an input, convolution patch and pool patch with sizes of $[t \times s]$, $[a \times b]$ and $[c \times d]$, respectively, are considered, the convolution patch is first applied to the input to extract features of interest, which generates a feature map of size $(t - a + 1) \times (s - b + 1)$. Then, the pooling process takes place, which in our case is max pooling. By taking a single desired value out of an area (of the feature map) defined by the size of the pool patch, this process generates a resulting feature map of size $(t - a + 1)/c \times (s - b + 1)/d$ for those cases in which the pooling process does not overlap. For our case $t = 28$, $s = 64$, a, b, c, d change depending on the patches being used. The resulting feature maps are then connected to the output layer in which classification takes place. While in the convolution process, the applied patches overlap, that is not normally the case for the pooling process (see Section 3.1.3 for details). In this study, we test also CNN structures that cause the pool patches to overlap. The convolution and pooling processes occur as many times as there are convolution layers in the CNN.

3.1. Proposed Structures

As with other neural network structures, there are several CNN parameters to be defined by the user that will directly impact CNN performance. In this study, we considered as variables the number of convolutional layers and the shape of convolutional and pool patches. However, the learning rate, experiment stopping conditions, pool stride and optimization method are always the same regardless of the structure being tested. By proposing variations to the above-mentioned parameters, we were able to evaluate 18 different CNN structures in terms of classification rate. Figure 2 shows the general structure of the CNN used in this study.

3.1.1. Number of Convolution Layers

We propose structures with one and two convolution layers. This is the biggest structural difference the proposed models could exhibit as it heavily affects the size of the resulting feature maps. Structures with more than two convolution layers are not advised for applications such as ours, as early tests showed that the input was over simplified and the classification rate highly affected in a negative way.

3.1.2. Shape of Convolution and Pool Patches

Each of the EEG electrodes experiences the presence of the P300 wave in different magnitudes, and there are certain regions that are more likely to show it. This has been reported by different studies [14,25]. However, in most studies that attempt to classify EEG data, the two-dimensional position of the channels along the scalp of the user is mapped, generating a one-dimensional array that positions channels from different regions of the brain next to one another.

Given that applying either of the kernels in a squared-shaped fashion like that demonstrated in Figure 2 will result in feature maps that mix data from both the space and time domains, it is advised [14] that patches be constructed such that they only consider information of one dimension and one channel at a time. In this study, we considered three different convolution and pool patch sizes, including one pool patch that considers data from two adjacent channels simultaneously in the one-dimensional array. These patch sizes were chosen as a result of preliminary tests, in which a wide number of options was analyzed using data from one subject. The different CNN structures that were tested for this study consist of combinations of the selected number of layers and sizes of convolution and pool patches, which are summarized in Table 1. For a given number of convolution layers, the nine possible combinations of convolution and pool patches are considered. For the CNN with two layers, the same combination of convolution and pool patches is used in each layer. For the pooling operation, max pooling is applied.

Table 1. Proposed and tested number of convolution layers, size of convolution patches and size of pool patches.

Convolution Layers	1	2	
Size of convolution patch	2×1	4×1	5×1
Size of pool patch	2×1	3×1	3×2

For easier identification within the text, we will use brackets to refer to the patches listed above, e.g., pool patch [3 × 2]. The convolution layers will only be referred to as layers in the following sections. In this study, 64 feature maps of the same size are generated after the convolution.

3.1.3. Pool Stride

For this study, a fixed pool stride of size 2 × 1 was considered for all 18 proposed CNN structures. Normally, the pool stride is the same size as the pool patch, which means that the pool process takes place in areas of the data that do not overlap. However, in early tests, that approach proved to be inadequate especially for the structures with two convolution layers. A fixed pool stride as the one proposed in this study implies that the area in which the pool kernels are applied overlap for the [3 × 1] and the [3 × 2] pool patches. The consequences of fixing the pool stride for the proposed pool patches can be seen in Figure 3, where the gray areas are those that the pool process has already considered, while the dark-colored ones correspond to those areas considered more than once (overlap) in the current application of the pool patches. Regardless of the pool patch size, their application occurs one space to the right of the previous one at a time and, when meeting the end of the structure, going back to the start, but spaced two spaces vertically. Although the consequences of the overlapping pooling process are still unknown in the application of CNN in BCI, this approach has successfully been used for image recognition [23]. With the selected size of the fixed pool stride and the proposed pool patches, we can account for CNN that do not experience overlapping in the case of pool patch [2 × 1], other ones that do experience overlap for the [3 × 1] patch and, finally, models that experience overlapping and also consider data from two channels simultaneously, which corresponds to the pool patch [3 × 2]. Depending on the pool strategy used, the size of the resulting feature map varies slightly.

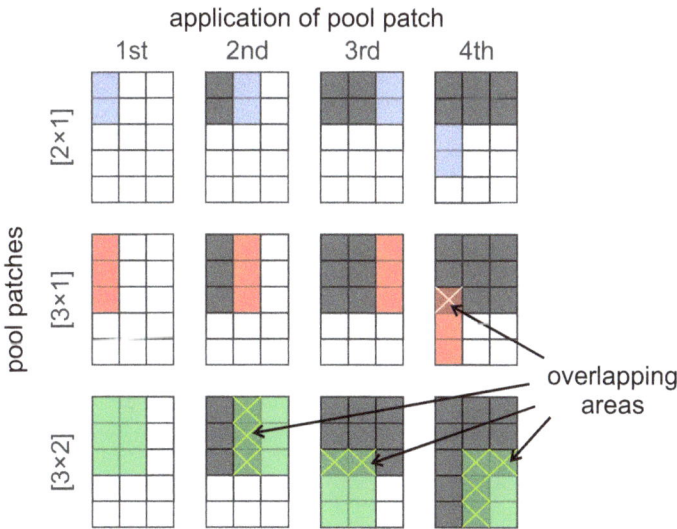

Figure 3. Proposed pooling patches applied on a 5 × 3 structure to show overlapping caused by the fixed pooling stride. Gray areas represent those in which the patch has been applied, and dark-colored areas are those in which the patch overlaps with previous iterations.

3.1.4. Learning Rate

The discussion towards learning rate usually goes in two directions: whether it is chosen based on how fast it is desired for the training to be finished or depending on the size of each example in the dataset. The learning rate used for this study is 0.008. Several other learning rate values ranging from 0.1 to 0.000001 were tested in preliminary tests with noticeable negative repercussions for CNN performance, either with respect to the time required to train or the overall classification rate. The value was chosen as it allows one to see gradual and meaningful changes in the accuracy rate evolution during both training and test phases.

3.1.5. Optimization Method

We used stochastic gradient descent (SGD) to minimize the error present during training. The work of [27] has demonstrated that this method is useful for training neural networks on large datasets. For this case, the error function $E(w)$ is given as a sum of terms for a set of independent observations $E_n(w)$, one for each example or batch of examples in the dataset being used in the form:

$$E(w) = \sum_{n=1}^{N} E_n(w). \tag{1}$$

Thus, making the weight updates based on one example or batch of examples at a time, such that:

$$w^{(\tau+1)} = w^{(\tau)} - \eta \nabla E_n(w^{(\tau)}) \tag{2}$$

where w is the weight and bias of the network grouped together (weight vector), τ is the number of iterations of the learning process in the neural network, η is the learning rate and n ranges from one to Q, which is the maximum number of examples or possible batches in the provided set depending on whether the batch approach is used or not. For this study, batches of 100 examples were used when training any of the proposed CNN structures.

3.1.6. Output Classification

We used a softmax function to evaluate the probability of the input x belonging to each of the possible classes. This is done by:

$$p(C_k|x) = \frac{p(x|C_k)p(C_k)}{\sum_j p(x|C_j)p(C_j)}, \tag{3}$$

where C_k is the current class being considered, and $j = 1, ..., L$, where L represents the maximum number of classes. After the probability is computed for each class, the highest value is forced to one and the rest to zero, forming a vector of the same size as the provided teaching vector (labels). The vectors are then compared to see if the suggested class is the same as the one given as the teaching vector.

3.1.7. Accuracy Rate

As the data-sets used for training and testing the different CNN contained examples for two classes of stimuli (target and non-target) in different amounts, the accuracy rate is defined by the expression:

$$accuracy = \sqrt{\frac{TP}{P} \times \frac{TN}{N}}, \tag{4}$$

which heavily penalizes poor individual classification performance in binary classification tasks. TP stands for true positives and is the number of correctly classified target examples, and TN, which stands for true negatives, is the number of correctly classified non-target examples. P and N represent the total number of examples of the target and non-target classes, respectively, for this case.

All the CNN structures were implemented using a GeForce GTX TITAN X GPU by NVIDIA in Python 2.7 using the work developed by [28].

4. Results for P300 Identification

In this section, we compare the obtained results from the 72 CNN models (18 for each of the four trial intervals) and group them in two different ways in order to facilitate the appreciation of patterns of interest. First, the obtained accuracy rates with fixed convolution patches as seen in Table 2 are discussed. Then, we describe the results of models with fixed pool patches, as shown in Table 3. These two ways of presenting the same results allows one to recognize some performance patterns linked to the convolution or pool patches used for each model. The results show the mean accuracy of each model for the nine subjects that took part in the experiment. At the same time, the results are the mean value obtained from a two-fold cross-validation, where the accuracy for each fold was calculated using Equation (4).

The highest accuracy rate obtained among all the tested models was 0.927 for trials 500 ms long in the model with one layer, convolution patch [4 × 1] and pool patch [3 × 2]. The lowest accuracy rate was 0.783 for trials 200 ms long in the model with two layers, convolution patch [5 × 1] and pool patch [2 × 1].

Table 2. Summarized results for the 18 convolutional neural networks (CNN) structures for all considered trial lengths with fixed convolution patches. PP = pool patch, CP= convolution patch, CL= convolution layers.

		500 ms		400 ms		300 ms		200 ms	
CP	PP	CL 1	CL 2	CL 1	CL 2	CL 1	CL 2	CL 1	CL 2
	[2 × 1]	0.882	0.915	0.860	0.914	0.850	0.903	0.842	0.880
[2 × 1]	[3 × 1]	0.920	0.910	0.899	0.906	0.865	0.883	0.880	0.891
	[3 × 2]	0.907	0.910	0.919	0.906	0.881	0.884	0.897	0.887
	[2 × 1]	0.855	0.869	0.867	0.880	0.796	0.809	0.837	0.814
[4 × 1]	[3 × 1]	0.880	0.884	0.901	0.864	0.858	0.855	0.838	0.832
	[3 × 2]	**0.927**	0.916	0.912	0.868	0.872	0.836	0.911	0.848
	[2 × 1]	0.869	0.840	0.855	0.867	0.805	0.827	0.841	0.783
[5 × 1]	[3 × 1]	0.896	0.847	0.880	0.868	0.874	0.839	0.841	0.826
	[3 × 2]	0.897	0.857	0.895	0.859	0.890	0.824	0.915	0.820

Table 3. Summarized results for the 18 CNN structures for all considered trial lengths with fixed pool patches. PP = pool patch, CP= convolution patch, CL= convolution layers.

		500 ms		400 ms		300 ms		200 ms	
PP	CP	CL 1	CL 2	CL 1	CL 2	CL 1	CL 2	CL 1	CL 2
	[2 × 1]	0.882	0.915	0.860	0.914	0.850	0.903	0.842	0.880
[2 × 1]	[4 × 1]	0.855	0.869	0.867	0.880	0.796	0.809	0.837	0.814
	[5 × 1]	0.869	0.840	0.855	0.867	0.805	0.827	0.841	0.783
	[2 × 1]	0.920	0.910	0.899	0.906	0.865	0.883	0.880	0.891
[3 × 1]	[4 × 1]	0.880	0.884	0.901	0.864	0.858	0.855	0.838	0.832
	[5 × 1]	0.896	0.847	0.880	0.868	0.874	0.839	0.841	0.826
	[2 × 1]	0.907	0.910	0.919	0.906	0.881	0.884	0.897	0.887
[3 × 2]	[4 × 1]	**0.927**	0.916	0.912	0.868	0.872	0.836	0.911	0.848
	[5 × 1]	0.897	0.857	0.895	0.859	0.890	0.824	0.915	0.820

4.1. Fixed Convolution Patches

Producing good results by mixing information from two adjacent channels in a mapped channel vector was considered with skepticism. However, if a direct comparison between the structures with different pool patches is considered (see Table 2), in all cases but one for the one-layered structures and considering all trial lengths, the best results were obtained by those models using the pool patch [3 × 2], which considers both spatial and temporal information. This behavior is not seen for models that use two layers. Additionally, regardless of the number of layers, 58.3% (42 models) of the time, the best results were from structures that used pool patch [3 × 2], 25% (18 models) of the time for when pool patch [3 × 1] was applied and 16.6% (12 models) of the time for structures that used pool patch [2 × 1].

With respect to the convolution patch [5 × 1] for trials 200 ms long, the lowest accuracy corresponded to the structure with one layer and pool patch [2 × 1]. In this condition, an accuracy rate of 0.915 was also achieved by another structure (one layer and pool patch [3 × 2]), thus representing the biggest accuracy rate gap (around 13%) among results produced for a trial of the same length and convolution patch.

4.2. Fixed Pool Patches

In Table 3, the results are now accommodated by fixing the pool patches. Comparing results between different convolution patches on the structures with one and two layers separately reveals a tendency for the convolution patch [2 × 1] to offer the best accuracy rates for 70.8% of the cases (51 models). As for the convolution patches [4 × 1] and [5 × 1], for 16.6% (12 models) and 12.5% (9 models) of the time, they produce the best results, respectively. If only the two-layer models are considered, the convolution patch [2 × 1] offers the best results for all cases except one (the model with $PP = [3 × 2]$ and $CP = [2 × 1]$), similar to the pattern for the one-layer models in Table 2 discussed before.

4.3. Mean Accuracy Rate for Fixed Patches

Given the fixed convolution or pool patches presented in Tables 2 and 3, the mean accuracy for all possible CNN models is presented in Figure 4 to show the differences in patch performance.

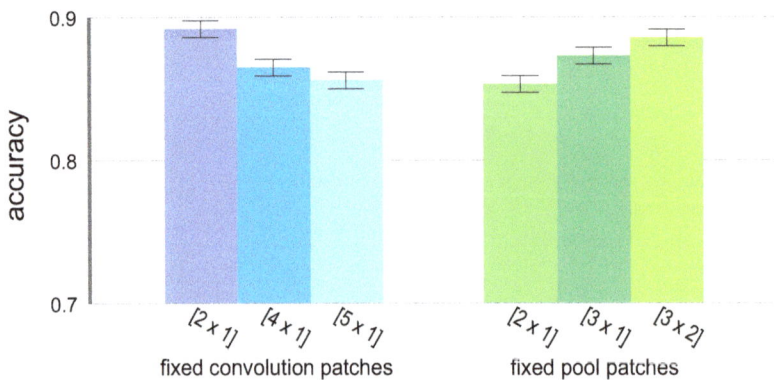

Figure 4. Mean accuracy, along with the accuracy's standard deviation, for all models with fixed convolution patches (left) and pool patches (right).

For the convolution patches, [2 × 1] achieved the overall highest accuracy (presented with the standard deviation), i.e., with $0.891 ± 0.021$, followed by [4 × 1] and [5 × 1], in that order, with $0.864 ± 0.034$ and $0.855 ± 0.032$, respectively. As for the pool patches, the approach in which the pool patch is the same size as the pool stride, i.e., no overlapping occurs, yielded the lowest overall accuracy at $0.853 ± 0.026$, followed by [3 × 1], the patch representing the overlapping pool process, with $0.872 ± 0.035$, and finally, by the pool patch that not only causes overlap, but also considers data from two adjacent channels in the mapped version of the channels, [3 × 2], with $0.885 ± 0.03$.

5. Discussion

In this study, we proposed and tested the efficacy of 18 different CNN for classifying the presence or absence of the P300 wave from the EEG readings of nine subjects in an auditory BCI with virtual direction sources. We approached the classification task by testing three pooling strategies and considering four different trial lengths for the presentation of the auditory stimuli. The implementation of the mentioned strategies is possible due to the fixed pooling stride explained in Section 3.1.3 and present in all of the CNN models. The fixed pooling stride also prevents the resulting feature maps from being oversimplified, as having a stride that matches the size of the pooling kernel might not be possible in all cases due to the down-sampling of the data.

5.1. Pooling Strategies and Other Studies

The first pooling strategy, represented by pool patch $[2 \times 1]$, is the most common approach used in CNN and consists of a pooling process in which the pool patch and stride are of the same size. In this study, this approach led to the lowest general accuracy rates.

The goal of the second strategy, represented by $[3 \times 1]$, is to cause overlapping of the pool patches. This strategy has been tested in previous work, although with data of a very different nature. While [24] reported no differences between performance for approaches with or without overlapping for speech tasks, we have found that, as in the work of [27], better CNN model performance can be achieved using an overlapping pool strategy.

The third strategy, represented by pool patch $[3 \times 2]$, showed that the performance of the overlapping strategy can be further enhanced by also considering data from two different adjacent channels simultaneously. This consideration is not applied to the original input, but rather to feature maps generated after the convolution patch is applied.

Past studies involving the classification of single-trial P300 includes the work of [14], in which results are reported for P300 identification using raw data from Dataset II from the third BCI competition [17]. Rather than changing the parameters of a CNN model, they presented the results of changing the way the input is constructed. In the best scenario, they achieve accuracy rates of 0.7038 and 0.7819 for each of two subjects, with a mean accuracy of 0.7428. In the work of [29], three experiments were conducted, which compared different classifiers for classification of single-trial ERPs for rapid serial visual presentation (RSVP) tasks. They found that the best performance is achieved by a CNN, with a mean accuracy and standard deviation of 0.86 ± 0.073. Another RSVP task is presented in [21] where CNNs are also used for classification. By applying the CNN classifier, they found that they could improve the results obtained in previous studies. These studies are well known, but were not focused on single trial P300 classification; however, they present approaches that inspired this work and provide a reference to what has normally been achieved in this context.

In [25], single trial P300 classification is reported as part of their results. By using support vector machines (SVM), they achieve a mean accuracy rate of approximately 0.70 for seven subjects when considering a reduced number of EEG channels. Another case of a single-trial identification attempt comes from [30], where Fisher's discriminant analysis (FDA) regularized parameters are searched for using particle swarm optimization, achieving an accuracy of 0.745 for single trials and no channel selection. These results can be fairly compared to ours (see Section 5.4), as the goal of these studies, their experimental setup and BCI approach are the same as the ones we present.

In this study, the highest mean accuracy rate for nine subjects was 0.927, and the lowest was 0.783. The mean accuracy rates for all the models for fixed trial intervals were 0.855 ± 0.036, 0.853 ± 0.031, 0.884 ± 0.021 and 0.888 ± 0.026 for the 200-, 300-, 400- and 500-ms trial interval, respectively.

5.2. Convolution Patches and Number of Layers

Although we approached this study expecting the pooling strategies to play the most relevant role performance-wise, we also observed patterns of improvement depending on the selected convolution patch (as presented in Figure 4) and the number of layers. Considering less information in the time domain for the convolution process leads to better mean accuracy rates. We found the difference between the highest and lowest mean accuracy rate in the convolution patch to be 0.036, which is slightly bigger than the difference between the lowest and highest mean accuracy rates between the tested pool strategies (0.032).

On a related matter, the models with only one layer outperformed those with two layers in 21 (58.3%) of the 36 cases. As each time the pool patch is applied, the size of the input is reduced significantly, a large number of layers might produce an oversimplification of the input. In our preliminary research, models with three and four layers were tested for different tasks using the datasets described in this work; however, they performed poorly in comparison to models with only one or two layers. This situation might be different if the input we used did not consist of down-sampled data, therefore not falling into the oversimplification problem with the proposed CNN models.

To analyze whether the down-sampling negatively affects the performance of the CNN, we used non-down-sampled data to test the model that achieved the highest accuracy as reported in Section 4. The results, which favor the down-sampled data, can be seen in Figure 5.

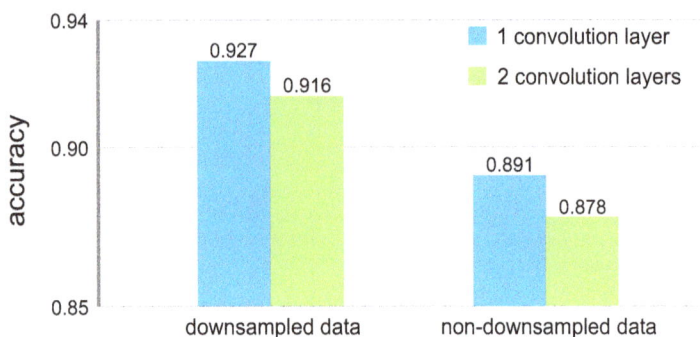

Figure 5. Difference in mean accuracy for the model with the highest performance using down-sampled and non-down-sampled data.

By using down-sampled data, we could not only boost the accuracy with respect to the non-down-sampled data, but also shorten the training/testing times inherent in the size of the input.

5.3. Alternative Training Approach

For the results presented in Section 4, we used the training approach 'single subject approach' described in Section 5.3.1 trying to achieve the best possible performance. However, the 'combined subjects approach' described in Section 5.3.2 is also a viable way to address CNN training. Next, we will discuss the differences between both approaches and offer results that support our decision to implement the former.

5.3.1. Single-Subject Approach

This approach consists of training one CNN using the data of a single subject at a time for each trial length. As the ability to correctly recognize irregular auditory cues varies from one subject to another, this approach allows some of the trained CNN models to perform particularly well if the data come from a subject that excelled in the recognition task. The drawback of using this approach is the large amount of time needed to obtain the mean accuracy of a single CNN model as each of the proposed structures is trained individually for each subject. Therefore, considering a single trial length, 9×18 CNN were trained. The mean accuracy rate obtained for all subjects is presented as the result for a single CNN model. The average time spent on training was about 20 min for the structures with one convolution layer and approximately 27 min for those with two layers.

5.3.2. Combined Subjects Approach

This approach consists of training only one CNN with examples from all of the subjects for each different trial length and then testing each subject individually on the trained CNN. This approach allows one to decrease the number of CNN to train in order to obtain the average accuracy rate for a single CNN model and therefore the time needed to analyze the results. Using this approach also means that the number of examples for training and testing will increase by the number of subjects. A major drawback of this approach is that subjects who fail to recognize the irregular auditory cues will produce examples that do not contain the P300 even if they are labeled otherwise, negatively affecting the CNN performance. For a single CNN model considering data from all subjects, the average time spent on training was about 32 min for the structures with one convolution layer and approximately 39 for those with two layers.

The model that exhibited the best performance in Section 4 was chosen to be tested using also the previously explained combined subjects approach to determine if it offered better performance than that of the currently used approach. Table 4 shows the comparison between the results for the single subject and combined subjects approaches for each subject considering the CNN structure with the overall highest accuracy rate (one layer, convolution patch $[4 \times 1]$, pool patch $[3 \times 2]$). The difference in the mean accuracy rate is about 6%, in favor of the single-subject approach. If subjects are compared in terms of the two approaches, the combined subjects approach is better only in one out of the nine cases. Subject 9, which produced the lowest accuracy in the single-subject approach, benefited slightly from the combined subjects approach. In the eight cases in which the single-subject approach obtains better results, the accuracy rates between both approaches varies between 0% and 11%, depending on the subject. Figure 6 shows the receiver operating characteristic (ROC) curves for each of the nine subjects for the CNN model with the highest accuracy under the single-subject approach. These curves can serve to better understand the results presented in Table 4 for such an approach.

Although there is a substantial difference between both approaches of about five times in terms of the amount of time each one required to produce the mean accuracy, the single-subject approach offered better results.

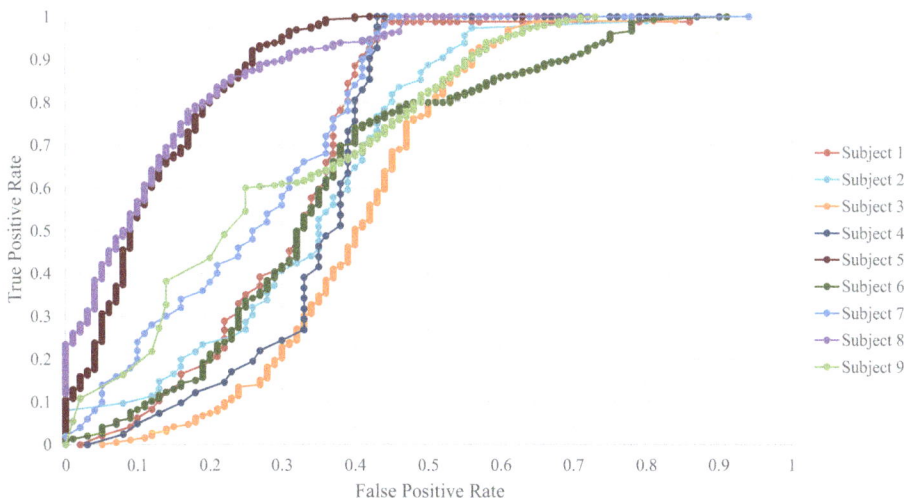

Figure 6. Receiver operating characteristic (ROC) curves of each subject for the CNN model with one layer, convolution patch $[4 \times 1]$ and pool patch $[3 \times 2]$ using the single-subject approach.

Table 4. Comparison between the accuracy rates obtained for the single-subject (SS) and combined subject (CS) approaches for the CNN model with 1 layer, convolution patch [4 × 1] and pool patch [3 × 2].

	500 ms	
Subject	SS	CS
1	0.926	0.858
2	0.893	0.817
3	0.914	0.912
4	0.948	0.877
5	0.969	0.851
6	0.953	0.854
7	0.969	0.915
8	0.916	0.836
9	0.854	**0.865**
Average accuracy	**0.927**	0.865

5.4. CNN and Other Classifiers

As presented in Section 5.1, there are many studies related somehow to the one we present now. We considered it appropriate to compare some of the classifiers those works present that are not CNN. To compare the obtained results from the CNN with other classifiers, we used support vector machines (SVM) and Fisher's discriminant analysis (FDA) for all of the available trial lengths. Details for how we implemented these two classifiers can be found in the Appendix. Table 5 shows the comparison between the accuracy, precision and recall results obtained from the SVM, FDA and the CNN model that achieved the highest accuracy rate in four different trial intervals: 200, 300, 400 and 500 ms.

Table 5. Comparison between the overall accuracy rates, precision and recall obtained for the CNN model with the highest accuracy rate, support vector machines (SVM) and Fisher's discriminant analysis (FDA).

	Accuracy			Precision			Recall		
	CNN	SVM	FDA	CNN	SVM	FDA	CNN	SVM	FDA
500 ms	**0.927**	0.709	0.745	**0.994**	0.37	0.43	**0.826**	0.70	0.71
400 ms	**0.912**	0.711	0.731	**0.987**	0.37	0.41	**0.836**	0.70	0.68
300 ms	**0.872**	0.691	0.707	**0.992**	0.35	0.39	**0.766**	0.68	0.65
200 ms	**0.911**	0.662	0.688	**1.00**	0.34	0.36	**0.833**	0.62	0.61

The results from CNN with the highest accuracy compared to those of the SVM or the FDA are clearly higher, and this difference decreases if we take into consideration the lowest accuracy obtained by one of our CNN models, which is 0.783, with a precision of 1.0 and a recall of 0.618 for the 200-ms trial interval.

5.5. Future Work

Like many previous studies, we used a mapped version of the EEG channels to create a two-dimensional input for CNN. However, EEG data, especially those recorded during experiments conducted using the oddball paradigm, exhibit areas where irregular events have more visible repercussions. For this reason, it is of interest in the future for EEG analysis that the input of the CNN is a three-dimensional structure. Images have this kind of topology, in which two dimensions are used for the position of pixels and three channels define the color of a given pixel. Thus, instead of analyzing one channel at a time to avoid mixing spatial and temporal information, larger two-dimensional patches could be used in both the convolution and pooling process to address a specific moment in the EEG readings. The result would be a map showing the complete brain activity in that particular point.

Embracing different training schemes to reduce the computation time needed for the results to be obtained should be considered. The presented accuracy rates were obtained by training and testing a CNN for each subject on each of the proposed structures and repeating that for the different trial lengths, resulting in long training sessions to obtain the mean accuracy rate of a single structure.

6. Conclusions

By proposing and testing CNN models to classify single-trial P300 waves, we obtained state of the art performances for CNN models using different pooling strategies in the form of mean accuracy rates for nine subjects. We proved that, in off-line classification, single-trial P300 examples could be correctly classified in the auditory BCI we proposed, which uses headphones to produce sound from six virtual directions, thus reducing the amount of hardware needed to implement the BCI in real life. While similar previous studies obtained accuracy rates varying from approximately 0.70 to 0.745, we found mean accuracy rates ranging from 0.855 to 0.888 depending on the trial interval and from 0.783 to 0.927 if individual models are considered. We achieved this by applying different pooling strategies that affect the performance of CNN models dealing with EEG data for classification purposes, as well as using a different number of convolution layers. We found that either of the approaches that overlap in the pooling process or also consider data from two adjacent channels performed better than the most common approach, which uses a pooling stride that is the same size as the pool patch and only considers data from one channel at a time. In most cases, models with simple structures (only one layer) perform better for this type of case and also offer faster training times. Other improvement patterns were also observed for the different convolution patches, as well as for how to approach the training and testing of CNN models.

Acknowledgments: This work was supported in part by Nagaoka University of Technology Presidential Grant and JSPS Kakenhi, Grants 24300051, 24650104 and 16K00182.

Author Contributions: Wada Yasuhiro, Isao Nambu and Miho Sugi conceived of and designed the experiments. Miho Sugi performed the experiments. Eduardo Carabez analyzed the data. Wada Yasuhiro, Isao Nambu and Eduardo Carabez wrote the paper.

Conflicts of Interest: The authors declare no conflict of interest.

Appendix A

SVM

Analysis was performed using LIBSVM software [31] and implemented in MATLAB (MathWorks, Natick, MA, USA). We used a weighted linear SVM [32] to compensate for imbalance in the target and non-target examples. Thus, we used a penalty parameter of C+ for the target and C-for the non-target examples. The penalty parameter for each class was searched in the range of 10^{-6} to 10^{-1} ($10^{-6} \leq 10^m \leq 10^{-1}$; m: $-6{:}0.5{:}-1$) within the training. We determined the best parameters as those that obtained the highest accuracy using 10-fold cross-validation for the training. Using the best penalty parameters, we constructed the SVM classifier using all training data and applied it to the test data.

FDA

We used a variant of the regularized Fisher discriminant analysis (FDA) as the classification algorithm [30]. In this algorithm, a regularized parameter for FDA is searched for by particle swarm optimization (for details, see [30]) within the training. In this study, we used all EEG channels without selection.

1. He, B.; Gao, S.; Yuan, H.; Wolpaw, J.R. Brain-computer interfaces. In *Neural Engineering*; Springer: Berlin/Heidelberg, Germany, 2013; pp. 87–151.
2. Nijboer, F.; Furdea, A.; Gunst, I.; Mellinger, J.; McFarland, D.J.; Birbaumer, N.; Kübler, A. An auditory brain-computer interface BCI. *J. Neurosci. Methods* **2008**, *167*, 43–50.
3. Vos, M.D.; Gandras, K.; Debener, S. Towards a truly mobile auditory brain-computer interface: Exploring the P300 to take away. *Int. J. Psychophysiol.* **2014**, *91*, 46–53.
4. Allison, B.Z.; McFarland, D.J.; Schalk, G.; Zheng, S.D.; Jackson, M.M.; Wolpaw, J.R. Towards an independent brain-computer interface using steady state visual evoked potentials. *Clin. Neurophysiol.* **2008**, *119*, 399–408.
5. Käthner, I.; Ruf, C.A.; Pasqualotto, E.; Braun, C.; Birbaumer, N.; Halder, S. A portable auditory P300 brain-computer interface with directional cues. *Clin. Neurophysiol.* **2013**, *124*, 327–338.
6. Citi, L.; Poli, R.; Cinel, C.; Sepulveda, F. L300-Based BCI Mouse With Genetically-Optimized Analogue Control. *IEEE Trans. Neural Syst. Rehabil. Eng.* **2008**, *16*, 51–61.
7. Rebsamen, B.; Burdet, E.; Guan, C.; Zhang, H.; Teo, C.L.; Zeng, Q.; Laugier, C.; Ang, M.H., Jr. Controlling a Wheelchair Indoors Using Thought. *IEEE Intell. Syst.* **2007**, *22*, 18–24.
8. Nijboer, F.; Birbaumer, N.; Kübler, A. The Influence of Psychological State and Motivation on Brain-Computer Interface Performance in Patients with Amyotrophic Lateral Sclerosis—A Longitudinal Study. *Front. Neurosci.* **2010**, *4*, doi:10.3389/fnins.2010.00055.
9. Baykara, E.; Ruf, C.A.; Fioravanti, C.; Käthner, I.; Simon, N.; Kleih, S.C.; Kübler, A.; Halder, S. Effects of training and motivation on auditory P300 brain-computer interface performance. *Clin. Neurophysiol.* **2016**, *127*, 379–387.
10. Sellers, E.W.; Donchin, E. A P300-based brain-computer interface: Initial tests by ALS patients. *Clin. Neurophysiol.* **2006**, *117*, 538–548.
11. Chang, M.; Nishikawa, N.; Struzik, Z.R.; Mori, K.; Makino, S.; Mandic, D.; Rutkowski, T.M. Comparison of P300 Responses in Auditory, Visual and Audiovisual Spatial Speller BCI Paradigms. *ArXiv* **2013**, arXiv:q-bio.NC/1301.6360.
12. Hoffmann, U.; Vesin, J.M.; Ebrahimi, T.; Diserens, K. An efficient P300-based brain-computer interface for disabled subjects. *J. Neurosci. Methods* **2008**, *167*, 115–125.
13. Nijboer, F.; Sellers, E.W.; Mellinger, J.; Jordan, M.A.; Matuz, T.; Furdea, A.; Halder, S.; Mochty, U.; Krusienski, D.J.; Vaughan, T.M.; et al. A P300-based brain-computer interface for people with amyotrophic lateral sclerosis. *Clin. Neurophysiol.* **2008**, *119*, 1909–1916.
14. Cecotti, H.; Gräser, A. Convolutional Neural Networks for P300 Detection with Application to Brain-Computer Interfaces. *IEEE Trans. Pattern Anal. Mach. Intell.* **2011**, *33*, 433–445.
15. Cecotti, H.; Gräser, A. Time Delay Neural Network with Fourier transform for multiple channel detection of Steady-State Visual Evoked Potentials for Brain-Computer Interfaces. In Proceedings of the 2008 16th European Signal Processing Conference, Lausanne, Switzerland, 25–29 August 2008; pp. 1–5.
16. Guler, I.; Ubeyli, E.D. Multiclass Support Vector Machines for EEG-Signals Classification. *IEEE Trans. Inf. Technol. Biomed.* **2007**, *11*, 117–126.
17. Kaper, M.; Meinicke, P.; Grossekathoefer, U.; Lingner, T.; Ritter, H. BCI competition 2003-data set IIb: Support vector machines for the P300 speller paradigm. *IEEE Trans. Biomed. Eng.* **2004**, *51*, 1073–1076.
18. Naseer, N.; Qureshi, N.K.; Noori, F.M.; Hong, K.S. Analysis of different classification techniques for two-class functional near-infrared spectroscopy-based brain-computer interface. *Comput. Intell. Neurosci.* **2016**, *2016*, doi:10.1155/2016/5480760.
19. Abdel-Hamid, O.; Deng, L.; Yu, D. Exploring convolutional neural network structures and optimization techniques for speech recognition. In Procedings of the 14th Annual Conference of the International Speech Communication Association, Lyon, France, 25–29 August 2013; pp. 3366–3370.
20. Simonyan, K.; Zisserman, A. Very Deep Convolutional Networks for Large-Scale Image Recognition. *arXiv* **2014**, arXiv:1409.1556.
21. Manor, R.; Geva, A.B. Convolutional Neural Network for Multi-Category Rapid Serial Visual Presentation BCI. *Front. Comput. Neurosci.* **2015**, *9*, doi:10.3389/fncom.2015.00146.
22. Bevilacqua, V.; Tattoli, G.; Buongiorno, D.; Loconsole, C.; Leonardis, D.; Barsotti, M.; Frisoli, A.; Bergamasco, M. A novel BCI-SSVEP based approach for control of walking in Virtual Environment using

a Convolutional Neural Network. In Proceedings of the 2014 International Joint Conference on Neural Networks (IJCNN), Beijing, China, 6–11 July 2014; pp. 4121–4128.

23. Krizhevsky, A.; Sutskever, I.; Hinton, G.E. ImageNet Classification with Deep Convolutional Neural Networks. In *Advances in Neural Information Processing Systems 25*; Pereira, F., Burges, C.J.C., Bottou, L., Weinberger, K.Q., Eds.; Curran Associates, Inc.: Red Hook, NY, USA, 2012; pp. 1097–1105.

24. Sainath, T.N.; Kingsbury, B.; Saon, G.; Soltau, H.; Mohamed, A.; Dahl, G.; Ramabhadran, B. Deep Convolutional Neural Networks for Large-scale Speech Tasks. *Neural Netw.* **2015**, *64*, 39–48.

25. Nambu, I.; Ebisawa, M.; Kogure, M.; Yano, S.; Hokari, H.; Wada, Y. Estimating the Intended Sound Direction of the User: Toward an Auditory Brain-Computer Interface Using Out-of-Head Sound Localization. *PLoS ONE* **2013**, *8*, 1–14.

26. Yano, S.; Hokari, H.; Shimada, S. A study on personal difference in the transfer functions of sound localization using stereo earphones. *IEICE Trans. Fundam. Electron. Commun. Comput. Sci.* **2000**, *83*, 877–887.

27. LeCun, Y.; Boser, B.; Denker, J.S.; Henderson, D.; Howard, R.E.; Hubbard, W.; Jackel, L.D. Backpropagation applied to handwritten zip code recognition. *Neural Comput.* **1989**, *1*, 541–551.

28. Goodfellow, I.J.; Warde-Farley, D.; Lamblin, P.; Dumoulin, V.; Mirza, M.; Pascanu, R.; Bergstra, J.; Bastien, F.; Bengio, Y. Pylearn2: A machine learning research library. *ArXiv* **2013**, arXiv:stat.ML/1308.4214.

29. Cecotti, H.; Eckstein, M.P.; Giesbrecht, B. Single-Trial Classification of Event-Related Potentials in Rapid Serial Visual Presentation Tasks Using Supervised Spatial Filtering. *IEEE Trans. Neural Netw. Learn. Syst.* **2014**, *25*, 2030–2042.

30. Gonzalez, A.; Nambu, I.; Hokari, H.; Wada, Y. EEG channel selection using particle swarm optimization for the classification of auditory event-related potentials. *Sci. World J.* **2014**, *2014*, doi:10.1155/2014/350270.

31. Chang, C.; Lin, C. LIBSVM: A library for support vector machines. *ACM Trans. Intell. Syst. Technol. (TIST)* **2011**, *2*, doi:10.1145/1961189.1961199.

32. Osuna, E.; Freund, R.; Girosi, F. An improved training algorithm for support vector machines. In Proceedings of the 1997 IEEE Signal Processing Society Workshop on Neural Networks for Signal Processing VII, Amelia Island, FL, USA, 24–26 September 1997; pp. 276–285.

applied sciences

MDPI

Article

Sound Synthesis of Objects Swinging through Air Using Physical Models †

Rod Selfridge [1,*], David Moffat [2] and Joshua D. Reiss [2]

[1] Media and Arts Technology, Electronic Engineering and Computer Science Department,
   Queen Mary University of London, Mile End Road, London E1 4NS, UK
[2] Centre for Digital Music, Electronic Engineering and Computer Science Department,
   Queen Mary University of London, Mile End Road, London E1 4NS, UK;
   d.j.moffat@qmul.ac.uk (D.M.); joshua.reiss@qmul.ac.uk (J.D.R.)
* Correspondence: r.selfridge@qmul.ac.uk; Tel.: +44-7773-609-280
† This paper is an extended version of our paper published in 14th Sound and Music Computing Conference,
   Espoo, Finland, 5–8 July 2017. Real-Time Physical Model for Synthesis of Sword Swing Sounds.

Academic Editor: Stefania Serafin
Received: 12 October 2017; Accepted: 10 November 2017; Published: 16 November 2017

Featured Application: A real-time physical model sound effect that can replicate the sound of a number of swinging objects, such as a sword, baseball bat and golf club, has great potential for dynamic environments within virtual reality or games. The properties exposed by the sound effects model could be automatically adjusted by a physics engine giving a wide corpus of sounds from one simple model, all based on fundamental fluid dynamics principles.

Abstract: A real-time physically-derived sound synthesis model is presented that replicates the sounds generated as an object swings through the air. Equations obtained from fluid dynamics are used to determine the sounds generated while exposing practical parameters for a user or game engine to vary. Listening tests reveal that for the majority of objects modelled, participants rated the sounds from our model as plausible as actual recordings. The sword sound effect performed worse than others, and it is speculated that one cause may be linked to the difference between expectations of a sound and the actual sound for a given object.

Keywords: sound synthesis; physical modelling; aeroacoustics; sound effects; real-time; game audio; virtual reality

---

1. Introduction

The sound of an object swinging through the air has a very distinctive swoosh sound. We expect this sound when watching a sword fight in a movie or playing a golfing game. This is a common sound within films, TV programmes and games covering genres like sports, material arts or a swashbuckling yarn. These distinct sounds are all generated by a similar physical process as the objects move through the air.

When sounds are added into media to replicate or emphasise original sounds, like a sword swoosh, they are classed as sound effects. A sound effect is usually implemented as a pre-recorded sample or from sound synthesis. Pre-recorded samples have a drawback in media like games and virtual reality as they are unable to change or evolve with the environment, but they are often viewed as more perceptually accurate than synthesised effects. Synthesised effects have the advantage of being based on algorithms and hence have the potential to adapt with their environments.

Being able to replicate these sounds within a single synthesis model offers the opportunity to cover a wide variety of objects travelling through the air. This potentially gives a programmer the ability to obtain the results required without having to find a sample within a sound effects library or record the sound themselves. It also provides an audio programmer the ability to integrate parameters of the model into a game engine. Thus, the synthesis model can evolve with the environment, increasing immersion within a game or virtual reality. A video illustrating the model being used to synthesise a sword swing within the Unity game engine is shown at https://www.youtube.com/watch?v=zVvNthqKQIk.

This article is a revised and extended version of [1], which won the best paper award at the 14th Sound and Music Computing Conference 2017. It presents a new sound synthesis method illustrating the design, implementation and analysis of a real-time physically-derived model that can be used to produce sounds similar to those of an object swooshing through the air. The objects examined were a metal sword, a wooden sword, a baseball bat, a golf club and a broom handle, which represent different object geometries commonly heard swinging through the air. To our knowledge, this is the first synthesis model that replicates a wide variety of objects swinging through the air by using bona fide fluid dynamics equations to calculate the sound output in real time.

Section 2 describes the state of the art and related work, while Section 3 gives a detailed description of our method. The implementation is given in Section 4 followed by both subjective and objective evaluations of our model in Section 5. A discussion of the work is presented in Section 6 followed by conclusions in Section 7.

2. Background and Related Work

Sound synthesis techniques can be split into two broad approaches, signal-based and physical models [2]. Signal-based models aim to replicate the sound properties; matching frequency components, replicating the time envelope or similar. Physical models aim to replicate the processes behind the natural sound creation by mathematical models.

The advantage of a signal-based model is that it is relatively computationally inexpensive to replicate the spectrum of a sound using established techniques such as additive synthesis or noise shaping. A drawback of this approach is that it is rarely possible to relate changes in signal properties to the physical processes creating the sound. For example, an increase in speed of a sword not only changes the fundamental tone frequency, but also the gain. Therefore, changing one signal-based property could lose realism in another.

Physical models aim to replicate the physics behind the sound generation process. Sounds generated by these models have the advantage of possessing greater authenticity in the generated sounds, especially in relation to parameter adjustments. A potential drawback is that the computational cost required to produce sounds is often high, and the physical models typically cannot adapt quickly to parameter adjustments, making real-time operation challenging and often not possible.

In the middle of these traditional techniques lay physically-inspired models. These hybrid approaches replicate the signal produced, but add characteristics of the physics that are behind the sound creation. For a simple sword model, this might be noise shaping with a bandpass filter with centre frequency proportional to the speed of the swing. A variety of examples of physically-inspired models was given in [3]; the model for whistling wires being exactly the bandpass filter mentioned.

Four different sword models were evaluated in [4]. Here, the application was for interactive gaming, and the evaluation was focused on perception and preference rather than accuracy of sound. The user was able to interact with the sound effect through the use of a Wii Controller. One model was a band-filtered noise signal with the centre frequency proportional to the acceleration of the controller. A physically-inspired model replicated the dominant frequency modes extracted from a recording of a bamboo stick swung through the air. The amplitude of the modes was mapped to the real-time acceleration data.

The other synthesis methods in [4] both mapped acceleration data from the Wii Controller to different parameters; one using the data to threshold between two audio samples, the other a granular synthesis method mapping acceleration to the playback speed of grains. Tests revealed that the granular synthesis was the preferred method for expression and perception. One possible reason that the physical model was less popular could be the lack of correlation between speed and frequency pitch, which the band-filtered noise had. This may also be present in the granular model.

A signal-based approach to a variety of environmental sound effects, including sword whoosh, waves and wind sounds, was undertaken in [2]. Analysis and synthesis occur in the frequency domain using a sub-band method to produce narrow band coloured noise. In [5], a rapier sword sound was replicated, but this focused on the impact rather than the swoosh when swung through the air.

A physical model of sword sounds was explored in [6]. Here, offline sound textures were generated based on the physical dimensions of the sword. The sound textures were then played back with speed proportional to the movement. The sound textures were generated using computational fluid dynamics software (CFD), solving the Navier–Stokes equations and used Lighthill's acoustic analogy [7] extended by Curle's method [8]. In this model [6], the sword was split into a number of compact sound sources (discussed in Section 3.2), spaced along the length of the sword. As the sword was swept thought the air, each source moved at a different speed; therefore, the sound texture for each source was adjusted accordingly. The sounds from each source were summed and output to the listener.

An overview of the different synthesis methods and parameters available to a user are presented in table form in Table 1. It can be seen that the only model offering real-time operation with instantaneous variability of physical parameter was [1]. Outputs from [4,6] were used within our listening test, Section 5, to represent alternative synthesis methods.

Table 1. Table highlighting different synthesis methods for swing sounds.

Reference	Synthesis Method	Parameters	Comments
[1]	Physically derived	Length, diameter, length of swing and speed of swing	Operates in real time
[2]	Frequency domain signal-based model	Amplitude control over analysis and synthesis filters	Operates in real time
[4]	Granular	Accelerometer speed	Mapped to playback speed
	Sample-based	Accelerometer speed	Triggered by threshold speeds
	Noise shaping	Accelerometer speed	Mapped to bandpass centre frequency
	Physically inspired	Accelerometer speed	Mapped to the amplitude of frequency modes
[6]	Computational fluid dynamics	Length, diameter and swing speed	Real-time operation, but requires initial offline computations

A Japanese katana sword was analysed in [9] by means of wind tunnel experiments. A number of harmonics from vortex shedding were observed along with additional harmonics from a cavity tone due to the shinogi or blood grooves in the profile of the sword.

3. Method

3.1. Aeroacoustics

When sound is generated by airflows or air interacting with objects, the process is labelled aeroacoustics. This falls under the wider body of research known as fluid dynamics, which describes the physical processes controlling the flow of fluids and enables the prediction of pressures, noises, strains on objects, etc. Understanding these processes enables better design of a wide number of objects including aircraft, cars, trains, ships, buildings, space vehicles and bridges.

Today, computers are able to solve the highly complex equations that govern these processes using techniques like finite difference or finite volume techniques and mapping out the domain of interest with complex mesh structures. Even with the advances in the computational power available, these processes can take hours and even days to complete depending on the level of detail required.

Prior to the availability of such processing power, engineers and scientists derived and defined simpler equations to allow them to calculate the approximate acoustic characteristics. These are labelled semi-empirical equations, where assumptions and generalisations have been made to simplify calculations or to yield results in accordance with observations. Although many of these equations may at first appear complicated, once all the relevant parameters are known, they produce exact results with errors only due to the approximations made during the equation derivation.

There is a number of fundamental aeroacoustic sounds that constitute the main focus of research. Figure 1 illustrates a number of these fundamental tones and gives examples of the types of objects that produce them. Each tone is generated by distinct fluid dynamics processes.

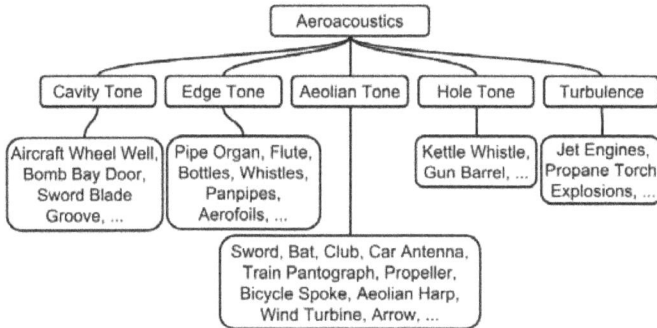

Figure 1. A simplified taxonomy of aeroacoustic sounds.

3.2. Aeolian Tone

It can be seen from Figure 1 that the Aeolian tone is the fundamental tone produced when an object like a sword or bat is swung through the air. A brief overview of the Aeolian tone characteristics will be given here, including a number of fundamental equations. For greater depth, the reader is directed to [10].

3.2.1. Tone Frequency

Strouhal (1878) defined a useful relationship between the tone frequency f_l, air speed $u(t)$ and cylinder diameter d, Equation (1). The variable S_t is known as the Strouhal number.

$$S_t = \frac{f_l\, d}{u(t)} \tag{1}$$

Rearranging to isolate the tone frequency gives:

$$f_l = \frac{S_t u(t)}{d} \tag{2}$$

As air flows around a cylinder, vortices are shed, causing a fluctuating lift force normal to the flow dominated by the fundamental frequency, f_l. Simultaneously, a side axial fluctuating drag force is present with frequency f_d, twice that of the lift frequency. It was noted in [11] that, "The amplitude of the fluctuating lift is approximately ten times greater than that of the fluctuating drag".

It was shown in [8] and confirmed in [12] that aeroacoustic sounds in low flow speed situations could be modelled by the summation of compact sound sources, namely monopoles, dipoles and quadrupoles. An acoustic monopole can be described as a pulsating sphere, much smaller than the acoustic wavelength. A dipole is equivalent to two monopoles separated by a small distance, but of opposite phase. Quadrupoles are two dipoles separated by a small distance with opposite phases. A longitudinal quadrupole has the dipole axes in the same line, while a lateral quadrupole can be considered as four monopoles at the corners of a rectangle [13]. Aeolian tones can be represented by dipole sources, one for the lift fundamental frequency and one for the drag; each source can include a number of harmonics.

The turbulence around the cylinder affects the frequency and the bandwidth of the tone produced. A measure of this turbulence is given by a dimensionless variable, the Reynolds number R_e, given by the relationship in Equation (3).

$$R_e = \frac{\rho_{air} \, d \, u(t)}{\mu_{air}} \tag{3}$$

where ρ_{air} and μ_{air} are the density and viscosity of air, respectively. An experimental study of the relationship between the Strouhal number and the Reynolds number was performed in [14], giving the following equation:

$$S_t = \lambda + \frac{\tau}{\sqrt{R_e}} \tag{4}$$

where λ and τ are constants and given in Table 1 of [14] (additional values were calculated in [10]). The different values represent the turbulence regions of the flow, starting at laminar up to sub-critical. With the Strouhal number obtained, diameter and air speed known, we can apply them to Equation (2) and obtain the fundamental frequency, f_l, of the Aeolian tone, generated by the lift force.

3.2.2. Source Gain

The time-averaged acoustic intensity $\overline{I_l}$ (W/m^2) of an Aeolian tone lift dipole source and the time-averaging period were given in [15]. The time-averaged acoustic intensity for low airspeeds is given as:

$$\overline{I_l} \approx \frac{\sqrt{2\pi}\kappa^2 S_t^2 l \, b \, \rho \, u(t)^6 \sin^2\theta \cos^2\varphi}{32c^3 r^2 (1 - M\cos\theta)^4} \left\{ exp\left[-\frac{1}{2}\left(\frac{2\pi M \, S_t \, l}{d}\right)^2 \sin^2\theta \sin^2\varphi \right] \right\} \tag{5}$$

where b is the cylinder length; M is the Mach number; $M = u(t)/c$, where c is the speed of sound. The elevation angle, azimuth angle and distance between listener and source are given by θ, φ and r, respectively. κ is a numerical constant that lies somewhere between 0.5 and 2 [15]. The correlation length, l, has dimensionless units of diameter d and indicates the span-wise length that the vortex shedding is in phase; after this, the vortices become decorrelated. The work in [15] states that the exponent of Equation (5) can be neglected at low Mach numbers, in accordance with [16]. The gain for the drag dipole is obtained from its relationship to the lift gain given in [11] and the lift dipole harmonics values from similar relationships published in [17].

3.2.3. Wake Noise

As the Reynolds number increases, the vortices diffuse rapidly and merge into a turbulent wake. The wake produces wide band noise modelled by lateral quadrupole sources whose intensities vary with $u(t)^8$ [18]. It was noted in [18] that there is very little noise content below the lift dipole fundamental frequency. Above the fundamental frequency, the roll off of the amplitude of the turbulent noise is $\frac{1}{f^2}$.

The sound generated by jet turbulence was examined in [15,19,20]. The work in [15] states that the radiated sound pattern is greatly influenced by a Doppler factor of $(1 - M\cos\theta)^{-5}$. The wake noise has less energy than a jet, and its intensity $\overline{I_w}$ has been approximated by the authors to capture this relationship as shown in Equation (6):

$$\overline{I_w} \sim \Gamma \frac{\sqrt{2\pi}\kappa^2 S_t^2 l \, b\rho \, u(t)^8}{16\pi^2 c^5 (1 - M\cos(\pi - \theta))^5 r^2} \left(1 + B\cos^4(\theta) - \frac{B+3}{4}\sin^2(2\theta)\sin^2(\varphi) \right) \tag{6}$$

where Γ is a scaling factor between wake noise and lift dipole noise and B is an empirical constant. A value of $B = 0.7$ was found in [19] to match measured values.

4. Implementation

Our model was built using Pure Data, a real-time graphical data flow programming language. This was chosen due to the open source nature of the code and ease of repeatability rather than high performance computations.

4.1. Discrete Compact Sound Source

4.1.1. Fundamental Frequency Calculation

A uniform sampling of the continuous air flow speed $u[n]$, along with the given diameter d set by the user, permits the calculation of the Reynolds number R_e from a discrete implementation of Equation (3). Using data published in [14] the discrete Strouhal number S_t was calculated, Equation (4). Thereafter, a discrete implementation of Equation (2) was used to obtain the lift fundamental frequency f_l.

4.1.2. Gain Calculations

The time-averaged intensity value $\overline{I_{l1}}$ calculated by Equation (5) pertains to the dipole associated with the fundamental lift frequency f_l. Previous theoretical research [16] has set the constant $\kappa = 1$ and neglected the exponent. We set $\kappa = 1$ matching conditions used by [16], likewise neglecting the exponent, which has a negligible effect due to the low Mach numbers used in this implementation [15]. The correlation length l was obtained from a graph published in [21] showing the ratio of correlation length to diameter, l/d, as a function of the Reynolds number. An equation replicating this relationship has been derived by the authors in Equation (7).

$$l = 10^{1.536} R_e^{-0.245} d \tag{7}$$

The discrete intensity value pertaining to the drag force $\overline{I_{d1}}$ was calculated using Equation (8).

$$\overline{I_{d1}} \sim 0.1 \frac{\sqrt{2\pi} S_t^2 \rho u[n]^6 l (\sin(\theta + \frac{\pi}{2}))^2 b (\cos \varphi)^2}{32 c^3 r^2 (1 - M \cos \theta)^4} \tag{8}$$

where constant $\frac{\pi}{2}$ was added to the value of θ due to the 90° phase difference between the lift and drag forces.

4.1.3. Harmonic Content Calculations

In [10], the Aeolian tone was presented with two harmonics for the lift dipole and one for the drag dipole. Due to the additional computational complexity this adds, multiplied by the number of sources in each swinging object, the number of harmonics was reduced down to the most perceptually significant; the first lift dipole harmonic at $3f_l$.

Hardin [17] stated that this value was 60% of the fundamental SPL. This was implemented as shown below:

$$\overline{I_{l3}} = 10^{0.6 \log_{10} \overline{I_{l1}}} \tag{9}$$

4.1.4. Tone Bandwidth Calculations

As stated in Section 3.2.1, there is a bandwidth around the tone, and this is related to the Reynolds number. Data available in [22] were limited to Reynolds numbers under 237,000. The relationship between the bandwidth and Reynolds number from 0–193,260 was found to be linear. This relationship was interpolated from the data as:

$$\frac{\Delta f}{f_l}(\%) = 4.624 \times 10^{-5} R_e + 0.9797 \tag{10}$$

where Δf is the tone bandwidth at -3 dB of the peak frequency. Above a Reynolds number of 193,260, a quadratic formula was found to fit the bandwidth data. This is shown in Equation (11).

$$\frac{\Delta f}{f_l}(\%) = 1.27 \times 10^{-10}R_e^2 - 8.552 \times 10^{-5}R_e + 16.5 \tag{11}$$

In signal processing, the relationship between the peak frequency and bandwidth is called the Q value, ($Q = f_l/\Delta f$), the reciprocal of the percentage value, obtained by an implementation of Equations (10) and (11).

4.1.5. Wake Calculations

A noise profile of $\frac{1}{f^2}$ is known as brown noise. This was approximated using white noise and the transfer function shown in Equation (12) [23].

$$H_{brown}(z) = \frac{1}{1 - \alpha z^{-1}} \tag{12}$$

In [23], α has a value of 1, but this proved unstable in our implementation. A value of 0.99 was chosen, giving a stable implementation while producing a virtually identical magnitude spectrum. The required noise profile was generated using the transfer function given in Equation (13):

$$B[z] = H_{brown}[z]W[z] \tag{13}$$

where $W[z]$ is a white noise source and the output $B[z]$ is a brown noise source. There is little wake contribution below the fundamental frequency [18]. Therefore, a high pass filter was applied to $B[z]$ with the filter cut-off set at the lift dipole fundamental frequency, f_l. This produces the turbulent noise profile required, $G[z]$:

$$G[z] = H_{hp}[z]B[z] \tag{14}$$

where $H_{hp}[z]$ is the high pass filter transfer function. The inverse Z-transform of $G[z]$ gives the wake output signal, $g[n]$. The wake gain was calculated by a discrete implementation of Equation (6). A value of $\Gamma = 0.2$ was set perceptually based on sounds generated from experiments (Section 5), giving $\overline{I_w}$.

4.1.6. Final Output

To generate the correct output sound for the fundamental lift dipole, we used a white noise source filtered by a bandpass filter. The centre frequency of the bandpass filter was set to f_l and the Q value as calculated in Section 4.1.4, giving the bandpass filter output $x_{l1}[n]$. The same process was applied in relation to the fundamental drag dipole, using f_d as a bandpass filter centre frequency, giving an output of $x_{d1}[n]$. The lift dipole harmonic $3f_l$ was computed in a similar way, giving output $x_{l3}[n]$.

The gain values for the lift and drag dipole outputs were obtained from Equations (5) and (8). The appropriate gain value for the lift dipole harmonic was given in Equation (9). Finally, the wake output $g[n]$ with gain $\overline{I_w}$ was added. Note that a single white noise source was used for all fundamental and harmonic dipoles and for the wake noise as they were all part of a single compact source.

Combining the outputs from the lift dipole, drag dipole, harmonic and wake, it is possible to define a final output, Equation (15):

$$y_{output}[n] = \chi\left[\overline{I_l}x_{l1}[n] + \overline{I_d}x_{d1}[n] + \overline{I_{l3}}x_{l3}[n] + \overline{I_w}g[n]\right] \tag{15}$$

where χ is an absolute gain value allowing the user to increase the overall sound level depending on artistic requirements.

4.2. Swinging Model

The basic concept of all the models was to line up a number of the compact sources to replicate the sounds created as a cylindrical object swings through the air. The intensities given in Equation (15) were time averaged, which caused an issue for our model due to the swing time being shorter than the averaging process. Thus, the intensity was implemented as an instantaneous value.

For each of our models, eight Aeolian tone compact sound sources were used to replicate the sounds. The distance between each source depends on the correlation length, the distance given in diameters before the vortices being shed go out of phase or become decorrelated.

To increase the flexibility and ease of use of our swinging objects model, two modes of operation were available; one allowing the user to adjust the diameter of the top and bottom of the object with a linear interpolation between them and the second with preset objects based on actual physical measurements. Both modes of operation allow the user to predefine the top speed of the tip, start and end position of the object being swung, as well as the position of the observer. These parameters can be easily mapped to graphics or animation to have an exact match with visuals. The coordinate system used for the model is shown in Figure 2.

For ease of calculation, the swing action throughout was made to be an arc of constant radius and hence always tracing a line on the surface of a sphere. This allowed us to calculate the distance between the start and end position of the swing using the Haversine formula [24]. This formula calculates the length of a great circle on a sphere and is shown in Equation (16) below:

$$\text{arc length} = 2r \arcsin\left(\sqrt{\sin^2\left(\frac{\phi_2 - \phi_1}{2}\right) + \cos(\phi_1)\cos(\phi_2)\sin^2\left(\frac{\lambda_2 - \lambda_1}{2}\right)}\right) \tag{16}$$

where ϕ_1 and ϕ_2 are the latitude of the start and finish points, respectively; λ_1 and λ_2 are the longitude values of the start and finish points. Latitude and longitude values are given in radians and determined from the start and end positions set by the user.

The radius r is the distance between the centre of the arc and tip of the object. This was set to be the length of the object with an additional 0.35 m to represent the length of the swinging arm. The top speed of the object being swung was set as the halfway distance of the arc, with linear acceleration and deceleration to and from rest.

In our implementation, the sword sweep created a two-dimensional plane in a three-dimensional environment with the observer taken as a point in that environment. Trigonometry identities were used to calculate the elevation and azimuth between each source and the observer.

Panning was included as the sound moves across the xy plane, as well as the Doppler effect. It was shown in [25] that the addition of the Doppler effect increases the natural perception. This effect was taken into account when the sword was moving towards or away from the observer and frequencies adjusted accordingly.

4.3. Variable Mode

This model gives the user the ability to vary the diameter of the object by setting the object diameter at the tip and the hilt. The user can also vary the length of the object. The position of the Aeolian tone compact sound sources depends on the choices made by the user when setting the diameter and length values.

Six of the eight compact sound sources were placed at the tip of the object. It is known from Equation (5) that the gain is proportional to $u(t)^6$, and the greatest speed will be at the tip of a sword, a golf club, etc., during a normal swing. The remaining sources were placed at the hilt and midway between the 6th source at the tip and the hilt. This is illustrated in Figure 2.

This positioning of the six sources at the tip was equivalent to each source having a set correlation length of $7d$; see Section 3.2.2. A range of correlation values from 17–$3d$ were given in [16] depending on the Reynolds number. A plot showing similar values was given in [21]. Since the position of the sources has to be chosen prior to calculation of the Reynolds number, the value 7*d* was chosen as a compromise, covering a reasonable length of the sword for a wide range of speeds (in [6], the correlation length of 3*d* was used; the number of sources set to match the length of the sword).

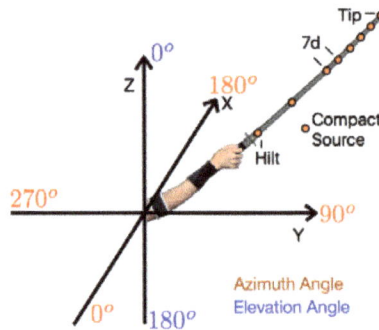

Figure 2. Position of 8 compact sources and coordinates used in the sword model.

4.4. Preset Mode

A number of actual objects were measured; a metal sword, wooden sword, baseball bat, 3–wood golf club, 7–iron golf club and a broom handle. Exact measurements gave us the opportunity to set the position and diameter of each of the compact sound sources individually, giving a more accurate model. The correlation length at the tip of all objects was set to 5*d* for all objects except the baseball bat, which had a reduced correlation length of 2*d* due to its thickness. The exact values of the source position from the base of the object to the tip and the corresponding object diameter are shown in Table 2.

Table 2. Diameter and radius of compact sound sources for the preset objects. All values in metres. Correlation length = 5*d* except for baseball bat, where correlation length = 2*d*.

Metal Sword		Wooden Sword		Baseball Bat		3–Wood Golf Club		Broom Handle	
Radius	Diameter	Radius	Diameter	Radius	Diameter	Radius	Diameter	Radius	Diameter
0	0.0046	0	0.0117	0	0.0237	0	0.0258	0	0.0270
0.418	0.0046	0.307	0.0111	0.159	0.0237	0.383	0.0124	0.313	0.0270
0.777	0.0046	0.370	0.0108	0.314	0.0246	0.767	0.0095	0.625	0.0270
0.780	0.0037	0.417	0.0105	0.371	0.0286	0.813	0.0092	0.760	0.0270
0.810	0.0029	0.465	0.0103	0.444	0.0366	0.857	0.0089	0.895	0.0270
0.821	0.0022	0.512	0.0100	0.549	0.0504	0.900	0.0086	1.030	0.0270
0.830	0.0017	0.560	0.0098	0.672	0.0637	1.050	0.0154	1.165	0.0270
0.836	0.0013	0.607	0.0095	0.804	0.0659	1.100	0.0388	1.300	0.0270

4.5. Grooved Profile

In [26], a physically-derived sound synthesis model of a cavity tone was presented. This covers a separate fundamental aeroacoustic sound with a different set of fluid dynamics equations governing the generation of the tone. In [9], the sound generated by a grooved sword was found to contain a number of discrete frequencies, including those from the cavity tone. Thus, we added in cavity tone compact sound sources at the same location as the Aeolian tone compact sources to our model.

5. Evaluations and Results

5.1. Subjective Evaluation

The subjective evaluation was split into two different tests, a listening test and an object recognition test. A total of 26 participants undertook the test, 18 males, 7 females and 1 preferring not to say. Participants were aged between 17 and 71 with a median of 28 and standard deviation of 13. The order of the listening test and object recognition was split to examine if the order had any influence on the results. Working models of both versions of the swinging object model are available at https://code.soundsoftware.ac.uk/projects/physicallyderivedswingingobjects, which includes a copy of all sounds used in our listening test.

5.1.1. Listening Tests

A double-blind listening test was carried out to evaluate the effectiveness of our synthesis model. The Web Audio Evaluation Tool [27] was used to build and run listening tests in the browser. This allowed test page order and samples on each page to be randomised. All samples were loudness normalised in accordance with [28]. Headphones were used to administer the sounds to participants. These were either AKG K553 Pro Closed-Back Studio Headphones or Beyerdynamic DT150 closed back Isolating Studio Headphones.

Each participant was presented with five test pages, one for each of the preset sound effects. The wooden sword, baseball bat, golf club and broom handle pages contained two real samples, two samples from our physical model (PM), two samples generated by spectral modelling synthesis (SMS) [29] from a recording and an anchor. The metal sword page included two real samples, one synthesis sample from [4], one synthesis sample from [6], one SMS sample, one sample from our physical model and a sample from the physical model with cavity tone compact sound sources added.

All the sampled recordings were captured by the authors within the Listening Room, Electronic Engineering and Computer Science Department, Queen Mary University of London. They were recorded on a Neumann U87 microphone placed approximately 20 cm from the midpoint of the swing and at 90 degrees to the plane of the swing. The impulse response of the room was captured and applied to all other sounds in the listening test so that the natural reverb of the room would not influence the results (except samples from [4,6]).

The anchors were created from a real-time browser-based synthesis effect (http://c4dm.eecs.qmul. ac.uk/audioengineering/RTSFX/app/main-panel/whoosh.html), to allow a thorough comparison of how plausible the synthesis method is compared to the recorded sample. It was expected that a low pass filtered sample, as used in the MUltiple Stimuli with Hidden Reference and Anchor (MUSHRA) standard, would still be considered plausible, whereas a low-quality anchor would encourage the full use of the scale and allow for better understanding as to the effectiveness of the synthesis method.

Rating the plausibility of sound from a physical model was the preferred judgement in [30], stating a plausible sound as one that listeners thought "was produced in some physical manner". Box plots for all five objects are shown in Figure 3. Our physical model outperforms the alternative synthesis methods on all of the objects except the metal sword. The metal sword performed poorly for plausibility in this test, with the model with added cavity tones performing slightly better.

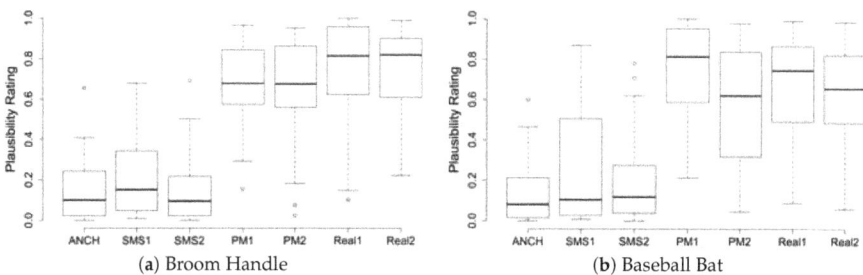

(a) Broom Handle   (b) Baseball Bat

Figure 3. *Cont.*

(c) Golf Club

(d) Wooden Sword

(e) Metal Sword

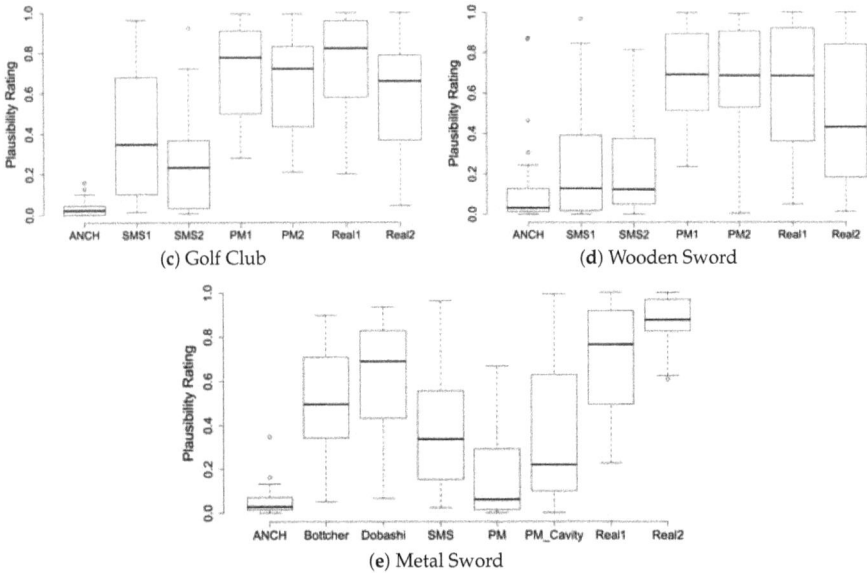

Figure 3. Box plots showing plausibility results for the preset objects. (ANCH, Anchor; SMS, Spectral Modelling Synthesis; PM, Physical Model; PMCavity, physical model including cavity tone; Real, recorded sample).

We performed the Shapiro–Wilk test for the plausibility ratings to examine the distribution of the ratings. The results are shown in Table 3, which indicate that 29 out of 36 tests were not normally distributed. To examine similarity between the ratings between each audio source in the listening test, we performed the Mann–Whitney U-test. Results of these are shown in Tables 4–8.

Table 3. Results for Shapiro–Wilk test for the plausibility ratings (**** $\Rightarrow p < 0.0001$, *** $\Rightarrow p < 0.001$, ** $\Rightarrow p < 0.01$, * $\Rightarrow p < 0.05$, - $\Rightarrow p \geq 0.05$). SMS, spectral modelling synthesis; PM, physical model; PMCavity, physical model including cavity tone; Real, recorded sample.

	Anchor	SMS1	SMS2	PM1	PM2	Real1	Real2
Broom Handle	***	**	***	-	**	***	***
Baseball Bat	***	***	**	*	-	*	-
Golf Club	***	*	**	*	*	*	-
Wooden Sword	****	***	***	-	*	*	**

	Anchor	SMS1	Bottcher	Dobashi	PM	PMCavity	Real1	Real2
Metal Sword	****	-	-	*	***	**	*	*

Table 4. The effect of different samples for a broom pole (**** $\Rightarrow p < 0.0001$, *** $\Rightarrow p < 0.001$, ** $\Rightarrow p < 0.01$, * $\Rightarrow p < 0.05$, - $\Rightarrow p \geq 0.05$). SMS, spectral modelling synthesis; PM, physical model; Real, recorded sample.

	Anchor	SMS1	SMS2	PM1	PM2	Real1	Real2
ANCH	.	-	-	****	****	****	****
SMS1		.	-	****	****	****	****
SMS2			.	****	****	****	****
PM1				.	-	-	-
PM2					.	*	-
Real1						.	-
Real2							.

Table 5. The effect of different samples for a baseball bat (**** $\Rightarrow p < 0.0001$, *** $\Rightarrow p < 0.001$, ** $\Rightarrow p < 0.01$, * $\Rightarrow p < 0.05$, - $\Rightarrow p \geq 0.05$). SMS, spectral modelling synthesis; PM, physical model; Real, recorded sample.

	Anchor	SMS1	SMS2	PM1	PM2	Real1	Real2
ANCH	.	-	-	****	****	****	****
SMS1		.	-	****	****	****	****
SMS2			.	****	****	****	****
PM1				.	*	-	*
PM2					.	-	-
Real1						.	-
Real2							.

Table 6. The effect of different samples for a golf club (**** $\Rightarrow p < 0.0001$, *** $\Rightarrow p < 0.001$, ** $\Rightarrow p < 0.01$, * $\Rightarrow p < 0.05$, - $\Rightarrow p \geq 0.05$). SMS, spectral modelling synthesis; PM, physical model; Real, recorded sample.

	Anchor	SMS1	SMS2	PM1	PM2	Real1	Real2
ANCH	.	****	****	****	****	****	****
SMS1		.	-	***	***	****	*
SMS2			.	****	****	****	****
PM1				.	-	-	-
PM2					.	-	-
Real1						.	-
Real2							.

Table 7. The effect of different sample for a wooden sword (**** $\Rightarrow p < 0.0001$, *** $\Rightarrow p < 0.001$, ** $\Rightarrow p < 0.01$, * $\Rightarrow p < 0.05$, - $\Rightarrow p \geq 0.05$). SMS, spectral modelling synthesis; PM, physical model; Real, recorded sample.

	Anchor	SMS1	SMS2	PM1	PM2	Real1	Real2
ANCH	.	-	*	****	****	****	****
SMS1		.	-	****	****	***	**
SMS2			.	****	****	***	**
PM1				.	-	-	*
PM2					.	-	*
Real1						.	-
Real2							.

Table 8. The effect of different samples for a metal sword (**** $\Rightarrow p < 0.0001$, *** $\Rightarrow p < 0.001$, ** $\Rightarrow p < 0.01$, * $\Rightarrow p < 0.05$, - $\Rightarrow p \geq 0.05$). SMS, spectral modelling synthesis; PM, physical model; PMCavity, physical model including cavity tone; Real, recorded sample.

	Anchor	SMS1	Bottcher	Dobashi	PM	PMCavity	Real1	Real2
ANCH	.	****	****	****	*	****	****	****
SMS1		.	-	**	**	-	***	****
Bottcher			.	-	****	**	**	****
Dobashi				.	****	***	-	****
PM					.	*	****	****
PMCavity						.	****	****
Real1							.	**
Real2								.

5.1.2. Object Recognition

For this test, participants were able to control the speed parameter of the physical model by use of a Wii controller and swinging the virtual object through the air. The five preset objects were presented in a pseudorandom order and the user asked to identify which object they were swinging from the list of presets. Fourteen participants completed the object recognition test prior to the listening test, and 12 completed it after the listening test. Each preset was presented twice giving 10 individual tests in total.

Tables 9 and 10 give the results of how often participants correctly identified the object being modelled by our physical model. A clear difference can be seen between participants who completed the object recognition test prior to the listening test compared to those who completed the object recognition after. It is reasonable to conclude that completing the listening test first provides some level of training for the object recognition.

Table 9. Objects identified from the Wii Controller; tested before the listening test.

Object	Correctly Guessed (%)
Wooden Sword	0
Metal Sword	36
Broom Handle	7
Baseball Bat	11
Golf Club	21

Table 10. Objects identified from Wii Controller; tested after the listening test.

Object	Correctly Guessed (%)
Wooden Sword	38
Metal Sword	63
Broom Handle	42
Baseball Bat	46
Golf Club	38

Results presented in Table 9 show that participants were far less able to identify the object being modelled by our synthesis model when having to choose before the listening test. In fact, it was more common to choose one of the other objects being modelled rather than the correct one. The wooden sword model was never correctly identified, while the metal sword object was correctly identified more than any other object, but still less than 50% of the time.

On examination of those who completed the object recognition test after the listening test, shown in Table 10, it can be seen that there was an increase for all objects being correctly identified. Similar to results shown in Table 9, the metal sword object was correctly identified more often than the other objects and on this occasion, more often than not. Although the results for the other objects are higher than those presented in Table 9, it was still more common for participants to choose one of the other objects being modelled rather than the one being replicated by our synthesis model.

5.2. Objective Evaluation

The sound produced by katana swords was examined in [9]. One sword examined had a profile with grooves on either side, which produced a cavity tone along with the Aeolian tone. To replicate this, we added a cavity tone model [26] to the sword model, which allowed a wider range of sword and object profiles to be modelled.

The sword in [9] had a thickness of 0.005 m, and the tones were measured in a wind tunnel with airspeed $u = 24$ ms^{-1}. Roger [9] observed a tone around 960 Hz due to vortex shedding (Aeolian tone) and a higher frequency sound around 6–9 kHz. The dimension of the groove in the sword was not published in [9], but it was possible for us to replicate this sound based on the published cavity tone peaks.

The magnitude spectrum output of a compact sound source, including the cavity tone, is shown in Figure 4. The parameters set were airspeed $u = 24$ ms^{-1}, diameter $d = 0.005$ m and cavity length $L = 0.00307$ m. The Aeolian tone frequency can be seen clearly at 969 Hz, with a harmonic at 2907 Hz. The cavity tone frequencies are seen at 3213 Hz, 7497 Hz, 11,780 Hz and 16,064 Hz. The length of the cavity was set to give the second cavity tone at 7497 Hz, approximately halfway between the 6 kHz and 9 kHz observed in [9].

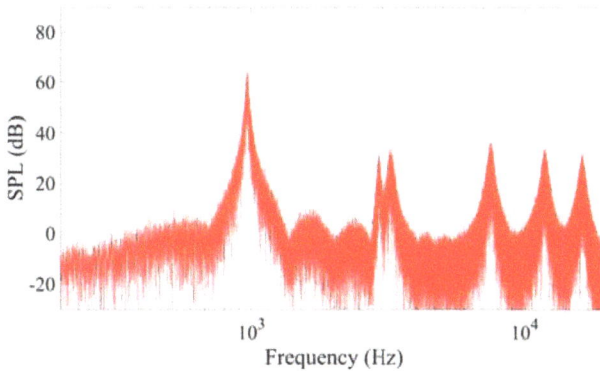

Figure 4. Magnitude spectrum of the physical model of the grooved sword.

The Aeolian tone and second cavity tone are very similar to the details published in [9]. The peaks around 3 kHz from the Aeolian tone harmonic and first cavity tone are at a greater magnitude in the synthesis model than in [9]. The published data do not cover frequencies as high as the third and fourth cavity tone.

It was noted in [9] that two oscillating motions around a sword with a groove will modulate each other. In [9], wind tunnel experiments were given where the airspeed was ramped from $u = 15$ ms^{-1} to $u = 30$ ms^{-1}. Under these circumstances there were a number of extra harmonics found. The magnitude of individual modulated frequencies varies with airspeed. Our model does not produce any harmonics that relate to the interaction between the two oscillating tones. The addition of these may increase the authenticity of our model and is a possible area of future work.

6. Discussion

The results from the listening test indicate that overall, our model performs well compared to other synthesis models. It has exceptional performance for the broom handle, baseball bat, golf club and wooden sword objects, where participants found sounds generated by our model to be as plausible as real recordings. The exception to this was the metal sword physical model sound effect, which actually performed worse in this test compared to our previously published test [1]. During the previous listening test [1], we did not have the physical dimensions of the sword samples. In this test, we had the dimensions, as well as the impulse response of the room in which the samples were recorded, thus enabling a fairer comparison.

One possible reason for the poorer performance of the metal sword physical model was that all the other modelled objects were thicker than the metal sword. Thicker objects have higher Reynolds numbers, which results in lower Q values. Spectral modelling synthesis analyses a recording and extracts sinusoidal components. Thinner objects produce sounds closer to pure tones and hence are better synthesised using SMS than thicker objects.

Table 8 shows that our physical model was significantly different from all other sounds, especially the real sounds and those synthesised by other methods. Since only one physical model sound was compared with a number of others, it is believed that a further listening test would be necessary to investigate if this result would be repeated over the range of sword dimensions and speeds. Results given in [1] indicated that the lower quality physical model sounds were rated as more plausible. These sounds had a fixed Q value that gave the impression of a thicker object. The diameter used to generate sounds in [6] was 0.01 m, substantially thicker than the sword we were modelling. It may be the case that listeners perceive a thicker sound as more plausible even if not physically accurate. This could be revealed in future perceptual evaluations.

In the original paper [1], the value of Γ in Equation (6) was set to 1×10^{-4}. This was set perceptually as no exact relationship between dipole and wake noise had been identified. During the design of the listening test for this article, the value was again set perceptually, but this time, all objects were considered, including sounds generated using the Wii Controller. This resulted in the value of Γ being set to 0.2, increasing the wake gain.

The broom handle, baseball bat and golf club objects were all cylindrical with thickness to width ratios of 1:1. For the wooden sword, this ratio decreases to approximately 0.37:1 and for a metal sword to approximately 0.14:1. The Aeolian tone model is designed around vortex shedding from cylindrical objects, and it is reasonable to assume that additional discrepancies may exist when there is a deviation from the thickness to width ratio of a cylinder.

Another possible reason for the poor rating of the metal sword object compared to the other objects is that the number of participants who have swung a real sword and heard the sound may well be less than those who have perhaps swung a golf club and the other objects. Memory plays an important role in perception [31]. If participants have heard a Foley sound effect for a sword more often than an actual sword sound, this may influence their perception of the physical model.

In contrast, it can be argued that participants will have more likely heard the actual sounds of a golf club at a live sporting event or within sporting broadcasts, and hence, their memory of these sounds would be closer to the physical model. Since all participants were from the U.K., the baseball bat would most likely not be as familiar to them as other objects, and hence, they might not have as strong a memory of the sound made by this object. This would make the difference between a memory of a Foley sound and an actual sound diminish.

It is clear from the object recognition that, with zero training, it was extremely difficult to identify an object from controlling the speed parameter from the swing of a Wii Controller. This is corroborated by the variation in results from those who did the object recognition test before the listening test to those who took the test after. Clearly, the listening tests provided participants with some form of informal training for the object recognition (it was found that the object recognition test provides negligible training for the listening test). This is in line with results from [32] where it was found that participant training was the dominant factor in determining whether or not similar tests produced significant results.

A common comment from participants when completing recognition tests was that they would like to have some visual stimulus to assist them with making their decision. It is anticipated that participants may have given more accurate choices if they were able to choose from pictures of the five objects being modelled rather than the names. The label of broom handle could produce a wide variety of images in the minds of participants, but a picture of the actual broom handle we were modelling would allow participants to focus on the same object. A further comment was that the participant would prefer a none of the above option when they believed the sound did not match any of the objects.

The use of a Wii Controller was an obvious interface for participants to swing and generate the sounds due to the sensors and ergonomic design. It was noted in a previous test as part of [1] that a participant would have liked some sense of weight in their hand to increase their sense of belief. This comment, along with the previous comment requesting visual stimulus, indicates that participants look for non-aural cues to assist in identifying sounds. Further research into which cues participants prefer and the effects on identification is required.

Since all sounds from the objects modelled were generated by the same physical model, it was understandable that there was some confusion between choices, possibly due to sonic similarities between the sound effects. The only differences between each synthesis model were the dimensions of the object being swung and the speed, either set as in the listening test or generated by each participant using the Wii Controller. A listening test that only provides a choice between a metal sword and a baseball bat would be expected to produce more clear-cut results.

The classification of different sound effects with sonic similarities was examined in [33] where nine categories of sound effects were identified. It is anticipated that objects modelled herein would be categorised into the same category in [33], but within weapons and sports in a traditional sound effect library.

Comparison with results published in [9] indicates that we have good agreement with the Aeolian tone frequency generated by vortex shedding. Wind tunnel results show the sword tested in [9] having an Aeolian tone peak at 960 Hz, while our model predicts the frequency at 969 Hz, a difference of 0.9%.

The inclusion of the cavity tone within the sword model provides the possibility to model more complex blade profiles. Listening tests indicate that it was found as plausible as the SMS sample, similar to Bottcher's sample, but not as plausible as Dobashi's sample and the real recordings. None of the other profiles are believed to include the cavity tone, but it was found that inclusion of it makes our model more plausible. It is difficult to draw overriding conclusions why this occurred, but it may be linked to Foley sword sound effects previously heard by participants.

Future research into the inclusion of the cavity tone compared to actual swords with known cavity profiles would be advantageous, enabling us to better judge how plausible the inclusion of this tone is in the generation of sword swoosh sounds. This would also assist in evaluating how the lack of modulation between the Aeolian tone and cavity tone in our model affects perception and if we need to extend our model to include this.

The range of sword profiles that we are able to model from using only the Aeolian tone and cavity tone is yet to be explored. Similarly, it is yet to be established if the sword material, bronze, steal, etc., plays an important role in the sound produced. It is known that when the vortex shedding frequency is approximately equal to a vibration frequency of the object, the sound is re-enforced. A physical model replicating this in the form of an Aeolian harp was given in [34]. Adding some of the physical properties implemented in this model would allow for consideration of the mass density of the metal and damping of the construction to be considered. Whether this would have an influence on perception is another area for further research.

Further objective evaluation would include obtaining exact velocity data for known object swings and comparing the physical model using these data and a recording of the swing from which these data were captured. This may involve wind tunnel measurements as in [9].

It is recognised that the swing sounds recorded for the listening tests were mono, and the output from the physical model includes basic stereo panning. The listener position within the virtual space of the physical model was set to replicate the microphone position when the other sounds were recorded. Although we believe this would not have a strong affect on plausible ratings, examination of spatialisation should be undertaken within future evaluation, and recording swing sounds binaurally would be preferred.

Additional models could be developed to replicate other sporting equipment, for example hockey sticks, cricket bats or even tennis racquets or lacrosse sticks, which have meshed faces. A physical model of a ball travelling through the air may also be possible although the fluid dynamics will differ from that of a cylinder, and the spinning of the ball may add other sounds not possible from our model. Authenticity may also be increased if the swinging arc of the objects was not restricted to great circles on the surface of a sphere. Normal swings often have the arms extending at the elbow, creating more elliptical arcs.

7. Conclusions

This article has presented a physically-derived synthesis model for objects swinging through the air. Adjustable parameters allow the user to approximate objects or to predefine the dimensions of objects. It is possible to match the object dimensions to graphics and for them to be morphed in real time.

Listening tests indicated that for all objects, except the very thinnest, participants found our model as plausible as real-world recordings. We have also highlighted that recognising an object from hearing the sound only was extremely difficult without any form of training.

An initial evaluation of extending the shape of profiles by adding the cavity tone has been carried out. Further evaluation is required in relation to this, examining the profiles of known objects that contain cavities and the interaction between the two fundamental aeroacoustic tones.

Acknowledgments: Rod Selfridge is funded by ESPRC, Grant EP/G03723X/1. David Moffat is funded by EPSRC Studentship—Award Reference No. 1513645.

Author Contributions: Rod Selfridge was responsible for researching relevant fluid dynamics equations, design and implementations of the synthesis model. He was also involved in devising and running the tests, analysing results and evaluation of the synthesis model. He drafted the article, critically reviewed and implemented revisions. David Moffat was involved in generating alternative synthesis sounds for comparison, devising and running of the tests, collating the data as well as analysis and evaluation of the synthesis model. He contributed to presentation of results, critical reviewing and redrafting this article. Joshua D. Reiss supervised all aspects of the research, implementation, testing and analysis. He critically reviewed and redrafted the article.

Conflicts of Interest: The authors declare no conflict of interest.

References

1. Selfridge, R.; Moffat, D.; Reiss, J.D. Real-time Physical Model for Synthesis of Sword Swing Sounds. In Proceedings of the 14th Sound and Music Computing: Best Paper Award, Espoo, Finland, 5–8 July 2017.
2. Marelli, D.; Aramaki, M.; Kronland-Martinet, R.; Verron, C. Time-frequency synthesis of noisy sounds with narrow spectral components. *IEEE Trans. Audio Speech Lang. Process.* **2010**, *18*, 1929–1940.
3. Farnell, A. *Designing Sound*; MIT Press: Cambridge, MA, USA, 2010.
4. Böttcher, N.; Serafin, S. Design and evaluation of physically inspired models of sound effects in computer games. In Proceedings of the 35th International Conference: Audio for Games, London, UK, 11–13 February 2009.
5. Mengual, L.; Moffat, D.; Reiss, J.D. Modal Synthesis of Weapon Sounds. In Proceedings of the 61st International Conference: Audio for Games, London, UK, 10–12 February 2016.
6. Dobashi, Y.; Yamamoto, T.; Nishita, T. Real-time rendering of aerodynamic sound using sound textures based on computational fluid dynamics. *ACM Trans. Graph.* **2003**, *22*, 732–740.
7. Lighthill, M.J. On sound generated aerodynamically. I. General theory. *Proc. R. Soc. Lond. Ser. A Math. Phys. Sci.* **1952**, *211*, 564–587.
8. Curle, N. The influence of solid boundaries upon aerodynamic sound. *Proc. R. Soc. Lond. Ser. A Math. Phys. Sci.* **1955**, *231*, 505–514.
9. Roger, M. Coupled oscillations in the aeroacoustics of a Katana blade. *J. Acoust. Soc. Am.* **2008**, *123*, 3023.
10. Selfridge, R.; Reiss, J.; Avital, E.; Tang, X. Physically Derived Synthesis Model of an Aeolian Tone. In Proceedings of the 141st Audio Engineering Society Convention: Best Student Paper Award, Audio Engineering Society, Los Angeles, CA, USA, 29 October–1 November 2016.
11. Cheong, C.; Joseph, P.; Park, Y.; Lee, S. Computation of aeolian tone from a circular cylinder using source models. *Appl. Acoust.* **2008**, *69*, 110–126.
12. Gerrard, J. Measurements of the sound from circular cylinders in an air stream. *Proc. Phys. Soc. Sect. B* **1955**, *68*, 453.
13. Russell, D.A.; Titlow, J.P.; Bemmen, Y.J. Acoustic monopoles, dipoles, and quadrupoles: An experiment revisited. *Am. J. Phys.* **1999**, *67*, 660–664.
14. Fey, U.; König, M.; Eckelmann, H. A new Strouhal-Reynolds-number relationship for the circular cylinder in the range $47 < \mathrm{Re} < 2 \times 10^5$. *Phys. Fluids* **1998**, *10*, 1547–1549.
15. Goldstein, M.E. *Aeroacoustics*; McGraw-Hill International Book Co.: New York, NY, USA, 1976; Volume 1.

16. Phillips, O.M. The intensity of Aeolian tones. *J. Fluid Mech.* **1956**, *1*, 607–624.
17. Hardin, J.C.; Lamkin, S.L. Aeroacoustic Computation of Cylinder Wake Flow. *AIAA J.* **1984**, *22*, 51–57.
18. Etkin, B.; Korbacher, G.; Keefe, R. Acoustic radiation from a stationary cylinder in a fluid stream (Aeolian tones). *J. Acoust. Soc. Am.* **1957**, *29*, 30–36.
19. Musafir, R. On The Sound Field of Organized Vorticity in Jet Flows. In Proceedings of the 13th International Congress on Acoustics, Belgrade, Serbia, 24–31 August 1989.
20. Avital, E.; Alonso, M.; Supontisky, V. Computational aeroacoustics: The low speed jet. *Aeronaut. J.* **2008**, *112*, 405–414.
21. Norberg, C. Flow around a circular cylinder: Aspects of fluctuating lift. *J. Fluids Struct.* **2001**, *15*, 459–469.
22. Norberg, C. *Effects of Reynolds Number and a Low-Intensity Freestream Turbulence on the Flow around a Circular Cylinder*; Chalmers University of Technology: Goteborg, Sweden, 1987.
23. Kasdin, N. Discrete simulation of colored noise and stochastic processes and $1/f\alpha$ power law noise generation. *Proc. IEEE* **1995**, *83*, 802–827.
24. Mahmoud, H.; Akkari, N. Shortest Path Calculation: A Comparative Study for Location-Based Recommender System. In Proceedings of the World Symposium on Computer Applications & Research, Cairo, Egypt, 12–14 March 2016.
25. Morrell, M.J.; Reiss, J.D. Inherent Doppler properties of spatial audio. In Proceedings of the 129th Audio Engineering Society Convention, San Francisco, CA, USA, 4–7 November 2010.
26. Selfridge, R.; Reiss, J.; Avital, E. Physically Derived Synthesis Model of a Cavity Tone. In Proceedings of the 20th Digital Audio Effects Conference, Edinburgh, UK, 5–9 September 2017.
27. Jillings, N.; De Man, B.; Moffat, D.; Reiss, J.D. Web Audio Evaluation Tool: A Browser-Based Listening Test Environment. In Proceedings of the 12th Sound and Music Computing, Maynooth, Ireland, 26 July–1 August 2015.
28. Recommendation ITU_R BS.1534-3. *Method for the Subjective Assessment of Intermediate Quality Level of Audio Systems*; Technical Report; International Telecommunication Union Radiocommunication Assembly: Geneva, Switzerland, 2015.
29. Amatriain, X.; Bonada, J.; Loscos, A.; Serra, X. Spectral processing. In *DAFX: Digital Audio Effects*; Wiley: Chichester, UK, 2002.
30. Castagné, N.; Cadoz, C. 10 criteria for evaluating physical modelling schemes for music creation. In Proceedings of the 6th Digital Audio Effects Conference, London, UK, 8–11 September 2003.
31. Gaver, W.W.; Norman, D.A. Everyday Listening and Auditory Icons. Ph.D. Thesis, University of California, San Diego, CA, USA, 1988.
32. Reiss, J.D. A Meta-Analysis of High Resolution Audio Perceptual Evaluation. *J. Audio Eng. Soc.* **2016**, *64*, 364–379.
33. Moffat, D.; Ronan, D.; Reiss, J.D. Unsupervised Taxonomy of Sound Effects. In Proceedings of the 20th Digital Audio Effects Conference, Edinburgh, UK, 5–9 September 2017.
34. Selfridge, R.; Moffat, D.; Reiss, J.D.; Avital, E. Real-Time Physical Model of an Aeolian Harp. In Proceedings of the 24th International Congress on Sound and Vibration, London, UK, 23–27 July 2017.

*applied
sciences*

MDPI

Article

SymCHM—An Unsupervised Approach for Pattern Discovery in Symbolic Music with a Compositional Hierarchical Model

Matevž Pesek [1,*], Aleš Leonardis [2] and Matija Marolt [1]

[1]   Faculty of Computer and Information Science, University of Ljubljana, Ljubljana 1000, Slovenia;
     matija.marolt@fri.uni-lj.si
[2]   School of Computer Science, University of Birmingham, Edgbaston, Birmingham B15 2TT, UK;
     a.leonardis@cs.bham.ac.uk
*   Correspondence: matevz.pesek@fri.uni-lj.si; Tel.: +386-1-4798-259

Academic Editor: Meinard Müller
Received: 13 September 2017; Accepted: 1 November 2017; Published: 4 November 2017

Abstract: This paper presents a compositional hierarchical model for pattern discovery in symbolic music. The model can be regarded as a deep architecture with a transparent structure. It can learn a set of repeated patterns within individual works or larger corpora in an unsupervised manner, relying on statistics of pattern occurrences, and robustly infer the learned patterns in new, unknown works. A learned model contains representations of patterns on different layers, from the simple short structures on lower layers to the longer and more complex music structures on higher layers. A pattern selection procedure can be used to extract the most frequent patterns from the model. We evaluate the model on the publicly available JKU Patterns Datasetsand compare the results to other approaches.

Keywords: music information retrieval; compositional modelling; pattern discovery; symbolic music representations

1. Introduction

In music, hierarchical representations are intuitive when one considers its spectral and temporal structures. In an analytical sense, the Generative Theory of Tonal Music (GTTM) by Lerdahl and Jackendoff [1] offers an approach of explicit hierarchical music modelling in musicology, well known in contemporary music theory. Although GTTM mostly relies on expert rules, the concept of hierarchical structuring seems reasonable, derived from the humans' search for structure in consciously perceived surroundings. There are several attempts to build a system capable of automatic analysis supported by the GTTM and Schenkerian analysis [2–4]. Several other rule-based models were also researched in Music Information Retrieval (MIR) and related fields [5,6]. Furthermore, the hierarchical models abound in analysis of music perception from the point of view of computational biology and neuroscience [7,8].

In parallel to explicit hierarchical representations, a variety of new approaches emerged under a common name of deep learning [9]. Several neural-network-based approaches have been proposed for melody transcription (e.g., [10]), genre classification (e.g., [11]), onset detection (e.g., [12]), drum pattern analysis (e.g., [13]) and chord estimation (e.g., [14]). The idea behind a deep learning algorithm is to construct multiple levels of data abstraction: a hierarchy of features. The high-level representations in the training data are reflected in the hierarchy. However, the encoded knowledge is implicit and is difficult to explain in a transparent (non black-box) way. Therefore, although deep learning enables unsupervised learning of features and achieves good results on a variety of tasks, it is not very appropriate for pattern discovery in music where explicit explanations of input are desired.

The discovery of repeated patterns is a known problem in different domains, including computer vision (e.g., [15]), bioinformatics (e.g., [16]) and music information retrieval (MIR). Although a common problem, its definition, as well as pattern discovery algorithms, significantly differs across these fields. In music, the importance of repetition has been addressed and discussed by a number of music theorists (e.g., [17]) and, more recently, also by researchers who develop algorithms for semi-automatic music analysis, such as one described by Marsden [4]. In the MIR field, an initiative for a common definition of different tasks was formalized into the Music Information Retrieval Evaluation eXchange (MIREX), in an attempt to compare different approaches. MIREX is a community-based framework for formal evaluation of algorithms and techniques related to MIR [18]. The MIREX community established several tasks dealing with patterns and structures in music, including structural segmentation, symbolic melodic similarity and pattern matching, and pattern discovery.

The aim of the discovery of repeated themes and sections task is to find repetitions which represent one of the more significant aspects of a music piece [19]. The MIREX task definition states "the algorithms take a piece of music as input, and output a list of patterns repeated within that piece" [20]. The task may also seem similar to the well-known pattern matching task [21], However, while a pattern matching algorithm aims to find the place of a searched pattern within a dataset and usually has a clear quantitative relation between a query and a match, a discovery of repeated patterns finds locations of multiple similar sequences of data in the dataset, without any information about the searched pattern. The definition of a pattern has been troubling researchers since the beginning; while a pattern may come as an intuitive representation with a repetitive substance, patterns in music are more difficult to define and are usually formalized using theoretical rules, specific to the music era and genre. In the discovery of repeated themes and sections task, a pattern is defined as "a set of on-time-pitch pairs that occurs at least twice (i.e., is repeated at least once) in a piece of music. The second, third, etc. occurrences of the pattern will likely be shifted in time and perhaps also transposed, relative to the first occurrence." [20]. As noted by Wang et al. [22], the pattern discovery task differs from the structural segmentation task, where segments cover the whole music piece and represent disjoint sets of events. In the pattern discovery task, patterns may partially overlap or be subsets of another pattern. However, some of the approaches mentioned in this section (e.g., [23,24]) perform pattern discovery by calculating a set of non-overlapping patterns.

A variety of approaches has been proposed for pattern discovery in music in the past years. Conklin and Anagnostopoulou [25] proposed a multiple viewpoint pattern discovery algorithm based on a suffix-tree. For a selected viewpoint (a transformation of a musical event into an abstract feature) the algorithm builds a suffix tree of viewpoint sequences (transformed music pieces). After selecting patterns which meet specified frequency and significance thresholds, the leafs of the suffix tree are reported as longest significant patterns in the corpus. Conklin and Bergeron [24] apply two algorithms, using viewpoints which represent abstract properties of musical notes for statistical modelling of melody [26]. A viewpoint is thus a function that computes values for events in a sequence; a pattern is a sequence of such feature sets, where the latter represent a logical conjunction of multiple viewpoints. The authors present a complete algorithm which can find all 'maximal frequent patterns' and an optimization algorithm using a faster heuristic approach, where the found patterns may not always be the maximal frequent patterns. The maximal frequent pattern represents a pattern whose component feature set cannot be further specialized without the pattern becoming infrequent. Rolland [27] presents the FlExPat (Flexible Extraction of Patterns) algorithm for extracting sequential patterns from sequences of data. The algorithm first identifies equipollent passage pairs and produces a similarity graph, representing the relations between each two passages; patterns are extracted from the similarity graph. The author evaluated the approach on a set of ten Charlie Parker solos from the subset of Owens' corpus [28] and reported a satisfactory pattern extraction of a large number of the annotated patterns. Cambouropoulos et al. [23] introduced an approach for extraction of patterns from abstract strings of symbols, allowing for a partial overlap of various abstract symbolic classes. They also focused on time complexity of their solution and addressed the problem of approximate pattern matching. Based on their previous work [29], they presented the PAT algorithm for segmentation based on maximal repeated patterns. Besides discovering the patterns, and subject and counter-subject entries in fugues, Meredith [30] described multiple point-set compression algorithms, including several COSIATEC and COSIATECCompress approaches and Forth's algorithm. The author evaluated these approaches on three music analysis tasks: the classification of folk song melodies into tune families, discovering entries of subjects and counter-subjects in fugues, and the discovery of repeated themes and sections in polyphonic works task. Meredith [31] also evaluated his SIATECCompressSegment algorithm for the task, which is a greedy compression algorithm based on the previously introduced SIATEC approach [19]. The algorithm evaluates patterns based on assumption that perceptually interesting patterns correspond to Maximal Translatable Patterns (MTP). The approach produces a compact encoding of a musical piece, defined by a point-set representation, in form of a set of Translational Equivalence Classes (TEC) of MTPs. The MTP with a defined particular vector is a set of points, which can be translated by that vector to give other points in the point-set representation. The authors observed that the MTPs often correspond to perceptually significant repeated patterns in music. The TEC defines a set of all patterns which are translationally equivalent to a pattern defining the specific TEC. The SIATECCompressSegment approach generates an ordered list of TECs which may overlap (in contrast to other related versions such as COSIATEC).

Recently, Velarde and Meredith [32] extended a previously introduced approach to melodic segmentation [33] for melodic classification and segmentation, where the symbolic input is first segmented, then compared and hierarchically clustered. Finally, the clusters are ranked, taking into account the cumulative length of all occurrences within each cluster. Based on their results, it can be assumed that the output is additionally filtered by a threshold defining the number of output patterns. Lartillot [34] introduced the PatMinr algorithm [35] which uses an incremental one-pass approach to identify pattern occurrences. To avoid redundancy, the author addresses two issues: closed pattern mining, which filters out the patterns that have more occurrences than their more specific patterns, thus providing more robust patterns, and pattern cyclicity, which removes redundant matches for successive occurrences of a single underlying pattern. The most recent approach submitted to the MIREX task by Ren [36] also employs a closed pattern approach commonly used in data mining. Nieto and FarBood [37] proposed the MotivesExtractor which obtains a harmonic representation of the audio or symbolic input and extracts patterns based on a produced self-similarity matrix. Using a score-based greedy algorithm ([38]) the approach extracts repeated segments, allowing the patterns to overlap. Finally, the segments are grouped into clusters and provided in the algorithm's output as patterns.

In contrast to the existing hierarchical and deep approaches, the Compositional Hierarchical Model (CHM) presented in this paper is a transparent deep architecture. The model provides an explicit (transparent) encoding of concepts, learned in an unsupervised manner, thus merging the benefits of explicit and deep hierarchical models in MIR. The CHM is built around the premise that the repetitive nature of patterns can be captured by observing statistics of occurrences of their sub-patterns, thus providing a hierarchy of the analysed symbolic music representation(s) [39]. Similar to other approaches that build a tree of patterns based on their subsumption (e.g., [25]), the CHM first extracts small atomic patterns and builds complex patterns as compositions of these atomic patterns. Its ability to concurrently provide multiple pattern hypotheses on several levels of complexity and their transparent descriptions makes it very suitable for pattern extraction, as patterns may overlap or be mutually included.

The compositional hierarchical model was first introduced by Pesek et al. [40] and was evaluated for several MIR tasks, including automated chord estimation and multiple fundamental frequency estimation [41]. In the paper, we present an adaptation of the model for analysis of Symbolic music (SymCHM) applied to the task of finding repeated patterns and sections. Instead of finding compositions in a frequency-magnitude audio representation, the adjusted model searches for compositions of symbolic events in the time-pitch-onset domain. The model learns a hierarchy of patterns; the transparent nature of the model allows the user to explore and analyse a music piece by observing the hierarchy of pattern occurrences. For the automatic discovery of repeated patterns, the patterns represented in the hierarchy are extracted. We analyse the model output and propose an extension of the model named SymCHMMerge, which refines the extracted patterns.

The contributions of this paper are as follows: the compositional hierarchical model for symbolic music analysis that can learn hierarchical melodic structures in an unsupervised manner is presented. An application of the model to the task of finding repeated patterns and sections is evaluated. The improved pattern extraction and merging approach from knowledge encoded in the model (SymCHMMerge) is proposed and analysed.

The paper is structured as follows: we present the SymCHM in Section 2, describe its application and extension to pattern extraction in Section 3 and present its evaluation and error analysis in Section 4. We conclude the paper with an overview of other possible applications of the presented model and outline future work in Section 5.

2. The Symbolic Compositional Hierarchical Model

The Symbolic Compositional Hierarchical Model (SymCHM) is derived from the CHM [40,41], which in turn was inspired by an approach for object categorization in computer vision, named the learned Hierarchy of Parts (lHoP) [42]. The SymCHM provides a hierarchical representation of a symbolic music piece, from individual notes on the lowest layer, up to complex musical patterns on higher layers. It is based on a hierarchical decomposition of music into atomic blocks, denoted as parts (not to be confused with 'voice' or 'vocal/instrumental part'). This denomination is used to retain the consistency in relation to the lHoP). According to their musical complexity, parts are structured across several layers, whereby parts on higher layers form compositions of parts on lower layers. A part can therefore describe a simple individual event as well as a complex composition of events. While events in the original compositional hierarchical model represent spectral audio features (frequencies, pitch partials and pitches), the SymCHM models notes and their compositions into melodic patterns.

2.1. Model Description

2.1.1. Compositional Layers

The SymCHM consists of an input layer $\mathcal{L}_0$ and several compositional layers $\{\mathcal{L}_1, \ldots, \mathcal{L}_N\}$. Each compositional layer $\mathcal{L}_n$ contains a set of parts $\{P_1^n, \ldots, P_{M_n}^n\}$, which are formed as compositions of parts from the previous layer $\mathcal{L}_{n-1}$. The parts on the layer $\mathcal{L}_{n-1}$ may form any number of compositions on the layer $\mathcal{L}_n$, which enables their effective reuse and thus learning of compact models, as shown later in this paper. A hierarchy of parts is illustrated in Figure 1.

The SymCHM retains part definitions of the original CHM model. The i-th composition on the layer $\mathcal{L}_n$, denoted P_i^n, is defined as:

$$P_i^n = \{P_{k_0}^{n-1}, \{P_{k_j}^{n-1}, (\mu_j, \sigma_j)\}_{j=1}^{K-1}\}. \tag{1}$$

P_i^n is a composition of K parts from the layer $\mathcal{L}_{n-1}$, called subparts. The composition is governed by parameters $\mu_{1,\ldots,K-1}$ and $\sigma_{1,\ldots,K-1}$, which model relationships between the subparts. In contrast to most existing hierarchical and deep approaches, the CHM encodes compositions in a relative rather than absolute manner. This is achieved by encoding the relative distance (offset) between each subpart $P_{k_j}^{n-1}$, from the first subpart $P_{k_0}^{n-1}$, called the central part. The offset is encoded as a Gaussian with parameters μ_j and σ_j. In SymCHM, offsets are modelled in semitones in the pitch domain (a semitone is the smallest musical interval commonly used in Western tonal music), thus a composition encodes the semitone distance between patterns represented by various subparts. Currently, the standard deviation σ_j is set to a small fixed value, which does not allow for deviations from the offset encoded by μ_j. In future work we may relax this condition to potentially achieve similar robustness as in chromatic to morphetic pitch translation [43]. As an example, the part P_2^3 in Figure 1 represents a composition of two subparts with offset 2 ($\mu = 2$), meaning its pattern is a concatenation of two sub-patterns spaced two semitones apart. All compositions and their parameters (μ, σ) are learned in an unsupervised manner as explained in Section 2.2.

Such relative encoding of knowledge enables the model to learn position-independent concepts, which in turn enables learning of compact models from small datasets, which still generalize well [41]. This is an advantage over most neural network deep approaches, which encode concepts in an absolute manner and therefore need very large datasets to train properly.

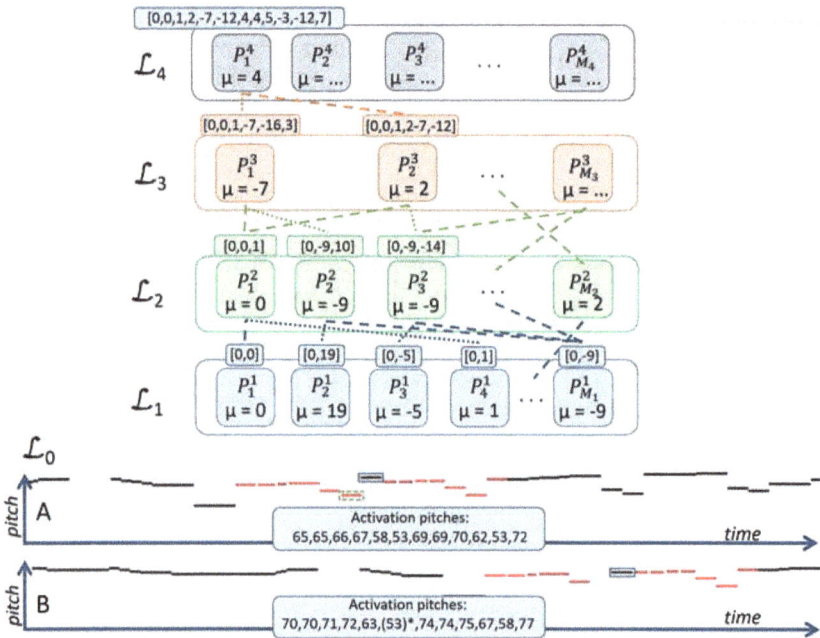

Figure 1. The symbolic compositional hierarchical model. The input layer corresponds to a symbolic music representation (a sequence of pitches). Parts on higher layers are compositions of lower-layer parts (depicted as connections between parts, the parameter μ is given in semitones). The structure of a part is displayed above each part in the figure, represented by a sequence of pitch values relative to the first subpart (e.g., [0,0,1] for the part P_1^2). A part may be contained in several compositions, e.g., $P_{M_1}^1$ is a part of compositions P_2^2 and P_3^2. The entire structure is transparent, thus we can observe the entire sub-tree of the part P_1^4. A part activates, when (a part of) the pattern it represents is found in the input. As an example, P_1^4 activates twice (Inputs A and B), however there are differences in the found patterns. Pattern A is positioned five semitones higher than B; Pattern B is missing one event (dotted green rectangle); and the pitch of one event (blue rectangle) differs between the two patterns.

2.1.2. Activations: Occurrences of Patterns

An activation of a part corresponds to the presence of the concept it encodes (melodic pattern in SymCHM) in the model input. An activation has three components: location and onset time, which map the relative pattern representation onto a specific MIDI (Musical Instrument Digital Interface technical standard) pitch and a time position within the input sequence of events (thus making it absolute) and magnitude, representing its strength.

A part will activate at a given location if all of its subparts are activated with magnitude greater than zero (this condition is relaxed with hallucination, which we introduce later in this section). A part can concurrently activate at different locations and times, which indicates multiple occurrences of its concept in the input representation. In terms of the repeated pattern discovery task, each activation of a part can be observed as a pattern occurrence: a repetition of the pattern encoded by the observed part.

More formally, the activation A is defined as a triplet $\langle A_L, A_T, A_M \rangle$ of location, time and magnitude. The activation location A_L and the time A_T of the part P_i^n are defined as:

$$
\begin{aligned}
A_L(P_i^n) &= A_L(P_{k_0}^{n-1}) \\
A_T(P_i^n) &= A_T(P_{k_0}^{n-1}).
\end{aligned}
\tag{2}
$$

The compositions therefore propagate their locations and onset times upwards through the hierarchy. Such propagation can be usefully employed as an indexing mechanism and allows for a top-down analysis of activations.

The activation magnitude represents the strength of the composition's match with the input and is defined as a weighted sum of subpart magnitudes:

$$
A_M(P_i^n) = \tanh\left(\tfrac{1}{K} \sum_{j=0}^{K-1} w_j A_M(P_{k_j}^{n-1}) \right),
\tag{3}
$$

where the weights w_j are defined by the match between the learned and the observed relative subpart pitch locations and bounded by the difference in their activation times:

$$
\begin{aligned}
w_j &= \begin{cases} 1: & j = 0 \\ \mathcal{N}(\delta_{Lj}, \mu_j, \sigma_j): & j > 0 \wedge \delta_{Tj} < \tau_W \\ 0: & \delta_{Tj} \geq \tau_W \end{cases} \\
\delta_{Lj} &= A_L(P_{k_j}^{n-1}) - A_L(P_{k_0}^{n-1}) \\
\delta_{Tj} &= A_T(P_{k_j}^{n-1}) - A_T(P_{k_0}^{n-1})
\end{aligned}
\tag{4}
$$

The motivation behind the usage of *tanh* function introduced in Equation (3) is retained from neural-network-based architectures: it provides a saturated output with the maximum limited to one. Any other function could be used to calculate the magnitude of the activation, but the hyperbolic tangent function possesses several interesting properties: it is a monotonically increasing function with a smooth gradient and has a value close to one as it approaches infinity. Since the activation magnitudes are directly used to calculate activations on a higher layer, the output of the function needs to be normalized.

The parameter τ_W represents the maximal difference between activation times of two subparts (time distance of two patterns) which still produces an activation. Such a limit must be imposed in order to avoid a combinatorial explosion of possible compositions. If subpart activations fall within this time window, their activation magnitude is calculated according to the match between their observed (δ_{Lj}) and their learned (μ_j, σ_j) relative pitch distances. A part will activate with maximal magnitude when its subparts activate at pitch distances according to the learned representation encoded by μ_j and σ_j. Note that onset times do not directly influence the activation magnitude. Thus, the activation strength of a pattern is not dependent on the temporal distance between its sub-patterns (within τ_W) and remains the same whether they are adjacent or separated by other events, allowing for gaps between sub-patterns.

2.1.3. The Input Representation and Input Layer

A symbolic music representation encoding note pitches and onset times represents input to the SymCHM. Any symbolic encoding that includes these values can be used, such as MusicXML, MIDI or text-based representations; the latter two are also available for the MIREX pattern discovery task.

We can thus define the input representation as a set of note onset (e.g., in seconds) and note pitch (e.g., MIDI pitch) tuples $\mathcal{S} = \{(N_o, N_p)\}$.

The input layer of SymCHM $\mathcal{L}_0$ models such a symbolic music representation. It consists of a single atomic part P_1^0, which activates for all note events as:

$$A = \langle A_L(P_1^0), A_T(P_1^0), A_M(P_1^0) \rangle \leftarrow \langle N_p, N_o, 1 \rangle \tag{5}$$

Thus, the activation locations A_L are equal to note pitches, the onset times A_T to note onsets, while the magnitude A_M is assumed to be 1 for all events (it can also represent note dynamics, if greater importance is to be put on accented notes).

An example of a learned hierarchy is shown in Figure 1. The part P_1^0 is activated for each input note event. The parts on the first layer represent intervals, e.g., P_4^1 represents a minor second (offset one semitone) and is activated for all such intervals in the input regardless of gaps, with notes spaced maximally τ_W apart. P_1^4 represents a sequence of note events defined by a series of offsets $[0,0,1,2,-7,-12,4,4,5,-3,-12,7]$ and is activated at MIDI locations 65 and 70.

2.2. Constructing a Hierarchy of Parts

The model is built layer-by-layer with unsupervised learning on a single or multiple musical pieces. In the 'intra-opus' pattern discovery task experiment described in this paper, we build a model for each musical piece separately.

The learning process is an optimization problem, where for each layer a set of all possible part compositions of the layer is searched for a minimal subset of compositions that covers a maximal amount of events in the training set. The learning process is driven by statistics of part activations that capture regularities in the input data. It consists of two main steps: (1) finding a set of all possible compositions, denoted candidate compositions, and (2) selecting compositions that explain a maximal amount of events in the training set.

To construct a new layer $\mathcal{L}_n$, a set of new candidate compositions $\mathcal{C}$, which will be considered for inclusion in the new layer, is first formed (Step 1). This set of candidate compositions is obtained by inferring the hierarchy with the training data and generating activations of parts layer-by-layer from $\mathcal{L}_0$ to $\mathcal{L}_{n-1}$, as explained in Section 2.3. The candidate compositions for layer $\mathcal{L}_n$ are generated from histograms of co-occurrences of $\mathcal{L}_{n-1}$ part activations within the time window τ_W (see also Equation (4)). Frequent co-occurrences indicate the presence of underlying patterns. New compositions are formed from combinations of $\mathcal{L}_{n-1}$ parts where the number of co-occurrences exceeds the learning threshold τ_C. The composition parameter μ is estimated from the corresponding histogram.

The $\mathcal{L}_1$ candidate compositions are thus constructed as a relative structure of two co-occurring $\mathcal{L}_0$ part activations, both occurring within the time window τ_W. This procedure is repeated on all consecutive layers, where activations of parts co-occurring within the time window on a previous layer $\mathcal{L}_{n-1}$ compose new part candidates on the next layer $\mathcal{L}_n$. Since the model allows for partial overlapping of the covered structure (e.g., P_1^2 in Figure 1), the structures on these layers represent 3–4 music events. Consequently, the $\mathcal{L}_N$ candidate compositions include all combinations of $\mathcal{L}_{N-1}$ part pairs representing structures of 2^{N-1}–2^N music events.

In the second step, a subset of compositions from $\mathcal{C}$ that covers a maximum number of events in the input data is selected. As the problem of selecting a set of compositions from $\mathcal{C}$ which optimally cover the input data is NP (nondeterministic polynomial time) complete, a greedy approach, which selects a subset of compositions and leaves a minimal amount of events in the input uncovered, was introduced in [41].

The composition selection uses part coverage as a measure of the part's suitability for selection. The coverage of the part P_i^n can be obtained by projecting its activations to the input layer and observing the covered events. For a single activation of the part P_i^n at the time T and the location L, coverage is defined as the union of coverages of its subparts:

$$C(A(P_i^n)) = \bigcup_{j=0}^{K-1} C(A(P_{k_j}^{n-1})). \tag{6}$$

When the input layer is reached, the coverage is defined by the presence of an event at the given location and time as:

$$C(A(P_1^0)) = \begin{cases} A_L(P_1^0) : & A_M(P_1^0) > 0 \\ \varnothing : & otherwise \end{cases}. \tag{7}$$

Based on coverage, the greedy composition selection approach is defined as follows:

- the coverage of each part from $\mathcal{C}$ is calculated as a union of events in the training data covered by all activations of the part,
- parts are iteratively added to the new layer $\mathcal{L}_n$ by choosing the part that adds most to the coverage of the entire training set in each iteration. This ensures that only compositions that provide enough coverage of new data with regard to the currently selected set of parts will be added,
- the algorithm stops when the additional coverage falls below the learning threshold τ_L.

The learning procedure is repeated for each layer until a desired number of layers is reached. The reader should note that the number of layers governs the maximal length of encoded patterns, as discussed in the evaluation.

2.3. Inferring Patterns

A learned model captures the repetitive patterns in the training data, which are relatively encoded and may be observed through an inspection of the model's parts on its various layers. When a trained model is presented with new input data, the learned patterns may be located in the input through the process of inference. Inference calculates part activations on the input data (and thus absolute pattern positions) according to Equations (2) and (3). They are calculated bottom-up layer-by-layer, whereby the input data activates the layer $\mathcal{L}_0$. As already mentioned, the activation of a part represents a specific occurrence of the pattern it represents in the input. An activation has three components: location and onset time, which map the relative pattern onto a specific set of pitches within the input sequence of events (thus making it absolute), and magnitude, representing its strength. A part can concurrently activate at different locations, which indicates multiple occurrences of the represented pattern in the input representation.

Inference may be exact or approximate, where in the latter case two additional mechanisms, hallucination and inhibition, enable the model to find patterns with deletions, changes or insertions, thus increasing its predictive power and robustness.

2.3.1. Hallucination

As described in Section 2.1, a part activation is produced only if all subparts activate with magnitude greater than zero at locations which approximately correspond to the structure encoded by the part. This conservative behaviour may be relaxed by hallucination. It enables a part to produce activations even when the structure it represents is incomplete or modified in the input (e.g., missing notes, added notes, changed pitch, changed note order). Hallucination is important, as it enables the model to find variations of patterns represented by individual parts. The missing information is obtained from knowledge acquired during learning and encoded in the model structure. Using hallucination, the model generates activations of parts most fittingly covering the input representation, where notes which are not present, but are encoded in the model, are hallucinated. It is implemented by changing the conditions under which a part may activate. With hallucination, a part may activate even if all of its subparts are activated, when the percentage of events it represents, covered in the input, exceeds a hallucination threshold τ_H. Thus, if we set τ_H to one, the default behaviour is obtained, while lowering its value leads to increased hallucination and tolerance to changes in patterns.

The hallucination threshold τ_H influences the number of discovered patterns and identified pattern occurrences. When lowered, the amount of activations increases, as parts may activate on incomplete matches, thus producing activations which would otherwise not be generated. Additionally, if used during learning, the number of parts on lower layers will decrease, as parts added to a layer will have higher coverage due to more activations.

2.3.2. Inhibition

Inhibition in our model is a hypothesis refinement mechanism, which reduces the amount of redundant activations. An activation of a part P_i^n is inhibited (removed) when one or multiple parts $P_{j_1}^n, \ldots, P_{j_K}^n$ cover a large part of the same events in the input, but with stronger magnitude. More formally, activation of the part P_i^n is inhibited when the following conditions are met:

$$\exists \{P_{j_1}^n \ldots P_{j_K}^n\} : \frac{|C(A(P_i^n)) \setminus \bigcup_{k=1}^K C(A(P_{j_k}^n))|}{|C(A(P_i^n))|} < \tau_I \tag{8}$$

and

$$\forall P_{j_k}^n \in \{P_{j_1}^n \ldots P_{j_K}^n\} : A_M(P_{j_k}^n) > A_M(P_i^n). \tag{9}$$

The $C(A)$ represents activation coverage (Equation (6)), A_M activation magnitude (Equation (3)) and τ_I controls the strength of inhibition. If τ_I is set to zero, no inhibition occurs; the larger its value, the more activations are inhibited and propagated less between model layers. Notably, only activations with magnitude larger than that of the part P_i^n are considered in the inhibition process.

Besides reducing the number of activations and output patterns, the inhibition mechanism can also be used for producing alternative explanations of the input. If activations of the strongest pattern which inhibits other competing hypotheses are removed from the model, the next best hypothesis is selected during inference, thus providing an alternative explanation with different pattern occurrences to appear in the model's output.

3. Pattern Selection with SymCHM

The SymCHM model can be trained on a single or multiple symbolic music representations. It learns a hierarchical representation of patterns occurring in the input, where patterns encoded by parts on higher layers are compositions of patterns on lower layers. The inference produces part activations which expose the learned patterns (and their variations) in the input data. Shorter and more trivial patterns naturally occur more frequently, longer patterns less frequently. On the other hand, longer patterns may entirely subsume shorter patterns. Occurrences of melodic patterns in a given piece are discovered by observing activations of the learned model's parts, where each activation of a part is interpreted as an occurrence of the pattern encoded by the part.

To use the model for the discovery of repeated patterns and sections task, we need to select which of the found patterns will be provided in the model's output. In this Section, we present two approaches for a pattern selection.

3.1. Basic Selection

In a basic pattern selection, we output all patterns of sufficient complexity, as encoded by parts starting from the layer L up to the highest layer N. First, we select all parts from the layers $\mathcal{L}_L \ldots \mathcal{L}_N$. Since parts on higher layers are compositions of parts on lower layers, we exclude all parts which are subparts of a composition on a higher layer to avoid redundancy. The final selection of parts can be formulated as:

$$\bigcup_{l=L}^{N} \{ P_i^l \in \mathcal{L}_l : (\neg \exists P_j^{l+1})[P_j^{l+1} \in \mathcal{L}_{l+1} \land P_i^l \in P_j^{l+1}] \} \tag{10}$$

Inference is then performed on a music piece and activations of the selected parts represent the found patterns and their locations in the piece. Hallucination and inhibition are applied during inference to provide balance between producing hypotheses which partially match the input representation (hallucination) and the amount of competitive hypotheses produced (inhibition).

3.2. SymCHMMerge: Improved Pattern Selection

An analysis of the basic pattern selection algorithm showed lack of diversity in the found patterns, as the patterns were often very similar and overlapping. We improved the algorithm by merging redundant patterns and adjusting the learning and inference parameters, and named the resulting model SymCHMMerge.

3.2.1. Merging Redundant Patterns

Since parts in our model are learned in an unsupervised manner, several parts may represent similar and overlapping patterns (e.g., patterns shifted by a few notes). Inhibition reduces redundant activations of such parts, however it is usually not enforced strongly, as it could overly reduce the number of activations and found patterns. To reduce the number of such overlapping patterns, we merge them into single, longer patterns.

Let $\pi(A(P_i^n))$ represent a pattern occurrence defined by the projection π of the activation A of the part P_i^n onto the layer $\mathcal{L}_0$. Ψ_i^n represents the set of all such pattern occurrences discovered by activations of the part:

$$\Psi_i^n = \bigcup_k \{ \pi(A_k(P_i^n)) \}. \tag{11}$$

Two pattern occurrences a_i and a_j, produced by the parts P_i^n and P_j^m, are taken to be redundant, if they overlap significantly. We express this by calculating the Jaccard similarity coefficient and compare it to a threshold τ_R:

$$a_i = \pi(A(P_i^n)), a_j = \pi(A(P_j^m))$$
$$J(a_i, a_j) = \frac{|a_i \cap a_j|}{|a_i \cup a_j|} > \tau_R. \tag{12}$$

We aim to merge redundant pattern occurrences of two parts if they frequently produce overlapping patterns. Therefore, we calculate the proportion of such patterns produced by the two parts as:

$$\frac{1}{|\Psi_i^n| + |\Psi_j^m|} \sum_{a_i \in \Psi_i^n} \sum_{a_j \in \Psi_j^m} |J(a_i, a_j) > \tau_R|. \tag{13}$$

If the proportion exceeds a threshold τ_M, all redundant pattern occurrences of the two parts are merged.

For evaluation, the thresholds τ_R and τ_M were both set to 0.5, meaning that pattern occurrences produced by two parts had to share at least 50% of events in the input layer and appear together in at least 50% of cases, to be merged.

3.2.2. Increasing Diversity

To address the problem of pattern diversity, we needed to increase the number of patterns found by the model. This was achieved with three simple adjustments. First, we lowered the candidate selection thresholds in the greedy phase of the learning process to add more parts to each layer (evaluation showed that on average 16% more parts were added). Second, more layers were considered when searching for pattern occurrences, and third, hallucination was increased during inference. All these modifications could also be made with the basic pattern selection approach; however, they would result in an even higher number of redundant patterns. With SymCHMMerge, redundant occurrences are merged and thus the diversity of the found patterns increases.

4. Evaluation

We evaluated the proposed model for the discovery of repeated themes and sections task in symbolic monophonic music pieces. Since we are searching for patterns within a given piece (and not across the entire corpus) the model was built independently for each piece and inferred on the same piece. All model parameters were kept constant during all evaluations and were not tuned to each specific case. The parameters were set to the values defined in Table 1. The τ_W parameter limiting the time span of activations was set to $\tau_W = 2^{n+2}$ events. The values and short descriptions of parameters are also listed in Table 1. The values for the τ_H and τ_I parameters are based on the stable performance achieved in the range around 0.5 for (see the Sensitivity to parameter values subsection. The τ_R and τ_M values were set to the majority thresholds of 50% and were not tuned. The τ_L parameter value was retained from the original spectral CHM where it was evaluated empirically.

Table 1. Model's parameter settings for the experiment.

Parameter	Description	Value
τ_H	Hallucination parameter retaining the activation of a part in an incomplete presence of the events in the input signal	0.5
τ_I	Inhibition parameter reducing the number of competing activations	0.4
τ_R	Redundancy parameter determining the the necessary amount of overlapping pattern occurrences in order for the occurrences to be merged	0.5
τ_M	Merging parameter determining the amount of redundant pattern occurrences needed for two patterns to be merged into one	0.5
τ_L	Learning threshold for added coverage which needs to be exceeded in order for a candidate composition to be retained while learning the model	0.005
τ_W	Window limiting the time span of activations, defined per layer $\mathcal{L}_n$	2^{n+2}

Table 2 shows the performance of SymCHM on the MIREX 2015 discovery of repeated themes and sections task. To compare SymCHM to SymCHMMerge, the Table 2 also includes the results of their evaluation on the publicly available JKU Patterns Development Dataset (PDD) [44]. Detailed results of SymCHMMerge on this dataset are shown in Table 3.

The JKU PDD dataset (the dataset is publicly available on this link: https://dl.dropbox.com/u/11997856/JKU/JKUPDD-Aug2013.zip) consists of five pieces:

- Bach's Prelude and Fugue in A minor (BWV(Bach-Werke-Verzeichnis) 889): 731 note events, 3 patterns, 21 pattern occurrences,
- Beethoven's Piano Sonata in F minor (Opus 2, No. 1), third movement: 638 note events, 7 patterns, 22 pattern occurrences,
- Chopin's Mazurka in B flat minor (Opus 24, No. 4): 747 note events, 4 patterns, 94 pattern occurrences,
- Gibbons' "The Silver Swan": 347 note events, 8 patterns, 33 pattern occurrences,
- Mozart's Piano Sonata in E flat major, K. 282-2nd movement: 923 note events, 9 patterns, 38 pattern occurrences.

Table 2. Evaluation of SymCHM, SymCHMMerge and Music Information Retrieval Evaluation eXchange (MIREX) results of other proposed approaches for the discovery of repeated themes and sections task on the JKU Patterns Development Dataset (PDD) and JKU Patterns Testing Dataset (PTD), denoted as MIREX 2015.

Algorithm	P_{est}	R_{est}	F_{1est}	$P_{occ(c=0.75)}$	$R_{occ(0.75)}$	$F_{1occ(c=0.75)}$
SymCHM MIREX 2015	53.36	41.40	42.32	81.34	59.84	67.92
NF1 MIREX 2014	50.06	54.42	50.22	59.72	32.88	40.86
DM1 MIREX 2013	52.28	60.86	54.80	56.70	75.14	62.42
OL1 MIREX 2015	61.66	56.10	49.76	87.90	75.98	80.66
VM2 MIREX 2015	65.14	63.14	62.74	60.06	58.44	57.00
SymCHM JKU PDD	67.92	45.36	51.01	93.90	82.72	86.85
SymCHMMerge JKU PDD	67.96	50.67	56.97	88.61	75.66	80.02

	TLF_1	$P_{occ(c=0.5)}$	$R_{occ(c=0.5)}$	$F_{1occ(c=0.5)}$	P	R	F_1
SymCHM MIREX 2015	37.78	73.34	62.48	67.24	10.64	6.50	5.12
NF1 MIREX 2014	33.28	54.98	33.40	40.80	1.54	5.00	2.36
DM1 MIREX 2013	43.28	47.20	74.46	56.94	2.66	4.50	3.24
OL1 MIREX 2015	42.72	78.78	71.08	74.50	16.0	23.74	12.36
VM2 MIREX 2015	42.20	46.14	60.98	51.52	6.20	6.50	6.2
SymCHM JKU PDD	51.75	78.53	72.99	75.41	25.00	13.89	17.18
SymCHMMerge JKU PDD	52.89	83.23	68.86	73.88	35.83	20.56	25.63

4.1. Evaluation Metrics

Evaluation metrics from the MIREX discovery of repeated themes and sections task were used for evaluation. This subsection provides a short description and formalization of the definitions found in the MIREX task definition [20]. The establishment measure (precision P_{est}, recall R_{est} and F score F_{1est}) evaluates the algorithm's ability to find at least one occurrence of each pattern shifted in time and pitch. Two occurrence measures F_{1occ} evaluate the extent of the model's ability to find all pattern occurrences, where the $c = \{0.5, 0.75\}$ factor represents the inexactness tolerance threshold. Meredith [30] proposed an additional three-layer metric (P_3, R_3, TLF_1) that provides balance between the establishment and the occurrence measures. The exact precision, recall and F score measures (P, R, F_1) show the algorithm's performance in matching the found patterns with the reference annotations in an exact manner.

The metrics are formally defined using the following set of symbols:

- n_P: the number of patterns in a ground truth
- $\Pi = \{\mathcal{P}_1, \mathcal{P}_2, \ldots, \mathcal{P}_{n_P}\}$: a set of ground truth patterns
- $P = \{P_1, P_2, \ldots, P_{m_P}\}$—occurrences of pattern $\mathcal{P}$
- n_Q: the number of patterns in the algorithm's output
- $\Xi = \{\mathcal{Q}_1, \mathcal{Q}_2, \ldots, \mathcal{Q}_{n_Q}\}$: a set of patterns returned by the algorithm
- $Q = \{Q_1, Q_2, \ldots, Q_{m_Q}\}$—occurrences of pattern $\mathcal{Q}$.
- k: the number of ground truth patterns identified by the algorithm

The standard precision is defined as $P = k/n_Q$, the recall as $R = k/n_P$, and the F_1 score as $F1 = 2 \times P \times R/(P + R)$. Due to the extreme difficulty of discovering strictly exact patterns, more robust versions of the metrics are provided: the occurrence and the establishment scores. First, the cardinality score is used to determine the music similarity between the annotated and the discovered patterns:

$$s_c(P_i, Q_j) : |P_i \cap Q_j| / \max\{|P_i|, |Q_j|\} \tag{14}$$

A score matrix is calculated based on the similarity as follows:

$$s(\mathcal{P}, \mathcal{Q}) = \begin{bmatrix} s(P_1, Q_1) & s(P_1, Q_2) & \cdots & s(P_1, Q_{m_Q}) \\ s(P_2, Q_1) & s(P_2, Q_2) & \cdots & s(P_2, Q_{m_Q}) \\ \vdots & \vdots & \ddots & \vdots \\ s(P_{m_P}, Q_1) & s(P_{m_P}, Q_2) & \cdots & s(P_{m_P}, Q_{m_Q}) \end{bmatrix} \tag{15}$$

Based on the score matrix, the establishment matrix is calculated from the set of annotated patterns Π and the set of algorithm's output patterns Ξ:

$$S(\Pi, \Xi) = \begin{bmatrix} S(\mathcal{P}_1, \mathcal{Q}_1) & S(\mathcal{P}_1, \mathcal{Q}_2) & \cdots & S(\mathcal{P}_1, \mathcal{Q}_{n_Q}) \\ S(\mathcal{P}_2, \mathcal{Q}_1) & S(\mathcal{P}_2, \mathcal{Q}_2) & \cdots & S(\mathcal{P}_2, \mathcal{Q}_{n_Q}) \\ \vdots & \vdots & \ddots & \vdots \\ S(\mathcal{P}_{n_P}, \mathcal{Q}_1) & S(\mathcal{P}_{n_P}, \mathcal{Q}_2) & \cdots & S(\mathcal{P}_{n_P}, \mathcal{Q}_{n_Q}) \end{bmatrix} \tag{16}$$

The establishment precision is thus defined as:

$$P_{est} = \frac{1}{n_Q} \sum_{j=1}^{n_Q} \max\{S(\mathcal{P}_i, \mathcal{Q}_j) | i = 1 \ldots n_P\} \tag{17}$$

The establishment recall is defined as:

$$R_{est} = \frac{1}{n_P} \sum_{j=1}^{n_P} \max\{S(\mathcal{P}_i, \mathcal{Q}_j) | i = 1 \ldots n_Q\} \tag{18}$$

Additionally, the establishment F_1 score is calculated as:

$$F1_{est} = 2 \times P_{est} \times R_{est} / (P_{est} + R_{est}) \tag{19}$$

The establishment metrics reward a single match between the annotated and algorithm's patterns. To counterbalance this bias, the occurrence metrics are used. The occurrence metrics reward the algorithm's ability to find all occurrences of a single pattern. To loosen the exactness, the found patterns may be inexact. This inexactness is implemented using a threshold c (default values used in the 0.5 and 0.75), The indices $\mathcal{I}$ of the establishment matrix with values greater than or equal this threshold c are considered discovered. The occurrence matrix $O(\Pi, \Xi)$ is calculated using the following approach, starting with an empty $n_P \times n_Q$ matrix and the establishment indices $\mathcal{I}$:

$$\forall (i, j) \in \mathcal{I} : O(\Pi, \Xi)[i, j] = s(\mathcal{P}_i, \mathcal{Q}_j). \tag{20}$$

The occurrence precision score is consequently calculated using the occurrence matrix as follows:

$$P_{occ} = \frac{1}{n_{col}} \sum_{j=1}^{n_Q} O(i, j) | i = 1 \ldots n_P, \tag{21}$$

where n_{col} represents the number of non-zero columns in occurrence matrix O. The occurrence recall score is analogously calculated as:

$$R_{occ} = \frac{1}{n_{row}} \sum_{j=1}^{n_P} S(i, j) | i = 1 \ldots n_Q, \tag{22}$$

where n_{row} represents the number of non-zero rows in the occurrence matrix O.

4.2. Performance

The SymCHM with the basic pattern selection algorithm was submitted to the MIREX 2015 discovery of repeated themes and sections task. The results are shown in Table 2. The submitted model learned a six layer hierarchy, where activations of parts on Layers 4–6 were output as the found pattern occurrences.

In the MIREX 2015 evaluation [20], the two state-of-the art approaches by Velarde and Meredith (VM2) [32] and Lartillot (OL1) [34] achieved better overall results. However, the SymCHM outperformed other algorithms on the first piece in the MIREX evaluation dataset and achieved better results than VM2 in pattern occurrence measures, which indicated the model's ability to robustly identify the occurrences of the identified patterns. Compared to other approaches proposed in previous MIREX evaluations, such as NF1'14 [37] and DM1'13 [45], SymCHM found more pattern occurrences, as well a higher number of exact matches. The SymCHM also achieved a higher TLF_1 score when compared to NF1'14 submission.

Table 3. A detailed list of JKU Patterns Development Dataset results for the SymCHMMerge approach. The n_P and n_Q columns represent the number of annotated patterns and the number of discovered patterns respectively. Song names are shortened, using a four letter abbreviation of the composer's name.

Piece	n_P	n_Q	P_{est}	R_{est}	F_{1est}	$P_{occ(c=0.75)}$	$R_{occ(c=0.75)}$	$F_{1occ(c=0.75)}$
bach	3	2	100.00	66.67	80.00	100.00	45.65	62.68
beet	7	7	65.81	60.02	62.78	80.71	80.71	80.71
chop	4	5	47.95	49.81	48.86	62.36	51.96	56.69
gbns	8	3	78.16	35.49	48.81	100.00	100.00	100.00
mzrt	9	8	47.88	41.39	44.40	100.00	100.00	100.00
Average	6.2	5	67.96	50.67	56.97	88.61	75.66	80.02

Piece	P_3	R_3	TLF_1	$P_{occ(c=0.5)}$	$R_{occ(c=0.5)}$	$F_{1occ(c=0.5)}$	P	R	F_1
bach	62.96	41.97	50.37	100.00	45.65	62.68	100.00	66.67	80.00
beet	77.38	64.95	70.62	79.24	72.44	75.69	0.00	0.00	0.00
chop	46.96	39.92	43.15	57.00	46.29	51.09	0.00	0.00	0.00
gbns	81.82	34.33	48.37	100.00	100.00	100.00	66.67	25.00	36.36
mzrt	57.21	47.54	51.93	79.92	79.92	79.92	12.50	11.11	11.77
Average	65.27	45.74	52.89	83.23	68.86	73.88	35.83	20.56	25.63

To increase diversity and decrease redundancy, we introduced the SymCHMMerge with an improved pattern selection algorithm. Activations of parts on Layers 2–6 were considered for finding pattern occurrences, where each layer included 16% more parts on average due to the more relaxed learning conditions.

A comparison between both models on the JKU PDD dataset showed that the SymCHMMerge achieved significantly better results (Friedman's test: $\chi^2 = 7.2, p < 0.01$). It mostly improved in establishment measures, which indicated an improvement of the algorithm's ability to discover at least one occurrence of a pattern, tolerating for time shift and transposition [20]. On the other hand, occurrence measures $F_{1occ(c=0.75)}$ and $F_{1occ(c=0.5)}$ which evaluated the algorithm's ability to find all occurrences of the established patterns, have dropped by 5%. We attribute this drop to a higher number of established patterns, for which the occurrence measure is calculated. Finally, the absolute precision, recall and F scores significantly increased due to the SymCHMMerge's pattern merging procedure and increased pattern diversity.

4.3. Sensitivity to Parameter Values

To assess the sensitivity of SymCHMMerge to changes of model parameters, we analysed its performance by varying the inhibition and hallucination parameters τ_I and τ_H, which affect inference.

We observed the behaviour of occurrence and establishment measures in order to estimate the balance between the two. Due to the large number of possible parameter combinations, we evaluated how changes in one parameter (set for all layers) affect performance when all other parameters are fixed.

4.3.1. Inhibition

The top part of Figure 2 shows how changes in the inhibition parameter τ_I affect the results. An increase of τ_I increases inhibition and removes activations which are only partially covered by others, while a decrease will allow for more overlapping activations to propagate to higher layers. The plots show that reduced inhibition has a positive effect on occurrence recall, which is expected, as more activations are produced. It is even more interesting that it also positively affects precision of found occurrences, which might be explained by the fact that overlapping activations are successfully merged by the merging algorithm of SymCHMMerge. For the establishment metrics, the effect of changes in inhibition is not so obvious, and apart from extreme values, performance is stable.

4.3.2. Hallucination

The bottom part of Figure 2 shows how changes in the hallucination parameter τ_H affect performance. As described in Section 2.3.1, larger τ_H values decrease hallucination and thus the number of activations. Decreased hallucination affects both occurrence and establishment of patterns, as there is little tolerance for pattern variations. With more hallucination, both measures increase and then remain stable; again, precision is not affected significantly, as the merging algorithm of SymCHMMerge reduces the growing number of activations on higher layers.

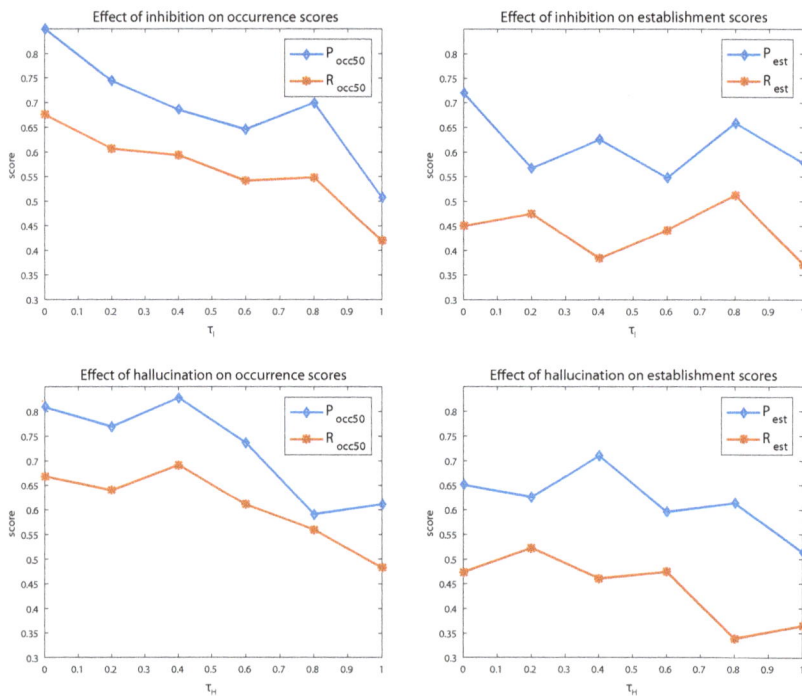

Figure 2. Sensitivity of the model to changes of the hallucination parameter τ_I (**top**) and the inhibition parameter τ_H (**bottom**). When one parameter was varied, all others remained fixed.

4.4. Error Analysis

To increase our understanding of the model's performance, we performed an analysis of its most common types of errors.

4.4.1. Incomplete Matches

We observed that the occurrence metrics increase when we allow for partially incomplete patterns to be discovered (hallucination), however, the exact F_1 scores do not always increase. After observing the pattern occurrences which do not contribute to the rise in F_1 score, we discovered that these patterns do not completely match the reference annotations, as shown in Figure 3.

The difference between a reference annotation and a model's proposed pattern usually presents itself at the edges of an occurrence, where the model assumes that one or more preceding or succeeding events belong to the pattern. These events frequently occur at the same locations (relative to the pattern), with similar time and pitch offsets. Thus, the model adds these events to the pattern occurrence, causing mismatch with the reference annotation. Such errors could be resolved by incorporating theoretical rules governing the beginnings and endings of patterns, e.g., gap rule ([46], p. 68) into the pattern selection algorithm.

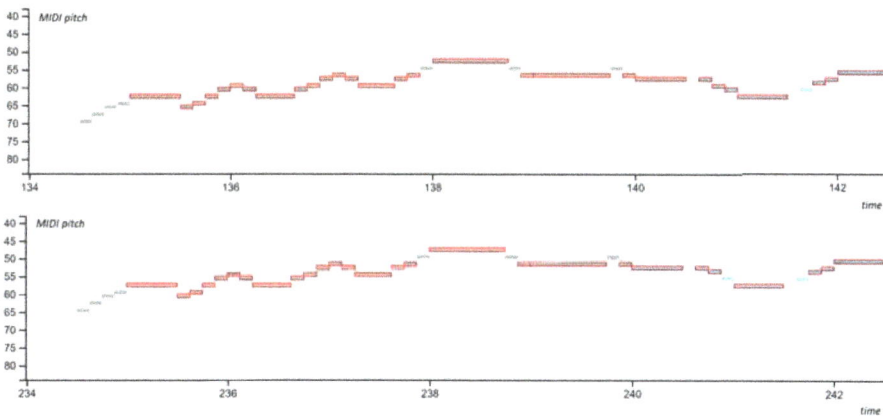

Figure 3. An incomplete pattern match of two pattern occurrences in Bach BWV 889 Fugue in A minor (from the JKU PDD dataset). Two pattern occurrences are presented in the figure (top and bottom). A piano roll representation is shown where the reference annotation is coloured in grey and the identified pattern occurrences outlined with red borders. Even though similar, events on the right side (shown in light blue) are not part of the reference annotations, however they are included in the model's patterns due to their co-occurrence with other events.

4.4.2. Unidentified Patterns

Patterns which were not identified by the model usually belong to one of two types: section patterns and short patterns.

Section patterns, such as in Mozart's Piano Sonata in E flat major, K. 282-2nd movement, remain unidentified. These section patterns represent large segments of music (50–137 events). The six layers in our model have the potential of encoding patterns of up to 64 events. While some of the reference patterns could be identified, the model did not contain a sufficient amount of layers to cover the largest patterns. We consequently focused on observing the absence of the shorter section patterns (between 50 and 64 events). While incomplete (often overlapping) matches of these patterns were found on the $\mathcal{L}_5$ and $\mathcal{L}_6$ layers (sub-patterns), there were no complete matches between the found patterns and the reference annotations. Furthermore, the overlap was not high enough that these sub-patterns would be merged during pattern merging.

The second subgroup—the short patterns—also frequently occur in evaluation datasets. These patterns are 4–5 events long. They are identified by the model on the layers $\mathcal{L}_2$ and $\mathcal{L}_3$, and also form compositions on higher layers. If such larger compositions are present, the pattern selection procedure excludes the short patterns from the final output.

The discovery of larger patterns could be improved by building additional compositional layers while learning the model, and by adjusting the merging rules for long patterns. To find more short patterns, we could add additional criteria that would counterbalance the promotion of longer patterns during pattern selection. For example, the event duration could be used when considering the importance of short events.

4.5. Drawbacks of the Evaluation

To establish the effectiveness of the proposed model in the symbolic domain, we evaluated the model for the pattern discovery task, where a comparison between the SymCHM and other approaches is based on the JKU PDD and JKU PTD datasets. To avoid diminishing the MIREX's position of being an evaluation exchange and not a benchmarking framework, we focused our evaluation on the two variants of the compositional model we developed, the SymCHM and SymCHMMerge, as shown in Table 2.

As thoroughly discussed by Meredith [30], this MIREX task possesses many drawbacks and thus might not be the optimal tool for an algorithm comparison. However, it is rather difficult to create an experiment which would provide a clearer evaluation of the algorithm's performance. First, a definition of a pattern is vague; there are several sources gathered in the JKU datasets. Some of the patterns in the ground truth represent themes, while others represent entire sections. Without any prior knowledge about the goal (length of pattern, perhaps a ratio between the length and the variation within the pattern occurrences), the metrics are logically leaning towards awarding the approach which finds most occurrences of the discovered pattern. It seems impossible to design an algorithm capable of finding a "pattern" when the definition of a pattern varies among the annotators. The three-layer F score proposed by Meredith is a step towards a metric which provides the balance between the establishment and the occurrence metrics otherwise provided. Second, the size of the dataset presents a limitation: the combined JKU PDD and JKU PTD datasets represent ten (classical) musical pieces in total. It is thus difficult to claim the datasets provide a representative sample of any kind of music or genre. However, we acknowledge the incredible effort put in the creation of the datasets and the tasks; we believe the size of the datasets is affected by the effort needed. Nevertheless, we believe the MIREX discovery of repeated themes and sections task is currently the best currently available approximation of a performance evaluation for the pattern discovery in music.

5. Conclusions

In the paper, we presented the compositional hierarchical model for pattern discovery in symbolic music representations. The model calculates a hierarchical representation of melodic patterns in a music corpus with a statistically-based learning algorithm. It can be viewed as a transparent deep architecture, combining the ability of unsupervised learning of multi-layer hierarchies with a transparent structure that enables insight into the learned concepts. The inference process with hallucination and inhibition mechanisms enables the search for pattern variations.

We evaluated the model in the MIREX evaluation campaign and its improved pattern selection algorithm on the JKU PDD dataset, where we show that we can obtain favourable results with the improved version of the model. We showed that the model can be used for finding patterns in symbolic music and that it can learn to extract patterns in an unsupervised manner without hard-coding the rules of music theory. We have also demonstrated the transfer of the model from classification tasks based on audio representations to pattern extraction in the symbolic domain. The results obtained by the model are not on par with the best two performing algorithms. Nevertheless, the proposed model performs better than several other proposed approaches. As discussed in Section 4.5, this evaluation contains many potential drawbacks, but it is currently the best approximation for pattern discovery evaluation. The definition of the 'pattern' itself is elusive and may contain many different explanations, varying from strictly music-theoretical, to mathematical formalization. The human perception of patterns in music itself is too difficult to explain and incorporate in a single formalized task. However, with the proposed model, we have demonstrated that a deep transparent architecture can tackle the pattern discovery by employing unsupervised learning and may thus better approximate how listeners recognize patterns than the rule-based systems. Due to its transparency, the model is not only applicable to tasks where a single output is provided, but can also be used for exploration and pattern discovery by an expert. The model produces multiple hypotheses on several layers, which can be used as reference points in a deeper semi-automatic music analysis. We believe this further strengthens the model's usefulness to the wider MIR community.

In our future work, we will focus on improving the model. We plan to include event duration into pattern selection and merging and adapt the model for polyphonic pattern discovery. We could also introduce pattern ranking, similar to [32], and add music theory rules, as discussed in Section 4.4. The model's output could further be optimized by supervised training of model parameters, especially the number of layers in the hierarchy and the layers in the model's output. However, a sufficiently large annotated dataset is needed for such an optimization, significantly larger than the datasets currently used to evaluate the pattern discovery task.

The proposed approach can also be applied to identify similar and inexact patterns across larger corpora. We plan on evaluating the model in an inter-opus pattern discovery task, aiding the current research in tune family identification and folk music analysis. To tackle classification tasks, the model can be observed as a feature generator; thus, its output can be employed as an input to tune family analysis, similarity comparison or composer identification.

Author Contributions: M.P., A.L. and M.M. conceived of and designed the experiments. M.P. performed the experiments. M.P. and M.M. analysed the data. M.P., A.L. and M.M. wrote the paper.

Conflicts of Interest: The authors declare no conflict of interest.

Abbreviations

The following abbreviations are used in this manuscript:

CHM   Compositional Hierarchical Model
SymCHM   Compositional Hierarchical model for Symbolic music representations
SymCHMMerge An extension of the SymCHM using a pattern merging technique

References

1. Lerdahl, F.; Jackendoff, R. *A Generative Theory of Tonal Music*; MIT Press: Cambridge, MA, USA, 1983.
2. Hamanaka, M.; Hirata, K.; Tojo, S. Implementing "A Generative Theory of Tonal Music". *J. New Music Res.* **2006**, *35*, 249–277.
3. Hirata, K.; Tojo, S.; Hamanaka, M. Techniques for Implementing the Generative Theory of Tonal Music. In Proceedings of the International Conference on Music Information Retrieval (ISMIR), Vienna, Austria, 23–30 September 2007.
4. Marsden, A. Schenkerian Analysis by Computer: A Proof of Concept. *J. New Music Res.* **2010**, *39*, 269–289.
5. Todd, N. A Model of Expressive Timing in Tonal Music. *Music Percept. Interdiscip. J.* **1985**, *3*, 33–57.
6. Farbood, M. Working memory and the perception of hierarchical tonal structures. In Proceedings of the International Conference of Music Perception and Cognition, Seattle, WA, USA, 23–27 August 2010; pp. 219–222.
7. Balaguer-Ballester, E.; Clark, N.R.; Coath, M.; Krumbholz, K.; Denham, S.L. Understanding Pitch Perception as a Hierarchical Process with Top-Down Modulation. *PLoS Comput. Biol.* **2009**, *4*, 1–15.

8. McDermott, J.H.; Oxenham, A.J. Music perception, pitch and the auditory system. *Curr. Opin. Neurobiol.* **2008**, *18*, 452–463.

9. Humphrey, E.J.; Bello, J.P.; LeCun, Y. Moving beyond feature design: Deep architectures and automatic feature learning in music informatics. In Proceedings of the 13th International Conference on Music Information Retrieval (ISMIR), Porto, Portugal, 8–12 October 2012.

10. Rigaud, F.; Radenen, M. Singing Voice Melody Transcription using Deep Neural Networks. In Proceedings of the International Conference on Music Information Retrieval (ISMIR), New York, NY, USA, 7–11 August 2016; pp. 737–743.

11. Jeong, I.Y.; Lee, K. Learning Temporal Features Using a Deep Neural Network and its Application to Music Genre Classification. In Proceedings of the International Conference on Music Information Retrieval (ISMIR), New York, NY, USA, 7–11 August 2016; pp. 434–440.

12. Schluter, J.; Bock, S. Musical Onset Detection with Convolutional Neural Networks. In Proceedings of the 6th International Workshop on Machine Learning and Music, held in Conjunction with the European Conference on Machine Learning and Principles and Practice of Knowledge Discovery in Databases, ECML/PKDD 2013, Prague, Czech Republic, 23–27 September 2013.

13. Battenberg, E.; Wessel, D. Analyzing Drum Patterns using Conditional Deep Belief Networks. In Proceedings of the International Conference on Music Information Retrieval (ISMIR), Porto, Portugal, 8–12 October 2012; pp. 37–42.

14. Deng, J.; Kwok, Y.K. A Hybrid Gaussian-Hmm-Deep-Learning Approach for Automatic Chord Estimation with Very Large Vocabulary. In Proceedings of the International Conference on Music Information Retrieval (ISMIR), New York, NY, USA, 7–11 August 2016; pp. 812–818.

15. Campilho, A.; Kamel, M. (Eds.) *Image Analysis and Recognition*; Lecture Notes in Computer Science; Springer: Berlin/Heidelberg, Gemany, 2012; Volume 7324.

16. Coward, E.; Drabløs, F. Detecting periodic patterns in biological sequences. *Bioinformatics* **1998**, *14*, 498–507.

17. Margulis, E.H. *On Repeat: How Music Plays the Mind*; Oxford University Press: Oxford, UK, 2014; p. 224.

18. Downie, J.S. The music information retrieval evaluation exchange (2005–2007): A window into music information retrieval research. *Acoust. Sci. Technol.* **2008**, *29*, 247–255.

19. Meredith, D.; Lemstrom, K.; Wiggins, G.A. Algorithms for discovering repeated patterns in multidimensional representations of polyphonic music. *J. New Music Res.* **2002**, *31*, 321–345.

20. The MIREX Discovery of Repeated Themes & Sections Task. Available online: http://www.music-ir.org/mirex/wiki/2015:Discovery_of_Repeated_Themes_%26_Sections (accessed on 19 June 2015)

21. Collins, T.; Thurlow, J.; Laney, R.; Willis, A.; Garthwaite, P.H. A Comparative Evaluation of Algorithms for Discovering Translational Patterns in Baroque Keyboard Works. In Proceedings of the International Conference on Music Information Retrieval (ISMIR), Utrecht, Netherlands, 9–13 August, 2010; pp. 3–8.

22. Wang, C.I.; Hsu, J.; Dubnov, S. Music Pattern Discovery with Variable Markov Oracle: A Unified Approach to Symbolic and Audio Representations. In Proceedings of the International Conference on Music Information Retrieval (ISMIR), Malaga, Spain, 26–30 October 2015; pp. 176–182.

23. Cambouropoulos, E.; Crochemore, M.; Iliopoulos, C.S.; Mohamed, M.; Sagot, M.F. A Pattern Extraction Algorithm for Abstract Melodic Representations that Allow Partial Overlapping of Intervallic Categories. In Proceedings of the International Conference on Music Information Retrieval (ISMIR), London, UK, 11–15 September 2005; pp. 167–174.

24. Conklin, D.; Bergeron, M. Feature Set Patterns in Music. *Comput. Music J.* **2008**, *32*, 60–70.

25. Conklin, D.; Anagnostopoulou, C. Representation and Discovery of Multiple Viewpoint Patterns. In Proceedings of the 2001 International Computer Music Conference, Havana, Cuba, 18–22 September 2001; pp. 479–485.

26. Conklin, D. Melodic analysis with segment classes. *Mach. Learn.* **2006**, *65*, 349–360.

27. Rolland, P.Y. Discovering Patterns in Musical Sequences. *J. New Music Res.* **1999**, *28*, 334–350.

28. Owens, T. *Charlie Parker: Techniques of Improvisation*; Number Let. 1 in Charlie Parker: Techniques of Improvisation; University of California: Los Angeles, CA, USA, 1974.

29. Cambouropoulos, E. Musical Parallelism and Melodic Segmentation. *Music Percept. Interdiscip. J.* **2006**, *23*, doi:10.1525/mp.2006.23.3.249.

30. Meredith, D. Music Analysis and Point-Set Compression. *J. New Music Res.* **2015**, *44*, 245–270.

31. Meredith, D. COSIATEC and SIATECCompress: Pattern Discovery by Geometric Compression. In Proceedings of the International Conference on Music Information Retrieval (ISMIR), Curitiba, Brazil, 4–8 November 2013; pp. 1–6.

32. Velarde, G.; Meredith, D. Submission to MIREX Discovery of Repeated Themes and Sections. In Proceedings of the 10th Annual Music Information Retrieval eXchange (MIREX'14), Taipei, Taiwan, 27–31 October 2014; pp. 1–3.

33. Velarde, G.; Weyde, T.; Meredith, D. An approach to melodic segmentation and classification based on filtering with the Haar-wavelet. *J. New Music Res.* **2013**, *42*, 325–345.

34. Lartillot, O. Submission to MIREX Discovery of Repeated Themes and Sections. In Proceedings of the 10th Annual Music Information Retrieval eXchange (MIREX'14), Taipei, Taiwan, 27–31 October 2014; pp. 1–3.

35. Lartillot, O. In-depth motivic analysis based on multiparametric closed pattern and cyclic sequence mining. In Proceedings of the International Conference on Music Information Retrieval (ISMIR), Taipei, Taiwan, 27–31 October 2014; pp. 361–366.

36. Ren, I.Y. Closed Patterns in Folk Music and Other Genres. In Proceedings of the 6th International Workshop on Folk Music Analysis, Dublin, Ireland, 15–17 June 2016, ; pp. 56–58.

37. Nieto, O.; Farbood, M. MIREX 2014 Entry: Music Segmentation Techniques And Greedy Path Finder Algorithm To Discover Musical Patterns. In Proceedings of the 10th Annual Music Information Retrieval eXchange (MIREX'14), Taipei, Taiwan, 27–31 October 2014; pp. 1–2.

38. Nieto, O.; Farbood, M.M. Identifying Polyphonic Patterns From Audio Recordings Using Music Segmentation Techniques. In Proceedings of the International Conference on Music Information Retrieval (ISMIR), Taipei, Taiwan, 27–31 October 2014; pp. 411–416.

39. Reber, A.S. *Implicit Learning and Tacit Knowledge : An Essay on the Cognitive Unconscious*; Oxford University Press: Oxford, UK, 1993.

40. Pesek, M.; Leonardis, A.; Marolt, M. A compositional hierarchical model for music information retrieval. In Proceedings of the International Conference on Music Information Retrieval (ISMIR), Taipei, Taiwan, 27–31 October 2014; pp. 131–136.

41. Pesek, M.; Leonardis, A.; Marolt, M. Robust Real-Time Music Transcription with a Compositional Hierarchical Model. *PLoS ONE* **2017**, *12*, doi:10.1371/journal.pone.0169411.

42. Fidler, S.; Boben, M.; Leonardis, A. Learning Hierarchical Compositional Representations of Object Structure. In *Object Categorization: Computer and Human Vision Perspectives*; Cambridge University Press: Cambridge, UK, 2009; pp. 196–215.

43. Meredith, D. *Method of Computing the Pitch Names of Notes in MIDI-Like Music Representations*; US Patent US 20040216586 A1, 4 November 2004.

44. Collins, T. *JKU Patterns Development Database*; August 2013. Available online: https://dl.dropbox.com/u/11997856/JKU/JKUPDD-Aug2013.zip (accessed on 13 September 2017)

45. Meredith, D. COSIATEC and SIATECCompress: Pattern Discovery by Geometric Compression. In Proceedings of the 9th Annual Music Information Retrieval eXchange (MIREX'13), Curitiba, Brazil, 4–8 November 2013.

46. Temperley, D. *The Cognition of Basic Musical Structures*; MIT Press: Cambridge, MA. USA, 2001.

applied
sciences

MDPI

Article

Supporting an Object-Oriented Approach to Unit Generator Development: The Csound Plugin Opcode Framework [†]

Victor Lazzarini

Music Department, Maynooth University, Maynooth W23 X021, Ireland; victor.lazzarini@mu.ie;
Tel.: +353-1-708-6936

† This article is a re-written and expanded version of "The Csound Plugin Opcode Framework", SMC 2017, Espoo, Finland, 5 July 2017.

Academic Editor: Stefania Serafin
Received: 31 July 2017; Accepted: 18 September 2017; Published: 21 September 2017

Abstract: This article presents a new framework for unit generator development for Csound, supporting a full object-oriented programming approach. It introduces the concept of unit generators and opcodes, and its centrality with regards to music programming languages in general, and Csound in specific. The layout of an opcode from the perspective of the Csound C-language API is presented, with some outline code examples. This is followed by a discussion which places the unit generator within the object-oriented paradigm and the motivation for a full C++ programming support, which is provided by the Csound Plugin Opcode Framework (CPOF). The design of CPOF is then explored in detail, supported by several opcode examples. The article concludes by discussing two key applications of object-orientation and their respective instances in the Csound code base.

Keywords: computer music languages; musical signal processing; csound; sound synthesis; object-oriented programming; C++; code re-use; opcodes; unit generators

1. Introduction

All modern music programming systems provide means for extending their capabilities [1]. In most cases, this extensibility applies to allowing new *unit generators* (UGs) to be added to the system. A UG is defined as a component of the system programming language responsible for processing input and/or generating some output [2]. These may take the form of control or audio signals, messages of some kind, or single values. UGs are core elements of any MUSIC N-type language and they are are responsible for much of the processing power provided by the system. While some languages allow the user to process signal using only primitive operations, UGs provide in most cases a more convenient and efficient way to implement a given algorithm. The efficiency gain can be of various orders of magnitude, depending on the system and the kind of operations involved [3]. In some cases, UGs are the only means possible to realise a given algorithm, as the language in question is not designed to process signals directly or does not provide the required primitive operations to implement it (on a sample-level basis).

1.1. Csound Unit Generators

Csound [4,5], is a MUSIC N-type system, which consists of an audio engine, a music programming language, and an application programming interface (API). UGs in Csound are more commonly called *opcodes* or *functions*. While the system does not distinguish between these two, we often reserve the latter name for UGs with no side effects, which are pure functions [6]. Structurally, however, there is no distinction between these as far as their implementation layout is concerned, and we will employ the term *opcode* more generally to denote a UG of any kind. Internally, all operations are implemented by opcodes/functions (including all primitive arithmetic and control-of-flow), which places them as a central pillar of the system. Some of these of course are going to be minimal and light-weight, and others can be of significant complexity.

Csound has a collection of internal or built-in opcodes that are compiled into the system, but also provides a mechanism for adding new opcodes [7]. This mechanism is made up by two separate components, namely

1. An interface for registering new opcodes, provided by Csound API.
2. Dynamic library loading is provided by the audio engine at startup. As part of this, a given directory is searched for suitable library files containing new opcodes.

Opcodes are usually written in C, although other languages producing a compatible binary format can be employed. Each opcode will be defined by a dataspace and up to three processing functions that are called by the engine at different times depending on the type of signals that are to be processed. An opcode is invoked by Csound code in units called *instruments* or *user-defined opcodes* (UDOs). The engine will instantiate it by allocating space for its data, and call its processing functions according to one or more of two action times (*passes*):

1. **initialisation time**: code is executed once, after instantiation. It can also be invoked again if a re-initialisation pass is requested.
2. **performance time**: code is executed repeatedly each control cycle to consume and/or produce a signal. There are two separate types of functions that can be defined for performance:

    (a) **control**: this function is invoked to deal with scalar inputs or outputs (e.g., processing a single sample at a time).
    (b) **audio**: code is expected to deal with audio signals, which are passed as vectors to it.

The three processing functions defined by an opcode are each connected to one of the processing cases listed above. They are also linked to the different types of variables supported by the Csound language, which the opcodes will be designed to operate on:

- i-type: these variables are floating-point scalars that can only change at initialisation or re-initialisation time.
- k-type: also floating-point scalars, these will only change at performance time, although they can also be initialised at i-time.
- a-type: this is a floating-point vector, which is modified at performance time. The length of this vector is defined by a ksmps variable that can assume local (instrument) values or can be set globally for the whole engine. Vectors can also be initialised at i-time. Audio-rate functions are designed to operate on these variables.
- S-type: character strings, which can be modified at i-time and perf-time, although it is more common to do so only at i-time.
- f-type: frequency-domain signals (fsigs), these contain self-describing spectral data (of different formats) that are processed at performance-time by control-rate functions.
- arrays: composite-type variables of each of the above types. A very important case is k and i arrays, for which there are various important applications.

An opcode is defined to operate on a given set of input and/or output argument types. Parameters passed to them have to match the pre-defined types. Multiple versions of the same opcode can be declared for different types, which is a case of *overloading*.

1.1.1. Opcode Layout

The Csound API mandates that opcodes should have the following components:

- A data structure declared with the following format. It always contains an OPDS member as its first component. This holds a set of elements common to all opcodes.

```
struct NAME {
OPDS h;
// output argument addresses
// input argument addresses
// dataspace members
};
```

where we need to declare one pointer variable for each output and input argument (in the order they should occur in the Csound language code). When the opcode is called, the output and input argument addresses are passed to Csound through these pointers. The C variable types for different Csound argument types are:

- MYFLT*: pointers to the internal floating-point data type (MYFLT is either defined as a 64 or a 32-bit float, depending on the engine version and platform) are used for all basic numeric arguments. (i, k, or a).
- STRINGDAT*: used for string arguments.
- PVSDAT*: holds an fsig argument
- ARRAYDAT*: for array arguments (of any fundamental type).

- A function with the signature

```
int func(CSOUND *, void *);
```

for each one of the required action times (init, control, and/or audio). The first argument is a pointer to the Csound engine that instantiated this opcode. The second argument receives a pointer to the allocated dataspace for a given instance of the opcode.

New opcodes are registered with the engine using the Csound C API function csoundAppendOpcode(),

```
int csoundAppendOpcode(CSOUND *csound, const char *opname,
                       int dsblksiz, int flags, int thread,
                       const char *outypes, const char *intypes,
                       int (*iopadr)(CSOUND *, void *),
                       int (*kopadr)(CSOUND *, void *),
                       int (*aopadr)(CSOUND *, void *));
```

This takes in the opcode name opname; the size of the dataspace dsblksiz in bytes; multithreading flags (normally 0); a thread code for the action times, which determines whether it should be active on i-time (1), k-rate (2), and/or a-rate (4); and the functions for i-time, k-rate, and a-rate (NULL if not needed), for an engine given by csound. Two strings, outtypes and intypes, define the output and input argument types expected by the opcode. Valid values are the characters i, k, a, S, f, with an added [] to indicate array arguments.

1.1.2. Plugin libraries

As discussed in Section 1.1, the Csound engine has a dynamic library loading mechanism that scans a given directory, the *opcode directory* (which can be defined by the OPCODE6DIR or OPCODE6DIR64 environment variables) and loads any suitable files containing opcodes. Alternatively, Csound can be passed the option --opcode-lib= to load a given plugin library file at startup.

To recognise a dynamic library as containing Csound code, the engine looks for the definition of three functions:

```
csoundModuleCreate(CSOUND *csound);
csoundModuleInit(CSOUND *csound);
csoundModuleDestroy(CSOUND *csound);
```

If these functions exist, the library is loaded and `csoundModuleCreate()` followed by `csoundModuleInit()`. One of these functions should contain the call to the opcode registration API function to append the new UGs to the list kept by the engine.

To enable libraries to be built without the need to link directly to the Csound library, an `AppendOpcode()` function exists in the `CSOUND` structure allowing code to call the API function indirectly via the engine pointer. For example,

```
PUBLIC int csoundModuleInit(CSOUND *csound) {
  csound->AppendOpcode(csound, "test", sizeof(struct OPTEST),
                            0, 1, "i", "i", test_init, NULL, NULL);
  return 0;
}
```

registers an opcode with the name `test`, running at init-time only, implemented by the function `test_init`. This function invokes the exact same code as the Csound API call, but it is more convenient for the purposes of a plugin library.

1.1.3. Discussion

Csound UG development relies on the conventions outlined in Section 1.1.1, plus a comprehensive set of API functions provided in the `CSOUND` structure (including, as we saw in Section 1.1.2, the opcode-registering function `AppendOpcode`). These support a range of facilities and access to the engine that allow a complete scope for the development of new unit generators. However, it is the case that this interface can be more complex and cumbersome than necessary, owing to the characteristics of the C language.

From an object-oriented (OO) perspective [8], we can observe that an opcode might be described as a class that inherits from `OPDS`. It will contain a variable number of argument objects $(0 - N)$, depending on its outputs and inputs. An opcode class can define up to three public methods, one for each action time required. These methods take as a parameter the Csound engine object, which has a large number of public methods that can be used for a variety of means: to retrieve engine attributes; to print messages; to handle MIDI data; to list and retrieve lists of arguments; to perform memory allocation and management; to access function tables; to perform FFTs and handle frequency-domain data; to access disk files; to generate random numbers; to spawn threads, manage locks and circular buffers; plus a number of other miscellaneous operations.

The Csound engine is responsible for constructing opcodes at instantiation. Effectively, this entails only the allocation of space for its data members, if no pre-allocated space exists already. Any further initialisation, if required, needs to be performed by the init-time method. When an instance of an opcode is no longer active, there is no automatic recovery of the memory space, and so new instances can take advantage of pre-existing space and skip allocation. However, memory can be recovered at certain stages explicitly if needed. Opcodes can also optionally register a destructor method with the engine, that will be called when an instance is no longer active.

From this analysis, we conclude that the structure of an opcode effectively takes an object-oriented form as far as it is possible under C. However, the language is not conducive to the application of techniques that would maximise code re-use and encapsulation, which would allow developers to concentrate directly on the implementation of their algorithms. Code re-use not only helps to save work in reproducing boilerplate lines, but also supports a more robust development, where a given functionality is implemented once and only once facilitating the task of ensuring correctness in the code. Encapsulation allows for certain fundamental components to be hidden away and manipulated through a logic interface, which greatly supports developers in concentrating on the task at hand.

An ideal candidate language to support this type of approach is C++ in its more modern incarnations [9]. Unlike other languages such as Java, it is possible to compile code to a binary form that can be taken directly by the Csound dynamic loading mechanism, and it does not require a virtual machine environment to run. It is also very closely integrated with C, especially if we regard it as its superset. Differently to Objective-C, there is no major syntactic chasm between C and C++ code, and its new constructs have evolved organically as extensions to the original language.

1.2. Unit Generators and Plugins in Other Systems

As outlined in Section 1.1.3, the original Csound opcode C API imposes effectively a C-based OOP idiom, with limitations determined by the choice of language. A similar arrangement for UG development is present under Pure Data (PD) [10], where an essentially object-oriented structure is implemented in C. The API for plugins provides functions to register a new class (an UG in PD), as well as to register methods to respond to various messages that the system provides. Due to this message-passing nature of the PD engine, it is slightly more complex to add a new UG than it is in Csound. However, there are a number of similarities, including the presence of an audio processing method (corresponding to the performance-time audio function in Csound) and an object constructor (with similarities) to the init-time function. New classes are registered with the system in a similar way to Csound, in a given plugin entry point function. Thus, a C++ treatment similar to what is described for Csound in Sections 2–4 is also possible for PD.

The case of SuperCollider [11] UGs (ugens) is a little more mixed. The synthesis engine is written in C++, adopting as we would expect, a full OO approach. However, the ugen API uses a mostly C-oriented approach (not unlike Csound), with heavy use of macro substitution and depending on C linkage (`extern "C"`) [12]. This is highly surprising, since we would expect that it could use a more up-to-date idiom, enabled as it is by the C++ language. Unlike Csound or PD, there would be no limitations as to what the engine might be able to handle, since there is no language barrier to speak of. Additionally, new ugen registration requires a class definition in the SuperCollider language to be provided separately, which should match the C++ code. This is, of course, not needed either in Csound or PD, and it is an aspect that does not compare very well with these systems. However, in general, the OOP approach explored in this paper could also potentially be implemented to aid SuperCollider ugen development, without some of constraints imposed by the C language as discussed in Section 2.1.

Finally, as an example of a fuller use of C++ for plugin development, we have the VST framework [13]. Under this model, we have the use of a C++ class inheritance mechanism to define new plugins, and a simple plugin registration process based on a single function call at the plugin entry point. Equally, there are programming libraries that are implemented in C++ and employ an OO approach, such as STK [14] and SndObj [15], which are based on earlier standards of the language, and AuLib [16], which takes advantage of the latest, C++14 [17]. However, VST(and audio programming libraries in general) is not fully comparable to Csound (PD or SuperCollider) as a system. Thus, we can conclude that an OOP C idiom predominates in UG development for the most important music programming environments. A more modern approach might be more conducive to better programming practices that can in particular support the implementation of signal processing code. The achievement of this result is one of the main objectives of the present work.

2. The Framework

In order to support object-oriented programming (OOP) for unit generator development, a new framework development is developed from the ground up. The main objective of the work to provide an environment that is conducive to modern programming practices discussed in Section 1.1.3. The remainder of this paper will concentrate on describing the design and implementation of the Csound Plugin Opcode Framework (CPOF (To be pronounced *see-pough* or *cipó*, *vine* in Portuguese, appropriated from the tupi-guarani word meaning "the hand of the branch")) [18], and the discussion of results. This framework provides an alternative for opcode programming that attempts to be thin, simple, complete, and that handles internal Csound resources in a safe way (using the same mechanisms provided by the underlying C interface).

An object-oriented *framework* is a type of library or API that supports the development of new classes through inheritance/sub-classing. This is set in contrast to a *toolkit* where existing classes are expected to be used through delegation or composition [19]. In this sense, CPOF is a framework with a small associated toolkit of support classes that encapsulate a number of key operations.

2.1. Design Fundamentals

The conditions in which the framework is supposed to work constrain significantly the scope of what is possible. In particular,

1. The main Csound engine is written in C. As we have noted, it instantiates opcodes and makes calls to opcode processing functions, but it does not support any C++ facilities.
2. Polymorphism [20] via virtual functions [21] is not possible (due to 1) since C does not provide dynamic dispatch. All function calls have to be resolved at compile time.
3. The process of registering an opcode with the engine requires that processing functions are defined as static. As we have seen, up to three different functions should be registered (for different action times).
4. In C, the sub-classing of the base structure (OPDS) is achieved by adding this as its first member variable.

In relation to the last item above, we will assume a similar behaviour in C++. While defining an opcode base class for the framework, we have the practical expectation that all C++ compilers place subclass members contiguously to their superclass object in memory. Although this layout is not imposed by the C++ standard, it is the standard practice [22]. CPOF assumes then that the following code

```
struct OPCD {
  OPDS h;
};
```

is binary equivalent to

```
struct OPCD : OPDS {
};
```

The absence of a virtual function mechanism for overriding base class methods can be overcome with different compile time strategies. One of the possibilities for designing a framework based on polymorphism without the use of dynamic binding is to employ a method called *curiously recurring template pattern* (CRTP) [23].

However, we can do better with a much simpler approach. Given the constraints in which the opcode classes are meant to operate, there is no need for a compile-time mimicking of the virtual-function mechanism. This is because it is not our objective to have a general purpose framework for C++ programs, where users would be instantiating their own objects and possibly using generic pointers and references that need to bind to the correct override.

Here the scope is much narrower: the developer supplies the code, but will not call it directly (the functions are effectively callbacks). Csound does the instantiation and the calls, so we can make the decision at compile time just by providing functions that *hide* rather than *override* (in the dynamic binding sense) the base class ones. In this case, hiding plays the same role as overriding, there is in practice no distinction between the two. A plugin base class can be defined from which we will inherit to create the actual opcodes. This class will inherit from OPDS (which is opaque to CPOF) and provide some extra members that are commonly used by all opcodes. It will also provide stub methods for the processing functions, which then can be *specialised* (hidden, in reality) by derived class methods. An initial design for an opcode base class would thus be

```
struct Plugin : OPDS {
  // dataspace
  ...
  // stub methods
  int init() { return OK; }
  int kperf() { return OK; }
  int aperf() { return OK; }
};
```

from which we would inherit our own class to implement the new opcode:

```
struct MyClass: Plugin {
...
};
```

Given that in any practical applications they will not ever be called, it would seem that these stubs are surplus to requirements. However, having these allows a considerable simplification in the plugin registration process. We can just register any plugin in the same way, even if it does only provide one or two of the required processing functions. The stubs play an important role to keep the compiler happy in this scheme, even if Csound will not take any notice of them.

This mechanism requires that we provide function templates for opcode registration. These get instantiated with exactly the derived types and are used to glue the C++ code into Csound. Each one of them is minimal: consisting just of a call to the instantiated class processing function:

```
template <typename T> int init(CSOUND *csound, T *p) {
  p->csound = (Csound *)csound;
  return p->init();
}

template <typename T> int kperf(CSOUND *csound, T *p) {
  p->csound = (Csound *)csound;
  return p->kperf();
}

template <typename T> int aperf(CSOUND *csound, T *p) {
  p->csound = (Csound *)csound;
  p->sa_offset();
  return p->aperf();
}
```

In this case, T is our derived class, which implements the new opcode. Registration can then be also implemented using a template function

```
template <typename T>
int plugin(CSOUND *cs, const char *name, const char *oargs,
           const char *iargs, uint32_t thread, uint32_t flags = 0) {
  return cs->AppendOpcode(cs, (char *)name, sizeof(T), flags, thread,
                          (char *)oargs, (char *)iargs, (SUBR)init<T>,
                          (SUBR)kperf<T>, (SUBR)aperf<T>);
}
```

In this design, a class can be registered by instantiating and invoking the template function,

```
Plugin<MyOpcode>(...);
```

This call will be resolved at compile time with the requested class argument (`MyOpcode`). When running, Csound calls the template functions for processing, which in their turn delegate directly to the ones defined in the opcode class in question (If they are not defined there, the call will default to the non-op stub). Note that this is all hidden from the framework user (in the header file `plugin.h`), who only needs to derive her classes and register them. As we will see in the following sections, this scheme enables significant economy, re-use and reduction in code verbosity (one of the issues with CRTP).

2.2. Opcode Arguments

To allow for a flexible handling of opcode output and input arguments, we can refine our earlier definition of an opcode base class. It is possible to take advantage of *non-type* (numeric) template variables to define the number of arguments for a class:

```
template <uint32_t N, uint32_t M> struct Plugin : OPDS { ... };
```

where N and M will define how many outputs and inputs an opcode will take, respectively, which are defined by the class declaration (template instantiation).

As we have seen in Section 1.1.1, opcode arguments can be of different pointer types, depending on the Csound variable types required. The `Param` class in CPOF is employed to encapsulate these, providing a general interface to arguments:

```
template <uint32_t N> class Param {
  MYFLT *ptrs[N];
  ...
};
```

In the base class, we declare two of these objects, `outargs` and `inargs` as its first two members (Figure 1):

```
template <uint32_t N, uint32_t M> struct Plugin : OPDS {
  Param<N> outargs;
  Param<M> inargs;
  ...
};
```

Note that this ensures a complete binary compatibility between the C-structure form of an opcode dataspace and the CPOF template base class (assuming the standard layout discussed in Section 2.1 is obeyed by the C++ compiler).

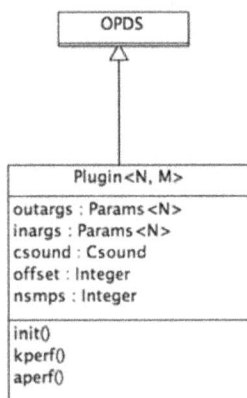

Figure 1. The Plugin template base class, derived from the opaque C structure OPDS.

2.3. The Base Class

In summary, this re-definition of the framework base class makes it a class template that needs to be instantiated by user code. To create a new opcode, we derive our own class by declaring the required number of output and inputs needs as template arguments (CPOF code uses the csnd namespace and is declared in the plugin.h header file):

```
#include <plugin.h>
struct MyPlug : csnd::Plugin<1,1> { };
```

The above lines will create a plugin opcode with one output (first template argument) and one input (second template argument). This class defines a complete opcode, but since it is only using the base class stubs, it is also fully non-op. It will inherit the following members from Plugin:

- outargs: a Params object holding output arguments.
- inargs: input arguments (Params).
- csound: a pointer to the Csound engine object.
- offset: the starting position of an audio vector (for audio opcodes only).
- nsmps: the size of an audio vector (also for audio opcodes only).
- init(), kperf() and aperf() non-op methods, to be reimplemented as needed.
- out_count() and $in_count()$: these functions return the number of arguments for output and input, respectively. They are useful for opcodes with variable number of arguments.
- sa_offset(): this method calculates the correct values for offset and nsmps. User called does not need to invoke it, as it is called implicitly by the aperf() template function before it delegates to the plugin code.

As we have outlined in Section 1.1.1, Csound has two basic passes for opcodes: init and perf-time. The former runs a processing routine once per instrument instantiation (and/or once again if a re-init is called for). Code for this is placed in the Plugin class init() function. Perf-time code runs in a loop and is called once every control (k-)cycle (also known as k-period). The other class methods kperf() and aperf() are called in this loop, for control (scalar) and audio (vectorial) processing. The following examples demonstrate the derivation of plugin classes for each one of these opcode types (i, k or a). Note that k and a opcodes can also use i-time functions if they require some sort of initialisation.

2.3.1. Initialisation-time Opcodes

For init-time opcodes, all we need to do is provide an implementation of the `init()` method, as shown in Listing 1.

Listing 1: `i`-time opcode example

```
struct Simplei : csnd::Plugin<1,1> {
  int init() {
    outargs[0] = inargs[0];
    return OK;
  }
};
```

In this simple example, we just copy the input arguments to the output once, at init-time. Each scalar input type can be accessed using array indexing. All numeric argument data is real, declared as `MYFLT`, which, as we have seen, is the internal floating-point type used in Csound.

2.3.2. Control-rate Opcodes

For opcodes running only at k-rate (no init-time operation), all we need to do is provide an implementation of the `kperf()` method, demonstrated by the code in Listing 2.

Listing 2: `k`-rate opcode example

```
struct Simplek : csnd::Plugin<1,1> {
  int kperf() {
    outargs[0] = inargs[0];
    return OK;
  }
};
```

Similarly, in this simple example, we just copy the input arguments to the output at each k-period.

2.3.3. Audio-Rate Opcodes

For opcodes running only at a-rate (and with no init-time operation), we need to do provide an implementation of the `aperf()` method to process an audio vector (Listing 3).

Listing 3: a-rate opcode example

```
struct Simplea : csnd::Plugin<1,1> {
 int aperf() {
  std::copy(inargs(0)+offset, inargs(0)+nsmps, outargs(0));
  return OK;
 }
};
```

Because audio arguments are `nsmps`-size vectors, we get these using the overloaded `operator()` for the `inargs` and `outargs` objects, which takes the argument number as input and returns a `MYFLT` pointer to the vector.

2.4. Registering Opcodes with Csound

We have discussed in Section 2.1 how the opcode registration mechanism is implement through the CPOF function template `plugin()`. In order to use it, we just have to instantiate and invoke it with the required parameters. It signature is:

```
template <typename T>
```

```
int plugin(Csound *csound,
           const char *name,
           const char *oargs,
           const char *iargs,
           uint32_t thread,
           uint32_t flags = 0)
```

where we have the following arguments:

- `csound`: a pointer to the `Csound` object to which we want to register our opcode.
- `name`: the opcode name as it will be used in Csound code.
- `oargs`: a string containing the output argument types, one identifier per argument.
- `iargs`: a string containing the input argument types, one identifier per argument.
- `thread`: a code to tell Csound when the opcode should be active.
- `flags`: multithread flags (generally 0 unless the opcode accesses global resources).

For the argument type identifiers, we have seen in Section 1.1.1 that the most common types are: a (audio), k (control), i (i-time), S (string) and f (fsig). The *thread* argument, which defines what methods will be called by the opcode, can be defined by the following constants:

- `thread::i`: indicates `init()`.
- `thread::k`: indicates `kperf()`.
- `thread::ik`: indicates `init()` and `kperf()`.
- `thread::a`: indicates `aperf()`.
- `thread::ia`: indicates `init()` and `aperf()`.
- `thread::ika`: indicates `init()`, `kperf()` and `aperf()`.

CPOF supports the symbol `on_load()` as its entry point (Declared in the header file `modload.h`). This function needs to implemented only **once** per plugin library, and it should contain the calls to one registration function for each opcode to be added to the engine. For example, the three opcodes defined in Section 2.3 can be registered as shown in Listing 4.

Listing 4: Registering opcodes with Csound

```
#include <modload.h>
void csnd::on_load(Csound *csound){
  csnd::plugin<Simplei>(csound, "simple", "i", "i",  csnd::thread::i);
  csnd::plugin<Simplek>(csound, "simple", "k", "k", csnd::thread::k);
  csnd::plugin<Simplea>(csound, "simple", "a", "a", csnd::thread::a);
  return CSOUND_OK;
}
```

These calls will register the `simple` *polymorphic* opcode, which can be used with i-, k- and a-rate variables. In each instantiation of the plugin registration template, the class name is passed as an argument to it, followed by the function call. If the class defines two specific static members, `otypes` and `itypes`, to hold the types for output and input arguments, declared as

```
struct MyPlug : csnd::Plugin<1,2> {
  static constexpr char const *otypes = "k";
  static constexpr char const *itypes = "ki";
  ...
};
```

then we can use a simpler overload of the plugin registration function:

```
template <typename T>
int plugin(Csound *csound,
           const char *name,
           uint32_t thread,
           uint32_t flags = 0)
```

For some classes, this might be a very convenient way to define the argument types. For other cases, where opcode polymorphism might be involved, we might re-use the same class for different argument types, in which case it is not desirable to define these statically in a class.

2.5. Constructing and Destroying Member Variables

As opcode classes are instantiated by the Csound engine through C-language code, constructors for member variables are not invoked automatically. For member variables of non-trivial types, this may pose an issue, especially if there are specific initialisation steps to be performed at construction. All classes in the toolkit have no such requirements and do not declare constructors, but external code from other libraries might do. For these, CPOF provides a mechanism to call the member variable constructor explicitly. This is based on the use of a placement new via a function template, which is used to access the class constructor for an object:

```
template <typename T, typename ... Types>
T *constr(T* p, Types ... args){
      return new(p) T(args ...);
}
```

As an example, consider an object of type A called obj, which needs to be constructed explicitly, using

```
A::A(int, float) { ... };
```

To invoke it, we should place the following call in the init() method of our opcode class:

```
csnd::constr(&obj, 10, 10.f);
```

where the arguments are the variable address, followed by any class constructor parameters. Again, given that the compiler knows that obj is of type A, it resolves the template without the need for an explicit type instantiation.

It is also important to note that if the object constructed in this form allocates any resources dynamically, we will need to free these. For this we are required to call the object destructor explicitly by using another function template, defined in CPOF as

```
template<typename T> void destr(T *p) {
  p->T::~T();
}
```

The call to the destructor should be issued at the point where we no longer need the object. For opcodes that run at perf-time, this is normally done in the opcoe deinit() method. For example, to clean up a member variable obj, we implement the following code:

```
int deinit() {
    csnd::destr(&obj);
    return OK;
}
```

3. The Engine Object

As noted in Section 1.1.1, the Csound API provides a large range of facilities to opcodes through several functions provided in the CSOUND data structure, which is made opaque to CPOF. In fact, since the API is designed to cater for a wider variety of applications beyond extending the language, many of these functions are not designed for use in UGs. For this reason, to provide a clearer interface for opcode programming, CPOF encapsulates the engine into an object that exposes only the relevant methods to the user.

All opcodes are run inside the Csound engine, represented by the `Csound` class (Figure 2). As we have seen above, the `Plugin` class holds a pointer of this type, which can be used to access the various utility methods provided by the engine. The following are the public methods of the Csound class in each category:

- Messaging:

    - `init_error()`: takes a string message and signals an initialisation error.
    - `perf_error()`: takes a string message, an instrument instance and signals a performance error.
    - `warning()`: warning messages.
    - `message()`: information messages.

- System parameters:

    - `sr()`: returns engine sampling rate.
    - `_0dbfs()`: returns max amplitude reference.
    - `_A4()`: returns A4 pitch reference.
    - `nchnls()`: return number of output channels for the engine.
    - `nchnls_i()`: same, for input channel numbers.
    - `current_time_samples()`: current engine time in samples.
    - `current_time_seconds()`: current engine time in seconds.
    - `is_asig()`: check for an audio signal argument.

- MIDI data access:

    - `midi_channel()`: midi channel assigned to this instrument.
    - `midi_note_num()`: midi note number (if the instrument was instantiated with a MIDI NOTE ON).
    - `midi_note_vel()`: same, for velocity.
    - `midi_chn_aftertouch()`: channel aftertouch.
    - `midi_chn_polytouch()`: polyphonic aftertouch.
    - `midi_chn_ctl()`: continuous control value.
    - `midi_chn_pitchbend()`: pitch bend data.
    - `midi_chn_list()`: list of active notes for this channel.

- FFT:

    - `fft_setup()`: FFT operation setup.
    - `rfft()`: real-to-complex, complex-to-real FFT.
    - `fft()`: complex-to-complex FFT.

- Memory allocation (Csound-managed heap):

    - `malloc()`: malloc-style memory allocation.
    - `calloc()`: calloc-style memory allocation.
    - `realloc()`: realloc-style memory allocation.
    - `strdup()`: string duplication.
    - `free()`: memory deallocation.

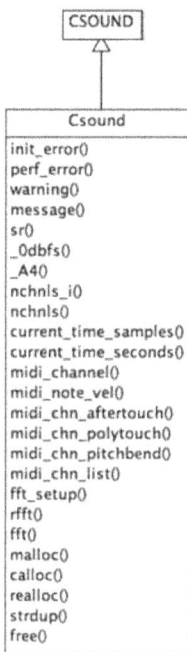

Figure 2. The Csound engine class, derived from the opaque C structure CSOUND.

In addition to these, the Csound class also holds a deinit method registration function template that can be used by opcodes to implement housekeeping tasks.

```
template <typename T> void plugin_deinit(T *p);
```

This is only required if the Plugin-derived class has allocated extra resources using mechanisms that require explicit clean-up. It is not need in most cases, as we will see in our examples. To use it, the plugin needs to declare and implement a deinit() method and then call the plugin_deinit() method passing itself (through a this pointer) in its own init() function:

```
csound->plugin_deinit(this);
```

Because of the presence of the opcode object as an argument, the compiler resolves the template instantiation without requiring an explicit template parameter.

4. Toolkit Classes

Plugins developed with CPOF can avail of a number of helper classes that compose its *toolkit*. These include the aforementioned Params class, as well as classes for resource allocation, input/output access/manipulation, threads, and support for constructing objects allocated in Csound's heap.

4.1. Parameters

As we have already discussed in Section 2.2, parameters passed to a Csound opcode instance are encapsulated by the Params class so that they can be conveniently accessed. The class has the following public methods:

- operator[](): array-style access to numeric (scalar) parameter values.
- begin(), cbegin(): begin iterators for the parameter list.

- `end()`, `cend()`: end iterators.
- `iterator` and `const_iterator`: iterator types for this class.
- `operator() ()`: function-style access to numeric (vector) parameter pointers.
- `data()`: same as the function operator, access to the parameter address.
- `str_data()`: access to parameter as a `STRINGDAT` reference (see Section 4.5).
- `fsig_data()`: access to parameter as a `Fsig` reference (fsig data class, see Section 5).
- `vector_data()`: access to parameter as a `Vector<T>` reference (Csound 1-D numeric array data, see Section 4.6).
- `myfltvec_data()`: access to parameter as a `myfltvec` reference (Csound 1-D numeric array, see Section 4.6).

As we can see, this is is a thin wrapper over the argument pointers, which translates between the original `MYFLT*` and the various argument types, and allows for iteration over the parameter lists.

4.2. Audio Signals

As outlined in Section 1.1.1, audio signal variables are vectors of `nsmps` samples and we can access them through raw `MYFLT` pointers from input and output parameters. While this works well in a C-language environment, it is possible to provide a better object-oriented support to the manipulation of vectors through encapsulation. The `AudioSig` class wraps audio signal vectors conveniently, providing iterators and subscript access:

- `operator[] ()`: array-style access to individual samples.
- `begin()`, `cbegin()`: begin iterators for the audio vector.
- `end()`, `cend()`: end iterators.
- `iterator` and `const_iterator`: iterator types for this class.
- `operator() ()`: function-style access to numeric (vector) parameter pointers.

Objects are constructed by passing the current plugin pointer (`this`) and the raw parameter pointer. The final parameter is flag for an optional resetting of the audio signal vector:

```
AudioSig(OPDS *p, MYFLT *s, bool res = false);
```

With this, we can re-write the `simple` audio example opcode to use this class and its iterators in a typical C++ idiom, as demonstrated in Listing 5.

Listing 5: a-rate opcode example, using `AudioSig` objects

```
struct Simplea : csnd::Plugin<1,1> {
  int aperf() {
    csnd::AudioSig in(this, inargs(0));
    csnd::AudioSig out(this, outargs(0));
    std::copy(in.begin(), in.end(), out.begin());
    return OK;
  }
};
```

4.3. Memory Allocation

Csound provides its own managed heap for dynamic memory allocation. The engine provides mechanisms to allocate space as required. This ensures that there are no leaks and that there is an efficient use of resources. When an opcode requires a certain amount of space that is not known at compile time, it can avail of this mechanism to get access to it.

It is not advisable for developers to employ any other memory allocation methods. In C, this means that standard library functions `malloc` etc should be avoided. In C++, we should avoid to use the new operator (`new`) and containers that employ it (for instance, `std::vector`). They are more difficult to integrate and use properly in this environment, especially given the fact that we are operating under in a hybrid space that includes a C platform supporting the C++ code.

The main mechanism for memory allocation in opcodes is provided by the `AuxAlloc()` function in the Csound API. This is encapsulated by the helper class `AuxMem` in CPOF, which allows an object-oriented approach to memory manipulation. This class has the following methods:

- `allocate()`: allocates new memory whenever required.
- `operator[]`: array-subscript access to the allocated memory.
- `data()`: returns a pointer to the data.
- `len()`: returns the length of the vector.
- `begin()`, `cbegin()`: begin iterators to the data memory.
- `end()`, `cend()`: end iterators.
- `iterator` and `const_iterator`: iterator types for this class.

In Listing 6, the `DelayLine` class implements a simple comb filter [24] with three parameters (audio input, i-time delay time, and k-rate feedback amount). It demonstrates the use of the `AuxMem` template class, which holds the delay memory for the opcode.

Listing 6: Delay line opcode example

```
struct DelayLine : csnd::Plugin<1,3> {
 static constexpr char const *otypes = "a";
 static constexpr char const *itypes = "aik";

 csnd::AuxMem<MYFLT> delay;
 csnd::AuxMem<MYFLT>::iterator iter;

 int init() {
  delay.allocate(csound, csound->sr()*inargs[1]);
  iter = delay.begin();
  return OK;
 }

 int aperf() {
  csnd::AudioSig in(this, inargs(0));
  csnd::AudioSig out(this, outargs(0));
  MYFLT g = inargs[2];

  std::transform(in.begin() ,in.end(), out.begin(), [this](MYFLT s) {
                 MYFLT o = *iter;
                 *iter = s + o*g;
                 if(++iter == delay.end()) iter = delay.begin();
                 return o;
                } );
  return OK;
 }
};
```

In this example, we use an `AuxMem` iterator to access the delay vector, in a typical C++ idiom. The delay line access is implemented via a lambda that captures he opcode dataspace and processes every sample of the input producing the output vector. While this example uses an iterator for convenience, it is equally possible to access each element with an array-style subscript. The memory allocated by this class is managed by Csound, so we do not need to be concerned about disposing of it. To register this opcode, we can use

```
csnd::plugin<DelayLine>(csound, "delayline", csnd::thread::ia);
```

because the output and input types have already been declared in the class as the compile-time constants `itypes` and `otypes`.

In addition to the automatic `AuxMen` mechanism, Csound also offers the more conventional `malloc`-style allocation. This is not generally used in opcode development, but it is accessible via methods of the `Csound` engine object. It requires explicit de-allocation of resources on clean-up.

4.4. Function Table Access

Function tables are used by Csound for a variety of applications. They hold an array of floating-point numbers that is normally created by one of the GEN routines offered by the system [4]. These can generate tables based on trigonometric functions, polynomials, envelopes, windows, and other various mathematical formulae. Function tables are essential for many opcodes, such as oscillators, granular generators, waveshapers, and various different types of processors.

The Csound C API offers access to function tables via a FUNC structure. In CPOF, this access is facilitated by a thin wrapper class that allows us to treat it as a vector object. This is provided by the Table class:

- `init()`: initialises a table object from an opcode argument pointer.
- `operator[]`: array-subscript access to the function table.
- `data()`: returns a pointer to the function table data.
- `len()`: returns the length of the table (excluding guard point).
- `begin()`, `cbegin()`: iterators to the beginning of the function table.
- `end()`, `cend()`: end iterators.
- `iterator` and `const_iterator`: iterator types for this class.

A typical usage example is given by the table-lookup oscillator algorithm [25]. In listing 7, the `Oscillator` class implements truncating lookup using a C++11 range-for facility.

Listing 7: Table-lookup oscillator opcode example

```
struct Oscillator : csnd::Plugin<1,3> {
  static constexpr char const *otypes = "a";
  static constexpr char const *itypes = "kki";
  csnd::Table tab;
  double scl;
  double x;

  int init() {
    tab.init(csound,inargs(2));
    scl = tab.len()/csound->sr();
    x = 0;
    return OK;
  }

  int aperf() {
    csnd::AudioSig out(this, outargs(0));
    MYFLT amp = inargs[0];
    MYFLT si = inargs[1] * scl;
    double ph = x;

    for(auto &s : out) {
      s = amp * tab[(uint32_t) ph];
```

```
      ph += si;
      while (ph < 0) ph += tab.len();
      while (ph >= tab.len()) ph -= tab.len();
    }
    x = ph;
    return OK;
  }
};
```

A table is initialised by passing the relevant argument pointer to it (using its `data()` method). This will hold the function table number that is passed to opcode. In this example, as we need a precise index value, it is more convenient to use array index access instead of iterators, although these are also available in the class. This opcode is registered by

```
csnd::plugin<Oscillator>(csound, "oscillator", csnd::thread::ia);
```

4.5. String Types

String variables in Csound are held in a `STRINGDAT` data structure, containing two members, a pointer to the address holding the zero-terminated string, and a size:

```
typedef struct {
    char *data;
    int size;
} STRINGDAT;
```

The size parameter contains the total space currently allocated for the string, which might be larger than the actual string. While CPOF does not wrap strings, it provides a translated access to string arguments through the argument objects `str_data()` function. This takes an argument index (similarly to `data()`) and returns a reference to the string variable, as demonstrated in this example:

```
struct Tprint : csnd::Plugin<0,1> {
  static constexpr char const *otypes = "";
  static constexpr char const *itypes = "S";
  int init() {
    char *s = inargs.str_data(0).data;
    csound->message(s);
    return OK;
  }
};
```

This opcode will print the string to the console. Note that we have no output arguments, so we set the first template parameter to 0. We register it using

```
csnd::plugin<Tprint>(csound, "tprint", csnd::thread::i);
```

4.6. Array Variables

Opcodes with array inputs or outputs use the data structure `ARRAYDAT` for parameters. Again, in order to facilitate access to these argument types, CPOF provides a template class, `Vector<typename T>`. This currently supports only one-dimensional arrays directly, but can be used with all basic Csound variable types.

This container class is derived from `ARRAYDAT` and wraps an array argument of a type defined by its template parameter. Input variables of these types are already properly initialised, but outputs need to be initialised with a given array size. The class has the following members:

- `init()`: initialises an output variable.
- `operator[]`: array-subscript access to the vector data.
- `data()`: returns a pointer to the vector data.
- `len()`: returns the length of the vector.
- `begin()`, `cbegin()`: iterators to the beginning and end of the vector.
- `end()`, `cend()`: end iterators.
- `iterator` and `const_iterator`: iterator types for this class.
- `data_array()`: returns a pointer to the vector data address.

In addition to this, the `inargs` and `outargs` objects in the `Plugin` class have a template method that can be used to get a `Vector` class reference. A trivial example is shown in Listing 8, implementing both i-time and k-rate array operations.

Listing 8: i-time and k-rate array opcode example

```
struct SimpleArray : csnd::Plugin<1, 1>{
  int init() {
    csnd::Vector<MYFLT> &out = outargs.vector_data<MYFLT>(0);
    csnd::Vector<MYFLT> &in = inargs.vector_data<MYFLT>(0);
    out.init(csound, in.len());
    std::copy(in.begin(), in.end(), out.begin());
    return OK;
  }

  int kperf() {
    csnd::Vector<MYFLT> &out = outargs.vector_data<MYFLT>(0);
    csnd::Vector<MYFLT> &in = inargs.vector_data<MYFLT>(0);
    std::copy(in.begin(), in.end(), out.begin());
    return OK;
  }
};
```

This opcode is registered using

```
csnd::plugin<SimpleArray>(csound, "simple", "i[]", "i[]", csnd::thread::i);
```

for i-time operation and

```
csnd::plugin<SimpleArray>(csound, "simple", "k[]", "k[]", csnd::thread::ik);
```

for perf-time processing. This is an example of an *overloaded* opcode, that can operate on i and k-type variables. To facilitate the manipulation of this more common type of array, based on MYFLT, CPOF defines the following type:

```
typedef Vector<MYFLT> myfltvec;
```

5. Streaming Spectral Processing

Csound has a very useful mechanism for *streaming* (as opposed to *batch*) spectral processing, which is based on its fsig data type [26]. This is a self-describing type, which holds a frame of spectral data in one of several formats, defined by the C data structure PVSDAT (This C structure is provided here for reference only, it is opaque to CPOF):

```
typedef struct pvsdat {
    int32   N;
    int32   sliding;
    int32   NB;
```

```
    int32   overlap;
    int32   winsize;
    int32    wintype;
    int32   format;
    uint32  framecount;
    AUXCH   frame;
} PVSDAT;
```

This structure contains information on the DFT size used in the analysis, the window and hop sizes, the window type, the data format, and the current frame count. It also sets the analysis mode (normal or sample-by-sample sliding, where the actual hopsize is 1). The actual frame data is stored in an AUXCH memory structure, which is managed by Csound.

An opcode implementing fsig processing will operate nominally at the control rate, but will actually compute new frames at a rate determined by the analysis hopsize. This is implemented by checking the fsig framecount and only proceeding to consume and produce new frames if this is greater than the opcode internal frame counter. For streaming spectral processing opcodes, CPOF provides a separate base class, FPlugin, derived from Plugin, with an extra member variable, framecount, used for this purpose.

To facilitate fsig manipulation, the toolkit provides the Fsig class, derived from PVSDAT, with the following methods:

- init(): initialises an fsig. There are two overloads: it is possible to initialise it from individual parameters (DFT size, hop size, etc.) or directly from another fsig. Initialisation is only needed for output variables.
- dft_size(): DFT size.
- hop_size(): hopsize.
- win_size(): window size.
- win_type(): returns the window type (Hamming = 0, von Hann =1, Kaiser = 2, custom = 3, Blackman = 4 and 5, Nutall = 6, Blackman-Harris = 7 and 8, rectangular = 9).
- nbins(): number of bins.
- count(): current frame count.
- isSliding(): checks for sliding mode.
- fsig_format(): returns the data format. This will vary depending on the source. To facilitate identification, CPOF defines the following constants:

  - fsig_format::pvs: standard phase vocoder frame, composed of bins containing amplitude and frequency pairs.There are $N/2 + 1$ bins (N is the DFT frame size), equally spaced between between 0 Hz and the Nyquist frequency (inclusive).
  - fsig_format::polar: similar to the pvs type, except that bins contain pairs of magnitude and phase data.
  - fsig_format::complex: as above, but with bins holding complex-format data (real, imaginary).
  - fsig_format::tracks: this format holds tracks of amplitude, frequency, phase, and track ID (used in partial tracking opcodes).

Phase Vocoder Data

The most common format used in Csound opcodes is phase vocoder amplitude-frequency [25,27]. To provide a convinient access access to bins, a container interface is provided by pv_frame (spv_frame for the sliding mode (pv_frame is a convenience typedef for Pvframe<pv_bin>), whereas spv_frame is Pvframe<spv_bin>). This is a class derived from Fsig that can hold a series of pv_bin (spv_bin for sliding (pv_bin is Pvbin<float> and spv_bin is Pvbin<MYFLT>)) objects, which have the following methods (Figure 3):

- amp(): returns the bin amplitude.

- `freq()`: returns the bin frequency.
- `amp(float a)`: sets the bin amplitude to a.
- `freq(float f)`: sets the bin frequency to f.
- `operator*(pv_bin f)`: multiply the amp of a pvs bin by `f.amp`.
- `operator*(MYFLT f)`: multiply the bin amp by `f`
- `operator*=()`: unary versions of the above.

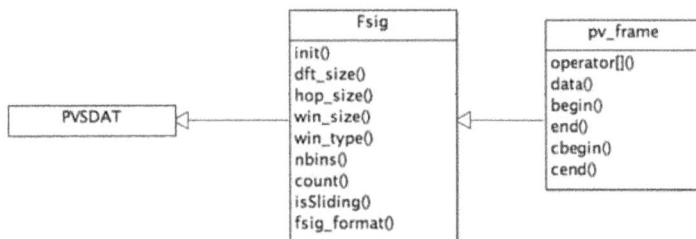

Figure 3. The `Fsig` and `pv_frame` classes and their relationship to the opaque C structure `PVSDAT`.

The `pv_bin` class can also be translated into a `std::complex<float>`, object if needed. This class is also fully compatible with the C complex type and an object obj can be cast into a float array consisting of two items (or a float pointer), using `reinterpret_cast<float(&)[2]>(obj)` or `reinterpret_cast<float*>(&obj)`. The `pv_frame` (or `spv_frame`) class contains the following members:

- `operator[]`: array-subscript access to the spectral frame
- `data()`: returns a pointer to the spectral frame data.
- `len()`: returns the length of the frame.
- `begin()`, `cbegin()` and `end()`, `cend()`: return iterators to the beginning and end of the data frame.
- `iterator` and `const_iterator`: iterator types for this class.

As noted above, fsig opcodes run at k-rate but will internally use an update rate based on the analysis hopsize. For this to work, a frame count is kept and checked to make sure we only process the input when new data is available. The example in Listing 9 shows a class implementing a simple gain scaler for fsigs:

Listing 9: Fsig opcode example

```
struct PVGain : csnd::FPlugin<1, 2> {
 static constexpr char const *otypes = "f";
 static constexpr char const *itypes = "fk";

 int init() {
  if(inargs.fsig_data(0).isSliding())
   return csound->init_error("sliding not supported");

  if(inargs.fsig_data(0).fsig_format() != csnd::fsig_format::pvs &&
     inargs.fsig_data(0).fsig_format() != csnd::fsig_format::polar){
    char *s = "format not supported";
    return csound->init_error(s);
  }
  csnd::Fsig &fout = outargs.fsig_data(0);
```

```
  fout.init(csound,  inargs.fsig_data(0));
  framecount = 0;
  return OK;
 }

 int kperf() {
   csnd::pv_frame &fin = inargs.fsig_data(0);
   csnd::pv_frame &fout = outargs.fsig_data(0);
   uint32_t i;

   if(framecount < fin.count()) {
     std::transform(fin.begin(), fin.end(), fout.begin(),
       [this](csnd::pv_bin f){ return f *= inargs[1]; });
     framecount = fout.count(fin.count());
   }
   return OK;
  }
};
```

The Params class offers a dedicated method that is used on the arguments objects to get references to the Fsig parameters. This can also be assigned directly to a pv_frame reference variable. At init-time, we initialise the output based on the input fsig format. At performance time, we check the input count and process the data if necessary. The facilities offered by CPOF allow us to use a standard library transform algorithm with a lambda object and implement the gain processing very compactly. This opcode is registered using

```
csnd::plugin<PVGain>(csound, "pvg", csnd::thread::ik);
```

6. Multithreading Support

The Csound API includes an interface for multithreading, which is implemented via pthreads [28] on POSIX systems, or other native threading libraries in non-POSIX platforms. To allow opcodes an object-oriented access to this C interface, CPOF provides the Thread pure virtual class. This is subclassed and instantiated to encapsulate a separate thread whose entry point is given by a run() method. The base class provides join() and get_thread() methods for joining a thread and getting its handle.

The example in Listing 10 illustrates the use of the Thread class. This implements a message printer on separate thread, picking up a string from the input and outputting it to the terminal. We derive a class that includes the run() method. To prevent data races, the class provides its own spin locks. A message is passed to it via the set_messsage() method.

Listing 10: Deriving a class from Thread

```
class PrintThread : public csnd::Thread {
  std::atomic_bool splock;
  std::atomic_bool on;
  std::string message;

  void lock() {
    bool tmp = false;
    while(!splock.compare_exchange_weak(tmp, true))
      tmp = false;
  }
```

```
    void unlock() {
      splock = false;
    }

    uintptr_t run() {
      std::string old;
      while(on) {
        lock();
        if(old.compare(message)) {
         csound->message(message.c_str());
         old = message;
        }
        unlock();
      }
      return 0;
    }

public:
  PrintThread(csnd::Csound *csound)
    : Thread(csound), splock(false), on(true), message("") {};

  ~PrintThread(){
    on = false;
    join();
  }

  void set_message(const char *m) {
    lock();
    message = m;
    unlock();
  }
};
```

The opcode class then is composed with it, as demonstrated in Listing 11. It will instantiate the object by calling its constructor and then pass messages to it from the performance method.

Listing 11: Threading opcode example

```
struct TPrint : csnd::Plugin<0, 1> {
  static constexpr char const *otypes = "";
  static constexpr char const *itypes = "S";
  PrintThread thread;

  int init() {
    csound->plugin_deinit(this);
    csnd::constr(&thread, csound);
    return OK;
  }

  int deinit() {
    csnd::destr(&thread);
```

```
    return OK;
  }

  int kperf() {
    thread.set_message(inargs.str_data(0).data);
    return OK;
  }
};
```

7. Results

As we have noted, CPOF supports a fully object-oriented approach to the development of new Csound opcodes, which is the main result of this work. In this section, we first oppose two versions of the same unit generator, the first using the original C API and the second based on CPOF. This demonstrates the advantages of applying the framework in the implementation of the opcode. To complement this, drawing from examples in the Csound codebase, we highlight two particular aspects of object-oriented programming that are enabled by CPOF: code re-use and the application of standard algorithms. Concluding this section, the overall contribution of this work to Csound and audio programming in general is discussed.

7.1. CPOF versus C API

As a way of comparing the CPOF approach with the original C API, we will focus on re-implementing an existing opcode using the framework. We have chose tone, a simple first-order lowpass filter, which provides a simple code that is easy to follow, but also demonstrates the compactness of the C++ approach. In Listing 12, we present the original code, taken from the Csound sources and adapted as a plugin. While the original code is actually an internal opcode, for the sake of making an exact comparison, we present it here in a modified version as an externally-loaded plugin. The only changes made to the original code are to do with opcode registration.

Listing 12: Tone opcode, original C API version

```
typedef struct {
    OPDS     h;
    MYFLT    *ar, *asig, *khp, *istor;
    double   c1, c2, yt1, prvhp;
} TONE;

int tonset(CSOUND *csound, TONE *p)
{
    double b;
    p->prvhp = (double)*p->khp;
    b = 2.0 - cos((double)(p->prvhp * TWOPI/csound->GetSr()));
    p->c2 = b - sqrt(b * b - 1.0);
    p->c1 = 1.0 - p->c2;

    if (LIKELY(!(*p->istor)))
      p->yt1 = 0.0;
    return OK;
}
```

```
int tone(CSOUND *csound, TONE *p)
{
    MYFLT        *ar, *asig;
    uint32_t offset = p->h.insdshead->ksmps_offset;
    uint32_t early  = p->h.insdshead->ksmps_no_end;
    uint32_t n, nsmps = CS_KSMPS;
    double       c1 = p->c1, c2 = p->c2;
    double       yt1 = p->yt1;

    if (*p->khp != (MYFLT)p->prvhp) {
      double b;
      p->prvhp = (double)*p->khp;
      b = 2.0 - cos((double)(p->prvhp * TWOPI/csound->GetSr()));
      p->c2 = c2 = b - sqrt(b * b - 1.0);
      p->c1 = c1 = 1.0 - c2;
    }
    ar = p->ar;
    asig = p->asig;
    if (UNLIKELY(offset)) memset(ar, '\0', offset*sizeof(MYFLT));
    if (UNLIKELY(early)) {
      nsmps -= early;
      memset(&ar[nsmps], '\0', early*sizeof(MYFLT));
    }
    for (n=offset; n<nsmps; n++) {
      yt1 = c1 * (double)(asig[n]) + c2 * yt1;
      ar[n] = (MYFLT)yt1;
    }
    p->yt1 = yt1;
    return OK;
}

int csoundModuleInit(CSOUND *csound) {
  csoundAppendOpcode(csound, "tone", sizeof(TONE), 0, 5, "a", "ako",
                     (SUBR) toneset, NULL, (SUBR) tone);
  return OK;
};

int csoundModuleCreate(CSOUND *csound) { return OK; };
int csoundModuleDestroy(CSOUND *csound) { return OK; };
```

This code example follows the straight C API opcode implementation: a data structure is provided, along with init-time and audio-rate perf-time functions. The opcode is registered using csoundAppendOpcode() called in csoundModuleInit(). Looking at the code, besides the typical C idiom, we see that there are a number of lines of code devoted to saving to the dataspace, setting sample-accurate parameters (offset, nsmps), and so on.

In contrast, the CPOF version, in Listing 13, is much more succinct. The functions have direct access to the dataspace, which allows them to update state directly. The use of iterators allow us to replace the loop with a single function call containing a lambda expression. The attention of the programmer is directed to the actual filter equation implemented therein. The code is made more compact also by replacing the filter update lines by inline calls to the update() method. All the boilerplate code present in the C version is hidden away by the framework. Finally, opcode registration is much simplified by the use of the function template.

Listing 13: Tone opcode, CPOF version

```
struct Tone : csnd::Plugin <1, 3> {
  static constexpr char const *otypes = "a";
  static constexpr char const *itypes = "aio";
  double c1;
  double c2;
  double yt1;
  double prvhp;

  void update() {
    prvhp = (double) inargs[1];
    double b = 2.0 - cos(prvhp*csnd::twopi/csound->sr());
    c2 = b - sqrt(b * b - 1.0);
    c1 = 1.0 - c2;
  }

  int init() {
    update();
    if (!inargs[2]) yt1 = 0.;
    return OK;
  }

  int aperf() {
    csnd::AudioSig in(this, inargs(0));
    csnd::AudioSig out(this, outargs(0));
    double y = yt1;
    if (prvhp != inargs[1]) update();
    std::transform(in.begin(), in.end(), out.begin(),
                   [this, &y](MYFLT x) {
                       return (y = c1 * x + c2 * y);
                   });
    yt1 = y;
    return OK;
  }
};

void csnd::on_load(Csound *csound) {
 csnd::plugin<Tone>(csound, "tone", csnd::thread::ia);
};
```

While detailed performance considerations are beyond the scope of this paper, it is nonetheless useful to observe how the two versions of this simple opcode compare in that respect. Using standard timing tests, it was found that the elapsed CPU times of both versions are very close, as illustrated in Table 1 for various vector size (ksmps) values.

Table 1. CPU time ratio C:C++ of versions of the opcode `tone`. These results are an average of five runs per test, each processing 600 seconds of audio at $f_s = 44100$ KHz. Tests were run on the x86_64 architecture under MacOS and the clang compiler.

ksmps	1	2	4	8	16	32	64	128
CPU time	0.996369	0.995616	0.982088	0.975774	0.966242	1.000936	0.960440	0.975930

7.2. Code Re-Use

With CPOF, massive code re-use can be applied to generate whole families of opcodes. For example, the following class template is used to generate a set of numeric array-variable operators for i-time and k-rate processing. This is based on creating a template opcode class that can be instantiated with different functions that have the same signature. The class in Listing 14 implements it for single argument operators.

Listing 14: Single-argument function class template

```
template <MYFLT (*op)(MYFLT)> struct ArrayOp : csnd::Plugin<1, 1> {
  int process(csnd::myfltvec &out, csnd::myfltvec &in) {
    std::transform(in.begin(), in.end(), out.begin(),
                   [](MYFLT f) { return op(f); });
    return OK;
  }

  int init() {
    csnd::myfltvec &out = outargs.myfltvec_data(0);
    csnd::myfltvec &in = inargs.myfltvec_data(0);
    out.init(csound, in.len());
    return process(out, in);
  }

  int kperf() {
    return process(outargs.myfltvec_data(0), inargs.myfltvec_data(0));
  }
};
```

We can instantiate it with a huge variety of one-in one-out functions from the standard library, thus creating over forty new opcodes. This is done by just registering each opcode using the same `ArrayOp` class with a different template parameter, as demonstrated in Listing 15.

Listing 15: Instantiating the array opcodes and registering them with Csound

```
void csnd::on_load(Csound *csound) {
  csnd::plugin<ArrayOp<std::ceil>>(csound, "ceil", "i[]", "i[]",
                                   csnd::thread::i);
  csnd::plugin<ArrayOp<std::ceil>>(csound, "ceil", "k[]", "k[]",
                                   csnd::thread::ik);
  csnd::plugin<ArrayOp<std::floor>>(csound, "floor", "i[]", "i[]",
                                    csnd::thread::i);
  csnd::plugin<ArrayOp<std::floor>>(csound, "floor", "k[]", "k[]",
                                    csnd::thread::ik);
  csnd::plugin<ArrayOp<std::round>>(csound, "round", "i[]", "i[]",
                                    csnd::thread::i);
  csnd::plugin<ArrayOp<std::round>>(csound, "round", "k[]", "k[]",
```

```
                                    csnd::thread::ik);
  csnd::plugin<ArrayOp<std::trunc>>(csound, "int", "i[]", "i[]",
                                    csnd::thread::i);
  csnd::plugin<ArrayOp<std::trunc>>(csound, "int", "k[]", "k[]",
                                    csnd::thread::i);
  csnd::plugin<ArrayOp<frac>>(csound, "frac", "i[]", "i[]",
                              csnd::thread::i);
  csnd::plugin<ArrayOp<frac>>(csound, "frac", "k[]", "k[]",
                              csnd::thread::ik);
  csnd::plugin<ArrayOp<std::exp2>>(csound, "powoftwo", "i[]", "i[]",
                                   csnd::thread::i);
  csnd::plugin<ArrayOp<std::exp2>>(csound, "powoftwo", "k[]", "k[]",
                                   csnd::thread::ik);
  csnd::plugin<ArrayOp<std::fabs>>(csound, "abs", "i[]", "i[]",
                                   csnd::thread::i);
  csnd::plugin<ArrayOp<std::fabs>>(csound, "abs", "k[]", "k[]",
                                   csnd::thread::ik);
  csnd::plugin<ArrayOp<std::log10>>(csound, "log2", "i[]", "i[]",
                                    csnd::thread::i);
  csnd::plugin<ArrayOp<std::log10>>(csound, "log2", "k[]", "k[]",
                                    csnd::thread::ik);
  csnd::plugin<ArrayOp<std::log10>>(csound, "log10", "i[]", "i[]",
                                    csnd::thread::i);
  csnd::plugin<ArrayOp<std::log10>>(csound, "log10", "k[]", "k[]",
                                    csnd::thread::ik);
  csnd::plugin<ArrayOp<std::log>>(csound, "log", "i[]", "i[]",
                                  csnd::thread::i);
  csnd::plugin<ArrayOp<std::log>>(csound, "log", "k[]", "k[]",
                                  csnd::thread::ik);
  csnd::plugin<ArrayOp<std::exp>>(csound, "exp", "i[]", "i[]",
                                  csnd::thread::i);
  csnd::plugin<ArrayOp<std::exp>>(csound, "exp", "k[]", "k[]",
                                  csnd::thread::ik);
  csnd::plugin<ArrayOp<std::sqrt>>(csound, "sqrt", "i[]", "i[]",
                                   csnd::thread::i);
  csnd::plugin<ArrayOp<std::sqrt>>(csound, "sqrt", "k[]", "k[]",
                                   csnd::thread::ik);
  csnd::plugin<ArrayOp<std::cos>>(csound, "cos", "i[]", "i[]",
                                  csnd::thread::i);
  csnd::plugin<ArrayOp<std::cos>>(csound, "cos", "k[]", "k[]",
                                  csnd::thread::ik);
  csnd::plugin<ArrayOp<std::sin>>(csound, "sin", "i[]", "i[]",
                                  csnd::thread::i);
  csnd::plugin<ArrayOp<std::sin>>(csound, "sin", "k[]", "k[]",
                                  csnd::thread::ik);
  csnd::plugin<ArrayOp<std::tan>>(csound, "tan", "i[]", "i[]",
                                  csnd::thread::i);
  csnd::plugin<ArrayOp<std::tan>>(csound, "tan", "k[]", "k[]",
                                  csnd::thread::ik);
  csnd::plugin<ArrayOp<std::acos>>(csound, "cosinv", "i[]", "i[]",
```

```
                                        csnd::thread::i);
  csnd::plugin<ArrayOp<std::acos>>(csound, "cosinv", "k[]", "k[]",
                                        csnd::thread::ik);
  csnd::plugin<ArrayOp<std::asin>>(csound, "sininv", "i[]", "i[]",
                                        csnd::thread::i);
  csnd::plugin<ArrayOp<std::asin>>(csound, "sininv", "k[]", "k[]",
                                        csnd::thread::ik);
  csnd::plugin<ArrayOp<std::atan>>(csound, "taninv", "i[]", "i[]",
                                        csnd::thread::i);
  csnd::plugin<ArrayOp<std::atan>>(csound, "taninv", "k[]", "k[]",
                                        csnd::thread::ik);
  csnd::plugin<ArrayOp<std::cosh>>(csound, "cosh", "i[]", "i[]",
                                        csnd::thread::i);
  csnd::plugin<ArrayOp<std::cosh>>(csound, "cosh", "k[]", "k[]",
                                        csnd::thread::ik);
  csnd::plugin<ArrayOp<std::sinh>>(csound, "sinh", "i[]", "i[]",
                                        csnd::thread::i);
  csnd::plugin<ArrayOp<std::sinh>>(csound, "sinh", "k[]", "k[]",
                                        csnd::thread::ik);
  csnd::plugin<ArrayOp<std::tanh>>(csound, "tanh", "i[]", "i[]",
                                        csnd::thread::i);
  csnd::plugin<ArrayOp<std::tanh>>(csound, "tanh", "k[]", "k[]",
                                        csnd::thread::ik);
  csnd::plugin<ArrayOp<std::cbrt>>(csound, "cbrt", "i[]", "i[]",
                                        csnd::thread::i);
  csnd::plugin<ArrayOp<std::cbrt>>(csound, "cbrt", "k[]", "k[]",
                                        csnd::thread::ik);
}
```

A similar approach can be used for functions of two arguments, yielding yet another large set of new opcodes. While this is a simple and indeed obvious example that can deliver re-use in a large scale, more generally the framework reduces the amount of code repetition significantly.

7.3. Standard Algorithms

Another example shows the use of standard algorithms in spectral processing. The opcode in Listing 16 implements *spectral tracing* [29], which retains only a given number of bins in each frame, according to their amplitude. To select the bins, we need to sort them to find out the ones we want to retain (the loudest N). For this we collect all amplitudes from the frame and then apply *nth element* sorting, placing the threshold amplitude in element n. Then we just filter the original frame according to this threshold. Here we have the performance code (amps is a dynamically allocated array belonging to the Plugin object).

Listing 16: Phase vocoder tracing opcode

```
struct PVTrace : csnd::FPlugin<1, 2> {
  csnd::AuxMem<float> amps;
  static constexpr char const *otypes = "f";
  static constexpr char const *itypes = "fk";

  int init() {
    if (inargs.fsig_data(0).isSliding())
      return csound->init_error(Str("sliding not supported"));
```

```cpp
    if (inargs.fsig_data(0).fsig_format() != csnd::fsig_format::pvs &&
        inargs.fsig_data(0).fsig_format() != csnd::fsig_format::polar)
      return csound->init_error(Str("fsig format not supported"));

    amps.allocate(csound, inargs.fsig_data(0).nbins());
    csnd::Fsig &fout = outargs.fsig_data(0);
    fout.init(csound, inargs.fsig_data(0));
    framecount = 0;
    return OK;
  }

  int kperf() {
    csnd::pv_frame &fin = inargs.fsig_data(0);
    csnd::pv_frame &fout = outargs.fsig_data(0);

    if (framecount < fin.count()) {
      int n = fin.len() - (int) inargs[1];
      float thrsh;
      std::transform(fin.begin(), fin.end(), amps.begin(),
                     [](csnd::pv_bin f) { return f.amp(); });
      std::nth_element(amps.begin(), amps.begin() + n, amps.end());
      thrsh = amps[n];
      std::transform(fin.begin(), fin.end(), fout.begin(),
                     [thrsh](csnd::pv_bin f) {
                       return f.amp() >= thrsh ? f : csnd::pv_bin();
                     });
      framecount = fout.count(fin.count());
    }
    return OK;
  }
};
```

7.4. Discussion

One of the main objectives of this work, as stated earlier, is to provide support for more modern C++ approaches to opcode development, which can, in particular, facilitate the writing of signal processing code. Under this point, we were able to show through the comparison of CPOF and C API code that such objective has been met. A cursory comparison of the two versions demonstrates a number of surface differences in the CPOF version: economy of expressions, code length (36 lines vs. 59), and clarity of context. A closer look will reveal how the semantics of the process (the application of a filter equation) is far better realised in the C++ version, without the intrusion of set-up and boilerplate code.

In fact, this comparative exercise points out to the fact that a lot of code re-use could actually replace the traditional approach to write components such as filters. We were able to demonstrate how this is possible in a simple case of stateless operators, but the principle is applicable to other types of opcodes, of varying levels of complexity. In the case of filters, by supplying different update() and filter lambdas to a template class, a whole family of such processors can be implemented. Beyond this extensive code re-use, we can also make avail of standard algorithms for these tasks, which not only simplifies some of the development stages, but also helps improve correctness of implementation.

We may conclude that the availability of CPOF in the Csound code allows for better engineering practices in the development of opcodes. A corollary of this conclusion is that CPOF provides Csound a much more advanced means of UG development than what is available in similar music programming environments. This is because, as we have noted in Section 1.2, a more traditional C-like OOP idiom is prevalent elsewhere (even where C++ is the implementation language).

Another important result of this work, which should be highlighted here is the integration of modern C++ practices into a pure-C environment. While this has been done elsewhere, the extent to which we were able to apply the C++11 standard and generate a completely C-compatible code is of note. Some of the techniques developed here, such as the use of hiding and templating as part of the polymorphism mechanism, while quite particular to the case, are original solutions introduced by this work. Some of these are applicable to other systems, including as noted in Section 1.2, PD, SuperCollider, and more generally should be of interest to developers of DSP applications.

8. Conclusions

Object-oriented programming is an established paradigm for systems implementation, which has been used in a variety of applications. In music programming systems, it is very well suited to the development of language components, unit generators, which themselves are loosely modelled on this approach. Csound originally provided through its API a C-language interface for the addition of new opcodes to the system, but the supports for object orientation under that language are incipient.

This paper described the motivations, design and implementation of a framework for opcode development in C++, CPOF, together with its support toolkit. We were able to demonstrate how it enables a range of idioms that are being adopted as standard for object-oriented programming under that language, especially following the advent of C++11 [30] standard. We have provided a detailed discussion of the framework code with several examples, complemented with the discussion of two specific cases of object-orientation that are well supported by CPOF.

This version of CPOF is already being adopted by Csound opcode developers; starting from version 6.09, a number of new UGs have appeared, which are based on the framework. While the design described here is solid and matches well the underlying opcode model, some additions and improvements might be considered in newer versions. One particular approach would be to consider a way to encapsulate the Csound type system in a more complete way under C++, representing not only the data formats (as we do now), but also the timing aspects of each type.

To build a plugin opcode library based on this framework, a C++ compiler supporting the C++11 standard (`-std=c++11`), and the Csound public headers. The Csound plugin mechanism does not depend on any particular link-time libraries. The opcode library should be built as a dynamic/shared module (e.g., `.so` on Linux, `.dylib` on OSX or `.dll` on Windows), and placed in the opcode directory (or, alternatively, it can be loaded with the `--opcode-lib=` flag).

All opcode examples in this paper, with the exception of the ones in Section 7, are provided in `opcodes.cpp`, found in the `examples/plugin` directory of the Csound source codebase (https://github.com/csound/csound). The cases discussed in Section 7 are part of the existing code base (in `Opcodes/arrayops.cpp` and `Opcodes/pvsops.cpp`). CPOF is part of Csound and is distributed alongside its public headers. Csound is free software, licensed by the Lesser GNU Public License.

Conflicts of Interest: The author declares no conflict of interest.

References

1. Lazzarini, V. The Development of Computer Music Programming Systems. *J. New Music Res.* **2013**, *1*, 97–110.
2. Dodge, C.; Jerse, T.A. *Computer Music: Synthesis, Composition and Performance*, 2nd ed.; Schirmer: New York, NY, USA, 1997.

3. Dannenberg, R.B.; Thompson, N. Real-Time Software Synthesis on Superscalar Architectures. *Comput. Music J.* **1997**, *21*, 83–94.

4. Lazzarini, V.; ffitch, J.; Yi, S.; Heintz, J.; Brandtsegg, Ø.; McCurdy, I. *Csound: A Sound and Music Computing System*; Springer: Berlin, Germany, 2016.

5. Boulanger, R. (Ed.) *The Csound Book*; MIT Press: Cambridge, MA, USA, 2000.

6. Sondergaard, H.; Sestoft, P. Referential Transparency, Definiteness and Unfoldability. *Acta Inf.* **1990**, *27*, 505–517.

7. Ffitch, J. Understanding an Opcode in Csound. In *The Audio Programming Book*; Boulanger, R., Lazzarini, V., Eds.; MIT Press: Cambridge, MA, USA, 2010; pp. 581–615.

8. Lieberman, H. Machine Tongues IX: Object Oriented Programming. In *The Well-tempered Object: Musical Applications of Object-Oriented Software Technology*; Pope, S., Ed.; MIT Press: Cambridge, MA, USA, 1991; pp. 18–31.

9. Stroustrup, B. *The C++ Programming Language*, 4th ed.; Addison-Wesley: Boston, MA, USA, 2013.

10. Puckette, M. *The Theory and Technique of Computer Music*; World Scientific Publishing: New York, NY, USA, 2007.

11. McCartney, J. Rethinking the Computer Music Language: Supercollider. *Comput. Music J.* **2002**, *26*, 61–68.

12. Stowell, D. Writing Unit Generator Plug-ins. In *The SuperCollider Book*; Wilson, S., Cottle, D., Collins, N., Eds.; MIT Press: Cambridge, MA, USA, 2010; pp. 692–720.

13. Boulanger, R.; Lazzarini, V. (Eds.) *The Audio Programming Book*; MIT Press: Cambridge, MA, USA, 2010.

14. Cook, P.; Scavone, G. The Synthesis Toolkit (STK). In Proceedings of the 1999 International Computer Music Conference, Beijing, China, 22–27 October 1999; Volume III, pp. 164–166.

15. Lazzarini, V. The SndObj Sound Object Library. *Organ. Sound* **2000**, *5*, 35–49.

16. Lazzarini, V. The Design of a Lightweight DSP Programming Library. In Proceedings of the 14th Sound and Music Computing Conference 2017, Aalto University, Espoo, Finland, 5–8 July 2017; pp. 5–12.

17. ISO/IEC. ISO International Standard ISO/IEC 14882:2014, Programming Language C++. 2014. Available online: https://www.iso.org/standard/64029.html (accessed on 25 July 2017).

18. Lazzarini, V. The Csound Plugin Opcode Framework. In Proceedings of the 14th Sound and Music Computing Conference 2017, Aalto University, Espoo, Finland, 5–8 July 2017; pp. 267–274.

19. Pope, S. Machine Tongues XI: Object-Oriented Software Design. In *The Well-tempered Object: Musical Applications of Object-oriented Software Technology*; Pope, S., Ed.; MIT Press: Cambridge, MA, USA, 1991; pp. 32–47.

20. Cardelli, L.; Wegner, P. On Understanding Types, Data Abstraction, and Polymorphism. *ACM Comput. Surv.* **1985**, *17*, 471–523.

21. Lippman, S.B. *Inside the C++ Object Model*; Addison Wesley Longman Publishing Co., Inc.: Redwood City, CA, USA, 1996.

22. Standard C++ Foundation. How to mix C and C++. Available online: https://isocpp.org/wiki/faq/mixing-c-and-cpp (accessed on 25 July 2017).

23. Abrahams, D.; Gurtovoy, A. *C++ Template Metaprogramming: Concepts, Tools, and Techniques from Boost and Beyond (C++ in Depth Series)*; Addison-Wesley Professional: Boston, MA, USA, 2004.

24. Lazzarini, V. Time-Domain Signal Processing. In *The Audio Programming Book*; Boulanger, R., Lazzarini, V., Eds.; MIT Press: Cambridge, MA, USA, 2010; pp. 463–512.

25. Moore, F.R. *Elements of Computer Music*; Prentice-Hall, Inc.: Upper Saddle River, NJ, USA, 1990.

26. Lazzarini, V. Spectral Opcodes. In *The Audio Programming Book*; Boulanger, R., Lazzarini, V., Eds.; MIT Press: Cambridge, MA, USA, 2010; pp. 617–626.

27. Dolson, M. The Phase Vocoder: A Tutorial. *Comput. Music J.* **1986**, *10*, 14–27.

28. IEEE/Open Group. The Open Group Base Specifications Issue 7, IEEE Std 1003.1-2008, 2016. Available online: http://pubs.opengroup.org/onlinepubs/9699919799/ (accessed on 25 July 2017).

29. Wishart, T. *Audible Design*; Orpheus The Pantomine Ltd.: North Yorkshire, UK, 1996.

30. ISO/IEC. ISO International Standard ISO/IEC 14882:2011, Programming Language C++, 2011. Available online: https://www.iso.org/standard/50372.html (accessed on 25 July 2017).

applied
sciences

MDPI

Article

A Two-Stage Approach to Note-Level Transcription of a Specific Piano

Qi Wang [1,2], **Ruohua Zhou** [1,2,*] **and Yonghong Yan** [1,2,3]

1   Key Laboratory of Speech Acoustics and Content Understanding, Institute of Acoustics, Chinese Academy of Sciences, Beijing 100190, China; wangqi@hccl.ioa.ac.cn (Q.W.); yanyonghong@hccl.ioa.ac.cn (Y.Y.)
2   University of Chinese Academy of Sciences, Beijing 100190, China
3   Xinjiang Laboratory of Minority Speech and Language Information Processing, Xinjiang Technical Institute of Physics and Chemistry, Chinese Academy of Sciences, Urumchi 830001, China
*   Correspondence: zhouruohua@hccl.ioa.ac.cn; Tel.: +86-010-8254-7570

Academic Editor: Tapio Lokki
Received: 22 July 2017 ; Accepted: 29 August 2017; Published: 2 September 2017

Abstract: This paper presents a two-stage transcription framework for a specific piano, which combines deep learning and spectrogram factorization techniques. In the first stage, two convolutional neural networks (CNNs) are adopted to recognize the notes of the piano preliminarily, and note verification for the specific individual is conducted in the second stage. The note recognition stage is independent of piano individual, in which one CNN is used to detect onsets and another is used to estimate the probabilities of pitches at each detected onset. Hence, candidate pitches at candidate onsets are obtained in the first stage. During the note verification, templates for the specific piano are generated to model the attack of note per pitch. Then, the spectrogram of the segment around candidate onset is factorized using attack templates of candidate pitches. In this way, not only the pitches are picked up by note activations, but the onsets are revised. Experiments show that CNN outperforms other types of neural networks in both onset detection and pitch estimation, and the combination of two CNNs yields better performance than a single CNN in note recognition. We also observe that note verification further improves the performance of transcription. In the transcription of a specific piano, the proposed system achieves 82% on note-wise F-measure, which outperforms the state-of-the-art.

Keywords: music information retrieval; piano transcription; note recognition; note verification; onset detection; multi-pitch estimation

1. Introduction

Automatic music transcription (AMT) is a process of transcribing a musical audio signal into a symbolic representation, such as a piano roll or music score. It has many applications in music information retrieval, composition, music education, and music visualization.

AMT has been researched for four decades (since 1977) [1,2], and it is still a challenging problem. While the transcription of monophonic music is considered solved, polyphonic AMT remains open because the signal is more complex. In polyphonic music, many notes overlap in the time domain and interact in the frequency domain. Additionally, the complexity of polyphony increases with the number of sound sources. For example, the concurrent notes in orchestral music come from instruments of different timbral properties, and the corresponding AMT performance is poor.

Note is the basic unit of music, and the main problem of transcription is to extract the information of every note in the music [3]. For each note, a set of information includes: pitch, onset, offset, loudness, and timbre. Pitch is a major attribute of auditory sensation, which can be reliably related to the fundamental frequency (F0). Onset refers to the beginning time of a note, and offset refers to the ending time. Loudness is the characteristic related to the amplitude of a sound. Timbre is that perceptual attribute in which a listener can judge that two sounds having the same loudness and pitch are dissimilar. In general, we only focus on which notes are played and when they appear in the music. Therefore, the pitch and onset time are necessary in the results of AMT.

The approaches to polyphonic transcription can be divided into frame-based methods and note-based methods [4]. The frame-based approaches estimate pitches in each time frame and form frame-level results in a post-processing stage. The most straightforward solution is to analyze the time–frequency representation of audio and compute the fundamental frequencies [5]. Short-time Fourier transform (STFT) [6,7] and constant Q transform (CQT) [8] are two widely used time–frequency analysis methods. Zhou proposed resonator time–frequency image (RTFI), in which a first-order complex resonator filter bank is adopted to the analysis of music [9]. Dressler used multi-resolution STFT, and the pitch was estimated by detecting peaks in the weighted spectrum [10]. Spectrogram factorization techniques are also very popular in AMT, such as non-negative matrix factorization (NMF) [11]. Probabilistic latent component analysis (PLCA) is another factorization technique, which aims to fit a latent variable probabilistic model to normalised spectrograms [12,13]. Apart from the discriminative approaches, deep neural networks have been used to identify pitches recently. Nam superimposed a support vector machine (SVM) on top of a deep belief network (DBN) to learn feature representations [14]. Sigtia compared the performance of neural networks and proposed a recurrent neural network (RNN) language model for music transcription [15]. Kelz utilized both a ConvNet and an AUNet in transcription, and investigated the glass ceiling effect of deep neural networks [16].

The note-based transcription approaches directly estimate notes, including pitches and onsets. One solution is combining the estimation of pitches and onsets into a single framework [17,18]. Kameoka [19] used harmonic temporal structured clustering to estimate the attributes of notes simultaneously. In [20], Böck used an RNN with bidirectional long short-term memory (LSTM) units. Similarly, Sigtia utilized three kinds of neural networks to transfer the input audio to a list of notes, along with the corresponding pitches and onset times [21]. Another solution is employing a separate onset detection stage and an additional pitch estimation stage. The approaches in this category often estimate the pitches using the segments between two successive onsets, and an accurate onset detection benefits the transcription. Marolt proposed a connectionist approach which contains a neural network of onset detection [22]. Costantini detected the onsets and estimated the pitches at the note attack using SVM [23]. However, little deep-learning-based research has been done in this category, to our knowledge.

Modeling the instrument being transcribed and learning the corresponding timbral properties is an efficient way to improve the AMT performance. Instrument-specific transcription research restricts the employed instrument models to a specific type. Depending on the timbral properties of different instruments, different sets of constraints are adopted in instrument-specific AMT systems [24–26]. As a typical multi-pitch instrument, the piano has been widely studied in AMT because its polyphony is challenging. The task of piano transcription has existed in MIREX (Music Information Retrieval Evaluation eXchange) since 2007, and it is competitive every year [27]. Figure 1 gives MIREX's annual best results for the note tracking of piano subset based on onset only over the past 10 years. The current state-of-the-art system won 82% on F-measure in MIREX 2016, which is employed as a baseline system to evaluate the performance of our proposed method [28].

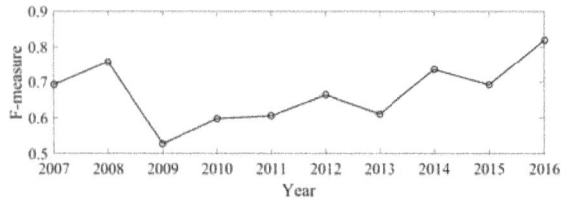

Figure 1. The 2007–2016 annual best results for piano transcription in MIREX (Music Information Retrieval Evaluation eXchange).

Individual-specific transcription is a new direction of AMT, which can make use of more characteristics of the individual piano. Cogliati and Duan modeled the temporal evolution of piano notes, and the spectrogram was factorized using the templates [29]. In the same context-dependent setting, they also employed convolutional sparse coding to transcribe the music from a specific piano in the specific environment [30]. In the supervised NMF, templates were usually formed by the isolated notes of the specific piano to be transcribed. Ewert employed spectro-temporal patterns to model the temporal evolution in NMF [31]. Cheng proposed a method to model the attack and decay of notes, and all the templates were trained by a Disklavier piano [32]. In the same transcription task, Gao combined the convolutional NMF with a differential spectrogram [33].

In this paper, we focus on the note-based polyphonic transcription for a specific piano. Deep learning technique is adopted to recognize notes preliminarily, and then the candidate notes are verified for the specific piano. In the stage of note recognition, a convolutional neural network (CNN) is used to detect onsets, and another CNN is used to estimate the probabilities of pitches at each detected onset. During the note verification, the spectrogram is factorized using attack templates of notes. Compared to existing AMT approaches, the proposed method has the following advantages:

(1) The note recognition stage yields a note-level transcription by estimating the pitch at each onset. Compared to existing deep-learning-based methods which use a single network, two consecutive CNNs yield better performance.

(2) An extra stage of note verification is conducted for the specific piano, in which the spectrogram factorization improves the precision of transcription. Compared with the traditional NMF, the proposed note verification stage could save computing time and storage space to a great extent.

(3) The proposed method achieves better performance in specific piano transcription compared to the state-of-the-art approach.

The outline of this paper is as follows. The proposed framework is described in Section 2. The transcription and comparison experiments are presented in Section 3. Finally, conclusions are drawn in Section 4.

2. Proposed Framework

The proposed transcription framework is shown in Figure 2, which comprises a note recognition module and a note verification module. In this section, we describe the two stages.

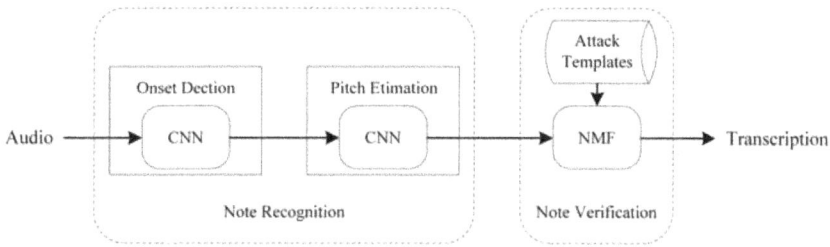

Figure 2. Diagram for the proposed framework. CNN: convolutional neural network; NMF: non-negative matrix factorization.

2.1. Note Recognition

Recently, convolutional learning has achieved great success in music signal processing, such as genre classification [34], artist classification [35], and chord detection [36]. In the task of AMT, CNNs have also been evaluated for onset detection and frame-based transcription, respectively. In the experiments of onset detection, Schlüter used CNNs of different architectures [37]. The results shows that a CNN with linear rectifier outperformed the state-of-the-art while requiring less manual preprocessing. Sigtia utilized a CNN to transcribe polyphonic piano music frame-by-frame, and the output was estimated pitches at each frame [21]. Although CNN yields the best performance on the frame-based metrics, an NMF method outperforms CNN on note-based metrics. So, it is promising for CNN to make use of the note onset and generate a note-based transcription. Here we train a CNN to detect onset and another CNN to estimate pitches at each detected onset.

CNNs are neural networks characterized by a convolutional structure. The convolutional layers are designed to preserve the spatial structure of the input. In each convolutional layer, a set of weights act on a fixed-size local region of the input. These weights are then repeatedly applied to the entire input to produce a feature map. After the convolution of input with shared weights, the output of the convolutional layer is obtained by adding a bias term and then applying a non-linear function. Each unit of out feature map in the convolutional layer can be computed as:

$$q_{j,m} = f(\sum_i^I \sum_n^N o_{i,n+m-1} w_{i,j,n} + b_j) \tag{1}$$

where $o_{i,m}$ is the mth unit of the ith input feature map, $q_{j,m}$ is the mth unit of the jth output feature map, $w_{i,j,n}$ is the nth element of the weight vector, b_j is the bias term added to the jth feature map, $f(\cdot)$ is the activation function. I is the number of input feature maps, and N is the size of weight filter. A convolutional layer is often followed by a pooling layer, which subsamples each feature map. For example, the most common max pooling only retains the maximum value in non-overlapping cells. When the max pooling function is used, the pooling layer is defined as:

$$p_{j,m} = \max_{k=1}^{K} q_{j,(m-1)\times s+k} \tag{2}$$

where K is the pooling size and s is the shift size of pooling windows. Here, $p_{j,m}$ is the mth unit of the jth output feature map. $q_{j,m}$ is the mth unit of the jth input feature map in this pooling layer, and it is also the corresponding unit of the output feature map in the last convolutional layer. Finally, the CNN ends in fully-connected layers that integrate the information of layers below. In audio signal processing, the input to the CNN is a window of feature frames centering around time t, whereas the output contains posterior probabilities of different categories at time t.

There are several motivations for applying CNNs to music transcription. Firstly, aggregating over several frames achieves better performance than processing a single frame. For example, the attack stage of notes can be modeled by applying a context window around the onset so that the onset will be detected more accurately. Secondly, the architecture of the CNN can learn features along both the time and frequency axes. CNN is proper for processing the harmonic structure in a spectrogram because of its shift invariance. Compared with deep neural network (DNN) and RNN, the weight sharing and pooling architecture leads to a reduction of parameters.

In the proposed note recognition stage, two CNNs are trained using a constant Q transform (CQT) of the music signal. The spectrogram of CQT is suited as time–frequency representation for music since its frequency bins are evenly spaced on a logarithmic axis. Additionally, the inter-harmonic spacings are constant for different pitches so that the CNN can learn pitch-invariant information. We trained a CNN of one output unit as an onset detector, giving binary labels to distinguish onsets from non-onsets. The architecture of this CNN is shown in Figure 3. The CNN takes a spectrogram slice of several frames as a single input, and each spectrogram excerpt centers on the frame to be detected. All of the spectrograms are extracted along the music signal, with a hop size of one frame. Feeding the spectrograms of the test signal to the network, we can obtain an onset activation function over time. The frame whose activation function is greater than the threshold is set to be the candidate onset.

The onset detector is followed by another CNN for multi-pitch estimation (MPE), which has the same architecture except for the output layer. Its input is a spectrogram slice centered at the onset frame. The CNN has 88 units in the output layer, corresponding the 88 pitches of piano. To make sure the multiple pitches can be estimated at the same time, all the outputs are transformed by a sigmoid function. For each training sample, the onset time is annotated accurately in advance. In the testing procedure, the input is a spectrogram slice centered at the candidate onset, which is detected by the previous CNN. A set of probabilities of 88 pitches is estimated through the network. Finally, the candidate pitches at candidate onsets are obtained by applying a threshold to the output.

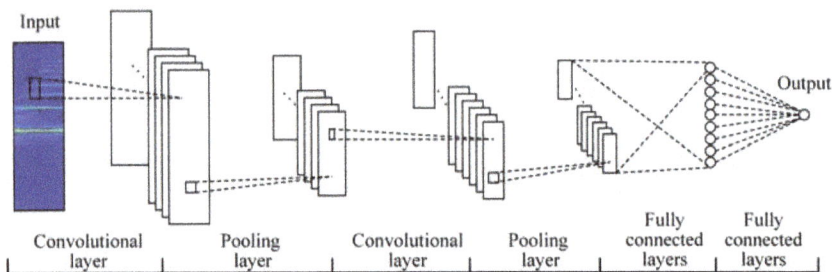

Figure 3. CNN architecture for onset detection.

2.2. Note Verification

Note verification for the specific piano is implemented through an NMF. As a frame-based approach, the traditional NMF factorises a spectrogram of a piano signal into 88 spectral bases and sparsity activations. Here the NMF only takes the candidate onsets and pitches into consideration and provides a note-wise representation. In the proposed framework, the sound to be transcribed is reconstructed by:

$$R_{t-T}^{t+T} = \sum_{k=1}^{K} W_k H_{t-T}^{t+T} \tag{3}$$

where R_{t-T}^{t+T} is the reconstructed spectrogram of $2T+1$ frames and t is the frame of candidate onset. W is the attack template for the specific piano, $k \in [1, K]$ is the index of candidate pitches, and H_{t-T}^{t+T} is the note activations. For the piano to be transcribed, 88 individual notes are pre-recorded and each template is obtained by computing the average spectrum over time frames. The attack template was calculated using the attack stage of each note rather than the whole duration. Note activations

H_{t-T}^{t+T} can be estimated by minimising the difference between the reconstruction R_{t-T}^{t+T} and the original spectrogram X_{t-T}^{t+T}. The spectrogram X_{t-T}^{t+T} is also the input being fed to the pitch estimation CNN. Finally, we verified the candidate notes from activations. Only the candidate pitches whose peaks in the activations exceed a threshold will be identified. Meanwhile, the time when activations exceed the threshold will be set as the onset. Compared with the traditional NMF, the proposed method can save computing time and storage space to a great extent.

An illustration of note verification is shown in Figure 4. Figure 4a is a spectrogram excerpt used for traditional NMF, in which a C4 note starts at 0.14 s and ends at 0.96 s. Additionally, a C#4 note fades away before the C4 note appears, and a A3 note is played at last. Here, we only present the factorization of note C4. The templates and activations are shown in Figure 4b,c, respectively. Compared with the traditional template (solid line), the attack template (dashed line) concentrates on the percussive stage of the note and shows a different characteristic. For example, both the high-order harmonics and components between harmonics have higher amplitude in the spectrum of the attack template. In Figure 4c, the solid line is the frame-wise activations for traditional NMF, and the dashed line corresponds to the attack activations for note verification. Both curves rise rapidly at the onset time, and a note C4 can be detected using a threshold of 3.0. However, another peak appears in the curve of traditional activations at the end of note C4, and a false positive will be detected using the threshold. Therefore, the NMF using attack templates are more suitable to be applied in note verification.

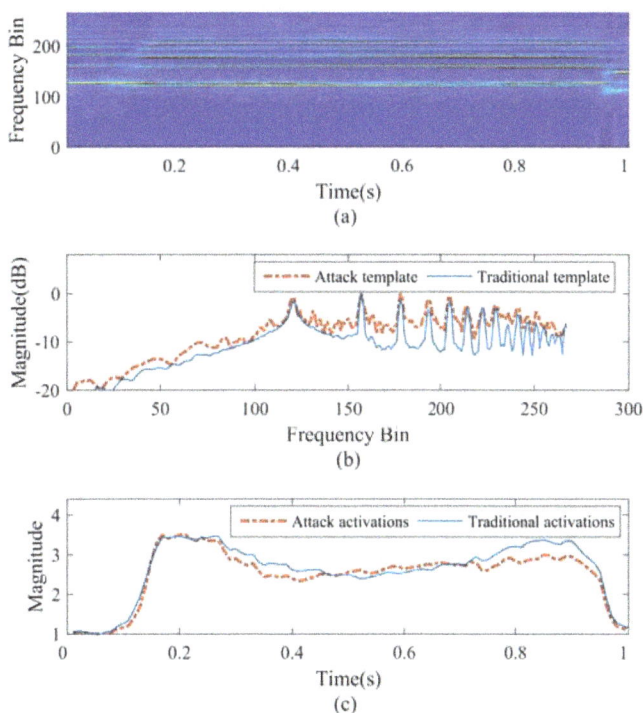

Figure 4. An illustration of note verification: (**a**) a spectrogram excerpt used for NMF; (**b**) the attack templates and traditional templates; (**c**) the attack activations and traditional activations in NMF.

In the stage of note verification, the effect of the dynamic level of templates is important. Even for a specific piano, the spectrograms of same pitch vary depending on different dynamics. Figure 5 shows the attack templates of note C4, played at three common dynamics: forte, mezzo-forte, and piano.

As shown in Figure 5, there are differences between the templates of three dynamics—especially for the higher partials. In the high-frequency range, the notes of louder dynamic have richer spectral content compared to notes of softer dynamics. This indicates that the louder dynamics excite more modes in the vibration of strings than softer dynamics, which is consistent with the assumption of [30]. If we factorize a forte note using piano templates, false positives may happen because the forte note contains some spectral content which is not present in the corresponding piano template. This error will not occur when we transcribe a note using attack templates of louder dynamics.

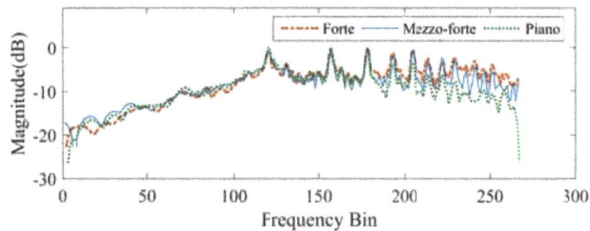

Figure 5. Attack templates of note C4 played at three dynamics: forte, mezzo-forte, and piano.

3. Experiments

In this section, we describe the dataset used in our experiments. Then, the experimental preprocessing, parameters, and metrics are introduced. Finally, we present the results from the different experiments and analyze the performance of the proposed approach.

3.1. Dataset

The transcription experiments were conducted on the MIDI aligned piano sounds (MAPS) dataset [38]. The MAPS dataset provides piano recordings, the related aligned MIDI files, and annotated text files. The overall size of MAPS is about 60 h of audio, and it is the largest database for piano transcription. There are nine categories of recordings corresponding to different piano types and recording conditions. Seven categories of audio are produced by software piano synthesizers, while two categories of recordings are obtained from a real Yamaha Disklavier upright piano. The dataset consists of isolated notes, chords, and 30 pieces of music in each category. For music pieces, the number of concurrent notes ranges from one to nine. Each music piece lasts more than 30 s, and all 270 pieces contain 18 h of audio signal.

We aim at the transcription of the Disklavier piano, which is in category "ENSTDkCl" of the MAPS dataset. For the real piano, the recording room was a studio with dimensions equal to about 4×5 m. The distance between the piano and the microphones was about 50 cm. MIDI files were created beforehand and were sent to the MIDI input of the Disklavier. Then, the audio was recorded using two omnidirectional microphones.

To build a universal model independent of the real individual, we trained the CNNs using 210 music pieces of synthesized pianos in the MAPS dataset. The training set contains 460,988 notes and the overall size is about 14 h. The proposed system was evaluated on the music pieces of the Disklavier piano. In the testing set, there are 30 music pieces, and only the first 30 s of each piece was used for transcription. The testing set contains 7345 notes in total. The setting is realistic because the training set and testing set are disjoint on piano types. During the note verification, the attack templates were obtained from the isolated notes produced by the same piano.

3.2. Experimental Settings

The proposed framework takes the spectrograms of CQT as input. The audio signal was segmented with a frame length of 100 ms and a hop-size of 10 ms. The CQTs cover 88 notes of piano, and there are 36 bins per octave. Hence, a 267-dimensional CQT vector is extracted for each frame. A context window of nine frames was applied to the CQTs so that we could obtain a spectrogram slice.

In the note recognition, architectures of these two CNNs were similar (as shown in Figure 3): two convolutional layers, two pooling layers, and two fully-connected layers. These two CNNs have the same structure, except for the final fully-connected layer. For the spectrogram slices of 267 dimensions by 9 frames, the first convolutional layer with 10 filters of size 16×2 computes 10 feature maps of size 252×8. The next layer performs max-pooling of 2×2, reducing the size of maps to 126×4. The second convolutional layer contains 20 filters of size 11×3, which generates 20 feature maps of 116×2. The max-pooling size of the second pooling layer was also set to 2×2, resulting in 20 maps of 58×1. The first fully-connected layer contains 256 units, and the number of units in the final layer changes with the task. In the CNN for onset detection, the final fully-connected layer has a single output unit. In the CNN for MPE, the final fully-connected layer has 88 output units. The two convolutional layers and the first fully-connected layer use the rectified linear unit (ReLU) activation function, and the final fully-connected layers use the sigmoid function. Appendix A shows more details about the CNNs.

The CNNs were trained using mini-batch gradient descent, and the size of a mini-batch was 256. The Adam algorithm [39] was also used in the training. An initial learning rate of 0.01 was decreased to 0 over 100 epochs. To prevent over-fitting, a dropout of 0.5 was applied to each network. We also used the method of early stopping, in which training was stopped if the cost (cross entropy) did not decrease for 20 epochs. The training of two CNNs was independent, whereas the CNNs were concatenated in the testing procedure. For the testing data, the first CNN estimates the candidate onset and the input of the second CNN is a spectrogram slice centered at the candidate onset.

During the note verification, we trained one attack template per pitch using the forte notes. The attack template was obtained by calculating the average of first 5-frame spectrogram followed by the onset. Each spectrum to be factorised is 267 dimension by 9 frames, and the central frame is the candidate onset detected by the first CNN.

Note-based metrics were employed to assess the performance of the proposed system [40]. A note event is regarded as right if its pitch is correct and its onset is within a ± 50 ms range of the ground truth onset. These measures are defined as:

$$P = \frac{N_{TP}}{N_{TP} + N_{FP}} \tag{4}$$

$$R = \frac{N_{TP}}{N_{TP} + N_{FN}} \tag{5}$$

$$F = \frac{2 \times P \times R}{P + R} \tag{6}$$

where P, R, F correspond to precision, recall, and F-measure, respectively, and N_{TP}, N_{FP}, and N_{FN} are the numbers of true positives, false positives, false negatives respectively.

3.3. Results

To evaluate the performance of proposed approach comprehensively, we present the results of each step. Firstly, we analyze the performance of two CNNs, which were trained for onset detection and pitch estimation, respectively. Additionally, the performance of the proposed note recognition module was evaluated on piano transcription. At last, we compared the proposed approach with a state-of-the-art method on individual-specific transcription.

3.3.1. Onset Detection

For comparative purposes, the DNN and RNN were used for onset detection. In the training of DNN and RNN, we performed a grid search over sets of parameters to find an architecture with the best performance. The uncertain parameters of neural networks are: number of layers $L \in \{1, 2, 3, 4\}$, number of hidden units $H \in \{32, 64, 128, 256, 512\}$. The hidden unit activation is a ReLU function and the output unit activation is sigmoid. In the architecture of RNN, LSTM [41] units are used, and the length of sequence was set to 10. The other parameters and methods in training are same as them in the CNN, such as dropout and early stopping.

All the results of onset detection are presented in Table 1. As shown in Table 1, the CNN performs best and the RNN outperforms DNN on all evaluation metrics. For example, the CNN yields a relative improvement of 2.84% over the RNN, and the RNN outperforms the DNN by 4.48% on F-measure. Both the CNN and RNN take a sequence of spectrums as input, which utilize the context information over time. Additionally, the spatial structure of the spectrogram is preserved by the CNN, which is useful for onset detection.

Table 1. Performance on onset detection using different neural networks. DNN: deep neural network; RNN: recurrent neural network.

Method	Recall	Precision	F-Measure
CNN	0.9731	0.9590	0.9660
DNN	0.9319	0.8683	0.8990
RNN	0.9530	0.9259	0.9393

Figure 6 shows the outputs of neural networks for a music excerpt along with the corresponding ground truth. The excerpt is the first 10 s of track MAPS_MUS-bk_xmas5_ENSTDkCl. It is a typical example for transcription, and it is analyzed in each of the following experiments. In the ground truth (Figure 6d), there are two values: zero represents non-onset, and one stands for onset. We can also observe that the onset is sparse in the excerpt's first 8.8 s, and it is dense in the last 1.2 s. As shown in Figure 6, the DNN's output is far away from the ground truth, which cannot detect the dense onset and bring many false positives. This example explains why the DNN yields low recall and precision in Table 1. RNN and CNN are more suitable for onset detection than DNN. This is largely due to the context information over time. The evolution of a note can be modeled using the sequence information, so the false positives will not be detected in the sustain or decay stage of the note. Compared to RNN, CNN's output is closer to the ground truth—especially for the dense onset. When two adjacent onsets have small time difference, their detection is difficult through change along the time axis. In this case, we can identify the onset using the pitch information. CNN is such a method, which learns a feature along both the time and frequency axes through its convolutional layers.

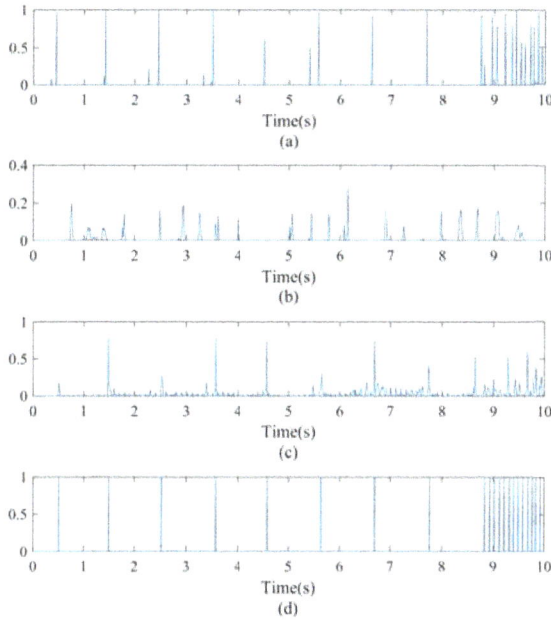

Figure 6. Results of onset detection: (a–c) the output of CNN, DNN, and RNN, respectively; (d) the corresponding ground truth.

3.3.2. Multi-Pitch Estimation

The DNN and RNN were also used as comparative methods for pitch estimation. The architecture and training parameters are the same as that in onset detection, except for the final layer. Each net has 88 units in the output layer, and the output unit activation is sigmoid. In the training and evaluation, all onset time was determined accurately in advance, and the pitch estimation was carried out at each onset.

The results of MPE are shown in Table 2. As shown in Table 2, the CNN outperformed other nets on all evaluation metrics. For example, the CNN yielded a relative improvement of 24.61% over the DNN and outperformed RNN by 15.91% on note-based F-measure. This is largely because the CNN can learn pitch-invariant features from the frames around the onset. We can also observe that the RNN outperformed the DNN on precision and F-measure, which indicates that the context information is helpful in pitch estimation. Therefore, the advantage of CNN is significant in the subtask of onset detection and MPE.

Table 2. Performance on pitch estimation using different neural networks.

Method	Recall	Precision	F-Measure
CNN	0.7810	0.8319	0.8056
DNN	0.6223	0.6727	0.6465
RNN	0.6020	0.8221	0.6950

Figure 7 shows the graphical representation of the outputs of neural networks for the music excerpt along with the corresponding ground truth piano roll. As shown in the ground truth (Figure 7d), the pitch estimation of this excerpt is challenging. The polyphony at each time instant is four in the excerpt's first 8.8 s, and the overlapping is serious. Additionally, the notes are much shorter in the excerpt's last 1.2 s. Compared to the posteriograms of CNN and RNN, DNN estimated more pitches, where many of them were false positives. This is because DNN's topology is simple and its input is just the spectrum at onset. Utilizing the note sequence information in piano music, RNN produced a higher-precision output. However, RNN's output seemed to be a result of monophonic pitch estimation, which yielded many false negatives and corresponded to low recall. In general, the CNN's output was much closer to the ground truth than DNN and RNN. Unlike RNN's input, the context information of CNN's input is from several frames around each onset. CNN can model the attack stage of each pitch through this information, such that the MPE at onset is more accurate. There are also some octave errors which require further effort in the CNN's posteriogram. For example, the MIDI pitch of 46 (about 116.54 Hz) was estimated to be MIDI pitch 58 (about 233.08 Hz) at the eighth onset.

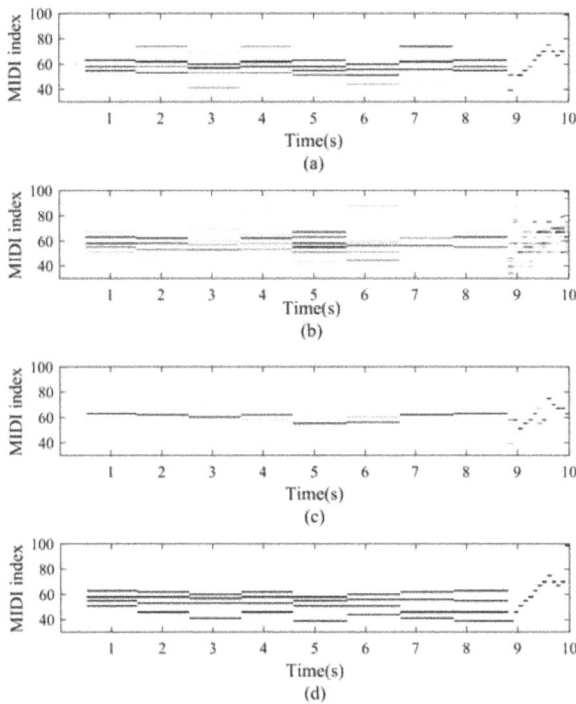

Figure 7. Results of multi-pitch estimation (MPE): (**a–c**) the output of CNN, DNN, and RNN, respectively; (**d**) the corresponding ground truth piano roll representation.

3.3.3. Note Recognition

To evaluate the performance of the proposed note recognition stage which contains two CNNs, another CNN system was used for comparison [21]. The system contained only a single CNN, which transcribes music frame-by-frame and returns a list of notes with pitches and onset. This system will be referred to as Sigtia. Actually, the note recognition stage can be treated as a piano transcription system, which takes no account of the individual to be transcribed. To make a comprehensive comparison, two state-of-the-art transcription methods were also used. Both were submitted to MIREX and evaluated in the task of piano tracking. Benetos's method uses a variable-Q transform representation as input and employs probabilistic latent component analysis in transcription [42].

Troxel's system is based on Microsoft's ResNet, and it has achieved the best performance in MIREX. For Sigtia's method, we trained a CNN using parameters he described in [21]. We have access to the code of Benetos's method, and the second baseline system was implemented by the code. For Troxel's system, the results were obtained from the transcription software named AnthemScore [43].

All of the note-based results of transcription are presented in Table 3. In general, the performance of the proposed note recognition stage is acceptable. Among these four methods, Benetos' approach performed the worst on each evaluation measure. This is because Benetos' model is trained for multiple instruments instead of piano, and the pre-shifted templates are not helpful for piano transcription. The proposed note recognition module outperformed Sigtia's method on all evaluation metrics, which indicates that two independent CNNs are superior to a single one in AMT. Troxel's method yielded the best performance, and it outperformed us by only 0.14% on F-measure. On the metrics of precision, our proposed note recognition stage was inferior to Troxel's system. Therefore, we can use a note verification stage to reduce the false positive notes and improve the precision of transcription.

Table 3. Performance on piano transcription.

Method	Recall	Precision	F-Measure
CNNs	0.7524	0.8593	0.8023
Sigtia	0.6786	0.8023	0.7353
Benetos	0.5857	0.6305	0.6073
Troxel	0.7477	0.8687	0.8037

Figure 8 shows the transcription of the MAPS_MUS-bk_xmas5_ENSTDkCl excerpt using the top two systems in Table 3. The corresponding ground truth has been shown in Figure 7d. Compared with the ground truth, the false positive notes are marked using red crosses and the false negative notes are marked using a blue dashed line. We can observe that the onset of notes in Figure 8a are detected more accurately than that in Figure 8b. This can be attributed to the CNN for onset detection in our system. In the excerpt's first 8.8 s, the transcription result of Troxel's system is better than that of our two consecutive CNNs. There are eight false negative errors and five false positive errors in Figure 8a. Correspondingly, there are only three false negative notes and two false positive notes in Figure 8b. One solution to reduce the false negative errors is to apply a small threshold to the output of the second CNN. This will bring more false positive notes, so an additional note verification stage is necessary. In the excerpt's last 1.2 s, the performance of our note recognition stage was much better than Troxel's system. As the duration of notes here are short, the accurate onset is essential for transcribing them. This also indicates the advantage of our CNNs on short-note transcription.

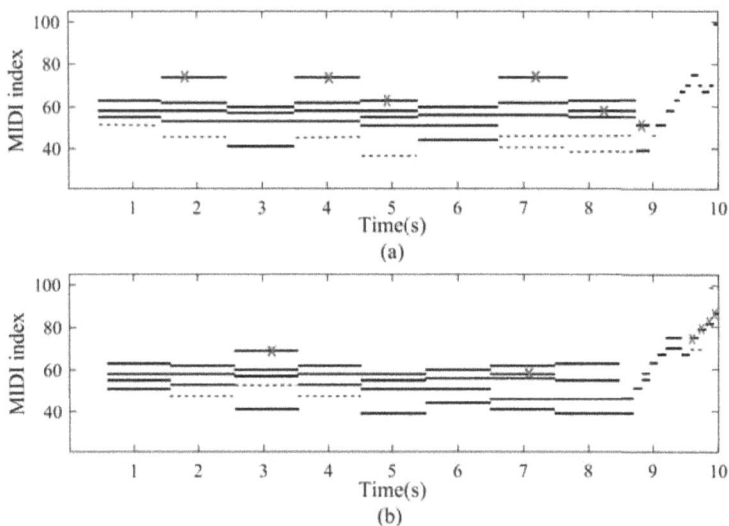

Figure 8. Results of piano transcription: (**a**) the transcription produced by CNNs in our proposed framwork; and (**b**) the transcription produced by Troxel's system.

3.3.4. Transcription for Specific Piano

In our proposed framework, the individual-specific transcription is conducted by feeding the output of note recognition into a note verification stage. For comparative purposes, two transcription systems were used to evaluate the performance of the proposed method. The first comparative approach was proposed by Cheng, which is the current state-of-the-art specific piano transcription method [32]. Cheng's method is implemented using a sparse NMF in AMT, and all the templates are extracted using the notes from "ENSTDkCl" of MAPS. Considering that the CNNs have shown advantages in the note recognition stage, the second comparative approach is based on them. Adding the specific individual's data to the training set, we got two adapted CNNs. To make a fair comparison, the newly-added training samples were isolated notes produced by the same piano.

The transcription results are shown in Table 4, and the proposed method performed best in general. Although they are based on the same note recognition module, the proposed system outperformed the adapted CNNs on all evaluation metrics. This illustrates the benefits of note verification. Another reason is that the CNNs cannot learn enough information about the specific individual through these limited isolated notes—especially the information of polyphony. The proposed system outperformed Cheng's system in terms of recall and F-measure. Our proposed method estimated 5511 notes correctly, whereas the number of true positive notes was 5421 for Cheng's method. This can be attributed to the use of note recognition, which achieved significant performance on recall through CNNs. Meanwhile, the preliminary results led to a limitation of note verification. Both the proposed method and Cheng's method achieved better performance than the adapted CNN on all evaluation metrics. One of the reasons may be that both of them use the templates of attack during the NMF.

Table 4. Performance comparison on specific piano transcription.

Method	Recall	Precision	F-Measure
Proposed	0.7503	0.9039	0.8200
Cheng	0.7381	0.9070	0.8139
Adapted CNNs	0.7458	0.8792	0.8070

In general, all of the specific piano transcription systems in Table 4 perform better than universal systems in Table 3. We can conclude that making use of the information of specific individual is promising in AMT. Compared with results in Table 3, The proposed system performed better on the metrics of precision and F-measure when the note verification stage was applied. Therefore, the effectiveness of note verification is validated again.

The results of the proposed method and the state-of-the-art method are compared concretely. Figure 9 shows the F-measure obtained by our proposed and Cheng's methods, which is along the different octaves of a piano. As shown in Figure 9, our proposed method outperformed Cheng's method for six octaves, except for the A5-Ab6 octave. Cheng's method achieved an F-measure of 0.4854 for A0-Ab1, which shows its poor performance in the transcription of low-pitch notes. The proposed method showed a more balanced result, with an F-measure of 0.5672 for the first octave. In general, the F-measure increased approximately along the increase of octaves for the two methods. This suggests the limitation of the time-domain approach, which brings a time–frequency resolution trade-off.

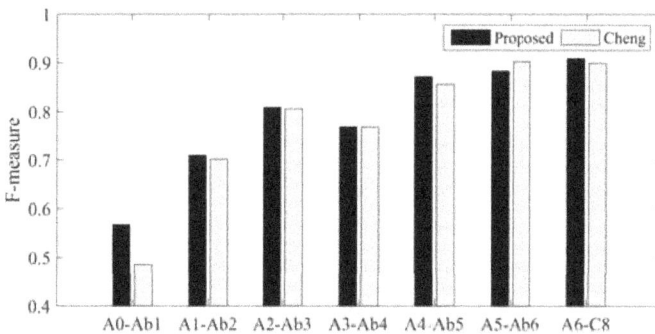

Figure 9. F-measure per octave achieved by our proposed system and Cheng's system.

Figure 10 shows the specific piano transcription of the MAPS_MUS-bk_xmas5_ENSTDkCl excerpt, which was produced by our proposed framework and Cheng's system. Compared with the ground truth in Figure 7d, the false positive notes are marked using red crosses, and the false negative notes are marked using a blue dashed line. The contrast between Figures 8a and 10a indicates that the note verification can improve the precision of transcription. As shown in Figure 10, Cheng's method estimated more correct pitches than our proposed method in the excerpt's first 8.8 s. This is due to a limitation in our proposed system. Although the note verification conducted on candidate notes can save computing time and storage space, it is limited because the candidate set is not complete. In the excerpt's last 1.2 s, our system yielded a better performance than Cheng's system. This indicates the advantage of our note recognition stage, which is good at transcribing short notes. Another reason is that modeling both the attack and decay stages in short duration is difficult for Cheng's system.

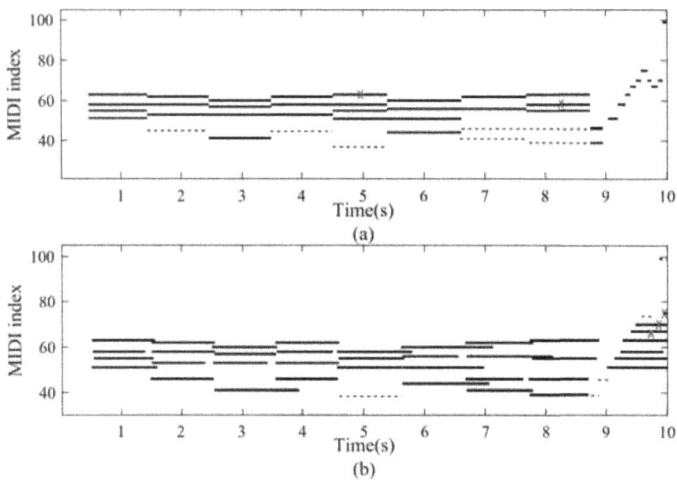

Figure 10. Results of specific piano transcription: (**a**) the transcription of our proposed system and (**b**) the transcription of Cheng's system.

4. Conclusions

We present a two-stage framework for note-level polyphonic piano music transcription, which comprises a note recognition stage and a note verification stage. In the note recognition, one CNN is trained for onset detection and another is trained for pitch estimation at each onset. To our knowledge, the combination of two CNNs has not been attempted before for AMT. The note verification for the specific piano is implemented using NMF. The factorization is conducted in the time slice around candidate onset, which only uses attack templates of the candidate pitches. Our experiments are carried out on the MAPS database and the performance of each module is discussed. The experiments demonstrate that CNN performs better than other types of neural networks in the subtasks of onset detection and pitch estimation, and the connection of two CNNs outperforms a single CNN in note recognition. We also observe that the performance of transcription is improved significantly when note verification is applied to the system, and our proposed system performs better than state-of-the-art systems in specific piano transcription.

There are some limitations of the proposed system. As the biggest dataset for piano AMT, the MAPS has only 270 solo pieces. So, the data may be not enough for training CNNs. Although training data and testing data are from synthesized pianos and a real piano, respectively, they contain overlaps in music pieces. The limited data and piece-dependent scheme led the CNNs to overfit. For the real pieces in the testing dataset, the recording environment was quiet and the distance between the piano and microphones was close. Therefore, one future research direction is to discuss whether the proposed method is robust to noise and reverberation. Additionally, the proposed method cannot estimate note offsets or loudness, which will be another research direction in the future.

Acknowledgments: This work is partially supported by the National Natural Science Foundation of China (Nos. 11461141004, 61271426, U1536117, 11504406, 11590770-4), the Strategic Priority Research Program of the Chinese Academy of Sciences (Nos. XDA06030100, XDA06030500, XDA06040603), the National 863 Program (No. 2015AA016306) and the National 973 Program (No. 2013CB329302).

Author Contributions: Qi Wang and Ruohua Zhou conceived of and designed the experiments; Qi Wang performed the experiments; Qi Wang and Ruohua Zhou analyzed the data; Yonghong Yan contributed analysis tools; Qi Wang wrote the paper.

Conflicts of Interest: The authors declare no conflict of interest.

Appendix A

```
#Builds the CNN. The code is based on the open source software library-TensorFlow.
import tensorflow as tf

def inference(images0):
    """Build the CNN model.
    Args: images0: Images placeholder, from inputs().
    Returns: sigmoid_linear: Output tensor with the computed probabilities.
    """
    images=tf.reshape(images0, [-1,267,9,1])

    # conv1
    with tf.variable_scope('conv1') as scope:
        weights = tf.Variable(tf.truncated_normal([16,2,1,10],stddev=0.1))
        conv = tf.nn.conv2d(images, weights, [1,1,1,1],padding='VALID')
        biases = tf.Variable(tf.constant(0.1,shape=[10]))
        pre_activation = tf.nn.bias_add(conv, biases)
        conv1 = tf.nn.relu(pre_activation, name=scope.name)

    # pool1
    pool1 = tf.nn.max_pool(conv1, ksize=[1, 2, 2, 1], strides=[1, 2, 2, 1], padding='SAME',
name='pool1')

    # conv2
    with tf.variable_scope('conv2') as scope:
        weights = tf.Variable(tf.truncated_normal([11,3,10,20],stddev=0.1))
        conv = tf.nn.conv2d(pool1, weights, [1, 1, 1, 1], padding='VALID')
        biases = tf.Variable(tf.constant(0.1,shape=[20]))
        pre_activation = tf.nn.bias_add(conv, biases)
        conv2 = tf.nn.relu(pre_activation, name=scope.name)

    # pool2
    pool2 = tf.nn.max_pool(conv2, ksize=[1, 2, 2, 1], strides=[1, 2, 2, 1], padding='SAME',
name='pool2')

    # fully-connected1
    with tf.variable_scope('fully-connected1') as scope:
        reshape = tf.reshape(pool2, [-1,58*1*20])
        weights = tf.Variable(tf.truncated_normal([58*1*20,256],stddev=0.1))
        biases = tf.Variable(tf.constant(0.1,shape=[256]))
        local = tf.nn.relu(tf.matmul(reshape, weights) + biases, name=scope.name)

    # fully-connected2
    with tf.variable_scope('fully-connected2') as scope:
        #dropout
        local3_drop =tf.nn.dropout(local, 0.5)
        weights = tf.Variable(tf.truncated_normal([256,num_classes],stddev=0.1))
        biases = tf.Variable(tf.constant(0.1,shape=[num_classes]))
        sigmoid_linear = tf.nn.sigmoid(tf.matmul(local3_drop, weights) + biases, name=scope.name)
```

```
    return sigmoid_linear

def loss(logits, labels):
    """Calculates the loss from the logits and the labels.
    Args:
    logits: Logits from inference(), float - [batch_size, num_classes].
    labels: Labels tensor, int32 - [batch_size, num_classes].
    Returns: cross_entropy: Loss tensor of type float.
    """

    cross_entropy = -tf.reduce_sum(labels*tf.log(logits+1e-10)+(1-labels)*tf.log(1-logits+1e-10))
    return cross_entropy

def evaluation(logits, labels, threshold):
    """Evaluate the quality of the logits at predicting the label.
    Args:
    logits: Logits from inference(), float - [batch_size, num_classes].
    labels: Labels tensor, int32 - [batch_size, num_classes].
    threshold: Threshold applied to the logits.
    Returns: accuracy: Compute precision of predicting.
    """

    pred=tf.cast(tf.greater(logits, threshold),"float")
    correct_prediction = tf.cast(tf.equal(pred, labels), "float")
    accuracy = tf.reduce_mean(correct_prediction)
    return accuracy

def training(loss, learning_rate):
    """Sets up the training Ops.
    Creates an optimizer and applies the gradients to all trainable variables.
    Args:
    loss: Loss tensor, from loss().
    learning_rate: The learning rate to use for gradient descent.
    Returns: train_op: The Op for training.
    """

    # Create the gradient descent optimizer with the given learning rate.
    optimizer = tf.train.AdamOptimizer(learning_rate)
    # Use the optimizer to apply the gradients that minimize the loss
    train_op = optimizer.minimize(loss)
    return train_op
```

Reference

1. Moorer, J.A. On the transcription of musical sound by computer. *Comput. Music J.* **1977**, *1*, 32–38.
2. Piszczalski, M.; Galler, B.A. Automatic music transcription. *Comput. Music J.* **1977**, *1*, 24–31.
3. Klapuri, A. Introduction to music transcription. In *Signal Processing Methods for Music Transcription*; Springer: Boston, MA, USA, 2006; pp. 3–20.
4. Cogliati, A.; Duan, Z.; Wohlberg, B. Piano transcription with convolutional sparse lateral inhibition. *IEEE Signal Process. Lett.* **2017**, *24*, 392–396.
5. Benetos, E.; Dixon, S.; Giannoulis, D.; Kirchhoff, H.; Klapuri, A. Automatic music transcription: Challenges and future directions. *J. Intell. Inf. Syst.* **2013**, *41*, 407–434.
6. Klapuri, A.P. Multiple fundamental frequency estimation based on harmonicity and spectral smoothness. *IEEE Trans. Speech Audio Process.* **2003**, *11*, 804–816.

7. Pertusa, A.; Inesta, J.M. Multiple fundamental frequency estimation using Gaussian smoothness. In Proceedings of the IEEE International Conference on Acoustics, Speech and Signal Processing (ICASSP), Las Vegas, NV, USA, 31 March–4 April 2008; pp. 105–108.

8. Brown, J.C. Calculation of a constant Q spectral transform. *J. Acoust. Soc. Am.* **1991**, *89*, 425–434.

9. Zhou, R.; Reiss, J.D. A real-time polyphonic music transcription system. In Proceedings of the 4th Music Information Retrieval Evaluation eXchange (MIREX), Philadelphia, PA, USA, 14–18 September 2008.

10. Dressler, K. Multiple fundamental frequency extraction for MIREX 2012. In Proceedings of the 8th Music Information Retrieval Evaluation eXchange (MIREX), Porto, Portugal, 8–12 October 2012.

11. Smaragdis, P.; Brown, J.C. Non-negative matrix factorization for polyphonic music transcription. In Proceedings of the IEEE Workshop on Applications of Signal Processing to Audio and Acoustics, New Paltz, NY, USA, 19–22 October 2003; pp. 177–180.

12. Smaragdis, P.; Raj, B.; Shashanka, M. A probabilistic latent variable model for acoustic modeling. *Adv. Models Acoust. Process.* **2006**, *148*, 1–8.

13. Benetos, E.; Dixon, S. A shift-invariant latent variable model for automatic music transcription. *Comput. Music J.* **2012**, *36*, 81–94.

14. Nam, J.; Ngiam, J.; Lee, H.; Slaney, M. A classification-based polyphonic piano transcription approach using learned feature representations. In Proceedings of the International Society for Music Information Retrieval Conference (ISMIR), Miami, FL, USA, 24–28 October 2011; pp. 175–180.

15. Sigtia, S.; Benetos, E.; Boulanger-Lewandowski, N.; Weyde, T.; Garcez, A.S.D.; Dixon, S. A hybrid recurrent neural network for music transcription. In Proceedings of the 2015 IEEE International Conference on Acoustics, Speech and Signal Processing (ICASSP), Brisbane, QLD, Australia, 19–24 April 2015; pp. 2061–2065.

16. Kelz, R.; Widmer, G. An experimental analysis of the entanglement problem in neural-network-based music transcription systems. In Proceedings of the AES Conference on Semantic Audio, Erlangen, Germany, 22–24 June 2017.

17. Berg-Kirkpatrick, T.; Andreas, J.; Klein, D. Unsupervised transcription of piano music. In Proceedings of the International Conference on Neural Information Processing Systems (NIPS), Montreal, QC, Canada, 8–13 Demcember 2014; pp. 1538–1546.

18. Ewert, S.; Plumbley, M.D.; Sandler, M. A dynamic programming variant of non-negative matrix deconvolution for the transcription of struck string instruments. In Proceedings of the IEEE International Conference on Acoustics, Speech and Signal Processing (ICASSP), Brisbane, QLD, Australia, 19-24 April 2015; pp. 569–573.

19. Kameoka, H.; Nishimoto, T.; Sagayama, S. A multipitch analyzer based on harmonic temporal structured clustering. *IEEE Trans. Audio Speech Lang. Process.* **2007**, *15*, 982–994.

20. Böck, S.; Schedl, M. Polyphonic piano note transcription with recurrent neural networks. In Proceedings of the IEEE International Conference on Acoustics, Speech and Signal Processing (ICASSP), Kyoto, Japan, 25–30 March 2012; pp. 121–124.

21. Sigtia, S.; Benetos, E.; Dixon, S. An end-to-end neural network for polyphonic piano music transcription. *IEEE/ACM Trans. Audio Speech Lang. Process.* **2016**, *24*, 927–939.

22. Marolt, M. A connectionist approach to automatic transcription of polyphonic piano music. *IEEE Trans. Multimed.* **2004**, *6*, 439–449.

23. Costantini, G.; Perfetti, R.; Todisco, M. Event based transcription system for polyphonic piano music. *Signal Process.* **2009**, *89*, 1798–1811.

24. Barbancho, I.; de la Bandera, C.; Barbancho, A.M.; Tardon, L.J. Transcription and expressiveness detection system for violin music. In Proceedings of the IEEE International Conference on Acoustics, Speech and Signal Processing (ICASSP), Taipei, Taiwan, 19–24 April 2009; pp. 189–192.

25. Marolt, M. Automatic transcription of bell chiming recordings. *IEEE Trans. Audio Speech Lang. Process.* **2012**, *20*, 844–853.

26. Barbancho, A.M.; Klapuri, A.; Tardón, L.J.; Barbancho, I. Automatic transcription of guitar chords and fingering from audio. *IEEE Trans. Audio Speech Lang. Process.* **2012**, *20*, 915–921.

27. Wan, Y.; Wang, X.; Zhou, R.; Yan, Y. Automatic Piano Music Transcription Using Audio-Visual Features. *Chin. J. Electron.* **2015**, *24*, 596–603.

28. 2016: Multiple Fundamental Frequency Estimation Tracking Results—MIREX Dataset. Available online: http://www.music-ir.org/mirex/wiki/2016:Multiple_Fundamental_Frequency_Estimation_%26_Tracking_Results_-_MIREX_Dataset (accessed on 15 October 2016).
29. Cogliati, A.; Duan, Z. Piano music transcription modeling note temporal evolution. In Proceedings of the IEEE International Conference on Acoustics, Speech and Signal Processing (ICASSP), Brisbane, QLD, Australia, 19–24 April 2015; pp. 429–433.
30. Cogliati, A.; Duan, Z.; Wohlberg, B. Context-dependent piano music transcription with convolutional sparse coding. *IEEE/ACM Trans. Audio Speech Lang. Process.* **2016**, *24*, 2218–2230.
31. Ewert, S.; Sandler, M. Piano transcription in the studio using an extensible alternating directions framework. *IEEE/ACM Trans. Audio Speech Lang. Process.* **2016**, *24*, 1983–1997.
32. Cheng, T.; Mauch, M.; Benetos, E.; Dixon, S. An attack/decay model for piano transcription. In Proceedings of the International Society for Music Information Retrieval Conference (ISMIR), New York, NY, USA, 7–11 August 2016.
33. Gao, L.; Su, L.; Yang, Y.H.; Lee, T. Polyphonic piano note transcription with non-negative matrix factorization of differential spectrogram. In Proceedings of the IEEE International Conference on Acoustics, Speech and Signal Processing (ICASSP), New Orleans, LA, USA, 5–9 March 2017; pp. 291–295.
34. Li, T.L.; Chan, A.B.; Chun, A. Automatic musical pattern feature extraction using convolutional neural network. In Proceedings of the International Multi Conference of Engineers and Computer Scientists, Hongkong, China, 17–19 March 2010.
35. Dieleman, S.; Brakel, P.; Schrauwen, B. Audio-based music classification with a pretrained convolutional network. In Proceedings of the International Society for Music Information Retrieval Conference (ISMIR), Miami, FL, USA, 24–28 October 2011; pp. 669–674.
36. Humphrey, E.J.; Bello, J.P. Rethinking automatic chord recognition with convolutional neural networks. In Proceedings of the International Conference on Machine Learning and Applications (ICMLA), Boca Raton, FL, USA, 12–15 December 2012; Volume 2, pp. 357–362.
37. Schluter, J.; Bock, S. Improved musical onset detection with convolutional neural networks. In Proceedings of the IEEE International Conference on Acoustics, Speech and Signal Processing (ICASSP), Florence, Italy, 4–9 May 2014; pp. 6979–6983.
38. Emiya, V.; Badeau, R.; David, B. Multipitch estimation of piano sounds using a new probabilistic spectral smoothness principle. *IEEE Trans. Audio Speech Lang. Process.* **2010**, *18*, 1643–1654.
39. Kingma, D.; Ba, J. Adam: A method for stochastic optimization. *arXiv* **2014**, arXiv:1412.6980.
40. Bay, M.; Ehmann, A.F.; Downie, J.S. Evaluation of Multiple-F0 Estimation and Tracking Systems. In Proceedings of the International Society for Music Information Retrieval Conference (ISMIR), Kobe, Japan, 26–30 October 2009; pp. 315–320.
41. Eyben, F.; Böck, S.; Schuller, B.W.; Graves, A. Universal Onset Detection with Bidirectional Long Short-Term Memory Neural Networks. In Proceedings of the International Society for Music Information Retrieval Conference (ISMIR), Utrecht, The Netherlands, 9–13 August 2010; pp. 589–594.
42. Benetos, E.; Weyde, T. An efficient temporally-constrained probabilistic model for multiple-instrument music transcription. In Proceedings of the International Society for Music Information Retrieval Conference (ISMIR), Malaga, Spain, 20–26 October 2015.
43. Troxel, D. Automatic Music Transcription Software. Available online: https://www.lunaverus.com/ (accessed on 9 December 2016).

![applied sciences logo] *applied* *sciences*

MDPI

Article

A Low Cost Wireless Acoustic Sensor for Ambient Assisted Living Systems

Miguel A. Quintana-Suárez [1,†], **David Sánchez-Rodríguez** [1,2,*,†], **Itziar Alonso-González** [1,2,†] and **Jesús B. Alonso-Hernández** [2,3,†]

[1]   Telematic Engineering Department, University of Las Palmas de Gran Canaria, Campus Universitario de Tafira, 35017 Las Palmas de Gran Canaria, Spain; mangel.quintana@ulpgc.es (M.A.Q.-S.); itziar.alonso@ulpgc.es (I.A.-G.)
[2]   Institute for Technological Development and Innovation in Communications, University of Las Palmas de Gran Canaria, Campus Universitario de Tafira, 35017 Las Palmas de Gran Canaria, Spain; jesus.alonso@ulpgc.es
[3]   Signal and Communications Department, University of Las Palmas de Gran Canaria, Campus Universitario de Tafira, 35017 Las Palmas de Gran Canaria, Spain
*   Correspondence: david.sanchez@ulpgc.es; Tel.: +34-928-458047; Fax: +34-928-451380
†   These authors contributed equally to this work.

Academic Editor: Tapio Lokki
Received: 31 July 2017; Accepted: 25 August 2017; Published: 27 August 2017

Abstract: Ambient Assisted Living (AAL) has become an attractive research topic due to growing interest in remote monitoring of older people. Development in sensor technologies and advances in wireless communications allows to remotely offer smart assistance and monitor those people at their own home, increasing their quality of life. In this context, Wireless Acoustic Sensor Networks (WASN) provide a suitable way for implementing AAL systems which can be used to infer hazardous situations via environmental sounds identification. Nevertheless, satisfying sensor solutions have not been found with the considerations of both low cost and high performance. In this paper, we report the design and implementation of a wireless acoustic sensor to be located at the edge of a WASN for recording and processing environmental sounds which can be applied to AAL systems for personal healthcare because it has the following significant advantages: low cost, small size, audio sampling and computation capabilities for audio processing. The proposed wireless acoustic sensor is able to record audio samples at least to 10 kHz sampling frequency and 12-bit resolution. Also, it is capable of doing audio signal processing without compromising the sample rate and the energy consumption by using a new microcontroller released at the last quarter of 2016. The proposed low cost wireless acoustic sensor has been verified using four randomness tests for doing statistical analysis and a classification system of the recorded sounds based on audio fingerprints.

Keywords: wireless acoustic sensor; ambient assisted living; internet of things; edge computing; low cost; ESP32

1. Introduction

As one of the fastest growing technologies in the emerging Internet of Things (IoT) environment, low power wireless sensor networks are expected to realize IoT applications and to provide connectivity for remote smart objects. The basic concept of IoT is that various smart objects can be automatically linked into a network for interacting with humans through perception and networking technologies [1]. Smart objects in the IoT have the ability to send information through the Internet to provide the interaction among multiple things and people. IoT is opening tremendous opportunities for novel applications that promise to improve the quality of people life.

The development of IoT technologies can be applied to a huge variety of applications, such as intelligent power grid [2], healthcare [3], environmental monitoring [4], localization [5], etc. In an AAL context where assisted living technologies are based on ambient intelligence, smart objects need to use wireless communications because of the requirements of supporting mobile applications and remote monitoring of people. AAL can be used for detecting and preventing distress situations, improving wellness and health conditions of older adults. AAL technologies can also provide more safety for the elderly, using mobile emergency response systems, detecting domestic accidents, monitoring activities of daily living, issuing reminders, as well as helping with mobility and automation, and, overall, improving their quality of life [6,7]. In fact, according [8], AAL should be understood as a system for extending the time people can live in their preferred environment by increasing their autonomy, self-confidence and mobility; supporting the preservation of health and functional capabilities of the elderly, promoting a better and healthier lifestyle for individuals at risk; enhancing security, preventing social isolation and supporting the preservation of the multifunctional network around the individual; supporting carers, families and care organizations; and increasing the efficiency and productivity of used resources in the ageing societies.

A survey of sensors in assisted living of older people is presented in [9], such as passive infrared (PIR) and vibration sensors, accelerometers, cameras, depth sensors, and microphones. These systems should satisfy some requirements as: low-cost, high accuracy, user acceptance and privacy. These can be connected to form a network for an intelligent home designed for elderly people. The data and decision results that the sensors produce can be processed and fused over a cloud or a fog. Authors expect that the IoT will lead to remote health monitoring and emergency notification AAL systems that will operate autonomously, without requiring user intervention. In this context, audio recognition is also a promising way to ensure more safety by contributing to detection of distress situations because of the interaction of each person with her environment may be identified. In fact, in [10] detection of distress situations and monitoring of activity and health are described as two challenges to address in AAL environments. On the one hand, the identification the sounds of everyday life can be particularly useful for detecting distress situations in which the person might be. For instance, the detection of a fall can be used to call an emergency number. On the other hand, audio processing can be quite useful for the monitoring of the person's activity and the assessment of some decline. For instance, an application might consist of recognising health related symptoms such as coughing, scraping throats and sniffles. Hence, the development of WASN with low power consumption and low cost are suitable for implementing AAL systems. In this research, we are focused in the development of a low cost wireless acoustic sensor with audio processing capabilities and network connectivity to be located at the edge of a WASN.

The WASN have been developed under the paradigms of both the Smart City and the IoT. In recent years, there has been a rapid evolution of WASN, and many works have been developed. To date, several authors have designed and deployed WASN for different purposes such as noise monitoring [11] or sound identification as road traffic, horns, and people [12]. For instance, in [13] the production and analysis of a real-life environmental audio database in two urban and suburban scenarios corresponding to the pilot areas of the DYNAMAP project was presented. The WASN of the DYNAMAP project is based on low cost acoustic sensor but it is connected to digital recorder for data saving. Hence, unlike our research, audio samples cannot be sent to a central server using wireless communications, and neither audio processing can not be carried out at node. Audio recordings have been categorized as road traffic noise, background city noise, and anomalous noise events. However, it was carried out offline with Audicity and Matlab software [14].

In [15], a distributed noise measurement system based on IoT technology was developed. The sensor node is based on a Raspberry Pi with an electret omnidirectional microphone and a sound card in order to record the audio. The data from WASN was interpolated for obtaining a spatial noise level in a small-sized city. However, the system was designed to measure, represent and predict urban noise levels, and not for audio processing and classification.

In [16], the design of low cost wireless sensor node for environmental noise measurement is described. The sensor node platform is built on ATmega128L with 4 kB RAM, and its internal 10-bit ADC can operate a peak sampling rate of 33 kHz. However, according the microcontroller specifications, the maximum sampling rate for 10-bit resolution is 9.8 kHz and not 33 kHz. In addition, only the effective sound pressure is sent, and an audio processing is not carried out.

Nevertheless, the WASN paradigm presents several challenges, ranging from those derived from the design and development of the wireless sensor network, such as energy harvesting and low cost hardware development and maintenance, to some specific challenges derived from the automation of the data collection and subsequent signal processing, such as to detect anomalous noise events [13]. In addition, the sensor of a WASN designed for AAL systems should process the enviromental sounds to rapidly infer hazardous situations instead to send the full audio record to a server for a centralized processing. Thus, processing data at the node can ensure shorter response time and better reliability. In this context, the use of devices with an increasing storage and computation capacity coins a new term: Edge or Fog Computing. This model extends Cloud computing and services to the edge of the network reducing network latency and offloading computation [17], as well as to avoid bottlenecks at remote server due to the throughput and volume of data to be collected and processed. Edge Computing has the potential to address the concerns of response time requirement, battery life constraint, bandwidth cost saving, as well as data safety and privacy [18]. This concept covers computational to be performed at the edge of the network and to exchange data from or to cloud IoT services. In [19], the design and deployment of a WASN at home, inspired by the Mobile Edge Computing paradigm [20] able to gather the data of all acoustic sensing nodes deployed to infer the audio events of interest in an AAL context is described. It follows a distributed implementation, where all digital signal processing is carried out in a concentrator offloading the sensor nodes and avoiding the use of the server to remotely process the data. This concentrator is based on a GPU embedded platform.

As has been discussed, many research using low cost sensors in a WASN have been developed. Nevertheless, those works have been designed to measure only noise levels and not for sound identification. On the other hand, research where audio processing is carried out are based on medium cost platforms, such as Raspberry or GPU, or using cloud services. Hence, the aim of this research is to solve these deficiencies designing a low cost acoustic sensor to do audio processing at the edge of network.

There is no doubt that significant progress has been made in the field of wireless acoustic sensor networks. However, an improvement to the actual sensors is needed because the main drawback of these recent WASN is that their power consumption and cost do not fit some of the critical requirements of AAL applications: power consumption, mobility, size and cost. In addition, humans are most sensitive to frequencies between 2 kHz and 5 kHz, and the speech and environmental sounds are often less than 5 kHz bandwidth. Therefore, the sensor has to be characterized by a spectrum with these frequencies. In this paper, a novel wireless acoustic sensor is proposed. The main novelty of this work comes from the fact that the proposed wireless acoustic sensor is able to record audio samples at least at 10 kHz sampling frequency (5 kHz bandwidth) and 12-bit resolution, and audio signal processing can be carried out at node without compromising the sample rate and the energy consumption. Furthermore, this sensor can be applied in AAL systems for personal healthcare because it has the following significant advantages: low cost, small size, wireless network connectivity, audio sampling and computation capabilities for audio processing. Thus, the identification of sounds for an AAL context, such as fall detection or health related symptoms, could be carried out at the edge of a WASN reducing network latency and improving response time and battery life of proposed sensor, enhancing quality of life and safety of older people.

The remainder of this paper is organized as follows. Section 2 describes the low cost proposed wireless acoustic sensor. Section 3 describes the used methods to validate and evaluate the proposed sensor. Section 4 shows the results of some experiments carried out to validate the designed sensor in this study. Finally, in Section 5 draws some conclusions and discusses some possible directions for future research.

2. Wireless Acoustic Sensor

In this section the proposed low cost wireless acoustic sensor is described which is formed by an audio sensor and a microcontroller based board. The main goal in designing the sensor was to obtain a product of small size, low cost, low consumption and versatile which allows to be used in permanent and remote monitoring in AAL systems.

2.1. Audio Sensor

The audio sensor is an electret microphone amplifier with adjustable gain [21]. It is based on an electret omnidirectional microphone, CMA-4544PF-W, and an op-amp specifically optimized for use as microphone preamplifiers, a Maxim MAX4466, Figure 1. They provide the ideal combination of an optimized gain bandwidth product with low voltage operation in ultra-small packages. Furthermore, it has an almost flat response in the frequency range between 50 Hz and 10 kHz, Figure 2. Therefore, the characterization of sensor is fulfilled whose operating frequency lies within the range 100 Hz–5 kHz.

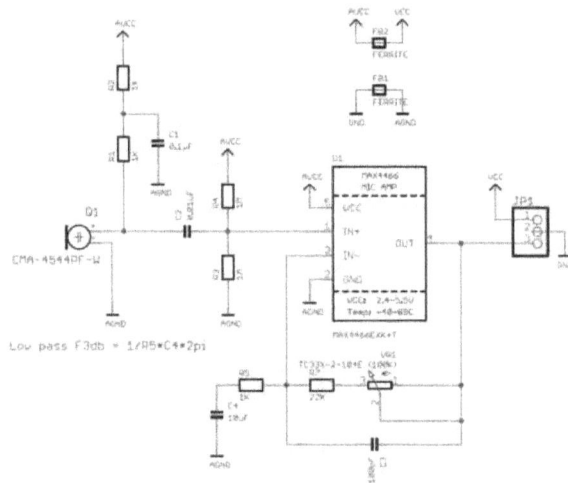

Figure 1. Schematic of audio sensor.

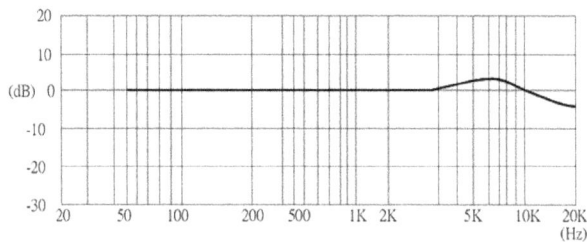

Figure 2. Frequency response curve of microphone.

2.2. Microcontroller Based Board

Three low cost microcontroller platforms were evaluated, joinly with the above audio sensor, to determine the best option for the proposed system: Libelium Waspmote platform [22], Espressif ESP8266 board [23], and Espressif ESP32 board [24]. Figure 3 shows the evaluated microcontroller boards.

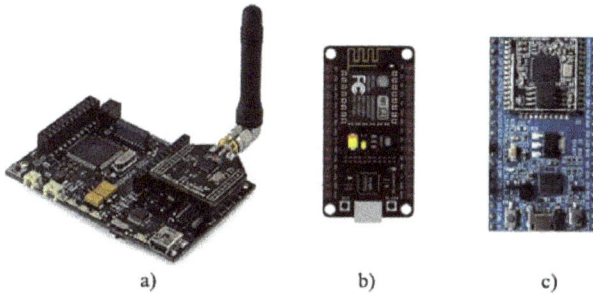

a) b) c)

Figure 3. Microcontroller boards: (**a**) Waspmote; (**b**) ESP8266; (**c**) ESP32.

Waspmote board is a modular device that allows us to install different sensors and different radio transceivers. Waspmote hardware architecture has been specially designed to be extremely low consumption. The Waspmote has an Atmega1281 microcontroller running at 14 MHz with programmable sleep modes. These sleep modes make Waspmote the lowest consumption sensor platform in the market (0.7 uA in hibernate mode and 55 uA in sleep mode). The whole set, formed by Waspmote and audio sensor, has a small size (85 × 75 × 35 mm, included battery). The ATmega1281 has 8 ADC channels with 10-bit resolution. Due to the microcontroller characteristics, the tested maximum ADC sampling frequency was 9.8 kHz. In addition, it has only 8 kB SRAM, and therefore, the audio recording is about a few tenths of a second maximum duration. Waspmote has an SD card and could be used to save the sampled data. Nevertheless, the sample rate of ADC converter must be fit to 8-bit resolution to carry out these extra operations needed.

ESP8266 board delivers a highly integrated Wi-Fi SoC solution for efficient power, with its complete and self-contained Wi-Fi networking capabilities. It integrates an enhanced version of Tensilica's L106 Diamond series 32-bit processor and on-chip SRAM with an ADC with 10-bit resolution, and can be interfaced with external sensors through the GPIOs, in low development cost at prototyped. One of the most common boards with the ESP8266 is NodeMCU, with ESP-12E module, Figure 3b. The whole set, formed by ESP8266 and audio sensor, has a very small size (50 × 30 × 20 mm) which is very useful to place at different positions in a discrete way. It can support up to 80 MHz frequency clock. It has a built-in SPI flash memory with 4MB capacity and the SRAM capacity available to users is about 36 kB. The tested maximum ADC sampling frequency was 10.6 kHz.

Espressif Systems announced the launch of ESP32 cloud on chip on September 6th, 2016. It is a Dual Core Wi-Fi + BT Combo MCU. Some of features of the ESP32 are the following: the CPU is an Xtensa Dual-Core 32-bit LX6 microprocessor, operating up to 240 MHz, 520 kB SRAM, 12-bit SAR ADC up to 18 channels and built-in Wi-Fi card, supporting IEEE 802.11 b/g/n standards, and Bluetooth v4.2 BR/EDR and BLE. Also, the ESP32 chip features 40 physical GPIO pads which can be used as a general purpose I/O, to connect new sensors, or can be connected to an internal peripheral signal. The most common development board is the ESP32S, with a ESP-WROOM-32 module and an SRAM capacity available to users about 300 kB. As previous board, the whole set, formed by ESP32 and audio sensor, has a very small size (55 × 30 × 20 mm), and is very useful to place at different positions in a discrete way. The tested maximum ADC sampling frequency was 100 kHz. It is enough for the system purposes. Furthermore, ADC is non-blocking, so the conversion process with other instructions execution can be overlapped.

Although all analyzed boards can fulfill for implementing a wireless acoustic sensor, ESP32 board was chosen because it has the biggest: ADC bit resolution, SRAM capacity, and microprocessor frequency. Furthermore, because of it has a dual core and the microcontroller's connectivity features and functionalities, audio samples are gathered while other operations can be simultaneously done, such as sending data to a server using IP protocol or data processing, thus promoting the edge computing idea. Figure 4 shows the wireless acoustic sensor based on ESP32 board. On the other hand, the cost of proposed sensor is about 10 Euros, being it very competitive for an AAL environment.

Figure 4. Wireless acoustic sensor based on ESP32 board.

Lastly, the principle of operation of the software implementation for the proposed wireless acoustic sensor is shown in Figure 5. First, a timer interruption is enabled for gathering the audio samples from ADC using *AnalogRead* function. Timer is set to 100 μs to obtain a sampling frequency at 10 kHz. Next, the data obtained from ADC are stored in an endless buffer. Finally, the raw data can be sent to a server for recording the sampled data in a wav format file or a suitable audio fingerprint to identify different sound events that can be used for detecting hazardous situations.

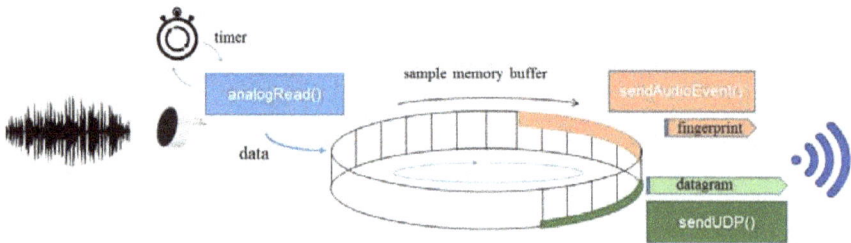

Figure 5. Graphical flow diagram implemented in the wireless acoustic sensor.

3. System Validation Methods

In order to validate the proposed wireless acoustic sensor a statistical analysis and an audio classification of recorded samples are carrying out. Thus, randomness tests and an audio fingerprint matching were the methods employed for system validation and are described in this section.

3.1. Randomness Tests

Randomness tests can be used to determine whether a dataset has a recognizable pattern, and therefore whether the process that generated it is significantly random. That is, it can be used to test the hypothesis that the elements of a sequence are mutually independent or not. Four randomness tests were used to demonstrate that recorded audio files by the proposed system have a recognizable pattern, and hence, the sampled audio information is not random. The following randomness tests were used: Bartels Test [25], Cox Stuart Test [26], Mann-Kendall Test [27] and Wald-Wolfowitz Test [28].

3.1.1. Bartels Test

This randomness test is the rank version of von Neumann's Ratio Test for Randomness [29].

3.1.2. Cox Stuart Test

In this test data are grouped in pairs with the *ith* observation of the first half paired with the *ith* observation of the second half of the time-ordered data. If the length of vector X is odd the middle observation is eliminated. The Cox Stuart test is then simply a sign test applied to these paired data.

3.1.3. Mann-Kendall Test

This randomness test is a non-parametric statistical test that analyzes difference in signs between earlier and later data points. The idea is that if a trend is present, the sign values will tend to increase constantly, or decrease constantly. Every value is compared to every value preceding it in the time series.

3.1.4. Wald-Wolfowitz Test

This randomness test is a non-parametric statistical test that transforms into a dichotomous vector according as each values is above or below a given threshold. Values equal to the level are removed from the sample. The default threshold value used in applications is the sample median.

3.2. Audio Fingerprint Matching

An audio fingerprint is a compact content-based signature that summarizes an audio recording. This technology has attracted attention since they allow the identification of audio independently of its format and without the need of meta-data or watermark embedding [30]. The main objective of an audio fingerprint mechanism is to efficiently compare the equality (or not) of two audio files, not by comparing the files themselves, but by comparing substantially smaller sets of information, referred to as audio fingerprints. Furthermore, audio fingerpint length is a lot less than the raw audio data. In order to validate the proposed wireless acoustic sensor, an open source application, termed Chromaprint [31], is used to generate the fingerprints of original and recorded audios. Then, to find an audio matching between original and recorded audios, the Hamming distance is evaluated using both fingerprints.

3.2.1. Chromaprint Process

Chromaprint converts the audio input to mono and downsampled to 11,025 Hz. The audio signal is converted to the frequency domain by performing a short-time Fourier Transform (STFT) with a frame size of 4096 samples (371 ms) and a 2/3 overlap (2731 samples). The resulting spectrum is converted to 12 bins representing the chroma of the signal. This information is called "chroma features". Each bin in the chromagram represents the energy that is present in a musical note. The 12 bins represent the 12 notes of the chromatic scale. In order to transform the bins in a more compact form to carry out the fingerprint matching, a 12-by-16 sliding window is moved over the chromagram one sample at a time. On each of them is applied a pre-defined set of 16 filters that capture intensity differences across musical notes and time. Each of the filters quantizes the energy value to a 2-bit number. The 2-bit value is encoded using Gray coding. The 2-bit hash values from each of the 16 filters are converted to a single 32-bit integer representing the subfingerprint of the 12-by-16 window. The window is advanced one sample to calculate the next subfingerprint. The full fingerprint is composed by the all subfinngerprints.

3.2.2. Hamming Distance

In order to find a simple audio matching for verifying and validating the proposed system, the Hamming distance is implemented because is performed at the bit-level and therefore, requires less computational complexity. The Hamming distance between two $(NFx32)$-bit binary fingerprint vectors fv_1 and fv_2 is computed as Equation (1):

$$H_d(fv_1, fv_2) = \sum_{i=1}^{NFx32} F(bit_{fv_1}(i) \neq bit_{fv_2}(i)) \tag{1}$$

where NF denotes the number of subfingerprints of vectors, $bit_{fv_1}(i)$ and bit_{fv_2} are the ith element of binary fingerprint vectors, and $F()$ is an function defined by Equation (2):

$$F(x) = \begin{cases} 1 & \text{if } x \text{ is true} \\ 0 & \text{otherwise} \end{cases} \tag{2}$$

3.2.3. Matching Algorithm

Algorithm 1 was used to evaluate the shortest Hamming distance between all original environmental sounds and each recorded audio. The shortest distance identifies and matches the recorded audio with the original sound.

Algorithm 1 Audio Fingerprint Matching

Require: Fingerprint of recorded audio, fv_1
Require: Fingerprint of all original audios, fv_{source}
  $L1 \leftarrow length(fv_1)$
  for $source \leftarrow 1$: all original fingerprints **do**

    $Lsource \leftarrow length(fv_{source})$
    for $i \leftarrow 1 : (Lsource - L1 - 1)$ **do**

      $distanceVector \leftarrow H_d(fv_1, fv_{source}(i : L1 + i - 1))$
    end for
    $distanceAllAudioVector \leftarrow min(distanceVector)$
  end for
  return $min(distanceAllAudioVector)$

---

4. Results and Discussion

This section describes the acoustic anechoic chamber where different audio samples from different environmental sounds were gathered and the dataset built to evaluate the validity of the proposed sensor. Afterwards, the results of aforementioned system evaluation methods, randomness tests and audio fingerprint matching, are presented and discussed. Also, the usefulness of the proposed sensor board in terms of energy consumption and audio processing capabilities are carried out.

4.1. Acoustic Anechoic Chamber

The acoustic anechoic chamber where experiments were carried out is located on the second floor of Institute for Technological Development and Innovation in Communications (IDeTIC) building at Las Palmas de Gran Canaria University, Spain. The chamber area is nearly 200 cm wide and 430 cm long, and it has a simple design to absorb reflections of sound waves and is also isolated from waves entering from its surroundings.

The soundproofing of the chamber is carried out using foam pyramidal panels which is a powerful sound absorber that dramatically reduces echo, reverberation and standing waves. For the acoustic insulation is used rock wool and polyurethane panels. Figure 6 shows acoustic anechoic chamber at IDeTIC and the foam pyramidal panels used.

Figure 6. Acoustic anechoic chamber at IDeTIC.

4.2. Dataset

A total of 48 audio records were gathered from fourteen different indoor and outdoor environmental sounds for performing the statistical analysis and audio classification. These environmental records have been downloaded from Freesound website [32]. Table 1 shows the dataset characteristics. Each environmental sound was recorded using a 10 kHz sampling frequency and 8-bit resolution during 10 s. In order to gather different samples, the start point of each recording was randomly established.

Table 1. Audio recordings dataset.

Environmental Sound	Duration (s)	Number of Records
S1—Traffic jam in a city	49	3
S2—People on a street without traffic	34	3
S3—Very strong traffic	70	5
S4—City park with children	33	3
S5—Pedestrian zone of a city with traffic	32	3
S6—Inside of a noisy room by traffic	60	4
S7—Ambulance passing with the siren	29	4
S8—Drilling machine in a city	18	3
S9—Police car passing with the siren	28	3
S10—Ambulance siren. Doppler effect	24	3
S11—Dense traffic in a city	72	5
S12—Indoor door slam	23	3
S13—Indoor gun shots	98	3
S14—Slicing vegetables in a kitchen	40	3

4.2.1. Randomness Tests

All recorded audios were evaluated with the four above randomness tests. The null hypothesis of randomness is tested against nonrandomness, and a *p*-value is calculated which is used in the context of null hypothesis testing in order to quantify the idea of statistical significance of evidence, that is, the probability of finding the observed results when the null hypothesis is true. In the tests, if the p-value is less than 0.05, the null hypothesis is rejected because a significant difference exists.

Table 2 shows the p-values of four randomness tests. As can be seen, all p-values for Bartels, Mann-Kendall and Wald-Wofowitz tests are less than 0.05. For Cox Stuart test, most results return a p-value less than 0.05, and only eight tests are slightly greater than this value. It is not significant, and hence the null hypothesis can be rejected. Thus, it can be considered that the recorded audio files by wireless acoustic sensor are not significantly random, and therefore, have a recognizable pattern.

Table 2. p-values of randomness tests.

Record	Bartels	Cox Stuart	Mann-Kendall	Wald-Wolfowitz
S1-R1	0	0.06	≈0	0
S1-R2	0	0.01	≈0	0
S1-R3	0	0.09	≈0	0
S2-R1	0	≈0	≈0	0
S2-R2	0	≈0	≈0	0
S2-R3	0	≈0	≈0	0
S3-R1	0	≈0	≈0	0
S3-R2	0	0.02	≈0	0
S3-R3	0	0.12	≈0	0
S3-R4	0	0.13	≈0	0
S3-R5	0	0.09	≈0	0
S4-R1	0	≈0	0	0
S4-R2	0	≈0	0	0
S4-R3	0	≈0	≈0	0
S5-R1	0	≈0	≈0	0
S5-R2	0	≈0	0	0
S5-R3	0	≈0	0	0
S6-R1	0	≈0	0	0
S6-R2	0	≈0	0	0
S6-R3	0	≈0	0	0
S6-R4	0	≈0	0	0
S7-R1	0	0.07	0	0
S7-R2	0	0.08	≈0	0
S7-R3	0	0.02	≈0	0
S7-R4	0	≈0	≈0	0
S8-R1	0	≈0	≈0	0
S8-R2	0	0.01	≈0	0
S8-R3	0	≈0	≈0	0
S9-R1	0	≈0	0	0
S9-R2	0	≈0	0	0
S9-R3	0	0.01	0	0
S10-R1	0	0.03	≈0	0
S10-R2	0	0.03	≈0	0
S10-R3	0	0.06	≈0	0
S11-R1	0	≈0	≈0	0
S11-R2	0	0.01	≈0	0
S11-R3	0	≈0	≈0	0
S11-R4	0	0.06	≈0	0
S11-R5	0	≈0	≈0	0
S12-R1	0	0.01	≈0	0
S12-R2	0	0.01	0	0
S12-R3	0	≈0	0	0
S13-R1	0	≈0	0	0
S13-R2	0	≈0	0	0
S13-R3	0	0.01	0	0
S14-R1	0	0.05	0	0
S14-R2	0	≈0	0	0
S14-R3	0	0.06	0	0

4.2.2. Audio Fingerprint Matching

To carry out the audio fingerprint matching, an audio fingerprint was computed both for each recorded audio and original environmental sounds. All recorded audios are 10 s long, therefore, each recorded audio fingerprint is 66 subfingerprints long. Thus, the vectors used in Equation (1) are 2112 (66 × 32) bits length. Afterwards, Algorithm 1 was used to evaluate the shortest Hamming distance between all original environmental sounds and each recorded audio. The shortest distance identifies and matches the recorded audio with the original sound.

Table 3 shows the results returned by Algorithm 1 when it was evaluated for each recorded audio. For each one the shortest Hamming Distance is bold marked. As can be seen, most of recorded audios match with its correspondent original sound, yielding an 85.4% accuracy, and there is not dependency with the kind of sound, indoor or outdoor. Furthermore, a 91.6% accuracy is reached when the three shortest Hamming distances are used, that is, a recorded audio is correctly classified and found in a set of three sounds with a 91.6% probability. On the other hand, all S10 recordings were classified as S3 original audio. It can be because doppler effect is perceived in both audios. In any case, the aim of this experiment is not to implement a robust classification system, but to demonstrate the validity of the proposed acoustic sensor. Taking in account the results, it can be asserted that the recorded audios have a high grade of similarity with its original sound, and hence, the proposed acoustic sensor can be validated.

Table 3. Audio fingerprints matching.

Record	S1	S2	S3	S4	S5	S6	S7	S8	S9	S10	S11	S12	S13	S14
S1-R1	**773**	1088	1075	1086	1132	1107	1249	1232	1096	1257	1111	1232	1060	1146
S1-R2	**1025**	1086	1047	1050	1111	1182	1323	1121	1146	1164	1053	1234	1040	1239
S1-R3	**883**	1012	1022	996	1151	1108	1227	1152	1111	1180	1101	1252	923	1114
S2-R1	1058	**351**	995	1017	1087	1100	1237	1118	1138	1208	997	1170	855	1156
S2-R2	1047	**336**	961	962	1083	1120	1231	1150	1161	1245	1022	1091	838	1127
S2-R3	1093	1088	1083	**1070**	1124	1224	1265	1162	1230	1288	1105	1204	1098	1194
S3-R1	1115	992	**549**	1002	1111	1037	1288	1212	1099	1298	1072	1150	890	1034
S3-R2	1144	995	**486**	1053	1121	1110	1253	1277	1127	1289	1102	1138	924	1116
S3-R3	1138	1037	**459**	1053	1223	1126	1225	1247	1178	1290	1097	1203	927	1144
S3-R4	1122	1036	**427**	1035	1117	1091	1303	1116	1179	1276	1081	1081	962	1060
S3-R5	1112	1066	**485**	1126	1180	1177	1247	1198	1168	1276	1061	1177	1023	1183
S4-R1	1032	939	1003	**370**	1133	1051	1250	1157	1160	1249	1064	1195	833	1159
S4-R2	1035	978	996	**396**	1107	1133	1187	1159	1135	1270	1061	1150	867	1142
S4-R3	1102	1051	1089	**313**	1122	1128	1280	1182	1218	1172	1116	1211	951	1164
S5-R1	1112	1112	1060	1134	**377**	1102	1215	1157	1178	1262	1088	1162	1081	1163
S5-R2	1095	1024	1033	1132	1053	1064	1246	1170	1119	1376	1058	1143	**973**	1121
S5-R3	1140	1080	1108	1182	**403**	1108	1236	1174	1204	1270	1135	1193	1102	1150
S6-R1	1209	1171	1153	1161	1131	**591**	1240	1198	1318	1306	1155	1188	1172	1010
S6-R2	1141	1019	1080	1141	1059	**598**	1248	1149	1191	1277	1099	1166	1022	1066
S6-R3	1043	1035	**972**	1051	1107	1040	1365	1106	1088	1152	1073	1153	1055	1059
S6-R4	1134	1051	1026	1164	1093	**485**	1289	1151	1172	1275	1152	1185	1007	1039
S7-R1	1167	1121	1089	1153	1150	1178	**1026**	1220	1196	1170	1145	1095	1059	1129
S7-R2	1208	1217	1091	1215	1247	1205	**991**	1226	1209	1234	1183	1179	1183	1169
S7-R3	1236	1192	1133	1256	1273	1225	**340**	1371	1283	1311	1166	1159	1335	1212
S7-R4	1201	1174	1136	1165	1216	1174	**351**	1246	1247	1240	1151	1196	1242	1227
S8-R1	981	1096	1155	1069	1096	1196	1265	**949**	1091	1079	977	1186	982	1218
S8-R2	1015	1111	1123	1047	1036	1161	1274	**340**	1095	1077	986	1181	986	1236
S8-R3	1007	1093	1136	1071	1081	1189	1289	**847**	1079	1027	975	1217	988	1238
S9-R1	1113	1138	1111	1133	1151	1036	1380	1056	**1040**	1103	1099	1167	1098	1018
S9-R2	1164	1195	1143	1245	1192	1061	1322	1156	**402**	1228	1128	1199	1106	1052
S9-R3	1140	1129	1142	1191	1199	1160	1314	1139	**445**	1195	1110	1231	1052	1193
S10-R1	1151	1118	**1065**	1160	1275	1201	1227	1355	1186	1380	1160	1197	1194	1097
S10-R2	1149	1061	**1060**	1124	1173	1105	1322	1250	1073	1305	1157	1219	1192	1128
S10-R3	1191	1140	**1092**	1095	1205	1200	1206	1218	1152	1299	1112	1237	1175	1148
S11-R1	1072	1031	1105	1071	1106	1147	1259	1121	1136	1334	**483**	1141	903	1125
S11-R2	1070	1084	1085	1059	1125	1112	1207	1090	1129	1287	**538**	1238	981	1162
S11-R3	997	1066	1121	1051	1080	1138	1280	1030	1111	1190	**483**	1175	922	1249
S11-R4	1144	1174	1161	1208	1132	1095	1187	1192	1179	1311	**358**	1167	1136	1156
S11-R5	1066	1022	1082	1038	1125	1086	1261	1163	1178	1340	**488**	1170	931	1068
S12-R1	1227	1222	1195	1219	1166	1140	1181	1263	1265	1297	1194	**495**	1288	1192
S12-R2	1094	1055	1051	1033	1125	1115	1252	1029	1102	1193	1080	1024	**953**	1183
S12-R3	1121	1095	1122	1094	1140	1156	1262	1122	1233	1228	1054	**1009**	1074	1149

Table 3. *Cont.*

Record	S1	S2	S3	S4	S5	S6	S7	S8	S9	S10	S11	S12	S13	S14
S13-R1	1039	1031	991	1019	1110	1076	1400	972	1108	1031	985	1169	758	1060
S13-R2	1029	966	956	942	1140	1036	1323	1036	1015	1128	986	1165	618	1058
S13-R3	953	954	994	860	1099	1072	1246	1045	1055	1142	937	1217	372	1180
S14-R1	1082	1052	1077	1111	1113	1148	1293	1071	1102	1140	1120	1158	985	758
S14-R2	1078	979	1022	1085	1106	1095	1295	1093	1122	1172	1117	1159	1001	768
S14-R3	1244	1158	1129	1228	1170	931	1381	1233	1110	1360	1219	1092	1156	249

4.3. Energy Consumption and Audio Proccesing Capabilities

In order to evaluate the energy consumption of the proposed sensor, three experiments with different processing capabilities were performed: (A) audio recording, (B) audio recording and Fast Fourier Transform (FFT) calculation, and (C) audio recording and UDP datagram sending via Wi-Fi connection. Audio recording was carried out in an infinite loop using a 10 kHz sampling frequency and 12-bit resolution. The ArduinoFFT library [33] was used to implement the FFT, and it was computed each 12.8 ms, that is, every FFT was run after 128 new samples were recorded. UDP datagram sending was carried out each 25.6 ms, therefore, each UDP datagram is sent when 256 new samples are gathered. The prototype sensor was powered by 5 V, and the current consumption was measurement for each experiment. Table 4 shows the results. As can be seen, the A and B experiments have similar energy consumption because all operations are carried out in an infinite loop and additional resources are not used, only the processor. However, in the C experiment, Wi-Fi module is periodiocally transmitting a datagram, and therefore, the energy consumption is higher. In any case, the maximun energy consumption is about 0.8 W, and the proposed sensor could be powered by battery for a long time.

Table 4. Energy consumption.

Experiment	Average Current (mA)	Energy Consumption (W)
A—Audio recording	139	0.695
B—Audio recording and FFT	141	0.705
C—Audio recording and UDP sending	165	0.825

On the other hand, audio processing capabilities were evaluated implementing the FFT with different number of samples and carrying out an audio recording using the same core. FFT was chosen because is an expensive computational algorithm in audio processing. Each experiment was performed 1000 times and the average execution time was computed. Table 5 shows the average execution time for each experiment. As can be seen, the execution time increases with the number of samples. In addition, using the same number of samples, the FFT execution time is slightly higher when the simultaneous sampling is carried out, that is, the FFT was computed while other samples were gathered. Nevertheless, about 20 ms is only spent to compute a 512 sample FFT. Hence, audio processing capabilities could be performed without compromising the sample rate and the energy consumption, and edge computing paradigm can be implemented in the proposed sensor.

Table 5. Fast Fourier Transform execution time.

Number of Samples	Simultaneous Sampling	Average Execution Time (ms)
128	No	3.32
128	Yes	4.45
256	No	7.04
256	Yes	9.43
512	No	14.92
512	Yes	19.97

5. Conclusions and Future Work

As was discussed in this paper, in recent years, many authors have experienced a growing interest in remote monitoring of older people, and several systems have been proposed in the literature. These systems are commonly termed as AAL systems, but the acoustic sensors were not designed with low cost and audio processing requisites. In this paper, we described the design of a low cost wireless acoustic sensor for AAL systems based on ESP32 board. In order to choose the best platform, three different low cost microcontroller boards were evaluated. It was given a detailed description of the hardware and the principle of operation of software implementation. The proposed sensor is capable of recording ambient sounds at least to 10 kHz sampling frequency and 12-bit resolution. Furthermore, the sensor board has computation capabilities to carry out audio signal processing and network communications without compromising the sample rate and the energy consumption. Hence, the proposed sensor can improve AAL solutions carrying out the audio identification for monitoring of activity and health, and the detection of distress situations at the edge of WASN. Thus, a shorter response time and better reliability is ensured enhancing quality of life and safety of older people. The acoustic sensor is very small in size, and therefore is very useful to be used in a discrete way for personal healthcare in AAL systems, and the cost of hardware platform is very competitive. The experiments on the proposed system showed that the system worked well. System validation is demonstrated by experimental results, which were statistically obtained analysing several tests of randomness and audio classification. Furthermore, evaluations of energy consumption and audio processing capabilities were carried out demonstrating the usefulness and low power, and that edge computing paradigm can be implemented at the proposed sensor.

In our ongoing work, we are planning to design a better sound classification system based on audio fingerprint to be implemented at each acoustic sensor. Moreover, we are also planning to deploy a WASN using the proposed acoustic sensor in this paper to evaluate the whole system and network performance.

Author Contributions: Miguel A. Quintana-Suárez designed and implemented the prototype of wireless acoustic sensor, and wrote part of manuscript. David Sánchez-Rodríguez, Itziar Alonso-González and Jesús B. Alonso-Hernández conceived and designed the experiments, analized the data, and wrote the rest of the manuscript. All authors have reviewed and approved the final manuscript.

Conflicts of Interest: The authors declare no conflict of interest.

References

1. Chang, C.Y.; Kuo, C.H.; Chen, J.C.; Wang, T.C. Design and implementation of an IoT access point for smart home. *Appl. Sci.* **2015**, *5*, 1882–1903.
2. Peruzzini, M.; Germani, M.; Papetti, A.; Capitanelli, A. Smart Home Information Management System for Energy-Efficient Networks. In *Collaborative Systems for Reindustrialization, Proceedings of the 14th IFIP WG 5.5 Working Conference on Virtual Enterprises, PRO-VE 2013 Dresden, Germany, 30 September–2 October 2013;* Springer: Heidelberg, Germany, 2013; pp. 393–401.
3. Amendola, S.; Lodato, R.; Manzari, S.; Occhiuzzi, C.; Marrocco, G. RFID technology for IoT-based personal healthcare in smart spaces. *IEEE Internet Things J.* **2014**, *1*, 144–152.
4. Sánchez-Rosario, F.; Sanchez-Rodriguez, D.; Alonso-Hernández, J.B.; Travieso-González, C.M.; Alonso-González, I.; Ley-Bosch, C.; Ramírez-Casañas, C.; Quintana-Suárez, M.A. A low consumption real time environmental monitoring system for smart cities based on ZigBee wireless sensor network. In Proceedings of the 2015 International Wireless Communications and Mobile Computing Conference (IWCMC), Dubrovnik, Croatia, 24–25 August 2015; IEEE: Washington, DC, USA, 2015; pp. 702–707.
5. Lin, K.; Chen, M.; Deng, J.; Hassan, M.M.; Fortino, G. Enhanced fingerprinting and trajectory prediction for IoT localization in smart buildings. *IEEE Trans. Autom. Sci. Eng.* **2016**, *13*, 1294–1307.
6. Memon, M.; Wagner, S.R.; Pedersen, C.F.; Beevi, F.H.A.; Hansen, F.O. Ambient assisted living healthcare frameworks, platforms, standards, and quality attributes. *Sensors* **2014**, *14*, 4312–4341.

7. Rashidi, P.; Mihailidis, A. A survey on ambient-assisted living tools for older adults. *IEEE J. Biomed. Health Inform.* **2013**, *17*, 579–590.
8. European Comission. Active and Assisted Living Programme. ICT for Ageing Well. Available online: http://www.aal-europe.eu/ (accessed on 1 June 2017).
9. Erden, F.; Velipasalar, S.; Alkar, A.Z.; Cetin, A.E. Sensors in Assisted Living: A survey of signal and image processing methods. *IEEE Signal Proc. Mag.* **2016**, *33*, 36–44.
10. Vacher, M.; Portet, F.; Fleury, A.; Noury, N. Development of audio sensing technology for ambient assisted living: Applications and challenges. *Int. J. E-Health and Medical Commun.* **2011**, *2*, 35–34.
11. Kivelä, I.; Gao, C.; Luomala, J.; Hakala, I. Design of Noise Measurement Sensor Network: Networking and Communication Part. In Proceedings of the Fifth International Conference on Sensor Technologies and Applications, Sensorcomm, French Riviera, France, 21–27 August 2011; pp. 280–287.
12. Paulo, J.; Fazenda, P.; Oliveira, T.; Carvalho, C.; Felix, M. Framework to Monitor Sound Events in the City Supported by the FIWARE platform. In Proceedings of the 46th Congreso Español de Acústica, Valencia, Spain, 21–23 October 2015; pp. 21–23.
13. Alías, F.; Socoró, J.C. Description of anomalous noise events for reliable dynamic traffic noise mapping in real-life urban and suburban soundscapes. *Appl. Sci.* **2017**, *7*, 146.
14. Sevillano, X.; Socoró, J.C.; Alías, F.; Bellucci, P.; Peruzzi, L.; Radaelli, S.; Coppi, P.; Nencini, L.; Cerniglia, A.; Bisceglie, A.; et al. DYNAMAP—Development of low cost sensors networks for real time noise mapping. *Noise Mapp.* **2016**, *3*, 172–189.
15. Segura Garcia, J.; Pérez Solano, J.J.; Cobos Serrano, M.; Navarro Camba, E.A.; Felici Castell, S.; Soriano Asensi, A.; Montes Suay, F. Spatial Statistical Analysis of Urban Noise Data from a WASN Gathered by an IoT System: Application to a Small City. *Appl. Sci.* **2016**, *6*, 380.
16. Hakala, I.; Kivela, I.; Ihalainen, J.; Luomala, J.; Gao, C. Design of Low-Cost Noise Measurement Sensor Network: Sensor Function Design. In Proceedings of the 2010 First International Conference on Sensor Device Technologies and Applications (SENSORDEVICES), Venice, Italy, 18–25 July 2010; IEEE: Washington, DC, USA, 2010; pp. 172–179.
17. Ahmed, A.; Ahmed, E. A Survey on Mobile Edge Computing. In Proceedings of the 2016 10th International Conference on Intelligent Systems and Control (ISCO), Coimbatore, India, 7–8 January 2016; IEEE: Washington, DC, USA, 2016; pp. 1–8.
18. Shi, W.; Cao, J.; Zhang, Q.; Li, Y.; Xu, L. Edge computing: Vision and challenges. *IEEE Internet Things J.* **2016**, *3*, 637–646.
19. Alsina-Pagès, R.M.; Navarro, J.; Alías, F.; Hervás, M. homeSound: Real-Time Audio Event Detection Based on High Performance Computing for Behaviour and Surveillance Remote Monitoring. *Sensors* **2017**, *17*, 854.
20. Mach, P.; Becvar, Z. Mobile edge computing: A survey on architecture and computation offloading. *IEEE Commun. Surv. Tutor.* **2017**, *19*, 1628–1656.
21. Adafruit. Electret Microphone Amplifier. Available online: https://www.adafruit.com/product/1063 (accessed on 15 July 2017).
22. Libelium. Waspmote Platform. Available online: http://www.libelium.com/products/waspmote/ (accessed on 3 May 2017).
23. Espressif. ESP8266 Specification. Available online: https://espressif.com/en/products/hardware/esp8266ex/overview (accessed on 15 July 2017).
24. Espressif. ESP32 Specification. Available online: https://espressif.com/en/products/hardware/esp32/overview (accessed on 15 July 2017).
25. Bartels, R. The rank version of von Neumann's ratio test for randomness. *J. Am. Stat. Assoc.* **1982**, *77*, 40–46.
26. Cox, D.R.; Stuart, A. Some quick sign tests for trend in location and dispersion. *Biometrika* **1955**, *42*, 80–95.
27. Kendall, M. *Rank Correlation Methods*; Oxford University Press: New York, NY, USA, 1990.
28. Wald, A.; Wolfowitz, J. On a test whether two samples are from the same population. *Ann. Math. Stat.* **1940**, *11*, 147–162.
29. Von Neumann, J. Distribution of the ratio of the mean square successive difference to the variance. *Ann. Math. Stat.* **1941**, *12*, 367–395.
30. Cano, P.; Batlle, E.; Kalker, T.; Haitsma, J. A review of audio fingerprinting. *J. VLSI Signal Proc. Syst. Signal Image Video Technol.* **2005**, *41*, 271–284.
31. Lukáš, L. Chromaprint. Available online: https://acoustid.org/chromaprint (accessed on 3 May 2017).

32. Freesound. Dataset of Audio Clips. Available online: http://freesound.org/ (accessed on 1 July 2017).
33. Condes, E. ArduinoFFT. Available online: https://github.com/kosme/arduinoFFT (accessed on 28 July 2017).

applied sciences

MDPI

Review

Room Response Equalization—A Review

Stefania Cecchi [1],*, Alberto Carini [2] and Sascha Spors [3]

[1] Department of Information Engineering, Università Politecnica delle Marche, 60121 Ancona, Italy; s.cecchi@univpm.it
[2] Department of Pure and Applied Sciences, University of Urbino Carlo Bo, 61029 Urbino, Italy; alberto.carini@uniurb.it
[3] Institute of Communications Engineering, University of Rostock, 18011 Rostock, Germany; sascha.spors@uni-rostock.de
* Correspondence: s.cecchi@univpm.it; Tel.: +39-071-2204-146

Academic Editor: Vesa Valimaki
Received: 3 November 2017; Accepted: 12 December 2017; Published: 23 December 2017

Abstract: Room response equalization aims at improving the sound reproduction in rooms by applying advanced digital signal processing techniques to design an equalizer on the basis of one or more measurements of the room response. This topic has been intensively studied in the last 40 years, resulting in a number of effective techniques facing different aspects of the problem. This review paper aims at giving an overview of the existing methods following their historical evolution, and discussing pros and cons of each approach with relation to the room characteristics, as well as instrumental and perceptual measures. The review is concluded by a discussion on emerging topics and new trends.

Keywords: room response equalization; single-point equalization; multi-point equalization; adaptive equalization; wave domain filtering

1. Introduction

When sound is reproduced by one or more loudspeakers, the perception of the desired auditory illusion is modified by the listening environment. To some extent this may be seen as positive, since spaciousness and depth is added, but the environment and the sound reproduction system can also introduce undesired artifacts. Excessive reflections or resonances within the listening environment may result in an undesired alteration of the auditory illusion. A non-ideal reproduction system may even add some artifacts (e.g., frequency band extension, nonlinearities) to the original sound.

Room response equalization (RRE) has been studied in theory and applied in practice for improving the quality of sound reproduction contrasting the detrimental effects of the room environment and reproduction system. In an RRE system, the room transfer function (RTF) characterizing the path from the sound reproduction system to the listener is equalized with a suitably designed equalizer that can be realized in several manners. The basic idea is to measure the room impulse response (RIR) using a microphone, and then obtain the equalizer through its inversion. However, several issues influence this method, and thus a wide variety of techniques have been developed over the last 40 years. The reader should be aware that many different names have been used in the literature for RRE, such as "room equalization", "room correction", "room compensation", "room inversion", "room dereverberation", "dereverberation", "reverberation reduction", and others. In this review, the collective term RRE is used to denote any technique that aims to design an equalizer from measurements of the RTF.

Borrowing the words of [1], there is a *"multidimensionality of alternatives for room inverse filter design"*. In particular, the inversion of the RIR can be performed considering a non-parametric approach such as least-mean-squares or direct inversion of the frequency response [1,2], a parametric approach such as autoregressive-moving average (ARMA) modeling [1,3], or temporal decay control at low frequency [1,4]. However, as reported in [1], this is not the only classification possible: RRE can also be classified into minimum- or mixed-phase. The former aims only at the equalization of RTF magnitude, while the latter also acts on the excess-phase RTF component.

In this review paper, a general classification is presented aiming at a broader view on the state-of-the-art in RRE. Figure 1 provides a conceptual scheme of this classification, clustering the various techniques that will be presented in the following. As shown in Figure 1, the RRE approaches are divided into single-point (single-input/single-output—SISO, multiple-input/single-output—MISO) and multi-point (single-input/multiple-output—SIMO, multiple-input/multiple-output—MIMO) room equalizers. A single position room equalizer estimates the equalization filter on the basis of the measurement in a single location of the RTF [5]. It is effective only in a limited zone around the measured point (of the size of a fraction of the acoustic wavelength). In reality, the RTF varies significantly with respect to the position in the room [6,7] and time [2], as the room can be considered a *"weakly non-stationary"* system [8]. To enlarge the equalized zone and to contrast the room response variations, multi-point equalizers have been proposed [9]. A multi-point room equalizer uses multiple measurements of the RTF at different locations in order to design the equalizer. These approaches can be used for fixed and adaptive equalization. The former is based on RTFs measured at fixed positions at a certain time. The latter is capable of tracking and adapting to changes in the room response due to its time varying nature resulting for instance from temperature changes or movement of people or other obstacles. Different pre-processing techniques are applied to contrast audible distortions caused by fixed equalization in scenarios where RTFs vary. Different equalizer design techniques can also be adopted (classified in the following as minimum-phase or mixed-phase). More recently [10], equalization in spatio-temporally transformed domains for the adaptive equalization of massive multichannel sound reproduction systems has been investigated, and is presently a topic of active research.

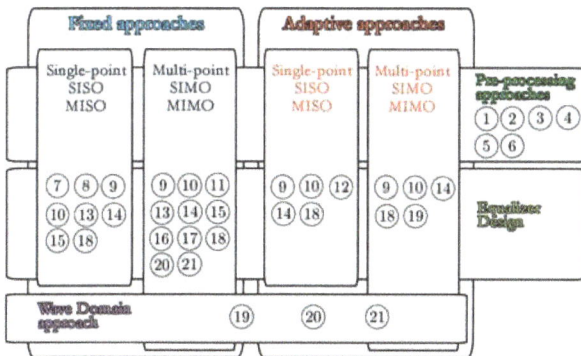

Figure 1. A general classification of room response equalization (RRE) systems. Possible approaches: (1) short filters, (2) complex smoothing, (3) frequency warping, (4) Kautz filters, (5) multirate approaches, (6) room impulse response (RIR) reshaping, (7) homomorphic filtering, (8) linear predictive coding analysis, (9) least-squares optimization techniques, (10) frequency domain deconvolution, (11) multiple-input/multiple-output inverse theorem (MINT) solutions, (12) average and weighted average methods, (13) clustering methods, (14) prototype approach, (15) common acoustical poles compensation, (16) modal equalization, (17) plane wave approach, (18) quasi-anechoic approach, (19) wave domain adaptive filtering, (20) transform domain approaches, (21) room geometry aware methods. MIMO: multiple-input/multiple-output; MISO: multiple-input/single-output; SIMO: single-input/multiple-output; SISO: single-input/single-output.

This paper aims to provide an up-to-date review on RRE, discussing the pros and the cons of each technique, following the historical evolution. It is worth underlining that the RRE problem is analyzed from the viewpoint of impulse response analysis. All approaches that are not directly based on RIR analysis (e.g., parametric or graphic equalizers) are not discussed. The reader is referred to [11] for a comprehensive review on this topic. Another research field related to RRE which is not addressed in this paper is sound spatialization. The reader is referred to [12] for a recent review.

This review article is organized as follows: Section 2 describes the characteristics of room impulse responses and its perception by the human auditory system. Section 3 introduces the basic concept of RRE, explaining the main challenges in inverting room responses. Section 4 describes the approaches used for equalizer design following their historical evolution. Section 5 discusses pre-processing techniques used to cope for RIR variations by exploiting human perception. Section 6 covers the evolution from single-point to multi-point equalization using multiple microphones placed within the room. Section 7 reports adaptive approaches for RRE in the framework of single-point and multi-point equalization. Section 8 introduces innovative approaches following a wave-theoretical view on the problem. Section 9 describes instrumental and perceptual measures used for state-of-the-art evaluation of RRE approaches. Section 10 reports emerging methods and new trends in the field. Finally, Section 11 concludes this review.

2. The Room Response and Its Perception

The characteristics of the room response in the time and frequency domain are related to the acoustic properties of the environment that influence human perception. Due to this aspect, it is sensible to shape the impulse response analysis in order to handle important issues that should be considered in the RRE procedure to reach a sound listening improvement. This includes knowledge on human perception and psychoacoustics to be exploited explicitly in the equalization procedure.

An impulse response, obtained from a sound source in a specific position of a real environment, can be divided into three parts [13]: (i) direct sound; (ii) early reflections, and (iii) late reflections, as reported in Figure 2a. The transition from early reflections to late reflections is given by the mixing time, estimating the time elapsed from early to late reflections. It can be estimated in several manners [14,15]. Direct sound and early reflections are fundamental for the localization of the sound source and perception of its timbre [16–18], while the late reverberation provides cues on the spaciousness of the room [19]. Studies on the perception of reflections and their influence on the timbre can be found in [19–25]. The spectral content of direct and reflected sound is different. Walls, drapes, and upholstery typically absorb the high frequencies of reflections. The effect is boosted by multiple reflections, with the late reverberation typically having a much lower energy in the high frequencies.

At low frequencies, the wavelength is comparable to typical room dimensions: standing waves may appear in a room for steady-state signals, resulting in well defined position-dependent maxima and minima of the magnitude response. At these frequencies, the room response has a smooth behavior characterized by well separated resonances and notches, as illustrated in Figure 2b. The resonances and notches are determined by interference patterns caused by the direct sound and reflections, with notches appearing when the path-length difference is an odd number of half wavelengths. The notches become increasingly dense with increasing frequency. For frequencies greater than the Schroeder frequency [13], the frequency response becomes extremely irregular. Spectral peaks are more audible than notches [20], but wide-bandwidth notches are also audible [26]. At high frequencies, the peaks and notches strongly depend on the position in the room and on factors like the room humidity and temperature [27–29] or obstacle movements [30–34]. It must be pointed out that these large variations in the response have little influence on the subjective impression of the listener [18]. It has been suggested [19,22,24] that the ear is more sensitive to signal onsets (i.e., to the full spectrum of the initial part of the RIR) and that it largely ignores the high-frequency components of the late reverb [35]. This aspect should be considered in the equalizer design.

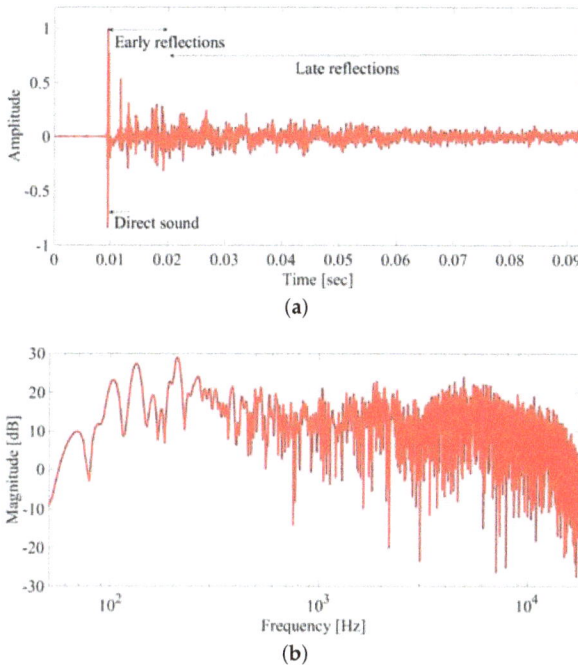

Figure 2. Real RIR behaviour (**a**) in the time domain and (**b**) in the frequency domain.

The perception of high frequencies is particularly affected by the frequency resolution of the human auditory system. The resolution of the ear is nonlinear and nonuniform with frequency, with an almost logarithmic dependency on frequency [36]. This aspect has led to the introduction of psychoacoustic frequency scales in the equalizer development with the aim of modifying the spectral content according to human perception. The mel scale [37], the Bark (critical band rate) scale [38], and the ERB (equivalent rectangular bandwidth) scale [39] are examples of psychoacoustic frequency scales that usually build on a filterbank model of hearing. The mel scale is a perceptual scale of pitches judged by listeners to be equal in distance from one another. The Bark scale is based on the critical bands (i.e., the bandwidth of the auditory filters modeling hearing and frequency masking at different frequencies). The ERB is also related to the Bark scale and to auditory filters, since the ERB filters pass the same amount of energy as the auditory filters they correspond to. It can be concluded that the logarithmic frequency scale of human hearing largely explains the low sensitivity to peaks and notches at high frequencies, and this aspect should be considered in the equalizer design.

The temporal integration and masking properties of the human auditory system also affect the perception of reflections. The ear perceives sounds by integrating them with a window of around 60 ms duration, having an equivalent rectangular duration of 5 ms [40]. The window is asymmetrical, with a slower rise and faster decay. The ear is insensitive to temporal events shorter than about 2 ms [41]. Masking indicates a condition where sounds which presented isolated would be audible are hidden by the presence of a higher level sound (the masker). We can have both simultaneous and non-simultaneous masking. Simultaneous masking depends on the frequency of the masker and the masked signal. It has its maximum effect when the two differ by less than a critical bandwidth. It diminishes quickly when the frequency of the masker is greater than the masked signal, while it diminishes more slowly when the frequency of masker is lower than the masked signal [42,43]. Non-simultaneous masking refers to situations where the masker and masked sound are separated in time. It is divided into backward masking, with the masked sound preceding the masker, and forward masking, with the masked sound following the masker. Backward masking is quite limited in time [43]: its effect disappears after 15–20 ms [44,45], with the most significant portion fading out after 5 ms [46]. Forward masking has a longer extension of 100–200 ms. Its behavior is similar to simultaneous masking, and it depends on the frequency relationship between masker and masked sound. According to [39], its effect starts as simultaneous masking and then fades out over time with a straight line in a graph representing the masking reduction in dB versus time [43]. An average forward masking curve has been introduced in [43,47]. For the first 10 ms, the curve has a constant value equal to -9 dB, which is the maximum level of masking in [19], and then it decays over time. This phenomenon can be exploited in the equalization procedure as discussed in the next sections.

The audibility of room reflections also depends on the direction of arrival of the direct sound and reflections with respect to the listener [48], on the loudness of the direct sound (reflections can be more easily perceived with louder direct sound), on the kind of signal [19], and on the spectral content of direct sound and reflections (masking has a stronger effect if the spectral content of direct sound and reflections coincides) [48].

In the following sections, different RRE techniques are discussed highlighting the problems following from the physical properties of the room response and how the characteristics of the human auditory system can be included.

3. Invertibility of the Room Response

The first research paper on RRE can be attributed to Neely and Allen in 1979 [5]. In their seminal paper, they studied the invertibility of the RTF and implications. Considering the RTF of a synthetic room, they showed that if the reflectivity of the wall is low (below 36%) the RTF is minimum-phase and thus invertible. On the contrary, with larger wall reflection coefficients, as those of typical rooms (in the range 70%–90%), the RIR is non-minimum-phase. However, it is still possible to equalize the minimum-phase part of the room response (i.e., the amplitude response and the minimum-phase part of the phase response) by factoring the RTF $H(z)$ into a product of a minimum-phase term $H_m(z)$ and a stable all-pass filter $A(z)$,

$$H(z) = H_m(z) \cdot A(z). \tag{1}$$

The equalization filter is simply computed by taking the inverse z-transform of the reciprocal of the spectrum of $H_m(z)$. By listening to the result of the minimum-phase equalization, Neely and Allen reported that "The room effect had been removed, but a tone, much like a bell chime, sounded in the background" [5].

The original approach of [5] is in reality affected by several problems, many discovered by researchers only in later studies. Following the chronological order in which these problems were addressed:

- When the room response is non-minimum-phase, an exact inverse cannot be implemented with a single sound source, since the inverse is either unstable or acausal.
- The exact equalization of the room response—or of its minimum-phase part—requires very long filters.
- The equalizer is affected by any imperfection in the measurement of the room response [6,34].
- The room response strongly depends on the location of the loudspeaker and the microphone used for the measurement [6,31–34,48,49].
- Exact equalization is possible only in one location, and the extent of the equalized zone is just a fraction of the acoustic wavelength [6]. At high frequencies, the equalized zone can be smaller than the inter-aural distance of the ears (around 18 cm).
- The notches of the room response—which are affected by the noise floor—are highly boosted by the equalizer with the generation of an often audible tone-like noise (the bell chime experienced by Neely and Allen) [50–52].

- The room response is actually slowly time-varying, affected by humidity and temperature [28,29] and by movement of people or other obstacles in the enclosure.
- The human ear is sensitive to the excess-phase of the RTF [53].
- The equalizer should preserve the natural roll-off of loudspeakers at low and high frequencies [54,55]. Amplifying these frequencies could cause an unnatural boost of the loudspeaker response, causing nonlinear effects, energy dissipation, and possible damages.

In what follows, we will discuss the different solutions that have been devised in order to contrast the above-mentioned problems. In particular, we will review the techniques used to design the equalizer, considering both minimum-phase and mixed-phase equalization strategies, and pre-processing techniques used to contrast the effects of the variations of the room response with position and time. As much as possible, we will try to follow the chronological order in which the techniques were proposed to illustrate the evolution of RRE.

4. Equalizer Design Techniques

In the techniques we discuss, the room response equalizer is designed on the basis of measurements of the RIR or RTF in one or more locations within the desired listening area. As we will see in Section 5, the room response is pre-processed in most cases in order to contrast some of the detrimental effects discussed in Section 3. In any case, a prototype room response is usually obtained and used for the equalizer design.

Most of the equalizer design techniques can be classified into the following five classes:

- Homomorphic filtering;
- Linear predictive coding (LPC) analysis;
- Least-squares (or other) optimization techniques;
- Frequency domain deconvolution;
- Multiple-input/multiple output inverse theorem (MINT) solutions.

The first two techniques are generally used for minimum-phase equalization, the latter three for mixed-phase equalization.

4.1. Homomorphic Filtering

Homomorphic filtering was already proposed for minimum phase equalization in the seminal paper of Neely and Allen [5], but many other authors introduced modified versions of the homomorphic technique [56–58]. In homomorphic filtering, the minimum phase part of the room response is extracted from the causal part of the complex cepstrum. A stable infinite impulse response (IIR) equalizer is then obtained by direct inversion of the minimum-phase part. Since the excess-phase part of the RTF was found to carry most of the reverberant energy [59], in [6,56] the homomorphic technique was also used for mixed-phase equalization. In particular, the minimum-phase equalizer was complemented with an excess-phase equalizer, designed with a least-squares technique. Another possibility for implementing an excess-phase equalizer is to use a matched filter—i.e., a filter having an impulse response that is the time-reversal of the impulse response of the excess-phase system [57]. However, mixed-phase equalization based on homomorphic technique was found to be oversensitive to errors in the initial homomorphic decomposition of the room response [56,60]. Improvements to the homomorphic technique were reported in [57] and [58]. In [57], an iterative homomorphic technique is proposed by iteratively flattening the RTF magnitude response. The technique overcomes potential numerical problems and *"provides more insight into subjective aspects of magnitude and phase equalization in the reduction of acoustic reverberation"* [57]. In [58], some of the low-frequency dominant poles of the filter transfer function are replaced by new ones with smaller magnitude before computing the inverse filter. The technique allows the extent of oscillations associated with these poles to be reduced. The main disadvantage of the homomorphic technique is the large length of the all-zero (finite impulse response) model of the room response and the high sensitivity of the model to *"changes in source/receiver placement"* [61]. From this point of view, the LPC analysis provides more robust results [61].

4.2. LPC Analysis

In LPC analysis, the room response is modeled with a minimum-phase all-pole filter and the equalizer is a finite impulse response (FIR) filter. The all-pole model can be obtained by different techniques, including the efficient Levison–Durbin algorithm [62]. LPC analysis has been one of the most successful approaches for minimum-phase equalization, and has been successfully used by many researchers [61,63–75]. An all-pole filter can adequately model the spectral peaks of the room response, while it provides a less accurate model of the notches. We should remember that in the human auditory system the spectral peaks are more audible than the notches [20]. Moreover, the room response varies significantly with respect to the position in correspondence to notches [49]. An all-pole equalizer can compensate the most audible parts of the room response (the spectral peaks), without boosting the notches, which is another desirable property of the equalizer.

The main limitation of LPC analysis is the fact that it can be used only for minimum-phase equalization, and it must be complemented with other techniques to equalize the excess-phase.

4.3. Least-Squares Optimization Methods

Mixed-phase equalization requires the approximation of the inverse of a non-minimum phase response, which is acausal. In order to approximate an acausal impulse response, it was proposed in [76] to add a delay in the response of the equalizer and to design the equalizer by minimizing a least-squares error criterion. The approach proposed in [76] was thereafter followed and improved upon by many researchers, for both single-position and multiple-position equalization [77] (see Section 6). Mixed-phase equalization requires the introduction of a delay in the equalizer. This delay should be kept as low as possible (on the order of a few milliseconds according to the backward masking characteristics of the ear [46]), since it can give rise to annoying artifacts in the form of pre-ringing or pre-echo effects. At the same time, the delay should be sufficiently long to obtain reasonable mixed-phase equalization. The least-squares optimization has been the key ingredient of many adaptive solutions, starting from the seminal paper of [77], as detailed in Section 7. Other least-squares optimization criteria considering further constraints have also been proposed; e.g., deconvolution with regularization [51], room response reshaping [78], Kautz filters [55], and short filters [79].

The main limitations of the least-squares methods are the high sensitivity to the peaks and notches of the room response, the non-uniform distribution of errors in the spectrum, and the possibility of pre-ringing or pre-echo artifacts caused by the equalizer delay.

4.4. Frequency Domain Deconvolution

Another technique used for the equalizer design is based on frequency domain deconvolution. As initially proposed in [80], the equalizer can be directly designed in the discrete Fourier transform (DFT) domain by considering the reciprocal of the room response. In [80], the technique is applied to the DFT of a windowed impulse response in order to correct only the early reflections of the room response that affect the perception of timbre and to obtain a short equalizer response. In general, this technique can be applied both for minimum-phase and mixed-phase equalization (adding an appropriate delay), but the room response must be properly pre-processed. In particular, the depth of the notches of the room response should be suitably limited to avoid excessive gains and long impulse responses of the equalizer, which could result in tonal artifacts [52]. In [50], the equalizer is designed by dividing the complex spectrum of a target response with the complex spectrum of the measured room response. To avoid the problem of notches, a positive bias is added to the measured room response. This technique is known as *"deconvolution with regularization"*, with the bias called *"regularization parameter"*. The concept was formalized by Kirkeby and colleagues in [51,81] by considering a least-squares optimization criterion with a *"penalty effort"*. It is also known as the *"Kirkeby algorithm"*. In [52], the technique was applied to RRE. The regularization parameter controls the longest time constant of the inverse filters [81] in practice. In order to ensure that the time constant is neither too long nor too short, the regularization parameter must be set appropriately [51]. In [82], the authors show how the poles of the deconvolution solution are influenced by the regularization parameter. In particular, for each zero close to the unit circle, a triplet of two poles and one zero is generated, with one of the poles outside the unit circle. This pole is responsible for an acausal response, and thus modeling delay should be introduced. In [43], an analysis of RRE based on the Kirkeby algorithm on the basis of psychoacoustic criteria is provided. In the considered conditions, it was shown that the "errors in the dereverberation process manifested themselves as extremely audible and annoying resonances. These arose from the presence of deep spectral notches in the transfer functions of loudspeaker–room combinations, which created tonal artifacts that occurred long before and after the direct-arrival sounds. Furthermore, an extreme sensitivity to changes in position was found, which prevented the optimization of dereverberation over practically sized listening areas. The quality of the dereverberation was found to degrade even further for larger acoustic spaces." Despite these limitations, deconvolution with regularization approaches has been successfully applied in combination with other techniques used to avoid perceivable distortions. For example, it has been combined with frequency warping [83,84], or used in wave field synthesis [85].

4.5. Multiple-Input/Multiple-Output Inverse Theorem Methods

A method for the exact inversion of the RIR—even when it is non-minimum phase—was proposed in [86,87]. The method is based on a principle called the multiple-input/multiple-output inverse theorem (MINT). With this method, the inverse is constructed from multiple FIR filters, by adding *"some extra acoustic signal-transmission channels produced by multiple loudspeakers or microphones."* In practice, the MINT states that it is possible to obtain an exact inversion of the room response if the number of loudspeakers is larger than the number of microphones (i.e., measurement points). Thus, the approach is intrinsically multi-channel. Let us consider the case of a system with two loudspeakers and one microphone. Let us indicate with $G_1(z)$ and $G_2(z)$ the transfer functions from the loudspeakers to the microphones, and with $H_1(z)$ and $H_2(z)$ the transfer functions of the equalizers associated to each loudspeaker. Then, for exact inversion of the room response, $H_1(z)$ and $H_2(z)$ must satisfy the following condition:

$$H_1(z)G_1(z) + H_2(z)G_2(z) = 1. \tag{2}$$

As shown in [87], the solution of Equation (2) exists if $G_1(z)$ and $G_2(z)$ are relatively prime (i.e., do not have common zeros), and when the solution exists the orders of $H_1(z)$ and $H_2(z)$ are lower than $G_2(z)$ and $G_1(z)$, respectively. The approach is very powerful because it allows the acausality problem of the equalizer to be overcome. However, the MINT approach also exhibits strong limitations. In [88], the MINT is analyzed under a numerical perspective, studying the condition number of the time domain matrix that is inverted. It is shown that the condition number of the time domain matrix is related to the singular values of the transfer matrix evaluated over frequency. The condition number decreases and the numerical performance is enhanced as the number of loudspeakers is increased. However, the condition number increases *"at the rate of approximately 1 bit"* (i.e., of approximately 6 dB) for each microphone added [88]. Moreover, an analysis of the MINT technique is also presented in [89], discussing the conditions which must be fulfilled for an exact inverse filter matrix to exist. Additionally, [89] demonstrates that the number of loudspeakers must exceed the number of microphones in a manner consistent with the findings of [87]. Moreover, an explicit formula is derived specifying the number of required inverse filter coefficients for the existence of an exact inverse. The paper also investigates the spatial extent of the zones of equalization produced by inverse filtering. It is shown that the equalized zone scales in size in accordance with the acoustic wavelength at the highest frequency of interest.

The low extent of the equalized zone and the numerical sensitivity to errors in the measured responses appear to be the main limitations of the MINT. An improvement of the method has been proposed in [90], where more control points are considered without increasing the number of inverse filters. Another improvement is discussed in [91], where an iterative method is applied to the MINT considering an optimally-stopped weighted conjugate gradient. To improve the computational efficiency of the MINT, an oversampled subband approach with decimation has been presented in [92].

4.6. Alternative Classification of Equalizers

As explained earlier, equalizers may be classified in several ways, and the above design techniques have already been classified into minimum-phase or mixed-phase. Another interesting classification of the equalizer design methods was provided in [55]. According to [55], the equalizer design methods can be classified into *"indirect"* and *"direct"* methods. As shown in Figure 3, indirect methods estimate a model of the room response—possibly processed—to obtain the equalizer by model inversion. Direct methods instead minimize the error between the equalized room response and a target response. From this point of view, homomorphic filtering, LPC analysis, and frequency domain deconvolution constitute indirect methods, while least-squares optimization constitutes a direct method. Multiple-input/multiple-output inverse theorem techniques can be classified as both direct and indirect methods, since they compute the equalizer considering the inversion of a matrix of room responses. However, according to Equation (2) they can also be estimated by minimizing the error with respect to an ideal response.

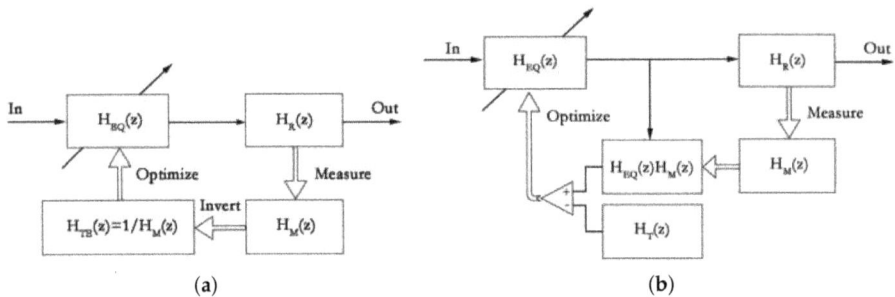

Figure 3. (a) Indirect and (b) direct equalizer design methods classification as reported in [55], where H_{EQ} represents the equalization filter, H_R is the reproduction channel, H_M is the measured impulse response, and H_T and H_{TE} are the target functions.

5. Pre-Processing Techniques

The main techniques that have been developed to overcome the limitations of RRE dictated by the characteristics of the room response, also taking advantage of the psychoacoustic properties of the ear, are discussed in the following. These approaches are capable of modifying the measured RIR and should be applied before the actual equalization procedure. They are suitable for both single-point and multi-point equalization.

The major problems of RRE that were addressed in the early approaches were the very long impulse responses of the equalizer, the limited region of space in which the RRE is effective, and the slow time variations of the room response. The very long impulse response of an exact equalizer is due to the spectral characteristics of the room response, as shown in Figure 2b, with many peaks and notches that increase their density towards high frequencies. The notches correspond to zeros close to the unit circle in the RTF. Thus, the inverse filter has poles close to the unit circle that determine the long impulse response. The notches at high frequencies are extremely variable with position and time, determining the small extent in space and time in which the equalizer is effective. Movements of listening position of 10 cm can cause variations of up to 20 dB in the room response [93], and a pre-processing technique is required to contrast these variations.

5.1. Short Filters

One of the first expedients to improve RRE resorted in using short equalization filters. By considering a coarse model of the room response which tries to capture and correct only the general trend of the room response, avoiding modeling the sharp peaks and notches, it is possible to reduce the temporal length of the equalizer impulse response. This solution is also beneficial for enlarging the extent of the equalized zone and to cope with the room response variations in time [2]. One of the most effective techniques for designing short equalization filters is that based on LPC analysis [61], which obtains a good modeling of the peaks of the room response, with a coarser modeling of notches.

5.2. Non-Uniform Frequency Resolution

To improve the accuracy and effectiveness of equalization, the equalizer should take advantage of the characteristics of the room response and the human ear. At low frequencies, the room frequency response is more regular and the peaks and notches are mostly insensitive to the position in the room. The resolution of the ear is nonuniform and nonlinear, with a logarithmic dependence on frequency. At high frequencies, the ear is rather insensitive to notches of the room response and to high-frequency reverberation. Accordingly, the equalizer should provide fine resolution at low frequencies and a coarser resolution at higher frequencies. Many techniques have been developed following this strategy:

- Complex smoothing,
- Frequency warping,
- Kautz filters and parallel IIR filters with fixed poles,
- Multirate approaches.

5.2.1. Complex Smoothing

Fractional octave-band smoothing of the power spectrum has been widely applied in audio processing. Its use can be traced back to analog equalizers (as for example the one-third-octave-band filterbank analyzers), and was later extended to digital spectrum analyzers. In [35,94], the authors extend the technique by introducing a methodology for smoothing the complex transfer function of the measured room response with fractional octave profiles. The technique can be implemented in the time or frequency domains. It is perceptually compliant since the spectral smoothing follows the frequency resolution of the ear, with a fine resolution at low frequencies and a lower resolution at high frequencies. As a result, in the time domain the application of complex smoothing can retain the initial high-frequency content of the early components (i.e., the transient behavior of the direct sound and of the first reflections) and then can progressively introduce a low-pass filtering of the later components (i.e., of room reverberation) [95]. This is also desirable from another psychoacoustic point of view. In dispersive room environments, the ear is very sensitive to the signal onsets (i.e., to the full frequency range of the first part of the RIR), while it is less sensitive to the high-frequency components of late reflections [19,22,24,35]. When the complex smoothed impulse responses are used for the design of an RRE, they allow the avoidance of compensating sharp notches at high frequencies in order to obtain a reduced length of the equalizer, and they provide a more robust equalization with lower sensitivity to possible changes in the listener position and to other variations in the room response [2,35]. Figure 4 shows the complex smoothing effect on an RTF for different resolutions. By introducing an appropriate delay in the equalizer, complex smoothing allows the mixed-phase equalization of a room response. As an alternative to complex smoothing, frequency-dependent signal windowing [96] or a separate smoothing of the magnitude and phase of the transfer function [97] have been proposed.

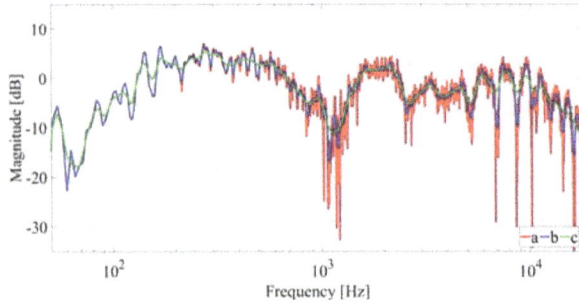

Figure 4. Complex smoothing of a measured room transfer function (RTF): (**a**) RTF; (**b**) 1/12 octave-band complex smoothing; (**c**) 1/3 octave-band complex smoothing.

5.2.2. Frequency Warping

Another technique that provides a nonuniform frequency resolution is *"frequency warping"* [98]. The original idea of frequency warping is presented in [99], where a nonuniform Fourier transform is introduced. The technique consists of replacing the unit delay z^{-1} of digital filters with a first-order all-pass filter,

$$D_1(z) = \frac{z^{-1} - \lambda}{1 - \lambda z^{-1}}, \tag{3}$$

thus obtaining a bilinear mapping of the unit circle on itself. The warping effect can be adjusted to approximate the spectral representation of the ear [100]. In [101], analytic expressions that approximate the Bark and ERB scale are provided. They allow for a very good approximation of the Bark scale and less accurate approximation of the ERB scale, due to the higher frequency resolution required, particularly at low frequencies. The effect of the frequency warping can be easily reversed by again replacing the unit delay z^{-1} with the all-pass filter

$$\tilde{D}_1(z) = \frac{z^{-1} + \lambda}{1 + \lambda z^{-1}}. \tag{4}$$

Figure 5 shows an example of the effect of frequency warping on an RTF for different values of λ. The reader should note the expansion of the low frequency range and the compression of the high frequencies obtained with positive values of the warping parameter λ.

Warped FIR and IIR filters can be obtained by replacing the tapped delay line with a chain of first-order all-pass filters, but while the implementation of warped FIR filters is immediate [66], warped IIR filters require appropriate structures to avoid delay-free loops [102]. Warped FIR filters are strictly related to the Laguerre filters [103], the only difference being the fact that in a Laguerre filter there is an additional prefilter placed before the all-pass chain [66]. A logarithmic frequency scale can also be approximated, but in this case the all-pass chain has to be replaced with a filterbank formed by all-pass filters.

Frequency warping has been exploited in many audio applications, from LPC analysis [100], audio equalization [104,105], loudspeaker equalization [106–108], and physical modeling of guitar bodies [109], to head-related transfer function (HRTF) filtering [66,109]. The reader is referred to [66] for a review of frequency warping techniques and their applications. In the context of RRE, frequency warping has been used by many researchers in minimum-phase equalization to improve the equalizer performance by expanding the resolution at low frequencies and compressing it at high frequencies. A psychoacoustically-motivated frequency scale—in most cases the Bark scale—is used. For example, in the approach of [70,71], a prototype room response is first frequency warped to an approximate Bark scale. Then, an all-pole model of the room response is obtained in the warped domain using

LPC analysis. Eventually, a minimum-phase equalizer is derived in the time domain by de-warping the inverse of the all-pole model with (4). The main disadvantage of this approach is represented by the high computational cost of the frequency warping operation. In [73,75], frequency warping was efficiently implemented by nonlinearly sampling a high-resolution fast Fourier transform (FFT) of the prototype room response.

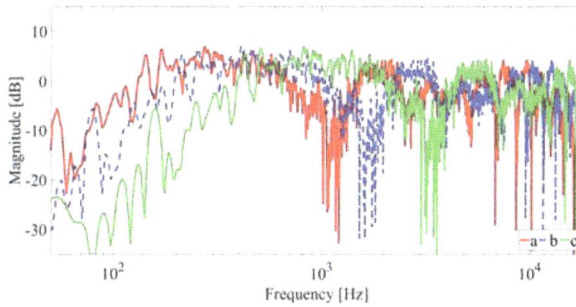

Figure 5. Frequency warping of a measured RTF: (**a**) RTF; (**b**) warped RIR with $\lambda = 0.2$; (**c**) warped RIR with $\lambda = 0.5$.

5.2.3. Kautz Filters and Parallel IIR Filters with Fixed Poles

Kautz filters are rational orthonormal filter structures. They are orthonormal since they have orthonormal impulse responses. Continuous-time rational transfer functions with orthonormal impulse responses were studied by Kautz in [110]. Discrete-time orthonormal transfer functions were later studied by Broome in [111], who named them *"discrete Kautz functions"*. Kautz filters can be considered as a generalization of warped FIR filters and Laguerre filters, where the chain of all-pass filters with equal poles is replaced by a chain of all-pass filters with individual poles, possibly complex [54]. Figure 6 shows the results of Kautz modelling a measured RTF.

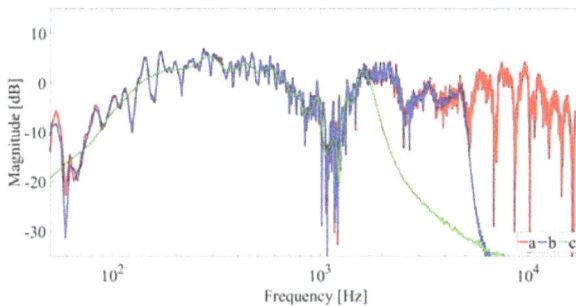

Figure 6. Kautz filter applied on a measured RTF: (**a**) RTF; (**b**) Kautz model of order 1000; (**c**) Kautz model of order 300.

By properly choosing the poles, it is possible to realize an arbitrary allocation of the frequency resolution of the designed filter. An approximation of the log-frequency scale resolution with Kautz filters can be found in [55]. The poles can be chosen a priori on the basis of the desired resolution, but they can also be tuned to the specific application by matching the pole frequency with the resonances of the system to be modeled [54]. In practice, fine tuning of the poles is necessary when designing low-order models for highly resonant systems [54]. Once the poles are chosen, system identification using Kautz filters can benefit from the orthogonality of the impulse responses. The reader is referred to [54] for a discussion about pole fitting and identification methods.

Kautz filters have been used for RRE, exploiting the nonuniform frequency resolution of these filters. They have been applied both for minimum-phase and mixed-phase equalization, using both fixed poles or tuned poles [54,55]. When a fixed pole approach is used, the Kautz filters can also be designed and implemented in the form of a filterbank of second-order sections [112–114], with advantages for the computational complexity. In [114,115], the theoretical equivalence of parallel filters and Kautz filters is shown, and formulas to convert the parameters of the two structures into each other are given. Figure 7 reports a parallel filter design example following the methodology of [114].

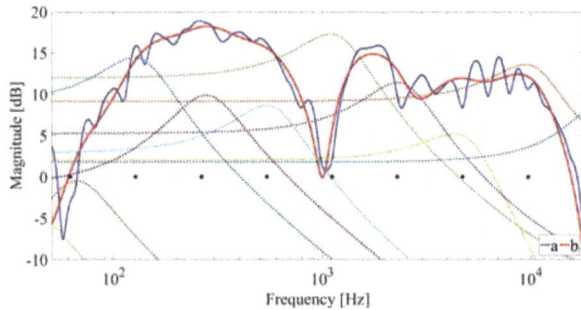

Figure 7. Bank's parallel filter design example: (**a**) RTF; (**b**) the resulting filter frequency response. The dotted lines represent the individual transfer functions of the 16 second-order sections, while the circles display the pole frequencies.

5.2.4. Multirate Approaches

Another possibility for achieving a nonuniform frequency resolution is given by multirate approaches. In these approaches the spectrum is divided into different bands, that are down-sampled and separately processed with filters of different length. In most of the proposed approaches, one of the filters covers the low frequencies [50,72,116–120] which is used for modal equalization and low resonances control (see Section 6.5) or for bass management. Generally, the low-frequency filters must compensate very long reverberation times, and thus the filters benefit from the high down-sampling at low frequencies. The filters used for mid and high frequencies generally use a lower resolution compared to the low frequencies, with strong computational savings.

For example, in [119] the authors propose a dual band equalization procedure. The low frequency channel is restricted approximately to the Schroeder frequency through down-sampling. An FFT-based technique with regularization is used to design a minimum-phase equalizer with homomorphic filtering. The upper band is also equalized with a minimum-phase equalizer designed with LPC analysis and warping techniques. In [120], the same authors have instead divided the spectrum into three bands: the low-, mid-, and high-frequency bands. The low band is again restricted approximately to the Schroeder frequency—specifically 150 Hz—through down-sampling, but the equalizer is now designed with the LPC technique. In the mid-frequency band from 150 Hz to 900 Hz, the equalizer is designed with a warped LPC technique to focus attention to the lower part of the band. Above 900 Hz, the high-frequency spectrum is smoothed to reduce sensitivity to position, and then the equalizer is found by inverting the smoothed spectrum, imposing a slightly decreasing target function. The authors have also combined this basic equalizer with an excess phase equalizer in the low-frequency band, and a pre-processing based on a deconvolution technique in the first 10 ms after the direct sound.

In [72], the authors have combined the multi-point fuzzy c-means clustering technique of [121] (see Section 6.2) with a dual-band multirate approach, separating the low-frequency band below 80 Hz from the high-frequency band beyond 80 Hz. The low-frequency band is decimated by a factor of 256 to work with small length room responses, prior to applying the fuzzy c-means clustering technique for designing the equalizer.

5.3. Room Impulse Response Reshaping

Another possibility for taking into account the psychoacoustic characteristics of the ear is that of reshaping the impulse response in such a way that the alteration of the room becomes inaudible. In RIR shortening, the attenuation of the original RIR is accelerated so that the reverberation effect is weakened. Different techniques have been proposed in the literature [78,122–130]. In what follows, we review the most relevant methods. Most of these methods are not RRE methods in a strict sense, but they could be easily combined with RRE techniques.

The first attempts of RIR reshaping [122,123] tried to adapt the concepts of channel shortening developed in the telecommunication area [131–134], applying least-squares optimization algorithms. By properly designing a reshaping filter, it is possible to maximize the energy of the equalized RIR in a desired time window, minimizing at the same time the tails of the room response in an undesired window. In this way, for example, it is possible to directly maximize the D50 measure for intelligibility of speech, which is the ratio of the energy within 50 ms after the first peak of an RIR versus the energy of the complete response. The least-squares optimization of the reshaping filter segregating the desired time window from the undesired window provided unsatisfactory results [122] in the form of audible late echoes or spectral distortions. These problems are caused by the strong separation imposed considering non-overlapping desired and undesired windows, and by the least-squares optimization that leads to a non-uniform error distribution. Thus, already in [122] the authors modified the channel shortening paradigm with the aim of shaping the desired impulse response to a shorter reverberation time, considering a gradual transition between the desired and undesired windows.

The approach was improved in [78,124,126,127]. These approaches exploit the psychoacoustic properties of the human auditory system, and in particular the forward-masking effect. They aim to obtain an equalized response that decays sufficiently quickly to avoid audible echoes, such that the reverberation time is masked by the direct sound according to the forward-masking effect of the human auditory system. The desired and undesired windows are here specified according to the average forward masking curve of [47] and [43]. Moreover, to avoid the problems due to least-squares optimization, infinity-norm and p-norm optimization (with large values of p) are proposed. The approach is also applied to multi-channel problems in [125,130].

No spectral requirement is imposed by any of the above-mentioned RIR reshaping approaches. In most cases, these approaches usually yield a flat overall frequency response, but with very long impulse responses they may lead to spectral distortions [128]. To contrast this problem, in [128] the objective function is modified to incorporate a p-norm-based regularization term in the frequency domain, thus imposing the joint optimization in time and frequency domains. In [135], the regularization term is replaced by an integrated spectral flatness measure, which allows the integration of the concept of auditory scales into the equalizer design. Thus, the approaches of [128,135] combine RIR reshaping with RRE.

6. From Single-Point to Multi-Point Equalization

Another classification of RRE is relative to the number of microphones or control points used. Classical approaches are based on the use of one RIR captured near the listener position (see Section 4), implying a specific sweet spot where the equalization is effective [136]. The objective of multi-point equalizers is to enlarge the equalized zone [137], also improving the robustness of the equalizer towards measurement errors and variations of the room response, implicitly exploiting the variation between the multiple measurements. In what follows, a review of multi-point equalization methods is given, taking into account that most of the techniques discussed in the previous sections have also been applied to multi-point RRE.

6.1. Average and Weighted Average Methods

One of the earliest multi-point approaches was proposed by Elliot and Nelson. In [77], the authors presented a method for designing an equalization filter for sound reproduction systems by adjusting the filter coefficients so as to minimize the sum of squared errors between the equalized responses at multiple points in the room and a delayed version of the original signal. The paper considers both fixed and adaptive equalizers based on filtered-x algorithms. The approach is effective and has also been applied in many other improved techniques [9,31,33,118]. The main limitation is given by the fact that the implicit averaging in the sum of squared errors cannot exploit the similarities in the room responses, nor can favorite equalization at certain positions. In the context of car audio equalization [118], the technique was improved by considering multi-point equalization with a weighted average of the errors. The solution provided improvements in the response at the selected location, *"without significant degradations at other points"* [118].

6.2. Clustering Methods

We can exploit the similarities between different spatially distributed room responses by clustering them according to a chosen distance measure. In [138], the *"extremely large set"* of possible RTFs within an enclosure was grouped together and equalized by a smaller number of equalizers. The RTFs were modeled with all-pole filters using LPC analysis, and thus minimum-phase equalizers were designed. Then, vector quantization was performed to optimally classify the all-pole filters. The classification can be used as a spatial equalization library, achieving reduction in reverberation over a wide range of positions within the enclosure, depending on the actual position of the listener. The main limitation of this method is the necessity to extract and memorize a large set of room responses and equalizers and to track the position of the listener.

A fuzzy c-means clustering method is applied in [30,70–72,121,139,140]. In the approach of [121,139], *"representative prototypical room responses"* are derived from several measured room responses that share similar characteristics using the fuzzy c-means unsupervised learning method. The prototypical responses are then combined to form *"a general point response"* based on the fuzzy standard additive model of Kosko [141,142]. The method employs a weighting according to *"the level of activation"* of a prototype, depending upon the degrees of assignment of the room responses to the cluster containing the prototype. The equalizer is then computed from the inverse of the general point response using LPC analysis, *"obtaining a significant improvement in equalization performance over the spatial averaging methods"* with the suppression of the peaks in the room magnitude spectra [139]. The method was further improved in [70,71,140] by applying the fuzzy c-means clustering to warped impulse responses, thus taking advantage of the perceptual properties of the ear. The approach was also combined with multirate filtering in [72] to allow effective filtering of the low frequency response at low sampling rates with computational savings.

The approach of [70,139] was later improved by applying frequency warping and fuzzy c-means clustering to the magnitude room responses [73,75], with a strong improvement in terms of computational complexity. A weighted fuzzy c-means clustering was also proposed in [143], where the RIR samples were weighted in a different manner to account for the different effect they have on RRE.

6.3. Prototype Approach

The fuzzy c-means clustering approach of [70,139] is also a first example of a *"multi-point prototype approach"*. These methods use measurements of the room response in different locations to extract a prototype response which is representative of the perceptual acoustic situation that has to be corrected. A single equalizer is then designed with indirect or direct methods [55], on the basis of this prototype response.

Different approaches for the determination of the prototype response were studied in [144]. In particular, the fuzzy c-means method was compared with the mean average, the median, the min-max, and the root-mean-square average, and applied to fractional octave complex-smoothed spectra. The equalizer was then derived by inversion or the Kirkeby algorithm [81], or LPC analysis, with minimum-phase equalization. In the considered conditions, the mean average gave the best results, with the other methods also providing similar performance. The prototype extraction approach based on mean average was also combined with the method of [73] and applied to room-response equalization [75,145]. Subjective listening tests confirmed the good results obtained with the approach [75]. The approach was further extended in [146,147] by also considering a group-delay equalization. In [146], after the determination of the minimum-phase equalizer, the smoothed phase responses measured at different positions are corrected with the phase response of the equalizer and are used to determine the group delay responses. A prototype group delay is computed by averaging the group delay at the different positions, and after spectral smoothing is used to extract an all-pass FIR group-delay equalizer. In [147], the prototype phase response used to determine the phase equalizer is extracted only from the early reflections, which represent the contribution of the direct sound, discarding the late reflections that represent the reverberation of the environment. The mixing time between early and late reflections is calculated using the approach presented in [147,148] based on Gaussianity estimators. The prototype function is truncated using the mixing time, and an FIR phase equalizer is obtained with the matched filter technique; i.e., time-reversing the all-pass impulse response. With this approach, pre-ringing artifacts are avoided, since only the early reflections are considered in the equalizer. In fact, taking into consideration only the first reflections, only the main characteristics of the room are considered and those parts of the impulse response which contain zeros that vary with the position and according to [149] produce the pre-ringing artifacts are avoided.

A prototype approach is also followed in [93,150–153]. According to the authors, *"part of the impact of a listening room is natural to the human ear and should not be removed by a room correction system"* [93]. In particular, sound reproduction in a room normally causes an increased sound pressure level at the lower frequencies, because of the lower absorption typically found at these frequencies. Since this effect is natural to the human ear, as it provides the sense of being in a room, room equalization systems should not be allowed to remove the smooth increase in level at low frequencies, also referred to as the *"room gain"*. The room gain describes how the room efficiency increases at low frequencies compared to high frequencies [152]. Moreover, the prototype response should preserve the basic characteristics of the loudspeaker; i.e., the equalizer should not try *"to make all loudspeakers sound alike"*. Thus, the developed system estimates the main characteristics of the loudspeaker: lower cut-off frequency and slope, sensitivity, directivity index, and upper cut-off for the treble driver. The equalizer is designed by acquiring information both of local properties at the listening position and on the acoustic power in the three-dimensional sound field. The RRE is based on measuring the sound pressure at the listening position and in at least three other randomly selected positions. The measurement in the listening position holds information on the perceived sound field, while the other room measurements hold information on the energy in the three-dimensional sound field. The information is then used to calculate lower and upper gain limits for the designed equalizer. The prototype response is automatically calculated based on the measurements. At low frequencies, the prototype response is designed to provide the same room gain of a listening room conforming to the IEC 268–13 standard [154], approximating a smooth room gain with a second-order shelving function, which adds 6 dB level smoothly below 120 Hz [152]. The equalizer is minimum phase and designed on basis of the homomorphic technique [151].

6.4. Common Acoustical Poles Compensation

At low frequencies strong resonances can appear in the room response. These resonances are often independent of the position and are associated with long slowly-decaying modes. Different techniques have been proposed to compensate the low-frequency response. Many of these techniques exploit multi-point measurements to determine the spectral properties of the resonances.

A model for a RTF using common acoustical poles corresponding to the resonances of a room is proposed in [63]. The common acoustical poles are estimated as the common pole values of many low-frequency RTFs estimated for different source and receiver positions. The poles are computed from an LPC model of the room response, estimated by two possible methods: (i) using a least-squares method, assuming all measured RTFs have the same LPC coefficients and (ii) averaging the LPC coefficients estimated from each measured RTF. The estimated poles correspond to the major resonance frequencies of the room. Then, using the estimated common poles, the method of [63] models the RTF with different moving average coefficients. The model is called by the authors the common acoustical pole and zero model, since it is a zero-pole model formed by the common acoustical poles and the zeros provided by the moving average coefficients. The approach was later expanded in [64,65]. In [64], a multi-point equalization filter using the common acoustical poles is proposed. The common acoustical poles are again estimated as common LPC coefficients from multiple measurement of the RTFs. The equalization is then achieved with an FIR filter having the inverse characteristics of the common acoustical pole function. As for the other all-pole models, the equalization filter is a minimum-phase equalizer that cannot compensate for the notches of the frequency response. Nevertheless, the filter can suppress the common peaks due to resonances in the multiple positions, and has low sensitivity to changes in the receiver position. In [65] a pre-conditioning stage is added to the common acoustical poles equalizer. The pre-conditioning stage suppresses low-Q resonances in the entire spectrum, while a second stage based on the common acoustical poles suppresses or minimizes the low-frequency resonances. In [155], an empirical technique to select an appropriate order for the common acoustical pole model is proposed. The technique selects the first order for which a further growth does not lead to an improvement in the modeling accuracy for at least one of the measured RIRs. The model order depends on the chosen maximum frequency of the modeled poles. The iterative algorithm of [156] is also based on the common-acoustical-pole and zero model. It designs biquadratic filters suitable for multi-point RRE.

The common acoustical poles compensation could also benefit from the filterbank technique based on second-order sections of [112–115], exploiting in particular the logarithmic frequency resolution and the ability to customize the pole positions.

6.5. Modal Equalization

Modal equalization has also been proposed at low frequencies [4,157]. Modal equalization aims to control excessively long decays in listening rooms caused by low-frequency modes, minimizing the audibility of these resonances. Modal equalization balances the rate of sound decay of the low-frequency modes to correspond to the reverberation time at mid and high frequencies. This is not an RRE technique by itself, but it can be used with conventional magnitude equalization to optimize the reproduced sound quality. In [157], two methods for implementing active modal equalization are proposed. The first approach considers a single loudspeaker and filters the sound such that the mode decay rates are controlled (e.g., using a filter with couples of zeros placed in correspondence to the poles responsible for the resonances). The second approach implements modal equalization by one or more secondary loudspeakers. A correction filter is considered for each secondary loudspeaker in order to produce a compensatory sound. The first approach was studied in depth, and different techniques for identifying the modes, estimating their parameters, and designing the equalizer are presented. Estimation of the modal decay parameters is based on the nonlinear optimization of the model for exponential decay plus stationary noise floor presented in [158].

6.6. Plane Wave Approach

Another possibility for equalizing the sound in the low-frequency region is that offered by the plane wave approach. In rectangular rooms with a symmetric arrangement of loudspeakers in two opposite walls, it has been shown in theory [159] and experiments [160] that equalization within the entire room can be achieved at low frequencies. The approach generates a plane wave that propagates from one wall to the opposite one, where it is absorbed by the loudspeakers. In the experiments of [160], the signals fed to the loudspeakers are determined with the RRE approach of [89]. The error sensors are positioned in two planes perpendicular to the direction of propagation of the simulated plane wave. The desired signal in the planes is a Dirac delta function with a delay corresponding to the time it takes the sound to travel the distance between the planes. A plane wave approach has also been studied by the authors of [161–165]. First, in [161], the authors developed an application based on finite-difference time domain approximation for studying low frequencies in audio reproduction. In particular, a rectangular room has been simulated by using a discrete model in time and space. Then, in [162] the application was used to study different configurations of loudspeakers in the room to reduce the effect of the acoustic modes. It is shown that by increasing the number of loudspeakers, the variation of the room response across positions is improved at the expense of an increment in the magnitude deviation at every position. The application has also been used to assess the effect of different equalization techniques, such as multi-point equalization and equalization of the acoustic radiation power of the loudspeaker. Eventually, a solution for equalizing the low-frequency sound field using multiple loudspeakers—named controlled acoustic bass system (CABS)—was proposed and studied in [163–165]. This solution creates a traveling plane wave in one side of the room and cancels it at the opposite wall using extra loudspeakers, with delayed and anti-phase response to remove back-wall reflections. Using the application of [161] and real measurements in rectangular rooms, the authors have shown that the CABS solution can produce a uniform acoustic field in the low-frequency range. In [166], the approach of [159,163] is further extended to rooms of arbitrary shape with multiple loudspeakers *"situated in more normal locations"* considering a 5.0 loudspeaker set-up. Additionally, [167] has addressed the problem of a non-rectangular room and of an asymmetric loudspeaker set-up. In [167] a multiple-input/multiple-output (MIMO) equalization technique that prescribes only the magnitude of the room response in the control points is proposed. The approach allows a smaller magnitude deviation to be obtained compared to the previous plane-wave approaches.

To improve equalization with plane waves, a control approach called effort variation regularization was proposed in [168]. In this approach, the conventional cost function of RRE of [169]—based on the minimization of the least-squares error in multiple control points—is modified by adding a regularization term proportional to the squared deviations between source strengths. The approach can be applied both in the frequency and time domain. Simulation results show that the technique can lead to smaller global reproduction errors and better equalization performance at listening positions away from the control points, than the Tikhonov regularization or the approach based on feeding the same signal to all loudspeakers placed on the same wall.

6.7. Other Low-Frequency RRE Approaches

At very low frequencies, instead of a plane wave, it is much more efficient to use a pressure-field chamber approach [170]. This approach is obtained by sending the same signal to all loudspeakers. This generates a standing wave pattern inside the room, which is homogeneous at wavelengths considerably larger than the room. For this reason, in [170] a hybrid-field playback approach is proposed which combines the efficiency of the pressure-field playback at the very low frequencies with the homogeneous sound-field obtained with the plane wave approach at higher frequencies.

In [79], the problem of multiple-loudspeaker low-frequency RRE for a wide listening area, with the equalized loudspeaker supported by the remaining ones, is addressed as a multipoint error minimization problem between the desired response and the synthesized magnitude response. The cost function is minimized, imposing physical and psychoacoustical criteria. In particular, to obtain short equalization filters, a temporal masking constraint is imposed on the equalization filters. To avoid perceivable echoes, a combination of delay and gain relative to the main loudspeaker is considered, with the auxiliary loudspeaker signals that should fall below the echo threshold [171]. To avoid modifications in the spatial perception, the delay of the auxiliary loudspeaker signals is enforced to be at least of 1 ms in order to exploit the precedence effect. To avoid boosting the notches, a maximum-gain is imposed on the equalizers. The room equalization filters are computed considering a convex optimization framework that takes all these constraints into account.

6.8. Quasi-Anechoic Approach

An approach that is complementary to the low-frequency techniques introduced in the previous sub-section is the quasi-anechoic approach of [172]. At mid- and high- frequencies, the timbre perception and localization is dominated by the direct sound. Thus, in [172], a quasi-anechoic loudspeaker response is obtained as a gated version (up to the first reflection) of the RIR and is used to design the equalizer in two steps. First, a mixed phase equalizer is derived from the quasi-anechoic RIR, computing the inverse filter with a least-squares approach. The quasi-anechoic loudspeaker response has a short length and the delay introduced by the equalizer is too short to produce pre-ringing artifacts. Then, a minimum-phase equalizer is used to correct the remaining part of the room response (i.e., the magnitude spectrum modifications caused by reverberation).

In [173], the quasi-anechoic approach is combined with the prototype approach described in Section 6.3. In particular, a novel prototype function is derived from the combination of quasi-anechoic impulse responses with the impulse responses recorded in the real environment to be equalized. The approach is used to equalize the direct sound only in the mid–high-frequency range, while applying full equalization in the modal frequency range. The approach is motivated again by the fact that at mid and high frequencies the timbre perception and localization is dominated by the direct sound. Thus, the measurable but mostly inaudible magnitude deviations due to reflections should not be equalized [174]. In [173], several experiments were conducted in order to validate the proposed approach, reporting objective measurements and subjective listening tests in comparison with approaches of the state-of-the-art. In this context, Figures 8 report the results of the equalization procedure. In particular, Figure 8a shows four impulse responses acquired in a real room—the prototype function and the equalizer obtained with the multi-point approach of [173] and the single-point equalizer derived as an inverse filter of the smoothed frequency response of IR1. Figure 8b shows the effect of the equalization procedure on the IRs applying the multi-point approach, while Figure 8c shows the effect of the single-point equalizer. It is evident that the performance of the single-point equalizer is very good only for IR1, while the multi-point equalizer exhibits flatter frequency responses compared to those obtained with the single-point approach.

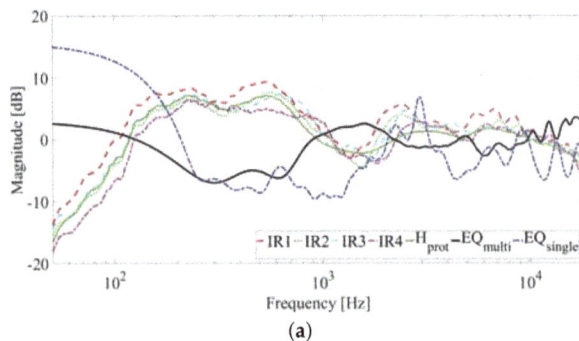

(a)

Figure 8. *Cont.*

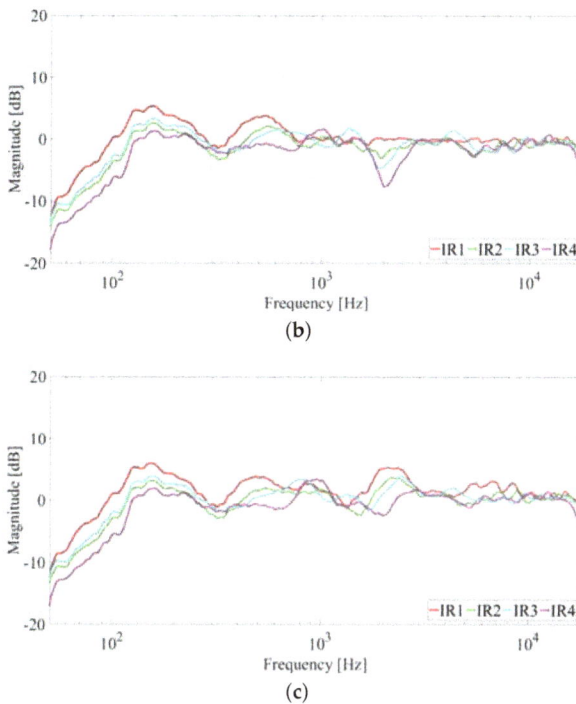

Figure 8. Equalization procedure: (**a**) impulse responses (IRs) with prototype function H_{prot}, multi-point equalizer EQ_{multi}, single-point equalizer EQ_{single}. Equalization results applying (**b**) single-point equalizer EQ_{single}; and (**c**) multi-point equalizer EQ_{multi}.

7. Adaptive Single-Point and Multi-Point Equalization

The room is generally a time-varying environment (a *"weakly non-stationary"* system as defined in [2]) that changes as a function of several parameters, such as the position of physical objects in the room, the opening of doors, as well as the movement of people and other obstacles in the enclosure [6,175]. Additionally, temperature variations can lead to large variations in the RIR, as reported by [176]. Furthermore, variations of the source and receiver positions, and of loudspeaker and microphone characteristics may occur as reported in [6]. Thus, adaptive solutions suitable to track and correct slow variations in the room response should be adopted. Different adaptive RRE techniques have been proposed in the literature. The approaches are here classified considering the number of input and output channels as SISO/SIMO, and MISO/MIMO, where input refers to the number of loudspeakers and output to the number of microphones, since these classes share similar problems in the identification procedure.

7.1. SISO/SIMO Approaches

These techniques can be classified into time domain and frequency domain approaches.

7.1.1. Time Domain Approaches

A first adaptive equalizer was proposed in [77], considering the variability of the environment from different points of view. The approach was based on a single-point technique, adaptively minimizing in the time domain the mean-squared error between the equalized response and a delayed single-channel version of the original signal using a filtered-x algorithm. The equalization was effective for the considered position, but a degradation in other points of the enclosure was introduced, as also described in Section 6. Therefore, a multi-point approach was also presented by the same authors in [77], where the equalizer was designed by adaptively minimizing the sum of squared errors between the equalized responses in several positions and a delayed version of the input signal. Unfortunately, the approach is very sensitive to peaks and notches in the room response and to room response variations at different positions. As a consequence, pre-echo problems can easily be experienced.

7.1.2. Frequency Domain Approaches

Working in the frequency domain, a single-point RRE technique was proposed in [177]. Here, the loudspeaker and microphone signals are split into subbands (a 20-band filterbank) and the equalization is achieved by adaptively updating the filter weights in these subbands. The approach is interesting because it combines simplicity, robustness towards peaks and notches of the room response, and the ability to track room response variations. It was improved in [178] by introducing a frequency-dependent step size. In this way, it is possible to optimize the adaptive equalization in each subband, improving the overall convergence speed. In [179], a further improvement of the previous methods [177,178] has also been presented to cope with the online identification of the impulse response. In particular, the room response estimation is obtained by means of inserting artificial test signals in such a way that they remain inaudible to listeners by exploiting frequency masking. The signal is then analyzed in the frequency domain to identify the test signal and to determine the RIR. In [180] the approach of [177,178] was elaborated and improved by developing a multi-point solution. After identification in frequency bands, a fractional octave smoothing is applied to the impulse responses, and a prototype filter is computed from the mean of the room magnitude responses. The obtained results have shown that the performance of this rather simple structure can be improved by considering a multi-point solution, which results in an increased width of the equalized zone. In [84], the approach of [180] was further elaborated considering frequency warping in the low-frequency region to improve perception. Specifically, the room responses at different positions in the zone to be equalized are estimated in the warped domain and the common trend of these responses is extracted as a prototype function. This allows the equalizer resolution to be increased at frequencies where the human auditory system is more sensitive. Adaptive versions of the filterbank techniques of [112–115] could also be used for the same purpose.

7.2. MISO/MIMO Approaches

The adaptive RRE techniques proposed in [84,178,180] (and many other papers) consider the equalization of a single sound source (i.e., of a single audio reproduction channel), due to the problem of estimating several impulse responses at the same time. If two or more channels are employed, the covariance matrix of a multichannel adaptive algorithm becomes ill-conditioned due to the correlations between the channels for typical reproduction techniques. The ill-conditioning generally causes convergence problems. This was shown, for instance, for stereophonic acoustic echo cancellation [181]. To cope with the non-uniqueness problem, a method to reduce the inter-channel coherence is usually exploited. In this context, many of the techniques used to reduce the channel cross-correlations often introduce significant distortions, which are unacceptable in high-quality sound reproduction systems [181,182]. Therefore, a suitable technique which is capable of decorrelating the loudspeaker signals and of preserving the audio quality must be considered. The approach in [183] introduces a multichannel solution which also considers the non-uniqueness problem. The room responses are estimated with good accuracy by reducing the inter-channel coherence using a technique that produces only a mild degradation of the sound quality. Specifically, the low-frequency region is decorrelated by exploiting the missing-fundamental phenomenon, while the high frequencies are decorrelated with a second-order time-varying all-pass filter combined with a multiple notch filter [184]. The equalizer is designed in the warped frequency domain to improve the equalization in the low-frequency region and, at the same time, to reduce the computational cost of the design. In [185], the adaptive multichannel and multi-position RRE system briefly introduced in [183] is fully detailed and extended, providing a real-time implementation in commercial Hi-Fi products.

To improve the convergence speed and the robustness of the adaptive identification algorithm in the presence of low signal-to-noise ratio, the use of a biased adaptive algorithm has recently been proposed in [186] for a MIMO system. In detail, the algorithm is based on the improved proportionate normalized least-mean squares algorithm (IPNLMS) within the conventional filtered-x scheme (IPNLMS-FX), previously introduced for active noise control (ANC) [187], and here extended towards multichannel equalization. However this method requires an a priori estimation of the impulse responses, which is not available in many practical applications. With the same purpose of improving convergence and robustness, a combination of block-based adaptive filters (also employing biased algorithms) was proposed in [188].

It is worth underlining that if a binaural system is considered, a natural decorrelation among stereo channels is obtained. A stereo representation of an adaptive RRE system can be achieved without channel decorrelation, as reported in [169,189]. An improvement of this technique is presented in [190], where a subband structure is proposed to reduce the computational complexity of the procedure.

8. Fixed and Adaptive Wave Domain Equalization

The equalization approaches reviewed so far considered the reproduced sound field at one or more points in space. These points should ideally coincide with a potential listener position or restricted listening area. A broader view of equalization can be gained by taking the entire reproduced sound field within the desired—potentially large—listening area into account. This can be achieved by taking the spatio-temporal character of the sound field instead of the sound pressure at a limited number of points into consideration. In order to lay the grounds, the background of equalization following such a field-centered view is reviewed in the next subsection. This is followed by a review of representative approaches in the subsequent subsections.

8.1. Physical Background

The Helmholtz integral equation (HIE) [191] provides the solution of the inhomogeneous wave equation with respect to homogeneous boundary conditions. This covers—among others—the sound field reproduced by a distribution of loudspeakers in a room. The HIE states that the sound pressure within a source and scatterer free volume V is uniquely determined by the sound pressure and its directional gradient at the boundary ∂V of the volume. This finding can be exploited for the analysis of sound fields as well as for their synthesis. For the analysis of sound fields, it is sufficient to capture the sound pressure and its gradient at the border of the volume of interest. The same holds for the synthesis of sound fields where placing loudspeakers around a listening area allows full control of the sound field within that area. However, in terms of technical complexity, it is generally not desirable to capture both the sound pressure and its directional gradient using two different types of microphones placed at the boundary of the listening area. The same also holds for the synthesis using loudspeakers. Here one would have to employ monopole and dipole loudspeakers. Microphones and loudspeakers with the properties of a monopole are desirable over their dipole counterparts. It has been shown [192] that the HIE can be reduced to a monopole-only variant under some practically feasible limitations. This lays the theoretical ground of RRE within an extended listening area. In summary, the sound field within the listening area can be analyzed and controlled by a continuous distribution of microphones and loudspeakers located on the boundary of the listening area. However, the solution of the underlying continuous problem requires the solution of integral equations derived from the HIE [193]. Operator theory provides a solution to this problem by expanding Green's function into orthogonal basis functions. A closer look onto this will be taken in the subsequent section on wave domain adaptive filtering.

For a practical implementation of the principles outlined above, only a finite number of microphones and loudspeakers can be used. Hence, the continuous distribution of microphones and loudspeakers must be sampled spatially. The geometry and sampling is illustrated in Figure 9.

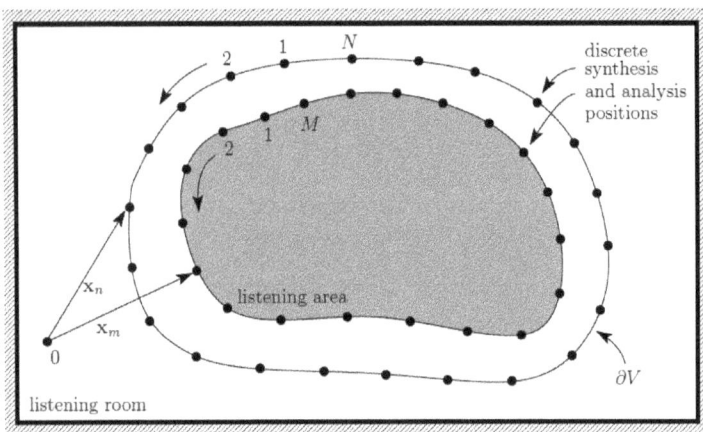

Figure 9. Application of the Helmholtz integral equation (HIE) to room compensation and spatial sampling of the loudspeaker and microphone contour.

The wave-theoretical view on RRE introduced above requires a sufficiently dense sampling of the loudspeaker and microphone contour. For typical systems, this calls for a high number of loudspeakers and microphones even when the upper frequency limit is quite low. Spatial sampling has been investigated intensively for different geometries and techniques [194–196]. The full three-dimensional coverage of the listening areas boundary by loudspeakers and microphone is often not feasible in practice. The limitations of considering only a planar listening area leveled with the listener's ears which is surround by loudspeakers are discussed in [197].

8.2. Wave-Domain Adaptive Filtering

An adaptive solution to the computation of RRE filters is desirable since the acoustic transfer paths may change, for instance due to people entering the room or due to temperature changes. As an example, the consequences of varying the room temperature on RRE using static filters are illustrated in [198]. A wide variety of adaptation algorithms have been developed in the past. Since RRE is an inverse problem, the class of filtered-x algorithms is well suited. The filters may be computed adaptively with the multichannel filtered-x recursive least-squares algorithm (X-RLS) [199]. However, in the context of multichannel RRE, an adaptive solution has three fundamental issues: (1) ill-conditioning; (2) non-uniqueness; and (3) numerical complexity. The first problem is related to the spatio-temporal correlation of typical loudspeaker signals, the second to the underlying optimization problem, and the third to the size of typical MIMO systems following the wave-theoretical view. A solution to the third problem—which also augments the other two issues—has been proposed by wave domain adaptive filtering (WDAF) [10,200]. Here the underlying MIMO system is decoupled by a set of spatio-temporal transforms, as illustrated in Figure 10.

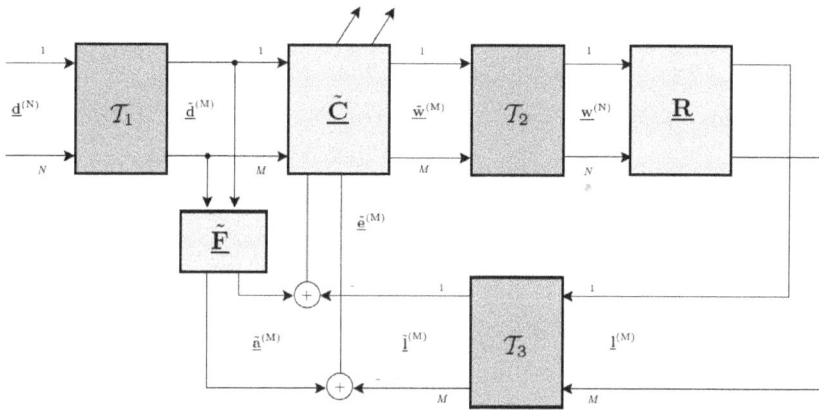

Figure 10. Block diagram illustrating the concept of wave domain adaptive filtering (WDAF)-based room equalization. The driving signals for the N loudspeakers—denoted by $\underline{\mathbf{d}}^{(N)}$—are transformed into the wave domain using the spatio-temporal transform $\mathcal{T}_1$, resulting in M transformed components $\tilde{\underline{\mathbf{d}}}^{(M)}$. These are filtered in the wave domain by the MIMO matrix $\tilde{\mathbf{C}}$ of equalization filters, resulting in the pre-filtered loudspeaker driving signals $\tilde{\underline{\mathbf{w}}}^{(M)}$, which are then transformed back by $\mathcal{T}_2$. The acoustic paths between the N loudspeakers and M control points (microphones) are combined into the MIMO room transfer matrix $\mathbf{R}$. The signals at the control points $\underline{\mathbf{l}}^{(M)}$ are transformed into the wave domain using transformation $\mathcal{T}_3$, resulting in the transformed control signals $\tilde{\underline{\mathbf{l}}}^{(M)}$. The desired free-field propagation is modeled in the wave domain by the MIMO matrix $\tilde{\mathbf{F}}$ of free-field transfer functions, resulting in the transformed desired signals $\tilde{\underline{\mathbf{a}}}^{(M)}$ at the control points. The error $\tilde{\underline{\mathbf{e}}}^{(M)}$ used for adaptation of the compensation filters is given by the difference of the transformed desired signals $\tilde{\underline{\mathbf{a}}}^{(M)}$ and the actual signals $\tilde{\underline{\mathbf{l}}}^{(M)}$ at the control points.

The transforms $\mathcal{T}_1$ through $\mathcal{T}_3$ are motivated by the physical background of the room equalization problem and its solution using orthogonal expansions, as outlined in the previous section. In terms of the underlying multichannel problem, this can be achieved by diagonalization of the MIMO systems using a generalized singular value decomposition (GSVD). This approach is known as eigenspace adaptive filtering (EAF) [193]. As a consequence, the adaptation problem is reduced to the adaptation of the main diagonal elements of the MIMO room equalization filter $\tilde{\mathbf{C}}$ in the transformed domain. In this way, the computational complexity is lowered significantly and the non-uniqueness problem is improved. However, EAF requires that the transfer paths from the loudspeakers to the microphones are known, which contradicts the idea of an adaptive computation of the equalization filters. Using analytic transformations which are based upon the free-field solutions of the wave equation, an approximate diagonalization of the MIMO system has been achieved [10,200].

The original approach focused on adapting only the diagonal paths in the transformed domain. In [201], this was extended towards a flexible adaptation framework also considering off-diagonal paths. The full adaptation of all paths in the transformed domain is investigated in [202]. Invertible transformations for WDAF have been introduced in [203], while a subband approach to WDAF has been published in [204]. Furthermore, strategies for the use of irregularly-spaced loudspeaker arrays have been proposed in [205].

8.3. Transform Domain Approaches

WDAF utilizes a set of transformations that transform the multichannel adaptive equalization problem into a transformed domain. This basic idea of applying a spatial transformation has also been applied to non-adaptive room equalization aiming at a large listening area. In [206], the sound field has been decomposed into circular/cylindrical basis functions for a concentric setup of loudspeaker and microphone array. This is essentially a two-dimensional problem. The equalization filters have been computed by least-squares optimization in the transformed domain. Room equalization has also been considered in the context of multizone synthesis by formulating the three-dimensional problem in the spherical harmonics domain (e.g., [207]).

A rather different approach is discussed in [208]. Here the original HIE is interpreted such that the sound field exterior to a spherical loudspeaker array is attenuated by the usage of variable directivity loudspeakers. The attenuation of the exterior sound field leads to less reflections traveling back into the listening area. Although such loudspeakers have not yet been realized, the simulation results look promising. The equalization problem is considered in the spherical harmonics domain, where the filters are computed by least-squares optimization.

8.4. Room Geometry-Aware Methods

The knowledge of the room geometry can be used to compute the resulting sound field in the room, for instance by the mirror image method. The control capabilities of a sound field synthesis system can then be used to cancel out the assumed contributions from the room. Methods which explicitly exploit knowledge on the room geometry can be seen as a specialization of the methods discussed so far, since they are based on a wave-theoretic view of the problem. A method for the equalization of early reflections for wave field synthesis (WFS) has been published in [209]. Here the mirror image sources are canceled out by anti-phase virtual point sources placed at the pre-computed positions of the image sources. A similar approach is presented in [210] for higher-order Ambisonics. An approach to room equalization for a linear loudspeaker array producing beams for a virtual surround system is discussed in [211]. The equalization of room reflections is achieved by accounting for the reflection of the beams in the room. The equalization filters are computed by solving the underlying least-squares problem in closed-form. In [212], a method is presented which is based on numerically simulating the impulse responses between the loudspeakers and control points. Only the early reflections are considered. The simulated impulse responses are fed into a MIMO solver for derivation of the equalization filters.

8.5. MIMO and SIMO Approaches

As an alternative to the wave-theoretic approach discussed so far, the acoustic paths between the loudspeakers and microphones can be interpreted as independent linear time-invariant systems. All resulting transfer functions can be combined together into a multiple-input/multiple-output (MIMO) system. MIMO room equalization approaches differ, amongst others, with respect to the loudspeaker and microphone positions (control points), and the particular technique used to compute the equalization filters. The difference between the wave-theoretic and the MIMO approaches discussed in the sequel is that the computation of equalization filters is not performed in a spatially transformed domain. Although the placement of the loudspeakers and microphones on the border of the listening area is motivated by the HIE, MIMO approaches may depart from this placement. As stated above, a sufficient number of loudspeakers and microphones must be used in order to synthesize and capture the entire sound field up to a given frequency. If the sampling is not dense enough, equalization may only be achieved at or in close vicinity to the microphone positions.

A non-adaptive MIMO approach which directly emerges from the discretization of the HIE is presented in [213]. The MIMO system is inverted in order to compute equalization filters for global equalization. As an alternative, a local solution is also discussed. A similar approach is followed in [85] for wave field synthesis. Channel shortening has also been investigated in the context of MIMO equalization [214,215] based on a least-squares solution.

The computation of equalization filters generally constitutes an inverse problem. Various algorithms have been proposed that improve the numerical and computational efficiency, as well as the numerical conditioning—for instance, a fast iterative MIMO inversion algorithm working in the DFT-domain [216], or a DFT-domain approximation of the MIMO filtered-x algorithm [217]. In [216], a steepest-descent and approximative Gauss–Newton iterative algorithm for the design of a MIMO equalizer is presented. In [218], a method for coping with the low conditioning of the transfer function matrix at some frequencies is proposed. The problem is amended by studying the structure of the MIMO transfer function matrix and replacing its inverse matrix by a pseudo-inverse that allows a range of acceptable solutions. Polynomial-based MIMO formulations of the room equalization problem are discussed in [219,220] with extensions towards explicitly controlling the number of active loudspeakers used for equalization [221].

There are also a number of specialized equalization approaches for specific scenarios. For instance, the equalization of multichannel stereophonic systems under the constraint that stereophonic pairs of loudspeakers should have similar transfer functions is discussed in [222–224]. The approach is split into two stages: (i) equalization of a single path also utilizing the other loudspeakers and (ii) similarity optimization between two channels that are used for stereophonic imaging. The room equalization in cars has been considered in various studies. A non-adaptive MIMO equalization approach utilizing IIR or FIR filters is presented in [225]. The optimization is performed in terms of the overall magnitude response to avoid coloration/tonal issues. A combined room equalization and cross-talk canceling approach for cars is discussed in [226].

Besides the MIMO approaches reviewed so far, single loudspeaker room equalization approaches have also been investigated which utilize multiple microphones. This constitutes a single-input/multiple-output (SIMO) problem. A non-adaptive polynomial multivariate control approach combined with a constrained mean squared error design and zero clustering is discussed in [149]. A statistic inferential method which considers the statistical variation between the different microphone positions for improved robustness and an enlarged listening area is presented in [227].

9. Evaluation Methods for RRE

One important aspect is the evaluation of RRE results, considering instrumental measures or subjective listening tests. The former aims at measures which are in relation with the goal of the procedure—for example, quantifying the similarity between the target function and the equalization result. However, an important role should be assigned to perceptual evaluation, since the final judgment is always performed by the human listener in the specific environment. In this section, we first analyze instrumental parameters used as a primary analysis stage of the obtained results. Then, a review of the most common listening test procedures is reported.

9.1. Instrumental Measures

In the following section, the most common instrumental measures for RRE evaluation are reviewed. Throughout the section, $h(n)$ denotes the RIR in the discrete time domain, while $H(e^{j\omega})$ denotes its discrete-time Fourier transform with ω being the normalized angular frequency.

9.1.1. Spectral Deviation Measures

The spectral deviation was first used for the evaluation of the RRE procedure in [76], and was then adopted in many other papers [92,228]. The spectral deviation, S_D, of a frequency response $E(e^{j\omega})$ can be expressed as

$$S_D = \sqrt{\frac{1}{Q_h - Q_l + 1} \sum_{i=Q_l}^{Q_h} \left(10 \log_{10} \left| E\left(e^{j\frac{2\Pi}{N} i} \right) \right| - D \right)^2}, \tag{5}$$

where

$$D = \frac{1}{Q_h - Q_l + 1} \sum_{i=Q_l}^{Q_h} \left(10 \log_{10} \left| E\left(e^{j\frac{2\Pi}{N} i} \right) \right| \right), \tag{6}$$

where Q_l and Q_h are the lowest and highest frequency indexes, respectively, of the equalized band.

Usually, the experimental results provide an initial spectral deviation $S_{D,in}$, calculated with $E(e^{j\omega}) = H(e^{j\omega})$, and a final spectral deviation $S_{D,fin}$, computed after equalization by considering $E(e^{j\omega}) = H(e^{j\omega}) \cdot H_{inv}(e^{j\omega})$, where $H_{inv}(e^{j\omega})$ represents the designed equalizer. Figure 11 shows the curves used for the S_D calculation. A Mean Spectral Deviation Measure (MSDM) that represents the mean value of the final spectral deviation measures over the entire set of measured RIRs has also been considered [74,146].

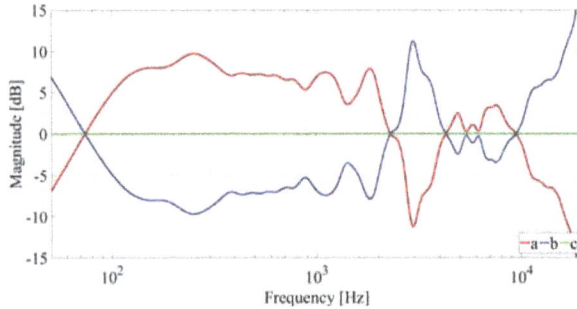

Figure 11. Equalization results: the spectral deviation measures the distance between the equalized response curve and the flat one: (red) $H(e^{j\omega})$, (blue) $H_{inv}(e^{j\omega})$, (green) $E(e^{j\omega}) = H(e^{j\omega}) \cdot H_{inv}(e^{j\omega})$.

In analogy to the mean spectral deviation measure, which gives a measure of the deviation of the magnitude frequency response from a flat one [228], a mean group delay deviation measure was introduced in [147,229] to quantify the average variation in terms of group delay:

$$GD_D = \frac{1}{M} \sum_{l=1}^{M} \sqrt{\frac{1}{Q_h - Q_l + 1} \sum_{i=Q_l}^{Q_h} (GD_l(i) - K_l)^2}, \tag{7}$$

where

$$K_l = \frac{1}{Q_h - Q_l + 1} \sum_{i=Q_l}^{Q_h} GD_l(i), \tag{8}$$

Q_l and Q_h are the lowest and the highest frequency indexes, respectively, of the equalized band, and $GD_l(i)$ is the group delay of the M RIRs for the i-th frequency index.

The objective of mixed-phase equalization is to achieve a linear phase, and therefore the group delay should be as flat as possible: using this parameter it is possible to quantify the distance of the obtained group delay from a constant delay.

9.1.2. Sammon Map

The Sammon map was introduced for the evaluation of RRE in [230]. It is a non-linear projection method that maps multidimensional data onto fewer dimensions (e.g., two or three). The main property of the Sammon map is that it retains the geometrical distances between signals in a multidimensional space in two or three dimensions. Given the M magnitude responses $|H_k(e^{j\omega})|$, $k = 1, \ldots, M$, of the measured RIRs, the Sammon map algorithm iteratively minimizes—by a gradient descent scheme—the cumulative sum of the differences between the Euclidean distances in the high and low dimensional space. The following objective function is minimized:

$$J_{Sammon} = \frac{1}{\sum\limits_{l=1}^{M} \sum\limits_{m=l+1}^{M} d_{lm}} \sum_{l=1}^{M} \sum_{m=l+1}^{M} \frac{(d_{lm} - d'_{lm})^2}{d_{lm}}, \tag{9}$$

with

$$d_{lm} = \sum_{w=1}^{W} \left(|H_l(e^{j\frac{2\Pi}{N}w})| - |H_m(e^{j\frac{2\Pi}{N}w})| \right)^2, \tag{10}$$

$$d'_{lm} = \sum_{i=1}^{L} |r_l(i) - r_m(i)|^2, \tag{11}$$

where W is the number of equally-spaced frequencies and L is the dimension of the Sammon map space. In the Sammon map, the point associated with $H_k(e^{j\omega})$ is represented as $(r_k(1), \ldots, r_k(L))$. Considering a two-dimensional mapping ($L = 2$), upon convergence, the points $(r_k(1), r_k(2))$ with $k = 1, \ldots, M$ are configured on a two-dimensional plane such that the relative distances between the different $H_k(e^{j\omega})$ are visually discernible. After equalization, the resulting performance can be determined from the size and shape of the region defined by the equalized frequency responses on the map. A circular shape around zeros indicates uniform equalization at all locations [230]. Figure 12 shows the results obtained using the Sammon map: it can be observed that for IRs without equalization (Figure 12a), the points are located far from the center of the map, while for IRs with equalization (Figure 12b), the points are uniformly distributed around the center of the map.

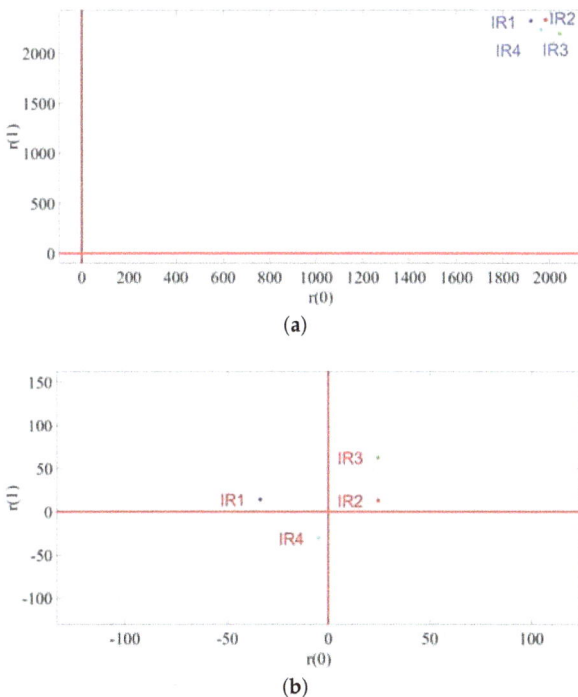

(a)

(b)

Figure 12. Sammon map results: (**a**) RIR without equalization; (**b**) RIR with equalization.

9.1.3. Energy Decay Reliefs

The effect of equalization can be evaluated considering the energy decay relief (EDR), which is a time-frequency generalization of the energy decay curve (EDC) used to calculate the reverberation time $T60$. Since room modes are characterized by peaks in the frequency response and extended ringing in the time domain, the EDR measure can help to understand the effect of the equalization procedure. The EDR is defined as the time-frequency representation of the RIR energy decay [231,232], and working in the continuous-time domain, it is calculated as follows:

$$EDR_h(t, f) = \int_t^{+\infty} \rho_h(\tau, f)\, d\tau, \tag{12}$$

where $\rho_h(\tau, f)$ is the energetic time-frequency representation of the RIR using a short-time Fourier transform (STFT) procedure applied with a rectangular analysis window.

Figure 13 shows the EDR calculated before and after the equalization procedure. Considering the temporal behavior, the plots show a reduction in decay times, while in the frequency domain, a reduction of the frequency peaks can be observed. Generally, after the equalization procedure a more uniform behavior is obtained, with a reduction of peaks and notches.

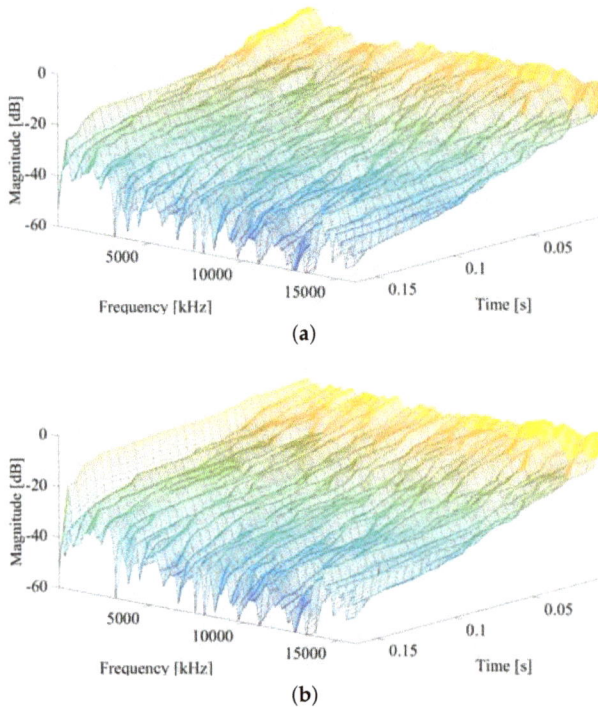

(a)

(b)

Figure 13. EDR results: (a) without equalization; (b) after equalization.

9.1.4. Acoustic Parameters

The quality of an audio signal can be evaluated considering some objective quality measures based on the RIR [233]. Acoustic parameters obtained using objective measures were first used for the assessment of RRE in [6]. The following acoustic parameters have been used in many papers about RRE:

- the definition index, which is defined as the percentage ratio of the energy of the first 50 ms or 80 ms after the main peak to the remaining energy of the RIR (D50 or D80) [13,234];
- the clarity index, which is defined as the logarithmic ratio of the energy of the first 50 ms or 80 ms after the main peak to the remaining energy of the RIR (C50 or C80) [13,235];
- the early decay time, which is defined as the time in which the first 10 dB fall of a decay process occurs, multiplied by a factor of 6 (EDT) [13,236];
- the direct-to-reverberation-ratio (DRR), also known as direct-to-reverberant-energy-ratio [233,237,238] is defined as the logarithmic ratio between the main peak and the remaining RIR;
- the central time (CT) [13,239] is the center of gravity of the energy of the RIR.

9.2. Perceptual Evaluation

To assess the audio quality, listening tests have to be performed following an appropriate procedure. Many proposals for the perceptual evaluation of an audio system can be found in the literature [240,241]. However, focusing on RRE and referring to the state-of-the-art, the perceptual assessment of RRE systems should adhere to the following standards:

- ITU-R BS.1116-1 [242]: "Methods for subjective assessment of small impairments in audio systems including multichannel sound systems",
- ITU-R BS.1534-1 [243]: "Method for the subjective assessment of intermediate quality level of coding systems",
- ITU-R BS.1284-1 [244]: "General methods for the subjective assessment of sound quality".

All these recommendations provide a description of the test methodology, test procedure, and statistical methods to elaborate the acquired data. However due to the broadness of this topic, the discussion will be focused only on the most relevant procedures that have been applied to RRE.

The ITU-R BS.1284-1 recommendation provides a guide to the general assessment of perceived audio quality, and has been applied in [75,145] for the assessment of RRE. It is worth noting that ITU-R BS.1284-1 is based on ITU-R BS.1116-1. According to the guidelines of [244], expert listeners should be preferred to *"give a better and a quicker indication of the likely results in the long term."* The subjective listening test is conceived as a comparison test, and the listeners should be instructed to provide a score using a seven-grade scale with a recommended resolution of 1 decimal place, as reported in [244]. The test is based on paired comparisons with references, and the score is given after listening to a repetition (four times consecutively) of the predetermined programme sequence. In the case of the assessment of an equalization procedure, the following sequence is considered in [75,145]:

1. reference sequence without equalization;
2. same sequence, equalized with one of the selected equalization techniques;
3. reference sequence without equalization (repeated);
4. same sequence, equalized with one of the selected equalization techniques (repeated).

As recommended in [244], the stimuli should never exceeded 20 s in length, thus lengths were limited between 15 to 20 s. Moreover, care was taken in order to guarantee that the tested musical items did not appear to be interrupted. For each reference signal, the presentation order of the different equalization methods was randomized and the listener did not know which equalization methodology was under test. Following the recommendation, before the listening test, a training set was subministrated to the listeners. As reported in [145], in order to familiarize with the test procedure, the test materials and the test environment, the subjects had the possibility to listen to each audio item in all conditions under evaluation. While the ITU-R BS.1284-1 recommendation suggests several attributes for characterizing the perceived sound quality, in [75,145] three attributes have been considered; i.e., "transparency" (all details of the performance can be clearly perceived), "timbre" (accurate portrayal of the different sound), "main impression" (the integrity of the total sound image and the interaction between other parameters). In order to test the room response equalizer using different spectral content, different music genres were considered as reference signals.

Finally, the obtained results were processed to derive the mean values and the confidence intervals. A significance level of 0.05 was considered for computing the confidence intervals.

10. Emerging Topics and New Trends

In this section, emerging topics and new trends related to RRE are analyzed. In particular, the necessity of improving the performance of the equalization algorithms combined with the increasing interest in new technologies have led to innovative applications and interesting developments.

10.1. Personal Sound Zones

In the last years, there is an increasing interest in the possibility of reproducing different content in adjacent spatially restricted zones for multiple listeners by reducing the interference between the zones. These approaches are known as personal sound zones [245], multi-zone synthesis, or multi-zone sound control. A recent review on this topic can be found in [246], and more details in the numerous published papers on the subject [247–274]. At the current state, the achievable suppression between the zones is limited by various acoustical and practical restrictions, resulting in a limited applicability to real-world scenarios. In [262], an overview is presented on the major challenges that have to be dealt with for multi-zone sound control in a reverberant environment. Interference mitigation and room compensation robust to changes and uncertainties in the acoustic environment remain as challenging problems. An approach to room equalization for sound pressure control over a region of space combined with a wave domain sound field representation is presented in [262]. The approach is reported to be robust at low frequencies, but ineffective at high frequencies where the reverberant sound field is diffuse, calling for a very high number of loudspeakers.

10.2. Portable Devices

In recent years the use of portable devices has increased enormously, reaching a very high level of expansion. However, due to the loudspeakers' characteristics and their interactions with the room environment, many of these devices are capable of satisfying just the basic audio requirements. This situation can be partially improved taking into consideration the acoustics of these devices and applying advanced audio techniques. In [275], a multi-point equalization procedure is introduced to improve the non-ideal response of a portable system such as the mobile phone. Objective measurements and subjective listening test results have confirmed the positive effect of the algorithms on personal portable devices. In [276], a static and an adaptive algorithm for frequency response linearization applied to mobile computers is reported. Subjective listening tests have underlined an improvement in the listener's perception, confirming the validity of such approaches.

10.3. Nonlinear Equalization

Sound reproduction systems can exhibit an undesirable behavior not only due to the room acoustics, but also due to loudspeaker and amplifier systems that can produce linear and nonlinear distortions. In order to remove the nonlinear effects, in [277–279] equalizers that involve Volterra filters to model the amplifier-loudspeaker-enclosure are used before driving the output signal through the loudspeakers. In this way it is possible to equalize not only the linear behavior of the system, but also its nonlinear behavior considering adaptive procedures.

10.4. Room Equalization with Moving Microphone

One of the main issues of multi-point equalization is the measurement of the RIRs, which requires a long time to achieve a certain spatial resolution inside the listening area. A solution to this problem can be found by using time-variant system identification techniques [280,281]. Here RIRs are measured by applying a dynamic method based on the use of one moving microphone instead of estimating the

RIRs independently. This procedure allows to obtain a dense grid of RIRs from one spatially continuous measurement that can be used in multi-point equalization to estimate the prototype function and equalization filters.

11. Conclusions

In this paper, following the historical path, a complete overview of the state-of-the-art has been presented. In order to underline the evolution and the potentiality of RRE, different classifications have been considered for the approaches. A first classification can be done considering the number of impulse responses used for the estimation of the equalization filter (i.e., single-point or multi-point equalizers). The former is effective only on a reduced zone around the measurement point, while the latter is capable of enlarging the equalized zone and contrasting the room response variations. The second classification can be performed considering an instantaneous or continuous measurement of the impulse responses (i.e., fixed or adaptive approaches). The former consists of a-priori measurement of the impulse responses, while the latter is based on a continuous update of the impulse responses and thus of the equalizer to cope with the temporal variations of the environment. Within this general classification, we must consider pre-processing techniques that are used to contrast the audible distortions caused by equalization errors due to the RIRs variations, minimum-phase and mixed-phase, direct and indirect approaches for different equalizer design techniques, and wave domain filters for the equalization of massive multichannel sound reproduction systems. Following this classification, different approaches have been described. Table 1 summarizes the state-of-the-art methods as function of classification criteria, i.e., pre-processing techniques, minimum phase and mixed phase technique, fixed and adaptive approaches, single-point and multi-point approaches, direct and indirect methods according to the definition of Section 4.6, and wave domain methods. It is evident that several methods can cover more than one aspect, extending the potential and the effectiveness of the methodology. In this context, the instrumental measurement and perceptual evaluation of the equalization results become crucial: some examples of the main approaches from the state-of-the-art in this field have been reported. Finally, a general discussion on emerging methodologies and new trends for RRE has been presented. It is evident that the increasing availability of personal devices will lead to an increased use of RRE techniques to enhance their performance.

Table 1. Comparison of the state-of-the-art approaches considering the aforementioned classification criteria. MINT: multiple-input/multiple-output inverse theorem; SISO: single-input/single-output; SIMO: single-input/multiple-output; MIMO: multiple-input/multiple-output.

	Pre-Processing	Minimum Phase	Mixed Phase	Fixed	Adaptive	Single-point	Multi-point	Direct	Indirect	Wave Domain	Section
Short filters [61]	✓	✓	✓					✓	✓		5.1
Complex smoothing [35,94]	✓	✓	✓					✓	✓		5.2.1
Frequency warping [98]	✓	✓	✓					✓	✓		5.2.2
Kautz filters [110]	✓	✓	✓					✓	✓		5.2.3
Multirate approaches [50,72,116–120]	✓	✓	✓					✓	✓		5.2.4
Room impulse response reshaping [78,122–130]	✓		✓						✓		5.3
Homomorphic filtering [5,56–58]		✓	✓	✓		✓			✓		4.1
Linear predictive coding analysis [61,63–75]		✓	✓	✓		✓			✓		4.2
Least-squares optimization techniques [76]			✓	✓		✓	✓	✓			4.3
Frequency domain deconvolution [51,80,81]		✓	✓	✓		✓			✓		4.4
MINT solutions [86,87]			✓	✓		✓	✓	✓	✓		4.5
Average and weighted average methods [77]			✓	✓	✓	✓		✓			6.1
Clustering methods [30,70–72,121,138–140]			✓	✓	✓	✓	✓		✓		6.2
Prototype approach [73,144,146,147]		✓	✓	✓	✓	✓	✓		✓		6.3
Common acoustical poles compensation [63–65]		✓	✓	✓		✓	✓		✓		6.4
Modal equalization [4,157]			✓	✓		✓	✓		✓		6.5
Plane wave approach [159–165]			✓	✓		✓	✓		✓		6.6
Quasi-anechoic approach [172,173]				✓	✓	✓	✓	✓			6.8
SISO/SIMO in time domain [77,169,189]			✓		✓	✓	✓	✓	✓		7.1.1
SISO/SIMO in frequency domain [177–180]			✓		✓	✓	✓	✓	✓		7.1.2
MIMO approaches [183,185,186]		✓	✓		✓	✓	✓	✓	✓		7.2
Wave domain adaptive filtering [10,200,201]			✓		✓	✓	✓	✓		✓	8
Transform domain approaches [206,208]				✓			✓	✓		✓	8
Room geometry aware methods [209–212]				✓			✓			✓	8

Author Contributions: The authors contributed equally to this work.

Conflicts of Interest: The authors declare no conflict of interest.

References

1. Karjalainen, M.; Paatero, T.; Mourjopoulos, J.N.; Hatziantoniou, P.D. About room response equalization and dereverberation. In Proceedings of the IEEE Workshop on Applications of Signal Processing to Audio and Acoustics (WASPAA), New Paltz, NY, USA, 14–16 October 2005; pp. 183–186.
2. Hatziantoniou, P.D.; Mourjopoulos, J.N. Errors in real-time room acoustics dereverberation. *J. Audio Eng. Soc.* **2004**, *52*, 883–899.
3. Karjalainen, M.; Esquef, P.A.A.; Antsalo, P.; Mäkivirta, A.; Välimäki, V. Frequency-zooming ARMA modeling of resonant and reverberant systems. *J. Audio Eng. Soc.* **2002**, *50*, 1012–1029.
4. Mäkivirta, A.; Antsalo, P.; Karjalainen, M.; Välimäki, V. Low-frequency modal equalization of loudspeaker-room responses. In Proceedings of the 111th Audio Engineering Society Convention, New York, NY, USA, 30 November–3 December 2001; Volume 5480.
5. Neely, S.T.; Allen, J.B. Invertibility of a room impulse response. *J. Acoust. Soc. Am.* **1979**, *66*, 165–169.
6. Mourjopoulos, J. On the variation and invertibility of room impulse response functions. *J. Sound Vib.* **1985**, *102*, 217–228.
7. Kyriakakis, C. Fundamental and technological limitations of immersive audio systems. *Proc. IEEE* **1998**, *86*, 941–951.
8. Mourjopoulos, J. Comments on "Analysis of traditional and reverberation-reducing methods of room equalization". *J. Audio Eng. Soc.* **2003**, *51*, 1186–1188.
9. Bharitkar, S.; Kyriakakis, C. *Immersive Audio Signal Processing*; Springer Science & Business Media: New York, NY, USA, 2006.
10. Spors, S.; Buchner, H.; Rabenstein, R.; Herbordt, W. Active listening room compensation for massive multichannel sound reproduction systems using wave-domain adaptive filtering. *J. Acoust. Soc. Am.* **2007**, *122*, 354–369.
11. Välimäki, V.; Reiss, J.D. All about audio equalization: Solutions and frontiers. *Appl. Sci.* **2016**, *6*, 129.
12. Hacihabiboglu, H.; De Sena, E.; Cvetkovic, V.; Johnston, J.; Smith III, J.O. Perceptual spatial audio recording, simulation, and rendering: An overview of spatial-audio techniques based on psychoacoustics. *IEEE Signal Process. Mag.* **2017**, *34*, 36–54.
13. Kuttruff, H. *Room Acoustics*; CRC Press: Boca Raton, FL, USA, 2016.
14. Defrance, G.; Daudet, L.; Polack, J.D. Using matching pursuit for estimating mixing time within room impulse responses. *Acta Acust. United Acust.* **2009**, *6*, 1071–1081.
15. Primavera, A.; Cecchi, S.; Li, J.; Piazza, F. Objective and Subjective Investigation on a Novel Method for Digital Reverberator Parameters Estimation. *IEEE/ACM Trans. Audio Speech Lang. Process.* **2014**, *22*, 441–452.
16. Wallach, H.; Newman, E.B.; Rosenzweig, M.R. A precedence effect in sound localization. *J. Acoust. Soc. Am.* **1949**, *21*, 468.
17. Haas, H. The influence of a single echo on the audibility of speech. *J. Audio Eng. Soc.* **1972**, *20*, 146–159.
18. Walker, R. Equalization of room acoustics and adaptive systems in the equalization of small room acoustics. In Proceedings of the 15th International Audio Engineering Society Conference: Audio, Acoustics & Small Spaces, Copenhagen, Denmark, 31 October– 2 November 1998.
19. Olive, S.E.; Toole, F.E. The detection of reflections in typical rooms. *J. Audio Eng. Soc.* **1989**, *37*, 539–553.
20. Toole, F.E.; Olive, S.E. The modification of timbre by resonances: Perception and measurement. *J. Audio Eng. Soc.* **1988**, *36*, 122–142.
21. Schuck, P.L.; Olive, S.E.; Ryan, J.G.; Toole, F.E.; Sally, S.; Bonneville, M.E.; Momtahan, K.L.; Verreault, E.S. Perception of perceived sound in rooms: some results of the Athena project. In Proceedings of the 12th International Audio Engineering Society Conference: The Perception of Reproduced Sound, Copenhagen, Denmark, 28–30 June 1993.
22. Bech, S. Perception of reproduced sound: audibility of individual reflections in a complete sound field, II. In Proceedings of the 99th Audio Engineering Society Convention, New York, NY, USA, 6–9 October 1995.
23. Bech, S. Timbral aspects of reproduced sound in small rooms. I. *J. Acoust. Soc. Am.* **1995**, *97*, 1717–1726.

24. Begault, D. Audible and inaudible early reflections: thresholds for auralization system design. In Proceedings of the 100th Audio Engineering Society Convention, Copenhagen, Denmark, 11–14 May 1996.
25. Bech, S. Perception of reproduced sound: Audibility of individual reflections in a complete sound field, III. In Proceedings of the 100th Audio Engineering Society Convention, Copenhagen, Denmark, 11–14 May 1996.
26. Bücklein, R. The audibility of frequency response irregularities. *J. Audio Eng. Soc.* **1981**, *29*, 126–131.
27. Hikichi, T. Time variation of room acoustic transfer functions and its effect on a multi-microphone dereverberation approach. In Proceedings of the Workshop on Microphone Arrays: Theory, Design & Application, Piscataway, NJ, USA, October 1994.
28. Omura, M.; Yada, M.; Saruwatari, H.; Kajita, S.; Takeda, K.; Itakura, F. Compensating of room acoustic transfer functions affected by change of room temperature. In Proceedings of the IEEE International Conference on Acoustics, Speech, and Signal Processing (ICASSP), Phoenix, AZ, USA, 15–19 March 1999; Volume 2, pp. 941–944.
29. Kajita, S.; Takeda, K.; Itakura, F. Compensation of room acoustic transfer functions affected by the change of room temperature. In Proceedings of the 107th Audio Engineering Society Convention, New York, NY, USA, 24–27 September 1999.
30. Bharitkar, S.; Kyriakakis, C. A classification scheme for acoustical room responses. In Proceedings of IEEE Sixth International, Symposium on Signal Processing and its Applications, Kuala Lumpur, Malaysia, 13–16 August 2001; Volume 2, pp. 671–674.
31. Bharitkar, S.; Hilmes, P.; Kyriakakis, C. Robustness of spatial averaging equalization methods: A statistical approach. In Proceedings of the IEEE Thirty-Sixth Asilomar Conference on Signals, Systems and Computers, Pacific Grove, CA, USA, 3–6 November 2002; Volume 1, pp. 184–187.
32. Bharitkar, S.; Hilmes, P.; Kyriakakis, C. Sensitivity of multichannel room equalization to listener position. In Proceedings of the International Conference on Multimedia and Expo (ICME), Baltimore, MD, USA, 6–9 July 2003; Volume 1, pp. I–721.
33. Bharitkar, S.; Hilmes, P.; Kyriakakis, C. Robustness of spatial average equalization: A statistical reverberation model approach. *J. Acoust. Soc. Am.* **2004**, *116*, 3491–3497.
34. Lundeby, A.; Vigran, T.E.; Bietz, H.; Vorländer, M. Uncertainties of measurements in room acoustics. *Acta Acust. United Acust.* **1995**, *81*, 344–355.
35. Hatziantoniou, P.D.; Mourjopoulos, J.N. Generalized fractional-octave smoothing of audio and acoustic responses. *J. Audio Eng. Soc.* **2000**, *48*, 259–280.
36. Greenwood, D. A cochlear frequency-position function for several species—29 years later. *J. Acoust. Soc. Am.* **1990**, *87*, 2592–2605.
37. Stevens, S.S.; Volkmann, J.; Newman, E.B. A scale for the measurement of the psychological magnitude pitch. *J. Acoust. Soc. Am.* **1937**, *8*, 185–190.
38. Scharf, B. Critical bands. *Foundations of Modern Auditory Theory*; Academic Press: New York, NY, USA, 1970; Volume 1.
39. Moore, B.C.; Peters, R.W.; Glasberg, B.R. Auditory filter shapes at low center frequencies. *J. Acoust. Soc. Am.* **1990**, *88*, 132–140.
40. Moore, B.C.; Glasberg, B.R.; Plack, C.; Biswas, A. The shape of the ear's temporal window. *J. Acoust. Soc. Am.* **1988**, *83*, 1102–1116.
41. Bilsen, F.A. The minimum integration time of the auditory system. In Proceedings of the 86th Audio Engineering Society Convention, Hamburg, Germany, 7–10 March 1989.
42. Moore, B. *An Introduction to the Psychology of Hearing*; Emerald: Bingley, UK, 2012.
43. Fielder, L.D. Analysis of traditional and reverberation-reducing methods of room equalization. *J. Audio Eng. Soc.* **2003**, *51*, 3–261.
44. Zwicker, E.; Fastl, H. *Psychoacoustics: Facts and models*; Springer Science & Business Media: New York, NY, USA, 2013; Volume 22.
45. Raab, D.H. Forward and backward masking between acoustic clicks. *J. Acoust. Soc. Am.* **1961**, *33*, 137–139.
46. Dolan, T.G.; Small, A.M., Jr. Frequency effects in backward masking. *J. Acoust. Soc. Am.* **1984**, *75*, 932–936.
47. Fielder, L.D. Practical limits for room equalization. In Proceedings of the 111th Audio Engineering Society Convention, New York, NY, USA, 30 November–3 December 2001.

48. Buchholz, J.; Mourjopoulos, J.; Blauert, J. Room masking: Understanding and modelling the masking of reflections in rooms. In Proceedings of the 110th Audio Engineering Society Convention, Amsterdam, The Netherlands, 12–15 May 2001.

49. Pedersen, J.A.; Hermansen, K.; Rubak, P. The distribution of the low frequency sound field and its relation to room equalization. In Proceedings of the 96th Audio Engineering Society Convention, Amsterdam, The Netherlands, 26 February–1 March 1994.

50. Bean, C.; Craven, P.G. Loudspeaker and room correction using digital signal processing. In Proceedings of the 86th Audio Engineering Society Convention, Hamburg, Germany, 7–10 March 1989.

51. Kirkeby, O.; Nelson, P.A.; Hamada, H.; Orduna-Bustamante, F. Fast deconvolution of multichannel systems using regularization. *IEEE Trans. Speech Audio Process.* **1998**, *6*, 189–194.

52. Kirkeby, O.; Nelson, P.A. Digital filter design for inversion problems in sound reproduction. *J. Audio Eng. Soc.* **1999**, *47*, 583–595.

53. Johansen, L.G.; Rubak, P. The excess phase in loudspeaker/room transfer functions: Can it be ignored in equalization tasks? In Proceedings of the 100th Audio Engineering Society Convention, Copenhagen, Denmark, 11–14 May 1996.

54. Paatero, T.; Karjalainen, M. Kautz filters and generalized frequency resolution: theory and audio applications. *J. Audio Eng. Soc.* **2003**, *51*, 27–44.

55. Karjalainen, M.; Paatero, T. Equalization of loudspeaker and room responses using Kautz filters: Direct least squares design. *EURASIP J. Appl. Signal Process.* **2006**, *2007*, 060949.

56. Mourjopoulos, J.; Clarkson, P.; Hammond, J. A comparative study of least-squares and homomorphic techniques for the inversion of mixed phase signals. In Proceedings of the IEEE International Conference on Acoustics, Speech, and Signal Processing (ICASSP), Paris, France, 3–5 May 1982; Volume 7, pp. 1858–1861.

57. Radlovic, B.D.; Kennedy, R.A. Nonminimum-phase equalization and its subjective importance in room acoustics. *IEEE Trans. Speech Audio Process.* **2000**, *8*, 728–737.

58. Maamar, A.; Kale, I.; Krukowski, A.; Daoud, B. Partial equalization of non-minimum-phase impulse responses. *EURASIP J. Adv. Signal Process.* **2006**, *2006*, 067467.

59. Hammond, J.; Mourjopoulos, J. Cepstral Methods applied to the Analysis of Room Impulse Response. In Proceedings of the Institute of Acoustics Autumn Conference, Windermere, UK, November 1980; pp. 51–54.

60. Mourjopoulos, J. The Removal of Room Reverberation from Signals; Ph.D. Thesis, University of Southamption, Southampton, UK, 1984.

61. Mourjopoulos, J.; Paraskevas, M. Pole and zero modeling of room transfer functions. *J. Sound Vib.* **1991**, *146*, 281–302.

62. Durbin, J. The fitting of time-series models. In *Revue de l'Institut International de Statistique*; International Statistical Institute (ISI): Haguethe, The Netherlands, 1960; pp. 233–244.

63. Haneda, Y.; Makino, S.; Kaneda, Y. Common acoustical pole and zero modeling of room transfer functions. *IEEE Trans. Speech Audio Process.* **1994**, *2*, 320–328.

64. Haneda, Y.; Makino, S.; Kaneda, Y. Multiple-point equalization of room transfer functions by using common acoustical poles. *IEEE Trans. Speech Audio Process.* **1997**, *5*, 325–333.

65. Fontana, F.; Gibin, L.; Rocchesso, D.; Ballan, O. Common pole equalization of small rooms using a two-step real-time digital equalizer. In Proceedings of the IEEE Workshop on Applications of Signal Processing to Audio and Acoustics (WASPAA), New Paltz, NY, USA, 20–20 October 1999; pp. 195–198.

66. Härmä, A.; Karjalainen, M.; Savioja, L.; Välimäki, V.; Laine, U.K.; Huopaniemi, J. Frequency-warped signal processing for audio applications. *J. Audio Eng. Soc.* **2000**, *48*, 1011–1031.

67. Härmä, A. Linear predictive coding with modified filter structures. *IEEE Trans. Speech Audio Process.* **2001**, *9*, 769–777.

68. Härmä, A. *Frequency-Warped Autoregressive Modeling and Filtering*; Helsinki University of Technology: Helsinki, Finland, 2001.

69. Harma, A.; Paatero, T. Discrete representation of signals on a logarithmic frequency scale. In Proceedings of the IEEE Workshop on the Applications of Signal Processing to Audio and Acoustics (WASPAA), New Paltz, NY, USA, 21–24 October 2001; pp. 39–42.

70. Bharitkar, S.; Kyriakakis, C. Perceptual multiple location equalization with clustering. In Proceedings of the IEEE Thirty-Sixth Asilomar Conference on Signals, Systems and Computers, Pacific Grove, CA, USA, 3–6 November 2002; Volume 1, pp. 179–183.

71. Bharitkar, S.; Kyriakakis, C. A comparison between multi-channel audio equalization filters using warping. In Proceedings of the IEEE Workshop on Applications of Signal Processing to Audio and Acoustics (WASPAA), New Paltz, NY, USA, 19–22 October 2003; pp. 63–66.

72. Bharitkar, S.; Kyriakakis, C. Multirate signal processing for multiple listener low frequency room acoustic equalization. In Proceedings of the IEEE Thirty-Eighth Asilomar Conference on Signals, Systems and Computers, Pacific Grove, CA, USA, 7–10 November 2004; Volume 1, pp. 263–267.

73. Omiciuolo, I.; Carini, A.; Sicuranza, G.L. Multiple position room response equalization with frequency domain fuzzy c-means prototype design. In Proceedings of the International Workshop on Acoustic Signal Enhancement (IWAENC), Seattle, WA, USA, 14–17 September 2008.

74. Carini, A.; Omiciuolo, I.; Sicuranza, G.L. Multiple position room response equalization: Frequency domain prototype design strategies. In Proceedings of the 6th International Symposium on Image and Signal Processing and Analysis (ISPA), Salzburg, Austria, 16–18 September 2009; pp. 638–643.

75. Carini, A.; Cecchi, S.; Piazza, F.; Omiciuolo, I.; Sicuranza, G.L. Multiple position room response equalization in frequency domain. *IEEE Trans. Audio Speech Lang. Process.* **2012**, *20*, 122–135.

76. Clarkson, P.M.; Mourjopoulos, J.; Hammond, J. Spectral, phase, and transient equalization for audio systems. *J. Audio Eng. Soc.* **1985**, *33*, 127–132.

77. Elliott, S.J.; Nelson, P.A. Multiple-point equalization in a room using adaptive digital filters. *J. Audio Eng. Soc.* **1989**, *37*, 899–907.

78. Mei, T.; Mertins, A.; Kallinger, M. Room impulse response reshaping/shortening based on least mean squares optimization with infinity norm constraint. In Proceedings of the 16th International Conference on Digital Signal Processing, Santorini-Hellas, Greece, 5–7 July 2009; pp. 1–6.

79. Kolundžija, M.; Faller, C.; Vetterli, M. Multi-channel low-frequency room equalization using perceptually motivated constrained optimization. In Proceedings of the IEEE International Conference on Acoustics, Speech and Signal Processing (ICASSP), Kyoto, Japan, 25–30 March 2012; pp. 533–536.

80. Kulp, B.D. Digital equalization using Fourier transform techniques. In Proceedings of the 85th Audio Engineering Society Convention, Los Angeles, CA, USA, 3–6 November 1988.

81. Kirkeby, O.; Nelson, P.A.; Hamada, H.; Orduna-Bustamante, F. Fast deconvolution of multi-channel systems using regularisation. *ISVR Tech. Rep.* **1996**, *18*, 2829–2832.

82. Kirkeby, O.; Rubak, P.; Farina, A. Analysis of ill-conditioning of multi-channel deconvolution problems. In Proceedings of the IEEE Workshop on Applications of signal processing to Audio and Acoustics (WASPAA), New Paltz, NY, USA, 20 October 1999; pp. 155–158.

83. Cecchi, S.; Palestini, L.; Peretti, P.; Piazza, F.; Carini, A. Multipoint equalization of digital car audio systems. In Proceedings of the 6th International Symposium on Image and Signal Processing and Analysis (ISPA), Salzburg, Austria, 16–18 September 2009; pp. 650–655.

84. Cecchi, S.; Carini, A.; Primavera, A.; Piazza, F. An adaptive multiple position room response equalizer in warped domain. In Proceedings of the 20th European Signal Processing Conference (EUSIPCO), Bucharest, Romania, 27–31 August 2012; pp. 1955–1959.

85. Spors, S.; Kuntz, A.; Rabenstein, R. Listening room compensation for wave field synthesis. In Proceedings of the IEEE International Conference on Multimedia and Expo (ICME), Baltimore, MD, USA, 6–9 July 2003; Volume 1, pp. I–725.

86. Miyoshi, M.; Kaneda, Y. Inverse control of room acoustics using multiple loudspeakers and/or microphones. In Proceedings of the IEEE International Conference on Acoustics, Speech, and Signal Processing (ICASSP), Tokyo, Japan, 7–11 April 1986; Volume 11, pp. 917–920.

87. Miyoshi, M.; Kaneda, Y. Inverse filtering of room acoustics. *IEEE Trans. Acoust. Speech Signal Process.* **1988**, *36*, 145–152.

88. Putnam, W.; Rocchesso, D.; Smith, J. A numerical investigation of the invertibility of room transfer functions. In Proceedings of the IEEE Workshop on Applications of Signal Processing to Audio and Acoustics (WASPAA), New Paltz, NY, USA, 15–18 October 1995; pp. 249–252.

89. Nelson, P.; Orduna-Bustamante, F.; Hamada, H. Inverse filter design and equalization zones in multichannel sound reproduction. *IEEE Trans. Speech Audio Process.* **1995**, *3*, 185–192.

90. Hirofumi, N.; Miyoshi, M.; Tohyama, M. Sound field control by indefinite MINT filters. *IEICE Trans. Fundam. Electron. Commun. Comput. Sci.* **1997**, *80*, 821–824.

91. Zhang, W.; Naylor, P.A. An iterative method for equalization of multichannel acoustic systems robust to system identification errors. In Proceedings of the International Workshop on Acoustic Signal Enhancement (IWAENC), Xi'an, China, 13–16 September 2016; pp. 1–5.

92. Gaubitch, N.D.; Naylor, P.A. Equalization of multichannel acoustic systems in oversampled subbands. *IEEE Trans. Audio Speech Lang. Process.* **2009**, *17*, 1061–1070.

93. Abildgaard Pedersen, J.; Thomsen, K. Fully automatic loudspeaker-room adaptation—the RoomPerfect system. In Proceedings of the 32nd International Conference of Audio Engineering Society: DSP for loudspeakers, Hillerød, Denmark, 21–23 September 2007.

94. Hatziantoniou, P.; Tsoukalas, D.; Mourjopoulos, J.; Salamouris, S. Time-frequency mapping based on non-uniform smoothed spectral representations. In Proceedings of the IEEE International Conference on Acoustics, Speech, and Signal Processing (ICASSP), Phoenix, AZ, USA, 15–19 March 1999; Volume 3, pp. 1425–1428.

95. Hatziantoniou, P.D.; Mourjopoulos, J.N. Results for room acoustics equalisation based on smoothed responses. In Proceedings of the 114th Audio Engineering Society Convention, Amsterdam, The Netherlands, 22–25 March 2003.

96. Karjalainen, M.; Paatero, T. Frequency-dependent signal windowing. In Proceedings of the IEEE Workshop on the Applications of Signal Processing to Audio and Acoustics (WASPAA), New Paltz, NY, USA, 24 October 2001; pp. 35–38.

97. Panzer, J.; Ferekidis, L. The use of continuous phase for interpolation, smoothing and forming mean values of complex frequency response curves. In Proceedings of the 116th Audio Engineering Society Convention, Berlin, Germany, 8–11 May 2004.

98. Karjalainen, M. Auditory interpretation and application of warped linear prediction. In Proceedings of the Consistent & Reliable Acoustic Cues for Sound Analysis, Aalborg, Denmark, 2 September 2001.

99. Oppenheim, A.; Johnson, D.; Steiglitz, K. Computation of spectra with unequal resolution using the fast Fourier transform. *Proc. IEEE* **1971**, *59*, 299–301.

100. Strube, H.W. Linear prediction on a warped frequency scale. *J. Acoust. Soc. Am.* **1980**, *68*, 1071–1076.

101. Smith, J.O.; Abel, J.S. Bark and ERB bilinear transforms. *IEEE Trans. Speech Audio Process.* **1999**, *7*, 697–708.

102. Härmä, A. Implementation of frequency-warped recursive filters. *Signal Process.* **2000**, *80*, 543–548.

103. Oliveira, T. Laguerre filters: An introduction. *Electrónica e Telecomunicações* **1995**, *1*, 237–248.

104. Moorer, J.A. The manifold joys of conformal mapping: Applications to digital filtering in the studio. *J. Audio Eng. Soc.* **1983**, *31*, 826–841.

105. Bank, B. Warped IIR filter design with custom warping profiles and its application to room response modeling and equalization. In Proceedings of the 130th Audio Engineering Society Convention, London, UK, 13–16 May 2011.

106. Tyril, M.; Pedersen, J.A.; Rubak, P. Digital filters for low-frequency equalization. *J. Audio Eng. Soc.* **2001**, *49*, 36–43.

107. López, J.J.; Pueo, B.; Ramos, G. Combination of warped and linear filter structures for loudspeaker equalization. In Proceedings of the 124th Audio Engineering Society Convention, Amsterdam, The Netherlands, 17–20 May 2008.

108. Ramos, G.; López, J.J.; Pueo, B. Cascaded warped-FIR and FIR filter structure for loudspeaker equalization with low computational cost requirements. *Digit. Signal Process.* **2009**, *19*, 393–409.

109. Karjalainen, M.; Harma, A.; Laine, U.K.; Huopaniemi, J. Warped filters and their audio applications. In Proceedings of the IEEE Workshop on Applications of Signal Processing to Audio and Acoustics (WASPAA), New Paltz, NY, USA, 19–22 October 1997; pp. 1–4.

110. Kautz, W. Transient synthesis in the time domain. *IEEE Trans. IRE Prof. Group Circuit Theory* **1954**, *1*, 29–39.

111. Broome, P.W. Discrete orthonormal sequences. *J. ACM (JACM)* **1965**, *12*, 151–168.

112. Bank, B. Direct design of parallel second-order filters for instrument body modeling. In Proceedings of the International Computer Music Conference (ICMC), Copenhagen, Denmark, August 2017; pp. 458–465.

113. Bank, B. Perceptually motivated audio equalization using fixed-pole parallel second-order filters. *IEEE Signal Process. Lett.* **2008**, *15*, 477–480.

114. Bank, B. Audio Equalization with Fixed-Pole Parallel Filters: An Efficient Alternative to Complex Smoothing. *J. Audio Eng. Soc.* **2013**, *61*, 39–49.

115. Bank, B. Loudspeaker and room response equalization Using parallel filters: comparison of pole positioning strategies. In Proceedings of the 51st International Audio Engineering Society Conference: Loudspeakers and Headphones, Helsinki, Finland, 21–24 August 2013.

116. Genereux, R.P. Adaptive loudspeaker systems: Correcting for the acoustic environment. In Proceedings of the 8th International Audio Engineering Society Conference: The Sound of Audio, Washington, WA, USA, 3–6 May 1990.

117. Craven, P.G.; Gerzon, M.A. Practical adaptive room and loudspeaker equaliser for HI-FI use. In Proceedings of the 7th Audio Engineering Society Conference: Digital Signal Processing (DSP), London, UK, 14–15 September 1992.

118. Elliott, S.J.; Bhatia, L.P.; Deghan, F.S.; Fu, A.H.; Stewart, M.S.; Wilson, D.W. Practical implementation of low-frequency equalization using adaptive digital filters. *J. Audio Eng. Soc.* **1994**, *42*, 988–998.

119. Rubak, P.; Johansen, L.G. Design and evaluation of digital filters applied to loudspeaker/room equalization. In Proceedings of the 108th Audio Engineering Society Convention, Paris, France, 19–22 February 2000.

120. Johansen, L.; Rubak, P. Listening test results from a new digital loudspeaker/room correction systems. In Proceedings of the 110th Audio Engineering Society Convention, Amsterdam, The Netherlands, 12–15 May 2001.

121. Bharitkar, S.; Kyriakakis, C. A cluster centroid method for room response equalization at multiple locations. In Proceedings of the IEEE Workshop on the Applications of Signal Processing to Audio and Acoustics, New Paltz, NY, USA, 21–24 October 2001; pp. 55–58.

122. Kallinger, M.; Mertins, A. Room impulse response shortening by channel shortening concepts. In Proceedings of the Asilomar Conf. Signals, Systems and Computers, Pacific Grove, CA, USA, 30 October–2 November 2005; pp. 898–902.

123. Kallinger, M.; Mertins, A. Impulse response shortening for acoustic listening room compensation. In Proceedings of the International Workshop on Acoustic Echo and Noise Control (IWAENC), Eindhoven, The Netherlands, 12–15 September 2005; pp. 197–200.

124. Mei, T.; Mertins, A.; Kallinger, M. Room impulse response shortening with infinity-norm optimization. In Proceedings of the IEEE International Conference on Acoustics, Speech and Signal Processing (ICASSP), Taipei, Taiwan, 19–24 April 2009; pp. 3745–3748.

125. Mei, T.; Mertins, A. On the robustness of room impulse response reshaping. In Proceedings of the International Workshop on Acoustic Echo and Noise control (IWAENC), Tel Aviv, Israel, 30 August–2 September 2010.

126. Mertins, A.; Mei, T.; Kallinger, M. Room impulse response shortening/reshaping with infinity-and *p*-norm optimization. *IEEE Trans. Audio Speech Lang. Process.* **2010**, *18*, 249–259.

127. Mazur, R.; Jungmann, J.O.; Mertins, A. On CUDA implementation of a multichannel room impulse response reshaping algorithm based on p-norm optimization. In Proceedings of the IEEE Workshop on Applications of Signal Processing to Audio and Acoustics (WASPAA), New Paltz, NY, USA, 16–19 October 2011; pp. 305–308.

128. Jungmann, J.O.; Mei, T.; Goetze, S.; Mertins, A. Room impulse response reshaping by joint optimization of multiple p-norm based criteria. In Proceedings of the 19th European Signal Processing Conference (EUSIPCO), Barcelona, Spain, 29 August–2 September 2011; pp. 1658–1662.

129. Mazur, R.; Jungmann, J.O.; Mertins, A. Optimized gradient calculation for room impulse response reshaping algorithm based on p-norm optimization. In Proceedings of the IEEE International Conference on Acoustics, Speech and Signal Processing (ICASSP), Kyoto, Japan, 25–30 March 2012; pp. 185–188.

130. Jungmann, J.O.; Mazur, R.; Kallinger, M.; Mei, T.; Mertins, A. Combined acoustic mimo channel crosstalk cancellation and room impulse response reshaping. *IEEE Trans. Audio Speech Lang. Process.* **2012**, *20*, 1829–1842.

131. Al-Dhahir, N.; Cioffi, J.M. Efficiently computed reduced-parameter input-aided MMSE equalizers for ML detection: A unified approach. *IEEE Trans. Inf. Theory* **1996**, *42*, 903–915.

132. Falconer, D.; Magee, F. Adaptive channel memory truncation for maximum likelihood sequence estimation. *Bell Labs Tech. J.* **1973**, *52*, 1541–1562.

133. Melsa, P.J.; Younce, R.C.; Rohrs, C.E. Impulse response shortening for discrete multitone transceivers. *IEEE Trans. Commun.* **1996**, *44*, 1662–1672.

134. Martin, R.; Ding, M.; Evans, B.L.; Johnson, C.R. Efficient channel shortening equalizer design. *EURASIP J. Adv. Signal Process.* **2003**, *2003*, 906491.

135. Jungmann, J.O.; Mazur, R.; Mertins, A. On the equalization and reshaping of room impulse responses. In Proceedings of the German Annual Conference on Acoustics (DAGA), Oldenburg, Germany, 10–13 March 2014; pp. 497–498.

136. Munshi, A. Equalizability of room acoustics. In Proceedings of the IEEE International Conference on Acoustics, Speech, and Signal Processing (ICASSP), San Francisco, CA, USA, 23–26 March 1992, Volume 2, pp. 217–220.

137. Howe, R.; Hawksford, M.J. Methods of local room equalization and their effect over the listening area. In Proceedings of the 91th Audio Engineering Society Convention, New York, NY, USA, 4–8 October 1991.

138. Mourjopoulos, J. Digital equalization of room acoustics. *J. Audio Eng. Soc.* **1994**, *42*, 884–900.

139. Kyriakakis, C.; Bharitkar, S. New factors in room equalization using a fuzzy logic approach. In Proceedings of the 111th Audio Engineering Society Convention, New York, NY, USA, 30 November–3 December 2001.

140. Bharitkar, S.; Kyriakakis, C. The influence of reverberation on multichannel equalization: an experimental comparison between methods. In Proceedings of the IEEE Thirty-Seventh Asilomar Conference on Signals, Systems and Computers, Pacific Grove, CA, USA, 9–12 November 2003; Volume 1, pp. 546–549.

141. Kosko, B. Fuzzy systems as universal approximators. *IEEE Trans. Comput.* **1994**, *43*, 1329–1333.

142. Kosko, B. Combining fuzzy systems. In Proceedings of the IEEE International Joint Conference of the Fourth International Conference on Fuzzy Systems and The Second International Fuzzy Engineering Symposium, Yokohama, Japan, 20–24 March 1995; Volume 4, pp. 1855–1863.

143. Turmchokkasam, S.; Mitaim, S. Effects of weights in weighted fuzzy c-means algorithm for room equalization at multiple locations. In Proceedings of the IEEE International Conference on Fuzzy Systems, Vancouver, BC, Canada, 16–21 July 2006; pp. 1468–1475.

144. Cecchi, S.; Palestini, L.; Peretti, P.; Romoli, L.; Piazza, F.; Carini, A. Evaluation of a multipoint equalization system based on impulse response prototype extraction. *J. Audio Eng. Soc.* **2011**, *59*, 110–123.

145. Cecchi, S.; Piazza, F.; Carini, A. Subjective evaluation of a multiple position room response equalization approach applied in frequency domain. In Proceedings of the IEEE 7th International Symposium on Image and Signal Processing and Analysis (ISPA), Dubrovnik, Croatia, 4–6 September 2011; pp. 84–89.

146. Carini, A.; Cecchi, S.; Romoli, L. Multipoint room response equalization with group delay compensation. In Proceedings of the International Workshop on Acoustic Echo and Noise Control (IWAENC), Tel Aviv, Israel, 30 August–2 September 2010.

147. Primavera, A.; Cecchi, S.; Piazza, F.; Carini, A. Mixed time-frequency approach for multipoint room response equalization. In Proceedings of the 45th International Audio Engineering Society Conference: Applications of Time-Frequency Processing in Audio, Helsinki, Finland, 1–4 March 2012.

148. Primavera, A.; Cecchi, S.; Romoli, L.; Peretti, P.; Piazza, F. An advanced implementation of a digital artificial reveberator. In Proceedings of the 130th Audio Engineering Society Convention, London, UK, 13–16 May 2011.

149. Brannmark, L.; Ahlén, A. Spatially robust audio compensation based on SIMO feedforward control. *IEEE Trans. Signal Process.* **2009**, *57*, 1689–1702.

150. Johansen, L.G.; Rubak, P. Target functions and preprocessing techniques in digital equalization design. In Proceedings of the 106th Audio Engineering Society Convention, Munich, Germany, 8–11 May 1999.

151. Abildgaard Pedersen, J. Loudspeaker-room adaptation for a specific listening position using information about the complete sound field. In Proceedings of the 121st Audio Engineering Society Convention, San Francisco, CA, USA, 5–8 October 2006.

152. Abildgaard Pedersen, J.; Mortensen, H.G. Natural timbre in room correction systems. In Proceedings of the 122nd Audio Engineering Society Convention, Vienna, Austria, 5–8 May 2007.

153. Abildgaard Pedersen, J.; El-Azm, F. Natural timbre in room correction systems (Part II). In Proceedings of the 32nd International Conference of Audio Engineering Society: DSP for loudspeakers, Hillerød, Denmark, 21–23 September 2007.

154. International Electrotechnical Commission (IEC). Sound system equipment-Part 13: Listening tests on loudspeakers. In Proceedings of the International Electrotechnical Commission, Geneva, Switzerland, 11 March 1988.

155. Bunkheila, G.; Parisi, R.; Uncini, A. Model order selection for estimation of common acoustical poles. In Proceedings of the IEEE International Symposium on Circuits and Systems (ISCAS), Seattle, WA, USA, 18–21 May 2008; pp. 1180–1183.

156. Lakhdhar, K.; Jaidane, M.; Shaiek, H.; Boucher, J. Iterative equalization of room transfer function using biquadratic filters. In Proceedings of the IEEE Instrumentation and Measurement Technology Conference (IMTC), Singapore, 5–7 May 2009; pp. 1463–1466.

157. Mäkivirta, A.; Antsalo, P.; Karjalainen, M.; Välimäki, V. Modal equalization of loudspeaker-room responses at low frequencies. *J. Audio Eng. Soc.* **2003**, *51*, 324–343.

158. Karjalainen, M.; Antsalo, P.; Mäkivirta, A.; Peltonen, T.; Välimäki, V. Estimation of modal decay parameters from noisy response measurements. *J. Audio Eng. Soc.* **2002**, *50*, 867–878.

159. Santillán, A.O. Spatially extended sound equalization in rectangular rooms. *J. Acoust. Soc. Am.* **2001**, *110*, 1989–1997.

160. Santillán, A.; Pedersen, C.S.; Lydolf, M. Experimental implementation of a low-frequency global sound equalization method based on free field propagation. *Appl. Acoust.* **2007**, *68*, 1063–1085.

161. Celestinos, A.; Birkedal Nielsen, S. Multi-Source low frequency room simulation using finite difference time domain approximations. In Proceedings of the 117th Audio Engineering Society Convention, San Francisco, CA, USA, 28–31 October 2004.

162. Birkedal Nielsen, S.; Celestinos, A. Optimizing placement and equalization of multiple low frequency loudspeakers in rooms. In Proceedings of the 119th Audio Engineering Society Convention, New York, NY, USA, 7–10 October 2005.

163. Celestinos, A.; Birkedal Nielsen, S. Low frequency sound field enhancement system for rectangular rooms using multiple low frequency loudspeakers. In Proceedings of the 120th Audio Engineering Society Convention, Paris, France, 20–23 May 2006.

164. Celestinos, A.; Birkedal Nielsen, S. Controlled acoustic bass system (CABS) A method to achieve uniform sound field distribution at low frequencies in rectangular rooms. *J. Audio Eng. Soc.* **2008**, *56*, 915–931.

165. Celestinos, A.; Birkedal Nielsen, S. Applications of a plane wave based room correction system for low frequencies using multiple loudspeakers. *J. Acoust. Soc. Am.* **2008**, *123*, 3091.

166. Vanderkooy, J. Multi-source room equalization: Reducing room resonances. In Proceedings of the 123rd Audio Engineering Society Convention, New York, NY, USA, 5–8 October 2007.

167. Bank, B. Full room equalization at low frequencies with asymmetric loudspeaker arrangements. In Proceedings of the 132nd Audio Engineering Society Convention, Budapest, Hungary, 26–29 April 2012.

168. Stefanakis, N.; Sarris, J.; Jacobsen, F. Regularization in global sound equalization based on effort variation. *J. Acoust. Soc. Am.* **2009**, *126*, 666–675.

169. Nelson, P.A.; Hamada, H.; Elliott, S.J. Adaptive inverse filters for stereophonic sound reproduction. *IEEE Trans. Signal Process.* **1992**, *40*, 1621–1632.

170. Pedersen, C.S.; Møller, H. Sound field control for a low-frequency test facility. In Proceedings of the 52nd International Audio Engineering Society Conference: Sound Field Control-Engineering and Perception, Guildford, UK, 2–4 September 2013.

171. Blauert, J. *Spatial Hearing: The Psychophysics of Human Sound Localization*; MIT Press: Cambridge, MA, USA, 1997.

172. Bank, B. Combined quasi-anechoic and in-room equalization of loudspeaker responses. In Proceedings of the 134th Audio Engineering Society Convention, Rome, Italy, 4–7 May 2013.

173. Cecchi, S.; Romoli, L.; Piazza, F.; Bank, B.; Carini, A. A novel approach for prototype extraction in a multipoint equalization procedure. In Proceedings of the 136th Audio Engineering Society Convention, Berlin, Germany, 26–29 April 2014.

174. Toole, F.E. Loudspeakers and rooms for sound reproduction — A scientific review. *J. Audio Eng. Soc.* **2006**, *54*, 451–476.

175. Ziomek, L. *Fundamentals of Acoustic Field Theory and Space-Time Signal Processing*; CRC Press: Boca Raton, FL, USA, 1995; pp. 651–662.

176. Elko, G.W.; Diethorn, E.; Gänsler, T. Room impulse response variation due to thermal fluctuation and its impact on acoustic echo cancellation. In Proceedings of the International Workshop on Acoustic Echo and Noise Control (IWAENC), Kyoto, Japan, September 2003; pp. 67–70.

177. Leitão, J.; Ferreira, A.J.; Fernandes, G. Adaptive room equalization in the frequency domain. In Proceedings of the 116th Audio Engineering Society Convention, Berlin, Germany, 8–11 May 2004.
178. Ferreira, A.; Leite, A. An improved adaptive room equalization in the frequency domain. In Proceedings of the 118th Audio Engineering Society Convention, Barcelona, Spain, 28–31 May 2005.
179. Ferreira, A.J.; Leite, A.; Pinto, F.; Rocha, A.F. Adaptive audio equalization of rooms based on a technique of transparent insertion of acoustic probe signals. In Proceedings of the 120th Audio Engineering Society Convention, Paris, France, 20–23 May 2006.
180. Cecchi, S.; Primavera, A.; Piazza, F.; Carini, A. An adaptive multiple position room response equalizer. In Proceedings of the 19th European Signal Processing Conference (EUSIPCO), Barcelona, Spain, 29 August–3 September 2011; pp. 1274–1278.
181. Benesty, J.; Morgan, D.; Sondhi, M. A better understanding and an improved solution to the specific problems of stereophonic acoustic echo cancellation. *IEEE Trans. Speech Audio Process.* **1998**, *6*, 156–165.
182. Ali, M. Stereophonic acoustic echo cancellation system using time-varying all-pass filtering for signal decorrelation. In Proceedings of the IEEE International Conference on Acoustics, Speech, and Signal Processing (ICASSP), Seattle, WA, USA, 15 May 1998; pp. 3689–3692.
183. Cecchi, S.; Romoli, L.; Piazza, F.; Carini, A. A multichannel and multiple position adaptive room response equalizer in warped domain. In Proceedings of the IEEE 8th International Symposium on Image and Signal Processing and Analysis (ISPA), Trieste, Italy, 4–6 September 2013; pp. 761–766.
184. Cecchi, S.; Romoli, L.; Peretti, P.; Piazza, F. Low-complexity implementation of a real-time decorrelation algorithm for stereophonic acoustic echo cancellation. *Signal Process.* **2012**, *92*, 2668–2675.
185. Cecchi, S.; Romoli, L.; Carini, A.; Piazza, F. A multichannel and multiple position adaptive room response equalizer in warped domain: Real-time implementation and performance evaluation. *Appl. Acoust.* **2014**, *82*, 28–37.
186. Fuster, L.; de Diego, M.; Ferrer, M.; Gonzalez, A.; Piñero, G. A biased multichannel adaptive algorithm for room equalization. In Proceedings of the 20th European Signal Processing Conference (EUSIPCO), Bucharest, Romania, 27–31 August 2012; pp. 1344–1348.
187. Arenas-Garcia, J.; de Diego, M.; Azpicueta-Ruiz, L.A.; Ferrer, M.; Gonzalez, A. Combinations of proportionate adaptive filters in acoustics: an application to ANC. In Proceedings of the 20th European Signal Processing Conference (EUSIPCO), Bucharest, Romania, 27–31 August 2012.
188. Fuster, L.; de Diego, M.; Azpicueta-Ruiz, L.A.; Ferrer, M. Adaptive filtered-x algorithms for room equalization based on block-based combination schemes. *IEEE/ACM Trans. Audio Speech Lang. Process.* **2016**, *24*, 1732–1745.
189. Nelson, P.; Orduna-Bustamante, F.; Hamada, H. Adaptive Audio Systems and Sound Reproduction Systems. U.S. Patent 5,949,894, 7 September 1999.
190. Weiss, S.; Rice, G.W.; Stewart, R.W. Multichannel equalization in subbands. In Proceedings of the IEEE Workshop on Applications of Signal Processing to Audio and Acoustics (WASPAA), New Paltz, NY, USA, 1999; pp. 203–206.
191. Williams, E. *Fourier Acoustics: Sound Radiation and Nearfield Acoustical Holography*; Academic Press: San Diego, CA, USA, 1999.
192. Zotter, F.; Spors, S. Is sound field control determined at all frequencies? How is it related to numerical acoustics? In Proceedings of the 52nd Conference on Sound Field Control—Engineering and Perception, Audio Engineering Society, Guildford, UK, 2–4 September 2013.
193. Spors, S. Active Listening Room Compensation for Spatial Sound Reproduction Systems. Ph.D. Thesis, University of Erlangen-Nuremberg, Nuremberg, Bavaria, Germany, 2006.
194. Spors, S.; Rabenstein, R. Spatial aliasing artifacts produced by linear and circular loudspeaker arrays used for wave field synthesis. In Proceedings of the 120th Convention of the Audio Engineering Society, Paris, France, 20–23 May 2006.
195. Ahrens, J.; Spors, S. A modal analysis of spatial discretization in spherical loudspeaker arrays used for sound field synthesis. *IEEE Trans. Audio Speech Lang. Process.* **2012**, *20*, 2564–2574.
196. Rafaely, B. *Fundamentals of Spherical Array Processing*; Springer: Berlin/Heidelberg, Germany, 2015.
197. Spors, S.; Renk, M.; Rabenstein, R. Limiting effects of active room compensation using Wave field synthesis. In Proceedings of the 118th Convention of the Audio Engineering Society, Barcelona, Spain, 28–31 May 2005.

198. Petrausch, S.; Spors, S.; Rabenstein, R. Simulation and visualization of room compensation for wave field synthesis with the functional transformation method. In Proceedings of the 119th Convention of the Audio Engineering Society, New York, NY, USA, 7–10 October 2005.

199. Bouchard, M.; Quednau, S. Multichannel recursive-least-squares algorithms and fast-transversal-filter Algorithms for Active Noise Control and Sound Reproduction Systems. *IEEE Trans. Speech Audio Process.* **2000**, *8*, 606–618.

200. Spors, S.; Buchner, H.; Rabenstein, R. Spatio-temporal adaptive inverse filtering in the wave domain. In *Speech and Audio Processing in Adverse Environments*; Haensler, E., Schmidt, G., Eds.; Springer: Berlin/Heidelberg, Germany, 2008; pp. 551–583.

201. Schneider, M.; Kellermann, W. Multichannel acoustic echo cancellation in the wave domain with increased robustness to nonuniqueness. *IEEE/ACM Trans. Audio Speech Lang. Process. (TASLP)* **2016**, *24*, 518–529.

202. Talagala, D.S.; Zhang, W.; Abhayapala, T.D. Efficient multi-channel adaptive room compensation for spatial soundfield reproduction using a modal decomposition. *IEEE/ACM Trans. Audio Speech Lang. Process. (TASLP)* **2014**, *22*, 1522–1532.

203. Schneider, M.; Kellermann, W. A Direct Derivation of Transforms for Wave-Domain Adaptive Filtering Based on Circular Harmonics. In Proceedings of the 20th European Signal Processing Conference (EUSIPCO), Bucharest, Romania, 27–31 August 2012.

204. Peretti, P.; Palestini, L.; Cecchi, S.; Piazza, F. A subband approach to wave domain adaptive filtering. In Proceedings of the Hands-Free Speech Communication and Microphone Arrays, Trento, Italy, 6–8 May 2008.

205. Hofmann, C.; Kellermann, W. Generalized wave-domain transforms for listening room equalization with azimuthally irregularly spaced loudspeaker arrays. In Proceedings of the IEEE International Conference on Acoustics, Speech and Signal Processing (ICASSP), Shanghai, China, 20–25 March 2016.

206. Betlehem, T.; Abhayapala, T.D. Theory and design of sound field reproduction in reverberant rooms. *J. Acoust. Soc. Am.* **2005**, *117*, 2100–2111.

207. Zha, M.f.; Bao, C.c.; Jia, M.s. 3D multizone soundfield reproduction in the reverberant room using a spherical loudspeaker array. In Proceedings of the Asia-Pacific Signal and Information Processing Association Annual Summit and Conference, Hong Kong, China, 16–19 December 2015.

208. Poletti, M.; Abhayapala, T.; Samarasinghe, P. Interior and exterior sound field control using two dimensional higher-order variable-directivity sources. *J. Acoust. Soc. Am.* **2012**, *131*, 3814–3823.

209. Corteel, E.; Nicol, R. Listening room compensation for wave field synthesis. What can be done? In Proceedings of the 23rd International Audio Engineering Society Conference: Signal Processing in Audio Recording and Reproduction, Helsingør, Denmark, 23–25 May 2003.

210. Sontacchi, A.; Hoeldrich, R. Enhanced 3D sound field synthesis and reproduction system by compensating interfering reflections. In Proceedings of the Conference on Digital Audio Effects (DAFX-00), Verona, Italy, December 2000.

211. Canclini, A.; Marković, D.; Antonacci, F.; Sarti, A.; Tubaro, S. A room-compensated virtual surround system exploiting early reflections in a reverberant room. In Proceedings of the 20th European Signal Processing Conference (EUSIPCO), Bucharest, Romania, 27–31 August 2012; pp. 1029–1033.

212. Canclini, C.; Markovic, D.; Bianchi, L.; Antonacci, F.; Sarti, A.; Tubaro, S. A geometrical approach to room compensation for sound field rendering applications. In Proceedings of the 22nd European Signal Processing Conference (EUSIPCO), Lisbon, Portugal, 1–5 September 2014.

213. Takane, S.; Toshio, S. A new theory for active suppression of reflected sound waves from the walls based on Kirchhoff-helmholtz boundary integral equation. *Acoust. Sci. Technol.* **2004**, *25*, 37–44.

214. Zhang, W.; Habets, E.A.; Naylor, P.A. On the use of channel shortening in multichannel acoustic system equalization. In Proceedings of the International Workshop on Acoustic Echo Noise Control (IWAENC), Tel Aviv, Israel, 30 August–2 September 2010.

215. Kallinger, M.; Mertins, A. Multi-channel room impulse response shaping-a study. In Proceedings of the IEEE International Conference on Acoustics, Speech and Signal Processing (ICASSP), Toulouse, France, 14–19 May 2006; Volume 5, pp. 101–104.

216. Guillaume, M.; Grenier, Y.; Richard, G. Iterative algorithms for multichannel equalization in sound reproduction systems. In Proceedings of the IEEE International Conference on Acoustics, Speech, and Signal Processing (ICASSP), Philadelphia, PA, USA, 23 March 2005; Volume 3, pp. 269–272.

217. Schneider, M.; Kellermann, W. Iterative DFT-domain inverse filter determination for adaptive listening room equalization. In Proceedings of the International Workshop on Acoustic Signal Enhancement (IWAENC), Aachen, Germany, 4–6 September 2012; pp. 1–4.

218. Korst-Fagundes, B.; Xie, J.; Snelgrove, M. Multipoint equalization with the condition number. In Proceedings of the 38th Midwest Symposium on Circuits and Systems, Rio de Janeiro, Brazil, 13–16 August 1995; Volume 2, pp. 807–810.

219. Brännmark, L.; Bahne, A.; Ahlen, A. Compensation of loudspeaker–room responses in a robust MIMO control framework. *IEEE Transa. Audio Speech Lang. Process.* **2013**, *21*, 1201–1216.

220. Brännmark, L.; Bahne, A.; Ahlén, A. Improved loudspeaker–room equalization using multiple loudspeakers and MIMO feedforward control. In Proceedings of the IEEE International Conference on Acoustics, Speech and Signal Processing (ICASSP), Kyoto, Japan, 25–30 March 2012; pp. 237–240.

221. Brännmark, L.; Ahlén, A. Multichannel room correction with focus control. *J. Audio Eng. Soc.* **2015**, *63*, 21–30.

222. Bahne, A.; Ahlén, A. Optimizing the Similarity of Loudspeaker–Room Responses in Multiple Listening Positions. *IEEE/ACM Trans. Audio Speech Lang. Process.* **2016**, *24*, 340–353.

223. Bahne, A.; Brännmark, L.; Ahlén, A. Symmetric loudspeaker-room equalization utilizing a pairwise channel similarity criterion. *IEEE Trans. Signal Process.* **2013**, *61*, 6276–6290.

224. Bahne, A.; Brännmark, L.; Ahlén, A. Improved loudspeaker-room equalization for stereo systems regarding channel similarity. In Proceedings of the International Conference on Audio, Language and Image Processing (ICALIP), Shanghai, China, 16–18 July 2012; pp. 254–259.

225. Johansson, M.; Brännmark, L.; Bahne, A.; Sternad, M. Sound field control using a limited number of loudspeakers. In Proceedings of the 36th International Audio Engineering Society Conference: Automotive Audio, Dearborn, MI, USA, 2–4 June 2009.

226. Farina, A.; Ugolotti, E. Spatial Equalization of sound systems in cars. In Proceedings of the 15th International Audio Engineering Society Conference: Audio, Acoustics & Small Spaces, Copenhagen, Denmark, 31 October–2 November 1998.

227. Lingvall, F.; Brännmark, L. Multiple-point statistical room correction for audio reproduction: Minimum mean squared error correction filtering. *J. Acoust. Soc. Am.* **2009**, *125*, 2121–2128.

228. Radlovic, B.D.; Williamson, R.C.; Kennedy, R.A. Equalization in an acoustic reverberant environment: Robustness results. *IEEE Trans. Speech Audio Process.* **2000**, *8*, 311–319.

229. Cecchi, S.; Romoli, L.; Piazza, F.; Carini, A. Multipoint room response equalization with group delay compensation: Subjective listening tests. In Proceedings of the IEEE 7th International Symposium on Image and Signal Processing and Analysis (ISPA), Dubrovnik, Croatia, 4–6 September 2011; pp. 57–62.

230. Bharitkar, S.; Kyriakakis, C. Visualization of multiple listener room acoustic equalization with the Sammon map. *IEEE Trans. Audio Speech Lang. Process.* **2007**, *15*, 542–551.

231. Jot, J. An analysis/synthesis approach to real-time artificial reverberation. In Proceedings of the IEEE International Conference on Acoustics, Speech and Signal Processing (ICASSP), San Francisco, CA, USA, 23–26 March 1992; pp. 221–224.

232. Jot, J.M.; Cerveau, L.; Warusfel, O. Analysis and synthesis of room reverberation based on a statistical time-frequency model. In Proceedings of the 103rd Audio Engineering Society Convention, New York, NY, USA, 26–29 September 1997.

233. Goetze, S.; Albertin, E.; Kallinger, M.; Mertins, A.; Kammeyer, K. Quality assessment for listening-room compensation algorithms. In Proceedings of the IEEE International Conference on Acoustics Speech and Signal Processing (ICASSP), Dallas, TX, USA, 14–19 March 2010; pp. 2450–2453.

234. Meyer, E.; Thiele, R. Raumakustische Untersuchungen in zahlreichen Konzertsälen und Rundfunkstudios unter Antoendung neuerer Meβverfahren. *Acta Acust. United Acust.* **1956**, *6*, 425–444.

235. Reichardt, W.; Alim, O.A.; Schmidt, W. Abhängigkeit der grenzen zwischen brauchbarer und unbrauchbarer durchsichtigkeit von der art des musikmotives, der nachhallzeit und der nachhalleinsatzzeit. *Appl. Acoust.* **1974**, *7*, 243–264.

236. Vilhelm Lassen, J. Acoustical criteria for auditoriums and their relation to model techniques. *J. Acoust. Soc. Am.* **1970**, *47*, 408–412.

237. Von Békésy, G. Über die Entstehung der Entfernungsempfindung beim Hören. *Akust. Z.* **1938**, *3*, 21–31.

238. Triki, M.; Slock, D. Iterated delay and predict equalization for blind speech dereverberation. In Proceedings of the International Workshop on Acoustic Echo and Noise Control (IWAENC), Paris, France, 12–14 September 2006; pp. 1–4.

239. Kürer, R. Zur Gewinnung von Einzahlkriterien bei Impulsmessung in der Raumakustik. *Acustica* **1969**, *21*, 370–372.

240. Bech, S.; Zacharov, N. *Perceptual Audio Evaluation-Theory, Method and Application*; Wiley: Hoboken, NJ, USA, 2006.

241. Campbell, D.; Jones, E.; Glavin, M. Audio quality assessment techniques—A review, and recent developments. *Signal Process.* **2009**, *89*, 1489–1500.

242. *Methods for the Subjective Assessment of Small Impairments in Audio System Including Multichannel Sound Systems*; ITU-R BS. 1116-1; 1997.

243. *Method for Subjective Listening Tests of Intermediate Audio Quality*; ITU-R BS. 1534; 2003.

244. *General Methods for the Subjective Assessment of Sound Quality*; ITU-R BS. 1284-1; 2003.

245. Rumsey, F. Sound field control: Personal sound zones and moving listeners. *J. Audio Eng. Soc.* **2016**, *64*, 808–813.

246. Zhang, W.; Samarasinghe, P.N.; Chen, H.; Abhayapala, T.D. Surround by sound: A review of spatial audio recording and reproduction. *Appl. Sci.* **2017**, *7*, 532.

247. Choi, J.W.; Kim, Y.H. Generation of an acoustically bright zone with an illuminated region using multiple sources. *J. Acoust. Soc. Am.* **2002**, *111*, 1695–1700.

248. Poletti, M. An investigation of 2-d multizone surround sound systems. In Proceedings of the 125th Audio Engineering Society Convention, San Francisco, CA, USA, 2–5 October 2008.

249. Chang, J.H.; Lee, C.-H.; Park, J.Y.; Kim, Y.H. A realization of sound focused personal audio system using acoustic contrast control. *J. Acoust. Soc. Am.* **2009**, *125*, 2091–2097.

250. Shin, M.; Lee, S.Q.; Fazi, F.M.; Nelson, P.A.; Kim, D.; Wang, S.; Park, K.H.; Seo, J. Maximization of acoustic energy difference between two spaces. *J. Acoust. Soc. Am.* **2010**, *128*, 121–131.

251. Wu, Y.J.; Abhayapala, T.D. Spatial multizone soundfield reproduction: Theory and design. *IEEE Trans. Audio Speech Lang. Process.* **2011**, *19*, 1711–1720.

252. Betlehem, T.; Withers, C. Sound field reproduction with energy constraint on loudspeaker weights. *IEEE Trans. Audio Speech Lang. Process.* **2012**, *20*, 2388–2392.

253. Chang, J.H.; Jacobsen, F. Sound field control with a circular double-layer array of loudspeakers. *J. Acoust. Soc. Am.* **2012**, *131*, 4518–4525.

254. Elliott, S.J.; Cheer, J.; Choi, J.W.; Kim, Y. Robustness and regularization of personal audio systems. *IEEE Trans. Audio Speech Lang. Process.* **2012**, *20*, 2123–2133.

255. Chang, J.H.; Jacobsen, F. Experimental validation of sound field control with a circular double-layer array of loudspeakers. *J. Acoust. Soc. Am.* **2013**, *133*, 2046–2054.

256. Cheer, J.; Elliott, S.J.; Gálvez, M.F.S. Design and implementation of a car cabin personal audio system. *J. Audio Eng. Soc.* **2013**, *61*, 412–424.

257. Radmanesh, N.; Burnett, I.S. Generation of isolated wideband sound fields using a combined two-stage lasso-ls algorithm. *IEEE Trans. Audio Speech Lang. Process.* **2013**, *21*, 378–387.

258. Cai, Y.; Wu, M.; Yang, J. Sound reproduction in personal audio systems using the least-squares approach with acoustic contrast control constraint. *J. Acoust. Soc. Am.* **2014**, *135*, 734–741.

259. Coleman, P.; Jackson, P.J.; Olik, M.; Abildgaard Pedersen, J. Personal audio with a planar bright zone. *J. Acoust. Soc. Am.* **2014**, *136*, 1725–1735.

260. Coleman, P.; Jackson, P.J.; Olik, M.; Møller, M.; Olsen, M.; Abildgaard Pedersen, J. Acoustic contrast, planarity and robustness of sound zone methods using a circular loudspeaker array a. *J. Acoust. Soc. Am.* **2014**, *135*, 1929–1940.

261. Helwani, K.; Spors, S.; Buchner, H. The synthesis of sound figures. *Multidimens. Syst. Signal Process.* **2014**, *25*, 379–403.

262. Betlehem, T.; Zhang, W.; Poletti, M.A.; Abhayapala, T.D. Personal sound zones: delivering interface-free audio to multiple listeners. *IEEE Signal Process. Mag* **2015**, *32*, 81–91.

263. Jin, W.; Kleijn, W.B. Theory and design of multizone soundfield reproduction using sparse methods. *IEEE/ACM Trans. Audio Speech Lang. Process. (TASLP)* **2015**, *23*, 2343–2355.

264. Menzies, D. Sound field synthesis with distributed modal constraints. *Acta Acust. United Acust.* **2012**, *98*, 15–27.

265. Poletti, M.A.; Fazi, F. An approach to generating two zones of silence with application to personal sound systems. *J. Acoust. Soc. Am.* **2015**, *137*, 598–605.

266. Simón G., Marcos, F.; Elliott, S.J.; Cheer, J. Time domain optimization of filters used in a loudspeaker array for personal audio. *IEEE/ACM Trans. Audio Speech Lang. Process. (TASLP)* **2015**, *23*, 1869–1878.

267. Antonello, N.; De Sena, E.; Moonen, M.; Naylor, P.A.; van Waterschoot, T. Sound field control in a reverberant room using the Finite Difference Time Domain method. In Proceedings of the 60th Audio Engineering Society International Conference: Dereverberation and Reverberation of Audio, Music, and Speech (DREAMS), Leuven, Belgium, 3–5 February 2016.

268. Poletti, M.A.; Fazi, F.M. Generation of half-space sound fields with application to personal sound systems. *J. Acoust. Soc. Am.* **2016**, *139*, 1294–1302.

269. Wakayama, K.; Takada, H. Spatial multi-zone sound field reproduction Using higher-order loudspeakers in reverberant rooms. In Proceedings of the 140th Audio Engineering Society Convention, Paris, France, 4–7 June 2016.

270. Zhang, W.; Abhayapala, T.D.; Betlehem, T.; Fazi, F.M. Analysis and control of multi-zone sound field reproduction using modal-domain approach. *J. Acoust. Soc. Am.* **2016**, *140*, 2134–2144.

271. Zhu, Q.; Coleman, P.; Wu, M.; Yang, J. Robust personal audio reproduction based on acoustic transfer function modelling. In Proceedings of the 2016 Audio Engineering Society International Conference on Sound Field Control, Guildford, UK, 18–20 July 2016.

272. Okamoto, T.; Sakaguchi, A. Experimental validation of spatial Fourier transform-based multiple sound zone generation with a linear loudspeaker array. *J. Acoust. Soc. Am.* **2017**, *141*, 1769–1780.

273. Piñero, G.; Botella, C.; de Diego, M.; Ferrer, M.; González, A. On the feasibility of personal audio systems over a network of distributed loudspeakers. In Proceedings of the 25th European Signal Processing Conference (EUSIPCO), Kos, Greece, 28 August–2 September 2017; pp. 2729–2733.

274. Zhu, Q.; Coleman, P.; Wu, M.; Yang, J. Robust reproduction of sound zones with local sound orientation. *J. Acoust. Soc. Am.* **2017**, *142*, EL118–EL122.

275. Cecchi, S.; Virgulti, M.; Primavera, A.; Piazza, F.; Bettarelli, F.; Li, J. Investigation on audio algorithms architecture for stereo portable devices. *J. Audio Eng. Soc.* **2016**, *64*, 75–88.

276. Czyzewski, A.; Ciarkowski, A.; Kostek, B.; Kotus, J.; Lopatka, K.; Suchomski, P. Adaptive personal tuning of sound in mobile computers. *J. Audio Eng. Soc.* **2016**, *64*, 405–428.

277. Fuster, L. Linear and Nonlinear Room Compensation of Audio Rendering Systems. Ph.D. Thesis, Universitat Politecnica de Valencia, Valencia, Spain, 2015.

278. Fuster, L.; de Diego, M.; Ferrer, M.; Gonzalez, A.; Pinero, G. Nonlinear filtered-X second-order adaptive voltcrra filters for listening-room compensation. In Proceedings of the 23rd European Signal Processing Conference (EUSIPCO), Nice, France, 31 August–4 September 2015; pp. 669–673.

279. Fuster, L.; Ferrer, M.; de Diego, M.; Gonzalez, A. Combination of filtered-x adaptive filters for nonlinear listening-room compensation. In Proceedings of the 24th European Signal Processing Conference (EUSIPCO), Budapest, Hungary, 29 August–2 September 2016; pp. 1773–1777.

280. Hahn, N.; Spors, S. Continuous measurement of impulse responses on a circle using a uniformly moving microphone. In Proceedings of the 23rd European Signal Processing Conference (EUSIPCO), Nice, France, 31 August–4 September 2015.

281. Mazur, R.; Katzberg, F.; Phan, H.; Mertins, A. Room equalization based on measurements with moving microphones. In Proceedings of the IEEE Hands-free Speech Communications and Microphone Arrays (HSCMA), San Francisco, CA, USA, 1–3 March 2017; pp. 121–125.

MDPI

St. Alban-Anlage 66

4052 Basel

Switzerland

Tel. +41 61 683 77 34

Fax +41 61 302 89 18

www.mdpi.com

Applied Sciences Editorial Office

E-mail: applsci@mdpi.com

www.mdpi.com/journal/applsci

www.ingramcontent.com/pod-product-compliance
Lightning Source LLC
Chambersburg PA
CBHW051700210326
41597CB00032B/5315